HANDBOOK OF SOLID WASTE MANAGEMENT

HANDBOOK OF SOLID WASTE MANAGEMENT

George Tchobanoglous

*Professor Emeritus of Civil and Environmental Engineering
University of California at Davis
Davis, California*

Frank Kreith

*Professor Emeritus of Engineering
University of Colorado
Boulder, Colorado*

Second Edition

McGRAW-HILL

New York   Chicago   San Francisco
Lisbon   London   Madrid   Mexico City   Milan
New Delhi   San Juan   Seoul   Singapore
Sydney   Toronto

Library of Congress Cataloging-in-Publication Data

Handbook of solid waste management / George Tchobanoglous, co-editor;
Frank Kreith, co-editor.—2nd ed.
    p. cm.
  Includes index.
  ISBN 0-07-135623-1
  1. Refuse and refuse disposal. I. Tchobanoglous, George. II. Kreith, Frank.

TD791 .H2723   2002
628.4'4'068—dc21   2002021284

McGraw-Hill

A Division of The McGraw·Hill Companies

ISBN 0-07-135623-1

*The sponsoring editor for this book was Ken McCombs and the production
supervisor was Pamela A. Pelton. It was set in TimesTen Roman by North
Market Street Graphics.*

Printed and bound by Quebecor/Martinsburg.

McGraw-Hill books are available at special quantity discounts to use as
premiums and sales promotions, or for use in corporate training programs.
For more information, please write to the Director of Special Sales,
McGraw-Hill, Two Penn Plaza, New York, NY 10121-2298. Or contact your
local bookstore.

This book is printed on recycled, acid-free paper containing a
minimum of 50% recycled, de-inked fiber.

CONTENTS

Chapter 9. Markets and Products for Recycled Material
Harold Leverenz and Frank Kreith   **9.1**

Chapter 10. Household Hazardous Wastes (HHW)
David E.B. Nightingale and Rachel Donnette   **10.1**

Chapter 11. Other Special Wastes
Part 11A. Batteries *Gary R. Brenniman, Stephen D. Casper,*
William H. Hallenbeck, and James M. Lyznicki   **11.1**

Part 11B. Used Oil *Stephen D. Casper, William H. Hallenbeck, and Gary R. Brenniman*

Part 11C. Scrap Tires *John K. Bell*

Part 11D. Construction and Demolition (C&D) Debris *George Tchobanoglous*

Part 13C. Emission Control *Floyd Hasselriis*

Chapter 14. Landfilling *Philip R. O'Leary and George Tchobanoglous*   **14.1**

Chapter 15. Siting Municipal Solid Waste Facilities *David Laws, Lawrence Susskind, and Jason Corburn*   **15.1**

Chapter 16. Financing and Life-Cycle Costing of Solid Waste Management Systems *Nicholas S. Artz, Jacob E. Beachey, and Philip R. O'Leary*   **16.1**

CONTRIBUTORS

Nicholas S. Artz *Franklin Associates, Ltd., 4121 W. 83rd Street, Suite 108, Prairie Village, KS 666208* (CHAP. 16).

Jacob E. Beachey *Franklin Associates, Ltd., 4121 W. 83rd Street, Suite 108, Prairie Village, KS 66208* (CHAP. 16).

John K. Bell *California Integrated Waste Management Board (CIWMB), 8800 Cal Center Drive, Sacramento, CA 95826* (CHAP. 11C).

Gary R. Brenniman *School of Public Health, University of Illinois, 2121 West Taylor Street, Chicago, IL 60612-7260* (CHAPS. 11A, 11B, 11E).

Calvin R. Brunner *Incinerator Consultant, Inc., 11204 Longwood Grove Drive, Reston, VA 22094* (CHAP. 13A).

Stephen D. Casper *School of Public Health, University of Illinois, 2121 West Taylor Street, Chicago, IL 60612-7260* (CHAPS. 11A, 11B).

Jason Corburn *Department of Urban Studies and Planning, Massachusetts Institute of Technology (MIT), 77 Massachusetts Ave., RM 3-411, Cambridge, MA 02139* (CHAP. 15).

Luis F. Diaz *CalRecovery, Inc., 1850 Gateway Boulevard, Suite 1060, Concord, CA 94520* (CHAP. 12).

Rachel Donnette *Thurston County Environmental Health, 2000 Lakeridge Drive SW, Olympia, WA 98502* (CHAP. 10).

Barbara Foster *National Conference of State Legislatures (NCSL), 1560 Broadway, Suite 700, Denver, CO 80202* (CHAP. 2).

Marjorie A. Franklin *Franklin Associates, Ltd., 4121 W. 83rd Street, Suite 108, Prairie Village, KS 66208* (CHAP. 5).

Ken Geiser *Toxics Use Reduction Institute, University of Lowell, 1 University Avenue, Lowell, MA 01854* (CHAP. 6B).

Jim Glenn *BioCycle, 419 State Avenue, Emmaus, PA 18049* (CHAP. 3).

Clarence G. Golueke *CalRecovery, Inc., 1850 Gateway Boulevard, Suite 1060, Concord, CA 94520* (CHAP. 12).

William H. Hallenbeck *1106 Maple Street, Western Springs, IL 60558* (CHAPS. 11A, 11B, 11E).

Floyd Hasselriis *Engineering Consultant, 52 Seasongood Road, Forest Hills Gardens, New York, NY 11375* (CHAPS. 13B, 13C).

Kelly Hill *105 Rosella Avenue, Fairbanks, AK 99701* (CHAP. 3).

Frank Kreith *Engineering Consultant, 1485 Sierra Drive, Boulder, CO 80302* (CHAPS. 1, 9, 13).

James E. Kundell *Vinson Institute of Government, University of Georgia, Athens, GA 30602* (CHAP. 4).

David Laws *Department of Urban Studies and Planning, Massachusetts Institute of Technology (MIT), 77 Massachusetts Ave., RM 3-411, Cambridge, MA 02139* (CHAP. 15).

Harold Leverenz *Department of Civil and Environmental Engineering, University of California, Davis, Davis, CA 95616* (CHAPS. 6A, 8, 9).

James M. Lyznicki   *School of Public Health, 2121 West Taylor Street, Chicago, IL 60612-7260* (CHAP. 11A).

David E. B. Nightingale   *Solid Waste and Financial Assistance Program, Washington State Department of Ecology, P.O. Box 47775, Olympia, WA 98504-7775* (CHAP. 10).

Philip R. O'Leary   *University of Wisconsin, 432 N. Lake Street, Madison, WI 53706* (CHAPS. 14, 16).

Edward W. Repa   *National Solid Waste Management Association (NSWMA), 1730 Rhode Island Avenue NW, Washington, DC 20036* (CHAP. 2).

Deanna K. Ruffer   *Vinson Institute of Government, University of Georgia, Athens, GA 30602* (CHAP. 4).

George M. Savage   *CalRecovery, Inc., 1850 Gateway Boulevard, Suite 1060, Concord, CA 94520* (CHAP. 12).

David B. Spencer   *WTE Corporation, 7 Alfred Circle, Bedford, MA 01730* (CHAP. 8).

Lawrence Susskind   *Department of Urban Studies and Planning, Massachusetts Institute of Technology, 77 Massachusetts Ave., RM 3-411, Cambridge, MA 02139* (CHAP. 15).

George Tchobanoglous   *Engineering Consultant, 662 Diego Place, Davis, CA 95616* (CHAPS. 1, 8, 11D).

Hilary Theisen   *Solid Waste Consultant, 2451 Palmira Place, San Ramon, CA 94583* (CHAP. 7).

Marcia E. Williams   *LECG, 333 South Grand Avenue, Los Angeles, CA 90071* (CHAP. 1).

PREFACE TO THE
SECOND EDITION

The first edition of this handbook was an outgrowth of a two-day conference on integrated solid waste management in June 1989, sponsored by the U.S. Environmental Protection Agency (EPA), the American Society of Mechanical Engineers (ASME), and the National Conference of State Legislatures (NCSL). At that time, the management of solid waste was considered a national crisis, because the number of available landfills was decreasing, there was a great deal of concern about the health risks associated with waste incineration, and there was growing opposition to siting new waste management facilities. The crisis mode was exacerbated by such incidents as the ship named *Mobro,* filled with waste, sailing from harbor to harbor and not being allowed to discharge its ever-more-fragrant cargo; a large number of landfills, built with insufficient environmental safeguards, that were placed on the Superfund List; and stories about the carcinogenic effects of emissions from incinerators creating fear among the population.

In the 12 years that have intervened between the time the first edition was written and the preparation of the second edition, solid waste management has achieved a maturity that has removed virtually all fear of it being a crisis. Although the number of landfills is diminishing, larger ones are being built with increased safeguards that prevent leaching or the emission of gases. Improved management of hazardous waste and the emergence of cost-effective integrated waste management systems, with greater emphasis on waste reduction and recycling, have reduced or eliminated most of the previous concerns and problems associated with solid waste management. Improved air pollution control devices on incinerators have proven to be effective, and a better understanding of hazardous materials found in solid waste has led to management options that are considered environmentally acceptable.

While there have been no revolutionary breakthroughs in waste management options, there has been a steady advance in the technologies necessary to handle solid waste materials safely and economically. Thus, the purpose of the second edition of this handbook is to bring the reader up to date on what these options are and how waste can be managed efficiently and cost-effectively. These new technologies have been incorporated in this edition to give the reader the tools necessary to plan and evaluate alternative solid waste management systems and/or programs. In addition to updating all of the chapters, new material has been added on (1) the characteristics of the solid waste stream as it exists now, and how it is likely to develop in the next 10 to 20 years; (2) the collection of solid waste; (3) the handling of construction and demolition wastes; (4) how a modern landfill should be built and managed; and (5) the cost of various waste management systems, so as to enable the reader to make reasonable estimates and comparisons of various waste management options.

The book has been reorganized slightly but has maintained the original sequence of topics, beginning with federal and state legislation in Chapters 2 and 3. Planning municipal solid waste (MSW) programs and the characterization of the solid waste stream are addressed in Chapters 4 and 5, respectively. Methods for reducing both the amount and toxicity of solid waste are discussed in Chapter 6. Chapter 7 is a new chapter dealing with the collection and transport of solid waste. Chapters 8 and 9, which deal with recycling and markets for recycled products, have been revised extensively. Household hazardous waste is discussed in Chapter

10. Special wastes are considered in Chapter 11, with new sections on construction and demolition and electronics and computer wastes. Composting, incineration, and landfilling are documented in Chapters 12, 13, and 14, respectively. Finally, siting and cost estimating of MSW facilities are discussed in Chapters 15 and 16, respectively. Many photographs have been added to the book to provide the reader with visual insights into various management strategies. To make the end-of-chapter references more accessible, they have been reorganized alphabetically. The glossary of terms, given in Appendix A, has been updated to reflect current practice, and conversion factors for transforming U.S. customary units to SI units have also been added.

George Tchobanoglous
Davis, CA

Frank Kreith
Boulder, CO

ABOUT THE EDITORS

George Tchobanoglous is a professor emeritus of civil and environmental engineering at the University of California at Davis. He received a B.S. degree in civil engineering from the University of the Pacific, an M.S. degree in sanitary engineering from the University of California at Berkeley, and a Ph.D. in environmental engineering from Stanford University. His principal research interests are in the areas of solid waste management, wastewater treatment, wastewater filtration, aquatic systems for wastewater treatment, and individual onsite treatment systems. He has taught courses on these subjects at UC Davis for the past 32 years. He has authored or coauthored over 350 technical publications including 12 textbooks and 3 reference books. He is the principal author of a textbook titled *Solid Waste Management: Engineering Principles and Management Issues,* published by McGraw-Hill. The textbooks are used in more than 200 colleges and universities throughout the United States, and they are also used extensively by practicing engineers in the United States and abroad.

Dr. Tchobanoglous is an active member of numerous professional societies. He is a corecipient of the Gordon Maskew Fair Medal and the Jack Edward McKee Medal from the Water Environment Federation. Professor Tchobanoglous serves nationally and internationally as a consultant to governmental agencies and private concerns. He is a past president of the Association of Environmental Engineering Professors. He is consulting editor for the McGraw-Hill book company series in Water Resources and Environmental Engineering. He has served as a member of the California Waste Management Board. He is a Diplomate of the American Academy of Environmental Engineers and a registered Civil Engineer in California.

Frank Kreith is a professor emeritus of engineering at the University of Colorado at Boulder, where he taught in the Mechanical and Chemical Engineering Departments from 1959 to 1978. For the past 13 years, Dr. Frank Kreith served as the American Society of Mechanical Engineers (ASME) legislative fellow at the National Conference of State Legislatures (NCSL), where he provided assistance on waste management, transportation, and energy issues to legislators in state governments. Prior to joining NCSL in 1988, Dr. Kreith was chief of thermal research at the Solar Energy Research Institute (SERI), now the National Renewable Energy Laboratory (NREL). During his tenure at SERI, he participated in the presidential domestic energy review and served as an advisor to the governor of Colorado. In 1983, he received SERI's first General Achievement Award. He has written more than a hundred peer-reviewed articles and authored or edited 12 books.

Dr. Kreith has served as a consultant and advisor all over the world. His assignments included consultancies to Vice Presidents Rockefeller and Gore, the U.S. Department of Energy, NATO, the U.S. Agency for National Development, and the United Nations. He is the recipient of numerous national awards, including the Charles Greeley Abbott Award from the American Solar Energy Society and the Max Jakob Award from ASME-AIChE. In 1992, he received the Ralph Coates Roe Medal for providing technical information to legislators about energy conservation, waste management, and environmental protection, and in 1998 he was the recipient of the prestigious Washington Award for "unselfish and preeminent service in advancing human progress."

CHAPTER 1
INTRODUCTION

George Tchobanoglous
Frank Kreith
Marcia E. Williams

Human activities generate waste materials that are often discarded because they are considered useless. These wastes are normally solid, and the word *waste* suggests that the material is useless and unwanted. However, many of these waste materials can be reused, and thus they can become a resource for industrial production or energy generation, if managed properly. Waste management has become one of the most significant problems of our time because the American way of life produces enormous amounts of waste, and most people want to preserve their lifestyle, while also protecting the environment and public health. Industry, private citizens, and state legislatures are searching for means to reduce the growing amount of waste that American homes and businesses discard and to reuse it or dispose of it safely and economically. In recent years, state legislatures have passed more laws dealing with solid waste management than with any other topic on their legislative agendas. The purpose of this chapter is to provide background material on the issues and challenges involved in the management of municipal solid waste (MSW) and to provide a foundation for the information on specific technologies and management options presented in the subsequent chapters. Appropriate references for the material covered in this chapter will be found in the chapters that follow.

1.1 WASTE GENERATION AND MANAGEMENT IN A TECHNOLOGICAL SOCIETY

Historically, waste management has been an engineering function. It is related to the evolution of a technological society, which, along with the benefits of mass production, has also created problems that require the disposal of solid wastes. The flow of materials in a technological society and the resulting waste generation are illustrated schematically in Fig. 1.1. Wastes are generated during the mining and production of raw materials, such as the tailings from a mine or the discarded husks from a cornfield. After the raw materials have been mined, harvested, or otherwise procured, more wastes are generated during subsequent steps of the processes that generate goods for consumption by society from these raw materials. It is apparent from the diagram in Fig. 1.1 that the most effective way to ameliorate the solid waste disposal problem is to reduce both the amount and the toxicity of waste that is generated, but as people search for a better life and a higher standard of living, they tend to consume more goods and generate more waste. Consequently, society is searching for improved methods of waste management and ways to reduce the amount of waste that needs to be landfilled.

Sources of solid wastes in a community are, in general, related to land use and zoning. Although any number of source classifications can be developed, the following categories have been found useful: (1) residential, (2) commercial, (3) institutional, (4) construction and demolition, (5) municipal services, (6) treatment plant sites, (7) industrial, and (8) agricultural. Typical facilities, activities, or locations associated with each of these sources of waste are reported in Table 1.1. As noted in Table 1.1, MSW is normally assumed to include all community wastes, with the exception of wastes generated by municipal services, water and waste-

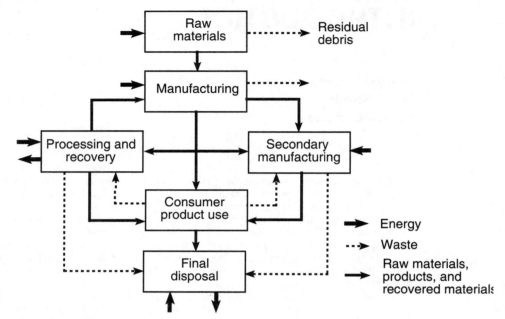

FIGURE 1.1   Flow of materials and waste in an industrial society.

water treatment plants, industrial processes, and agricultural operations. It is important to be aware that the definitions of terms and the classifications of solid waste vary greatly in the literature and in the profession. Consequently, the use of published data requires considerable care, judgment, and common sense.

Solid waste management is a complex process because it involves many technologies and disciplines. These include technologies associated with the control of generation, handling, storage, collection, transfer, transportation, processing, and disposal of solid wastes (see Table 1.2 and Fig. 1.2). All of these processes have to be carried out within existing legal and social guidelines that protect the public health and the environment and are aesthetically and economically acceptable. For the disposal process to be responsive to public attitudes, the disciplines that must be considered include administrative, financial, legal, architectural, planning, and engineering functions. All these disciplines must communicate and interact with each other in a positive interdisciplinary relationship for an integrated solid waste management plan to be successful. This handbook is devoted to facilitating this process.

1.2   ISSUES IN SOLID WASTE MANAGEMENT

The following major issues must be considered in discussing the management of solid wastes: (1) increasing waste quantities; (2) wastes not reported in the national MSW totals; (3) lack of clear definitions for solid waste management terms and functions; (4) lack of quality data, (5) need for clear roles and leadership in federal, state, and local government; (6) need for even and predictable enforcement regulations and standards, and (7) resolution of intercounty, interstate, and intercountry waste issues for MSW and its components. These topics are considered briefly in this section and in the subsequent chapters of this handbook.

TABLE 1.1   Sources of Solid Wastes in a Community

Source	Typical facilities, activities, or locations where wastes are generated	Types of solid wastes
Residential	Single-family and multifamily dwellings; low-, medium-, and high-density apartments; etc.	Food wastes, paper, cardboard, plastics, textiles, leather, yard wastes, wood, glass, tin cans, aluminum, other metal, ashes, street leaves, special wastes (including bulky items, consumer electronics, white goods, yard wastes collected separately, batteries, oil, and tires), and household hazardous wastes
Commercial	Stores, restaurants, markets, office buildings, hotels, motels, print shops, service stations, auto repair shops, etc.	Paper, cardboard, plastics, wood, food wastes, glass, metal wastes, ashes, special wastes (see preceding), hazardous wastes, etc.
Institutional	Schools, hospitals, prisons, governmental centers, etc.	Same as for commercial
Industrial (nonprocess wastes)	Construction, fabrication, light and heavy manufacturing, refineries, chemical plants, power plants, demolition, etc.	Paper, cardboard, plastics, wood, food wastes, glass, metal wastes, ashes, special wastes (see preceding), hazardous wastes, etc.
Municipal solid waste*	All of the preceding	All of the preceding
Construction and demolition	New construction sites, road repair, renovation sites, razing of buildings, broken pavement, etc.	Wood, steel, concrete, dirt, etc.
Municipal services (excluding treatment facilities)	Street cleaning, landscaping, catch-basin cleaning, parks and beaches, other recreational areas, etc.	Special wastes, rubbish, street sweepings, landscape and tree trimmings, catch-basin debris; general wastes from parks, beaches, and recreational areas
Treatment facilities	Water, wastewater, industrial treatment processes, etc.	Treatment plant wastes, principally composed of residual sludges and other residual materials
Industrial	Construction, fabrication, light and heavy manufacturing, refineries, chemical plants, power plants, demolition, etc.	Industrial process wastes, scrap materials, etc.; nonindustrial waste including food wastes, rubbish, ashes, demolition and construction wastes, special wastes, and hazardous waste
Agricultural	Field and row crops, orchards, vineyards, dairies, feedlots, farms, etc.	Spoiled food wastes, agricultural wastes, rubbish, and hazardous wastes

* The term *municipal solid waste* (MSW) is normally assumed to include all of the wastes generated in a community, with the exception of waste generated by municipal services, treatment plants, and industrial and agricultural processes.

Increasing Waste Quantities

As of 2000, about 226 million tons of MSW were generated each year in the United States. This total works out to be over 1600 lb per year per person (4.5 lb per person per day). The amount of MSW generated each year has continued to increase on both a per capita basis and a total generation rate basis. In 1960, per capita generation was about 2.7 lb per person per day and 88 million tons per year. By 1986, per capita generation jumped to 4.2 lb per person per day. The waste generation rate is expected to continue to increase over the current level to a

TABLE 1.2   Functional Elements of a Solid Waste Management System

Functional element	Description
Waste generation	Waste generation encompasses those activities in which materials are identified as no longer being of value and are either thrown away or gathered together for disposal. What is important in waste generation is to note that there is an identification step and that this step varies with each individual. Waste generation is, at present, an activity that is not very controllable.
Waste handling and separation, storage, and processing at the source	Waste handling and separation involve the activities associated with managing wastes until they are placed in storage containers for collection. Handling also encompasses the movement of loaded containers to the point of collection. Separation of waste components is an important step in the handling and storage of solid waste at the source. On-site storage is of primary importance because of public health concerns and aesthetic considerations.
Collection	Collection includes both the gathering of solid wastes and recyclable materials and the transport of these materials, after collection, to the location where the collection vehicle is emptied, such as a materials-processing facility, a transfer station, or a landfill.
Transfer and transport	The functional element of transfer and transport involves two steps: (1) the transfer of wastes from the smaller collection vehicle to the larger transport equipment, and (2) the subsequent transport of the wastes, usually over long distances, to a processing or disposal site. The transfer usually takes place at a transfer station. Although motor vehicle transport is most common, rail cars and barges are also used to transport wastes.
Separation, processing, and transformation of solid waste	The means and facilities that are now used for the recovery of waste materials that have been separated at the source include curbside collection and drop-off and buyback centers. The separation and processing of wastes that have been separated at the source and the separation of commingled wastes usually occurs at materials recovery facilities, transfer stations, combustion facilities, and disposal sites.
	Transformation processes are used to reduce the volume and weight of waste requiring disposal and to recover conversion products and energy. The organic fraction of MSW can be transformed by a variety of chemical and biological processes. The most commonly used chemical transformation process is combustion, used in conjunction with the recovery of energy. The most commonly used biological transformation process is aerobic composting.
Disposal	Today, disposal by landfilling or landspreading is the ultimate fate of all solid wastes, whether they are residential wastes collected and transported directly to a landfill site, residual materials from MRFs, residue from the combustion of solid waste, compost, or other substances from various solid waste processing facilities. A modern sanitary landfill is not a dump. It is a method of disposing of solid wastes on land or within the earth's mantel without creating public health hazards or nuisances.

per capita rate of about 4.6 lb per person per day and an overall rate of 240 million tons per year by 2005. While waste reduction and recycling now play an important part in management, these management options alone cannot solve the solid waste problem. Assuming it were possible to reach a recycling (diversion) rate of about 50 percent, more than 120 million tons of solid waste would still have to be treated by other means, such as combustion (waste-to-energy) and landfilling.

FIGURE 1.2 Views of the functional activities that comprise a solid waste management system: (*a*) waste generation; (*b*) waste handling and separation, storage, and processing at the source; (*c*) collection; (*d*) separation, processing, and transformation of solid waste; (*e*) transfer and transport; and (*f*) disposal.

Waste Not Reported in the National MSW Totals

In addition to the large volumes of MSW that are generated and reported nationally, larger quantities of solid waste are not included in the national totals. For example, in some states waste materials not classified as MSW are processed in the same facilities used for MSW. These wastes may include construction and demolition wastes, agricultural waste, municipal sludge, combustion ash (including cement kiln dust and boiler ash), medical waste, contaminated soil, mining wastes, oil and gas wastes, and industrial process wastes that are not classified as hazardous waste. The national volume of these wastes is extremely high and has been estimated at 7 to 10 billion tons per year. Most of these wastes are managed at the site of generation. However, if even 1 or 2 percent of these wastes are managed in MSW facilities, it can dramatically affect MSW capacity. One or two percent is probably a reasonable estimate.

Lack of Clear Definitions

To date, the lack of clear definitions in the field of solid waste management (SWM) has been a significant impediment to the development of sound waste management strategies. At a fundamental level, it has resulted in confusion as to what constitutes MSW and what processing capacity exists to manage it. Consistent definitions form the basis for a defensible measurement system. They allow an entity to track progress and to compare its progress with other entities. They facilitate quality dialogue with all affected and interested parties. Moreover, what is measured is managed, so if waste materials are not measured they are unlikely to receive careful management attention. Waste management decision makers must give significant attention to definitions at the front end of the planning process. Because all future legislation, regulations, and public dialogue will depend on these definitions, decision makers should consider an open public comment process to establish appropriate definitions early in the strategy development (planning) process.

Lack of Quality Data

It is difficult to develop sound integrated MSW management strategies without good data. It is even more difficult to engage the public in a dialogue about the choice of an optimal strategy without these data. While the federal government and some states have focused on collecting better waste generation and capacity data, these data are still weaker than they should be. Creative waste management strategies often require knowledge of who generates the waste, not just what volumes are generated.

The environmental, health, and safety (EHS) impacts and the costs of alternatives to landfilling and combustion are another data weakness. Landfilling and combustion have been studied in depth, although risks and costs are usually highly site-specific. Source reduction, recycling, and composting have received much less attention. While these activities can often result in reduced EHS impacts compared to landfilling, they do not always. Again, the answer is often site- and/or commodity-specific.

MSW management strategies developed without quality data on the risks and costs of all available options under consideration are not likely to optimize decision making and may, in some cases, result in unsound decisions. Because data are often costly and difficult to obtain, decision makers should plan for an active data collection stage before making critical strategy choices. While this approach may appear to result in slower progress in the short term, it will result in true long-term progress characterized by cost-effective and environmentally sound strategies.

Need for Clear Roles and Leadership in Federal, State, and Local Government

Historically, MSW has been considered a local government issue. That status has become increasingly confused over the past 10 years as EHS concerns have increased and more waste has moved outside the localities where it is generated. At the present time, federal, state, and local governments are developing location, design, and operating standards for waste management facilities. State and local governments are controlling facility permits for a range of issues including air emissions, stormwater runoff, and surface and groundwater discharges in addition to solid waste management. These requirements often result in the involvement of multiple agencies and multiple permits. While product labeling and product design have traditionally been regulated at the federal level, state and local governments have looked increasingly to product labeling and design as they attempt to reduce source generation and increase recycling of municipal waste.

Understandably, the current regulatory situation is becoming increasingly less efficient, and unless there is increased cooperation among all levels of government, the current trends will continue. However, a more rational and cost-effective waste management framework can result if roles are clarified and leadership is embraced. In particular, federal leadership on product labeling and product requirements is important. It will become increasingly unrealistic for multinational manufacturers to develop products for each state. The impact will be particularly severe on small states and on small businesses operating nationally. Along with the federal leadership on products, state leadership will be crucial in permit streamlining. The cost of facility permitting is severely impacted by the time-consuming nature of the permitting process, although a long process does nothing for increased environmental protection. Moreover, the best waste management strategies become obsolete and unimplementable if waste management facilities and facilities using secondary materials as feedstocks cannot be built or expanded. Even source reduction initiatives often depend on major permit modifications for existing manufacturing facilities.

Need for Even and Predictable Enforcement of Regulations and Standards

The public continues to distrust both the individuals who operate waste facilities and the regulators who enforce proper operation of those facilities. One key contributor to this phenomenon is the fact that state and federal enforcement programs are perceived as being understaffed or weak. Thus, even if a strong permit is written, the public lacks confidence that it will be enforced. Concern is also expressed that governments are reluctant to enforce regulations against other government-owned or -operated facilities. Whether these perceptions are true, they are the crucial ones to address if consensus on a sound waste management strategy is to be achieved.

There are multiple approaches which decision makers can consider. They can develop internally staffed state-of-the-art enforcement programs designed to provide a level playing field for all facilities, regardless of type, size, or ownership. If decision makers involve the public in the overall design of the enforcement program and report on inspections and results, public trust will increase. If internal resources are constrained, decision makers can examine more innovative approaches, including use of third-party inspectors, public disclosure requirements for facilities, or separate contracts on performance assurance between the host community and the facility.

Resolution of Intercounty, Interstate, and Intercountry Waste Issues for MSW and Its Components

The movement of wastes across jurisdictional boundaries (e.g., township, county, and state) has been a continuous issue over the past few years, as communities without sufficient local capacity ship their wastes to other locations. While a few receiving communities have welcomed the waste because it has resulted in a significant income source, most receiving communities have felt quite differently. These communities have wanted to preserve their existing capacity, knowing they will also find it difficult to site new capacity. Moreover, they do not want to become dumping grounds for other communities' waste, because they believe the adverse environmental impacts of the materials outweigh any short-term financial benefit.

This dilemma has resulted in the adoption of many restrictive ordinances, with subsequent court challenges. While the current federal legislative framework, embodied in the interstate commerce clause, makes it difficult for any state or local official to uphold state and local ordinances that prevent the inflow of nonlocal waste, the federal legislative playing field can be changed. At this writing, it is still expected that Congress will address the issue in the near future. However, this is a difficult issue in part because of the following concerns:

- Most communities and states export some of their wastes (e.g., medical wastes, hazardous wastes, and radioactive wastes).
- New state-of-the-art waste facilities are costly to build and operate, and they require larger volumes of waste than can typically be provided by the local community in order to cover their costs.
- Waste facilities are often similar in environmental effects to recycling facilities and manufacturing facilities. If one community will not manage wastes from another community, why should one community have to make chemicals or other products which are ultimately used by another community?
- While long-distance transport of MSW (over 200 mi) usually indicates the failure to develop a local waste management strategy, shorter interstate movements (less than 50 mi) may provide the foundation for a sound waste management strategy. Congress should be careful to avoid overrestricting options.

1.3   INTEGRATED WASTE MANAGEMENT

Integrated waste management (IWM) can be defined as the selection and application of suitable techniques, technologies, and management programs to achieve specific waste management objectives and goals. Because numerous state and federal laws have been adopted, IWM is also evolving in response to the regulations developed to implement the various laws. The U.S. Environmental Protection Agency (EPA) has identified four basic management options (strategies) for IWM: (1) source reduction, (2) recycling and composting, (3) combustion (waste-to-energy facilities), and (4) landfills. As proposed by the U.S. EPA, these strategies are meant to be interactive, as illustrated in Fig. 1.3a. It should be noted that the state of California has chosen to consider the management options in a hierarchical order (see Fig. 1.3b). For example, recycling can be considered only after all that can be done to reduce the quantity of waste at the source has been done. Similarly, waste transformation is considered only after the maximum amount of recycling has been achieved. Further, the combustion (waste-to-energy) option has been replaced by waste transformation in California and other states. Interpretation of the IWM hierarchy will, most likely, continue to vary by state. The management options that comprise the IWM are considered in the following discussion. The implementation of integrated waste management options is considered in the following three sections. Typical costs for solid waste management options are presented in Sec. 1.5.

Source Reduction

Source reduction focuses on reducing the volume and/or toxicity of generated waste. Source reduction includes the switch to reusable products and packaging, the most familiar example being returnable bottles. However, bottle bill legislation results in source reduction only if bottles are reused once they are returned. Other good examples of source reduction are grass clippings that are left on the lawn and never picked up and modified yard plantings that do not result in leaf and yard waste. The time to consider source reduction is at the product or process design phase.

Source reduction can be practiced by everybody. Consumers can participate by buying less or using products more efficiently. The public sector (government entities at all levels: local, state, and federal) and the private sector can also be more efficient consumers. They can reevaluate procedures which needlessly distribute paper (multiple copies of documents

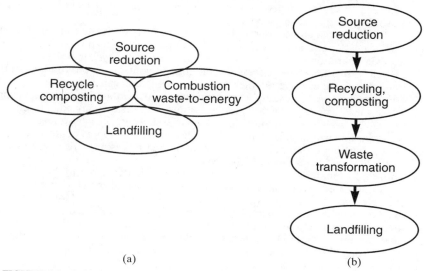

(a)   (b)

FIGURE 1.3 Relationships between the management options comprising integrated waste management: (*a*) interactive, and (*b*) hierarchical.

can be cut back), initiate procedures which require the purchase of products with longer life spans, and cut down on the purchase of disposable products. The private sector can redesign its manufacturing processes to reduce the amount of waste generated in manufacturing. Reducing the amount of waste may require the use of closed-loop manufacturing processes, different raw materials, and/or different production processes. Finally, the private sector can redesign products by increasing their durability, substituting less toxic materials, or increasing product effectiveness. However, while everybody can participate in source reduction, doing so digs deeply into how people go about their business—something that is difficult to mandate through regulation without getting mired in the tremendous complexity of commerce.

Source reduction is best encouraged by making sure that the cost of waste management is fully internalized. *Cost internalization* means pricing the service so that all of the costs are reflected. For waste management, the costs that need to be internalized include pickup and transport, site and construction, administrative and salary, and environmental controls and monitoring. It is important to note that these costs must be considered whether the product is ultimately managed in a landfill, combustion, recycling, or composting facility. Regulation can aid cost internalization by requiring product manufacturers to provide public disclosure of the costs associated with these aspects of product use and development.

Recycling and Composting

Recycling is perhaps the most positively perceived and doable of all the waste management practices. Recycling will return raw materials to market by separating reusable products from the rest of the municipal waste stream. The benefits of recycling are many. Recycling saves precious finite resources; lessens the need for mining of virgin materials, which lowers the environmental impact for mining and processing; and reduces the amount of energy consumed. Moreover, recycling can help stretch landfill capacity. Recycling can also improve the efficiency and ash quality of incinerators and composting facilities by removing noncombustible materials, such as metals and glass.

Recycling can also cause problems if it is not done in an environmentally responsible manner. Many Superfund sites are what is left of poorly managed recycling operations. Examples include operations for newsprint deinking, waste-oil recycling, solvent recycling, and metal recycling. In all of these processes, toxic contaminants that need to be properly managed are removed. Composting is another area of recycling that can cause problems without adequate location controls. For example, groundwater can be contaminated if grass clippings, leaves, or other yard wastes that contain pesticide or fertilizer residues are composted on sandy or other permeable soils. Air contamination by volatile substances can also result.

Recycling will flourish where economic conditions support it, not where it is merely mandated. For this to happen, the cost of landfilling or resource recovery must reflect its true cost—at least \$40 per ton or higher. Successful recycling programs also require stable markets for recycled materials. Examples of problems in this area are not hard to come by; a glut of paper occurred in Germany in 1984 to 1986 due to a mismatch between the grades of paper collected and the grades required by the German paper mills. Government had not worked with enough private industries to find out whether the mills had the capacity and equipment needed to deal with low-grade household newspaper. In the United States, similar losses of markets have occurred for paper, especially during the period from 1994 through 1997. Prices have dropped to the point at which it actually costs money to dispose of collected newspaper in some parts of the country.

Stable markets also require that stable supplies are generated. This supply-side problem has been troublesome in certain areas of recycling, including metals and plastics. Government and industry must work together to address the market situation. It is crucial to make sure that mandated recycling programs do not get too far ahead of the markets.

Even with a good market situation, recycling and composting will flourish only if they are made convenient. Examples include curbside pickup for residences on a frequent schedule and easy drop-off centers with convenient hours for rural communities and for more specialized products. Product mail-back programs have also worked for certain appliances and electronic components.

Even with stable markets and convenient programs, public education is a crucial component for increasing the amount of recycling. At this point, the United States must develop a conservation, rather than a throwaway, ethic, as was done during the energy crisis of the 1970s. Recycling presents the next opportunity for cultural change. It will require moving beyond a mere willingness to collect discarded materials for recycling. That cultural change will require consumers to purchase recyclable products and products made with recycled content. It will require businesses to utilize secondary materials in product manufacturing and to design new products for easy disassembly and separation of component materials.

Combustion (Waste-to-Energy)

The third of the IWM options (see Fig. 1.2) is combustion (waste-to-energy). Combustion facilities are attractive because they do one thing very well—they reduce the volume of waste dramatically, up to ninefold. Combustion facilities can also recover useful energy, either in the form of steam or in the form of electricity. Depending on the economics of energy in the region, this can be anywhere from profitable to unjustified. Volume reduction alone can make the high capital cost of incinerators attractive when landfill space is at a premium, or when the landfill is distant from the point of generation. For many major metropolitan areas, new landfills must be located increasingly far away from the center of the population. Moreover, incinerator bottom ash has promise for reuse as a building material. Those who make products from cement or concrete may be able to utilize incinerator ash.

The major constraints on incinerators are their cost, the relatively high degree of sophistication needed to operate them safely and economically, and the fact that the public is very skeptical concerning their safety. The public is concerned about both stack emissions from incinerators and the toxicity of ash produced by incinerators. The U.S. EPA has addressed both of these concerns through the development of new regulations for solid waste combustion waste-to-energy plants and improved landfill requirements for ash. These regulations will ensure that well-designed, well-built, and well-operated facilities will be fully protective from the health and environmental standpoints.

Landfills

Landfills are the one form of waste management that nobody wants but everybody needs. There are simply no combinations of waste management techniques that do not require landfilling to make them work. Of the four basic management options, landfilling is the only management technique that is both necessary and sufficient. Some wastes are simply not recyclable, because they eventually reach a point at which their intrinsic value is dissipated completely, so they no longer can be recovered, and recycling itself produces residuals.

The technology and operation of a modern landfill can ensure protection of human health and the environment. The challenge is to ensure that all operating landfills are designed properly and are monitored once they are closed. It is crucial to recognize that today's modern landfills do not look like the old landfills that are on the current Superfund list. Today's operating landfills do not continue to take hazardous wastes. In addition, they do not receive bulk liquids. They have gas-control systems, liners, leachate collection systems, extensive groundwater monitoring systems, and perhaps most important, they are better sited and located in the first place to take advantage of natural geological conditions.

Landfills can also turn into a resource. Methane gas recovery is occurring at many landfills today and carbon dioxide recovery is being considered. After closure, landfills can be used for recreation areas such as parks, golf courses, or ski areas. Some agencies and entrepreneurs are looking at landfills as repositories of resources for the future—in other words, today's landfills might be mined at some time in the future when economic conditions warrant. This could be particularly true for *monofills*, which focus on one kind of waste material, such as combustion ash or shredded tires.

Status of Integrated Waste Management

The U.S. EPA has set a national voluntary goal of reducing the quantity of MSW by 25 percent through source reduction and recycling. It should be noted that several states have set higher recycling (diversion) goals. For example, California set goals of 25 percent by the year 1995 and 50 percent by the year 2000. It is estimated that source reduction currently accounts for from 2 to 6 percent of the waste reduction that has occurred. There is no uniformly accepted definition of what constitutes recycling, and estimates of the percentage of MSW that is recycled vary significantly. The U.S. EPA and the Office of Technology Assessment (OTA) have published estimates ranging from 15 to 20 percent. It is estimated that about 5 to 10 percent of the total waste stream is now composted. Today, 50 to 70 percent of MSW is landfilled. Landfill gas is recovered for energy in more than 100 of the nation's larger landfills and most of it is burned with energy recovery.

1.4   IMPLEMENTING INTEGRATED WASTE MANAGEMENT STRATEGIES

The implementation of IWM for residential solid waste, as illustrated in Fig. 1.4, typically involves the use of several technologies and all of the management options discussed previously and identified in Fig. 1.2. At present, most communities use two or more of the MSW management options to dispose of their waste, but there have been only a few instances in

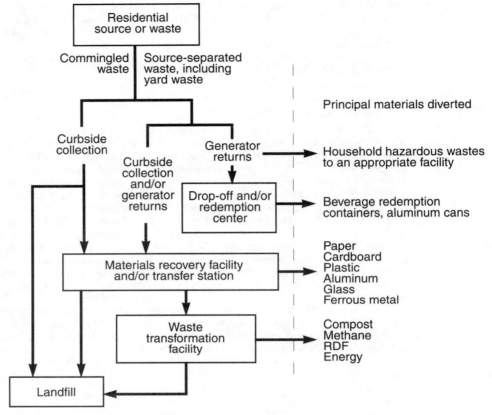

FIGURE 1.4   Flow diagram for residential integrated waste management.

which a truly integrated and optimized waste management plan has been developed. To achieve an integrated strategy for handling municipal waste, an optimization analysis combining all of the available options should be conducted. However, at present, there is no proven methodology for performing such an optimization analysis.

The most common combinations of technologies used to accomplish IWM are illustrated in Fig. 1.5. The most common in the United States is probably Strategy 4, consisting of curbside recycling and landfilling the remaining waste. In rural communities, Strategy 3, consisting of composting and landfilling, is prevalent. In large cities, where tipping fees for landfilling sometimes reach and exceed $100 per ton, Strategy 5, consisting of curbside recycling with the help of a materials recovery facility (MRF), followed by mass burning or combustion at a refuse-derived fuel (RDF) facility and landfilling of the nonrecyclable materials from the MRF and ash from the incinerator, is the most prevalent combination. However, as mentioned previously, each situation should be analyzed individually, and the combination of management options and technologies which fits the situation best should be selected. As a guide to the potential effect of any of the nine strategies in Fig. 1.5 on the landfill space and its lifetime, the required volume of landfill per ton of MSW generated for each of the nine combinations of options is displayed in Fig. 1.6. Apart from availability of landfill volume and space, the cost of the option combinations is of primary concern to the planning of an integrated waste management scheme. Costs are discussed in the following section.

1.5   *TYPICAL COSTS FOR MAJOR WASTE MANAGEMENT OPTIONS*

This section presents typical cost information for the various waste management technologies. More detailed cost information, including the cost of individual components, labor, land, and financing, is presented in Chap. 16. At the outset, it should be noted that the only reliable way to compare the costs of waste management options is to obtain site-specific quotations from experienced contractors. It is often necessary to make some preliminary estimates in the early stages of designing an integrated waste management system.

To assist in such preliminary costing, cost data from the literature for many parts of the country were examined, and published estimates of the capital costs and operating costs for the most common municipal solid waste options (materials recycling, composting, waste-to-energy combustion, and landfilling) were correlated. All of the cost data for the individual options were converted to January 2002 dollars to provide a consistent basis for cost comparisons. The cost data were adjusted using an Engineering News Record Construction Cost Index (ENRCCI) value of 6500.

In addition to the externalized costs presented in this chapter, there are also social costs associated with each of the waste management options. For example, recycling will generate

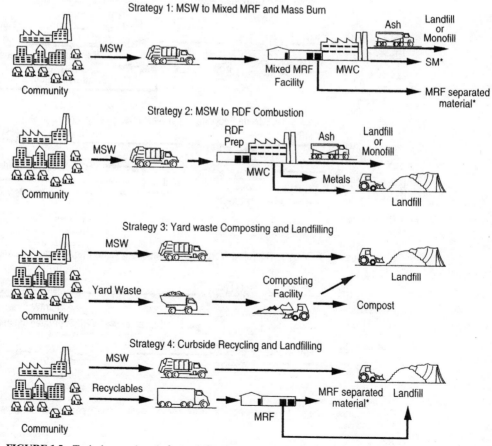

FIGURE 1.5   Typical examples of waste management options for a community.

Strategy 5: Curbside MRF Mass Burn and Ash Landfilling

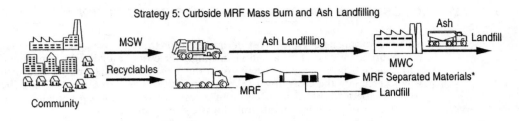

Strategy 6: Curbside Recycling RDF Combustion and Ash Landfilling

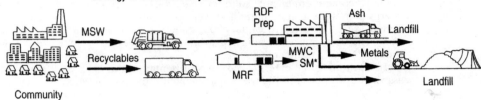

Strategy 7: Curbside Recycling, RDF Composting, and Landfilling

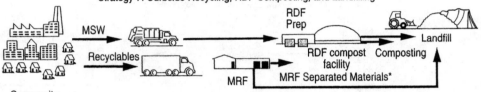

Strategy 8: Curbside Recycling, Yard Waste Composting, and Landfilling

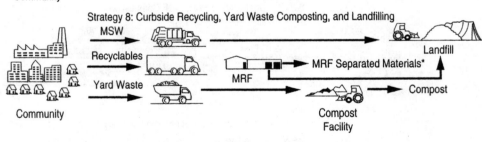

Strategy 9: Curbside Recycling, Yard Waste Composting, Mass Burning and Ash Landfilling

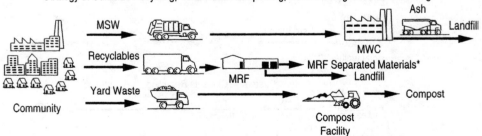

*MRF separates materials for processing by industry. These materials include paper, cardboard, glass, aluminum, steel, and plastics.

FIGURE 1.5   *(Continued)*

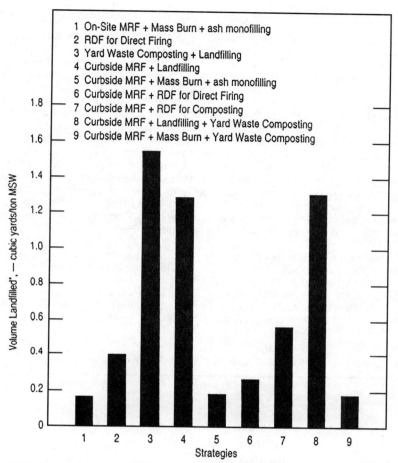

1 On-Site MRF + Mass Burn + ash monofilling
2 RDF for Direct Firing
3 Yard Waste Composting + Landfilling
4 Curbside MRF + Landfilling
5 Curbside MRF + Mass Burn + ash monofilling
6 Curbside MRF + RDF for Direct Firing
7 Curbside MRF + RDF for Composting
8 Curbside MRF + Landfilling + Yard Waste Composting
9 Curbside MRF + Mass Burn + Yard Waste Composting

*Excludes volume required for residue from remanufacturing recyclables.

FIGURE 1.6   Landfill volume required per ton of MSW generated from the waste management options illustrated in Fig. 1.5.

air pollution from the trucks used to pick up, collect, and distribute the materials to be recycled. Many steps in recycling processes, such as deinking newspaper, create pollution whose cost must be borne by society, since it is not a part of the recycling cost. Waste-to-energy combustion creates air pollution from stack emissions and water pollution from the disposal of ash, particularly if heavy metals are present. Landfilling has environmental costs due to leakage of leachates into aquifers and the generation of methane and other gases from the landfill. It has been estimated that 60 to 110 lb of methane will be formed per ton of wet municipal waste during the first 20 years of operation of a landfill. About 9 to 16 lb of that gas will not be recovered, but will leak into the atmosphere because of limitations in the collection system and the permeability of the cover. The U.S. EPA has estimated that about 12 million tons of methane are released from landfills per year in the United States. New regulations, however, will reduce the environmental impact of landfilling in the future.

Capital Costs

It should be noted that capital cost data available in the literature vary in quality, detail, and reliability. As a result, the range of the cost data is broad. Factors which will affect the costs reported are the year when a facility was built, the interest rate paid for the capital, the regulations in force at the time of construction, the manner in which a project was funded (privately or publicly), and the location in which the facility is located. Also, costs associated with ancillary activities such as road improvements, pollution control, and land acquisition greatly affect the results. Cost data on separation, recycling, and composting are scarce and, in many cases, unreliable. Therefore, it is recommended when comparing various strategies to manage MSW, costs for all systems should be built up from system components, using a consistent set of assumptions and realistic cost estimates at the time and place of operation. The most extensive and reliable data available appear to be those for the combustion option. Combustion is a controlled process that is completed within a short period of time and for which there is a good deal of recorded experience. Also, inputs and outputs can be measured effectively with techniques that have previously been used for fossil fuel combustion plants. Typical capital costs for collection vehicles and materials recovery facilities, and for composting, waste-to-energy combustion, and landfilling are presented in Tables 1.3 and 1.4, respectively.

Collection.   Capital costs for collection vehicles are presented in Table 1.3. As reported, vehicle costs will vary from $100,000 to $140,000, depending on the functions and capacity of the vehicle.

Materials Recovery Facilities (MRFs).   The range of capital costs for existing low-tech and high-tech MRFs that sort reusable materials, whether mixed or source-separated, varies from about $10,000 to $40,000 per ton of design capacity per day.

TABLE 1.3   Typical Capital Costs for Waste Collection Vehicles and Materials Recovery Systems

System	Major system components	Cost basis	Cost,* dollars
Waste collection			
Commingled waste	Right-hand stand-up-drive collection vehicle	$/truck	100,000–140,000
	Mechanically loaded collection vehicle	$/truck	115,000–140,000
Source-separated waste	Right-hand stand-up-drive collection vehicle equipped with four separate compartments	$/truck	120,000–140,000
Materials recovery			
Low-mechanical intensity[†]	Processing of source-separated materials only; enclosed building, concrete floors, 1st stage hand-picking stations and conveyor belts, storage for separated and prepared materials for 1 month, support facilities for the workers	$/ton of capacity per day	10,000–20,000
High-mechanical intensity[‡]	Processing of commingled materials or MSW; same facilities as the low-end system plus mechanical bag breakers, magnets, shredders, screens, and storage for up to 3 months; also includes a 2d stage picking line	$/ton of capacity per day	20,000–40,000

* All cost data have been adjusted to an Engineering News-Record Construction Cost Index of 6500.
[†] Low-end systems contain equipment to perform basic material separation and densification functions.
[‡] High-end systems contain equipment to perform multiple functions for material separation, preparation of feedstock, and densification.

TABLE 1.4 Typical Capital Costs for Composting Facilities, Combustion Facilities, and Landfills

System	Major system components	Cost basis	Cost,* dollars
Composting			
Low-end system	Source-separated yard waste feedstock only; cleared, level ground with equipment to turn windrows	$/ton of capacity per day	10,000–20,000
High-end system	Feedstock derived from processing of commingled wastes; enclosed building with concrete floors, MRF processing equipment, and in-vessel composting; enclosed building for curing of compost product	$/ton of capacity per day	25,000–50,000
Waste-to-energy			
Mass burn, field-erected	Integrated system of a receiving pit, furnace, boiler, energy recovery unit, and air discharge cleanup	$/ton of capacity per day	80,000–120,000
Mass burn, modular	Integrated system of a receiving pit, furnace, boiler, energy recovery unit, and air discharge cleanup	$/ton of capacity per day	80,000–120,000
RDF production	Production of fluff and densified refuse-derived fuel (RDF from processed MSW)	$/ton of capacity per day	20,000–30,000
Landfilling			
Commingled waste	Disposal of commingled waste in a modern landfill with double liner and gas recovery system	$/ton of capacity per day	25,000–40,000
Monofill	Disposal of single waste in a modern landfill with double liner and gas recovery system, if required	$/ton of capacity per day	10,000–25,000

* All cost data have been adjusted to an Engineering News-Record Construction Cost Index of 6500.

Composting.   Published capital cost data for MSW composting facilities are limited. As reported in Table 1.4, capital costs for MSW composting facilities are in the range of $10,000 to $50,000 per ton of daily capacity. Further, investment costs show no scale effects (i.e., investment is a linear function of capacity within the capacity range of 10 to 1000 ton/d).

Mass Burn: Field-Erected.   Most field-erected mass burn plants are used to generate electricity. The average size for which useful data are available is 1200 tons/day of design capacity (with a range of 750 to 3000 ton/d). The range of capital cost varies from $80,000 to $120,000 per ton per day. The mass burn facilities were not differentiated by the form of energy produced.

Mass-Burn: Modular.   Modular mass-burn steam and electricity generating plants are typically in the range of 100 to 300 ton/d. The range of capital costs is from $80,000 to $120,000 per ton per day.

Refuse-Derived Fuel (RDF) Facilities.   The range of capital costs for operating RDF production facilities with a processing capacity in the range of 100 to 300 ton/d varies from $20,000 to $30,000 per ton per day (see Table 1.4).

Landfilling.   Landfilling capital costs are difficult to come by, because construction often continues throughout the life of the landfill instead of being completed at the beginning of operations. Consequently, capital costs are combined and reported with operating costs. Capital and operating costs of landfills can be estimated by using cost models, but such models are valid only for a particular region. The range of costs reported in Table 1.4 represents the start-up costs for a new modern landfill that meets all current federal regulations, with a capacity greater than 100 tons/day.

Operation And Maintenance (O&M) Costs.   Along with capital investment, operation and maintenance (O&M) costs are important in making an analysis of integrated waste management systems. Once again, it should be noted that the O&M cost data show large variations. For a reliable estimate, a study of the conditions in the time and place of the project must be made. Operating costs are affected by local differences in labor rates, labor contracts, safety rules, and crew sizes. Accounting systems, especially those used by cities and private owners, and the age of landfills or incinerators can greatly affect O&M costs. Typical O&M costs for collection vehicles and materials recovery facilities, and for composting, combustion, and landfilling are presented in Tables 1.5 and 1.6, respectively.

Collection O&M Costs.   Collection O&M costs, expressed in dollars per ton, are affected by both the number of stops made and the tonnage collected. Typical O&M costs for the collection of commingled wastes with no source separation range from $50 to $70 per ton. Typical O&M costs for the collection of the commingled wastes remaining after source separation of recyclable materials range from $60 to $100 per ton. Costs for curbside collection of source-separated materials vary from $100 to $140 per ton.

MRF O&M Costs.   O&M costs for MRFs range from $20 to $60 per ton of material separated, with a typical value in the range of $40 to $50/ton. The large variation in O&M costs is due, in large part, to inconsistencies in the methods of reporting cost data and not on predictable variations based on the type of technology or the size of the facility. In general, low-technology MRFs have higher operating costs than high-technology MRFs, because of the greater labor intensity of the former.

Composting O&M Costs.   The range of O&M costs for composting processed MSW varies from $30 to $70 per ton. While the capital costs show little or no effect with scale, O&M costs show some decline with plant capacity, but the correlation is quite poor.

Mass Burn: Field-Erected O&M Costs.   Typical O&M cost estimates for field-erected mass burn combustion facilities reported in Table 1.6 are for electricity-only mass burn plants.

TABLE 1.5   Typical Operation and Maintenance Costs for Waste Collection Vehicles and Materials Recovery Systems

System	Major system components	Cost basis	Cost,* dollars
Waste collection			
Commingled waste	Right-hand stand-up-drive collection vehicle	$/ton	60–80
	Mechanically loaded collection vehicle	$/ton	50–70
Source-separated waste	Right-hand stand-up-drive collection vehicle equipped with four separate compartments	$/ton	100–140
Materials recovery			
Low-mechanical intensity[†]	Processing of source-separated materials only; enclosed building, concrete floors, 1st stage hand-picking stations and conveyor belts, storage for separated and prepared materials for 1 month, support facilities for the workers	$/ton	20–40
High-mechanical intensity[‡]	Processing of commingled materials or MSW; same facilities as the low-end system plus mechanical bag breakers, magnets, shredders, screens, and storage for up to 3 months; also includes a 2d stage picking line	$/ton	30–60

* All cost data have been adjusted to an Engineering News-Record Construction Cost Index of 6500.
† Low-end systems contain equipment to perform basic material separation and densification functions.
‡ High-end systems contain equipment to perform multiple functions for material separation, preparation of feedstock, and densification.

TABLE 1.6 Typical Operation and Maintenance Costs for Composting Facilities, Combustion Facilities, and Landfills

System	Major system components	Cost basis	Cost,* dollars
Composting			
Low-end system	Source-separated yard waste feedstock only; cleared, level ground with equipment to turn windrows	$/ton	20–40
High-end system	Feedstock derived from processing of commingled wastes; enclosed building with concrete floors, MRF processing equipment, and in-vessel composting; enclosed building for curing of compost product	$/ton	30–50
Waste-to-energy			
Mass burn, field-erected	Integrated system of a receiving pit, furnace, boiler, energy recovery unit, and air discharge cleanup	$/ton	40–80
Mass burn, modular	Integrated system of a receiving pit, furnace, boiler, energy recovery unit, and air discharge cleanup	$/ton	40–80
RDF production	Production of fluff and densified refuse-derived fuel (RDF from processed MSW)	$/ton	20–40
Landfilling			
Commingled waste	Disposal of commingled waste in a modern landfill with double liner and gas recovery system	$/ton	10–120
Monofill	Disposal of single waste in a modern landfill with double liner and gas recovery system, if required	$/ton	10–80

* All cost data have been adjusted to an Engineering News-Record Construction Cost Index of 6500.

O&M costs range from $60 to $80 per ton. O&M costs for plants producing steam and electricity are about the same.

Mass Burn: Modular O&M Costs.   Typical O&M costs for modular mass burn combustion range from $40 to $80 per ton. Although the capital costs are sometimes lower for the steam-only plants, the O&M costs are not. Typical tipping fees for the steam-and-electricity plants and for steam-only plants range from $50 to $60/ton and $40 to $50/ton, respectively.

RDF Facility O&M Costs.   Typical O&M costs for RDF facilities is about $40 per ton of MSW processed, with a range of $20 to $40 per ton. Note that the averages cited previously are based on wide ranges and the number of data points is small. Hence, these averages are only a rough estimate of future RDF facility costs.

Landfilling O&M Costs.   Available data are few and indicate wide variability in landfill costs as a result of local conditions. Some cost data reflect capital recovery costs that others do not. The O&M costs for MSW landfills range from $10 to over $120 per ton. The cost range for monofills varies from $10 to $80 per ton of ash.

1.6 FRAMEWORK FOR DECISION MAKING

The preceding sections present information on the four waste management options—source reduction, recycling, waste-to-energy, and landfilling. With that material as a background, we must map out a framework for making decisions. In a world without economic constraints, the tools for waste management could be ordered by their degree of apparent environmental desirability. Source reduction would clearly be at the top, as it prevents waste from having to be managed at all. Recycling, including composting, would be the next-best management tool, because it can return resources to commerce after the original product no longer serves its intended purpose. Waste-to-energy follows because it is able to retrieve energy that otherwise would be buried and wasted. Finally, landfilling, while often listed last, is really not any better or worse than incineration, as it too can recover energy. Moreover, waste-to-energy facilities still require landfills to manage their ash.

In reality, every community and region will have to customize its integrated management system to suit its environmental situation and its economic constraints. A small, remote community such as Nome, Alaska, has little choice but to rely solely on a well-designed and -operated landfill. At the other extreme, New York City can easily and effectively draw on some combination of all the elements of the waste management hierarchy. Communities that rely heavily on groundwater that is vulnerable, such as Long Island, New York, and many Florida communities, usually need to minimize landfilling and look at incineration, recycling, and residual disposal in regions where groundwater is less vulnerable. Communities that have problems with air quality usually avoid incineration to minimize more atmospheric pollutants. Sometimes these communities can take extra steps to ensure that incineration is acceptable by first removing metals and other bad actors out of the waste stream. In all communities, the viability of recycling certain components of the waste stream is linked to volumes, collection costs, available markets, and the environmental consequences of the recycling and the reuse operations.

Planning for Solid Waste Management

Long-term planning at the local, state, and even regional level is the only way to come up with a good mix of management tools. It must address both environmental concerns and economic constraints. As discussed earlier, planning requires good data. This fact has long been recognized in fields such as transportation and health-care planning. However, until recently, databases for solid waste planning were not available, and, even now, they are weak.

There are a number of guidelines that planners should embrace. First, it is crucial to look to the long term. The volatility of today's spot market prices is a symptom of the crisis conditions in which new facilities are simply not being sited. Examples already exist of locations where current prices are significantly reduced from their highs as new capacity options have emerged.

Second, planners must make sure that all costs are reflected in each option. Municipal accounting practices sometimes hide costs. For example, the transportation department may purchase vehicles while another department may pay for real estate, and so on. Accurate accounting is essential.

Third, skimping on environmental controls brings short-term cost savings with potentially greater liability down the road. It is always better to do it right the first time, especially for recycling and composting facilities, as well as for incineration facilities and landfills.

Fourth, planners should account for the volatility of markets for recyclables. The question becomes: In a given location for a given commodity, can a recycling program survive the peaks and valleys of recycling markets without going broke in between?

Fifth, planners must consider the availability of efficient facility permitting and siting for waste facilities using recycled material inputs, and for facilities which need permit changes to implement source reduction.

Finally, planners should look beyond strictly local options. When political boundaries are not considered, different management combinations may become possible at reasonable costs. Potential savings can occur in the areas of procurement, environmental protection, financing, administration, and ease of implementation. Regional approaches include public authorities, nonprofit public corporations, special districts, and multicommunity cooperatives.

Formulating an Integrated Solid Management Waste Strategy

The process of formulating a good integrated solid waste management strategy is time-consuming and difficult. Ultimately, the system must be holistic; each of its parts must have its own purpose and work in tandem with all the other pieces like a finely crafted, highly efficient piece of machinery. Like a piece of machinery, it is unlikely that an efficient and well-functioning output is achieved unless a single design team, understanding its objective and working with suppliers and customers, develops the design. The successful integrated waste plan drives legislation; it is not driven by legislation. More legislation does not necessarily lead to more source reduction and recycling. In fact, disparate pieces of legislation or regulation can work at cross purposes. Moreover, the free market system works best when there is some sense of stability and certainty, which encourages risk taking because it is easier to predict expected market response. The faster a holistic framework for waste management is stabilized, the more likely public decision making will obtain needed corporate investment.

The first stage of planning involves carefully defining terminology, including what wastes are covered, what wastes are not covered, and what activities constitute recycling and composting. It also requires the articulation of clear policy goals for the overall waste management strategy. Is the goal to achieve the most cost-effective strategy that is environmentally protective or to maximize diversion from landfills? There are no absolutely right or wrong answers. However decision makers should share the definitions, key assumptions, and goals with the public for their review and comment.

The second stage involves identification of the full range of possible options and the methodical collection of environmental risks and costs associated with each option. Data collection is best done before any strategy has been selected. The cost estimates for recycling and composting can be highly variable depending on what assumptions are made about market demand and what actions are taken to stimulate markets. These differing assumptions about markets can also impact the assumptions on environmental risks, since some types of reuse scenarios have more severe environmental impacts than others. The stringency of the regulatory permitting and enforcement programs that set and enforce standards for each type of waste management facility, including recycling facilities and facilities that use recycled material inputs in the manufacturing process, will also impact the costs and environmental risks associated with various options. Finally, the existence of product standards for recycled materials will impact the costs and risks of various recycling and composting strategies. The costs of all management strategies will be volume-dependent. Once this information is collected, the public should have the opportunity for meaningful input on the accuracy of the assumptions. Acceptance by the public at this stage can foster a smoother and faster process in the long run.

The final step involves examining the tradeoffs between available options so that an option or package of options can be selected. At the core, these tradeoffs involve risk and cost comparisons. However, they also involve careful consideration of implementation issues such as financing, waste volumes, enforcement, permit time frames, siting issues, and likely future behavior changes.

Some examples of implementation issues are useful. Pay-by-the-bag disposal programs may result in less garbage because people really cut back on their waste generation when they can save money. On the other hand, there has been some indication that pay-by-the-bag systems have actually resulted in the same amount of garbage generated, but an increase in burning at home or illegal dumping. Another example is the need to assess the real effect of bottle bills. Bottle bills may be very effective if collected bottles are reused, or if markets exist so that the collected bottles can be recycled. However, in some locations bottle bills result in a double payment—once to collect the bottles and then again to landfill the bottles because no viable market strategy is in place. A final example concerns flow control. Flow control promises a way to ensure that each of the various solid waste facilities has enough waste to run efficiently. On the other hand, if governments use flow control to send a private generator's waste to a poorly designed or operated solid waste facility, then the government may be tampering with the generator's Superfund liability, or the government may increase the amount of waste that is going to an environmentally inferior facility.

Some computerized decision models have been developed which compare the costs of various strategies. However, these often require considerable tailoring before they accurately fit a local situation. It is often useful to develop a final strategy in an iterative manner by first selecting one or two likely approaches and then setting the exact parameters of the selected approach in a second iteration. Public involvement is critical throughout the selection process.

It may also be useful to develop an integrated waste management strategy by formulating a series of generator-specific strategies. Residential generators are one group of MSW generators. Another important group includes the public sector, including municipalities and counties, who generate their own waste streams. Finally, there are numerous specific industry groups such as the hotel industry, the restaurant industry, petrochemical firms, the pulp and paper industry, and the grocery industry. In each case, the character of the solid waste generated will vary. For some groups, all the waste will fall into the broad category of MSWs. For other groups, much of the waste will include industrial, agricultural, or other non-MSW waste. In some cases, the variation within the generator category will be significant. In other cases, the within-group waste characterization is likely to be relatively uniform. Industry-by-industry strategies, focusing on the largest waste generator categories, may result in more implementable and cost-effective strategies.

1.7 KEY FACTORS FOR SUCCESS

Arriving at successful solid waste management solutions requires more than just good planning. The best technical solution may fail if politicians and government officials do not consider a series of other important points. This section attempts to identify some of these points.

Credibility for Decision Makers

It is absolutely crucial to work to protect the credibility of those individuals who must ultimately make the difficult siting and permitting decisions. Proper environmental standards for all types of facilities, including recycling, can help give decision makers necessary support. Credible enforcement that operates on a level playing field is also crucial. Operator certification programs, company-run environmental audit programs, company-run environmental excellence programs, government award programs for outstanding facilities, and financial assurance provisions can also increase the public's level of comfort with solid waste management facilities. Finally, clear-cut siting procedures and dispute resolution processes can provide decision makers with a crucial support system.

Efficient Implementation Mechanisms Including Market Incentives

A number of things can be done to help facilitate program implementation. Expedited permitting approaches for new facilities and expedited permit modification approaches for existing facilities can be helpful. Approaches such as class permits or differential requirements based on the complexity of the facilities are examples. Pilot programs can be particularly helpful in determining whether a program which looks good on paper will work well in real life.

Much of today's federal and state legislation and regulation has focused on a command-and-control strategy. Such a strategy relies on specified mandates that cover all parties with the same requirements. These requirements are developed independent of market concepts and other basic business incentives, and, as a result, these approaches are often slower and more expensive to implement for both the regulated and the regulators.

Some of the most efficient implementation mechanisms involve the consideration of market incentives. Market approaches can significantly cut the cost of achieving a fixed amount of environmental protection, energy conservation, or resource conservation when compared with a traditional command-and-control approach. The concept behind this approach is simple. Determine what total goal is needed. Then, let those who can achieve the goal most cost-effectively do so. They can sell extra credits to those who have a more difficult time meeting the goal. Other market approaches rely on using market pricing to strongly encourage desirable behaviors.

Another incentive-type program which has achieved major environmental benefits in a cost-effective way has been the implementation of the Federal Emergency Planning and Community Right-to-Know Act (1986) emissions reporting program. The law does not mandate specific reductions in emissions to air, water, and land. However, it does require affected facilities to publicly report quantities of chemicals released. The mere fact of having to publicly report has resulted in a dramatic lowering of emissions.

The types of programs which decision makers at the state or federal level could examine include the following:

- An overall program to reduce average per capita waste generation rates through the use of a marketable permit program. There would be several ways to implement this type of program. A fixed per capita figure could be established throughout a state. Whichever municipalities (or counties) could achieve it most efficiently could sell extra credits to other affected municipalities. Other alternatives would set the per capita rate by size of municipality or require all municipalities to achieve a fixed percent reduction from established baseline rates.

- A marketable permit program to implement recycling goals. Rather than require all townships, municipalities, and counties to achieve the same recycling rates, let those municipalities and counties that can achieve the recycling rates most cost-effectively sell any extra credits to other affected parties.

- A program that would develop differential business tax rates based on the amount of recycling (or source reduction) which the company achieves. The tax rates could be based on fixed rate standards (for example, source reduction of 10 percent or recycling of 25 percent) or percentage improvements over a baseline year.

- A program that would develop differential property taxes for homes that recycle or reduce their disposed waste by a given percentage. The percentage could be increased gradually each year in order to maintain the tax break.

- Product and service procurement preferences for those companies who have high overall recycling rates or who utilize a high percentage of secondary materials.

- Differential business tax rates or permit priority for companies who use recycled material inputs in production processes or who buy large quantities of recycled materials for consumption.

- Differential water rates for companies who use large volumes of compost or who reduce their green waste.

- Information disclosure requirements that require certain types and sizes of businesses to provide the public with information on their waste generation rates, their recycling rates, their procurement of secondary materials, and their waste management methods. (Good examples would be hotels and other types of consumer businesses.) The state could also compile state average values by industry group and require that these rates be posted along with the company-specific rates.

Significant Attention on Recycling Markets

Recycling will not be sustainable in the long term unless it is market-driven, so that there is a market demand for secondary materials. The market incentive discussion provides some ideas as to how market incentives can be utilized broadly to drive desirable integrated waste strategies by influencing the behavior of affected entities. Some of these behaviors may lead to the creation of market demand for specific secondary materials. However, it is also important to examine secondary material markets on a commodity-specific basis, particularly in the subset of materials that compose a large fraction of the MSW stream.

There is a wide range of policy choices that can impact market demand. These include commodity-specific procurement standards, entity-specific procurement plans, equipment tax credits, tax credits for users of secondary materials, mandated use of secondary materials for certain government-controlled activities (such as landfill cover or mine reclamation projects), use of market development mechanisms in enforcement settlements, recycled content requirements for certain commodities, manufacturer take-back systems, virgin material fees, and labeling requirements. Whether any of these actions are needed, and if so, which ones, can be determined only after a careful analysis of each commodity.

If such actions are needed, two cautions are in order. First, it is often better to discuss the need for market strengthening with affected parties before mandating a specific result. If, after a fixed time frame, the market does not improve, a regulated outcome can be automatically implemented. That hammer often provides the needed impetus for action without regulatory involvement. Second, while the first six program examples of market demand approaches can be implemented at the federal or state level, the last four examples of market demand approaches are best implemented at the federal level.

Public Involvement

As mentioned previously, the best technical solution is unlikely to work unless the public is active in helping to reach the final choice of options. Public involvement must be just that—it cannot be a one-way street; rather, the public must be involved in two-way discussions. There must be a give and take on the final solution. Included in this dialogue must be a serious discussion about the tradeoff between risk reductions and cost. This public involvement is best done with multiple opportunities for both formal and informal inputs.

Continuous Commitment to High-Quality Operations for All Facilities

Today's solid waste solutions require a commitment to high-quality operations. In the past, solid waste management, as with many other government services, was often awarded to the lowest bidder. This approach needs to be seriously reconsidered, given the environmental liabilities associated with poorly managed solid waste.

Evaluation of the Effectiveness of the Chosen Strategy

In developing specific legislation and regulations, it is important that the full impact of individual legislative or regulatory provisions be monitored after the program has been implemented. MSW planning is a process, not a project. That process must continually ensure that the plan mirrors reality and that implementation obstacles are addressed expeditiously.

1.8   PHILOSOPHY AND ORGANIZATION OF THIS HANDBOOK

The philosophy of this handbook is that the integrated waste management approach is not a hierarchical scheme, but is integrative in nature, as shown in Fig. 1.3a. In other words, an appropriate design of a toxicity reduction and/or recycling program which removes heavy metals from the waste stream—in particular lead, mercury, and cadmium—should not be considered merely a reduction or recycling function, because it also assists the waste-to-energy incineration function that benefits from the absence of heavy metals and batteries. Recycling is not a complete process unless the legal and institutional framework can create markets for the recycled products that can beneficially utilize the materials picked up from the curb. The technical and engineering aspects of waste management cannot function in a vacuum; decision makers must be aware of the political and social ramifications of their action.

Another philosophic underpinning of this book is that there is no single prescription for an integrated waste management program that will work successfully in every instance. Each situation must be analyzed on its own merit—an appropriate integrated waste management plan must be developed from hard data, social attitudes, and the legal framework that must be taken into account. The waste management disposal field is in a constant state of flux, and appropriate solutions should be innovative, as well as technically and economically sound.

The organization of this handbook reflects the realities of the situation, as well as the philosophy of its editorship. Chapter 2, "Federal Role in Municipal Solid Waste Management," deals with federal laws and regulations that impact the different solid waste management schemes. It should be noted, however, that the federal role has diminished over the last few years, because the federal government has passed authority and responsibility for waste management to the states; many states, in turn, have passed their responsibilities on to municipalities. Chapter 3, "Solid Waste State Legislation," provides an overview of the state legislation within the framework of which any waste management plan needs to be devised. Chapter 4, "Planning Municipal Solid Waste Management Programs," contains a discussion of how to plan an integrated municipal waste management program. Chapter 5, "Solid Waste Stream Characteristics," contains background data and information on what constitutes waste today and a projection of what it will consist of later in the twenty-first century.

Chapters 6 to 14 are devoted to the major technologies for an integrated waste management scheme: source reduction in Chap. 6, collection and transport of solid waste in Chap. 7, recycling in Chap. 8, products and markets for recyclable materials in Chap. 9, household hazardous wastes in Chap. 10, other special wastes in Chap. 11, composting in Chap. 12, waste-to-energy combustion in Chap. 13, and landfilling in Chap. 14.

It is clear, though, that irrespective of what combination of technologies is employed in a waste management scheme, new facilities will have to be sited and financed. The old approach, when technical experts determined the best location for a waste management site, then announced their decision and defended it to the public, is no longer accepted or acceptable. The confrontational results of the "decide, announce, and defend" strategy must be replaced by an interactive procedure in which the public participates in the siting process as a full partner. Therefore, Chap. 15 is devoted to a recommended procedure for siting of MSW facilities. Chapter 16 considers financing and life-cycle analysis for waste management facilities.

Appendixes contain a glossary of terms, conversion factors, a list of organizations active in solid waste management, and a list of state offices responsible for waste disposal in each state. Lists of companies that can provide technical help and equipment in the development and operation of an integrated solid waste management scheme are integrated into the chapters on specific technologies.

1.9   CONCLUDING REMARKS

The technologies to handle solid waste economically and safely are available today. This handbook describes each of them and provides a framework to coordinate them into an integrated system. However, the approach takes cognizance that the responsible disposal of solid waste is not merely an engineering problem, but involves sociological and political factors. The public supports waste reduction, reuse, recycling, and, sometimes, composting, on the assumption that these measures are environmentally benign, economical, energy conserving, and capable of solving the waste management problem. There is often, however, a lack of understanding of the limitations of these measures and the meaning of the terms. Waste reduction is not merely a matter of using less or reducing the amount of packaging. For example, doubling the life of the tires on automobiles cuts the number of tires that need to be disposed of in half. Deposit laws on bottles and batteries can be used to encourage reuse and recycling, as well as waste reduction. The public usually ignores the fact that recycling has technical and market limitations, becomes more expensive as the percentage of the waste recycled increases, and also has adverse environmental impacts.

There is general agreement that waste reduction, reuse, and recycling should be supported within their respective technical and economic limits. However, even if current source reduction and recycling efforts are successful, the amount of waste that must be disposed of in the year 2010 will be as much as or more than that today, as a consequence of increased waste generation and growth. Consequently, the need to site additional facilities for composting, waste-to-energy combustion, and landfilling will continue. But these technically obvious requirements for waste disposal often create opposition from the public that is politically difficult to resolve. These are factors that cannot be treated adequately in a technical handbook, but they must be kept in mind when devising an IWM strategy.

It should also be pointed out that the technologies for waste disposal are in a state of flux. New and more efficient methods are being introduced. Better equipment to control air and water pollution is being developed. Materials that increase the life of a product and thereby reduce the production of waste are becoming available. More economical methods of recycling and composting are being tried in pilot projects. For the waste management professional it is, therefore, important to keep up with the current literature. Table 1.7 lists some professional publications that describe the state of the art and present new developments.

To establish responsible IWM systems, public education programs must be developed to convince everyone to accept responsibility. Once the problem becomes a shared responsibility, people can find solutions by working together. Failure to site new facilities, more than any-

TABLE 1.7 Some Professional Journals for Solid Waste Management Issues

Solid Waste and Power HCI Publications 910 Archibald Street Kansas City, MO 64111-3046 (816) 931-1311 Fax (816) 931-2015	*Solid Waste Management* UIC, School of Public Health 2121 West Taylor Street Chicago, IL 60612-7260 (312) 996-8944	*Journal of Air and Waste Management Association* P.O. Box 2861 Pittsburgh, PA 15230 (412) 232-3444 Fax (412) 232-3450
Resource Recycling P.O. Box 10540 Portland, OR 97210 (800) 227-1424 Fax (503) 227-6135	*Waste Tech News* 131 Madison St. Denver, CO 80206 (303) 394-2905 Fax (303) 394-3011	*Energy from Biomass & Waste* U.S. Department of Energy, OSTI P.O. Box 62 Oak Ridge, TN 37831 (615) 576-1168
Bio-Cycle P.O. Box 37 Marysville, PA 17053 (717) 957-4195	*MSW Management* 216 East Gutierrez Santa Barbara, CA 93101 (805) 899-3355 Fax (805) 899-3350	*Public Works* P.O. Box 688 Ridgewood, NJ 07451 (201) 445-5800 Fax (201) 445-5170
Waste Age 1730 Rhode Island Avenue, NW Washington, DC 20036 (202) 861-0708 Fax (202) 659-0925	*Household Hazardous Waste Management News* 16 Haverville St. Andover, MA 01810 (508) 470-3044	*Resource Recovery Report* 5313 38th St. NW Washington, DC 20015 (202) 362-6034

thing else, can create a crisis situation that often leads to the implementation of less than optimal solutions, with serious later consequences. Engineers and regulators must work with the public to find acceptable sites for MSW facilities. The public must understand that waste-to-energy combustion produces electric power with less environmental impact than fossil fuel power plants and, at the same time, reduces the amount of waste that needs to be landfilled. Technically trained people should build continuous program evaluation into waste management plans and share information to improve the process in the future. Engineers, politicians, and the waste management industry will have to work to win the confidence of the public, so that technical solutions will be accepted and implemented. Because technologies for managing waste safely are now available, this handbook provides the information and data needed to implement programs to manage the wastes generated by an industrial society in a manner that protects public health and safety and the environment.

CHAPTER 2
FEDERAL ROLE IN MUNICIPAL SOLID WASTE MANAGEMENT

Barbara Foster
Edward W. Repa

There is a plethora of regulations at the federal level that impact the management and disposal of municipal solid waste. Many of these regulations have been in effect for years and further updates to this chapter will be needed as the regulations are updated. This chapter covers only the more important regulations that have been developed over the past few years that affect solid waste management. Readers are encouraged to refer to the environmental statutes for a complete list of laws and regulations affecting the industry or contact any of the organizations listed in App. D.

These regulations have been authorized by numerous pieces of legislation and include the following:

- Resource Conservation and Recovery Act (RCRA)
- Clean Air Act (CAA)
- Clean Water Act (CWA)
- Federal Aviation Administration (FAA) guidelines
- Flow control implications (court cases)

2.1 RESOURCE CONSERVATION AND RECOVERY ACT

On October 21, 1976, Congress passed the Resource Conservation and Recovery Act (RCRA), which has since been amended numerous times. RCRA for the first time divided the management of waste into two main categories: (1) Subtitle C—Hazardous Waste, and (2) Subtitle D—Non-Hazardous Waste. RCRA directed the U.S. Environmental Protection Agency (EPA) to promulgate criteria within a year for determining which facilities should be classified as sanitary landfills and which should be classified as open dumps.

- *Section 239.* Specifies requirements that state permit programs must meet to be determined adequate by the EPA. The section also specifies the procedure that EPA will follow in determining the adequacy of state Subtitle D permit programs or other systems of prior approval and conditions required to be adopted and implemented by states under RCRA Sec. 4005(c)(1)(B). Nothing in the section precludes a state from adopting or enforcing requirements that are more stringent or more extensive than those required under Sec. 239 or from operating a permit program or other system of prior approval and conditions with more stringent requirements or a broader scope of coverage that are required under Sec. 239. If EPA determines that a state Subtitle D permit program is inadequate, EPA will have the authority to enforce the Subtitle D federal revised criteria on the RCRA Sec. 4010(c) regulated facilities under the state's jurisdiction. The state must have compliance monitoring authority and enforcement authority and must provide for the intervention in civil enforcement proceedings.

- *Section 240.*   Specifies guidelines for the thermal processing of solid wastes in terms of defining the solid wastes accepted, the solid wastes excluded, site selection and general design, water quality, air quality, vectors, aesthetics, residue, safety, general operations, and record keeping.
- *Section 243.*   Provides guidelines for the storage and collection of residential, commercial, and institutional solid waste. The guidelines cover storage, design, safety, collection equipment, collection frequency, and collection management.
- *Section 244.*   Provides guidelines for solid waste management for beverage containers.
- *Section 246.*   Provides guidelines for source separation and materials recovery.
- *Section 247.*   Provides comprehensive procurement guidelines for products containing recovered materials.
- *Section 254.*   Specifies requirements for prior notice of citizen suits.
- *Section 255.*   Contains identification of regions and agencies for solid waste management.
- *Section 256.*   Provides guidelines for development and implementation of state solid waste regulations.

40 CFR Part 257 Regulations

The regulations promulgated on September 13, 1979, by EPA are contained in 40 CFR Part 257, "Criteria for Classification of Solid Waste Disposal Facilities and Practices." The act was later amended by the Hazardous and Solid Waste Amendments of 1984 (HSWA). Under HSWA, EPA was directed to develop minimum criteria for solid waste management facilities that may receive household hazardous waste or small quantity hazardous waste exempted from the Subtitle C requirements. The EPA promulgated these criteria on October 9, 1991, under 40 CFR Part 258, "Criteria for Municipal Solid Waste Landfills."

The Part 257 criteria, developed in 1979, remain in effect for all the disposal facilities that accept nonhazardous waste except for municipal solid waste landfills (MSWLFs) subject to the revised criteria contained in Part 258. A MSWLF is defined under RCRA as:

> A discrete area of land or an excavation that receives household waste, and that is not a land application unit, surface impoundment, injection well, or waste pile, as those terms are defined in this section. A MSWLF unit also may receive other types of RCRA Subtitle D wastes, such as commercial solid waste, nonhazardous sludge, and industrial solid waste. Such a landfill may be owned publicly or privately. A MSWLF may be a new MSWLF unit or a lateral expansion.

Furthermore, RCRA defines household waste as follows:

> Any solid waste (including garbage, trash, and sanitary waste in septic tanks) derived from households (including single and multiple residences, hotels and motels, bunkhouses, ranger stations, crew quarters, campgrounds, picnic grounds, and day-use recreation areas).

These definitions are important because they form the basis of the Subtitle D program. The origin of the waste, whether it was derived from a household or not, will determine the applicable regulations that must be complied with when the waste is landfilled.

The disposal of most nonhazardous solid waste occurs in landfills that meet the Part 257 criteria. These criteria apply to all nonhazardous waste streams except the following:

- Household wastes (as previously defined)
- Sewage sludge
- Municipal solid waste incinerator ash

- Agricultural wastes
- Overburden resulting from mining operations
- Nuclear wastes

The federal criteria contained in this part are minimal and encompass seven general areas. The requirements of each of these areas is described here.

Floodplains.   Facilities or practices in floodplains must not restrict the flow of the base flood, reduce the temporary water storage capacity of the floodplain, or result in the washout of solid waste that could pose a hazard to human life, wildlife, or land or water resources.

Endangered Species.   Facilities or practices must not cause or contribute to the taking of any endangered or threatened species or result in the destruction or adverse modification of the critical habitat of endangered or threatened species.

Surface Water.   Facilities or practices must not cause a point-source or non-point-source discharge of pollutants into waters of the United States in violation of the Clean Water Act. Also, facilities are not allowed to cause a discharge of dredged materials or fill materials into waters in violation of the CWA.

Groundwater.   A facility or practice is prohibited from contaminating an underground drinking water source beyond the solid waste boundary or an alternative boundary established in court. An alternative boundary can be established only if the change would not result in contamination of groundwater that may be needed or used for human consumption.

Disease Vectors.   The facility or practice shall not exist or occur unless the on-site propagation of disease vectors is minimized through the periodic application of cover material or other techniques as appropriate to protect public health. Specific areas of concern are the land application of sewage sludge and septic tank sludge.

Air.   Facilities are prohibited from engaging in the open burning of solid waste. The requirement does not apply to the infrequent burning of agricultural wastes, silvicultural wastes from forest management practices, land clearing debris, diseased trees, debris from emergency cleanup operations, and ordnance. Also, a facility is not permitted to violate any applicable requirements of a State Implementation Plan (SIP) developed under the Clean Air Act.

Safety.   The areas covered by the safety requirements include the control of explosive gases, prevention of fires, control of bird hazards, and control of public access. Facilities are required to control the concentration of explosive gases generated by the operation so that they do not exceed 25 percent of the lower explosive limit (LEL) for the gas in facility structures and LEL at the property boundary. Fires are to be controlled so that they do not pose a hazard to the safety of persons or property.

Facilities disposing of putrescible wastes that may attract birds and are located within 10,000 ft of an airport used by turbojet aircraft or 5000 ft of an airport used by piston-type aircraft must be operated in a manner so as not to pose a bird hazard to aircraft. Finally, the facility must not allow uncontrolled public access, which would expose the public to potential health and safety hazards at the disposal site.

Facilities or practices that fail to meet the requirements established in Part 257 are considered to be open dumps under RCRA. States are responsible for developing and enforcing programs within their jurisdiction to ensure that all applicable facilities are complying with the criteria. Operations not in compliance can be sued under the citizen suit provisions of RCRA.

40 CFR Part 258 Regulations

On October 9, 1991, EPA promulgated its long-awaited regulations for municipal solid waste landfills (MSWLFs). These regulations are contained in 40 CFR Part 258. Both existing and new MSWLFs were affected by the rules that became effective on October 9, 1993. Subtitle D set forth *performance-based* minimum criteria in the following six areas:

- Location restrictions
- Operations
- Design
- Groundwater monitoring and corrective action
- Closure and postclosure care
- Financial assurance

Each of these criteria is summarized in detail in the following sections.

Structure of Rule.   The structure of Subtitle D allows for the requirements to be self-implementing or for states to receive approval from EPA to implement and enforce its provisions. In those states that do not receive or seek approval, the owner or operator is responsible for ensuring that the facility is in compliance with all provisions of the rule. Owners and operators must document compliance and make this documentation available to the state on request. Enforcement of the Subtitle D criteria in unapproved states can occur through the citizen suit provisions of RCRA, or, if EPA finds the state program wholly inadequate, the agency itself may provide enforcement.

To encourage states to adopt Subtitle D of Part 258, EPA added greater flexibility for approved states to specify alternative requirements and schedules. These flexibilities are discussed in further detail in later sections as they occur in context of the rule.

Applicability.   The revised Subtitle D criteria apply to any landfill that accepts household waste (as previously defined), including those that receive sewage sludge or municipal waste combustion ash. The criteria in Part 258 do not affect landfills that accept only industrial and special waste and construction and demolition waste.

The requirements vary, depending on the landfill's closure date, according to the following breakdown:

- The rule does not apply to any MSWLF that ceased receipt of waste prior to October 9, 1991.
- Only the final cover requirements apply to any MSWLF that ceases receipt of waste after October 9, 1991, but before October 9, 1993.
- The entire Subtitle D requirements apply to any MSWLF that receives waste on or after October 9, 1993.

In addition, a small landfill exemption is available in two cases for owners and operators of landfills disposing of less than 20 ton/d and where there is no evidence of existing groundwater contamination. These exemptions are as follows:

Alaska provision.   Landfills located in areas where there is an annual interruption of at least three consecutive months of surface transportation that prevents access to a regional facility.

Arid provision.   Landfills located in areas that annually receive less than 25 in of precipitation and where no practicable waste management alternative exists. In addition, a landfill must dispose of less than 20 tons of solid waste that is actually buried, based on an annual average.

MSWLFs meeting the preceding requirements can be exempted from the design and corrective action requirements contained in these criteria. However, if an owner or operator has been exempted and receives knowledge of groundwater contamination, the exemption no longer applies and the landfill must comply with the design and corrective action requirements.

Location Restrictions.   The criteria establish restrictions or bans on locating and operating new and existing MSWLFs in six unsuitable areas. Restrictions regarding airports and floodplains existed in the Part 257 criteria, while the remainder are new to Part 258. The restricted areas in the final rule include:

- *Airports.*   New, existing, and lateral expansions to existing MSWLFs that are located within 10,000 ft of an airport runway end used by turbojet aircraft or within 5000 ft of a runway end used by piston-type aircraft must demonstrate that they are designed and operated so as not to pose a bird hazard to aircraft. Also, new MSWLFs and lateral expansions at existing MSWLFs within a 5-mi radius of any airport runway end must notify the affected airport and the Federal Aviation Administration (FAA).

- *Floodplains.*   New, existing, and lateral expansions at MSWLFs located in the 100-year floodplain must demonstrate that they will not restrict the flow of the 100-year flood, reduce the temporary water storage capacity of the floodplain, or result in washout of solid waste so as to pose a hazard to human health and the environment.

- *Wetlands.*   New and lateral expansions of MSWLFs into wetlands are prohibited in unapproved states. However, an approved state may allow siting of a landfill, provided it can be demonstrated that a practicable alternative is not available; construction and operation will not violate any other local, state, or federal law or cause or contribute to significant degradation of the wetland; and steps have been taken to achieve no net loss of wetlands. Other concerns are that a landfill must not cause or contribute to violations of any applicable state water quality standard; violate any applicable toxic effluent standard or prohibition under Sec. 307 of the Clean Water Act; jeopardize the continued existence of endangered or threatened species or result in the destruction or adverse modification of a critical habitat protected under the Endangered Species Act of 1973; or violate any requirement under the Marine Protection Research and Sanctuaries Act of 1972 for the protection of a marine sanctuary. The MSWLF must address the erosion, stability and migration potential of native wetland soils, moods, and deposits used to support the MSWLF unit; erosion stability and migration of potential of dredge fill materials used to support the MSWLF unit; the volume and chemical nature of the waste managed in the MSWLF unit; impacts on fish, wildlife, and other aquatic resources and their habitat from release of the solid waste; the potential effects of catastrophic release of waste to the wetland and the resulting impacts on the environment; and any additional factors as necessary to demonstrate that ecological resources in the wetland are protected sufficiently.

- *Fault areas.*   New and lateral expansions to MSWLFs are prohibited within 200 ft of a fault that has had displacement in Holocene time (i.e., last 9000 years). Approved states may allow an alternative setback of less than 200 ft if it can be demonstrated that the structural integrity of the MSWLF will not be damaged and it will be protective of human health and the environment.

- *Seismic impact zones.*   New and lateral expansions of MSWLFs are prohibited in seismic impact zones, unless it can be demonstrated to an approved state or tribe that all containment structures, including liners, leachate collection systems, and surface water control systems, are designed to resist the maximum horizontal acceleration in lithified earth material for the site.

- *Unstable areas.*   New, existing, and lateral expansions of MSWLFs located in unstable areas are required to demonstrate engineering measures that have been incorporated into the MSWLF design that ensure that the integrity of the structural components will not be disrupted. Unstable areas can include poor foundation conditions, areas susceptible to mass movements, and karst terranes. *Unstable area, structural components, poor foundation conditions, areas susceptible to mass movement,* and *karst terranes* are all defined.

Closure of Existing Municipal Solid Waste Landfill Units.   Existing MSWLFs that cannot make the demonstration pertaining to airports, floodplains, or unstable areas are required to close (according to the closure provisions) within 3 years of the effective date (i.e., October 9, 1996). This deadline can be extended up to 2 years in approved states if the facility can demonstrate that there is no alternative disposal capacity and no immediate threat to human health and the environment. Owners and operators of MSWLFs should be aware that a state in which their landfill is located or is to be located may have adopted a state wellhead protection program in accordance with Sec. 1428 of the Safe Drinking Water Act. Such state wellhead protection programs may impose requirements on owners or operators of MSWLFs in addition to those set forth in 40 CFR Chap. 1.

Operating Criteria.   The revised Subtitle D criteria imposed 10 operating requirements on MSWLFs. Of the 10, five were carryovers from the Part 257 criteria. The operating criteria are as follows:

1. *Procedures for excluding the receipt of hazardous waste.*   Landfill owners and operators are required to implement a program for detecting and preventing the disposal of regulated hazardous wastes and PCBs at their facilities. The program must include random inspections of incoming loads, maintaining records of inspections, training facility personnel to recognize regulated hazardous wastes and PCB wastes, and notification procedures to the regulatory authority if such wastes are discovered.

2. *Cover material requirements.*   MSWLFs must be covered with 6 in of earthen material at the end of each operating day, or more frequently, to control disease vectors, fires, odor, blowing litter, and scavenging. States with approved programs may allow alternative materials that meet or exceed the performance standard and may grant temporary waivers from the requirements when extreme seasonal climatic conditions make the requirements impractical.

3. *Disease vector control.*   Landfills are required to prevent or control on-site populations of disease vectors using techniques appropriate for the protection of human health and the environment.

4. *Explosive gas control.*   Landfill owners and operators must ensure through a routine methane monitoring program that the concentration of methane gas generated by the facility does not exceed 25 percent of the lower explosive limit in facility structures and does not exceed the lower explosive limit at the property boundary. If the landfill exceeds these limits, a remediation program must be implemented immediately.

5. *Air criteria.*   Landfills are required to meet applicable requirements developed under a State Implementation Plan pursuant to Sec. 110 of the Clean Air Act. Also, open burning of solid waste is prohibited except for the infrequent burning of agricultural wastes, silvicultural wastes, land-clearing debris, diseased trees, and debris from emergency cleanup operations.

6. *Access requirements.*   Owners and operators of MSWLFs must control public access and prevent unauthorized traffic and illegal dumping of wastes through the use of artificial barriers, natural barriers, or both.

7. *Runoff/run-on controls.*   MSWLFs must design, construct and maintain a run-on and runoff control system on the active portion of the landfill capable of handling the peak discharge from a 24-hour, 25-year storm event.

8. *Surface water requirements.*   MSWLFs must not cause a discharge of pollutants into waters, including wetlands, violating any requirements of the Clean Water Act, including the National Pollutant Discharge Elimination System (NPDES), or cause a nonpoint source of pollution to waters that violates any requirement of an areawide or statewide water quality management plan approved under Secs. 208 or 319 of the Clean Water Act.

9. *Liquid restrictions.* MSWLFs are prohibited from accepting bulk and noncontainerized liquid wastes unless the waste is a household waste other than septic water waste. Leachate and gas condensate derived from a MSWLF unit can be recirculated back into that unit as long as the unit is designed with a composite liner.

10. *Record keeping requirements.* Owners and operators are required to record and retain any location restriction demonstrations; inspection records, training procedures, and notification procedures; gas monitoring results; design documentation for gas condensate and leachate recirculation; monitoring, testing, and analytical data; and closure and postclosure care plans. They are also required to keep any cost estimates and financial assurance documentation required by Subpart G and any information used to demonstrate compliance with small community exemptions as required by Part 258.1(f)(2).

Design Criteria. The design criteria establish a specific engineering design for those states that are not approved and an alternative design standard based on performance that approved states can allow. The Part 258 design criteria are as follows:

- *Unapproved states.* A composite liner with a leachate collection system that is capable of maintaining less than 30-cm (12-in) depth of leachate over the liner must be installed. The composite liner system must consist of two components: an upper component composed of a flexible membrane liner (FML) at least 30 mil thick and a lower component composed of at least 2 ft of compacted soil with a hydraulic conductivity of no more than 1×10^{-7} cm/s. If the FML component consists of high-density polyethylene (HDPE), its thickness must be at least 60 mil. Alternative designs meeting the performance standard below are allowed in unapproved states but require that a demonstration be made to the unapproved state and that the state review and petition EPA for concurrence.

- *Approved states.* A design that ensures the concentration of 24 organic and inorganic constituents in the uppermost aquifer at the point of compliance does not exceed maximum contaminant levels (MCLs). The point of compliance can range from the waste management unit boundary to 150 m from the boundary, depending on local hydrogeologic conditions.

These new design standards apply to new MSWLFs and to new units and lateral expansions to existing MSWLFs. Existing units are not required to be retrofitted with liners and leachate collection systems.

Groundwater Monitoring and Corrective Action. The groundwater monitoring program requirements apply to all MSWLF units unless the owner or operator can demonstrate that no potential for migration of hazardous constituents from the unit to the uppermost aquifer exists during the active life of the unit and the postclosure care period.

As with the design criteria, the compliance schedule for the groundwater monitoring provisions of the rule are or can be different, depending on whether the state has an approved program. The following compliance schedule applies to existing units and lateral expansions in unapproved states:

- Units less than 1 mi from a drinking water intake (surface or subsurface)—October 9, 1994
- Units greater than 1 mi but less than 2 mi from a drinking water intake—October 9, 1995.
- Units greater than 2 mi from a drinking water intake—October 9, 1996

New MSWLF units are required to be in compliance before waste can be placed in the unit.

States with approved programs can adopt an alternative schedule. This schedule must ensure that 50 percent of all existing MSWLF units and lateral expansions are in compliance by October 9, 1994, and all units are in compliance by October 9, 1996.

The groundwater monitoring program for an MSWLF consists of four steps:

Groundwater Monitoring System.   The first step is the establishment of a groundwater monitoring system. The regulations require that a system be installed that has a sufficient number of wells, installed at appropriate locations and depths, to yield groundwater samples from the uppermost aquifer. The system must include wells that represent the quality of background groundwater that has not been affected by leakage from the unit, as well as groundwater that has passed under and is downgradient of the unit.

Detection Monitoring Program.   The second step is the establishment of a detection monitoring program. The monitoring program consists of semiannual monitoring of wells for both water quality and head levels during the active life of the facility through the postclosure care period. The minimum detection monitoring program includes the monitoring of 15 heavy metals and 47 volatile organics (Table 2.1). A minimum of four independent samples from each well must be collected and analyzed during the first semiannual sampling event. During subsequent sampling events, at least one sample from each well must be collected and analyzed for the preceding parameters.

Approved states are permitted to establish an alternative list of inorganic indicator parameters (i.e., in lieu of some or all of the heavy metals) as long as it provides a reliable indication of inorganic releases from the landfill to the groundwater. Also, an approved state can specify an appropriate alternative frequency for repeated sampling and analysis during the active life and postclosure care period. However, the sampling frequency cannot be less than annual during the active life and closure period.

Results of sampling events must be analyzed statistically to determine if a statistically significant increase over background has occurred for one or more of the monitored constituents. A number of statistical methods are permitted to analyze the data collected. If a statistically significant increase occurs over background, the owner or operator must establish an assessment monitoring program (i.e., the next step in the sequence) and demonstrate that a source other than the landfill is the cause, or show that the increase is the result of sampling, analysis, or statistical error or natural variation.

Assessment Monitoring Program.   An assessment monitoring program is required whenever a statistically significant increase over background has been detected for one or more of the constituents in the detection monitoring program. Within 90 days of triggering an assessment monitoring program, and annually thereafter, an owner or operator is required to sample and analyze groundwater for some 213 organic and inorganic constituents (Table 2.2). A minimum of one sample from each downgradient well must be analyzed during this sampling event. For any constituents detected, a minimum of four independent samples from each well are required to be collected and analyzed to establish background for these constituents.

After the initial sampling, owners and operators must sample all wells twice a year for detection monitoring parameters and those assessment monitoring parameters that were detected in the first assessment sampling. Also, all 213 assessment monitoring parameters must be sampled and analyzed annually.

The owner or operator is required to establish a groundwater protection standard for any constituents found in the assessment monitoring program. The standard must be based on an MCL, background concentration established during the assessment monitoring program for those constituents for which an MCL has not been promulgated, or background concentration for constituents where the background level is higher than the MCL.

States with approved programs are permitted to establish a number of alternative standards in the assessment monitoring program. Also, approved states are allowed to specify an appropriate subset of wells to be sampled and analyzed, delete any monitoring parameters that are not reasonably expected to be derived from the waste contained in the unit, specify an appropriate alternative frequency for repeated sampling and analysis, and establish an alternative groundwater protection standard based on an appropriate health-based level.

TABLE 2.1 Constituents for Detection Monitoring*

Common name[†]	CAS RN[‡]	Common name[†]	CAS RN[‡]
Inorganic constituents:		35. 1,1-Dichloroethylene;	
1. Antimony	Total	1,1-dichloroethene; vinylidene	
2. Arsenic	Total	chloride	75-35-4
3. Barium	Total	36. cis-1,2-Dichloroethylene;	
4. Beryllium	Total	cis-1,2-dichloroethene	156-59-2
5. Cadmium	Total	37. trans-1,2-Dichloroethylene;	
6. Chromium	Total	trans-1,2-dichloroethene	156-60-5
7. Cobalt	Total	38. 1,2-Dichloropropane;	
8. Copper	Total	propylene dichloride	78-87-5
9. Lead	Total	39. cis-1,3-Dichloropropene	10061-01-5
10. Nickel	Total	40. trans-1,3-Dichloropropene	10061-02-6
11. Selenium	Total	41. Ethylbenzene	100-41-4
12. Silver	Total	42. 2-Hexanone; methyl	
13. Thallium	Total	butyl ketone	591-78-6
14. Vanadium	Total	43. Methyl bromide;	
15. Zinc	Total	bromomethane	74-83-9
Organic constituents:		44. Methyl chloride;	
16. Acetone	67-64-1	chloromethane	74-87-3
17. Acrylonitrile	107-13-1	45. Methylene bromide;	
18. Benzene	71-43-2	dibromomethane	74-95-3
19. Bromochloromethane	74-97-5	46. Methylene chloride;	
20. Bromodichloromethane	75-27-4	dichloromethane	75-09-2
21. Bromoform; tribromomethane	75-25-2	47. Methyl ethyl ketone;	
22. Carbon disulfide	75-15-0	MEK; 2-butanone	78-93-3
23. Carbon tetrachloride	56-23-5	48. Methyl iodide; iodomethane	74-88-4
24. Chlorobenzene	108-90-7	49. 4-Methyl-2-pentanone;	
25. Chloroethane;		methyl isobutyl ketone	108-10-1
ethyl chloride	75-00-3	50. Styrene	100-42-5
26. Chloroform;		51. 1,1,1,2-Tetrachloroethane	630-20-6
trichloromethane	67-66-3	52. 1,1,2,2-Tetrachloroethane	79-34-5
27. Dibromochloromethane;		53. Tetrachloroethylene; tetra-	
chlorodibromomethane	124-48-1	chloroethene; perchloro-	
28. 1,2-Dibromo-3-chloropro-		ethylene	127-18-4
pane; DBCP	96-12-8	54. Toluene	108-88-3
29. 1,2-Dibromoethane;		55. 1,1,1-Trichloroethane;	
ethylene dibromide; EDB	106-93-4	methylchloroform	71-55-6
30. o-Dichlorobenzene; 1,2-		56. 1,1,2-Trichloroethane	79-00-5
dichlorobenzene	95-50-1	57. Trichloroethylene; trichloro-	
31. p-Dichlorobenzene;		ethene	79-01-6
1,4-dichlorobenzene	106-46-7	58. Trichlorofluoromethane;	
32. trans-1,4-Dichloro-2-butene	110-57-6	CFC-11	75-69-4
33. 1,1-Dichloroethane;		59. 1,2,3-Trichloropropane	96-18-4
ethylidene chloride	75-34-3	60. Vinyl acetate	108-05-4
34. 1,2-Dichloroethane;		61. Vinyl chloride	75-01-4
ethylene dichloride	107-06-2	62. Xylenes	1330-20-7

* This list contains 47 volatile organics for which possible analytical procedures are provided in EPA Report SW-846 "Test Methods for Evaluating Solid Waste," November 1986, as revised December 1987, includes Method 8260; and 15 metals for which SW-846 provides either Method 6010 or a method from the 7000 series of methods.
† Common names are those widely used in government regulations, scientific publications, and commerce; synonyms exist for many chemicals.
‡ Chemical Abstracts Service registry number. Where "Total" is entered, all species in the ground water that contain this element are included.

TABLE 2.2 List of Hazardous Inorganic and Organic Constituents[1]

Common name[2]	CAS RN[3]	Chemical abstracts service index name[4]	Suggested methods[5]	POL,[6] µg/L
Acenaphthene	83-32-9	Acenaphthylene, 1,2-dihydro-	8100	200
			8270	10
Acenaphthylene	208-96-8	Acenaphthylene	8100	200
			8270	10
Acetone	67-64-1	2-Propanone	8260	100
Acetonitrile; methyl cyanide	75-05-8	Acetonitrile	8015	100
Acetophenone	98-86-2	Ethanone, 1-phenyl-	8270	10
2-Acetylaminofluorene; 2-AAF	53-96-3	Acetamide, N-9H-fluoren-2-yl-	8270	20
Acrolein	107-02-8	2-Propenal	8030	5
			8260	100
Acrylonitrile	107-13-1	2-Propenenitrile	8030	5
			8260	200
Aldrin	309-00-2	1,4:5,8-Dimethanonaphthalene, 1,2,3,4,10,10-hexachloro-1,4,4a,5,8,8a-hexahydro-(1α,4α,4aβ,5α,8α,8aβ)-	8080	0.05
			8270	10
Allyl chloride	107-05-1	1-Propene, 3-chloro-	8010	5
			8260	10
4-Aminobiphenyl	92-67-1	[1,1'-Biphenyl]-4-amine	8270	20
Anthracene	120-12-7	Anthracene	8100	200
			8270	10
Antimony	Total	Antimony	6010	300
			7040	2000
			7041	30
Arsenic	Total	Arsenic	6010	500
			7060	10
			7061	20
Barium	Total	Barium	6010	20
			7080	1000
Benzene	71-43-2	Benzene	8020	2
			8021	0.1
			8260	5
Benz[a]anthracene; benzanthracene	56-55-3	Benz[a]anthracene	8100	200
			8270	10
Benzo[b]fluoranthene	205-99-2	Benz[e]acephenanthrylene	8100	200
			8270	10
Benzo[k]fluoranthene	207-08-9	Benzo[k]fluoranthene	8100	200
			8270	10
Benzo[ghi]perylene	191-24-2	Benzo[ghi]perylene	8100	200
			8270	10
Benzo[a]pyrene	50-32-8	Benzo[a]pyrene	8100	200
			8270	10
Benzyl alcohol	100-51-6	Benzenemethanol	8270	20
Beryllium	Total	Beryllium	6010	3
			7090	50
			7091	2

Common name	Chemical name	CAS No.	Method	Value
beta-BHC	Cyclohexane, 1,2,3,4,5,6-hexachloro-, (1α,2β,3α,4β,5α,6β)-	319-85-7	8270	10
			8080	0.05
delta-BHC	Cyclohexane, 1,2,3,4,5,6-hexachloro-, (1α,2α,3α,4β,5α,6β)-	319-86-8	8270	20
			8080	0.1
gamma-BHC; lindane	Cyclohexane, 1,2,3,4,5,6-hexachloro-, (1α,2α,3β,4α,5α,6β)-	58-89-9	8270	20
			8080	0.05
Bis[2-chloroethoxy]methane	Ethane, 1,1'-[methylenebis(oxy)]bis(2 chloro-	111-91-1	8270	20
			8110	5
Bis(2-chloroethyl) ether; dichloroethyl ether	Ethane, 1,1'-oxybis[2-chloro-	111-44-4	8270	10
			8110	3
Bis-(2 chloro-1-methylethyl) ether; 2,2'-dichlorodiisopropyl ether; DCIP[7]	Propane, 2,2'-oxybis[1-chloro-	108-60-1	8270	10
			8110	10
Bis(2-ethylhexyl) phthalate	1,2-Benzenedicarboxylic acid, bis(2-ethylhexyl) ester	117-81-7	8270	10
			8060	20
Bromochloromethane; chlorobromomethane	Methane, bromochloro-	74-97-5	8021	0.1
			8260	5
Bromodichloromethane; dibromochloromethane	Methane, bromodichloro-	75-27-4	8010	1
			8021	0.2
			8260	5
Bromoform; tribromomethane	Methane; tribromo-	75-25-2	8010	5
			8021	2
			8260	15
4-Bromophenyl phenyl ether	Benzene, 1-bromo-4-phenoxy-	101-55-3	8110	25
			8270	10
Butyl benzyl phthlate; benzyl butyl phthalate	1,2-Benzenedicarboxylic acid, butyl phenylmethl ester	85-68-7	8060	5
			8270	10
Cadmium	Cadmium	Total	6010	40
			7130	50
			7131	1
Carbon disulfide	Carbon disulfide	75-15-0	8260	100
Carbon tetrachloride	Methane, tetrachloro-	56-23-5	8010	1
			8021	0.1
			8260	10
Chlordane	4,7-Methano-1H-indene, 1,2,4,5,6,7,8,8-octochloro-2,3,3a,4,7,7a-hexahydro-	See Note 8	8080	0.1
			8270	50
p-Chloroaniline	Benzenamine, 4-chloro-	106-47-8	8270	20
Chlorobenzene	Benzene, chloro-	108-90-7	8010	2
			8020	2
			8021	0.1
			8260	5
Chlorobenzilate	Benzeneacetic acid, 4-chloro-α-(4-chlorophenyl-α-hydroxy-, ethyl ester	510-15-6	8270	10
p-Chloro-m-cresol; 4-chloro-3-methylphenol	Phenol, 4-chloro-3-methyl-	59-50-7	8040	5
			8270	20
Chloroethane; ethyl chloride	Ethane, chloro-	75-00-3	8010	5
			8021	1
			8260	10

TABLE 2.2 List of Hazardous Inorganic and Organic Constituents[1] (Continued)

Common name[2]	CAS RN[3]	Chemical abstracts service index name[4]	Suggested methods[5]	PQL,[6] µg/L
Chloroform; trichloromethane	67-66-3	Methane, trichloro-	8010	0.5
			8021	0.2
			8260	5
2-Chloronaphthalene	91-58-7	Naphthalene, 2-chloro	8120	10
			8270	10
2-Chlorophenol	95-57-8	Phenol, 2-chloro-	8040	5
			8270	10
4-Chlorophenyl phenyl ether	7005-72-3	Benzene, 1-chloro-4-phenoxy-	8110	40
			8270	10
Chloroprene	126-99-8	1,3-Butadiene, 2-chloro-	8010	50
			8260	20
Chromium	Total	Chromium	6010	70
			7190	500
			7191	10
Chrysene	218-01-9	Chrysene	8100	200
			8270	10
Cobalt	Total	Cobalt	6010	70
			7200	500
			7201	10
Copper	Total	Copper	6010	60
			7210	200
			7211	10
m-Cresol, 3-methylphenol	108-39-4	Phenol, 3-methyl-	8270	10
o-Cresol, 2-methylphenol	95-48-7	Phenol, 2-methyl-	8270	10
p-Cresol, 4-methylphenol	106-44-5	Phenol, 4-methyl-	8270	10
Cyanide	57-12-5	Cyanide	9010	200
2,4-D; 2,4-dichlorophenoxyacetic acid	94-75-7	Acetic acid, (2,4-dichlorophenoxy)-	8150	10
4,4'-DDD	72-54-8	Benzene 1,1'-(2,2-dichloroethylidene)bis-[4-chloro-	8080	0.1
			8270	10
4,4'-DDE	72-55-9	Benzene, 1,1'-(dichloroethyenylidene)bis[4-chloro-	8080	0.05
			8270	10
4,4'-DDT	50-29-3	Benzene, 1,1'-(2,2,2-trichloroethylidene)bis[4-chloro-	8080	0.1
			8270	10
Diallate	2303-16-4	Carbamothioic acid, bis(1-methylethyl)-,S-(2,3-dichloro-2-propenyl) ester	8270	10
Dibenz[a,h]anthracene	53-70-3	Dibenz[a,h]anthracene	8100	200
			8270	10
Dibenzofuran	132-64-9	Dibenzofuran	8270	10
Dibromochloromethane; chlorodibromomethane	124-48-1	Methane, dibromochloro-	8010	1
			8021	0.3
			8260	5
1,2-Dibromo-3-chloropropane; DBCP	96-12-8	Propane, 1,2-dibromo-3-chloro-	8011	0.1
			8021	30
			8260	25

Name	CAS No.	Synonym	Method	Detection limit
Di-n-butyl phthlate	84-74-2	1,2-Benzenedicarboxylic acid, dibutyl ester	8021	10
			8260	5
			8060	5
			8270	10
o-Dichlorobenzene; 1,2-dichlorobenzene	95-50-1	Benzene, 1,2-dichloro-	8010	2
			8020	5
			8021	0.5
			8120	10
			8260	5
			8270	10
m-Dichlorobenzene; 1,3-dichlorobenzene	541-73-1	Benzene, 1,3-dichloro-	8010	5
			8020	5
			8021	0.2
			8120	10
			8260	5
			8270	10
p-Dichlorobenzene; 1,4-dichlorobenzene	106-46-7	Benzene, 1,4-dichloro-	8010	2
			8020	5
			8021	0.1
			8120	15
			8260	5
			8270	10
3,3¹-Dichlorobenzidine	91-94-1	[1,1¹-Biphenyl]-4,4¹-diamine, 3,3¹-dichloro-	8270	20
trans-1,4-Dichloro-2-butene	110-57-6	2-Butene, 1,4-dichloro, (E)-	8260	100
Dichlorodifluoromethane; CFC 12	75-71-8	Methane, dichlorodifluoro-	8021	0.5
			8260	5
1,1-Dichloroethane; ethyldidene chloride	75-34-3	Ethane, 1,1-dichloro-	8010	1
			8021	0.5
			8260	5
1,2-Dichloroethane; ethylene dichloride	107-06-2	Ethane, 1,1-dichloro-	8010	0.5
			8021	0.3
			8260	5
1,1-Dichloroethylene; 1,1-dichloroethene; vinylidene chloride	75-35-4	Ethene, 1,1-dichloro-	8010	1
			8021	0.5
			8260	5
cis-1,2-Dichloroethylene; cis-1,2-dichloroethene	156-59-2	Ethene, 1,2-dichloro-, (Z)-	8021	0.2
			8260	5
trans-1,2-Dichloroethylene trans-1,2-dichloroethene	156-60-5	Ethene, 1,2-dichloro-, (E)-	8010	1
			8021	0.5
			8260	5
2,4-Dichlorophenol	120-83-2	Phenol, 2,4-dichloro-	8040	5
			8270	10
2,6-Dichlorophenol	87-65-0	Phenol, 2,6-dichloro-	8270	10
1,2-Dichloropropane; propylene dichloride	78-87-5	Propane, 1,2-dichloro-	8010	0.5
			8021	0.05
			8260	5

TABLE 2.2 List of Hazardous Inorganic and Organic Constituents[1] (*Continued*)

Common name[2]	CAS RN[3]	Chemical abstracts service index name[4]	Suggested methods[5]	PQL,[6] µg/L
1,3-Dichloropropane; trimethylene dichloride	142-28-9	Propane, 1,3-dichloro-	8021 8260	0.3 5
2,2-Dichloropropane; isopropylidene chloride	594-20-7	Propene, 2,2-dichloro-	8021 8260	0.5 15
1,1-Dichloropropene	563-58-6	1-Propene, 1,1-dichloro-	8021 8260	0.2 5
cis-1,3-Dichloropropene	10061-01-5	1-Propene, 1,3-dichloro-, (Z)	8010 8260	20 10
trans-1,3-Dichloropropene	10061-02-6	1-Propene, 1,3-dichloro-, (E)-	8010 8260	5 10
Dieldrin	60-57-1	2,7,3,6-Dimethanonaphth[2,3-b]oxirene,3,4,5,6,9,9-hexa-chloro-1a,2,2a,3,6,6a,7,7a-octahydro-, (1aα,2β,2aα,3β,6β,6aα,7β,7aα)-	8080 8270	0.05 10
Diethyl phthlate	84-66-2	1,2-Benzenedicarboxylic acid, diethyl ester	8060 8270	5 10
0,0-Diethyl 0-2-pyrazinyl phosphoro-thioate; thionazin	297-97-2	Phosphorothioic acid, 0,0-diethyl 0-pyrazinyl ester	8141 8270	5 20
Dimethoate	60-51-5	Phosphorodithioic acid, 0,0-dimethyl S-[2-(methylamino)-2-oxyethyl] ester	8141 8270	3 20
p-(Dimethylaminc)azobenzene	60-11-7	Benzenamine, N,N-dimethyl-4-(phenylazo)-	8270	10
7,12-Dimethylbenz[a]anthracene	57-97-6	Benz[a]anthracene, 7,12-dimethyl	8270	10
3,3'-Dimethylbenzidene	119-93-7	[1,1'-Biphenyl]-4,4'-diamine, 3,3,'-dimethyl-	8270	10
2,4-Dimethylphenol; m-xylenol	105-67-9	Phenol, 2,4-dimethyl	8040 8270	5 10
Dimethyl phthalate	131-11-3	1,2-Benzenedicarboxylic acid, dimethyl ester	8060 8270	5 10
m-Dinitrobenzene	99-65-0	Benzene, 1,3-dinitro-	8270	20
4,6-Dinitro-o-cresol 4,6-dinitro-2-methylphenol	534-52-1	Phenol, 2-methyl-4,6-dinitro	8040 8270	150 50
2,4-Dinitrophenol	51-28-5	Phenol, 2,4-dinitro-	8040 8270	150 50
2,4-Dinitrotoluene	121-14-2	Benzene, 1-methyl-2,4-dinitro-	8090 8270	0.2 10
2,6-Dinitrotoluene	606-20-2	Benzene, 2-methyl-1,3-dinitro-	8090 8270	0.1 10
Dinoseb; DNBP; 2-sec-butyl-4,6-dinitrophenol	88-85-7	Phenol, 2-(1-methylpropyl)-4,6-dinitro	8150 8270	1 20
Di-n-octyl phthalate	117-84-0	1,2-Benzenedicarboxylic acid, dioctyl ester	8060 8270	30 10
Diphenylamine	122-39-4	Benzenamine, N-phenyl-	8270	10
Disulfoton	298-04-4	Phosphorodithioic acid, 0,0-diethyl S-[2-(ethylthio)ethyl] ester	8140 8141 8270	2 0.5 10

Compound	Chemical name	CAS No.	Method	Value
Endosulfan sulfate	chloro-1,5,5a,6,9,9a-hexahydro-, 3 oxide, (3α,5aα,6β,9β,9aα)- 6,9-Methano-2,4,3-benzodioxathiepin, 6,7,8,9,10,10-hexachloro-1,5,5a,6,9,9a-hexahydro-3,3-dioxide	1031-07-8	8270	20
			8080	0.5
Endrin	2,7:3,6-Dimethanonaphth[2,3-b]oxirene, 3,4,5,6,9,9-hexachloro-1a,2,2a,3,6,6a,7,7a-octahydro-, (1aα,2β,2aβ,3α,6α,6aβ,7β,7aα)-	72-20-8	8270	10
			8080	0.1
			8270	20
Endrin aldehyde	1,2,4-Methenocyclopenta[cd]pentalene-5-carboxaldehyde, 2,2a,3,3,4,7-hexachlorodecahydro-, (1α,2β,2aβ,4β,4aβ,5β,6aβ,6bβ,7R*)-	7421-93-4	8080	0.2
			8270	10
Ethylbenzene	Benzene, ethyl-	100-41-4	8020	2
			8221	0.05
			8260	5
Ethyl methacrylate	2-Propenoic acid, 2-methyl-, ethyl ester	97-63-2	8015	5
			8260	10
			8270	10
Ethyl methanesulfonate	Methanesulfonic acid, ethyl ester	62-50-0	8270	20
Famphur	Phosphorothioic acid, O-[4-[(dimethylamino)sulfonyl]phenyl] O,O-dimethyl ester	52-85-7	8270	20
Fluoranthene	Fluoranthene	206-44-0	8100	200
			8270	10
Fluorene	9H-Fluorene	86-73-7	8100	200
			8270	10
Heptachlor	4,7-Methano-1H-indene, 1,4,5,6,7,8,8-heptachloro-3a,4,7,7a-tetrahydro-	76-44-8	8080	0.05
			8270	10
Heptachlor epoxide	2,5-Methano-2H-indeno[1,2-b]oxirene, 2,3,4,5,6,7,7-heptachloro-1a,1b,5,5a,6,6a-hexahydro-, (1aα,1bβ,2α,α,5aβ, 6β,6aα)	1024-57-3	8080	1
			8270	10
Hexachlorobenzene	Benzene, hexachloro-	118-74-1	8120	0.5
			8270	10
Hexachlorobutadiene	1,3-Butadiene, 1,1,2,3,4,4-hexachloro-	87-68-3	8021	0.5
			8120	5
			8260	10
			8270	10
Hexachlorocyclopentadiene	1,3-Cyclopentadiene, 1,2,3,4,5,5-hexachloro-	77-47-4	8120	5
			8270	10
Hexachloroethane	Ethane, hexachloro-	67-72-1	8120	0.5
			8260	10
Hexachloropropene	1-Propene, 1,1,2,3,3,3-hexachloro-	1888-71-7	8270	10
2-Hexanone; methyl butyl ketone	2-Hexanone	591-78-6	8270	10
			8260	50
Indeno(1,2,3-cd)pyrene	Indeno(1,2,3-cd)pyrene	193-39-5	8100	200
			8270	10
Isobutyl alcohol	1-Propanol, 2-methyl-	78-83-1	8015	50
			8240	100
Isodrin	1,4,5,8-Dimethanonaphthalene, 1,2,3,4,10,10-, hexachloro-1,4,4a,5,8,8a hexahydro-(1α,4α,4aβ,5β,8β,8aβ)-	465-73-6	8270	20
Isophorone	2-Cyclohexen-1-one, 3,5,5-trimethyl-	78-59-1	8270	10
			8090	60

TABLE 2.2 List of Hazardous Inorganic and Organic Constituents[1] *(Continued)*

Common name[2]	CAS RN[3]	Chemical abstracts service index name[4]	Suggested methods[5]	PQL,[6] µg/L
Isosafrole	120-58-1	1,3-Benzodioxole,5-(1-propenyl)-	8270	10
Kepone	143-50-0	1,3,4-Metheno-2H-cyclobuta[cd]pentalen-2-one 1,1a,3,3a,4,5,5a,5b,6-decachlorooctahydro-	8270	10
			8270	20
Lead	Total	Lead	6010	400
			7420	1000
			7421	10
Mercury	Total	Mercury	7470	10
Methacrylonitrile	126-98-7	2-Propenenitrile, 2-methyl-	8015	2
			8260	5
Methapyrilene	91-80-5	1,2-Ethanediamine, N,N-dimethyl-N¹-2-pyridinyl-N¹-thienyl-methyl)-	8270	100
Methoxychlor	72-43-5	Benzene,1,1¹-(2,2,2,trichloroethylidene)bis[4-methoxy-	8080	2
			8270	10
Methyl bromide; bromomethane	74-83-9	Methane, bromo-	8010	20
			8021	10
Methyl chloride; chloromethane	74-87-3	Methane, chloro-	8010	1
			8021	0.3
3-Methylcholanthrene	56-49-5	Benz[j]aceanthrylene, 1,2-dihydro-3-methyl-	8270	10
Methyl ethyl ketone; MEK; 2-butanone	78-93-3	2-Butanone	8015	10
			8260	100
Methyl iodide; iodomethane	74-88-4	Methane, iodo-	8010	40
			8260	10
Methyl methacrylate	80-62-6	2-Propenoic acid, 2-methyl-, methyl ester	8015	2
			8260	30
Methyl methanesulfonate	66-27-3	Methanesulfonic acid, methyl ester	8270	10
2-Methylnaphthalene	91-57-6	Naphthalene, 2-methyl-	8270	10
Methyl parathion; parathion methyl	298-00-0	Phosphorothioic acid, 0,0-dimethyl 0-(4-nitrophenyl) ester	8140	0.5
			8141	1
			8270	10
4-Methyl-2-pentanone; methyl isobutyl ketone	108-10-1	2-Pentanone, 4-methyl-	8015	5
			8260	100
Methylene bromide; dibromomethane	74-95-3	Methane, dibromo-	8010	15
			8021	20
			8260	10
Methylene chloride; dichloromethane	75-09-2	Methane, dichloro-	8010	5
			8021	0.2
			8260	10
Naphthalene	91-20-3	Naphthalene	8021	0.5
			8100	200
			8260	5
			8270	10
1,4-Naphthoquinone	130-15-4	1,4-Naphthalenedione	8270	10

Constituent	CAS No.	Chemical name	Method	Value
Nickel		Nickel	6010	150
Nickel	Total		7520	400
o-Nitroaniline; 2-nitroaniline	88-74-4	Benzenamine, 2-nitro-	8270	50
m-Nitroaniline; 3-nitroaniline	99-09-2	Benzenamine, 3-nitro-	8270	50
p-Nitroaniline; 4-nitroaniline	100-01-6	Benzenamine, 4-nitro-	8270	20
Nitrobenzene	98-95-3	Benzene, nitro-	8090	40
			8270	10
o-Nitrophenol; 2-nitrophenol	88-75-5	Phenol, 2-nitro-	8040	5
			8270	10
p-Nitrophenol; 4-nitrophenol	100-02-7	Phenol, 4-nitro-	8040	10
			8270	50
N-Nitrosodi-n-butylamine	924-16-3	1-Butanamine, N-butyl-N-nitroso-	8270	10
N-Nitrosodiethylamine	55-18-5	Ethanamine, N-ethyl-N-nitroso-	8270	20
N-Nitrosodimethylamine	62-75-9	Methanamine, N-methyl-N-nitroso-	8070	2
N-Nitrosodiphenylamine	86-30-6	Benzenamine, N-nitroso-N-phenyl-	8070	5
N-Nitrosodipropylamine; N-nitroso-N-dipropylamine; Di-n-propylnitrosamine	621-64-7	1-Propanamine, N-nitroso-N-propyl-	8070	10
N-Nitrosomethylethalamine	10595-95-6	Ethanamine, N-methyl-N-nitroso-	8270	10
N-Nitrosopiperidine	100-75-4	Piperidine, 1-nitroso	8270	20
N-Nitrosopyrrolidine	930-55-2	Pyrrolidine, 1-nitroso-	8270	40
5-Nitro-o-toluidine	99-55-8	Benzenamine, 2-methyl-5-nitro-	8270	10
Parathion	56-38-2	Phosphorothioic acid, 0,0-diethyl 0-(4-nitrophenyl) ester	8141	0.5
			8270	10
Pentachlorobenzene	608-93-5	Benzene, pentachloro-	8270	10
Pentachloronitrobenzene	82-68-8	Benzene, pentachloronitro-	8270	20
Pentachlorophenol	87-86-5	Phenol, pentachloro-	8040	5
			8270	50
Phenacetin	62-44-2	Acetamide, N-(4-ethoxyphenl)	8270	20
Phenanthrene	85-01-8	Phenanthrene	8100	200
			8270	10
Phenol	108-95-2	Phenol	8040	1
p-Phenylenediamine	106-50-3	1,4-Benzenediamine	8270	10
Phorate	298-02-2	Phosphorodithioic acid, 0,0-diethyl S-[(ethylthio)methyl] ester	8140	2
			8141	0.5
			8270	10
Polychlorinated biphenyls; PCBs; aroclors	See Note 9	1,1'-Biphenyl, chloro derivatives	8080	50
			8270	200
Pronamide	23950-58-5	Benzamide, 3,5-dichloro-N-(1,1-dimethyl-2-propynyl)-	8270	10
Propionitrile Ethyl cyanide	107-12-0	Propanenitrile	8015	60
			8280	150
Pyrene	129-00-0	Pyrene	8100	200
			8270	10
Safrole	94-59-1	1,3-Benzodioxole, 5-(2-propenyl)-	8270	10
Selenium	Total	Selenium	8010	750
			7740	20
			7741	20

TABLE 2.2 List of Hazardous Inorganic and Organic Constituents[1] (*Continued*)

Common name[2]	CAS RN[3]	Chemical abstracts service index name[4]	Suggested methods[5]	PQL,[6] µg/L
Silver	Total	Silver	6010	70
			7760	100
			7761	10
Silvex: 2,4,5-TP	93-72-1	Propanoic acid,2-(2,4,5-trichlorophenoxy)-	8150	2
Styrene	100-42-5	Benzene, ethenyl-	8020	1
			8021	0.1
			8260	10
Sulfide	18496-25-8	Sulfide	9030	4000
2,4,5-T: 2,4,5-trichlorophenoxyacetic acid	93-76-5	Acetic acid (2,4,5-trichlorophenoxy)-	8150	2
1,2,4,5-Tetrachlorobenzene	95-94-3	Benzene, 1,2,4,5-tetrachloro-	8270	10
1,1,1,2-Tetrachloroethane	630-20-6	Ethane, 1,1,1,2-tetrachloro-	8010	5
			8021	0.05
			8260	5
1,1,2,2-Tetrachloroethane	79-34-5	Ethane, 1,1,2,2-tetrachloro-	8010	0.5
			8021	0.1
			8260	5
Tetrachloroethylene; tetrachloroethene; perchloroethylene	127-18-4	Ethene, tetrachloro-	8010	0.5
			8021	0.5
			8260	5
2,3,4,6-Tetrachlorophenol	58-90-2	Phenol, 2,3,4,6-tetrachloro-	8270	10
Thallium	Total	Thallium	6010	400
			7840	1000
			7841	10
Tin	Total	Tin	6010	40
Toluene	108-88-3	Benzene, methyl-	8020	2
			8021	0.1
			8260	5
o-Toluidine	95-53-4	Benzenamine, 2-methyl-	8270	10
Toxaphene	See Note 10	Toxaphene	8080	2
1,2,4-Trichlorobenzene	120-82-1	Benzene, 1,2,4-trichloro-	8021	0.3
			8120	0.5
			8260	10
			8270	10
1,1,1-Trichloroethane; methylchloroform	71-55-6	Ethane, 1,1,1-trichloro-	8010	0.3
			8021	0.3
			8260	5
1,1,2-Trichloroethane	79-00-5	Ethane, 1,1,2-trichloro-	8010	0.2
			8260	5
Trichloroethylene; trichloroethene	79-01-6	Ethene, trichloro-	8010	1
			8021	0.2
			8260	5
Trichlorofluoromethane; CFC-11	75-69-4	Methane, trichlorofluoro-	8010	10
			8021	0.3
			8260	5

Substance	CAS RN	Common name	Method	PQL
1,2,3-Trichloropropane	96-18-4	Propane, 1,2,3-trichloro-	8270	10
			8010	10
			8021	5
0,0,0-Triethyl phosphorothioate	126-68-1	Phosphorothioic acid, 0,0,0-triethylester	8260	15
sym-Trinitrobenzene	99-35-4	Benzene, 1,3,5-trinitro-	8270	10
Vanadium	Total	Vanadium	8270	10
			6010	80
			7910	2000
Vinyl acetate	106-05-4	Acetic acid, ethenyl ester	7911	40
			8260	50
Vinyl chloride, chloroethene	75-01-4	Ethene, chloro-	8010	2
			8021	0.4
Xylene (total)	See Note 11	Benzene, dimethyl-	8260	10
			8020	5
			8021	0.2
Zinc	Total	Zinc	8260	5
			6010	20
			7950	50
			7951	0.5

[1] The regulatory requirements pertain only to the list of substances; the right-hand columns (Methods and PQL) are given for informational purposes only. See also footnotes 5 and 6.

[2] Common names are those widely used in government regulations, scientific publications, and commerce; synonyms exist for many chemicals.

[3] Chemical Abstracts Service registry number. Where "Total" is entered, all species in the groundwater that contain this element are included.

[4] CAS index names are those used in the 9th Collective Index.

[5] Suggested methods refer to analytical procedure numbers used in EPA Report SW-846 "Test Methods for Evaluating Solid Waste," 3d ed., November 1986, as revised, December 1987. Analytical details can be found in SW-846 and in documentation on file at the agency. Caution: The methods listed are representative SW-846 procedures and may not always be the most suitable method(s) for monitoring an analyte under the regulations.

[6] Practical quantitation limits (PQLs) are the lowest concentrations of analytes in groundwaters that can be reliably determined within specified limits of precision and accuracy by the indicated methods under routine laboratory operating conditions. The PQLs listed are generally stated to one significant figure. PQLs are based on 5-mL samples for volatile organics and 1-L samples for semivolatile organics. Caution: The PQL values in many cases are based only on a general estimate for the method and not on a determination for individual compounds; PQLs are not a part of the regulation.

[7] This substance is often called bis(2-chloroisopropyl) ether, the name Chemical Abstracts Service applies to its non-commercial isomer, propane,2,2"-oxybis[2-chloro-(CAS RN 39638-32-9).

[8] Chlordane: This entry includes alpha-chlordane (CAS RN 5103-71-9), beta-chlordane (CAS RN 5103-74-2), gamma-chlordane (CAS RN 5566-34-7), and constituents of chlordane (CAS RN 57-74-9 and CAS RN 12789-03-6). PQL shown is for technical chlordane. PQLs of specific isomers are about 20 µg/L by method 8270.

[9] Polychlorinated biphenyls (CAS RN 1336-36-3); this category contains congener chemicals, including constituents of Aroclor 1016 (CAS RN 12674-11-2), Aroclor 1221 (CAS RN 11104-28-2), Aroclor 1232 (CAS RN 11141-16-5), Aroclor 1242 (CAS RN 53469-21-9), Aroclor 1248 (CAS RN 12672-29-6), Aroclor 1254 (CAS RN 11097-69-1), and Aroclor 1260 (CAS RN 11096-82-5). The PQL shown is an average value for PCB congeners.

[10] Toxaphene: This entry includes congener chemicals contained in technical toxaphene (CAS RN 8001-35-2), i.e., chlorinated camphene.

[11] Xylene (total): This entry includes o-xylene (CAS RN 96-47-6), m-xylene (CAS RN 108-38-3), p-xylene (CAS RN 106-42-3), and unspecified xylenes (dimethylbenzenes) (CAS RN 1330-20-7). PQLs for method 8021 are 0.2 for o-xylene and 0.1 for m- or p-xylene. The PQL for m-xylene is 2.0 µg/L by method 8020 or 8260.

Results of the sampling must be analyzed statistically using an appropriate procedure. On the basis of the statistical analysis, the owner/operator:

- May return to detection monitoring if a demonstration can be made that a source other than the MSWLF caused the contamination, or that the increase resulted from an error in sampling, analysis, statistical evaluation, or natural variation in groundwater quality
- May return to detection monitoring if the concentration of all assessment monitoring constituents are shown to be at or below background values for two consecutive sampling events
- Must continue assessment monitoring if the constituent concentrations are above background values, but are below the groundwater protection standard
- Must initiate a corrective action program if one or more of constituents are detected at statistically significant levels above the groundwater protection standard

Corrective Action Program.   The corrective action program requires that owners and operators characterize the nature and extent of any release, assess the corrective action measures, select an appropriate corrective action, and implement a remedy. The major components of each of these steps is summarized here:

1. Characterize the nature and extent of a release by installing additional monitoring wells as necessary to characterize the release fully and notify all persons who own the land or reside on the land that directly overlies any part of the plume if contaminants have migrated off-site.
2. Assess appropriate corrective measures within 90 days of finding a statistically significant increase exceeding the groundwater protection standard.
3. Select a remedy that is protective of human health and the environment, attains the groundwater protection standard, controls the source of releases, and complies with RCRA standards for waste management.
4. Implement the selected corrective action remedy and take any interim measures necessary to ensure the protection of human health and the environment.

An approved state can determine that remediation of a release from a MSWLF is not necessary if the owner or operator can demonstrate that:

- Groundwater is additionally contaminated by substances originating from another source and cleanup of the MSWLF releases would not provide a significant reduction in risk to actual or potential receptors caused by such substances.
- The contaminants are present in groundwater that is not currently or reasonably expected to be a source of water, and not hydraulically connected with waters where contaminant migration would exceed the groundwater protection standard.
- Remediation of the release is technically impracticable.
- Remediation will result in unacceptable cross-media impacts.

Closure and Postclosure Care.   The final closure and postclosure care requirements impose significant new requirements on landfill owners and operators.
Closure Criteria.   The closure criteria require owners and operators to install a final cover system that is designed to minimize infiltration and erosion, prepare a written closure plan and place it in the operating records, notify the state when closure is to occur, and make a notation on the deed to the landfill that landfilling has occurred on the property. The final cover system must comprise an erosion layer underlaid by an infiltration layer meeting the following specifications:

- An erosion layer of a minimum of 6 in of earthen material that is capable of sustaining native plant growth.

- An infiltration layer of a minimum of 18 in of earthen material that has a permeability less than or equal to the permeability of any bottom liner system, natural soils present, or a permeability not greater than 1×10^{-5} cm/s, whichever is less.

An owner or operator is required to begin closure activities of each MSWLF unit not later than 30 days after the date the unit receives its last known final receipt of waste.

However, if the unit has remaining capacity and there is a reasonable likelihood that the unit will receive additional waste, the unit may close no later than 1 year after the most recent receipt of wastes.

In approved states, alternative final cover designs may be permitted if the design provides equivalent infiltration reduction and erosion protection. Also, an extension beyond the 1 year deadline for beginning closure may be granted in approved states if the owner or operator demonstrates that the unit has additional capacity and has taken all necessary steps to prevent threats to human health and the environment.

Postclosure Care Requirements.   Following closure of the unit, the owner or operator must conduct postclosure care for 30 years. Postclosure care must be performed in accordance with a prepared postclosure care plan. The postclosure care requirements include the following:

- Maintaining the integrity and effectiveness of any final cover systems
- Maintaining and operating the leachate collection system
- Monitoring groundwater and maintaining the groundwater monitoring system
- Maintaining and operating the gas monitoring system

In approved states, an owner or operator may be allowed to stop managing leachate if a demonstration can be made that leachate no longer poses a threat to human health and the environment. Also, an approved state can decrease or increase the postclosure care period as appropriate to protect human health and the environment.

Financial Assurance Criteria.   Owners and operators of MSWLFs are required to show financial assurance for closure, postclosure care, and known corrective actions. The requirement applies to all owners and operators except state and federal government entities whose debts and liabilities are the debts and liabilities of a state or the United States. The requirements for financial assurance are effective 30 months after promulgation of the rule.

The rule requires that the owner or operator have a detailed written estimate, in current dollars, of the cost of hiring a third party to perform closure, postclosure care, and any known corrective action. The cost estimates must be based on a worst-case analysis (i.e., most costly) and be adjusted annually. The owner or operator must increase or may decrease the amount of financial assurance on the basis of these estimates.

The allowable mechanisms for demonstrating financial assurance for closure, postclosure care, and known corrective actions are as follows:

- Trust fund
- Surety bond guaranteeing payment or performance
- Letter of credit
- Insurance
- Corporate financial test
- Local government financial test (effective date: April 9, 1997)
- Corporate guarantee
- Local government guarantee

On April 10, 1998, EPA amended RCRA to increase the flexibility available to owners and operators by adding two mechanisms: a financial test for use by private owners and operators, and a corporate guarantee that allows companies to guarantee the costs for another owner or operator.

Status of State Adequacy Determinations Under Part 258.   Section 4005(c)(1) of the Resource Conservation and Recovery Act requires states to develop and implement permit programs to ensure that municipal solid waste landfills are in compliance with the revised federal criteria contained in Part 258. These permit programs were required to be in place not later than 18 months after the promulgation date of the criteria (i.e., by April 9, 1993). Also, under Sec. 4005(c)(1), EPA is required to determine whether states have an adequate permit program.

On a national level, the number of solid waste landfills in the United States decreased from 8000 in 1988 to 2400 in 1996.

RCRA Reauthorization

The 102d Congress (1991–1992 session) began work on the reauthorization of RCRA, but it did not complete its work before the session ended. However, both houses' authorizing committees undertook activities that would have resulted in significant change for the waste management industry.

Bills were introduced and discussed that addressed the following:

- Restrictions on the interstate transportation and disposal of municipal solid waste
- State solid waste management planning
- Municipal solid waste recycling requirements
- Management of batteries and scrap tires
- Industrial nonhazardous waste reporting requirements
- Marketing claims pertaining to the environment

Only Senate Bill 2877 cleared the floor, and it contained a narrow set of provisions for restricting the interstate movement of municipal solid waste. The House never called the bill for a floor vote because House members wanted to pass a comprehensive RCRA bill, not pieces of bills.

With the start of the 103d Congress (1993–1994 session), indications were that a comprehensive RCRA reauthorization bill would not pass in 1993 because other environmental issues had priority (e.g., Superfund and Clean Water Act reauthorization). However, there appeared to be enough support in Congress to break out the interstate provisions of RCRA and pass a stand-alone bill on this issue. Bills authorizing restrictions on the movement of interstate waste were introduced in early 1993 in both the Senate and the House.

In 1996 the 104th Congress further amended RCRA with the Land Disposal Program Flexibility Act (P.L. 104-119), which exempts hazardous waste from RCRA regulation if it is treated to a point where it no longer exhibits the characteristic that makes it hazardous and is subsequently disposed in a facility regulated under the CWA or in a Class I deep injection well regulated under the Safe Drinking Water Act. It also exempts small landfills in arid or remote areas from groundwater monitoring requirements if there is no evidence of groundwater contamination. Approved states with any landfill that receives 20 tons or less of municipal solid waste are also provided additional flexibility.

2.2   CLEAN AIR ACT*

The regulation of gas emissions from solid waste management facilities was ignored, for all practical purposes, until the late 1970s. However, with growing public concern in the United

---

* (40 CFR 60, 62)

States over solid waste disposal and widespread nonattainment of national ambient air quality standards, the U.S. Congress and the U.S. Environmental Protection Agency have initiated a number of programs to control gas emissions, both inorganic and organic, at municipal solid waste landfills and combustors. This section summarizes future regulatory or legislative actions that are pending at the federal level.

Guidelines for Control of Existing Sources and Standards of Performance for New Stationary Sources

The EPA proposed new source performance standards (NSPSs) for new MSWLFs under Sec. III(b) of the Clean Air Act Amendments (CAAA) and emission guidelines for existing MSWLFs under Sec. III(d) on May 30, 1991. This action was in response to EPA's findings that MSWLFs can be a major source of air pollution which contributes to ambient ozone problems, airborne toxic gas concerns, global warming, and potential explosion hazards. The regulations were scheduled for promulgation in late 1993.

As part of its regulatory analysis, EPA developed a baseline emission estimate for nonmethane organic compounds (NMOCs) and methane using the Scholl Canyon model. To make the predictions, EPA needed to estimate the values for the methane generation rate constant k, potential methane generation capacity of the refuse LO_2, and NMOCs. These values were estimated from information in three publicly available sources.

The concentrations of NMOCs reported in the data vary widely, from 237 to 14,294 parts per million (ppm). EPA was not able to develop any apparent correlation between landfill gas composition and site-specific factors. However, EPA concluded that the highest NMOC concentrations were measured at sites with a known history of codisposing hazardous waste (i.e., prior to EPA restrictions on such activity).

The compounds occurring most frequently in landfill gas included trichlorofluoromethane, trichloroethylene, benzene, vinyl chloride, toluene, and perchloroethylene. Four of these compounds are known or suspected carcinogens. However, the compounds with highest average concentration included toluene, ethylbenzene, propane, methylene chloride, and total xylenes. Only one of these, methylene chloride, is a known or suspected carcinogen, and it is also a possible laboratory contaminant.

EPA's data on uncontrolled air emissions at MSWLFs were limited to seven landfills using active gas collection systems. The NMOC mass emission rates (based on inlet flow measurements) ranged from 43 to 1853 Mg/year with no apparent correlation between landfill design and operation parameters.

From its database and other assumptions, EPA estimated that the baseline (1987) emissions from the 7124 existing landfills in the United States was 300,000 Mg/year of NMOCs and 15 million Mg/year of methane. These predictions did not include emissions from some 32,000 landfills closed prior to 1987.

Regulatory Approach.   Because of the air emissions outlined previously, EPA proposed regulations to control gas emissions from MSWLFs under Sec. III of the CAAA. The development of the regulations according to EPA will respond to several health and welfare concerns, including the following:

- Contribution of NMOC emissions in the formation of ozone
- Contribution of methane to possible global warming
- Cancer and other potential health effects of individual compounds emitted
- Odor nuisance associated with emissions
- Fire and explosion hazard concerns

Because of all these concerns, EPA is considering the regulation of municipal landfill gas emissions in total, rather than regulation of the individual pollutants or class of pollutants emitted. Regulating total emissions has several advantages, according to EPA, including control of all air emissions with the same control technology, less expense for the regulated community, and easier enforcement and implementation for the regulatory agencies.

The regulatory approach proposed by EPA is twofold: a maximum landfill design capacity value coupled with a landfill-specific emission rate greater than a designated level. This format requires emission controls to be installed when NMOC emissions exceed a designated mass emission rate. The mass emission rate approach has a number of advantages, according to EPA, including the following:

- Control of landfills with the greatest emissions
- A high level of cost efficiency in terms of national cost per ton of NMOC emission reduction
- Relative ease to understand, implement, and enforce

EPA will be setting the stringency level for controlling emissions at MSWLFs when the regulations are finalized. However, the maximum design capacity level being considered is 111,000 tons and an emission rate set at 167 tons of nonmethane organics per year. At this level, EPA expects that 621 existing landfills will be affected. The methane and NMOC emission reductions expected are 255 million and 10.6 million Mg, respectively.

The emission guidelines require the MSWLF owner or operator to use the tiered calculation described in 40 CFR 60.754 to determine the eventual need for controls. The procedure involves the calculation of the NMOC emission rate from a landfill if the emission rate equals or exceeds a specified threshold (50 Mg NMOC per year), the landfill owner or operator must install a gas collection and control system. The first tier of the tiered calculation is conservative to ensure that landfill emissions are controlled. Tiers 2 and 3 allow site-specific measurement to determine emissions more accurately. However, if landfill owners or operators want to use an alternative, more accurate method, they can seek approval from the administrator. For state inventories and related state programs such as Title V permitting and new source review, a state may use its own procedures. Tier 1 default values are not recommended for inventories because they tend to overestimate emissions from many landfills.

The emission controls installed at a landfill that exceeds the stringency levels will be required to achieve a 98 percent reduction in collected NMOC emissions. These control devices must be operated until all of the following conditions are met:

- The landfill is no longer accepting waste and is closed permanently.
- The collection and control system has been in continuous operation for a minimum of 15 years.
- The calculated NMOC emission rate has been less than 167 ton/year on three successive test dates that must be no closer together than 3 months but no longer than 6 months apart.

Clean Air Act Amendments of 1990

The Clean Air Act Amendments of 1990 required EPA to promulgate rules regulating air emissions from a variety of sources, including many associated with solid waste management (e.g., air emissions from landfills and transportation of waste). Title 1, the General Provisions for Non-attainment Areas, expands the requirements for State Implementation Plans to include pollutants primarily responsible for urban air pollution problems. The primary causes of this air pollution include ozone, carbon monoxide, and particulate matter.

In August 2001 there were 75 areas that did not meet the minimum ozone health standard and were required to be classified into one of five categories: marginal, moderate, serious, severe, and extreme. The requirements for revised SIPs could include the following:

- Mandatory transportation controls—for example, programs for inspection and maintenance of vehicle emission control systems, programs to limit or restrict vehicle use in downtown areas, and employer requirements to reduce employee work-trip-related vehicle emissions

- Specified reductions in ozone levels—for example, controlling volatile organic compounds (VOC) emissions from stationary sources
- Offsets and/or reductions of existing sources prior to allowing new or modified sources in an area
- Regulation of small sources as "major" sources in certain categories (i.e., for the extreme category a major source could be defined as one emitting 10 tons of VOCs per year)
- Improved emission inventories

The CAA Amendments of 1990 also required the implementation of reasonably achievable control technology (RACT) at all major sources.

These new requirements could affect the ability of an industry (e.g., a landfill) to operate or expand in certain areas and may require the installation of extensive systems for the control of gas emissions (e.g., VOCs). In addition, the requirements could restrict the access of collection vehicles in nonattainment areas to certain times of the day or certain roadways. The development of the CAA requirements under Title I could dramatically affect solid waste management operations as the states revise SIPs and reduce and control emissions in nonattainment areas.

Title III (Air Toxics) of the Clean Air Act Amendments of 1990 also affects the management of solid waste. Under this title, requirements are set forth for hazardous air pollutants, air emissions from municipal waste combustion, and ash management and disposal.

Under Title III, the CAA amendments set forth a new regulatory program for hazardous air pollutants from major stationary and area sources and establish a list of 189 regulated pollutants (Table 2.3). Major stationary sources are those operations that emit:

- 10 ton/year of any single hazardous air pollutant
- 25 ton/year of any combination of hazardous air pollutants

Area sources are considered smaller sources that had not been previously controlled under the act.

On June 21, 1991, EPA proposed a preliminary draft list of categories of major and area sources of hazardous air pollutants by industry (those sources considered small sources that had not been previously controlled under the act). Waste treatment and disposal was listed as an industry group, and the categories listed under the group include the following:

Solid waste disposal—open burning

Sewage sludge incineration

Municipal landfills

Groundwater cleaning

Hazardous waste incineration

Tire burning

Tire pyrolysis

Cooling water chlorination for steam electric generators

Wastewater treatment systems

Water treatment purification

Water treatment for boilers

For these listed sources, the CAA Amendments require EPA to establish a maximum achievable control technology (MACT) for each category. MACT standards can differ, depending on whether the source is an existing source or a new source:

- *New sources.* MACT must be established as the degree of emission reduction that is achievable by the best controlled similar source and may be more stringent where feasible

TABLE 2.3 List of Hazardous Air Pollutants

CAS number	Chemical name
75070	Acetaldehyde
60355	Acetamide
75058	Acetonitrile
98862	Acetophenone
53963	2-Acetylaminofluorene
107028	Acrolein
79061	Acrylamide
79107	Acrylic acid
107131	Acrylonitrile
107051	Allyl chloride
92671	4-Aminobiphenyl
62533	Aniline
90040	o-Anisidine
1332214	Asbestos
71432	Benzene (including benzene from gasoline)
92875	Benzidine
98077	Benzotrichloride
100447	Benzyl Chloride
92524	Biphenyl
117817	Bis(2-ethylhexyl)phthalate (DEHP)
542881	Bix(chloromethyl)ether
75252	Bromoform
106990	1,3-Butadiene
156627	Calcium cyanamide
105602	Caprolactam
133062	Captan
63252	Carbaryl
75150	Carbon disulfide
56235	Carbon tetrachloride
463581	Carbonyl sulfide
120809	Catechol
133904	Chloramben
57749	Chlordane
7782505	Chlorine
79118	Chloroacetic acid
532274	2-Chloroacetophenone
108907	Chlorobenzene
510156	Chlorobenzilate
67663	Chloroform
107302	Chloromethyl methyl ether
126998	Chloroprene
1319773	Cresols/cresylic acid (isomers and mixture)
95487	o-Cresol
108394	m-Cresol
106445	p-Cresol
98828	Cumene
94757	2,4-D, salts and esters
3547044	DDE
334883	Diazomethane
132649	Dibenzofurans
96128	1,2-Dibromo-3-chloropane
84742	Dibutylphthalate
106467	1,4-Dichlorobenzene(p)
91941	3,3-Dichlorobenzidene

TABLE 2.3   List of Hazardous Air Pollutants *(Continued)*

CAS number	Chemical name
111444	Dichloroethyl ether [bis(2-chloroethyl)ether]
542756	1,3-Dichloropropene
62737	Dichlorvos
111422	Diethanolamine
121697	N,N-Diethyl aniline (N,N-dimethylaniline)
64675	Diethyl sulfate
119904	3,3-Dimethoxybenzidine
60117	Dimethyl aminoazobenzene
119937	3,3'-Dimethyl benzidine
79447	Dimethyl carbamoyl chloride
68122	Dimethyl formamide
57147	1,1-Dimethyl hydrazine
131113	Dimethyl phthalate
77781	Dimethyl sulfate
534521	4,6-Dinitro-o-cresol, and salts
51285	2,4-Dinitrophenol
121142	2,4-Dinitrotoluene
123911	1,4-Dioxane (1,4-Diethyleneoxide)
122667	1,2-Diphenylhydrazine
106898	Epichlorohydrin (1-chloro-2,3-epoxypropane)
106887	1,2-Epoxybutane
140885	Ethyl acrylate
100414	Ethyl benzene
51796	Ethyl carbamate (urethane)
75003	Ethyl chloride (chloroethane)
106934	Ethylene dibromide (dibromoethane)
107062	Ethylene dichloride (1,2-dichloroethane)
107211	Ethylene glycol
151564	Ethylene imine (aziridine)
75218	Ethylene oxide
96457	Ethylene thiourea
75343	Ethylidene dichloride (1,1-dichloroethane)
50000	Formaldehyde
76448	Heptachlor
118741	Hexachlorobenzene
87683	Hexachlorobutadiene
77474	Hexachlorocyclopentadiene
67721	Hexachloroethane
822060	Hexamethylene-1,6-diisocyanate
680319	Hexamethylphosphoramide
110543	Hexane
302012	Hydrazine
7647010	Hydrochloric acid
7664393	Hydrogen fluoride (hydrofluoric acid)
123319	Hydroquinone
78591	Isophorone
58899	Lindane (all isomers)
108316	Maleic anhydride
67561	Methanol
72435	Methoxychlor
74839	Methyl bromide (bromomethane)
74873	Methyl chloride (chloromethane)
71556	Methyl chloroform (1,1,1-trichloroethane)
78933	Methyl ethyl ketone (2-butanone)

TABLE 2.3 List of Hazardous Air Pollutants *(Continued)*

CAS number	Chemical name
60344	Methyl hydrazine
74884	Methyl iodide (iodomethane)
108101	Methyl isobutyl ketone (hexone)
624839	Methyl isocyanate
80626	Methyl methacrylate
1634044	Methyl tert butyl ether
101144	4,4′-Methylene bis(2-chloroaniline)
75092	Methylene chloride (dichloromethane)
101688	Methylene diphenyl diisocyanate (MDI)
101779	4,4-Methylenedianiline
91203	Naphathalene
98953	Nitrobenzene
92933	4-Nitrobiphenyl
100027	4-Nitrophenol
79469	2-Nitropropane
684935	N-Nitroso-N-methylurea
62759	N-Nitrosodimethylamine
59892	N-Nitrosomorpholine
56382	Parathion
82688	Pentachloronitrobenzene (quintobenzene)
87865	Pentachlorophenol
108952	Phenol
106503	p-Phenylenediamine
75445	Phosgene
7803512	Phosphine
7723140	Phosphorus
85449	Phthalic anhydride
1336363	Polychlorinated biphenyls (aroclors)
1120714	1,3-Propane sultone
57578	beta-Propiolactone
123386	Propionaldehyde
114261	Propoxur (Baygon)
78875	Propylene dichloride (1,2-dichloropropane)
75569	Propylene oxide
75558	1,2-Propylenimine (2-methyl aziridine)
91225	Quinoline
106514	Quinone
100425	Styrene
96093	Styrene oxide
1746016	2,3,7,8-Tetrachlorodibenzo-p-dioxin
79345	1,1,2,2-Tetrachloroethane
127184	Tetrachloroethylene (perchloroethylene)
7550450	Titanium tetrachloride
108883	Toluene
95807	2,4-Toluene diamine
584849	2,4-Toluene diisocyanate
95534	o-Toluidine
8001352	Toxaphene (chlorinated camphene)
120821	1,2,4-Trichlorobenzene
79005	1,1,2-Trichloroethane
79016	Trichloroethylene
95954	2,4,5-Trichlorophenol
88062	2,4,6-Trichlorophenol
121448	Triethylamine

TABLE 2.3 List of Hazardous Air Pollutants *(Continued)*

CAS number	Chemical name
1582098	Trifluralin
540841	2,2,4-Trimethylpentane
108054	Vinyl acetate
593602	Vinyl bromide
75014	Vinyl chloride
75354	Vinylidene chloride (1,1-dichloroethylene)
1330207	Xylenes (isomers and mixture)
97576	p-Xylenes
108383	m-Xylenes
106423	p-Xylenes
0	Antimony compounds
0	Arsenic compounds (inorganic including arsine)
0	Beryllium compounds
0	Cadmium compounds
0	Chromium compounds
0	Cobalt compounds
0	Coke oven emissions
0	Cyanide compounds[1]
0	Glycol ethers[2]
0	Lead compounds
0	Manganese compounds
0	Mercury compounds
0	Fine mineral fibers[3]
0	Nickel compounds
0	Polycyclic organic matter[4]
0	Radionuclides (including radon)[5]
0	Selenium compounds

Note: For all listings which contain the word *compounds* and for glycol ethers, the following applies: unless otherwise specified, these listings are defined as including any unique chemical substance that contains the named chemical (i.e., antimony, arsenic, etc.) as part of that chemical's infrastructure.

[1] $X'CN$ where $X = H'$ or any other group where a formal dissociation may occur. For example KCN or $Cal(CN)_2$.

[2] Includes mono- and diethers of ethylene glycol, diethylene glycol, and triethylene glycol $R-(OCH2CH2)_n-OR'$ where $n = 1, 2,$ or 3

R = alkyl or aryl groups

$R' = R, H,$ or groups which, when removed, yield glycol ethers with the structure: $R-(OCH2CH)_n-OH$. Polymers are excluded from the glycol category.

[3] Includes mineral fiber emissions from facilities manufacturing or processing glass, rock, or slag fibers (or other mineral-derived fibers) of average diameter 1 μm or less.

[4] Includes organic compounds with more than one benzene ring, and which have a boiling point greater than or equal to 100°C.

[5] A type of atom which spontaneously undergoes radioactive decay.

- *Existing sources.*   MACT can be less stringent than standards for new sources; however, it must be at least as stringent as:

  The average emission limitations achieved by the best-performing 12 percent of the existing sources where there are 30 or more sources

  The average emission limitation achieved by the best-performing three sources where there are fewer than 30 sources

  On September 24, 1992, EPA proposed a schedule for the promulgation of emission standards for categories and subcategories of hazardous air pollutants:

Hazardous waste incineration	November 15, 2000
Municipal landfills	November 15, 1997
Public owned treatment works (POTW) emissions	November 15, 1995
Sewage sludge incineration	November 15, 1997
Site remediation	November 15, 2000
Treatment, storage, and disposal facilities (TSDFs)	November 15, 1994

The new standards, once promulgated, would result in emission reductions of approximately 80 to 90 percent below current levels.

Section 129 of the act required EPA to revise the new source performance standards and emission guidelines (EGs) for new and existing municipal waste combustion (MWC) facilities. The revised NSPS will eventually apply to all solid waste combustion facilities according to category:

- Solid waste incineration units with capacity greater than 250 ton/d
- Solid waste incineration units with capacity equal to or less than 250 ton/d or combusting hospital waste, medical waste, or infectious waste
- Solid waste incineration units combusting commercial or industrial waste
- All other categories of solid waste incineration

The revised standards require the greatest degree of emission reduction achievable through the application of best available control technologies and procedures that have been achieved in practice, or are contained in a state or local regulation or permit by a solid waste incineration unit in the same category, whichever is more stringent. The performance standards established are required to specify opacity and numerical emission limitations for the following substances or mixtures:

Particulate matter (total and fine)

Sulfur dioxide

Hydrogen chloride

Oxides of nitrogen

Carbon monoxide

Lead

Cadmium

Mercury

Dioxins and dibenzofurans

The EPA is allowed to promulgate numerical emissions limitations, or provide for the monitoring of postcombustion concentrations of surrogate substances, parameters, or periods of residence times.

For existing facilities, EPA is required to promulgate guidelines that include:

- Emission limitations (as previously defined)
- Monitoring of emissions and incineration and pollution control technology performance
- Source-separation, recycling, and ash management requirements
- Operator training requirements

On June 10, 1999, EPA finalized a final federal plan (40 CFR Part 62) to implement emission guidelines for existing municipal solid waste (MSW) landfills. The effective date for the federal plan was January 7, 2000. The emission guidelines apply to existing landfills that handle everyday household waste and were in operation from November 8, 1987, to May 30, 1991, or have capacity available for future waste deposition. Landfills constructed on or after May 30, 1991, or which undergo changes in design capacities on or after May 30, 1991, are subject to EPA's new source performance standards, and not the federal plan. The federal plan applies to any existing MSW landfill that is not covered by an approved and effective state or tribal plan. Following implementation of an approved state or tribal program, EPA's federal plan is rescinded.

The federal plan is projected to affect approximately 3837 MSW landfills in 28 states, 5 territories, and 1 municipality.

Despite the preceding requirements, Title V does not restrict states from incorporating their own standards or emission limitations prior to the time a permit is issued. States may also use permits to impose new standards and emission limitations. None of the provisions of the Clean Air Act may have as great an impact on the waste management industry as the general operating permit provisions of Title V.

Title V establishes a program for issuing operating permits to all major sources (and certain other sources) of air pollutants in the United States. These permits will collect in one place all applicable requirements, limitations, and conditions governing regulated air emissions. Whereas in the past, air regulations governed specific air emission sources, beginning in November 1993, the law required states and localities to regulate emissions from all major stationary sources that directly emit or have the potential to emit 100 tons or more of any pollutant, 10 tons or more of a single hazardous air pollutant, or 25 tons or more of two or more hazardous air pollutants.

The applicability of these provisions is to major sources, which are defined variously in Secs. 112 and 302 and Part D of Title I of the act. The generally accepted definition is one having "the potential to emit." Such sources will be defined in the same way EPA has defined major sources under the Prevention of Significant Deterioration of Air Quality (PSD) and nonattainment New Source Review (NSR) permit programs. The term *potential to emit* means:

> . . . the maximum capacity of a stationary source to emit any air pollutant under its physical and operational design. Any physical or operational limitation on the capacity of a source to emit an air pollutant, including air pollution control equipment and restrictions on hours of operation or on the type or amount of material combusted, stored, or processed, shall be treated as part of its design if the limitation is enforceable by the administrator.

The Title V permit program must satisfy certain federal standards (40 CFR Part 70, 57 *Fed. Reg.* 32249, July 21, 1992), and will be administered by state and local air pollution control authorities. Under the terms of Title V, state and local authorities are required to submit their own operating permit programs to EPA for review and approval by November 15, 1993. If such authorities fail to submit and implement an approvable permit program by this date, Title V directs EPA to impose severe sanctions on states, including the withdrawal of federal highway assistance funds (80 percent of state highway budgets comes from federal highway assistance funds), and the imposition of a minimum 2-to-1 offset ratio for emissions from new or modified sources in certain nonattainment areas. In addition, Title V directs EPA to establish and administer a federal permit program where state and local programs are deemed to be inadequate. States and localities will have 1 year after the submittal of their programs to EPA to issue permits.

The immediate impact of Title V is that any source fitting the preceding description must apply for an operating permit. Facilities that are the least bit uncertain about their status should conduct a sitewide air emission inventory for those substances listed in Secs. 111 and 112 of the act. For landfills, substances identified in Secs. 111(b) and (d) and identified in EPA's New Source Performance Standards must be inventoried. MSW combustors, recycling centers, materials recovery facilities (MRFs), transfer stations, hazardous waste depots, and treatment and disposal facilities that emit substances listed under Sec. 112 are also likely to be subject to this title and to various sections under the act whether or not they are currently subject to specific regulations.

For those facilities that are required to obtain an operating permit, Sec. 70.3(c)(1) states that a permit for a major source must contain all applicable requirements for each of the source's regulated emission units. Therefore, if a source is listed as a major source for a single pollutant—say, methane—all other emissions from the site are subject to regulation under the permit. Section 504(a) requires the permit to include all applicable implementation plans (e.g., state, tribal, or federal implementation plans), and, where applicable, monitoring, compliance plans and reports, and information that is necessary to allow states to calculate permit fees.

Included in the implementation plans are National Ambient Air Quality Standards (NAAQS), which deal with non-emission-related control strategies, such as collection access limitations, roadway access limitations, odd-even day operation requirements, etc. Thus, as states move forward to implement the CAA requirements under Title I (nonattainment), solid waste management operations could be severely impacted.

In addition, all permits are judicially reviewable in state court and are to be made available for public review. Each state's program must include civil and criminal enforcement provisions, including fines for unauthorized emissions. Fines are to begin at a rate of $10,000 per day per violation. A failure to submit a "timely and complete" permit application is subject to civil penalties. A complete application is one that includes information "sufficient to evaluate the subject source and its application, and to determine all applicable requirements." Because EPA did not adopt a standard application form, the requirements of each state in which a facility is located will have to be fulfilled in order to satisfy this provision. Section 70.5(c) of the act lists the minimum requirements that are to be included in permit applications, including the following:

- Company information (e.g., name, address, phone numbers including those for emergencies, nature of business)
- A plant description (e.g., size, throughput, special characteristics, emission sources and emission rates, as well as emission control equipment)
- A description of applicable Clean Air Act requirements and test methods for determining compliance
- Information necessary to allow states to calculate permit fees, which are anticipated to be $25 per ton of actual emissions of regulated pollutants

Permits must include requirements for emission limitations and standards, monitoring, record keeping, reporting, and inspection and entry to ensure compliance with applicable emission limitations. Section 70.6(a)(3) states that, where periodic emissions monitoring is not required by applicable emission standards, the permit itself must provide for "periodic monitoring sufficient to yield reliable data for the relevant time period that are representative of the source's compliance with the permit." However, EPA's rules indicate that in some cases record keeping may be sufficient to satisfy this requirement. In addition, monitoring reports are to be submitted at least every 6 months and must be maintained for at least 5 years. Sources deemed to be out of compliance with any applicable provision of the act must also submit semiannual progress reports. Finally, the permit must contain a certificate of compliance which must be signed by a responsible official.

Last, Sec. 608 of the CAA required EPA to develop a regulatory program to reduce chlorofluorocarbon (CFC) and hydrochlorofluorocarbon (HCFC) emissions from all refrigeration and air-conditioning sources to the "lowest achievable level." Also, this section prohibits individuals from knowingly releasing ozone-depleting compounds into the atmosphere. Penalties for violating this prohibition on venting CFCs and HCFCs can be assessed up to $25,000 per day per violation by EPA. These fines can be levied against any person in the waste management process if a refrigeration unit's charge is not intact and the possessor cannot verify that the refrigerants were removed in accordance with these regulations. Haulers, recyclers, and landfill owners can face significant penalties for noncompliance with these rules.

On May 14, 1993, EPA promulgated final regulations that established:

- Restrictions on the sale of refrigerants to only certified technicians
- Service practices for the maintenance and repair of refrigerant-containing equipment
- Certification requirements for service technicians and equipment and reclaimers
- Disposal requirements to ensure that refrigerants were removed from equipment prior to disposal

The final regulations establish safe disposal requirements to ensure the recovery of the refrigerants from equipment that is disposed with an intact charge; however, they do not require that the recovery take place at any specific point along the disposal route. The recovery of the refrigerant can be done at the place of use prior to disposal (e.g., a consumer's home), at an intermediate processing facility, or at the final disposal site. To ensure that the refrigerant is removed properly, the regulation requires the "final processor" (e.g., landfills) to (1) verify that the refrigerant has been removed or (2) remove the refrigerant themselves prior to disposal. EPA final rules do not establish any type of specific markings to be placed on equipment that has had its refrigerant recovered or removed, although they are recommended. However, the regulations require some form of verification. Verification may include the following:

- A signed statement with the name and address of the person delivering the equipment and the date the refrigerant was removed
- The establishment of contracts for removal with suppliers such as those presently used for polychlorinated biphenyl (PCB) removal

Regardless of who supplies the service, the service provider must recover at least 90 percent of the refrigerant in the unit and must register the removal equipment with the appropriate EPA regional office.

Clean Air Act Amendments of 1996

On March 12, 1996, EPA issued a final rule, Standards of Performance for New Stationary Sources and Guidelines for Control of Existing Sources: Municipal Solid Waste Landfills. Certain new and existing municipal solid waste landfills must install landfill gas collection and control systems. Landfills with more than 2.5 million metric tons of waste in place and annual emissions of nonmethane volatile organic compounds (NMVOCs) exceeding 50 metric tons must collect and combust their landfill gas. The product of combustion may be flared or used as an energy resource. According to EPA estimates of landfill sizes and NMVOC emissions, this regulation should affect about 300 of the nation's largest landfills, doubling the number recovering methane. For example, the large landfills that emit nonmethane organic compounds (NMOCs) in excess of 50 Mg (55.1 tons) per year must control emissions by constructing collection systems or routing the gas to suitable energy recovery or combustion devices. The rule will affect landfills that have a lifetime design capacity of greater than 2.75 million tons and received waste on or after November 8, 1987.

New landfills are those that started construction or began waste acceptance on or after May 30, 1991. They must monitor the surface concentrations of methane on a quarterly basis. If methane is detected at levels greater than 500 parts per million, installation of a landfill gas collection system and gas utilization or disposal system that achieves a 98 percent reduction of NMOCs is necessary. A MSW landfill for which a NMOC emission rate of greater than 50 Mg/year has been calculated must install and operate a gas collection and control system at the landfill.

Existing landfills are those whose construction, modification or reconstruction began before May 30, 1991. The requirements of the emissions guidelines are almost identical to those of the NSPS, but include flexibility for state-implemented emission standards. For each affected landfill, planning, award of contracts, and installation of controls must be implemented within 30 months of the effective date of issuance of state standards.

On February 12, 1998, EPA published the final rule, Tribal Authority Rule (63 *Fed. Reg.* 7524). If a tribe develops a clean air program, it will be called a Tribal Implementation Plan (TIP). The final rule sets forth the CAA provisions under which it is appropriate to treat Indian tribes in the same manner as states, establishes the requirements that tribes must meet if they choose to seek such treatment, and provides for awards of federal financial assistance to tribes to address air quality problems.

Global Warming

Global warming, its cause, effects, and prevention, is one of the major environmental concerns of the decade. One of the sources commonly listed and recommended for control is methane emissions from MSWLFs.

There is growing consensus in the scientific community that changes and increases in the atmospheric concentrations of the "greenhouse" gases (carbon dioxide, methane, chlorofluorocarbons, nitrous oxide, and others) will alter the global climate by increasing world temperatures. The atmospheric greenhouse gases naturally absorb heat radiated from the earth's surface and emit part of the energy as heat back toward the earth, warming the climate. Increased concentrations of these gases on a global basis can intensify the greenhouse effect.

The specific rate and magnitude of future changes to the global climate caused by human activities is hard to predict. However, EPA predicts that if nothing is done, global temperatures may increase as much as 10°C by the year 2100. Reportedly, global warming of just a few degrees would present an enormous change in climate. For example, the difference in mean annual temperature between Boston and Washington, D.C., is only 3.3°C, and the difference between Chicago and Atlanta is 6.7°C. The total global warming since the peak of the last ice age (18,000 years ago) was only about 5°C, a change that shifted the Atlantic Ocean inland by about 100 mi, created the Great Lakes, and changed the composition of forests throughout the continent.

Many human activities contribute to the greenhouse gases currently accumulating in the atmosphere. The most important gas is carbon dioxide (CO_2), followed by methane (CH_4), chlorofluorocarbons, and nitrous oxide (N_2O).

Carbon dioxide is a primary by-product of burning fossil fuels such as coal, oil, and gas, and is also released as a result of deforestation. The largest source of methane is organic matter decaying in the absence of oxygen. CFCs are predominantly produced by the chemical industry. Nitrous oxide sources are not well characterized but are assumed to be related to soil processes such as nitrogenous fertilizer use.

The sources of methane emissions to the atmosphere can be broken into six broad categories: natural resources, rice production, domestic animals, fossil fuel production, biomass burning, and landfills.

The largest source is naturally occurring and is derived from the decomposition of organics in environments such as swamps and bogs. Rice production contributions result from the anaerobic decomposition of organics that occur when rice fields are flooded. The top three rice-producing countries are India, China, and Bangladesh.

Domestic animals, such as beef and dairy cattle, produce methane as a by-product of enteric fermentation, a digestive process in which grasses are broken down by microorganisms in the animal's stomach. The top three countries or regions in domestic animal utilization are India, the countries of the former Soviet Union, and Brazil. Methane releases through fossil fuel production are primarily related to the mining of coal. The United States, the countries of the former Soviet Union, and China are the largest coal-producing countries or regions.

Biomass burning results in the production of methane through the burning related to deforestation and shifting cultivation, burning agricultural waste, and fuelwood use.

The smallest source, landfills, generates methane through the decomposition of organic refuse. EPA predicts that landfilling will not increase very much in countries such as the United States in the future, but can be expected to increase dramatically in developing countries.

The total contribution to the global warming problem that is directly attributed to MSWLFs is less than 2 percent (i.e., 18 percent attributable to methane × 8 percent of the methane attributable to MSWLFs = 1.4 percent of the total greenhouse gases attributable to MSWLFs). The actual amount of methane attributable to landfills in the United States in 1998 was estimated to be 58.8 million metric tons of carbon emissions (MMTCE). Based on these estimates, the contribution to global warming by U.S. landfills is about 0.6 percent.

The contribution of landfill-generated methane to the overall greenhouse gases is relatively small, but landfills are one of the few sources that potentially can be controlled. The question remains as to whether control of such a small source of emissions is economically justifiable. The answer to this question will likely be the subject of discussion and debate in many hearings in Congress as it addresses the issue of global warming and how it should be controlled.

A June 2001 National Academy of Sciences (NAS) report, *Climate Change Science,* makes four points relevant to solid waste management:

1. The climate of the globe is getting warmer.
2. Greenhouse gases contribute to the increase in temperature.
3. Human activity contributes to the increase in greenhouse gases.
4. Climate change's greatest effects are on large land masses in higher latitudes.

Although there have been climate changes in the past, they were ascribed to natural causes such as volcanic eruptions or El Niño effects. In this century there are still uncertainties about natural variability, but the amounts of greenhouse gases that are the result of human activity can be evaluated.

Methane traps heat more effectively than carbon dioxide. Landfills are the chief source of methane. According to EPA, in spite of the increase in landfill waste, the small percentage increase in methane, 58.2 percent to 58.8 percent, was achieved by increasing the collection and combustion of methane gases.

2.3   CLEAN WATER ACT*

The U.S. Army Corps of Engineers (ACOE) and the EPA are the two federal authorities responsible for implementing the Clean Water Act (Title 33 U.S.C. Chap. 26). The purpose of the Clean Water Act is to protect the surface waters of the United States by eliminating the discharge of pollutants into navigable waters. Subchapter IV, for point-source discharges, establishes the permits and licenses programs, such as the National Pollution Discharge Elimination System (NPDES) EPA permit program for controlling storm water and other pollution discharge and the Sec. 404 (ACOE) permit program for wetlands protection. EPA may delegate its NPDES authority to states with approved programs.

As of 2001, there are six states that do not have approved EPA-delegated state NPDES permit programs: Alaska, Arizona, Idaho, Massachusetts, New Hampshire, and New Mexico. EPA retains regulatory authority in those states and over land subject to Indian tribe jurisdiction in Maine.

Until recently, the Clean Water Act only required waste management facilities to obtain permits for their point-source discharges under the NPDES. These regulations have been extensively described in older texts, and readers should refer to them for greater detail.

---

* (33 U.S.C. s/s 1251 et seq. 1977)

On November 16, 1990, the EPA promulgated regulations requiring an NPDES permit for storm water discharges associated with industrial activities. The EPA defined *storm water discharge* as a discharge from any conveyance which is used for collecting and conveying storm water. The word *conveyance* has a very broad meaning and includes almost any natural or human-made depression that carries storm water runoff, snowmelt runoff, and surface runoff and drainage (i.e., not process wastewater). These conveyances are required to be permitted and must achieve CWA 301 best available technology/best control technology (BAT/BCT) and water-quality-based limitations.

The types of waste management activities covered by the regulations include the following:

- Transportation facilities [Standard Industrial Classifications (SIC) 40, 41, 42 (except 4221–4225), 43, 44, 45, and 5171)] which have vehicle maintenance shops, equipment cleaning operations, and refueling and lubrication operations. These classifications cover most haulers/transporters of solid and hazardous waste.
- Material recycling facilities classified under SIC 5015 and 5093.
- Landfills, land application sites, and open dumps that receive any industrial wastes, where industrial wastes are defined very broadly.
- Steam electric power generating facilities such as waste-to-energy facilities.
- Hazardous waste treatment, storage, and disposal facilities.

The only exempted facilities are those that hold an NPDES permit that incorporates storm water runoff, have no storm water runoff that is carried through a conveyance (i.e., all sheet flow), or discharge all runoff to a sewage treatment facility.

The regulations allowed existing regulated industrial activities to apply for permits through one of three methods:

1. A general permit is the most efficient for most industrial facilities. Where EPA is the NPDES permitting authority, the multisector general permit (MSGP) is the general permit currently available to facility operators. Other types of general permits may be available in NPDES authorized states.

2. The multisector general permit (MSGP-2000, *Federal Register,* October 30, 2000) allows for group permits that are tailor-made industry specific permits. New facilities within the regulated industrial sectors must obtain permit coverage under MSGP-2000. The MSGP-2000 is effective in areas in EPA Regions 1,2,3,4,6,8,9, and 10 where EPA is the permitting authority, with a few exceptions.

3. If circumstances are such that a general permit is not available or not applicable to a specific facility, the operator must obtain coverage under an individual permit that the NPDES permitting authority will develop with requirements specific to the facility.

Because the NPDES program is a federal permit program that states can seek delegation under, many states have adopted different programs for application or modified the federal program. States that presently have delegated authority for all or part of the NPDES permit program are listed in Table 2.4. Facilities seeking storm water discharge permits should check with a delegated state to ascertain the availability of permits.

On January 9, 2001, the U.S. Supreme Court, in a 5 to 4 decision (*Solid Waste Agency of Northern Cook County v. U.S. Army Corps of Engineers,* 99-1178), ruled that the Clean Water Act (CWA) does not enable the Army Corps of Engineers to protect migratory bird habitats in intrastate nonnavigable waters. The decision allows a group of municipalities in Northern Cook County, Illinois, to locate a landfill on a former quarry that had become a pond and wetlands used by migratory birds. The Chief Justice wrote that the abandoned gravel pits are a "far cry" from the large and navigable waters that Congress intended to protect under the Clean Water Act.

TABLE 2.4 NPDES General Permitting Authorities

State	NPDES states With general permitting authority	NPDES states Without general permitting authority	Non-NPDES states*
Alabama	X		
Alaska		X	Region 10
Arizona		X	Region 9
Arkansas	X		
California	X		
Colorado	X		
Connecticut	X		
Delaware	X		
District of Columbia	X		
Florida	X		
Georgia	X		
Hawaii	X		
Idaho		X	Region 10
Illinois	X		
Indiana	X		
Iowa	X		
Kansas	X		
Kentucky	X		
Louisiana	X		
Maine	X		
Maryland	X		
Massachusetts		X	Region 1
Michigan	X		
Minnesota	X		
Mississippi	X		
Missouri	X		
Montana	X		
Nebraska	X		
Nevada	X		
New Hampshire		X	Region 1
New Jersey	X		
New Mexico		X	Region 6
New York	X		
North Carolina	X		
North Dakota	X		
Ohio	X		
Oklahoma	X		
Oregon	X		
Pennsylvania	X		
Rhode Island	X		
South Carolina	X		
South Dakota	X		
Tennessee	X		
Texas	X		
Utah	X		
Vermont	X		
Virginia	X		
Washington	X		
West Virginia	X		
Wisconsin	X		
Wyoming	X		

TABLE 2.4   NPDES General Permitting Authorities *(Continued)*

| | NPDES states | | |
State	With general permitting authority	Without general permitting authority	Non-NPDES states*
American Samoa		X	Region 9
Guam		X	Region 9
Northern Mariana Islands		X	Region 9
Puerto Rico		X	Region 2
Virgin Islands	X		

* Permitting in non-NPDES states is done by the EPA regional office indicated.

Section 404 of the Clean Water Act requires a permit from the Army Corps of Engineers to "dredge and fill" wetlands. However, the Corps can no longer rely on the Migratory Bird Act alone to assert CWA jurisdiction. States may choose to enact legislation to protect similar isolated and intrastate wetlands.

2.4   FEDERAL AVIATION ADMINISTRATION GUIDELINES

Federal Aviation Administration (FAA) Advisory Circular (AC) 150/5200-34 (August 8, 2000) establishes guidance concerning the siting, construction, and operation of municipal solid waste facilities (i.e., landfills, recycling facilities, and transfer stations) on or in the vicinity of FAA-regulated airports. The directive reflects the intent of Congress to place further limitations on the construction of MSWLFs near certain smaller airports, especially those landfills that attract birds. Bird-aircraft collisions are dangerous. If a new landfill (constructed or established after April 5, 2000) is intended to be located within 6 mi of an airport, either it should be relocated or the proponents should apply to the appropriate state agency for an exemption before starting construction. The airports are considered to be nonhub, nonprimary commercial services that are recipients of federal grants under 49 U.S.C. 4701. Other specifics apply. The advisory does not apply to Alaska.

2.5   FLOW CONTROL IMPLICATIONS

The theory of flow control is that states control the flow of solid waste to the extent of being able to restrict the import of waste from other states. This concept of flow control was challenged in several courts with the following results.

The Supreme Court decision, *Carbone v. Clarkstown* (1994), prohibits states from discrimination, in violation of the commerce clause and in the absence of authorizing Congressional legislation, by directing solid waste to a specific facility and/or excluding out-of-state waste. Relief may be gained if "the local government demonstrates under rigorous scrutiny that it has no other means to advance a legitimate local interest."

Local planning units cannot plan for the disposal of solid waste in terms of recycling or waste-to-energy because garbage is now protected by the Interstate Commerce Commission (ICC). (There are proposals bills in Congress as of this writing to override the Carbone decision). Local governments cannot manage their own solid waste to the exclusion of out-of-state waste.

On June 4, 2001, four Virginia laws restricting out-of-state trash were struck down by a federal appeals court. The court upheld the lower judge's decision that under the interstate commerce clause of the Constitution, states cannot stop the import of waste to their landfills (*State Recycling Laws Update,* June 2001).

In Virginia, a solution to protect the electricity producing waste-to-energy facilities was achieved by reducing the tipping fees at the landfill to encourage the haulers to maintain their delivery schedules. But the waste-to-energy facility bonds had to be refinanced.

Some of the obvious conclusions were the following:

- Solid waste haulers cannot be prohibited from taking waste to cheaper landfills in other states

- Landfills or waste-to-energy facilities that were built with the expectation of receiving specified wastes will not have those resources. In fact, a number of the waste-to-energy facilities have had their bond ratings lowered.

- There may be more land transportation miles involving solid waste.

- There may be more air pollution as a result of the increase in transportation.

- Increased transportation will increase fuel use.

In New Jersey, *Atlantic Coast II* [921 F. Supp. At 351, P23 (1997)] discussed possible ramifications of the Carbone decision.

The court stated "we disagree with the State's presumption that its problems are insurmountable." The district court listed several alternatives by which the state could lift its flow control laws yet ensure the financial integrity of the local government entities. In particular, the court suggested that the state: (1) issue new bonds to refinance its in-state solid waste disposal facilities, (2) implement "user charges" for those who use the facilities or a "system benefit charge" to make up for lost funds, (3) issue a statewide solid waste tax (or assessment) on all waste generated in-state regardless of where it was sent (in or out of state) for disposal, (4) have the municipalities establish long-term contracts with solid waste facilities (assuming, of course that out-of state facilities could compete with in-state facilities on equal footing), or (5) fund the system through a combination of municipal, county, or state "general revenues" (i.e., taxes).

In a recent case, *A.G.G. Enterprises, Inc. v. Washington County, Oregon* (9th Circuit Court of Appeals), discussed in *Municipal Solid Waste* (March/April 2001, pp. 91–92) the question was: Is garbage property and are franchise agreements valid?

The ICC regulates the interstate transportation of property, something that is owned and that has "economic value." Garbage can be owned, but it is questionable that it has any positive economic value unless it consists of recyclable materials in commercial quantities to attract a buyer and the hauler is in fact carrying segregated recycled materials to a processing facility. Curbside pickup of recylables may not reach the level of "commercial" quantity. Congress used the Federal Aviation Administration Act (FAAA) to deregulate trucking and "to prevent nonfederal interference with deregulation. The FAAA included language that prevents state and local governments from enacting or enforcing any 'law, regulation, or other provision having the force or effect of law to the price, route, service of any motor carrier . . . with respect to the transportation of property.' " The question then is, are haulers who have a franchise agreement to haul or process solid waste, including curbside recylables, hauling property? If they are, their franchise agreements may be invalidated.

CHAPTER 3
SOLID WASTE STATE LEGISLATION

Kelly Hill
Jim Glenn

3.1 INTRODUCTION

Since the 1960s the level of sophistication of solid waste management laws has grown significantly. Requirements on where disposal and processing facilities can be located, how they must be constructed, and how they are to be operated are becoming increasingly stringent. Beyond placing restrictions on the actual facilities, state solid waste management legislation has also mandated that municipalities and counties start planning for the proper disposal of their solid waste.

Following the lead set by the 1976 Federal Resource Conservation and Recovery Act (RCRA), states began to see municipal solid waste in a broader context. Rather than just viewing what is thrown away as waste to be disposed of, it has increasingly been recognized that waste contains resources that can be utilized if they are put back into the economy. In the late 1970s and early 1980s the principal direction of this trend was toward waste-to-energy projects. However, the passage of Oregon's "Opportunity to Recycle" legislation in 1983 ushered in an era of waste reduction legislation that focuses on recycling, composting, and source reduction.

With the majority of state waste reduction goals set by the early 1990s, the latter part of the decade proved to be a time for states to evaluate the effectiveness of their waste management programs and to adjust their expectations based on an evaluation of their progress. For example, some states increased their recycling goals in light of the success of their recycling programs. After New Jersey reached its goal of diverting 25 percent of its waste from disposal, the state revised its goal to divert 65 percent of the waste by the year 2000.

3.2 TRENDS IN MUNICIPAL WASTE GENERATION AND MANAGEMENT

According to the U.S. Environmental Protection Agency (U.S. EPA), 217 million tons of municipal solid waste were generated in this country in 1997, compared with 195.7 million tons in 1990. This increase reflects a steady increase in the annual amount of MSW generated since 1960, when the figure was 88 million tons. By the year 2000, the U.S. EPA estimates that the amount generated had increased to 222 million tons (U.S. EPA, 1999).

In terms of per capita generation, in 1960 the rate was 2.7 lb per person per day of MSW. By 1997, that rate had increased to 4.4 lb per person per day. By 2000, the U.S. EPA estimated the rate to be 4.5 lb per person per day.

Another study of the amount of waste generated throughout the United States based on information provided by solid waste officials in all the states and the District of Columbia typically exhibits MSW figures larger than the U.S. EPA estimates (Glenn and Riggle, 1989a).

TABLE 3.1 Waste Generation and Disposal and Methods of Disposal (by State)

State	Solid waste, tons/yr	Recycled, %	Incinerated, %	Landfilled, %
Alabama*	5,630,000	23	5	72
Alaska*	675,000	7	15	78
Arizona*	5,142,000	17	0	83
Arkansas*	3,316,000	36	<1	67
California*	56,000,000	33	0	67
Colorado*	5,085,000	18	<1	82
Connecticut	3,047,000	24	64	12
Delaware	825,000	22	30	48
Dist. Of Columbia	250,000	8	92	0
Florida*	23,770,000	39	16	45
Georgia*	10,745,000	33	1	66
Hawaii	1,950,000	24	30	46
Idaho	987,000	n/a	n/a	n/a
Illinois*	13,300,000	28	1	71
Indiana*	5,876,000	23	10	67
Iowa*	2,518,000	34	<1	66
Kansas	2,380,000	13	<1	87
Kentucky*	6,320,000	32	0	68
Louisiana*	4,100,000	19	0	81
Maine*	1,635,000	42	40	18
Maryland*	5,700,000	30	25	47
Massachusetts*	7,360,000	34	45	21
Michigan	19,500,000	25	5	70
Minnesota*	5,010,000	45	28	27
Mississippi*	3,070,000	14	5	81
Missouri*	7,950,000	30	<1	70
Montana*	1,001,000	5	2	93
Nebraska*	2,000,000	29	0	71
Nevada*	2,800,000	14	0	86
New Hampshire*	880,000	26	20	54
New Jersey*	7,800,000	43	20	37
New Mexico*	2,640,000	10	0	90
New York*	30,200,000	43	11	46
North Carolina*	12,575,000	32	1	67
North Dakota	501,000	26	0	74
Ohio*	12,335,000	17	0	83
Oklahoma*	3,545,000	12	10	78
Oregon*	4,100,000	30	11	59
Pennsylvania*	9,200,000	26	22	52
Rhode Island	420,000	27	0	23
South Carolina*	10,010,000	42	2	56
South Dakota*	510,000	42	0	58
Tennessee*	9,513,000	35	10	55
Texas*	33,750,000	35	<1	65
Utah*	3,490,000	22	3	75
Vermont*	550,000	30	7	63
Virginia	10,000,000	40	18	42
Washington*	6,540,000	33	8	59
West Virginia*	2,000,000	20	0	80
Wisconsin*	5,600,000	36	3	61
Wyoming*	530,000	5	0	95
Total	374,631,000	31.5	7.5	61

* Figures include some industrial waste

Source: Adapted from Robert Steutville and Nora Goldstein (1993).

There are several reasons for the difference in MSW generation rates in these two studies that may be important to future legislation. The U.S. EPA definition for MSW includes "wastes such as durable goods, non-durable goods, containers and packaging, food scraps, yard trimmings, and miscellaneous inorganic wastes from residential, commercial, institutional, and industrial sources." It does not include wastes from other sources, such as construction and demolition wastes, municipal sludges, combustion ash, and industrial process wastes that might also be disposed of in municipal landfills or incinerators. However, the definition used by some states to characterize MSW includes items such as construction and demolition waste and municipal sewage sludge. Another reason is that several states base the rate on disposal facility records, which also receive non-MSW wastes. Consequently, according to state-provided data the amount of MSW generated in the United States was approximately 250 million tons in 1988 (Glenn and Riggle, 1989a). By 1998 the state-provided estimate had risen to 374.6 million tons (see Table 3.1) (Glenn, 1998a).

By either measure, a significant amount of MSW must be managed. Landfilling is still the way most MSW is managed. According to U.S. EPA figures, in 1960 approximately 62 percent of all MSW was landfilled. That percentage increased until 1980, when it reached roughly 81 percent; by 1997 landfilled MSW had declined to 55 percent (U.S. EPA, 1999).

The steady decline in the reliance on landfills is also consistent with data generated from the *BioCycle* surveys. The 1989 survey found that in 1988 approximately 85 percent of the MSW waste was landfilled. By the end of 1998, that number had declined to 61 percent (U.S. EPA, 1999).

Stimulated by state waste reduction legislation, the landfilling rate has declined and the use of alternatives such as recycling, composting, and incineration has increased correspondingly. In 1989, the sum of recycling and yard waste composting rate was approximately 7 percent and the incineration rate was 8 percent. By the end of 1997, the recycling and composting rate was estimated to total 28 percent, while the incineration rate had decreased to 7.5 percent (U.S. EPA, 1999).

3.3 THE WASTE REDUCTION LEGISLATION MOVEMENT

One of the factors stimulating waste reduction legislation was the perception that available landfill space was dwindling. While this assertion is unfounded, it is clear that the number of landfills in this country continues to decline. At the end of 1988, there were at least 7924 landfills operating in this country (Glenn and Riggle, 1998a). By the end of 1998, that figure had dropped to 2314 (see Table 3.2).

Although the number of landfills is decreasing, it does not necessarily follow that there is a comparable decline in disposal capacity. In Pennsylvania, for example, at the end of 1988 the state had 75 landfills that could accept municipal solid waste, while by the end of 1998 there were only 51. But, although the number of landfills declined, the disposal capacity in the state rose from something less than 5 years to approximately 10 to 15 years (Glenn, 1999).

At the end of 1998, approximately 21 years of landfill capacity remained in the country. Regionally, of those states reporting, the mid-Atlantic states (Maryland, New Jersey, New York, Rhode Island, Pennsylvania, and West Virginia) have an average of 12 years of remaining capacity. They are followed by the Great Lakes states (Illinois, Indiana, Michigan, Minnesota, Ohio, and Wisconsin) and New England (Delaware, Connecticut, Maine, Massachusetts, New Hampshire, and Vermont) with an average of 14 years each. The South (Alabama, Arkansas, Florida, Georgia, Kentucky, Louisiana, North Carolina, South Carolina, Tennessee, and Virginia), and the Midwest (Iowa, Kansas, Mississippi, Missouri, Nebraska, North Dakota, Oklahoma, South Dakota, and Texas) have average remaining capacities of 15 and 17 years, respectfully. The western half of the country seems to be in the best shape, with more than 40 years of remaining capacity.

Another factor stimulating waste reduction legislation has been the escalation of disposal prices. The average tipping fee was $33.60/ton in 1999, an increase of $7 from the $26.50 fee

TABLE 3.2   Disposal Capacity, Number of Facilities, and Tipping Fees (by State)

State	Landfills			Incinerators		
	Number	Average tipping fee, $	Remaining capacity, years	Number	Average tipping fee, $	Capacity, tons/day
Alabama	30	33	10	1	40	700
Alaska	322	50	n/a	4	80	210
Arizona	54	22	n/a	2	n/a	n/a
Arkansas	23	27	20	1	n/a	n/a
California	188	39	28	3	34	6,440
Colorado	68	33	50	1	n/a	<20
Connecticut	3	n/a	n/a	6	64	6,500
Delaware	3	58.50	20	0	—	—
D.C.	0	—	0	0	—	—
Florida	95	43	n/a	13	55	18,996
Georgia	76	28	20	1	n/a	480
Hawaii	8	24	n/a	1	n/a	2,000
Idaho	27	21	n/a	0	—	—
Illinois	56	28	15	1	n/a	1,600
Indiana	45	30	n/a	1	27.50	2,175
Iowa	60	30.50	12	1	n/a	100
Kansas	53	23	n/a	1	n/a	n/a
Kentucky	26	25	19	0	—	—
Louisiana	25	23	n/a	0	—	—
Maine	8	n/a	18	4	47	2,850
Maryland	22	48	10+	3	51	3,860
Massachusetts	47	n/a	n/a	8	n/a	8,621
Michigan	58	n/a	15–20	5	n/a	3,700
Minnesota	26	50	9	9	50	4,681
Mississippi	19	18	10	1	n/a	150
Missouri	26	27	9	0	—	—
Montana	33	32	20	1	65	n/a
Nebraska	23	25	n/a	0	—	—
Nevada	25	23	75	0	—	—
New Hampshire	19	55	11	2	55	700
New Jersey	11	60	11	5	51	6,491
New Mexico	55	23	20	0	—	—
New York	28	n/a	n/a	10	n/a	10,350
North Carolina	35	31	5	1	n/a	540
North Dakota	15	25	35+	0	—	—
Ohio	52	30	20	2	n/a	n/a
Oklahoma	41	18	n/a	2	n/a	1,200
Oregon	33	25	40+	2	67	600
Pennsylvania	51	49	10–15	6	69	8,952
Rhode Island	4	35	5+	0	—	—
South Carolina	19	29	16	1	n/a	255
South Dakota	15	32	10+	0	—	—
Tennessee	34	35	10	2	43	1,250
Texas	181	n/a	30	4	n/a	n/a
Utah	45	n/a	20+	2	n/a	340
Vermont	5	65	5–10	0	—	—
Virginia	70	35	20	5	n/a	n/a
Washington	21	n/a	37	5	n/a	n/a

Source: Adapted from Robert Steutville and Nora Goldstein (1993).

average in 1991. The highest tipping fees in the country were in the New England states, where they averaged $59.50 per ton, and in the mid-Atlantic, where they averaged $48 per ton. The lowest average fees are in the Rocky Mountain states (Arizona, Colorado, Idaho, Montana, New Mexico, Utah, and Wyoming), where they average $23.50 per ton, and the Midwest ($24.81 per ton).

The highest average tipping fee in the country is in Vermont ($65/ton). Alaska, Delaware, Minnesota, New Hampshire, and New Jersey all have tipping fees of $50 or more (see Table 3.2) (Glenn, 1999). Generally, the states most likely to pursue waste reduction have been those with high tipping fees.

With the exception of Oregon's 1983 legislation, the states that initially developed waste reduction laws, such as Florida, New Jersey, New York, Pennsylvania, and Rhode Island, were ones which had relatively high disposal fees. In most cases, these same states also had limited amounts of disposal capacity remaining. For example, at the time of the legislative initiatives, Pennsylvania and Rhode Island both had an average of less than five years of remaining capacity, while New York's was less than 10 years.

Because of both disposal cost and limited capacity, the first wave of waste reduction legislation tended to be concentrated in the mid-Atlantic and New England states. But since then, the pattern has not been clear-cut. As just illustrated, most of the states in the South do not have high tipping fees at the present, but are facing a lack of disposal capacity. In the Middle West and Rocky Mountain states, the tipping fees are not high and there is no lack of capacity. However, while most of the Middle Western states have passed some form of legislation, many of the Rocky Mountain states (such as Colorado and Wyoming) have yet to pass comprehensive waste management laws.

3.4   THE EFFECT OF LEGISLATION

The passage of legislation is just the first step in developing a statewide solid waste management strategy. The evidence that progress is being made lies in the amount of material that goes to alternative uses. At the end of 1998, 35 states landfilled less than 75 percent of their waste. Twelve of those—Connecticut, Delaware, Florida, Hawaii, Maine, Maryland, Massachusetts, Minnesota, New Jersey, New York, Rhode Island, and Virginia—landfilled less than 50 percent (see Table 3.1). That is a considerable change from 1992, when only six states (Connecticut, Delaware, Florida, Maine, Massachusetts, and Minnesota) could make that claim (Glenn, 1999).

In 1998, 30 states estimated their recycling/waste reduction rate to be 25 percent or more. Arkansas, California, Florida, Georgia, Iowa, Kentucky, Maine, Maryland, Massachusetts, Minnesota, Missouri, New Jersey, New York, North Carolina, Oregon, South Carolina, South Dakota, Tennessee, Vermont, Virginia, Washington, and Wisconsin all estimated their waste reduction rates at 30 percent or more. This rate contrasts significantly with the waste reduction levels recorded by states in 1988, when legislation in this area was just beginning to take effect. In 1988, no state had a waste reduction rate of 25 percent or higher; the highest was Washington, with 22 percent (Glenn and Riggle, 1989a). The eight states that had waste-to-energy incineration rates of 25 percent or greater in 1998 remained constant with the 1992 rate. In 1988, only half of those had incineration rates above 25 percent (Stoutville and Goldstein, 1993).

Another measure of the effect of legislation is the number of projects that have been developed over the last several years. While it is difficult to gauge how many waste reduction projects have become part of the country's solid waste management system, some indicators are relatively easy to track.

One way to chart the trend in recycling is to keep track of the most visible of the various collection techniques—curbside recycling. In 1981, there were fewer than 300 known curbside recycling projects in the United States (Glenn and Riggle, 1989a). By the end of 1988, there were an estimated 1042 curbside programs collecting recyclables. In 1998 that figure had

TABLE 3.3 Curbside Recycling Programs By State*

State	Curbside programs	Curbside population served	Percentage of population served
Alabama	38	1,020,000	23
Alaska	1	10,000	2
Arizona	32	1,810,000	39
Arkansas	41	40,000	16
California	511	18,000,000	55
Colorado	70	700,000	18
Connecticut	169	3,270,000	100
Delaware	3	5,000	1
D.C.	1	250,000	48
Florida	315	11,070,000	74
Georgia	179	3,988,000	52
Hawaii	0	0	0
Idaho	6	200,000	16
Illinois	450	6,000,000	50
Indiana	169	4,133,000	70
Iowa	574	1,500,000	52
Kansas	101	n/a	n/a
Kentucky	43	n/a	n/a
Louisiana	33	2,000,000	46
Maine	84	400,000	32
Maryland	100	4,004,000	78
Massachusetts	156	4,758,000	77
Michigan	200	2,500,000	25
Minnesota	771	3,600,000	76
Mississippi	15	420,000	15
Missouri	197	2,100,000	39
Montana	6	8,000	1
Nebraska	15	425,000	26
Nevada	8	397,000	23
New Hampshire	38	433,000	37
New Jersey	510	7,300,000	90
New Mexico	3	400,000	23
New York	1,472	17,230,000	95
North Carolina	271	3,500,000	46
North Dakota	25	90,000	14
Ohio	372	6,600,000	59
Oklahoma	8	639,000	19
Oregon	122	1,830,000	56
Pennsylvania	879	8,800,000	73
Rhode Island	26	860,000	71
South Carolina	186	507,000	13
South Dakota	3	158,000	21
Tennessee	35	n/a	n/a
Texas	159	4,700,000	24
Utah	14	265,000	13
Vermont	80	111,000	19
Virginia	79	4,500,000	66
Washington	102	5,000,000	88
West Virginia	75	500,000	28
Wisconsin	600	3,000,000	57
Wyoming	2	24,000	5
Total	9,349	139,415,000	54

* Municipal, county, and other curbside recycling programs

Source: Adapted from Robert Steutville and Nora Goldstein (1993).

increased to 9349 (Glenn, 1999). In 1998, 23 states had at least 100 programs in operation (see Table 3.3), while in 1988 only three states (New Jersey, Pennsylvania, and Oregon) had 100 functioning programs (Glenn and Riggle, 1989a). Early in 1998, New Jersey, which had just initiated its mandatory recycling legislation, had 439 curbside recycling programs—more than three times the next closest state. By the end of 1998 New York, with 1,472 programs, topped the list. Rounding out the top five are Pennsylvania (879 programs), Minnesota (771 programs), Wisconsin (600 programs), and Iowa (574 programs) (Glenn, 1998b).

By the end of 1998, curbside recycling programs served more than 139 million people in the United States (Glenn, 1998b). Twenty-five states have programs that serve in excess of one million people (see Table 3.3). In 1988, only four states (California, New Jersey, Oregon, and Pennsylvania) had programs serving at least that many people (Glenn and Riggle, 1989a). California estimated that in 1998, 18 million of its residents had access to curbside service, followed closely by New York (17.2 million), Florida (11 million), Pennsylvania (8.8 million), and New Jersey (7.3 million) (see Table 3.3). The common thread among the programs is that all of these states have implemented waste reduction and recycling programs.

By the end of 1998, 3807 facilities were composting some part of the yard waste stream in the United States—which is more than five times the 651 sites operating in 1988 (Steutville and Goldstein, 1993). Twelve states have at least 100 sites composting yard waste. Ohio has 458, followed by Minnesota (433), Pennsylvania (329), Massachusetts (250), New York (200), Wisconsin (176), New Jersey (171), Florida (169), Texas (166), North Carolina and Michigan (120 programs each), and New Hampshire (103) (see Table 3.4). All these states, with the

TABLE 3.4 Yard Waste Composting Programs (by State)

State	Number	State	Number
Alabama	20	Montana	32
Alaska	0	Nebraska	5
Arizona	23	Nevada	1
Arkansas	22	New Hampshire	103
California	74	New Jersey	171
Colorado	11	New Mexico	5
Connecticut	65	New York	200
Delaware	3	North Carolina	120
District of Columbia	0	North Dakota	50
Florida	35	Ohio	458
Georgia	169	Oklahoma	4
Hawaii	9	Oregon	50
Idaho	7	Pennsylvania	329
Illinois	55	Rhode Island	21
Indiana	51	South Carolina	69
Iowa	57	South Dakota	10
Kansas	70	Tennessee	46
Kentucky	37	Texas	166
Louisiana	21	Utah	14
Maine	50	Vermont	14
Maryland	17	Virginia	11
Massachusetts	250	Washington	17
Michigan	120	West Virginia	22
Minnesota	433	Wisconsin	176
Mississippi	9	Wyoming	8
Missouri	97		
Total	3,807		

Source: Adapted from Glenn (1999).

exceptions of New York and Texas, have passed legislation or regulations banning yard waste from disposal facilities.

Beyond sheer numbers, the most dramatic change in yard waste composting over the past several years has been the number of facilities that are composting brush and grass as well as leaves. Of the 22 states banning yard waste from landfill disposal, only Connecticut, Indiana, Pennsylvania, Maryland, and New Jersey limit the types of waste accepted at their sites.

3.5   STATE MUNICIPAL SOLID WASTE LEGISLATION

The approach to municipal solid waste management legislation varies significantly from state to state. Some states, such as Minnesota and Illinois, pass MSW-related legislation annually. In some cases, several laws are passed each year. For instance, in 1991, 94 solid-waste-related bills were introduced in the Illinois legislature. At least 15 of those bills became law, including ones that dealt with procurement of recycled products, household hazardous waste collection, and establishment of a tire recycling fund (Glenn, 1992).

In other cases, states work on solid waste legislation periodically, revisiting the issue every 5 to 10 years. As an example, Pennsylvania passed omnibus solid waste legislation in 1968. Twelve years later, with Act 97 of 1980, that legislation was updated, with particular emphasis on hazardous waste management. In 1988, the state legislature passed Act 101, the Municipal Waste Planning and Waste Reduction Act, which, among other things, mandated some municipalities to establish recycling programs.

Waste management issues arise, throughout the country, during every legislative session. In 1999, more than 1300 bills that dealt with solid waste management in some way were introduced. Legislation addressing recycling issues numbered more than 400 bills. Bills that passed include California's S.B. 332, which expands the state's Beverage Container Recycling and Litter Reduction Act to include carbonated/noncarbonated water and sport drink containers, among other things, and Hawaii's H.B. 1350, which requires government procurement processes to give preference to recycled oils.

3.6   STATE PLANNING PROVISIONS

MSW management-planning provisions in legislation generally direct that planning be conducted on two levels. Numerous laws, including those in Alabama, Minnesota, Montana, and Washington, direct the responsible state agency to develop a state solid waste management plan. Beyond this statewide planning, which often serves as a guide for local governments, laws also include provisions that local governments or counties develop solid waste management plans on a periodic basis.

A state planning requirement that is representative of most of those passed in the late 1980s and early 1990s is contained in New Mexico's Solid Waste Act of 1990. That law required that a comprehensive and integrated solid waste management plan for the state be developed by December 31, 1991. The plan had to rank management techniques, placing source reduction and recycling first, environmentally safe transformation second, and landfilling third. As laid out in the law, the basis for developing the plan was information provided by each county and municipality.

The plan was required to establish a goal of diverting 50 percent of all solid waste from disposal facilities by July 1, 2000. Other elements of the plan include waste characterization, source reduction, recycling and composting, facility capacity, education and public information, funding, special waste, and siting.

The content of a state solid waste management plan often dictates what type of solid waste management planning activities are undertaken by local and county governments. For instance, in South Carolina, each local solid waste management plan has to be designed to achieve the recycling and waste reduction goals established in the state plan. This type of "top-down" planning is not always the case. In North Dakota, local plans are being used to formulate the comprehensive statewide solid waste management plan.

Like state plans, local planning requirements established by legislation follow the same general pattern. Plan contents typically include a description of the current solid waste management situation, both physical and institutional, and how adequate processing and disposal capacity will be made available over a 10- to 20-year period. Most plans now are required to have a waste reduction element that is aimed at achieving the state's reduction goal.

Most states place the primary responsibility for planning on counties, although there are other approaches. For instance, Connecticut requires all of its municipalities to submit 20-year plans. Nevada also requires municipalities to plan. In Alabama, counties are given the planning responsibility, unless the local municipality chooses to retain it. In several states, including North Dakota, Ohio, and Vermont, laws require that separate solid waste management districts be formed to plan and implement solid waste programs.

3.7   PERMITTING AND REGULATION REQUIREMENTS

The permitting and regulations provisions of state laws vary significantly from state to state. At the most fundamental level, laws simply direct that a state agency develops a means of permitting and regulating municipal waste management activities. In other cases, the law establishes a regulatory framework. At the extreme, lawmakers actually write into law the requirements that facilities must meet. For example, when the Illinois legislature amended the Solid Waste Management Act in 1988, it put into law requirements on how yard waste composting facilities were to be sited and operated. This strategy necessitates that any adjustments to requirements be formulated, debated, and passed by the legislature. For instance, in 1991, the legislature mandated an increase in yard waste composting facility setback from 200 to 660 feet from the nearest residence. Thus, the setback could not be determined by the Illinois Environmental Protection Agency, the state regulatory agency.

The breadth of legislative direction in permitting and regulations can be illustrated by looking at laws passed, respectively, in North Dakota and South Carolina. H.B. 1060, passed in North Dakota during 1991, directs that the Department of Health and Consolidated Laboratories, "adopt and enforce rules governing solid waste management." Additionally, the agencies are to adopt rules to establish standards and requirements for various categories of solid waste management facilities, establish financial assurance requirements, and conduct an environmental compliance background review of any permit applicant.

South Carolina's Solid Waste Policy Management Act of 1991 goes into much more detail on how solid waste facilities are to be permitted and regulated. How the permitting process is to take place and what minimum requirements must be met by different types of facilities, including landfills, incinerators, processing facilities and land application facilities, is spelled out in the law.

While it's beyond the scope of this chapter to detail the solid waste regulations in individual states, information on those rules can be obtained from the state solid waste programs. A listing of the appropriate state agencies may be found in the Appendix at the end of this chapter.

3.8   WASTE REDUCTION LEGISLATION

While most changes in solid waste management laws over the last 20 to 30 years can be described as evolutionary, waste reduction provisions are better characterized as revolutionary. In 1980, the most far-reaching waste reduction initiatives being implemented were mandatory deposit legislation on certain beverage containers. Ten years later, essentially every legislature in the country was at least giving serious consideration to bills targeting 25 to 50 percent of the municipal waste stream for reduction. With those goals in place throughout the 1990s, during the next decade states are likely to be devoting time to evaluating the success in reaching those goals and in revising their waste management plans accordingly.

The approaches states have taken are not uniform. Some states, like Pennsylvania and New Jersey, have opted for laws that ultimately require waste generators to participate. Others, like Arizona and Washington, require only that recycling programs be made available to citizens, while other methods, like those used in Florida and Iowa, require local governments to reach a certain goal (U.S. EPA, 1999).

The strategies do not stop with the various ways of developing waste reduction programs. Most states have banned outright the disposal of some materials such as yard waste, oil, and white goods, and are taxing disposal of items such as tires. Increasingly states are requiring manufacturers and retailers to take responsibility for disposal of their products.

3.9   ESTABLISHING WASTE REDUCTION GOALS

Perhaps the most fundamental provision in any solid waste legislation relates to the establishment of a statewide waste reduction goal. By the end of 1996, 42 states had put some type of waste reduction goals on the books (see Table 3.5). By comparison only eight states (Connecticut, Florida, Maine, Maryland, New Jersey, New York, Pennsylvania, and Rhode Island) had waste reduction goals in 1988 (Starkey and Hill, 1996).

When states first started to establish goals, they were primarily recycling goals. New Jersey was one of the first states to put a goal of 25 percent in legislation in 1987. The state did not even allow the composting of leaves to count for part of the 25 percent. New Jersey has since set a waste reduction goal of 65 percent, which includes material that is composted. In Florida and North Carolina no more than half of the waste reduction goal can be met by yard waste composting. No more than 50 percent of South Carolina's 25-percent recycling goal can be met by yard waste, land clearing debris, white goods, tires, and construction and demolition debris (Glenn, 1992).

Since the first implementation of state recycling goals, their focus has changed from strictly recycling and yard waste composting to overall waste reduction, which may also include other forms of composting and source reduction. While South Carolina has a 25 percent recycling goal, its overall goal is to reduce by 30 percent the amount of solid waste received at MSW landfills and incinerators. In West Virginia, the year 2010 goal is to reduce the disposal of MSW by 50 percent (Glenn, 1992).

As the focus of the goals has expanded, so has the amount of waste expected to be diverted. Recycling goals, as in Connecticut, Illinois, Maryland, and Rhode Island, typically ranged from 15 to 25 percent. Most waste reduction laws have rates from 25 to 70 percent. States at the lower end of the scale include Alabama, Louisiana, and Ohio. The high end includes New Jersey and Rhode Island. It should be noted that most states with lower goals placed deadlines for achieving them sooner than did states with higher goals. The deadlines for states with 25 percent goals ranged from 1991 to 1996. Deadlines for 50 to 70 percent goals usually stretch from the year 2000 and beyond (see Table 3.5).

Although a majority of states have established some form of waste reduction goals, for some it is just that. Thus far, only 10 of 21 states have been able to meet their goals by the legislated deadline. Two of those states, New Jersey and Pennsylvania, met their goals of 25 percent before their legislatures raised their goals. Florida has surpassed its goal of 30 percent. Four of the 11 states that did not meet their waste reduction goals had goals of 25 percent (Idaho, Louisiana, Mississippi, and Montana). The remaining states have goals ranging between 40 to 50 percent (Glenn, 1998b). Generally speaking, state agencies, regional solid waste districts, and counties and municipalities responsible for achieving these waste reduction rates

TABLE 3.5   Statewide Solid Waste Management Goals

State	Goal (%)	Recovery rate (%)	Deadline	Mandated
Alabama	25	23	—	No
Arkansas	40	36	2000	No
California	50	33	2000	Yes
Connecticut	40	24	2000	Yes
Delaware	25	22	2000	No
D.C.	45	8	1994	No
Florida	30	39	1994	No
Georgia	25	33	1996	No
Hawaii	50	24	2000	No
Idaho	25	n/a	1995	No
Illinois	25	28	2001	Yes
Indiana	50	23	2000	No
Iowa	50	34	2000	No
Kentucky	25	32	1997	No
Louisiana	25	19	1992	No
Maine	50	42	1998	No
Maryland	20	30	1994	Yes
Massachusetts	46	34	2000	No
Michigan	50	25	2005	No
Minnesota	50	45	1996	No
Mississippi	25	14	1996	No
Missouri	40	30	1998	No
Montana	25	5	1996	No
Nebraska	40	29	1999	No
Nevada	25	14	1995	No
New Hampshire	40	26	2000	No
New Jersey	65	43	2000	Yes
New Mexico	50	10	2000	No
New York	50	43	1997	No
North Carolina	40	32	2001	No
North Dakota	40	26	2000	No
Ohio	25	17	2000	No
Oregon	50	30	2000	Yes
Pennsylvania	35	26	2003	No
Rhode Island	70	27	—	No
South Carolina	30	42	1997	No
South Dakota	50	42	2001	No
Tennessee	25	35	1995	Yes
Texas	40	35	1994	No
Vermont	40	30	2000	No
Virginia	25	40	1997	No
Washington	50	33	2000	No
West Virginia	50	20	2010	No

Source: Adapted from Glenn (1999).

do not face penalties for failing to meet the stated goals. Only seven of the 43 states with goals have laws in place enforcing the requirements on local governments. For example, in California jurisdictions may face fines of up to $10,000 a day (Glenn, 1998b). Other states use the "carrot and the stick" approach to meeting goals. Any county in South Carolina that meets the state goal is to be rewarded for that effort by sharing in a special bonus grant program.

3.10 LEGISLATING LOCAL GOVERNMENT RESPONSIBILITY

During the 1980s and 1990s waste reduction legislation centered on requiring municipalities to develop recycling programs. In general, two approaches have been utilized. One type of law mandates that municipalities require generators to separate recyclables, and in some cases compostables, for further processing. The most prevalent form of legislation mandating municipal involvement is that which requires local governments to reach specified goals. The other model is legislation that requires local governments to provide some form of waste reduction system to its citizens.

Mandatory Recycling Laws

So-called mandatory recycling laws are employed by six states (see Table 3.6). In four of those states (Connecticut, New Jersey, New York, and Rhode Island) every municipality must pass an ordinance requiring municipal waste generators to recycle certain materials. For instance, in New Jersey, municipalities must collect a minimum of three recyclables. Connecticut and Rhode Island both require collection of a more extensive list of recyclables including, among other things, newspapers, glass containers, metal cans, and some plastic bottles. Connecticut also includes leaves in its list of recyclables that must be collected.

TABLE 3.6 States with Legislation Requiring Municipalities to Pass Mandatory Ordinances

State	Municipalities involved	Deadline	Act ID
Connecticut	All	1/91	PA 90-220
New Jersey	All	8/88	P.L. 1987, C. 107
New York	All	9/92	Chap. 70-1988
Pennsylvania	Population of 5000 or greater*	9/26/91	101-1988
Rhode Island	All	†	23-18—1986
West Virginia	Population of 10,000 or greater	10/93	S.B. 18—1991

* All municipalities with 10,000 and above must pass mandatory ordinances. For municipalities with populations between 5000 and 10,000, only those that have a population density of 300 people per square mile must pass ordinances.
† Deadline based on the implementation schedule for each municipality.

Source: Revised from Kreith (1994)

Original Source: Glenn (1992)

The two remaining states, Pennsylvania and West Virginia, limit the number of municipalities that are required to pass ordinances. In Pennsylvania, initially, only those municipalities with a population of 10,000 or greater needed to comply. As of September 1991, those municipalities with a population between 5000 and 10,000 and a population density of 300 or more people per square mile were also required to comply. West Virginia limits its mandate to those municipalities with a population of 10,000 or greater. In both cases, waste generators have to recycle at least three materials.

In addition to requiring municipalities to pass ordinances, these laws compel them to establish recycling programs that meet certain criteria. For instance, West Virginia municipalities must establish curbside programs that collect at least on a monthly basis. The programs must also include a comprehensive public information and education element.

"Opportunity to Recycle" Laws

The prototype of recycling legislation in this country was Oregon's "Opportunity to Recycle" Act. In this type of legislative scheme, municipalities are required to provide some form of recycling program, but the municipalities are not required to pass ordinances requiring participation by waste generators.

In Oregon's first recycling law, curbside recycling programs (which collected at least monthly) had to be put in place in every municipality with a population of 4000 or more. Additionally, every MSW disposal facility had to provide a drop-off program. In 1991, the Oregon legislature saw fit to modify this earlier legislation. Some of the improvements in this update include requirements for weekly collection of recyclables, distribution of home storage containers for recycling, and an expanded education and promotion program.

Besides Oregon, 12 other states have passed legislation requiring local government to develop recycling programs (see Table 3.7). In six of those states (Alabama, Arkansas, Maryland, Minnesota, Nevada, and South Carolina), counties are charged with the responsibility. In Arkansas, counties and municipalities can join together and form a single sanitation authority. Nevada's law applies only to counties with populations of more than 10,000. Three states

TABLE 3.7 State Legislation Requiring Local Government Units to Develop Recycling Programs

State	Local government units involved	Deadline	Act ID
Alabama	Counties[a]	5/92	824 1989
Arizona	Cities & Counties	Not set	H.B. 2574-1990
Arkansas	Sanitation Authorities	7/92	Chapt. 14-233
California	Cities & Counties[b]	1995	A.B. 939-1989
Maryland	Counties	1/94	H.B. 714 1988
Minnesota	Counties	10/90	115A 1989
Nevada	Counties over 10,000	Not set	A.B. 449—1995
North Carolina	C	7/91	S.B. 111—1989
Oregon[d]	Municipality	7/92	S.B. 66—1991
South Carolina	Counties	Not set	H.B. 388—1991
Vermont	SW management districts	Not set	78 1987
Virginia	Regional SW Planning Unit	1997	H.B. 1750—1995
Washington	Cities & Counties	1994	E.S.H.B. 1671—1989

[a] Local municipalities can develop programs on their own if they choose.
[b] Cities and counties can combine to form regional agencies to carry out this task.
[c] Designated local government, of which 90 are counties and 15 are municipalities
[d] Oregon's original legislation (S.B. 405) was effective July 1, 1986.

Source: Revised from Kreith (1994).
Original Source: Glenn (1992).

(Arizona, California, and Washington) targeted both cities and counties. In addition to Oregon, only Virginia's legislation puts the requirement at the municipal level where local and regional solid waste planning units have been created. In Vermont, solid waste management districts, which are generally groups of municipalities, are responsible, while in North Carolina, it is "designated local governments," of which 90 are counties and 15 are municipalities.

Oregon's law to the contrary, most "Opportunity to Recycle" legislation was constructed so that local governments could establish programs that were right for them. In Nevada, designated counties are required to make available a program for the "separation at the source of recyclable material from other solid waste originating from the residential premises where services for collection of solid waste are provided." Additionally, those counties are required to establish a recycling center if none are already available. In South Carolina, the legislation states only that counties may include curbside collection, drop-offs, or multifamily systems in their recycling programs.

Arkansas' statute defines the opportunity to recycle as the "availability of curbside pick-up or collections centers for recyclable materials at sites that are convenient for persons to use." It is up to the county or regional solid waste board to determine the number and type of facilities needed and what type of recyclables are to be collected. However, each board must develop a public education program and establish a yard-waste composting program.

Required Goals

Beyond the fact that most legislatures generally feel that it isn't prudent to dictate what type of recycling program will work best in a particular locality, one reason most states do not require a certain type of program be established is that they also specified in the law a goal which the program must reach. In fact, 10 of the 13 states that have "Opportunity to Recycle" laws also require local governments to reach certain waste reduction goals. These include Alabama, California, Maryland, Minnesota, Nevada, North Carolina, Oregon, Vermont, and Virginia. Additionally, Connecticut, New Jersey, Rhode Island, and South Carolina, which have mandatory recycling laws, also have goal requirements. Of the 21 states that put goal requirements on local governments and regional solid waste authorities, only eight (Florida, Georgia, Hawaii, Illinois, Iowa, Louisiana, Ohio, and Tennessee) do not combine them with some other form.

In most cases, the goal a local government must reach is identical to the state goal established in the law. The major exception to that is Oregon, where different groups of counties had different goals to meet, ranging from 15 to 45 percent by 1995. Each county's goal was revised after 1995 to assure that Oregon would meet its statewide goal of 50 percent by 2000. Other legislation has allowances for local governments that cannot meet a goal, either by providing it additional time to comply, as is the case in Tennessee, or by allowing the state agency that oversees the program to reduce or modify the goal if circumstances warrant.

What happens if a local government does not meet a goal varies widely. In some cases, it is not clear if anything will occur. In Oregon, if a recovery rate is not achieved, the municipality must take steps to upgrade its program. In Tennessee, fines can ultimately be levied for not complying with the law.

Disposal Bans

There is probably no more direct approach to waste reduction than banning specific types of waste from disposal facilities. For MSW, bans were first ushered in back in 1984 when Minnesota passed legislation banning the disposal of tires from landfills (Glenn and Riggle, 1989b). Since then, 47 additional states have passed bans on one or more waste materials (see Table 3.8).

Over the years, lawmakers have focused particularly on materials coming from vehicles, such as batteries, tires, and oil. Oregon goes so far as to ban the disposal of discarded vehicles. The most popular product ban is vehicle batteries. By the end of 1997, 43 states had passed such restrictions. Thirty-eight states have banned the disposal of, at least, whole tires, although Missouri allows landfill operators to use rubber chips for landfill cover (H.B. 783, 1999). Thirty states ban motor oil from landfills. Of those states that ban the disposal of tires, at least three (Minnesota, Vermont, and Wisconsin), ban any form of tire from being landfilled. Another, Ohio, has banned the disposal of tires in MSW landfills, but will permit them to be buried in tire "monofills."

The ban which can have the greatest effect on reducing the amount of waste being disposed of is a ban on yard waste. In all, 22 states have put yard waste bans on the books. For 20 of those states the ban is on all types of yard waste. In the remaining two it applies only to a

TABLE 3.8 Disposal Bans for Selected Waste Materials

State	Vehicle batteries	Tires	Yard trimmings	Motor oil	White goods	Others
Alabama	X	—	—	X	—	—
Arizona	X	X	—	X	X[1]	—
Arkansas	X	X	—	—	—	X[3]
California	X	X	—	X	X	—
Connecticut	X	—	X[2]	—	—	X[3]
Delaware	—	X	—	—	X[1]	—
Florida	X	X	X	X	X	X[4]
Georgia	X	X	X	—	—	—
Hawaii	X	X[5]	X[6]	X	X	X[7]
Idaho	X	X	—	—	—	—
Illinois	X	X[5]	X	X	X[1]	—
Indiana	X	X[8]	X[9]	—	—	—
Iowa	X	X[5]	X	X	—	X[10]
Kansas	X	X[5]	—	—	—	—
Kentucky	X	X[5]	—	X	—	—
Louisiana	X	X[5]	—	X	X	—
Maine	X	X	—	X	—	X[11]
Maryland	—	X	X[12]	X	—	—
Massachusetts	X	X	X	X	X	X[13]
Michigan	X	—	X	X	—	—
Minnesota	X	X	X	X	X	X[14]
Mississippi	X	X[5]	—	—	—	—
Missouri	X	X[5]	X	X	X	—
Nebraska	X	X	X	X	X	—
Nevada	X	—	—	X	—	—
New Hampshire	X	X	X	—	—	X[3]
New Jersey	X	—	X[15]	—	—	X[3]
New Mexico	X	—	—	X	—	—
New York	X	X[5]	—	—	—	—
North Carolina	X	X	X	X	X	X[16]
North Dakota	X	—	—	X	X	—
Ohio	—	X	X	—	—	—
Oklahoma	—	X	—	—	—	—
Oregon	X	X	—	X	X	X[7]
Pennsylvania	X	X	X[17]	—	—	—
Rhode Island	X	X	—	—	—	—
South Carolina	X	X	X	X	X	—
South Dakota	X	X	X	X	X	X[18]
Tennessee	X	X	—	X	—	—
Texas	X	X	—	X	—	—
Utah	X	X	—	X	—	—
Vermont	X	X	—	X	X	X[19]
Virginia	X	—	—	—	—	—
Washington	X	X	—	X	—	—
West Virginia	X	X	X	X	—	—
Wisconsin	X	X	X	X	X	X[18]
Wyoming	X	—	—	—	—	—

[1] White goods containing CFC gases, mercury switches, and PCBs; [2] Grass clippings; [3] Mercury batteries; [4] Disposal ban on demolition debris, devices containing mercury banned from incinerators; [5] Whole tires; [6] Landfills must divert 75 percent of commercial and 50 percent of residential green waste, or face a ban; [7] Scrap automobiles; [8] 1996 legislation allows incidental disposal of amounts of whole tires; [9] Leaves and woody vegetation greater than 3 feet in length; [10] Nondegradable grocery bags, carbonated beverage containers and liquor bottles with deposits; [11] NiCad, mercuric-oxide batteries; [12] Separately collected loads of yard trimmings; [13] Glass and metal containers, recyclable paper and single polymer plastics; [14] NiCad batteries, telephone books and sources of mercury, motor vehicle fluids and filters; [15] Leaves; [16] Antifreeze; [17] Leaves and brush; [18] Old newsprint, corrugated and paperboard, glass, steel and aluminum containers; [19] Various dry cell and NiCad batteries, paint.

Source: Adapted from Glenn (1999).

portion of the yard waste stream. In New Jersey (which in 1987 was the first state to pass a yard waste ban) and in Pennsylvania the prohibition against disposal does not include grass. There are instances where the bans are not absolute. In Pennsylvania and South Carolina, the bans pertain only to loads that are primarily yard waste. In both Florida and North Carolina, yard waste is banned from incinerators and certain classes of landfills.

Another ban that can have significant effect on the amount of disposal is white goods. By the end of 1997, white goods had been banned in 18 states (U.S. EPA, 1999). In 15 of the 18 states, the ban applies to all the appliances that are discarded. However, in Arizona, Delaware, and Illinois, it applies only to those appliances that have not had CFC gases, mercury switches, and/or PCBs removed.

Since 1991, another item that has become the target of bans is dry cell batteries. In H.B. 7216, Connecticut put a disposal ban on mercury oxide batteries. Minnesota prohibits the disposal of rechargeable nickel-cadmium batteries. The Vermont legislature passed a ban on mercuric oxide, silver oxide, nickel-cadmium, and sealed lead acid batteries used in commercial applications. Additionally, the Vermont law bans the disposal of retail nickel-cadmium batteries and bans alkaline batteries from incinerators. By the end of 1997, six states had passed bans on at least some household batteries (U.S. EPA, 1999).

While many states have legislatively banned multiple materials from disposal, none have developed as extensive a list as Wisconsin. In its 1989 recycling act, the legislature banned from disposal: appliances, waste oil, automotive batteries, yard waste, cardboard boxes, glass containers, newspapers, plastic bottles, office paper, magazines, steel cans, tires, aluminum cans, and foam polystyrene packaging (waste oil, yard waste, and tires are allowed to be incinerated if there is energy recovery). Wisconsin estimates that those items account for about 60 percent of the discarded waste in the state. Other states with extensive lists of banned items include Massachusetts, Oregon, and South Dakota (see Table 3.8).

3.11   MAKING PRODUCERS AND RETAILERS RESPONSIBLE FOR WASTE

Several states have attempted to shift some of the burden of waste disposal and recovery of materials back to the manufacturers of products. To accomplish this shift, one approach is to require manufacturers to take back products or packages after their useful life has expired.

Beverage Container Deposits

The first attempts at making manufacturers responsible for products came in the 1970s with the passage of mandatory deposits on selected beverage containers. The first mandatory deposit law was passed in Oregon in 1971. Since then, another eight states have passed mandatory deposit legislation, the last being New York in 1983.

The principal focus of these laws is the packages that are used for soft drink and beer containers, and to a lesser extent, mineral water and liquor. Maine has the most extensive list of containers. It includes the four mentioned previously as well as juice, water, and tea. In nine states the legislation includes glass, steel, aluminum, and plastic containers. Delaware has exempted aluminum cans from the deposits (see Table 3.9).

Another approach similar to mandatory deposits for beverage containers was taken in California in 1987. California's law is different from the other nine in that no actual deposit is paid by the consumers; instead distributors pay either two or four cents per container (based on size) into a state-administered fund. Consumers returning the containers to state-approved redemption centers receive 2.5 cents for each container under 24 oz and five cents for those over 24 oz.

Deposit requirements have been applied to other products, most notably auto vehicle batteries. The first vehicle battery deposit law was passed by the Rhode Island legislature in 1987.

TABLE 3.9 State Mandatory Deposits and Take-Back Laws

State	Type of product	Deposit or take back	Act ID	Effective year	Effective date
Arizona	Vehicle batteries	Take-back	H.B. 2012	1990	9/90
Arkansas	Vehicle batteries	Deposit*	HB1170	1991	7/92
Connecticut	Beverage containers		Sec. 22A 243-246	1978	1980
Connecticut	Mercury oxide batteries	Take-back	H.B. 7216	1991	1/92
Delaware[†]	Beverage containers	Deposit	Title 7, Chap. 60	1979	1982
Florida	Ni-cad batteries	Take-back	H.B. 461	1993	10/95
Idaho	Vehicle batteries	Deposit	H.B. 122	1991	7/91
Illinois	Vehicle batteries	Deposit	PA86-723	1989	9/90
Iowa	Beverage containers	Deposit	Chap. 445C	1978	1979
Louisiana	Vehicle batteries	Take-back	185	1989	8/89
Maine	Beverage containers	Deposit	P.L. 1975, C. 739 (as amended)	1975	1978
Massachusetts	Beverage containers	Deposit	301 CMR 4.00	1981	1983
Michigan	Beverage containers	Deposit	M.C.L. 445.571-576	1976	1978
Michigan	Vehicle batteries	Deposit	P.A. 20	1990	1/93
Minnesota	Vehicle batteries	Take-back	325E.115		
Mississippi	Vehicle batteries	Take-back	S.B. 2985	1991	7/91
Missouri	Vehicle batteries	Take-back	S.B. 530	1990	1/91
Nevada	Tires	Take-back	A.B. 320	1991	1/92
New Jersey	Vehicle batteries	Take-back	S.B. 2700	1991	10/91
New York	Beverage containers	Deposit	Title 10, C. 200	1982	1983
New York	Vehicle batteries	Deposit	Chapt. 152	1990	1991
North Carolina	Vehicle batteries	Take-back	H.B. 620	1991	10/91
North Dakota	Vehicle batteries	Take-back	H.B. 1060	1991	1/92
Oregon	Beverage containers	Deposit	O.R.S. 459.810-.890	1971	1972
Oregon	Vehicle batteries	Take-back	H.B. 3305	1989	1/90
Pennsylvania	Vehicle batteries	Take-back	101	1988	9/88
Rhode Island	Vehicle batteries	Deposit	23-60-1	1987	7/89
South Carolina	Vehicle batteries	Deposit	S.B. 366	1991	5/92
Texas	Vehicle batteries	Take-back	S.B. 1340	1991	9/91
Utah	Vehicle batteries	Take-back	H.B. 146	1991	1/92
Vermont	Beverage containers	Deposit	Title 10, C. 53	1972	1973
Vermont	Vehicle batteries	Take back	Title 10, C. 6622	1991	7/93
Washington	Vehicle batteries	Deposit	E.S.H.B. 1671	1989	8/89
Wisconsin	Vehicle batteries	‡	335	1990	1/91
Wyoming	Vehicle batteries	Take-back	W.S. 35-11-509-513	1989	6/89

* Retailers must take back lead-acid batteries.
† Any container that holds a carbonated beverage, except aluminum cans.
‡ Retailers are required to accept old lead acid batteries when a person purchases a new one and may place up to a $5 deposit on a battery which is sold.

Source: Revised from Kreith (1994).

Original Source: Steutville et al. (1993).

Since then, five other states (Arkansas, Idaho, Michigan, South Carolina, and Washington) have put deposits on auto batteries.

Beyond putting an added value on the batteries in question, these laws also require that retailers take back at least as many old batteries as customers buy new ones and also require that wholesalers accept the old batteries from the retailers.

Take-Back Provisions

What has become even more prominent than deposits on batteries is requiring retailers and then wholesalers in turn to take back products. As of the end of 1998 a total of 17 states had laws requiring retailers to accept batteries from consumers (see Table 3.9). These actions have come about since 1988, when Pennsylvania passed a "take-back" provision without including a deposit.

But the first state to require retailers to accept what they sell was Minnesota. In 1985, it applied that concept to tires. Arizona, Indiana, Kansas, and Nevada have since passed similar requirements. Additionally, Connecticut and Vermont passed legislation in 1991 mandating retailers to take back mercuric oxide batteries and lead acid auto batteries respectively.

Mandating Manufacturer Responsibility

In 1991, a number of the state legislatures developed another approach to requiring manufacturers to become responsible for their products. Vermont's H.B. 124 requires that manufacturers of dry cell batteries containing mercuric oxide electrode, silver oxide electrode, nickel-cadmium, or sealed lead, "ensure that a system for the proper collection, transportation and processing" be put in place. As part of that system, a manufacturer has to develop a link between the consumer and itself, and accept the batteries it produces at its manufacturing facility (Glenn, 1992). In 1992, two other states applied the same type of requirements to other products. Maryland passed S.B. 37, which required manufacturers of mercury oxide batteries to set up a collection program and Minnesota passed a law requiring telephone directory manufacturers to set up recycling programs (Stoutville and Goldstein, 1993).

3.12 ADVANCED DISPOSAL FEES

Advanced disposal fees (ADF), where the cost of disposal is at least partially paid for upfront, have been considered for a wide variety of products and packages. To date, states have put fees on only a limited range of products. All 28 of the states with some form of ADF legislation have put them on tires (see Table 3.10).

Only four states have put advanced disposal fees on multiple products. In Rhode Island there is a 50-cent fee on new tires, a 5-cent/quart charge on motor oil, a 10-cent/gallon fee on antifreeze, and a $0.0025/gallon charge on organic solvents. Other states with fees on multiple products include Maine (tires and vehicle batteries), South Carolina (tires, motor oil, white goods, and vehicle batteries), and North Carolina (tires and white goods). Florida had a $2 per ton fee for newsprint not made with at least 60 percent recycled paper fiber, but since the newsprint mandate was met, the fee has been effectively rescinded.

Twenty-one states require that the fees on tires be collected at the retail level. All but two of those place flat fees on each new tire sold. Those fees on tires range from 25 cents per tire in California to $2 per tire in South Carolina. Arizona and North Carolina's are 2 percent per tire. In Rhode Island, in addition to the fees placed on individual items, a $3 fee is put on each new car sold to cover the cost of materials that are hard to dispose of.

Michigan and Minnesota, the two states that do not collect the tire fee at the retail level, place it on the title transfer when a vehicle is sold. Michigan charges 50 cents and Minnesota

TABLE 3.10   States That Have Enacted Packaging and Product Taxes and Fees

State	Type of product or package	Type of fee or tax	Act ID	Effective year	Effective date
Arizona	Tires	2% new tire*	H.B. 2687	1990	9/90
Arkansas	Tires	$1.50/new tire	8-9-404.	1991	7/91
California	Tires	$0.25/new tire	A.B. 1843	1989	7/90
Florida	Tires	$1/new tire	S.B. 1192	1988	1/90
	Newsprint‡	$2/ton	S.B. 1192	1988	1/99
Georgia	Tires	$1/new tire	H.B. 1385	1992	1/93§
Illinois	Tires	$1/new tire	H.B. 989	1991	7/92
Indiana	Tires	$0.25/new tire	H.B. 1427	1993	1/94
Kansas	Tires	$0.50/new tire	S.B. 310	1990	7/90
Louisiana	Tires	*	Act 185	1989	1/93
Maine	Tires	$1/new tire	Chapt. 585	1989	7/90
	Vehicle batteries	$1 each	(as amended)		7/90
Maryland	Tires	$1/new tire	H.B. 12.02	1991	2/92
Michigan	Tires	$0.50/title transfer	Act 133	1990	1/91
Minnesota	Tires	$4/title transfer	Chapt. 654	1984	9/84
Mississippi	Tires	$1/new tire	S.B. 2985	1991	1/92
Missouri	Tires	$0.50/new tire	S.B. 530	1990	1/91
Nebraska	Tires	$1/new tire	L.B. 163	1990	10/90
Nevada	Tires	$1/new tire	A 320	1991	1/92
North Carolina	Tires	2% sales tax	S.B. 111	1989	1/90
	White goods	$10/CFC-containing appliance ($5/non)	S.B. 60	1993	1/94
Oklahoma	Tires	$1/new tire	H.B. 1532	1989	7/89
Rhode Island	Tires	$0.50/new tire	H 5504†	1989	1/90
	Motor Oil	$0.05/quart			1/90
	Antifreeze	$0.10/gallon			1/90
	Organic Solvents	$0.0025/gallon			1/90
Ohio	Tires	$0.50/tire	S.B. 165	1993	10/94
South Carolina	Tires	$2/new tire	S.B. 388	1991	11/91
	Motor oil	$0.08/gallon			11/91
	White goods	$2 each			11/91
	Vehicle batteries	$5 each			11/91
Texas	Motor oil	$0.02/quart	S.B. 1340	1991	9/91
Utah	Tires	$1/new tire	H.B. 34	1990	1/91
Virginia	Tires	$0.50/new tire	H.B. 1745	1989	1/90

* Tax cannot exceed $2 per tire.
† $3 for each new vehicle purchased to cover all materials that are hard to dispose of.
‡ Florida's newsprint fee has been rescinded because the 60% recycled content requirement for 1999 has been met.
§ This law sunset on June 30, 2000.
Source: Revised from Kreith (1994).
Original Source: Steutville et al. (1993).

charges $4 per transfer. In South Carolina and Ohio, the states have opted to collect all the fees they charge at the wholesale level. While this approach hides the cost of the fee from the consumer, it makes collection easier.

Perhaps one reason for the popularity of ADFs is that they provide a substantial funding source. In most cases, tire fees are utilized to fund tire cleanup and recovery programs. In others, such as South Carolina, the funds will be used to help finance all state waste reduction efforts.

Although ADFs on tires and white goods have appeared to be an effective way to finance the disposal of tires and white goods, it has not been as successful with other products. Florida's 1992 legislation that placed ADFs on glass, metal, and plastic containers sunset in 1995. Despite Florida's need to manage the high level of container waste associated with its tourism industry there is no indication that a new version of the legislation will be introduced.

3.13   SPECIAL WASTE LEGISLATION

Beyond waste reduction, another trend in state legislation is an increasing awareness that certain waste products need special consideration when it comes to developing a legislative package. For example, special legislation is required for such products as scrap tires, used motor oil, and household hazardous waste. States are going beyond simply taxing a material and/or banning it from disposal sites and are developing comprehensive management programs. For instance, 1991 saw Arkansas pass laws which developed a permit program for waste tire facilities, required solid waste management districts to establish collection sites, and provided grants for a variety of public education and collection activities. Pennsylvania has established a grant program for municipalities to establish hazardous waste collection programs.

Tires

The disposal of used tires has been a vexing problem for years. Whole tires cause operational problems at landfills (they tend to work their way to the surface after being buried). And with the exception of burning the tires as a fuel source, there has been only limited success at utilizing scrap tires. The result is thousands of tire piles, some numbering in the millions, throughout the United States.

The tire management laws in most states do not ban disposal outright. In all, 38 states have bans (see Table 3.8). However, of those states with tire bans, nine states (Hawaii, Illinois, Iowa, Kansas, Kentucky, Louisiana, Mississippi, Missouri, and New York) allow disposal if the tires have been shredded, chipped, or halved. To keep track of who is collecting and transporting tires some states also have put permitting or registration requirements into legislation. For instance, Florida and Iowa register haulers. Iowa requires haulers to be bonded. In Washington, legislation requires that all used tire haulers be licensed and pay a $250/year fee. Perhaps most important, haulers have to document where the tires were delivered. Georgia has established a manifest-and-tracking system for scrap tires as well.

Disposal and processing facilities have also come under legislative oversight. In Kentucky, H.B. 32, which passed in 1990, required that piles with more than 100 tires had to be registered with the Department of Environmental Protection. Missouri dictates that anyone storing more than 500 tires for longer than 30 days must be permitted by the state.

Other states are taking scrap tire management one step further. To ensure that collection and disposal facilities are available, some are making it the responsibility of local governments. Counties in North Carolina have had to provide collection sites, as do regional solid waste management authorities in Arkansas. A.B. 1843 in California includes provisions for a system of designated landfills that will accept and store shredded tires.

Another element of state tire management programs is the provision of grants to perform a variety of tasks. Arizona's tire disposal law provides that grant monies be used for counties and private companies to establish tire processing facilities, and for counties to establish collection centers and contract for hauling and processing services. A number of states, including Kentucky, Michigan, Minnesota, and Oklahoma use grant monies to help with the cleanup of existing disposal sites. To help develop utilization programs, some states such as

Georgia and Illinois use monies to fund innovative technology development. In 1996, Iowa started a Waste Tire Management Fund to appropriate $15 million by 2002 to foster the creation of new markets for used tires. Companies throughout the country have found markets for using used tires as crumb rubber, as a substitute for sand and gravel in landfill leachate control, running tracks and playground covers. These and other disposal technologies are treated in later chapters.

Used Oil

A second special waste to be tackled by state legislators is used motor oil. Unlike tires, there are not significant accumulations of used oil around the country. But that is not to say that poor used-oil disposal practices do not cause environmental harm. Less than two-thirds of the oil used in this country is accounted for. The speculation is that much of the remainder gets dumped in the drain or onto the ground.

Much of state used-oil legislation is directed at providing "do-it-yourselfers" with sufficient collection alternatives, and ensuring that once collected, the oil is properly handled and processed. In states such as Texas, the program consists of the voluntary establishment of used oil collection sites by private industry and local government. South Carolina has a similar provision, although its Department of Highways and Public Transportation is ultimately responsible for ensuring that at least one collection facility exists in each county. Additionally, both statutes require, as many other states do, that retailers post signs about a citizen's responsibilities and where to get additional information. In other states, like Arkansas, New York, and Wisconsin, retailers of motor oil must establish collection sites for used oil. Tennessee law requires each county has to provide a collection for not only used oil, but other automotive fluids such as antifreeze, as well as scrap tires and batteries.

But today states are going far beyond simply encouraging the development of collection programs. As with scrap tires, a growing number of states are seeking to control who collects used oil and what is done with it. South Carolina requires that used oil transporters register with the Department of Health and Environmental Control. They must submit annual reports to the Department and have liability insurance. Used-oil recycling facilities in the state must be permitted. In Arizona, used-oil transporters have to register as a hazardous waste transporter. The transporter must also manifest any used-oil shipments.

Household Hazardous Waste

The first state effort to manage household hazardous waste (HHW) was probably in Florida. In the mid-1980s, it provided a series of temporary collection opportunities throughout the state. Since then, other states have begun to develop comprehensive HHW management programs. For instance, in 1988, Pennsylvania's Act 101 established the "Right-Way-to-Throw-Away" program. In addition to initiating a grant program of local governments wanting to develop HHW collection programs, it also required the Department of Environmental Resources to register the programs, establish operational guidelines, and inspect sites.

3.14   MARKET DEVELOPMENT INITIATIVES

States have come to recognize that developing recycling programs has to encompass more than the supply side of the equation. Markets hold the key to sustain growth in recycling. Therefore, states have included market development initiatives in waste reduction legislation. The earliest market development efforts were directed at providing financial incentives to companies willing to convert to the use of recycled feedstock. In the mid-1970s, Oregon instituted a series of tax credits that companies could use if they made recycling-related capital improvements.

New Jersey was the first state to incorporate a full range of financial incentives in legislation. Its 1987 omnibus recycling law included a market development package that contained tax credits, loans, grants, and sales tax exemptions (see Table 3.11). Since then, numerous states have followed suit. Twenty-six states have tax credit programs. Most of the state legislative activity promoting tax credits took place in 1992 when six states—Arizona, Iowa, Kansas, New York, Pennsylvania, and Virginia—passed such legislation (Stoutville et al., 1993). The amount of credit given on income tax typically ranges from Colorado's 20 percent to Louisiana's 50 percent.

Montana put an innovative piece of tax legislation on the books in 1991. In addition to providing for a 25 percent income tax credit, the state now gives a tax deduction to encourage businesses to purchase recycled goods. The law, S.B. 111, allows for a deduction of 5 percent "of the taxpayer's expenditures for the purchase of recycled material that was otherwise deductible by the taxpayer as a business-related expense." Montana extended this legislation in 1997 (S.B. 336).

Although the bulk of state tax credit programs were developed in the early 1990s, states continued to implement programs throughout the decade. In 1997 Virginia H.B. 544 provided a 10 percent tax credit on equipment purchases used in facilities to manufacture items from recyclable material. H.B. 595, of that same year, provides a tax exemption for certified recycling equipment.

Because the effectiveness of tax credits diminishes when they are applied to firms just starting, some states also give out low-interest loans and grants. In all, 33 states have grants and loan programs (Goldstein and Glenn, 1997).

In addition to financial and market incentives, a number of states recognize that advances in market development require the coordinated action of many players in both the public and private sectors. To help in that coordination, numerous states have put together market development councils. For instance, South Carolina's Solid Waste Policy and Management Act established a council that was to analyze existing and potential markets and make recommendations on how to increase the demand for recovered materials. In Tennessee, not only was a markets council established, but its Department of Economic and Community Development had to set up an office of cooperative marketing.

Minimum Content Standards

Another approach states have taken to assist in the development of markets is to directly intervene in the market. In 1989, California and Connecticut passed statutes that required newspaper publishers to utilize print made with recycled paper or face fines (see Table 3.12). Since then another 11 states have passed newsprint "minimum content" legislation, and newspaper publishers in another 11 states have voluntarily agreed to increase purchases of newsprint containing recycled fiber (Glenn and Riggle, 1989a). As a result of these actions, numerous paper mills in both the United States and Canada have converted to the use of deinked fiber.

The content standards vary significantly in the statutes, ranging from Oregon's 7.5 percent of postconsumer fiber to West Virginia's requirement that 80 percent of the newsprint used by the newspaper publishers contain the highest postconsumer recycled paper content practicable. For the most part the standards were to be phased in through the year 2000.

Now that content standards for newsprint have started to take hold around the country, lawmakers are beginning to utilize them to tackle other products as well. In 1991 Maryland and Oregon passed laws that require phone directories to have recycled content of 40 percent and 25 percent respectively. Oregon is also pushing recycled content standards for plastic and glass containers. In the case of glass containers, the Oregon law (S.B. 66) requires that each glass container manufacturer use a minimum of 50 percent recycled glass by January 1, 2000.

TABLE 3.11 State Financial Incentives to Produce Goods Made with Recycled Materials

State	Tax credits	Loans	Grants	Other
Alabama	Yes	—	Yes	—
Arizona	Yes	—	Yes	—
Arkansas	Yes	—	Yes	—
California	—	Yes	Yes	—
Colorado	—	Yes	Yes	—
Connecticut	—	Yes	Yes	—
Delaware	Yes	—	—	—
Florida	Yes	—	—	—
Georgia	—	—	Yes	—
Hawaii	Yes	Yes	—	—
Idaho	—	—	—	Property tax exemption
Illinois	—	Yes	Yes	Property tax abatement
Indiana	—	Yes	Yes	—
Iowa	—	Yes	Yes	Sales & property tax exemptions
Kansas	—	Yes	Yes	—
Kentucky	Yes	—	—	—
Louisiana	Yes	—	—	—
Maine	—	Yes	Yes	—
Maryland	Yes	Yes	Yes	—
Massachusetts	—	Yes	Yes	—
Minnesota	Yes	Yes	Yes	—
Mississippi	—	Yes	Yes	—
Missouri	—	Yes	Yes	Tax exemption
Montana	Yes	Yes	—	—
Nebraska	—	Yes	Yes	—
Nevada	—	—	—	Property tax exemptions
New Jersey	—	—	—	Tax exemptions
New Mexico	Yes	Yes	Yes	—
New York	Yes	Yes	Yes	—
North Carolina	—	—	Yes	Property tax abatement
Ohio	—	Yes	Yes	—
Oklahoma	Yes	—	—	—
Oregon	Yes	Yes	Yes	—
Pennsylvania	Yes	Yes	Yes	—
South Carolina	Yes	—	Yes	—
South Dakota	—	Yes	Yes	—
Tennessee	—	Yes	Yes	—
Texas	Yes	—	—	—
Utah	Yes	—	—	—
Vermont	—	—	Yes	—
Virginia	Yes	—	Yes	—
West Virginia	—	Yes	Yes	—
Wisconsin	Yes	Yes	Yes	—
Wyoming	—	Yes	Yes	—

Source: Revised from Kreith (1994).
Original Source: Adapted from Goldstein and Glenn (1997).

TABLE 3.12 Recycled Content Standards

State	Material	Deadline	Act
Arizona	Newsprint	2000	
California	Newsprint	2000	
	Glass containers	2005	
Connecticut	Newsprint	2000	
Illinois	Newsprint	2000	
Kentucky	Newsprint	2004	H.B. 282
Maryland	Newsprint	2000	H.B. 1148/H.B. 629
	Phone directories	2000	H.B. 1148
Missouri	Newsprint	2000	
North Carolina	Newsprint	1999	H.B. 1224/H.B. 1055
Oregon	Newsprint	1995	S.B. 66
	Phone directories	1995	S.B. 66
	Glass containers	2000	S.B. 66
	Plastic containers	1995	S.B. 66
Rhode Island	Newsprint	2001	H.B. 5638
Texas	Newsprint	2001	S.B. 1340
West Virginia	Newsprint	1997	S.B. 18
Wisconsin	Newsprint	2000	

Source: Revised from Kreith (1994).
Original Source: Steutville et al. (1993).

Rigid plastic containers are to either contain at least 25 percent recycled content, be recycled at a 25 percent rate by the same date, or be a reusable package.

Procurement Provisions

Using the purchasing power of a state is another market development tool. Procurement initiatives go back well before when most states began to seriously consider developing comprehensive recycling programs. Initially, procurement provisions were directed at paper products, but more recently they have begun to be used in conjunction with a wide variety of products from plastics to compost.

Over the years, virtually every state in the country has passed some form of legislation encouraging the governmental purchase of products made from recycled materials. Legislation tends to focus on two things: eliminating any biases against recycled products, and price preferences, particularly for paper and paper products. Additionally, some states have begun to direct their agencies to make specific purchases of recycled products.

One such law (H.B. 2020), passed in Illinois in 1991, required that by July 1, 2000, 50 percent of the "total dollar value of paper and paper products" must be recycled. In Arkansas, H.B. 1170 established a progressive goal which aimed to reach 60 percent of paper purchases by calendar year 2000. Oregon requires that 35 percent of their paper purchases be made from recycled paper, while West Virginia requires that figure to be 40 percent.

Procurement requirements are going far beyond paper these days. In addition to merely telling procurement agencies they have to give a preference to recycled products, states are now targeting what materials have to be procured. For instance, Oregon's S.B. 66 requires the purchase of re-refined oil by both state and other public agencies. Illinois mandates that recycled cellulose insulation be used in weatherization projects done with state funds. Texas now can grant a 15 percent life-cycle price preference for rubberized asphalt. Main passed a bill that requires compost to be used on all public land maintenance and landfill closures that use state funds.

3.15 STATE FUNDING

Where states come up with the money to administer their solid waste management programs has changed significantly over the years. Traditionally, the vast majority of the funds have come from an appropriation from the legislature. However, as programs have become more diverse, so too have the funding sources. In addition to advanced disposal fees discussed previously, one of the most popular sources of funds is the landfill-tipping surcharge. By 1997, 17 states had a disposal surcharge. The lowest rate in the country is Arizona's 25 cents per ton. Iowa's $4.25 per ton is the highest. Most are in the $1 to $2 per ton range (Glenn and Riggle, 1989a).

3.16 FLOW CONTROL LEGISLATION: INTERSTATE MOVEMENT OF UNPROCESSED AND PROCESSED SOLID WASTE

Although many states' concerns about reduced landfill capacity have diminished since the passage of their comprehensive state solid waste plans in the late 1980s and early 1990s, there are still some states—particularly in the mid-Atlantic region—that are fighting vigorously over solid waste disposal rights. This issue has come to be known as "flow control." Flow control measures are those whose goal is to limit the disposal of out-of-state waste (see Table 3.13).

Flow control became a national issue in 1994 when the Supreme Court ruled against it in *C&A Carbone, Inc. v. Town of Clarkstown* stating that it was unconstitutional under the Com-

TABLE 3.13 Municipal Solid Waste Imports and Exports (44 States Reporting)

State	Imported, tons	Exported, tons	State	Imported, tons	Exported, tons
Alabama	210,000	n/a	Missouri	74,690	1,570,000
Alaska	n/a	20,000	Montana	37,850	0
Arizona	226,274	n/a	Nevada	214,680	0
Arkansas	n/a	151,000	New Hampshire	715,000	148,000
California	14,000	490,000	New Jersey	600,000	1,600,000
Connecticut	412,548	261,482	New Mexico	112,160	n/a
Delaware	0	248,000	New York	n/a	4,600,000
District of Columbia	n/a	230,000	North Carolina	148,209	632,044
Georgia	193,819	n/a	North Dakota	50,000	30,000
Hawaii	0	0	Ohio	1,014,716	709,788
Illinois	4,300,000	n/a	Oregon	1,048,188	20,709
Indiana	2,800,000	230,000	Pennsylvania	9,800,000	300,000
Iowa	304,486	170,289	South Carolina	673,275	0
Kansas	1,000,000	n/a	Tennessee	297,140	64,037
Kentucky	473,500	n/a	Texas	44,813	n/a
Louisiana	58,500	n/a	Utah	3,400	n/a
Maine	138,000	130,000	Vermont	800	151,000
Maryland	60,000	762,000	Virginia	2,800,000	100,000
Massachusetts	542,000	860,000	Washington	213,322	785,741
Michigan	1,900,000	n/a	West Virginia	300,000	200,000
Minnesota	0	435,000	Wisconsin	1,200,000	n/a
Mississippi	856,000	n/a	Wyoming	0	0
Total	32,837,370	14,899,090			

Source: Adapted from Glenn (1999).

merce Clause of the U.S. Constitution. The Supreme Court made its decisions based on the following considerations:

- Whether the action controlling waste flow regulates evenhandedly with only "incidental" effects on interstate commerce, or discriminates against interstate commerce either on its face or in practical effect
- Whether the action serves a legitimate purpose, and if so
- Whether alternative means could promote this local purpose as well without discriminating against interstate commerce (Starkey and Hill, 1996)

In addition to these considerations, the Supreme Court also ruled that local flow control does fit within the limits of the Commerce Clause if it serves a legitimate local purpose. These purposes are limited to health and safety effects, not economic impacts (Kundell et al., 1993). Thus, a disposal facility owned and operated by a state or local government can refuse to accept waste if health and safety criteria are not met.

There have been several Supreme Court rulings on solid waste during the 1990s. The 1992 ruling on *Fort Gratiot Sanitary Landfill, Inc. v. Michigan DNR* invalidated a state municipal solid waste management law that banned importation of out-of-state waste. In 1994, differential disposal fees were struck down in *Oregon Waste Systems, Inc. v. Department of Environmental Quality of Oregon*. These decisions were based on the court's rulings in the 1978 case of *Philadelphia v. New Jersey,* where the court held that state and local governments could not hoard an item of interstate commerce (in this case landfill space or disposal capacity) for the benefits of their own residents at the expense of out-of-state residents. Such acts have been determined by the court to be economic protectionism (Parker, 1994).

Despite the court rulings, states are still arguing over where the waste should go. The most visible battle is between the states of New York, New Jersey, and Virginia. In 2001 New York closed its Fresh Kills landfill, and had to find a place to dispose of the 13,000 tons of garbage per day the landfill accepted in the past. The landfills in closest proximity were in New Jersey and Virginia, the governments of both of those states opposed bringing out of state waste to their landfills. For example, New Jersey's governor, Christine Todd Whitman, said she would not permit garbage barges to travel past New Jersey State beaches (Ewel, 1999). In a similar response, Virginia passed a law (S.B. 1308, 1999) to regulate the commercial transport of certain types of solid wastes by ship, barge, or other vessel on the Rappahannock, James, and York Rivers.

The constitutionality of these and other attempts at flow control will likely have to be decided by the Supreme Court unless Congress addresses the issue directly.

REFERENCES

Ewel, D. (1999) "Garbage: Can't Keep It Out, Can't Keep It In," *BioCycle,* vol. 40, no. 3, pp. 34–37.

Glenn, J. (1992) "The State of Garbage in America, Part II," *BioCycle,* vol. 33, no. 5, pp. 30–37.

Glenn, J. (1998a) "The State Of Garbage in America, Part II," *BioCycle,* vol. 39, no. 5, pp. 48–52.

Glenn, J. (1998b) "The State of Garbage in America," *BioCycle,* vol. 39, no. 4, pp. 60–71.

Glenn, J. (1999) "The State of Garbage in America," *BioCycle,* vol. 40, no. 4, pp. 60–71.

Glenn, J., and D. Riggle (1989a) "Where Does the Waste Go?" *BioCycle,* vol. 30, no. 4, pp. 34–39.

Glenn, J., and D. Riggle (1989b) "How States Make Recycling Work," *BioCycle,* vol. 30, no. 5, pp. 47–49.

Goldstein, N., and J. Glenn (1997) "The State of Garbage in America, Part II," *BioCycle,* vol. 38, no. 5, pp. 71–75.

Kreith, F. (ed.) (1994) *Handbook of Solid Waste Management,* McGraw-Hill, Inc., New York, NY.

Kundell, J. E., D. L. Ruffer, and S. Thompson (March, 1993) "Solid Waste Flow Control (Designation)," Paper presented at the Conference of Southern County Associations Regional Solid Waste/Environmental Network, pp. 3.

Parker, B. (1994) "Supreme Court Finds Clarkstown Flow Control Law Unconstitutional," *Waste Age,* vol. 25, no. 6, pp. 21.

Starkey, D., and K. Hill (1996) *A Legislator's Guide to Municipal Solid Waste Management.* National Conference of State Legislatures: Denver, CO.

Steutville, R., and N. Goldstein (1993) "The State of Garbage in America," Part I, *BioCycle,* vol. 34, no. 5, pp. 42–50.

Steutville, R., K. Grotz, and N. Goldstein (1993) "The State of Garbage in America," Part II, *BioCycle,* vol. 34, no. 6, pp. 42–50.

U.S. EPA (1999) *Characterization of Municipal Solid Waste in the United States: 1998 Update,* EPA/530-R-99-021, U.S. Environmental Protection Agency, Washington, DC.

APPENDIX: STATE SOLID WASTE REGULATORY AGENCIES

Alabama
Engineering Services Branch
Department of Environmental Management
1751 Dickerson Dr.
Montgomery, AL 36130
205-275-7735

Alaska
Hazardous and Solid Waste Management
Department of Environmental Conservation
410 Willoughby
Juneau, AK 99801
907-465-5133

Arizona
Arizona Department of Environmental Quality
Recycling Unit
3033 N. Central Dr.
Phoenix, AZ 85012
602-207-4173

Arkansas
Solid Waste Division
Department of Pollution Control & Ecology
P.O. Box 8913
Little Rock, AR 72219
501-562-7444

California
Integrated Waste Management Board
8800 Cal Center Dr.
Sacramento, CA 95826
916-255-2182

Colorado
Office of Energy Conservation
1675 Broadway, Suite 1300
Denver, CO 80202
303-668-5445

Connecticut
Department of Environmental Protection
79 Elm St.
Hartford, CT 06106
860-424-3237

Delaware
Solid Waste Authority
P.O. Box 455
Dover, DE 19903
302-739-5361

District of Columbia
Solid Waste Management Administration
Department of Public Works
2000 14th St. NW
6th Floor
Washington, DC 20001

Florida
Bureau of Solid Waste Management
Department of Environmental Regulation
2600 Blairstone Rd.
Tallahassee, FL 32399
904-922-0300

Georgia
Environmental Protection Division
Department of Natural Resources
4244 International Parkway
Atlanta, GA 30354
404-679-4922

Hawaii
Solid & Hazardous Waste Management
  Branch
Department of Health
919 Ala Moana Blvd.
Honolulu, HI 96814
808-586-4240

Idaho
Department of Environmental Quality
1410 N. Hilton St.
Boise, ID 83706
208-334-5879

Illinois
Land Pollution Control
Environmental Protection Agency
P.O. Box 19278
Springfield, IL 62704
217-782-6760

Indiana
Recycling Bureau
Office of Pollution Prevention and
  Technical Assistance
P.O. Box 6015
Indianapolis, IN 46206
317-233-5431

Iowa
Natural Resources Department
Waste Reduction Bureau
900 E. Grand Ave.
Des Moines, IA 50319
515-281-8176

Kansas
Solid Waste Management Section
Department of Health & Environment
Forbes Field Building
Topeka, KS 66620
913-296-1595

Kentucky
Waste Management
Resource Conservation Section
14 Reilly Rd.
Frankfort, KY 40601
502-564-6716

Louisiana
Department of Environmental Quality
Solid Waste Division
P.O. Box 82178
Baton Rouge, LA 70884
504-765-0249

Maine
Waste Management Agency
State House, Station 154
Augusta, ME 04333
207-289-5300

Maryland
Hazardous & Solid Waste Management
Department of the Environment
2500 Broening Hwy
Baltimore, MD 21224
301-631-3304

Massachusetts
Department of Environmental Protection
1 Winter St.
4th Floor
Boston, MA 02108
617-556-1021

Michigan
Waste Management Division
Department of Natural Resources
P.O. Box 30241
Lansing, MI 48909
517-373-4743

Minnesota
Office of Environmental Assistance
520 Lafayette Rd.
2nd Floor
St. Paul, MN 55155
612-215-0198

Mississippi
Waste Reduction and Minimization
 Division
Department of Environmental Quality
2380 Highway 80 West
Jackson, MS 39204
601-961-5241

Missouri
Waste Management Program
Department of Natural Resources
P.O. Box 176

Jefferson City, MO 65102
314-751-5401

Montana
Solid and Hazardous Waste Division
Department of Environmental Quality
P.O. Box 200901
2209 Phoenix Dr.
Helena, MT 59620
406-444-1430

Nebraska
Land Quality Division
Department of Environmental Control
P.O. Box 98922
Lincoln, NB 68509
402-471-2186

Nevada
Environmental Protection Division
Department of Conservation and Natural
 Resources
Capitol Complex
Carson City, NV 89710
702-687-5872

New Hampshire
Waste Management Division
Department of Environmental Services
6 Hazen Dr.
Concord, NH 03301
603-271-3712

New Jersey
Department of Environmental Protection
120 South Stockton St.
3rd Floor
Trenton, NJ 08625
609-984-3438

New Mexico
Solid Waste Bureau
1190 St. Francis Dr.
Santa Fe, NM 87502
502-827-2883

New York
Division of Solid Waste
Department of Environmental Conservation
50 Wolf Rd.
Albany, NY 12233
518-457-7337

North Carolina
Office of Waste Reduction
Department of Environmental Health and Natural
 Resources
P.O. Box 27626-9569
Raleigh, NC 27626
919-715-6512

North Dakota
Department of Waste Reduction
P.O. Box 5520
Bismarck, ND 58506
701-328-5166

Ohio
Division of Litter Prevention and Recycling
Environmental Protection Agency
1889 Fountain Square
Building F2
Columbus, Ohio 43224
614-265-7069

Oklahoma
Department of Environmental Quality
1000 NE 10th St.
Oklahoma City, OK 73117
405-271-7353

Oregon
Hazardous & Solid Waste Division
Department of Environmental Quality
811 SW 6th Ave.
Portland, OR 97204
503-229-5356

Pennsylvania
Bureau of Land, Recycling and Waste
  Management
Department of Environmental Resources
P.O. Box 8472
Harrisburg, PA 17105
717-787-7382

Rhode Island
Department of Environmental Management
83 Park St.
Providence, RI 02903
401-277-3434

South Carolina
Department of Health and Environmental Control
Office of Solid Waste Planning and Recycling
2600 Bull St.
Columbia, SC 29201
803-896-4000

South Dakota
Division of Environmental Regulation
Office of Solid Waste Management
523 Capitol
Pierre, SD 57501
605-773-3153

Tennessee
Division of Solid Waste Assistance
Department of Environment and Conservation
401 Church St.

14th Floor, Land C Tower
Nashville, TN 37243
615-532-0082

Texas
Division of Pollution Prevention and Recycling
Natural Resources Conservation Commission
P.O. Box 13087
Austin, TX 78711
512-239-6741

Utah
Office of Planning and Public Affairs
168 N. 1950 West
P.O. Box 144810
Salt Lake City, UT 84114
801-536-4477

Vermont
Department of Environmental Conservation
103 S. Main St.
Waterbury, VT 05671
802-241-3449

Virginia
Department of Environmental Quality
629 E. Main St.
P.O. Box 10009
Richmond, VA 23240
804-698-4000

Washington
Department of Ecology
P.O. Box 44600
Olympia, WA 98504
360-407-6097

West Virginia
Division of Natural Resources
Capitol Complex
Building 3, Room 732
1900 Kanawha Blvd. East
Charleston, WV 25305
304-558-3370

Wisconsin
Division of Environmental Quality
Department of Natural Resources
P.O. Box 7921 AD/5
Madison, WI 53703
608-266-2121

Wyoming
Solid Waste Management Program
Department of Environmental Quality
122 W. 25th St.
Cheyenne, WY 82002
307-777-7752

CHAPTER 4

PLANNING FOR MUNICIPAL SOLID WASTE MANAGEMENT PROGRAMS

James E. Kundell
Deanna L. Ruffer

Planning for the management of municipal solid waste becomes increasingly important as the complexity of management needs expands, the tools and procedures for addressing these needs require greater sophistication, and competition increases. In addition, as the roles and responsibilities of states and their subdivisions in the management of solid waste have evolved, both state and local or regional solid waste planning is required.

4.1 STATE SOLID WASTE MANAGEMENT PLANNING

Some states may have planned for solid waste management needs in the past, but there is little evidence that such planning occurred prior to the passage of the federal Solid Waste Disposal Act of 1965. The federal focus for solid waste management planning has been at the state level, and the form and substance of state solid waste planning has been responsive to federal directives. Although some local governments were provided planning grants for demonstration projects, the plans called for under federal legislation were designed to show the U.S. Environmental Protection Agency (EPA) that states had the authority and capability to oversee the management of solid waste within their borders. In recent years, however, states and local governments have found solid waste management planning necessary without any federal directives or incentives to carry out such planning.

Historic Perspective: State Planning

The federal Solid Waste Disposal Act of 1965, like other environmental laws passed in the 1960s, did not establish a federal permit requirement for solid waste management facilities, but focused initially on the provision of "financial and technical assistance and leadership in the development, demonstration, and application of new and improved methods and processes to reduce the amount of waste and unsalvageable materials and to provide for proper and economical solid waste disposal practices." [P.L. 89-272, Sec 202(6)] In addition to supporting research and demonstration projects and efforts toward regional solid waste management solutions, the federal law identified planning for solid waste disposal as an important component. The Secretary of Health, Education and Welfare was directed to encourage regional solid waste management planning (P.L. 89-272, Sec 205) and to provide 50 percent matching grants to states to make surveys of solid waste disposal practices and problems within their jurisdictions and to develop solid waste disposal plans (P.L. 89-272, Sec. 206). To consider all aspects essential to statewide planning for the proper and effective disposal of solid waste, the law identified factors to be considered in planning such as "population growth, urban and metropolitan development, land use planning, water pollution control, air pollution control, and the feasibility of regional disposal programs" [P.L. 89-272, Sec. 204(2)].

The emphasis of the 1965 law was, in part, to generate a database for existing solid waste disposal problems and efforts. It must be remembered that this law predated the creation of the EPA and state environmental agencies. Consequently, most states had not assigned the responsibility of overseeing solid waste disposal practices to a state agency. Because permits and reports were not routinely required by states, states were not in a position of knowing what they were dealing with. Surveys to build a database and a better understanding of practices and problems thus became an important first step for state solid waste management planning. Beyond generating information, the plans were not very useful by today's standards (Lewis, 1992).

The Solid Waste Disposal Act was amended in 1970 with the passage of the Resource Recovery Act (P.L. 91-512). This law provided funds for planning and development of resource recovery facilities and other solid waste disposal programs. States were eligible for 75 percent federal/25 percent state grants for conducting surveys of solid waste disposal practices and problems and for "developing and revising solid waste disposal plans as part of regional environmental protection systems for such areas, providing for recycling or recovery of materials from wastes whenever possible and including planning for the reuse of solid waste disposal areas and studies of the effect and relationship of solid waste disposal practices on areas adjacent to waste disposal sites" (P.L. 91-512, Sec. 207). In addition, funds were allotted to plan for the removal and processing of abandoned motor vehicle hulks.

Grants were also allowed for planning and demonstration of resource recovery systems or for construction of new or improved solid waste disposal facilities. Interestingly, mass burn steam and electric power generating systems were not considered by federal officials at the time to be resource recovery systems. Instead, resource recovery meant refuse-derived fuel (RDF) systems involving front-end automated materials recovery followed by combustion (Lewis, 1992). Most of this grant money was used to fund resource recovery facility demonstration projects.

This early solid waste legislation was replaced by the passage of the Resource Conservation and Recovery Act (RCRA) in 1976 (P.L. 94-580). The thrust of RCRA, primarily through Subtitle C of the act, was to remove most of the hazardous waste, principally industrial chemical waste, from the solid waste stream and to establish a separate management program for the hazardous waste. The principal solid waste section of RCRA, Subtitle D, for the first time provided legislative guidance on the preparation of state solid waste management plans. Ten factors to be considered in state planning included:

1. Geologic, hydrologic, and climatic circumstances, and the protection of ground and surface waters
2. Collection, storage, processing, and disposal methods
3. Methods for closing dumps
4. Transportation
5. Profile of industries
6. Waste composition and quantity
7. Political, economic, organizational, financial, and management issues
8. Regulatory powers
9. Types of waste management systems
10. Markets for recovered materials and energy

In addition to this guidance, requirements for plan approval were also established (P.L. 94-580, Sec. 2). To be approved by EPA, each state plan had to comply with the following six requirements:

1. Identify state, local, and regional authorities responsible for plan implementation.
2. Prohibit the establishment of new dumps.
3. Provide for the closing or upgrading of existing dumps.
4. Provide for the establishment of state regulatory powers.
5. Allow for long-term contracts to be entered into for the supply of solid waste to resource recovery facilities.
6. Provide for resource conservation or recovery and for disposal of solid waste in environmentally sound facilities such as sanitary landfills.

When RCRA was amended in 1984, (P.L. 98-616), these provisions were not altered and, as a result, this is the latest guidance provided by Congress for state solid waste management planning.

State-Initiated Solid Waste Management Planning

State solid waste management planning reemerged in the late 1980s and early 1990s as a result of state initiatives rather than federal directives. With the 1984 amendments to RCRA, Congress directed EPA to develop environmentally protective landfill standards. These Subtitle D standards were released in draft form in 1988 and in final form in 1991. The message sent by the draft regulations was that, although greater assurance would be provided that landfills would not result in environmental degradation, the cost of landfilling would increase dramatically. It was this perceived cost increase plus the difficulty in siting new disposal facilities that caused local government officials to turn to their state legislatures for help. Between 1988 and 1991, states across the country enacted legislation to help resolve the solid waste problems facing local governments. Of note is that, although these laws were in part the result of federal action (Subtitle D regulations), there was no new federal legislative guidance for states to address their overall solid waste management concerns. As a result, the legislation enacted by states varied but, due to commonalities in problems and alternative solutions and interplay among the states, similarities emerged as state after state enacted comprehensive solid waste management legislation.

One common theme identified was the need for state and local/regional solid waste management planning. One analysis found this to be less evident in those states that were among the first to enact their legislation, but common among those states that were able to build on the experience of other states (Kundell, 1991). Also, the nature and extent of the planning and requirements varied considerably from state to state. State policy makers understood the need for planning, but the type of planning needed was more nebulous. As a result, states adopted planning provisions either built on historic solid waste management planning guidance as identified in 1976 by RCRA, or developed planning requirements tailored to meet perceived needs.

There are four reasons why states have undertaken solid waste management planning:

1. To meet federal solid waste management planning requirements
2. To inventory and assess the solid waste management facilities and procedures in the state to determine future capacity needs
3. To provide guidance to local governments and the private sector on solid waste management matters
4. To set forth the state's policies and strategy for managing solid waste

All of these are valid reasons for states to plan, but the emphasis has varied with time and from state to state. All current state solid waste management plans contain three components (facility and program inventory and assessments, provision of guidance, and formation of state policy and strategy), but considerable variation exists in the emphasis placed on each component. For this reason, it is possible to categorize state solid waste management plans based on their emphasis.

Inventory and assessment documents tend to follow the historic model for state solid waste management plans. They attempt to quantify the status of programs and facilities in the state and identify problems that must be addressed. The plans prepared in Alabama and Rhode Island are of this type. Plans taking the form of technical assistance documents are generally designed to identify problem areas with solid waste management and to provide guidance to local officials (and others) on how to address the problems. The plans developed in Indiana and Tennessee emphasize this approach. The third type of plan, appearing more recently, takes the form of policy documents that set forth the state strategy for reducing and managing solid waste and present the strategy for doing so. The plans developed for Georgia and New York are policy and strategy documents.

Inventory and assessment has been a big part of state solid waste management planning since the 1960s. Consequently, it is not surprising that some states continued this approach. It is interesting, however, that state plans have appeared that vary from the historic model in that they are designed to meet identified state needs rather than federal directives. Plans that provide technical assistance to local governments are addressing perceived needs. Since local governments are the ones that have been faced with financing increasingly expensive solid waste facilities, attempting to site facilities that no one wants near them, and responding to the concerns of irate citizens, it is not surprising that states would use the state solid waste management plan as a mechanism to provide local governments with guidance. Consequently, the use of state solid waste management plans as vehicles for providing technical assistance to local governments has increased in recent years.

The approach that differs most from the traditional model for state solid waste management plans, however, is using the plan as a policy and strategy document. The historic role of the state in solid waste management has been to provide guidance and technical assistance and to regulate disposal activities. These responsibilities were generally assigned to one agency and had little direct impact on other units of state government. With the increased complexity of integrated solid waste management, however, multiple state agency involvement is now the norm. Although one agency still retains regulatory authority, other agencies may be involved in planning, recycling programs, market development efforts, procurement of products made from recovered materials, education, enforcement, and so forth. With this complexity comes the need to clearly articulate the policies and goals of the state, and to set forth a strategy that assigns responsibilities to agencies and identifies the actions necessary for goal attainment.

The appearance of these new solid waste management planning efforts underscores the recognition of the complexity and interrelatedness of efforts to reduce and effectively manage solid waste. Integrated solid waste management is multifaceted, and decisions made to address one matter will likely affect other components of the system. Thus a systems management orientation is emerging that requires a continuous loop of planning and feedback. The result is a stronger commitment to solid waste management planning by state and local officials.

Emergence of a New Model for State Solid Waste Management Planning.   The *North Carolina Recycling and Solid Waste Management Plan* (State of North Carolina, 1992) exemplifies the type of planning that emerged to address state needs. It is composed of three volumes: Vol. 1 is an assessment of local and regional infrastructure and resources; Vol. 2, a policy document that identifies state goals and the actions necessary to achieve those goals, is the state strategy for reducing and managing solid waste; and Vol. 3 provides guidance and technical assistance to local governments. Some of the elements of the strategy are derived directly from existing legislative mandates, and some were developed as a result of the research conducted as part of the planning process. Each section of the strategy discusses a certain aspect of solid waste management, including:

- Solid waste reduction, reuse, recycling, and composting
- Waste processing and disposal
- Illegal disposal of solid waste
- Education and technical assistance
- Planning and reporting
- Resources

Since the intent of the strategy is to forge a clear path to meet state and local solid waste management needs, it clearly states goals and the actions necessary to meet those goals. A total of 29 goals were identified. For example, nine goals were presented for solid waste reduction, five for waste processing and disposal, and so forth. Each goal statement focuses on a major effort required to effectively implement a principal component of the state solid waste strategy. Generally, the goals are based on policies established in legislation or that require actions by the General Assembly to implement or alter them. Goals are ordered to be consistent with the hierarchy of decision making in an integrated solid waste management program, not necessarily in the order of priority for implementation.

A total of 185 specific actions were identified to implement the 29 goals. Implementation actions identify the specific state agencies responsible for taking the action. Since time had elapsed between the enactment of the comprehensive state legislation and the development of the state plan, considerable effort had been exerted to implement portions of the act. Consequently, progress toward implementing each goal is also presented. Progress made reflects the resources available and the priorities of the implementation agencies.

Once a goal has been established, actions necessary to reach that goal identified, and attempts made to implement the identified actions, problems and issues associated with the goal or the actions may become apparent. To provide policy makers and agency personnel with insights into how these potential problems might be addressed, for each goal there is a section on future issues and guidance.

The plan recognizes that it was not possible for the state to achieve all 29 goals or implement all 185 actions at one time. Consequently, it was necessary to prioritize the goals and actions so that the most important ones would be achieved first, and less important ones would be implemented when resources became available. Priorities were set both for goals and for actions, based on four criteria:

1. Protection of public health and environment
2. Waste reduction
3. Promotion of integrated solid waste management
4. Formalization of organizational arrangements and responsibilities

The forcing mechanism for effectively reducing and managing municipal solid waste is the need to protect public health and the environment. Therefore, in setting priorities, those goals and actions designed to protect public health and the environment were given greater preference. Most of the major requirements to ensure that disposal facilities are environmentally benign, however, had already been adopted through rules for the design, construction, operation, closure, and postclosure care of disposal facilities. As a result, the environmental protection goals and actions included in the plan, although important, were in some cases of lower priority than other goals and actions.

Waste reduction is one way of reducing environmental problems from disposal. If the waste is never generated, it cannot pose an environmental threat when disposed. Second priority was given to those goals and actions designed to reduce the amount of waste being disposed through source reduction, reuse, recycling, and composting.

Third priority was given to those goals and actions that support an integrated approach to solid waste management. It is through an integrated approach that local governments will be better able to avoid actions that have unforeseen consequences and to balance priorities of goals and actions.

Fourth priority was given to those goals and actions designed to formalize organizational arrangements and responsibilities. The early solid waste management legislation generally assigned all related responsibilities to one agency. As solid waste reduction and management have become more complex and integrated, this is no longer possible. Many agencies have roles to play in the solid waste arena relating to in-house recycling and waste reduction, public education, market development, curriculum development, and so forth. It is important that agency roles and responsibilities be formalized through the use of memoranda of agreements (MOAs) and other mechanisms so that each agency understands its specific functions and its working relationship with other agencies, local governments, associations, industry, and the public.

The drafters of the plan also found that it was necessary to separate goals and actions in order to prioritize them. The goals are more general policy statements that may differ from the specific actions when viewed in light of the criteria for prioritizing. For example, it may be that the greatest return on investment can be achieved by taking a specific action, but it may not relate to the highest-priority goal. Thus, both goals and actions were separately prioritized. In prioritizing implementation actions, it was found that they were often interconnected. Sometimes it is difficult to proceed on one action without another one being done (e.g., even though formalizing organizational arrangements is the fourth priority, MOAs may be needed before an agency is assured of its role and/or ability to take other actions).

From this discussion of the North Carolina solid waste management plan it is apparent that the form and substance of such planning is quite different from what was proposed in the Resource Conservation and Recovery Act of 1976. Such planning efforts were undertaken without federal directives or financial assistance. It is this type of experimentation that leads to planning being relevant and of value to the states.

Revising and Updating State Solid Waste Management Plans.   A major indirect impact of the 1984 RCRA amendments was forcing states and local governments to more comprehensively plan how they would reduce and manage solid waste. The first round of plans developed by states tended to be comprehensive in nature. This was important because of the complexity and interrelatedness of management options. As states moved to implement their plans, the need for planning continued, but the form of the planning changed to be more strategic and targeted toward specific concerns.

In a telephone survey conducted of the 48 contiguous states during 2000, 39 states responded that they had a state solid waste management plan in place, and 26 states responded that they had updated their plans within the past 5 years (Adams and Kundell, 2000). As might be expected, those states with more severe solid waste management challenges were the ones more likely to be involved in solid waste management planning. In particular, Connecticut, Rhode Island, and other northeastern states, faced with limited disposal capacity within their borders and the loss of flow control as a means to direct waste to specific facilities, were focusing their planning efforts on waste reduction or diversion and capacity expansion, including long-haul options.

States in other regions of the country are also involved in solid waste planning. It is more difficult to categorize planning efforts in the other regions of the country due to variation among states within each region. Southern and some western states, however, seem to place greater emphasis on compiling and overseeing local and regional plans because local governments historically have been more involved in direct service provision in these regions, a characteristic that is changing as more local governments privatize their solid waste services. In addition, the more sparsely populated plains states placed less emphasis, in general, on the need to develop solid waste management plans.

Almost all states provide oversight and, at least some, technical and financial assistance for the development of local and regional plans. A 1998 survey of state solid waste management concerns and efforts showed that states were placing less emphasis on local and regional planning than they did in the late 1980s to the early 1990s (Kundell et al., 1998).

One lesson learned by those involved in solid waste management is that the public should be involved in the process. The desire is to structure the process so that the involvement can be positive and constructive rather than negative. Mechanisms for incorporating public involvement in the planning process are well established in the current round of planning efforts. Kansas and North Carolina are using public meetings and discussion groups to help formulate their plans. Ohio and Oregon have advisory committees or councils to assess progress and to help develop plans.

The biggest change in state solid waste management planning, however, relates to the nature of the planning. In the late 1980s and early 1990s, states developed comprehensive solid waste management plans. Revisions and updates to these plans, however, are more strategic in nature, identifying key areas of emphasis and focusing on them. Georgia, for example, updated its state plan in 1997 (State of Georgia, 1997). The revisions focused on three major areas. First, the plan compiled and presented data on what had been achieved since the passage of the comprehensive law. Second, the plan identified strategic areas where greater emphasis was needed (i.e., how the role of local governments was changing in solid waste management, how to achieve greater reduction from the commercial and industrial sectors, and how to better measure progress toward reducing the waste stream). Third, the plan set out a five-year work program for the multiple agencies involved in solid waste reduction and management efforts (i.e., Georgia Department of Natural Resources, Georgia Department of Community Affairs, and the Georgia Environmental Facilities Authority). By developing this work program, Georgia emphasized the need for interagency coordination and cooperation to meet its goals.

Current planning efforts are also focusing on needed policy changes to better address solid waste management concerns. Both Minnesota and Wisconsin, for example, produce policy reports and recommendations for consideration by their state legislatures.

As this review of current state solid waste management activities suggests, the process and nature of the state planning is changing to meet the new and emerging needs of each state. Of the four reasons previously mentioned for states to undertake solid waste management planning (to meet federal solid waste management planning requirements; to inventory and assess the solid waste management facilities and procedures in the state to determine capacity needs; to provide guidance to local governments and the private sector on solid waste management matters; and to set forth the state's policies and strategy for managing solid waste), compiling and presenting data to track progress in meeting goals (i.e., benchmarking) and setting forth the state's policies and strategies seem to be the major focus of current state solid waste management planning efforts.

4.2   LOCAL AND REGIONAL SOLID WASTE MANAGEMENT PLANNING

At the local and regional level, integrated solid waste management planning involves a wide variety of programs, facilities, strategies, procedures, and practices (elements) which together, in varying combinations, constitute a complete system of management. Beginning with the federal Solid Waste Disposal Act of 1965, it was envisioned that very detailed and comprehensive plans would be prepared, with many of the federal planning requirements placed on states finding their way into the guidelines for the preparation of local plans. Yet, until the late 1980s and early 1990s few local and regional plans actually met these expectations. It is interesting to note that current planning efforts are now attempting to do what was called for in the 1965 Solid Waste Disposal Act (i.e., defining methods for reducing the amount of waste being discarded and effectively disposing of the remainder).

In the early 1970s, as efforts were instituted to move from open dumps to sanitary landfills, emphasis was placed on increasing recycling and waste reduction. It was determined, however, that landfilling was still the least expensive alternative, and recycling efforts declined. With the adoption of more stringent federal standards for landfills, interest again focused on waste reduction and recycling. The major difference now is that there is a greater understanding of what it will take to reduce the waste stream. Central to this effort was sound solid waste management planning. The planning efforts that have occurred over the past decade have provided a greater understanding of the value of such planning.

Today, solid waste management planning efforts are being influenced by state requirements and the need for many solid waste management systems to become more cost-effective and competitive—in other words, to operate in a businesslike fashion, not simply as a government or public service. In many instances, this change in management focus is having a significant effect on how and what type of planning is being done. It is also further enhancing the understanding of the value of such planning and changing the emphasis of planning activities from the more traditional assessment of needs to the strategic analysis of needs and opportunities.

Historic Perspective: Local and Regional Planning

In the 1960s and early 1970s, local solid waste management planning was primarily a community exercise in learning to understand and view solid waste collection and disposal practices and facilities as a total system. However, all too often local government leaders gave little priority to these planning efforts, and actual decision making was seldom incorporated into or preceded by the planning process. Thus, these planning efforts were little more than academic exercises or project plans used to define and justify the development of a specific program or facility.

In the 1980s, due to the increasing complexity and interrelatedness of integrated solid waste management, local and regional planning took on renewed importance in the management of solid waste. No longer was one type of program or facility adequate, or acceptable to all parties, to manage the entire waste stream. While one part of the waste stream was suitable for recycling, another was more suited to composting or energy recovery, and still others needed to be landfilled.

As a result, planning for the management of solid waste had to take into consideration the commonalities, differences, and interrelationships between the various programs, facilities, and procedures to be used. For example, specialized handling, processing, or segregation of materials may be required with some management approaches. Other aspects of the management system may require the establishment of ordinances and fee structures or the development of educational programs. Thus, planning for the management of solid waste no longer involves a simple comparison of technical options and costs, but includes consideration of how multiple waste streams can be handled, the interrelationship between management practices, as well as consideration of business risks and requirements, public policy, and social impacts of decisions. Furthermore, since an integrated solid waste management system contains multiple facilities, processes, programs, and procedures, it is unlikely that all aspects of the management system can be developed at one time. More likely, the system would be developed over several years. As a result, provisions needed to be made within the plan and the planning process for periodic reviews, updates, and—as necessary—modification.

In the late 1980s and early 1990s, it became increasingly common for states to require local governments to plan for solid waste management. In many instances, state planning requirements focused on defining how local governments should accomplish specific state objectives such as regionalization, the provision of adequate disposal capacity, or waste reduction and recycling. Typically, the state also dictated the format and content of the plan.

In Ohio, local governments were required to form solid waste districts consisting of a population of at least 120,000 people. Each solid waste management district had to develop and adopt a solid waste management plan that described its existing facilities and its ability to accommodate the area's solid waste (Mishkin, 1989). In comparison, Pennsylvania's Municipal Waste Planning, Recycling and Waste Reduction Act required each county to develop a plan for municipal waste generated within its boundaries, with emphasis on integrating recycling into existing disposal activities. The approach taken by the state of Georgia was to develop very specific planning standards and procedures to be used by local governments in demonstrating how they intended to meet the two overall objectives of ensuring 10-year disposal capacity and reducing by 25 percent the amount of waste (on a per capita basis) requiring disposal.

The federal government provided implicit guidance on the objectives and priorities to be used in solid waste management planning (Lewis, 1992) when it released its report, *The Solid Waste Dilemma: An Agenda for Action* (U.S. EPA, 1989). In this document, EPA stated that the elements of integrated solid waste management should be prioritized as follows:

1. Reduce the generation of solid waste.
2. Recycle (including composting) for productive reuse as much as is practicable.
3. Combust and recover energy for productive use.
4. Landfill the remainder.

EPA later revised its position and stated that the third and fourth priorities were in fact equal in priority. This statement of priorities, as simple a concept as it is, enabled states, local governments, planners, and citizens alike to focus their efforts and thus develop plans and strategies that are based on a rationale that can be clearly stated and defended.

On May 16, 1994, the U.S. Supreme Court ruling on what has come to be referred to as "The Carbone" case (*C&A Carbone v. Town of Clarkstown*) forced local governments and solid waste managers to reassess the manner by which solid waste management services are funded. This court ruling has profoundly affected solid waste management planning. Prior to the Supreme Court ruling, many local governments relied on their assumed right to control the flow of solid waste generated within their community to direct solid waste to specific solid waste management facilities at which a tipping fee was charged. This tipping fee covered all or part of the costs of the solid waste management system. By controlling the flow of the solid waste, the local government was assured a source of revenue for the system.

As a result of the Carbone decision, local governments may no longer rely on flow control to direct solid waste to designated facilities. Consequently, they may no longer be able to rely on tipping fees charged at the designated management facility to fully cover the costs of the solid waste management system. They may also face competition in the provision of services from private facilities or from other publicly owned facilities. This dilemma has resulted in a significant change in the economic analysis required as part of any solid waste management planning effort. It has also increased the importance of such planning (Ruffer, 1997).

The Carbone decision has forced many solid waste management systems to become more cost-effective and competitive. In many instances, this change in management focus is having a significant effect on how and what type of planning is being done.

Planning Responsibility

The local entity responsible for planning has varied from state to state, depending on the state's planning objectives and the regulatory assignment of responsibility for solid waste management within the state. At least one state requires that regional solid waste districts be formed if a minimum population is not present within one local government (Mishkin, 1989). Other states place the responsibility for planning with those units of government that own or operate solid waste management facilities.

Who is responsible for preparing the plan influences what topics are addressed, how planning and implementation responsibilities are allocated, and the relationships that are established with other organizations and individuals involved in solid waste management in the planning area. At a minimum, solid waste management should be of significance to the entity responsible for preparing the plan, staff should have the capability to conduct the necessary analysis, and there should be a clear understanding of the authority of the entity both during plan preparation and implementation.

Individual local governments must assess and define how solid waste planning and management should be accomplished based on applicable state requirements and their own unique local circumstances. However, in determining who will be responsible for planning and how solid waste will be managed, local and regional entities must consider the public expectations for a safe, reliable, and cost-effective solid waste management system. At a minimum, local government must exercise overall responsibility for planning for municipal solid waste management and for the provision of municipal solid waste management services (SWANA, 1990).

Regardless of who is given the specific responsibility for planning, the following four concepts should be embraced by those developing the plan (McDowell, 1986).

1. *Understanding needs.* The idea is to learn more about current problems and needs and future prospects before deciding on a course of action to accomplish objectives. By this means, decisions become more rational, more objective, and based on more reliable information.

2. *Commitment to solid waste management.* Some local governments make a decision to plan for solid waste management because they are committed to addressing the issue in a logical and comprehensive manner. Others simply develop plans because they are required. A plan can easily be written down on paper, but for a plan to work, the local government must be as committed to decision making and implementation as they are to planning. This includes paying enough attention to the planning process to ensure that the plan can be implemented.

3. *Leadership.* More often than not there is a single jurisdiction, agency, or individual that is deeply committed to seeing the planning process through to fruition and in many cases carrying on to lead implementation. When interest begins to taper off, need is considered to be less critical, or tough decisions need to be made, these leaders push on.

4. *Public involvement.* A successful planning process not only defines programs but also opens up lines of communication, often among parties that rarely spoke to one another before the process. This communication results in consensus building. As a result, it helps define what management practices are really needed and which are most likely to succeed. Effective public involvement in integrated solid waste management planning and program development provides the mechanism for addressing public concerns and values at each stage of the planning and decision-making process.

The Planning Process

The development of the local or regional integrated solid waste management plan should follow a clearly defined, rational process. This process should evolve through a sequence of analysis from the definition of goals and objectives to decision making on how the goals and objectives will be achieved. The steps in this process need to allow for continuous information flow, feedback, and adjustments to the planning process. The following six-step planning process accomplishes these objectives.

1. *Goals and objectives.* The first step in the planning process should be to identify and prioritize goals and objectives for solid waste management in the planning service area. A goal statement should specify the direction and desired outcome of the solid waste management system as defined by the philosophies, values, ideals, and constraints of the community. Goal setting gives an overall, explicit purpose to the system and specific programs, facilities, and management practices in terms of the desired end result. Objectives, on the other hand, provide ways in which progress toward meeting solid waste management goals can be assessed. Monitoring objectives furnishes incremental information, or milestones/benchmarks, for gauging how well the system and specific programs are attaining the stated goals.

Goals and objectives serve as the foundation of the plan and management system. When possible, goals and objectives should be developed in a public process. Goals and objectives should be realistic and achievable, but also challenging. It may be necessary to reassess the goals and objectives at various points during the development of the plan. In addition, regular evaluation of the goals and objectives after plan adoption should be a routine part of the evaluation of the management system.

2. *Inventory and assessment.* The foundation of the plan is the inventory of what resources are currently available and the assessment of the sufficiency of these resources to meet federal, state, and local goals for a projected period of time. The inventory and assessment should consider all aspects of the existing solid waste management infrastructure as well as both public and private resources. It may be helpful to organize the inventory and assessments around the principal functional components of an integrated solid waste management system, which include waste characteristics, collection (i.e., recyclables and compostables as well as solid waste), reduction, disposal, administration, education, and financing. At minimum, a cursory assessment should also be made of the infrastructure and resources that may be available beyond the planning area borders. The inventory and assessment should evaluate factors that affect waste stream generation and existing management practices in the planning area. These will vary from area to area but may include such factors as population, economic conditions, competition, major industries, and tourism.

3. *Identifying needs.* Based on the inventory and assessment, a determination should be made of what is needed to meet federal, state, and local management goals and objectives. For example, if existing waste reduction efforts have reduced disposal needs by 8 percent but a state goal requires a 25 percent reduction in disposal, one need would be to increase reduction efforts to achieve an additional 17 percent reduction in disposal. Each local government or regional entity should develop a list of solid waste management needs before beginning to define its desired solid waste management system. Often, such an explicit process will bring out needs that would otherwise be overlooked.

4. *Evaluating management options.* For each of the defined needs, options to meet the need should be identified and evaluated. The feasibility of each option should be evaluated on technical, environmental, managerial, and economic grounds. Each option may have a number of components or a combination of components, and will have impacts on other aspects of the management system that must be taken into consideration in the evaluation.

5. *Defining the recommended management system.* Once options have been evaluated, a series of options can be selected to form the basis of the solid waste management system. Ideally, the option selected can be integrated and will meet all the needs set forth, based on the inventory and assessment. If this does not occur, it may be necessary to revisit some aspects of the evaluation of options before finalizing the management system.

6. *Developing an implementation strategy.* Once the management system is selected, an implementation strategy can be developed. The implementation strategy is the road map of actions to be taken and the measuring stick by which progress can be evaluated. It defines who is going to take what action and when. This strategy must take into consideration the process to be followed for procurement, development of facilities, funding, administration and operation, and decision making.

Consideration of how each program (existing and planned) affects other aspects of solid waste management is the essence of integrated solid waste management planning. Changing a management practice in one area almost always affects some other aspect of solid waste management. In its most basic sense, this is exemplified by the planning priorities set by EPA. The point of reducing, reusing, and recycling is to reduce the amount of solid waste requiring disposal. Integrated solid waste management requires that an examination be made of how the flow of waste will be altered by implementing certain options. In addition, consideration must be given to the impacts of waste management strategies on finances, personnel, public participation, and other areas.

Regionalization

In many instances, intergovernmental cooperation during the planning stage can be instrumental for the development of regional management systems. Through such cooperation, local government decision makers can evaluate the potential for implementing solid waste management programs on a regional level. This can include:

- Examining geographic patterns of waste generation, waste management activities, material flow, and markets for recovered materials
- Identifying areas providing or planning for duplicative or competitive services
- Identifying available resources
- Evaluating alternative strategies for allocating responsibilities through existing governmental units and/or new entities.

The criteria to be used in evaluating the potential for regionalization will vary from situation to situation. It is likely that economics will be a major criterion utilized in evaluating most regional options. Other factors that may be considered include:

- Geographic pattern of need compared to the location and capacity of proposed services
- Level and consistency of service provided
- Availability and condition of transportation routes
- Presence of physical and natural barriers
- Presence of a population center
- Need for new facilities versus the ability to utilize existing facilities
- Institutional/legislative/regulatory requirements and time frame
- Consistency with short- and long-term management objectives, individually and collectively

It is logical that these factors be considered during the planning process to determine if and how regional solid waste management options should be implemented.

Privatization

As solid waste management has become more complex, and expectations for government services have changed, political leaders have searched for different ways to provide public services without straining the capabilities of government. In some cases this has led to treating local government services and departments more like private sector businesses. With increasing frequency, the search has led to the privatization of solid waste management services. This shift in the provision of service has been aggressively pursued by the private sector over the past decade.

When privatization is considered, there are two key issues that must be understood.

1. There is no right answer to solid waste management. Although privatization can be a valuable means of management, it is by no means the only approach available. The decision to privatize depends heavily on the needs of the community, the type of service to be provided, the capabilities and availability of private firms, and long-term as well as short-term financial consequences.

2. Privatization does not eliminate the ultimate responsibility of local government. Although private involvement can help carry out services, the ultimate responsibility for the welfare of the community remains in the hands of government. Even if the local government develops a relationship in which all management activities are handled privately, it must, at a minimum, ensure that the services are meeting community needs and are cost-effective.

If there is an interest in considering privatization, this option should be taken into consideration during the inventory and assessment and the identification of needs steps in the planning process. Integrating this type of managerial decision into the planning process ensures that all waste streams and management practices and needs are accounted for in a comprehensive manner. The planning process brings each element of the solid waste management system into perspective, including reduction, collection, disposal, education, administration, and costs. The analysis of management options can examine different scenarios for ownership and operation of facilities and programs, taking into consideration the protection of human health and the environment, public opinion, and financial considerations.

Plan Implementation

An annual program of monitoring and evaluation should be established to ensure that the strategy laid out in the plan is being accomplished. As part of this annual process, it may be necessary to reassess program goals and objectives to adapt to the evaluation of what has been accomplished and what is still required to achieve the goals and objectives. It is extremely challenging to tailor activities to existing conditions as more is learned about what works well and what does not.

Once implementation of the plan begins, the problems and complexities seem to grow. Since an integrated solid waste management system includes multiple facilities, processes, programs, and procedures, it is unlikely that the entire system can be developed and put into place at the same time. Elements of the system will probably be developed and implemented over a period of many years. This means that the complete, integrated system as envisioned in the plan may not be in place for several years after plan adoption, if indeed it ever is. Since the waste stream is changing over time, markets for recovered materials fluctuate, and processing technologies continue to evolve, it is highly likely that by the time some elements of the plan are implemented some changes will be required to the programs, facilities, and procedures defined in that or other elements of the plan. This natural evolution of the management system will increase the importance of annual updates to the plan to ensure its continued usefulness as a management tool.

4.3   CONCLUSIONS

As solid waste reduction and management become more complex, and management options become more sophisticated, planning becomes increasingly important. This planning is evolving over time to truly meet the solid waste reduction and management needs of both state and local governments.

Integrated solid waste management planning is not simple. It involves what seems to be an infinite number of combinations and interactions of programs, all of which are changing constantly. Because of this, it may be more appropriate to refer to the product of the process as an integrated solid waste management *strategy* rather than an integrated solid waste management *plan*. Indeed, it is important to recognize the importance of the planning process itself in the development of a worthwhile, broadly acceptable, and implementable plan. It must also be recognized that there must be a clear, explicit, and logical rationale for the approaches, actions, and strategies that the plan proposes. At the same time, the decision-making environment in which most solid waste management planning is done makes it a complex political process. Thus, success is often not simply a function of the clarity, completeness, and quality of the technical analysis. Rather, the planning process and the plan itself must produce relevant products that policy makers can use within a context that may be much broader than the solid waste management system itself.

REFERENCES

Adams, L., and J. E. Kundell (2000) A telephone survey was conducted during May–June 2000 for updating this chapter, Vinson Institute of Government, The University of Georgia, Atlanta, GA.

Kundell, J. E. (1991) "Ten Commandments for Developing State Solid Waste Policies," Conference of Southern County Associations, Athens, GA.

Kundell, J. E., H. Lu, and D. Dudley (1998) "State Progress and Programs Since Subtitle D," *Solid Waste Technologies,* vol. 12, no. 1/2.

Lewis, S. G. (March, 1992) "Integrated Waste Management Planning (Chinese Checkers Planning)," Roy F. Weston, Inc., West Chester, PA.

McDowell, B. D. (1986) "Approaches to Planning," in F. S. So, I. Hand, and B. D. McDowell, eds., *The Practice of State and Regional Planning,* American Planning Association, pp. 3–22.

Mishkin, A. E. (1989) "Recent Federal and State Mandates and Regulatory Initiatives and Their Effects on Solid Waste Management Planning for Local Governments," in *1989 National Solid Waste Forum on Integrated Municipal Waste Management,* Association of State and Territorial Solid Waste Management Officials, pp. 259–270.

P.L. 89-272, Sec. 202 (6).

P.L. 89-272, Sec. 205.

P.L. 89-272, Sec. 206.

P.L. 89-272, Sec. 206 (2).

P.L. 91-512.

P.L. 91-512, Sec. 207.

P.L. 94-580.

P.L. 95-580, Sec. 2.

P.L. 98-616.

Ruffer, D. L. (1997) "Life After Flow Control, Assuring the Economic Viability of Local Government Solid Waste Management Systems," *Waste Age,* vol. 28, No. 1.

State of Georgia (1997) *Georgia Solid Waste Management Plan,* State of Georgia, Georgia Department of Community Affairs, Georgia Department of Natural Resources, Georgia Environmental Facilities Authority, Athens, GA.

State of North Carolina (1992) *North Carolina Recycling and Solid Waste Management Plan,* State of North Carolina, Department of Environmental Health and Natural Resources, Raleigh, NC.

SWANA (August, 1990) "The Role of the Public Sector in the Management of Municipal Solid Waste," *The Solid Waste Association of North America,* Technical Policy Position, August 1990.

U.S. EPA (1989) *The Solid Waste Dilemma: An Agenda for Action,* U.S. Environmental Protection Agency, EPA/530-SW-80-019, Washington, DC.

CHAPTER 5

SOLID WASTE STREAM CHARACTERISTICS

Marjorie A. Franklin

No matter what method of solid waste management is being considered or implemented, an understanding of the characteristics of the waste stream is a must. Good planning goes beyond developing a snapshot of current waste composition; the long-term trends in waste stream characteristics are also important. If future quantities and components of the waste stream are under- or overestimated, then facilities may be over- or undersized, and project revenues and costs can be affected.

Most of the data presented in this chapter are taken from a report published by the U.S. Environmental Protection Agency (EPA), *Characterization of Municipal Solid Waste in the United States: 1998 Update* (Franklin Associates, 1999). This report, the latest in a series published periodically since the 1970s, includes nearly 40 years of historical data on municipal solid waste generation and management, with projections to the year 2005. An additional update of 1998 municipal solid waste data prepared for the EPA in 1999 was also used (Franklin Associates, 2000).

5.1 MUNICIPAL SOLID WASTE DEFINED

The definition of municipal solid waste (MSW) used in this chapter is the same as that used in the EPA reports. This definition states that MSW includes wastes from residential, commercial, institutional, and some industrial sources.

Three other definitions are important in this chapter:

Generation refers to the amount of materials and products in MSW as they enter the waste stream before any materials recovery, composting, or combustion take place.

Recovery refers to removal of materials from the waste stream for recycling or composting. Recovery does not automatically equal recycling.

Discards refers to the MSW remaining after recovery. The discards are generally combusted or landfilled, but they could be littered, stored, or disposed on-site, particularly in rural areas.

Types of Wastes Included

The sources of municipal solid waste are as follows:

Source	Examples
Residential	Single-family homes, duplexes, town houses, apartments
Commercial	Office buildings, shopping malls, warehouses, hotels, airports, restaurants

Institutional	Schools, medical facilities, prisons
Industrial	Packaging of components, office wastes, lunchroom and restroom wastes (but not industrial process wastes)

The wastes from these sources are categorized into durable goods, nondurable goods, containers and packaging, and other wastes. These categories are defined in detail in the following sections.

Types of Wastes Excluded

As defined in this chapter, municipal solid waste does *not* include a wide variety of other nonhazardous wastes that often are landfilled along with MSW. Examples of these other wastes are municipal sludges, combustion ash, nonhazardous industrial process wastes, construction and demolition wastes, and automobile bodies.

5.2   METHODS OF CHARACTERIZING MUNICIPAL SOLID WASTE

There are two basic methods for characterizing MSW—sampling and the material flows methodology used to produce the data referenced in this chapter. Each method has merits and drawbacks, as shown:

Material Flows	**Sampling**
Characterizes residential, commercial, institutional, and some industrial wastes	Characterizes wastes received at the sampling facility
Characterizes MSW nationwide	Is site-specific
Characterizes MSW generation as well as discards	Usually characterizes only discards as received
Characterizes MSW on an as-generated moisture basis	Usually characterizes wastes after they have been mixed and moisture transferred
Provides data on long-term trends	Provides only one point in time (unless multiple samples are taken over a long period of time)
Characterizes MSW on an annual basis	Provides data on seasonal fluctuations (if enough samples are taken)
Does not account for regional differences	Can provide data on regional differences

Not mentioned in this comparison is the fact that on-site sampling can be very expensive, especially if done with large enough samples and with the frequency required for reasonable accuracy.

To date, only the material flows method has been used to characterize the MSW stream nationwide. The idea for this methodology was developed at the EPA in the early 1970s. The following methodology has been further developed and refined for EPA and other organizations over the past two decades.

Data on domestic production of the materials and products in municipal solid waste provide the basis of the material flows methodology. Every effort is made to obtain data series that are consistent from year to year rather than a single point in time. This allows the methodology to provide meaningful historical data that can be used for establishing time trends. Data sources include publications of the U.S. Department of Commerce and statistical reports published by various trade associations. Numerous adjustments are made to the raw data, as follows:

- Deductions are made for converting/fabrication scrap, which is classified as industrial process waste rather than MSW.

- Where imports and/or exports are a significant portion of the products being characterized, adjustments are made, usually using U.S. Department of Commerce data. For example, more than half of the newsprint consumed in the United States is imported.

- Adjustments are made for various diversions of products from disposal as MSW. Examples include toilet tissue, which goes into sewer systems rather than solid waste, and paperboard used in automobiles, which are not classified as MSW when disposed.

- Adjustments are made for product lifetimes. It is assumed that all containers and packaging and most nondurables are disposed of the same year they are produced. Durable products such as appliances and tires, however, are assigned product lifetimes and "lagged" before they are assumed to be discarded. Thus, a refrigerator is assumed to be discarded 20 years after production.

While the basis of the material flows methodology is adjusted production data, it is necessary to use the results of sampling studies to determine the generation of food wastes, yard trimmings, and some miscellaneous inorganic wastes. A wide variety of sampling studies from all regions of the country over a long time period have been scrutinized to determine the relative percentages of these latter wastes in MSW. Since production data are as-generated rather than as-disposed, data on food, yard, and miscellaneous inorganic wastes are adjusted to account for the moisture transfer that occurs before these wastes are sampled.

5.3 MATERIALS IN MUNICIPAL SOLID WASTE BY WEIGHT

The materials generated in municipal solid waste over a nearly 40-year period (1960 through 1998) are shown in Table 5.1 (by weight), Table 5.2 (by percentage), and in Figs. 5.1 and 5.2. Projections are shown for the year 2005. The projections are based on trend line analysis and on data about the industries involved in production of the materials and products.

Generation of MSW in the United States has grown from about 88 million tons in 1960 to about 220 million tons in 1998.* Generation is projected to grow to 240 million tons in the year 2005. (Management of this tonnage of MSW by source reduction, recycling, composting, combustion, and landfilling is discussed elsewhere in this chapter and this handbook.)

Data on the most prominent materials in MSW are summarized in the following sections.

Paper and Paperboard

For the entire historical period documented, paper and paperboard have been the largest component of the municipal solid waste stream, always comprising more than one-third of total generation. In 1998, paper and paperboard were 39.6 percent of total generation.

Paper and paperboard are found in a wide variety of products in two categories of MSW—nondurable goods and containers and packaging (Table 5.3). In the nondurables category, newspapers comprise the largest portion at about 6 percent of total MSW generation. Other

---

* Units of weight in this chapter are expressed in short tons. SI units can be obtained by multiplying short tons by 907.2 to obtain kilograms.

TABLE 5.1   Materials Generated* in the Municipal Waste Stream, 1960 to 2005
(In millions of tons)

Materials	1960	1965	1970	1975	1980	1985	1990	1995	1998	2005[†]
Paper and paperboard	30.0	38.1	44.3	43.2	55.2	62.8	72.7	81.7	84.1	94.8
Glass	6.7	8.7	12.7	13.6	15.1	13.2	13.1	12.8	12.5	11.2
Metals:										
Ferrous	10.3	11.1	12.4	12.3	12.6	11.4	12.6	11.6	12.4	13.6
Aluminum	0.3	0.5	0.8	1.1	1.7	2.2	2.8	3.0	3.1	3.8
Other nonferrous										
metals	0.2	0.6	0.7	0.9	1.2	1.1	1.1	1.3	1.4	1.3
Total metals	10.8	12.2	13.8	14.3	15.5	14.6	16.6	15.9	16.8	18.7
Plastics	0.4	1.5	2.9	4.3	6.8	11.1	17.1	18.9	22.4	26.7
Rubber and leather	1.8	2.4	3.0	3.8	4.2	4.6	5.8	6.0	6.9	7.7
Textiles	1.8	1.9	2.0	2.1	2.5	2.9	5.8	7.4	8.6	10.2
Wood	3.0	3.4	3.7	4.3	7.0	8.4	12.2	10.4	11.9	15.8
Other	0.1	0.4	0.8	1.7	2.5	3.0	3.2	3.7	3.9	4.3
Total materials in										
products	54.6	68.5	83.3	87.2	108.8	120.6	146.5	156.8	167.1	189.4
Other wastes:										
Food wastes	12.2	12.7	12.8	13.4	13.0	13.2	20.8	21.7	22.1	23.5
Yard trimmings	20.0	21.6	23.2	25.2	27.5	30.0	35.0	29.7	27.7	23.0
Miscellaneous										
inorganic wastes	1.3	1.6	1.8	2.0	2.3	2.5	2.9	3.2	3.3	3.7
Total other wastes	33.5	35.9	37.8	40.6	42.8	45.7	58.7	54.6	53.2	50.1
Total MSW generated	88.1	104.4	121.1	127.8	151.6	166.3	205.2	211.4	220.2	239.5

* Generation before materials recovery or combustion. Details may not add to totals due to rounding.
[†] Projected data.
Source: Adapted from U.S. EPA (1999) and unpublished data developed for the U.S. EPA.

important contributions in this category come from office papers, third-class mail[†] (including catalogs), and other commercial printing, which includes advertising inserts in newspapers, reports, brochures, and the like. Some paper components of MSW that attract considerable attention, such as telephone books and disposable plates and cups, amount to less than 1 percent of total MSW generation.

In the containers and packaging category, corrugated boxes are by far the largest contributor. Indeed, corrugated boxes, at 13.5 percent of total MSW generation, are the largest single product category for all materials. Other paper and paperboard contributors to MSW include folding cartons (e.g., cereal boxes), paper bags and sacks, and other kinds of packaging.

Glass

Glass in MSW is found primarily in glass containers, although a small portion is found in durable goods (Table 5.4). The glass containers are used for beer and soft drinks, wine and liquor, food products, toiletries, and a variety of other products. Glass containers were 5 per-

---

† Now called Standard (A) Mail by the U.S. Postal Service.

TABLE 5.2   Materials Generated* in the Municipal Waste Stream, 1960 to 2005
(In percent of total generation)

Materials	1960	1965	1970	1975	1980	1985	1990	1995	1998	2005†
Paper and paperboard	34.0	36.5	36.6	33.8	36.4	37.8	35.4	38.6	38.2	39.6
Glass	7.6	8.3	10.5	10.6	10.0	7.9	6.4	6.1	5.7	4.7
Metals:										
Ferrous	11.7	10.6	10.2	9.6	8.3	6.8	6.2	5.5	5.6	5.7
Aluminum	0.4	0.5	0.7	0.9	1.1	1.3	1.4	1.4	1.4	1.6
Other nonferrous metals	0.2	0.5	0.6	0.7	0.8	0.6	0.5	0.6	0.6	0.6
Total metals	12.3	11.6	11.4	11.2	10.2	8.8	8.1	7.5	7.6	7.8
Plastics	0.5	1.4	2.4	3.4	4.5	6.7	8.3	8.9	10.2	11.2
Rubber and leather	2.1	2.3	2.5	3.0	2.8	2.8	2.8	2.9	3.1	3.2
Textiles	2.0	1.8	1.7	1.7	1.7	1.7	2.8	3.5	3.9	4.3
Wood	3.4	3.2	3.1	3.3	4.6	5.0	5.9	4.9	5.4	6.6
Other	0.1	0.4	0.7	1.3	1.7	1.8	1.5	1.7	1.8	1.8
Total materials in products	62.0	65.6	68.8	68.2	71.8	72.5	71.4	74.2	75.9	79.1
Other wastes:										
Food wastes	13.9	12.2	10.6	10.5	8.6	7.9	10.1	10.3	10.0	9.8
Yard trimmings	22.7	20.7	19.2	19.7	18.1	18.0	17.1	14.0	12.6	9.6
Miscellaneous inorganic wastes	1.5	1.5	1.5	1.6	1.5	1.5	1.4	1.5	1.5	1.5
Total other wastes	38.0	34.4	31.2	31.8	28.2	27.5	28.6	25.8	24.1	20.9
Total MSW generated	100.0	100.0	100.0	100.0	100.0	100.0	100.0	100.0	100.0	100.0

* Generation before materials recovery or combustion. Details may not add to totals due to rounding.
† Projected data.
Source: Adapted from U.S. EPA (1999) and unpublished data developed for the U.S. EPA.

cent of total MSW generation in 1998, with the total contribution of glass at less than 6 percent.

As a percentage of MSW generation, glass containers grew throughout the 1960s and into the 1970s. In the 1970s, however, first aluminum cans and then plastic bottles encroached on the markets for glass bottles. As a result, glass as a percentage of MSW has dropped in the 1980s and 1990s, and this trend is projected to continue.

Ferrous Metals

Overall, ferrous metals (steel and iron) made up 5.6 percent of MSW generation in 1998 (Table 5.5). The most significant sources of ferrous metals in MSW generation are durable goods, including major appliances, furniture, tires, and miscellaneous items such as small appliances. These sources contributed enough ferrous metals to make up 4.3 percent of MSW generation in 1998.

Steel cans contribute the remainder of ferrous metals in MSW, about 1.3 percent of total MSW in 1998. The steel cans mainly package food, with small amounts of steel found in beverage cans and other packaging such as strapping. The pattern of steel in containers and packaging is much like that of glass. Steel was commonly used in beverage cans until the 1970s, when aluminum cans became popular. Steel has also been displaced in many applications by plastics.

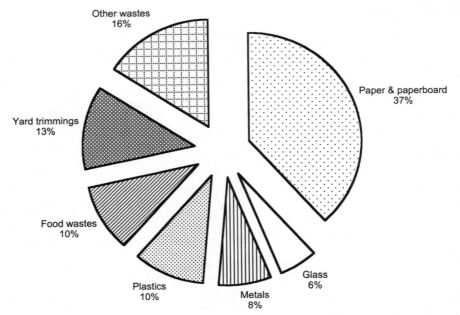

FIGURE 5.1   Materials generated in municipal solid waste, 1998. (*Unpublished data developed for the U.S. EPA, 1999.*)

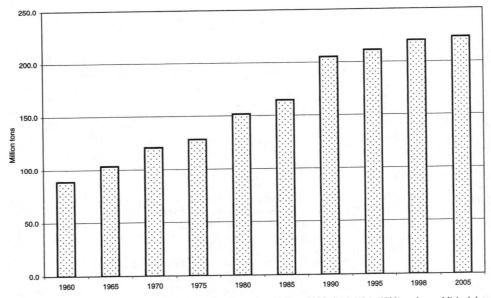

FIGURE 5.2   Materials generated in municipal solid waste, 1960 to 2005. (*U.S. EPA (1999) and unpublished data developed for the U.S. EPA.*)

TABLE 5.3   Paper and Paperboard Products in Municipal Solid Waste, 1998

Product category	Generation, millions of tons	Percent
Nondurable goods:		
Newspapers	13.6	6.2
Books	1.1	0.5
Magazines	2.3	1.0
Office papers*	7.0	3.2
Directories	0.7	0.3
Standard (A) mail[†]	5.2	2.4
Other commercial printing	6.6	3.0
Tissue paper and towels	3.1	1.4
Paper plates and cups	0.9	0.4
Other nonpackaging paper[‡]	4.4	2.0
Total paper and paperboard nondurable goods	45.0	20.4
Containers and packaging:		
Corrugated boxes	29.8	13.5
Milk cartons	0.5	0.2
Folding cartons	5.5	2.5
Other paperboard packaging	0.2	0.1
Bags and sacks	1.7	0.8
Other paper packaging	1.4	0.6
Total paper and paperboard containers and packaging	39.1	17.8
Total paper and paperboard	84.1	38.2
Total MSW generation	220.2	100.0

* Includes high-grade papers and an adjustment for files removed from storage.
[†] Formerly called Third Class Mail by the U.S. Postal Service.
[‡] Includes tissue in disposable diapers, paper in games and novelties, cards, etc. Details may not add to totals due to rounding.
Source: Unpublished data developed for the U.S. EPA.

TABLE 5.4   Glass Products in Municipal Solid Waste, 1998

Product category	Generation, millions of tons	Percent
Durable goods*	1.5	0.7
Containers and packaging:		
Beer and soft drink bottles	5.3	2.4
Wine and liquor bottles	1.8	0.8
Food and other bottles and jars	3.9	1.8
Total glass containers	11.0	5.0
Total glass	12.5	5.7
Total MSW generation	220.2	100.0

* Glass as a component of appliances, furniture, consumer electronics, etc. Details may not add to totals due to rounding.
Source: Based on unpublished data prepared for the U.S. EPA.

TABLE 5.5 Metal Products in Municipal Solid Waste, 1998

Product category	Generation, millions of tons	Percent
Durable goods:		
Ferrous metals*	9.4	4.3
Aluminum[†]	0.9	0.4
Batteries (lead only)	1.0	0.4
Other nonferrous metals[‡]	0.4	0.2
Total metals in durable goods	11.7	5.3
Nondurable goods:		
Aluminum	0.2	0.1
Containers and packaging:		
Steel:		
Food and other cans	2.7	1.2
Other steel packaging	0.2	0.1
Total steel packaging	2.9	1.3
Aluminum:		
Beer and soft drink cans	1.5	0.7
Food and other cans	0.1	0.0
Foil and closures	0.4	0.2
Total aluminum packaging	2.0	0.9
Total metals in containers and packaging	4.9	2.2
Total metals	16.8	7.6
Total MSW generation	220.2	100.0

* Ferrous metals in appliances, furniture, tires, and miscellaneous durables.
† Aluminum in appliances, furniture, and miscellaneous durables.
‡ Other nonferrous metals in appliances, lead-acid batteries, and miscellaneous durables. Details may not add to totals due to rounding.
Source: Based on unpublished data prepared for the U.S. EPA.

Aluminum

Most aluminum in MSW is found in containers and packaging, primarily in beverage cans (Table 5.5). Some aluminum is also found in durable and nondurable goods. Overall, aluminum amounted to an estimated 1.4 percent of MSW generation in 1998.

Other Nonferrous Metals

Other nonferrous metals (lead, copper, zinc) are found in MSW, primarily in durable goods. These metals have totaled less than 1 percent of MSW over the entire period quantified. The major source of nonferrous metals in MSW is lead in automotive batteries. In 1998, this lead amounted to an estimated 0.4 percent of all MSW generated (Table 5.5).

Plastics

Plastics are used very widely in the products found in municipal solid waste; in 1998, plastics comprised over 10 percent of MSW generation (Table 5.6). Use of plastics has grown rapidly, with plastics in MSW increasing from less than 1 percent in 1960. This growth is projected to continue, with plastics making up about 11 percent of MSW in 2005.

TABLE 5.6   Plastic Products in Municipal Solid Waste, 1998

Product category	Generation, millions of tons	Percent
Durable goods*	6.9	3.1
Nondurable goods:		
Plastic plates and cups	0.9	0.4
Trash bags	0.8	0.4
Disposable diapers†	0.4	0.2
Clothing and footwear	0.4	0.2
Other misc. nondurables‡	3.1	1.4
Total plastics nondurable goods	5.6	2.5
Containers and packaging:		
Soft drink bottles	0.8	0.4
Milk and water bottles	0.7	0.3
Other containers	2.3	
Bags and sacks	1.5	0.7
Wraps	2.0	0.9
Other plastic packaging	2.6	1.2
Total plastics containers and packaging	9.9	4.5
Total plastics	22.4	10.2
Total MSW generation	220.2	100.0

* Plastics as a component of appliances, furniture, lead-acid batteries, and miscellaneous durables. Adjustments have been made for lifetimes of products.
† Does not include other materials in diapers.
‡ Trash bags, eating utensils and straws, shower curtains, etc. Details may not add to totals due to rounding.

Source: Based on unpublished data prepared for the U.S. EPA.

Because plastics are relatively light, no one plastic product makes up a large portion of MSW. In 1998, plastics in durable goods (mainly appliances, carpeting, and furniture) amounted to about 3 percent of MSW generation. Plastics in nondurable goods made up 2.5 percent of MSW generation in 1998. The plastics in nondurables are found in plates and cups, trash bags, and many other products.

The largest source of plastics in MSW is containers and packaging, where plastics amounted to an estimated 4.5 percent of MSW generation in 1998. Containers for soft drinks, milk, water, food, and other products were the largest portion of plastics in containers and packaging. The remainder of plastic packaging is found in bags, sacks, wraps, closures, and other miscellaneous packaging products.

Other Materials in Products

In addition to the materials in products previously summarized, other materials making up lesser percentages of MSW generation are described in this section.

Rubber and Leather.   In 1998, rubber and leather made up an estimated 3.1 percent of MSW generation (Table 5.2). Most of the rubber and some of the leather was found in the durable goods category in products such as tires, furniture and furnishings, and carpets. Both rubber and leather were found in clothing and footwear in MSW.

Textiles.   Textiles comprised almost 4 percent of MSW generation in 1998 (Table 5.2). The primary sources of textiles in MSW are clothing and household items such as sheets and towels. However, textiles are also found in such items as tires, furniture, and footwear.

Wood.   Wood is a surprisingly important component of MSW, amounting to over 5 percent of MSW generation in 1998 (Table 5.2). The wood is found in durable goods such as furniture and cabinets for electronic goods, and in the containers and packaging category in shipping pallets and boxes.

Other Materials.   Since the material flows methodology is essentially a materials balance, some materials that cannot be classified into one of the basic material categories of MSW are put into an "Other" category in order to account for all the components associated with a product. This category amounted to 1.8 percent of MSW generation in 1998 (Table 5.2).

Most of the materials in this category are associated with disposable diapers, including the fluff (wood) pulp used in the diapers as well as the feces and urine that are disposed along with the diapers. The electrolyte in automotive batteries is also included in this category.

Food Wastes

Food wastes in MSW include uneaten food and food preparation wastes from residences, commercial establishments (e.g., restaurants), institutions (e.g., schools and hospitals), and some industrial sources (e.g., factory cafeterias or lunchrooms). In 1998, food wastes made up an estimated 10 percent of MSW generation (Table 5.2).

As described, the only source of data on food wastes is sampling studies conducted around the country. These studies show that a declining percentage of MSW is composed of food wastes; the decline has been from about 14 percent of the total in 1960 to the more recent amount of 10 percent of total MSW in 1998.

Yard Trimmings

Yard trimmings include grass, leaves, and tree and brush trimmings from residential, commercial, and institutional sources. About 17 percent of MSW generation was yard trimmings in 1990 (Table 5.2). Like food wastes, yard trimmings in MSW are estimated based on sampling study data. Due to increased emphasis on management of yard trimmings through composting or grasscycling (leaving grass clippings on the lawn), the percentage of yard trimmings in 1998 was estimated to have decreased to 12.6 percent of total MSW generation.

Miscellaneous Inorganic Wastes

This relatively small category, which includes soil, bits of stone and concrete, and the like, was estimated to be 1.5 percent of MSW generation in 1998 (Table 5.2). The estimates are derived from sampling studies, where the items in the category would usually be classified as fines.

Municipal Solid Waste Generation on a Per Capita Basis

For planning purposes, municipal solid waste generation on a per person, per day basis is often important. Some of the MSW generation data in Table 5.1 is converted to a per capita basis in Table 5.7. This table reveals some interesting trends. Overall, MSW generation per person increased over the 30 years for which historical data are available. Generation grew from 2.7 lb per person per day in 1960 to about 4.5 lb per person per day in 1998, with a pro-

TABLE 5.7   Materials Generated* in the Municipal Waste Stream, 1960 to 2005 *(In pounds per person per day)*

Materials	1960	1970	1980	1990	1998	2005[†]
Paper and paperboard	0.9	1.2	1.3	1.6	1.7	1.8
Glass	0.2	0.3	0.4	0.3	0.3	0.2
Metals	0.3	0.4	0.4	0.4	0.3	0.4
Plastics	Neg.	0.1	0.2	0.4	0.5	0.5
Rubber and leather	0.1	0.1	0.1	0.1	0.1	0.1
Textiles	0.1	0.1	0.1	0.1	0.2	0.2
Wood	0.1	0.1	0.2	0.3	0.2	0.3
Other	Neg.	Neg.	0.1	0.1	0.1	0.1
Total materials in products	1.7	2.2	2.6	3.2	3.4	3.6
Food wastes	0.4	0.3	0.3	0.5	0.4	0.4
Yard trimmings	0.6	0.6	0.7	0.8	0.6	0.4
Miscellaneous inorganic wastes	Neg.	Neg.	0.1	0.1	0.1	0.1
Total MSW generated	2.7	3.3	3.7	4.5	4.5	4.6

* Generation before materials recovery or combustion. Details may not add to totals due to rounding.
[†] Projected data.
Neg. = Negligible (less than 0.05 pounds per person per day).
Source: U.S. EPA (1999) and unpublished data developed for the U.S. EPA.

jected increase to 4.6 lb per person per day in 2005. In the decade of the 1990s, however, per capita generation of MSW stabilized, largely due to improved management of yard trimmings.

5.4   PRODUCTS IN MUNICIPAL SOLID WASTE BY WEIGHT

The materials in municipal solid waste are found in products that are used and discarded. These products can be classified as durable goods, nondurable goods, and containers and packaging. To these categories food wastes, yard trimmings, and miscellaneous inorganic wastes are added to obtain total MSW. (Note that these totals are by definition the same as the totals of all materials in MSW.) Products in MSW are summarized in Table 5.8 (by weight), in Table 5.9 (by percentage), and in Figs. 5.3 and 5.4.

In 1998, containers and packaging accounted for 72.4 million tons of MSW, or almost 33 percent of generation. Nondurable goods in MSW generation weighed over 60 millions tons, or over 27 percent of generation. Durable goods were over 34 million tons, or almost 16 percent of generation. The remainder of MSW was food wastes, yard trimmings, and other miscellaneous wastes, as previously discussed.

Durable Goods

Durable goods are generally defined as products having lifetimes of 3 years or more. This category includes major appliances, small appliances, furniture and furnishings, carpets and rugs, rubber tires, lead-acid automotive batteries, and miscellaneous durables such as consumer electronics and sporting goods. Historical and projected data on durable goods in MSW are shown in Tables 5.10 and 5.11.

TABLE 5.8   Products Generated* in the Municipal Waste Stream, 1960 to 2005
(In millions of tons)

Products	1960	1965	1970	1975	1980	1985	1990	1995	1998	2005[†]
Durable goods	9.9	12.4	14.7	17.4	21.8	23.5	29.8	31.1	34.4	39.0
Nondurable goods	17.3	21.9	25.1	25.3	34.4	42.2	52.2	57.3	60.3	67.7
Containers and packaging	27.4	34.2	43.6	44.5	52.7	54.9	64.5	68.4	72.4	82.8
Total product[‡] wastes	54.6	68.5	83.3	87.2	108.9	120.6	146.5	156.8	167.1	189.4
Other wastes:										
Food wastes	12.2	12.7	12.8	13.4	13.0	13.2	20.8	21.7	22.1	23.5
Yard trimmings	20.0	21.6	23.2	25.2	27.5	30.0	35.0	29.7	27.7	23.0
Miscellaneous inorganic wastes	1.3	1.6	1.8	2.0	2.3	2.5	2.9	3.2	3.3	3.7
Total other wastes	33.5	35.9	37.8	40.6	42.8	45.7	58.7	54.6	53.2	50.1
Total MSW generated	88.1	104.4	121.1	127.8	151.6	166.3	205.2	211.4	220.2	239.5

* Generation before materials recovery or combustion. Details may not add to totals due to rounding.
[†] Projected data.
[‡] Other than food products.
Source: Adapted from U.S. EPA (1999) and unpublished data developed for the U.S. EPA.

TABLE 5.9   Products Generated* in the Municipal Waste Stream, 1960 to 2005
(In percent of total generation)

Products	1960	1965	1970	1975	1980	1985	1990	1995	1998	2005[†]
Durable goods	11.3	11.8	12.1	13.6	14.4	14.1	14.5	14.7	15.6	16.3
Nondurable goods	19.7	21.0	20.7	19.8	22.7	25.4	25.4	27.1	27.4	28.3
Containers and packaging	31.1	32.8	36.0	34.8	34.7	33.0	31.4	32.4	32.9	34.6
Total product[‡] wastes	62.0	65.6	68.8	68.2	71.8	72.5	71.4	74.2	75.9	79.1
Other wastes:										
Food wastes	13.8	12.2	10.6	10.5	8.6	7.9	10.1	10.3	10.0	9.8
Yard trimmings	22.7	20.7	19.2	19.7	18.1	18.0	17.1	14.0	12.6	9.6
Miscellaneous inorganic wastes	1.5	1.5	1.5	1.6	1.5	1.5	1.4	1.5	1.5	1.5
Total other wastes	38.0	34.4	31.2	31.8	28.2	27.5	28.6	25.8	24.1	20.9
Total MSW generated	100.0	100.0	100.0	100.0	100.0	100.0	100.0	100.0	100.0	100.0

* Generation before materials recovery or combustion. Details may not add to totals due to rounding.
[†] Projected data.
[‡] Other than food products.
Source: Adapted from U.S. EPA (1999) and unpublished data developed for the U.S. EPA.

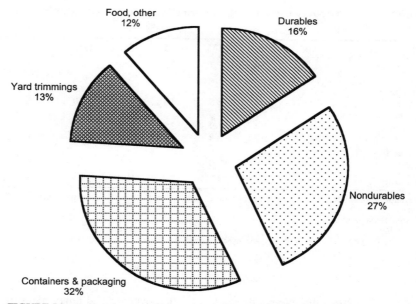

FIGURE 5.3   Products generated in municipal solid waste, 1998. (*Unpublished data developed for the U.S. EPA, 1999.*)

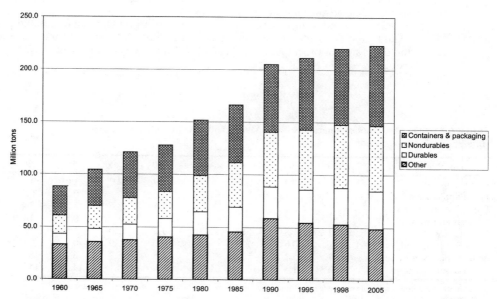

FIGURE 5.4   Products generated in municipal solid waste, 1960 to 2005. *[U.S. EPA (1999) and unpublished data developed for the U.S. EPA (1999).]*

TABLE 5.10   Durable Goods Generated* in the Municipal Waste Stream, 1960 to 2005
(In millions of tons)

Durable goods	1960	1965	1970	1975	1980	1985	1990	1995	1998	2005[†]
Major appliances	1.6	2.0	2.2	2.5	3.0	3.3	3.3	3.4	3.7	3.5
Small appliances[‡]							0.5	0.7	0.9	1.3
Furniture and furnishings	2.2	2.4	2.8	3.7	4.8	6.0	6.8	7.2	7.6	9.4
Carpets and rugs[‡]							1.7	2.2	2.4	3.2
Rubber tires	1.1	1.4	1.9	2.5	2.7	3.3	3.6	3.8	4.5	4.6
Batteries, lead-acid	Neg.	0.7	0.8	1.2	1.5	1.5	1.5	1.8	1.9	1.9
Miscellaneous durables	5.0	6.0	7.0	7.6	9.9	9.4	12.5	12.0	13.4	15.1
Total durable goods	9.9	12.4	14.7	17.4	21.8	23.5	29.8	31.1	34.4	39.0

* Generation before materials recovery or combustion. Details may not add to totals due to rounding.
[†] Projected data.
[‡] Not estimated separately in earlier years.
Neg. = Negligible.
Source: Adapted from U.S. EPA (1999) and unpublished data developed for the U.S. EPA.

Most durable goods would be called oversize and bulky items by solid waste managers. They typically would not be counted in a sampling survey, but they must nevertheless be managed, though perhaps in a somewhat different manner from other wastes.

On a weight basis, the miscellaneous durables are the largest line item in the durable goods category, at an estimated 13.4 million tons generated in 1998. This category includes electronics goods such as televisions and personal computers, as well as a variety of other products such as luggage and sporting goods.

Furniture and furnishings comprise the second largest segment of durable goods, at an estimated 7.6 million tons generated in 1998. Furniture from both residences and commercial buildings such as offices is counted as MSW. Items such as mattresses are also included in this category.

TABLE 5.11   Durable Goods Generated* in the Municipal Waste Stream, 1960 to 2005
(In percent of total generation)

Durable goods	1960	1965	1970	1975	1980	1985	1990	1995	1998	2005[†]
Major appliances	1.8	1.9	1.8	1.9	1.9	2.0	1.6	1.6	1.7	1.5
Small appliances[‡]							0.2	0.3	0.4	0.5
Furniture and furnishings	2.4	2.3	2.3	2.9	3.1	3.6	3.3	3.4	3.5	3.9
Carpets and rugs[‡]							0.8	1.1	1.1	1.4
Rubber tires	1.2	1.3	1.6	2.0	1.8	2.0	1.8	1.8	2.0	1.9
Batteries, lead-acid	Neg.	0.6	0.7	0.9	1.0	0.9	0.7	0.9	0.9	0.8
Miscellaneous durables	5.7	5.7	5.7	7.6	6.5	5.7	6.1	5.7	6.1	6.3
Total durable goods	11.3	11.8	12.1	15.3	14.4	14.1	14.5	14.7	15.6	16.3

* Generation before materials recovery or combustion. Details may not add to totals due to rounding.
[†] Projected data.
[‡] Not estimated separately in earlier years.
Neg. = Negligible.
Source: Adapted from U.S. EPA (1999) and unpublished data developed for the U.S. EPA.

Rubber tires contributed 4.5 million tons to MSW in 1998, and major appliances ("white goods") weighed 3.7 million tons. These appliances include refrigerators, washing machines, stoves, etc. They have long lifetimes before they are finally discarded.

Finally, carpets and rugs contributed 2.4 million tons, and lead-acid batteries were 1.9 million tons in 1998.

Nondurable Goods

Nondurable goods are generally defined as those having lifetimes of less than 3 years. The majority of these products are, however, discarded the same year they are manufactured. Paper products account for a large portion of nondurables, with plastics and textiles accounting for most of the remainder (Tables 5.12 and 5.13).

Newspapers, at 13.6 million tons generated in 1998, are the largest single item in nondurables, with office papers second at 7 million tons generated in 1998, and other commercial printing third at 6.6 million tons.

Some disposable nondurable products that are highly visible and consequently attract much attention are actually fairly minor constituents of the municipal waste stream. Thus Standard (A) mail (formerly called Third Class mail by the U.S. Postal Service, and often

TABLE 5.12   Nondurable Goods Generated* in the Municipal Waste Stream, 1960 to 2005
(In millions of tons)

Nondurable goods	1960	1965	1970	1975	1980	1985	1990	1995	1998	2005†
Newspapers	7.1	8.3	9.5	8.8	11.1	12.4	13.4	13.1	13.6	13.8
Books and magazines	1.9	2.2	2.5	2.3	3.4	4.7				
Books‡							1.0	1.2	1.1	1.4
Magazines‡							2.8	2.5	2.3	3.1
Office papers	1.5	2.2	2.7	2.6	4.0	5.8	6.4	6.6	7.0	8.0
Directories‡							0.6	0.5	0.7	0.6
Standard (A) mail‡							3.8	4.6	5.2	5.5
Other commercial printing	1.3	1.8	2.1	2.1	3.1	3.2	4.5	6.8	6.6	7.5
Tissue paper and towels	1.1	1.5	2.1	2.1	2.3	2.7	3.0	3.0	3.1	3.4
Paper plates and cups	0.3	0.3	0.4	0.4	0.6	0.6	0.7	1.0	0.9	1.0
Plastic plates and cups§					0.2	0.3	0.7	0.8	0.9	1.3
Trash bags‡							0.8	0.8	0.8	1.0
Disposable diapers	Neg.	Neg.	0.4	1.1	1.9	2.5	2.7	3.0	3.2	3.6
Other nonpackaging paper	2.7	3.9	3.6	3.5	4.2	4.7	3.8	4.3	4.4	5.0
Clothing and footwear	1.4	1.5	1.6	1.7	2.2	2.4	4.0	5.1	6.0	7.3
Towels, sheets, and pillowcases‡							0.7	0.7	0.8	0.8
Other misc. nondurables	0.1	0.2	0.2	0.6	1.4	3.0	3.3	3.3	3.6	4.5
Total nondurable goods	17.3	21.9	25.1	25.3	34.4	42.2	52.2	57.3	60.3	67.7

* Generation before materials recovery or combustion.
† Projected data.
‡ Not estimated separately in earlier years. Some categories used in previous years have been reallocated.
§ Not estimated prior to 1980.
Neg. = Negligible (less than 50,000 tons).
Source: Adapted from U.S. EPA (1999) and unpublished data developed for the U.S. EPA.

TABLE 5.13   Nondurable Goods Generated* in the Municipal Waste Stream, 1960 to 2005
(In percent of total generation)

Nondurable goods	1960	1965	1970	1975	1980	1985	1990	1995	1998	2005[†]
Newspapers	8.1	7.9	7.9	6.9	7.3	7.4	6.5	6.2	6.2	5.7
Books and magazines	2.2	2.1	2.0	1.8	2.2	2.8	0.0	0.0	0.0	0.0
Books[‡]							0.5	0.5	0.5	0.6
Magazines[‡]							1.4	1.2	1.0	1.3
Office papers	1.7	2.1	2.2	2.1	2.6	3.5	3.1	3.1	3.2	3.3
Directories[‡]							0.3	0.2	0.3	0.2
Standard (A) mail[‡]							1.9	2.2	2.4	2.3
Other commercial printing	1.4	1.7	1.8	1.7	2.0	1.9	2.2	3.2	3.0	3.1
Tissue paper and towels	1.2	1.4	1.7	1.7	1.5	1.6	1.4	1.4	1.4	1.4
Paper plates and cups	0.3	0.3	0.3	0.3	0.4	0.4	0.3	0.5	0.4	0.4
Plastic plates and cups[§]					0.1	0.2	0.3	0.4	0.4	0.5
Trash bags[‡]							0.4	0.4	0.4	0.4
Disposable diapers	Neg.	Neg.	0.3	0.9	1.3	1.5	1.3	1.4	1.5	1.5
Other nonpackaging paper	3.1	3.7	3.0	2.8	2.8	2.8	1.9	2.0	2.0	2.1
Clothing and footwear	1.5	1.4	1.3	1.4	1.4	1.4	1.9	2.4	2.7	3.0
Towels, sheets and pillowcases[‡]							0.4	0.4	0.3	0.3
Other miscellaneous nondurables	0.1	0.2	0.2	0.4	0.9	1.8	1.6	1.6	1.6	1.9
Total nondurable goods	19.7	21.0	20.7	19.8	22.7	25.4	25.4	27.1	27.4	28.3

* Generation before materials recovery or combustion. Details may not add to totals due to rounding.
[†] Projected data.
[‡] Not estimated separately in earlier years. Some categories used in previous years have been reallocated.
[§] Not estimated prior to 1980.
Neg. = Negligible (less than 50,000 tons).
Source: Adapted from U.S. EPA (1999) and unpublished data developed for the U.S. EPA.

called junk mail by others) is about 2.4 percent of MSW generation, paper and plastic plates and cups combined (including the "clamshells" used in fast food restaurants) are less than 1 percent of MSW, and disposable diapers (including the waste products contained within them) are about 1.5 percent of MSW generation.

Containers and Packaging

The containers and packaging category includes both primary packaging (the containers that directly hold food, beverages, toiletries, and a host of other products) and secondary and tertiary packaging, which contain the packaged products for shipping and display. By definition, it is assumed that all containers and packaging are discarded the same year they are manufactured (with a few exceptions, such as reusable wood pallets). Containers and packaging generation is shown in Table 5.14 (by weight) and Table 5.15 (in percentage).

By far the dominant material in this category is paper and paperboard, which accounted for about 63 percent of the weight of containers and packaging generated in 1998. Corrugated boxes, at nearly 30 million tons generated in 1998, are the single largest product line item in MSW.

TABLE 5.14 Containers and Packaging Generated* in the Municipal Waste Stream, 1960 to 2005
(In millions of tons)

Containers and packaging	1960	1965	1970	1975	1980	1985	1990	1995	1998	2005†
Glass packaging:										
Beer and soft drink bottles	1.4	2.6	5.6	6.3	6.7	5.6	5.6	5.1	5.4	4.2
Wine and liquor bottles	1.1	1.4	1.9	2.0	2.5	2.2	2.0	1.8	1.8	1.4
Food and other bottles and jars	3.7	4.2	4.4	4.4	4.8	4.3	4.2	4.6	3.9	4.0
Total glass packaging	6.2	8.1	11.9	12.7	14.0	12.1	11.8	11.5	11.0	9.6
Steel packaging:										
Beer and soft drink bottles	0.6	0.9	1.6	1.3	0.5	0.1	0.2	Neg.	Neg.	Neg.
Food and other cans	3.8	3.6	3.5	3.4	2.9	2.6	2.5	2.7	2.7	2.9
Other steel packaging	0.3	0.3	0.3	0.2	0.2	0.2	0.2	0.2	0.3	0.2
Total steel packaging	4.7	4.8	5.4	4.9	3.6	2.9	2.9	2.9	2.9	3.1
Aluminum packaging:										
Beer and soft drink bottles	Neg.	Neg.	0.1	0.4	0.9	1.3	1.6	1.6	1.5	2.1
Other cans	Neg.	Neg.	0.1	Neg.	Neg.	Neg.	Neg.	Neg.	0.1	Neg.
Foil and closures	0.2	0.3	0.4	0.4	0.4	0.3	0.3	0.4	0.4	0.4
Total aluminum packaging	0.2	0.3	0.6	0.8	1.3	1.6	1.9	2.0	2.0	2.5
Paper and paperboard packaging:										
Corrugated boxes	7.3	10.0	12.8	13.6	17.1	19.2	24.0	28.8	29.8	35.8
Milk cartons‡					0.8	0.5	0.5	0.5	0.5	0.5
Folding cartons‡					3.8	4.1	4.3	5.3	5.6	5.7
Other paperboard packaging	3.8	4.5	4.8	4.4	0.2	0.2	0.3	0.3	0.2	0.2
Bags and sacks‡					3.4	3.1	2.4	2.0	1.7	1.6
Wrapping papers‡					0.2	0.1	0.1	0.1		
Other paper packaging	2.9	3.4	3.8	3.3	0.9	1.3	1.0	1.2	1.4	1.5
Total paper and board packaging	14.1	17.9	21.4	21.3	26.3	28.6	32.7	38.1	39.1	45.5
Plastics packaging:										
Soft drink bottles‡					0.3	0.4	0.4	0.7	0.8	0.9
Milk bottles‡					0.2	0.3	0.5	0.6	0.7	0.9
Other containers	0.1	0.3	0.9	1.3	0.9	1.2	1.4	1.2	2.3	1.8
Bags and sacks‡					0.4	0.6	0.9	1.2	1.5	1.9
Wraps‡					0.8	1.0	1.5	1.7	2.0	2.7
Other plastics packaging	0.1	0.7	1.2	1.4	0.8	1.2	2.0	2.2	2.6	3.6
Total plastics packaging	0.1	1.0	2.1	2.8	3.4	4.6	6.9	7.6	9.9	11.7
Wood packaging	2.0	2.1	2.1	2.0	3.9	4.9	8.2	6.2	7.3	10.2
Other miscellaneous packaging	0.1	0.1	0.1	0.1	0.2	0.1	0.2	0.2	0.2	0.2
Total containers and packaging	27.4	34.2	43.6	44.5	52.7	54.9	64.5	68.4	72.4	82.8

* Generation before materials recovery or combustion. Details may not add to totals due to rounding.
† Projected data.
‡ Not estimated prior to 1980. Paper wrapping papers not estimated separately after 1996.
Neg. = Negligible (less than 50,000 tons).

Source: Adapted from U.S. EPA (1999) and unpublished data developed for the U.S. EPA.

TABLE 5.15 Containers and Packaging Generated* in the Municipal Waste Stream, 1960 to 2005
(In percent of total generation)

Containers and packaging	1960	1965	1970	1975	1980	1985	1990	1995	1998	2005[†]
Glass packaging:										
Beer and soft drink bottles	1.6	2.5	4.6	4.9	4.4	3.4	2.7	2.4	2.4	1.7
Wine and liquor bottles	1.2	1.3	1.6	1.6	1.6	1.3	1.0	0.8	0.8	0.6
Food and other bottles and jars	4.2	4.0	3.6	3.4	3.2	2.6	2.0	2.2	1.8	1.6
Total glass packaging	7.0	7.7	9.8	9.9	9.2	7.3	5.8	5.5	5.0	4.0
Steel packaging:										
Beer and soft drink bottles	0.7	0.9	1.3	1.0	0.3	0.1	0.1	Neg.	Neg.	Neg.
Food and other cans	4.3	3.4	2.9	2.7	1.9	1.6	1.2	1.3	1.2	1.2
Other steel packaging	0.3	0.3	0.2	0.2	0.1	0.1	0.1	0.1	0.1	0.1
Total steel packaging	5.3	4.6	4.4	3.8	2.4	1.7	1.4	1.4	1.3	1.3
Aluminum packaging:										
Beer and soft drink bottles	Neg.	Neg.	0.1	0.3	0.6	0.8	0.8	0.8	0.7	0.9
Other cans	Neg.	Neg.	0.0	Neg.	Neg.	Neg.	Neg.	Neg.	0.0	Neg.
Foil and closures	0.2	0.3	0.3	0.3	0.3	0.2	0.2	0.2	0.2	0.1
Total aluminum packaging	0.2	0.3	0.5	0.6	0.8	1.0	0.9	0.9	0.9	1.0
Paper and paperboard packaging:										
Corrugated boxes	8.3	9.6	10.5	10.6	11.3	11.5	11.7	13.6	13.5	15.0
Milk cartons‡					0.5	0.3	0.2	0.2	0.2	0.2
Folding cartons‡					2.5	2.5	2.1	2.5	2.5	2.4
Other paperboard packaging	4.4	4.3	4.0	3.4	0.2	0.1	0.1	0.1	0.1	0.1
Bags and sacks‡					2.2	1.9	1.2	0.9	0.8	0.7
Wrapping papers‡					0.1	0.1	0.1	0.0		
Other paper packaging	3.3	3.2	3.1	2.6	0.6	0.8	0.5	0.5	0.6	0.6
Total paper and board packaging	16.0	17.2	17.7	16.7	17.4	17.2	15.9	18.0	17.8	19.0
Plastics packaging:										
Soft drink bottles‡					0.2	0.2	0.2	0.3	0.4	0.4
Milk bottles‡					0.2	0.2	0.3	0.3	0.3	0.4
Other containers	0.1	0.3	0.7	1.0	0.6	0.7	0.7	0.6	1.1	0.8
Bags and sacks‡					0.3	0.3	0.5	0.6	0.7	0.8
Wraps‡					0.6	0.6	0.7	0.8	0.9	1.1
Other plastics packaging	0.1	0.7	1.0	1.1	0.5	0.7	1.0	1.1	1.2	1.5
Total plastics packaging	0.1	1.0	1.7	2.2	2.2	2.8	3.4	3.6	4.5	4.9
Wood packaging	2.3	2.0	1.7	1.6	2.6	2.9	4.0	2.9	3.3	4.2
Other miscellaneous packaging	0.1	0.1	0.1	0.1	0.1	0.1	0.1	0.1	0.1	0.1
Total containers and packaging	31.1	32.8	36.0	34.8	34.7	33.0	31.4	32.4	32.9	34.6

* Generation before materials recovery or combustion. Details may not add to totals due to rounding.
† Projected data.
‡ Not estimated prior to 1980. Paper wrapping papers not estimated separately after 1996.
Neg. = Negligible (less than 50,000 tons).
Source: Adapted from U.S. EPA (1999) and unpublished data developed for the U.S. EPA.

The second largest material category in containers and packaging, by weight, was glass bottles and jars, at 11 million tons generated in 1998. Plastics packaging was the third largest category, at nearly 10 million tons in 1998, while wood was in fourth place at over 7 million tons generated in 1998. Steel and aluminum cans and other packaging occupy relatively minor positions in MSW generation.

Other Wastes

Food wastes, yard trimmings, and miscellaneous inorganic wastes are added to the product categories of durable goods, nondurable goods, and containers and packaging to obtain total MSW generation. These other wastes were discussed in Sec. 5.3.

5.5 MUNICIPAL SOLID WASTE MANAGEMENT

Once municipal solid waste is generated, it must be managed somehow. In the United States, the usual management alternatives are recovery for recycling or composting, combustion, or landfilling. (The source reduction alternative is discussed elsewhere in this handbook. Chapter 3 deals with MSW as generated, after any source reduction measures have been applied.)

Recovery for Recycling and Composting

Recovery of MSW for recycling and composting in 1998 is estimated by weight and by percent of generation in Tables 5.16 through 5.19. (Note that the estimates are for *recovery*. The EPA data source referenced in this chapter does not attempt to determine whether materials recovered are actually recycled; in fact, some materials collected for recycling or composting are unrecyclable and become residues that must be disposed.)

Materials Recovery.   Of all materials recovered from MSW (Table 5.16), paper and paperboard comprise by far the largest tonnage—35 million tons out of a total of 62.2 million tons recovered in 1998. Yard trimmings represented the second highest tonnage recovered, at 12.6 million tons, with ferrous metals third at 4.3 million tons.

In terms of percentage of generation recovered, other nonferrous metals were the highest, at over 67 percent recovered in 1998. This is almost entirely due to the high rate of recovery of lead in automotive batteries. Recovery of yard trimmings ranked second, at over 45 percent of generation recovered. Recovery of paper and paperboard ranked third, at nearly 42 percent of generation recovered. Aluminum had the next highest recovery percentage in 1998 at 30 percent of generation, due to the relatively high recovery rate of aluminum beverage cans.

Durable Goods Recovery.   Because it is really products in MSW that are recovered, these estimates are more enlightening. Recovery of durable goods is shown in Table 5.17. Recovery of lead from lead-acid automotive batteries was estimated to be at nearly a 97 percent level in 1998, the highest recovery rate of all products in MSW. The other significant recovery in this category is ferrous metals from major appliances. The estimated ferrous recovery was estimated to be about 53 percent of the total weight of the appliances. Rubber was also recovered from rubber tires—about 23.5 percent of their weight in 1998.

Nondurable Goods Recovery.   Recovery of nondurable goods in 1998 is shown in Table 5.18. Newspapers have a long history of recovery, and they were recovered at an estimated rate of over 56 percent of generation in 1998 (7.7 million tons recovered). High-grade office papers were recovered at an estimated rate of over 50 percent, with other paper products recovered at lower, but yet significant, rates. The only other significant recovery identified was

TABLE 5.16 Materials Generated, Recovered, and Discarded* in the Municipal Solid Waste Stream, 1998
(In millions of tons and percent)

Materials	Generation, million tons	Recovery, million tons	Recovery, % of generation	Discards,* million tons	Discards, % of total
Paper and paperboard	84.1	35.0	41.6	49.2	31.1
Glass	12.5	3.2	25.5	9.3	5.9
Metals:					
Ferrous	12.4	4.3	35.1	8.0	5.1
Aluminum	3.1	0.9	27.9	2.2	1.4
Other nonferrous	1.4	0.9	67.4	0.5	0.3
Total metals	16.8	6.1	36.4	10.7	6.8
Plastics	22.4	1.2	5.4	21.2	13.4
Rubber and leather	6.9	0.9	12.5	6.0	3.8
Textiles	8.6	1.1	12.9	7.5	4.7
Wood	11.9	0.7	6.0	11.2	7.1
Other	3.9	0.9†	22.1	3.0	1.9
Total materials in products	167.1	49.0	29.3	118.1	74.7
Other wastes:					
Food wastes	22.1	0.6	2.6	21.6	13.6
Yard trimmings	27.7	12.6	45.3	15.2	9.6
Miscellaneous inorganic wastes	3.3	Neg.	Neg.	3.3	2.1
Total other wastes	53.2	13.1	24.7	40.0	25.3
Total MSW generated	220.2	62.2	28.2	158.1	100.0

* Discards after recovery for recycling or composting. Details may not add to totals due to rounding.
† Recovery of electrolytes in batteries. May not be recycled.
Neg. = negligible.
Source: Based on unpublished work prepared for the U.S. EPA.

TABLE 5.17 Durable Goods Generated, Recovered, and Discarded* in the Municipal Waste Stream, 1998

Durable goods	Generation, million tons	Recovery, million tons	Recovery, % of generation	Discards,* million tons	Discards, % of MSW total
Major appliances	3.7	1.9	53.2	1.7	1.1
Small appliances	0.9	Neg.	1.1	0.9	0.6
Furniture and furnishings	7.6	Neg.	Neg.	7.6	4.8
Carpets and rugs	2.4	Neg.	1.2	2.4	1.5
Rubber tires	4.5	1.1	23.5	3.5	2.2
Batteries, lead-acid	1.9	1.9	96.9	0.1	0.0
Miscellaneous durables	13.4	0.8	6.1	12.6	7.9
Total durable goods	34.4	5.7	16.7	28.6	18.1

* Discards after recovery for recycling or composting. Details may not add to totals due to rounding.
Neg. = Negligible.
Source: Based on unpublished data prepared for the U.S. EPA.

TABLE 5.18 Nondurable Goods Generated, Recovered, and Discarded* in the Municipal Waste Stream, 1998

Nondurable goods	Generation, million tons	Recovery, million tons	Recovery, % of generation	Discards,* million tons	Discards, % of MSW total
Newspapers	13.6	7.7	56.4	5.9	3.8
Books	1.1	0.2	14.0	1.0	0.6
Magazines	2.3	0.5	20.8	1.8	1.1
Office papers	7.0	3.6	50.4	3.5	2.2
Directories	0.7	0.1	13.5	0.6	0.4
Standard (A) mail	5.2	1.0	18.8	4.2	2.7
Other commercial printing	6.6	0.4	6.4	6.2	3.9
Tissue paper and towels	3.1	Neg.	Neg.	3.1	2.0
Paper plates and cups	0.9	Neg.	Neg.	0.9	0.6
Plastic plates and cups	0.9	Neg.	Neg.	0.9	0.6
Trash bags	0.8	Neg.	Neg.	0.8	0.5
Disposable diapers	3.2	Neg.	Neg.	3.2	2.0
Other nonpackaging paper	4.4	Neg.	Neg.	4.4	2.8
Clothing and footwear	6.0	0.8	13.2	5.2	3.3
Towels, sheets, and pillowcases	0.8	0.1	17.3	0.6	0.4
Other miscellaneous nondurables	3.6	Neg.	Neg.	3.6	2.3
Total nondurable goods	60.3	14.3	23.7	46.0	29.1

* Discards after recovery for recycling or composting. Details may not add to totals due to rounding.
Neg. = Negligible (less than 50,000 tons).
Source: Based on unpublished data prepared for the U.S. EPA.

recovery of textiles for export. (Reuse of textile products, e.g., clothing, in the United States was not considered to be recycling.)

Containers and Packaging Recovery.   Recovery of containers and packaging (29 million tons in 1998) comprised about 47 percent of all MSW recovery in that year, largely due to recovery of corrugated boxes at over 70 percent, or nearly 21 million tons. The significance of this level of recovery can be illustrated by the point that corrugated containers were 13.5 percent of MSW *generation* in 1998 (Table 5.15), but their discards after recovery were less than 6 percent of total *discards* (Table 5.19).

Steel packaging (mostly steel cans) was recovered at a rate of 57 percent in 1998. Aluminum beverage cans were recovered at a rate of about 54 percent, plastic soda bottles were recovered at over 35 percent, plastic milk and water bottles at over 31 percent, and glass bottles and jars at about 29 percent of generation. Other paper and plastic packaging products were also recovered to make total recovery at 40 percent of containers and packaging generation in 1998.

Combustion

In 1998 about 37 million tons of MSW were combusted, with most of that amount sent to energy recovery facilities. Thus combustion was the management of choice for about 17 percent of MSW generated, or about 23.5 percent of MSW discarded after recovery (Franklin Associates, 2000).

TABLE 5.19 Containers and Packaging Generated, Recovered, and Discarded* in the Municipal Waste Stream, 1998

Containers and packaging	Generation, million tons	Recovery, million tons	Recovery, % of generation	Discards,* million tons	Discards, % of MSW total
Glass packaging:					
Beer and soft drink bottles	5.4	1.7	31.4	3.7	2.3
Wine and liquor bottles	1.8	0.5	27.1	1.3	0.8
Food and other bottles and jars	3.9	1.0	26.3	2.9	1.8
Total glass packaging	11.0	3.2	28.9	7.8	4.9
Steel packaging:					
Beer and soft drink cans	Neg.	Neg.	Neg.	Neg.	Neg.
Food and other cans	2.7	1.5	56.1	1.2	0.7
Other steel packaging	0.3	0.2	68.0	0.1	0.1
Total steel packaging	2.9	1.7	57.1	1.3	0.8
Aluminum packaging:					
Beer and soft drink cans	1.5	0.8	53.9	0.7	0.4
Other cans	Neg.	Neg.	Neg.	Neg.	Neg.
Foil and closures	0.4	Neg.	Neg.	0.3	0.2
Total aluminum packaging	2.0	0.9	43.9	1.1	0.7
Paper and paperboard packaging:					
Corrugated boxes	29.8	20.9	70.%	8.8	5.6
Milk cartons	0.5	Neg.	Neg.	0.5	0.3
Folding cartons	5.6	0.4	7.0	5.2	3.3
Other paperboard packaging	0.2	Neg.	Neg.	0.2	0.1
Bags and sacks	1.7	0.3	17.3	1.4	0.9
Wrapping papers[†] Other paper packaging	1.4	Neg.	Neg.	1.4	0.9
Total paper and board packaging	39.1	21.6	55.2	17.5	11.1
Plastics packaging:					
Soft drink bottles	0.8	0.3	35.4	0.5	0.3
Milk bottles	0.7	0.2	31.4	0.5	0.3
Other containers	2.3	0.3	10.7	2.1	1.3
Bags and sacks	1.5	Neg.	0.7	1.5	0.9
Wraps	2.0	0.1	6.1	1.9	1.2
Other plastics packaging	2.6	0.1	2.7	2.5	1.6
Total plastics packaging	9.9	1.0	9.7	8.9	5.6
Wood packaging	7.3	0.7	9.8	6.6	4.2
Other miscellaneous packaging	0.2	Neg.	Neg.	0.2	0.1
Total containers and packaging	72.4	29.0	40.0	43.4	27.5

* Discards after recovery for recycling or composting. Details may not add to totals due to rounding.
† Not reported separately after 1996.
Neg. = Negligible (less than 50,000 tons or 0.05 percent).
Source: Based on unpublished data prepared for the U.S. EPA.

Landfilling

Over 140 million tons of MSW were landfilled in 1990, according to the EPA datasource (Franklin Associates, 1999). This was over 68 percent of MSW generated, or about 82 percent of MSW discarded after recovery. (Note that some of the amount of MSW assumed to be landfilled may in fact be littered, self-disposed, or otherwise disposed. These amounts are not estimated, but are thought to be relatively small.)

Trends in MSW Management

When nearly 40 years of data from the EPA database are combined with projections made for EPA, some interesting trends are demonstrated (Tables 5.20 and 5.21 and Fig. 5.5). Generation of MSW has increased steadily over the entire period except for some recession years. Recovery for recycling and composting was quite modest until the late 1980s, when the level of activity increased markedly, reaching 28 percent of generation in 1998.

Combustion of MSW in the United States exhibits a different pattern. In 1960 an estimated 30 percent of MSW generated in the U.S. was combusted, mostly in old-fashioned incinerators without energy recovery. When pollution controls began to be required, the old incinerators that did not meet the new standards were phased out. By 1985, it was estimated that only about 7 percent of MSW was combusted. Since then, there has been a steady increase in the amount of MSW going to combustion units—about 17 percent in 1998. (It should be noted, however, that the increase in tonnage was very slow in the late 1990s.)

With combustion declining in the 1960s and recycling still at relatively low levels, discards of MSW to landfills grew rapidly in the 1970s and 1980s. These discards appeared to peak around 1985 at over 136 million tons landfilled, or about 83 percent of generation that year.

TABLE 5.20 Management of Municipal Solid Waste, 1960 to 2005
(In millions of tons)

Materials	1960	1965	1970	1975	1980	1985	1990	1995	1997	1998	2005[‡]
Generation	88.1	104.4	121.1	127.8	151.6	166.3	205.2	211.4	216.4	220.2	239.5
Recovery for recycling	5.6	6.8	8.0	9.9	14.5	16.4	29.0	45.3	47.3	49.0	61.8
Recovery for composting	0.0	0.0	0.0	0.0	0.0	0.0	4.2	9.6	12.1	13.1	14.9
Total recovery	5.6	6.8	8.0	9.9	14.5	16.4	33.2	54.9	59.4	62.2	76.7
Discards after recovery*	82.5	97.6	113.0	117.9	137.1	148.1	172.0	156.5	157.0	158.1	162.8
Combustion with energy recovery	0.0	0.2	0.4	0.7	2.7	7.6	29.7	35.54	36.7	37.0	38
Combustion without energy recovery	27.0	26.8	24.7	17.8	11.0	4.1	2.2	0.0	0.0	0.0	0.0
Total combustion	27.0	27.0	25.1	18.5	13.7	11.7	31.9	35.5	36.7	37.0	38.0
Discards to landfill, other disposal[†]	55.5	70.6	87.9	99.4	123.4	136.4	140.1	120.9	120.3	121.1	124.8

* Does not include residues from recycling/composting processes.
[†] Does not include residues from recycling, composting, or combustion processes.
[‡] Total recovery of 32 percent assumed.
Details may not add to totals due to rounding.
Source: Adapted from U.S. EPA (1999) and unpublished data developed for the U.S. EPA.

TABLE 5.21 Management of Municipal Solid Waste, 1960 to 2005
(In percent of total generation)

Materials	1960	1965	1970	1975	1980	1985	1990	1995	1997	1998	2005[‡]
Generation	100.0	100.0	100.0	100.0	100.0	100.0	100.0	100.0	100.0	100.0	100.0
Recovery for recycling	6.4	6.5	6.6	7.7	9.6	9.9	14.2	21.5	21.9	22.3	25.8
Recovery for composting	Neg.	Neg.	Neg.	Neg.	Neg.	Neg.	2.0	4.5	5.6	6.0	6.2
Total recovery	6.4	6.5	6.6	7.7	9.6	9.9	16.2	26.0	27.4	28.2	32.0
Discards after recovery*	93.6	93.5	93.4	92.3	90.4	89.1	83.8	74.0	72.6	71.8	68.0
Combustion with energy recovery	Neg.	0.2	0.3	0.5	1.8	4.6	14.5	16.8	17.0	16.8	15.9
Combustion without energy recovery	30.6	25.7	20.4	13.9	7.3	2.5	1.1	Neg.	Neg.	Neg.	Neg.
Total combustion	30.6	25.9	20.7	14.5	9.0	7.0	15.5	16.8	17.0	16.8	15.9
Discards to landfill, other disposal[†]	63.0	67.6	72.6	77.8	81.4	82.0	68.3	57.2	55.6	55.0	52.1

* Does not include residues from recycling/composting processes.
[†] Does not include residues from recycling, composting, or combustion processes.
[‡] Total recovery of 32 percent assumed.
Details may not add to totals due to rounding.
Neg. = Negligible.
Source: Adapted from U.S. EPA (1999) and unpublished data developed for the U.S. EPA.

Since then, increased recovery and combustion of MSW have caused a decline in the estimated tonnage landfilled. In 1998, landfilled tonnage was about 121 million tons, or 55 percent of generation.

The EPA characterization report projects continued growth of MSW generation. Recovery of MSW was projected in scenarios, with 30 to 32 percent recovery in 2005. Combustion of MSW was also projected to grow very little—to about 38 million tons in 2005. Landfilling would go up somewhat in tonnage, but would go down to about 52 percent of generation in 2005.

5.6 DISCARDS OF MUNICIPAL SOLID WASTE BY VOLUME

Since weighing on a set of scales is quick and convenient, MSW is most often quantified in tons. Also, the material flows methodology for characterizing weight nationwide relies on data that are most often expressed in tons. Measurement of MSW by volume (cubic yards in the United States) is also relevant, however, since space occupied by the waste is an important consideration.

Unfortunately, density factors for MSW have been difficult to obtain. A 1990 study (Hunt, 1990) did provide a uniform set of experimental data for landfill densities of many of the products in MSW (Table 5.22), and these factors have been used to calculate the relative amounts of MSW discards as landfilled for the 1998 EPA characterization report (Table 5.23). Note that the data reported by EPA are for MSW *discards* after recovery (not generation), as these are the quantities corresponding most closely to MSW landfilled.

The information in Table 5.23 is primarily useful as an indicator of the *relative* density of materials in a landfill. In the real world, the materials in a landfill are mixed together before they

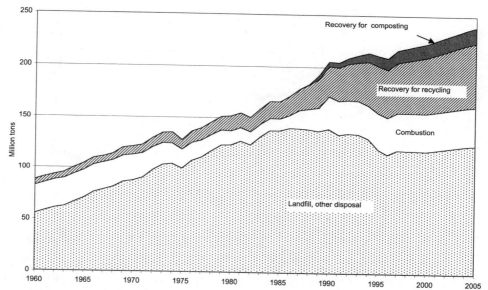

FIGURE 5.5   Generation and management of municipal solid waste, 1960 to 2005. *[U.S. EPA (1999) and data developed for the U.S. EPA.]*

are compacted, with the effect that air spaces are filled by small objects. Thus, the overall density in a particular landfill is probably greater than the sum of the individual components measured separately, and the total volume is probably less than the sum of the individual components.

A relative comparison of the weights and volumes of materials can be obtained by taking the ratio of volume percentage to weight percentage (the right-hand column in Table 5.23). A ratio of 1.0 means that the material occupies the same proportion of volume as weight in the landfill. Paper and paperboard exhibit this characteristic. A ratio greater than 1.0 shows that the material occupies a larger proportion by volume than by weight. Four materials have ratios of approximately 2.0 or higher: plastics, rubber and leather, textiles, and aluminum. Materials that are relatively dense and occupy proportionately less volume compared to their weight include glass, food wastes, and yard trimmings.

5.7   *THE VARIABILITY OF MUNICIPAL SOLID WASTE GENERATION*

The data presented in this chapter relate to MSW generation and management in the United States as a whole, and these data can provide useful guidelines for local use. Local planners should take care, however, when adapting general data for local planning. Some of the factors affecting variability in the waste stream are discussed in this section.

Commercial vs. Residential Waste

In general, people are most conscious of the wastes coming from their own homes, whether single-family residences, apartment buildings, or other residential options. Large amounts of waste, however, are also generated where people work, shop, travel, attend classes, or engage in

TABLE 5.22 Summary of Density Factors for
Landfilled Materials
(In pounds per cubic yard)

Products	Density
Durable goods*	475
Nondurable goods:	
Nondurable paper	800
Nondurable plastic	315
Disposable diapers:	
Diaper materials	795
Urine and feces	1350
Rubber	345
Textiles	435
Miscellaneous nondurables	
(mostly plastics)	390
Packaging:	
Glass containers:	
Beer and soft drink bottles	2800
Other containers	2800
Steel containers:	
Beer and soft drink cans	560
Food cans	560
Other packaging	560
Aluminum:	
Beer and soft drink cans	250
Other packaging	550
Paper and paperboard:	
Corrugated	750
Other paperboard	820
Paper packaging	740
Plastics:	
Film	670
Rigid containers	355
Other packaging	185
Wood packaging	800
Other miscellaneous packaging	1015
Food wastes	2000
Yard trimmings	1500

* No measurements were taken for durable goods or plastic
coatings.

    Source: U.S. EPA (1999).

other activities. These latter wastes are generally classified as commercial. To add to the confusion, waste haulers often classify wastes collected from apartment buildings as commercial, although the nature of the wastes may be very similar to that from single-family residences.

    The EPA report used as a source for much of this chapter includes a classification of MSW into residential and commercial fractions. The range of residential wastes is estimated to be between 55 and 65 percent of MSW generation, with commercial wastes estimated to range between 35 and 45 percent of generation. (MSW from multifamily residences was classified as residential, not commercial.)

TABLE 5.23   Volume of Materials Discarded in Municipal Solid Waste, 1997

Materials	1997 discards,* million tons	Weight,* % of MSW total	Landfill density,[†] lb/yd³	Landfill volume,[‡] million yd³	Volume, % of MSW total	Ratio, vol %/ wt %
Paper and paperboard	48.9	31.3	795	123.1	29.2	0.9
Plastics	20.4	13.0	370	110.0	26.1	2.0
Yard trimmings	16.2	10.4	1,500	21.7	5.1	0.5
Ferrous metals	7.6	4.9	570	26.7	6.3	1.3
Rubber and leather	5.8	3.7	355	32.8	7.8	2.1
Textiles	7.2	4.6	410	35.0	8.3	1.8
Wood	11.0	7.0	850	25.8	6.1	0.9
Food wastes	21.3	13.6	2,000	21.3	5.1	0.4
Other[†]	6.7	4.3	2,100	6.4	1.5	0.4
Aluminum	2.1	1.3	380	10.9	2.6	2.0
Glass	9.1	5.8	2,500	7.3	1.7	0.3
Totals	156.3	100.0	743	420.9[§]	100.0[§]	1.0

* From Table 5.16. Discards after materials recovery and landfilling, before combustion and landfilling.
[†] Composite factors derived by Franklin Associates for the source report.
[‡] This assumes that all waste is landfilled, but some is combusted and otherwise disposed.
[§] This density factor and volume are derived by adding the individual factors. Actual landfill density may be considerably higher (see discussion in text).

Source: U.S. EPA (1999).

Local/Regional Variability

Municipal solid waste managers generally agree that there are variations in the amount and characteristics of MSW around the country, although it is not easy to generalize with any degree of reliability. Some observations based on experience can be made, however.

First, there is some agreement that residential wastes vary less from location to location than do commercial wastes (Hunt, 1990). People across the country tend to buy much the same kinds of goods, whether they live in rural or urban areas or in different climates. Exceptions to this generalization include:

- *Yard trimmings.* Yard trimmings tend to be much more plentiful in warmer, moister parts of the country. Also, there are marked differences in how yard trimmings are managed. In rural areas and small towns, yard trimmings often are not hauled to landfills or compost facilities, while in suburban and urban areas, they usually are handled off-site. In addition, some states have banned landfilling of yard trimmings, which forces more on-site management by householders plus more community composting projects.

- *Food wastes.* Discards of food wastes in MSW will vary according to the prevalence of food disposers, which put the food wastes into the wastewater treatment system. Use of food disposers may not be allowed (e.g., in New York City). Also, food disposers typically cannot be used in rural areas that are not on a municipal sewer system.

- *Newspapers.* Newspapers, which are mostly discarded from residences, vary greatly in size, and thus contribute to regional and urban/rural variations in MSW generation. As an example, annual per capita generation of daily newspapers varies from about 120 lb per person in states like California, Massachusetts, and Florida, to 30 or 40 lb per person in less densely populated states like Wyoming and South Dakota.

Generation of MSW in a particular locality will be strongly influenced by commercial activity in the area. A concentration of office buildings will produce office papers and other wastes. Shopping malls, warehouses, and factories generate large amounts of corrugated containers and other wastes as well. Schools, hospitals, airports, train and bus stations, hotels and motels, and sports facilities all contribute to the commercial waste stream. Thus, small towns and rural areas without concentrations of commercial activities will typically generate less MSW per person than urban areas.

Seasonal Variations

Another well-known phenomenon in municipal waste management is seasonal variations in waste generation. Yard trimmings are generally the important variable for most communities, with seasonal cleanup of yards and garages often contributing to peak generation weeks. Late spring and autumn are peak generation periods in many communities, while generation of yard trimmings may approach zero in winter months in cold climates. As a rule of thumb, MSW generation may vary around 30 percent above or below the average in many communities.

Changes over Time

Municipal solid waste generation has increased in the United States, both in tonnage and in per capita generation. This does not mean, however, that generation of each material and product in MSW has grown at the same rate. In fact, generation of some materials and products has grown rapidly, while others have had slow growth or an actual decline. An understanding of this phenomenon is especially important in making projections of MSW generation and in planning waste management facilities.

Some factors tending to increase MSW generation are:

- *Increasing population.* Obviously, more people use and throw away more things. One preliminary analysis indicates that about one-half of the growth of MSW generation over a 15-year period can be attributed to population growth (Franklin Associates, 1992).

- *Increasing levels of affluence.* There is a rather strong correlation between generation of MSW and economic activity, as measured by gross domestic product (GDP) or personal consumption expenditures (PCE). Generation of paper and paperboard products is especially sensitive to economic activity. As an example, a plot of paper and paperboard generation (Fig. 5.6) shows declines in recession years such as 1975, 1982, and 1991. The reasons are obvious: when orders for goods go down, fewer boxes and other packaging are ordered for shipping. Also, advertising in newspapers and magazines declines during a recession.

- *Changes in lifestyles.* Changes in lifestyles are somewhat related to affluence. The United States has increasing numbers of individuals living alone, families with two wage earners, and single-parent families. People in these situations tend to buy more prepackaged food and to eat out more, often at fast-food establishments using disposable packaging. They may also do more shopping through catalogs, which increases the amounts of mail received and discarded at home. Also, each new household, however small, must have some appliances and furnishings.

The explosion of information and shopping opportunities through on-line electronic communications is causing changes in waste generation that are not yet fully understood. For example, readership of newspapers is declining, but people with computers at home may generate more office-type paper as they print out information and e-mail communications.

FIGURE 5.6 Generation of paper and paperboard in municipal solid waste, 1960 to 1998. *[U.S. EPA (1999) and unpublished data developed for the U.S. EPA.]*

- *Changes in work patterns.* Over a 15-year period, the number of office workers increased 72 percent, while manufacturing jobs declined.[4] At the same time, offices added personal computers, high-speed copiers, and facsimile machines, resulting in an increase in office papers generated.

- *New products.* New products may increase the amounts of MSW generated. Disposable diapers are an example of this phenomenon.

While the overall pattern has been an increase, some factors tend to decrease MSW generation. Some of these factors include:

- *Redesign of products.* Some products in MSW have actually grown lighter over the years. Appliances such as refrigerators are one example, due largely to changes in insulation and use of more lightweight plastics. Another example is rubber tires, which have not only been made smaller but last longer. Newsprint used to publish newspapers has been made lighter in weight, and sometimes page size has been decreased. Also, many kinds of packaging have been lightweighted over the years, often to save on transportation costs.

- *Materials substitution.* Especially in packaging, there has been a tendency to substitute lighter materials in many applications. Thus, aluminum cans have replaced steel cans in beverage packaging, and plastic bottles have been substituted for glass. This is reflected in declining or "flat" generation of steel and glass packaging, while aluminum and plastics have shown rapid growth. Plastics have also substituted for paper in many applications. For example, even though generation of paper packaging has grown overall, generation of paper bags and sacks has declined (Fig. 5.7). The decline is primarily due to increased use of plastic bags, which are much lighter.

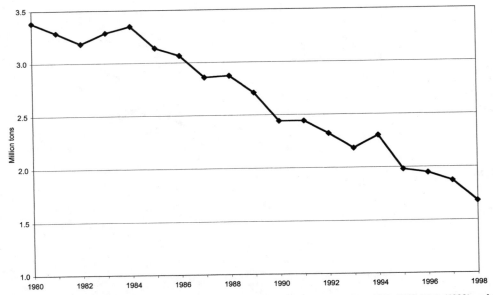

FIGURE 5.7   Generation of paper bags and sacks in municipal solid waste, 1960 to 1998. *[U.S. EPA (1999) and unpublished data developed for the U.S. EPA.]*

Trends in MSW generation are thus quite complex and difficult to quantify. Planners need to look at what is happening in their communities and nationwide when making projections affecting waste management facilities.

REFERENCES

Franklin Associates (2000) unpublished data developed for the U.S. Environmental Protection Agency.

Franklin Associates, Ltd. (1992) *Analysis of Trends in Municipal Solid Waste Generation, 1972 to 1987,* The Procter & Gamble Company, Browning-Ferris Industries, General Mills, and Sears.

Hunt, R. G., et al., (1990) *Estimates of the Volume of MSW and Selected Components in Trash Cans and Landfills.* Franklin Associates, Ltd., with The Garbage Project for the Council for Solid Waste Solutions.

U.S. EPA (1999) *Characterization of Municipal Solid Waste in the United States: 1998 Update,* EPA/530-R-99-021, U.S. Environmental Protection Agency, Washington, DC, Report prepared for the U.S. EPA by Franklin Associates.

CHAPTER 6
SOURCE REDUCTION: QUANTITY AND TOXICITY
Part 6A.   Quantity Reduction

Harold Leverenz

Whenever a consumer or establishment takes part in an activity that reduces the amount and/or toxicity of waste which otherwise would have been generated, they are participating in *source reduction*. Because of the economic and environmental advantages associated with generating less waste, the U.S. Environmental Protection Agency (EPA) has recognized source reduction as one of the most important approaches to deal with the increasing waste disposal and pollution problems in the United States (U.S. EPA, 1989). The purpose of this chapter is: (1) to define source reduction and the relevant terminology, (2) to provide a description of source reduction efforts and their potential impact on solid waste management, (3) to present a framework for developing a source reduction program, and (4) to describe strategies for source reduction.

6A.1   INTRODUCTION

According to the U.S. EPA (1999a), per capita waste generation rates in the United States have risen from 2.68 lb per person per day in 1960 to 4.44 lb per person per day in 1997. While the waste generation rate continues to increase, the recycling rate has increased to what appears to be a plateau of about 30 percent. Without some government intervention, it is unlikely that recycling rates will get much higher given the economic and environmental costs of solid waste management and the enormous quantities of waste generated in our society.

In recent years source reduction, also known as waste prevention, has been gaining more attention in the United States and around the world. The goal of a source reduction program is to decrease the amount and toxicity of material that must be managed by preventing its generation in the first place. Thus, source reduction is distinguished from other forms of solid waste management, such as recycling and yard waste collection, because it eliminates and/or facilitates the need to manage waste.

In general, the primary routes of source reduction are:

- Decreasing or eliminating the amount or toxicity of material used in the manufacture and packaging of products
- Redesigning products for increased life span, reusability, and repairability
- Changing purchasing decisions to favor those products that have minimized residual toxicity and waste associated with them
- Modifying patterns of consumption and material use in a way that reduces the amount and toxicity of waste generated

The terminology relevant to source reduction is presented in Table 6A.1. Additional information on source reduction can be obtained from the web sites listed in Table 6A.2. Strategies for source reduction are discussed further in Sec. 6A.5.

TABLE 6A.1   Definitions of Terms Relevant to Source Reduction

Term	Description
Waste	A material that the possessor considers to not have sufficient value to retain (Tchobanoglous, 1993).
Source reduction (also known as waste prevention)	Any change in the design, manufacturing, purchase, or use of materials or products (including packaging) to reduce the amount or toxicity before they become MSW. Source reduction also refers to the reuse of materials (U.S. EPA, 1999b).
Waste reduction and minimization	Activities that reduce the amount of waste that needs to be disposed of in landfills or incinerated, such as recycling, off-site composting, reuse, reprocessing, and remanufacture. However, waste reduction does not reduce the amount of waste generated.
Reuse and refurbishing	A source reduction activity involving the recovery and reapplication of a package, used product, or material in a manner that retains its original form or identity, such as refillable glass bottles, reusable plastic storage containers, or refurbished wood pallets (U.S. EPA, 1999b).
Lightweighting packaging	Reducing the amount of a particular material used to package a unit volume of product.
Source expansion	The increased generation of a waste material, effectively the opposite of source reduction (U.S. EPA, 1999b).
Functional product groupings	Considering items serving a similar purpose together. Allows for the quantification of source reduction activity due to material substitution.

6A.2   EFFECTS OF SOURCE REDUCTION

There are both economic and environmental advantages to source reduction, primarily the reduction in pollution and cost of solid waste management and disposal. In addition, source reduction activities can result in changes to the composition of solid waste.

Economic

The total cost of a solid waste management system is associated with collection, processing, and disposal of materials. Source reduction can reduce the costs of solid waste management in several ways, primarily by reducing the quantity of waste to be managed, avoided purchasing costs, and collecting revenues from resale of items.

In a study of solid waste management in New York City (Clarke et al., 1999) it was found that $300 million was spent for waste collection per year, $50 million for disposal, and slightly less for recycling, while only $1 to 2 million was invested in waste prevention programs. However, a 9 percent reduction in the solid waste stream would save an estimated $90 million in collection and disposal costs annually, along with other environmental benefits such as:

- Reduced pollution from trucks and disposal
- Less resource depletion from excess packaging not generated
- Economic development of New York reuse and repair industries
- Reduced need for landfill capacity

Choosing to refurbish, reuse, and repair an item can represent a substantial savings over disposal. Choosing products for reusability also has long-term cost benefits; for example, when a restaurant or cafeteria switches to reusable utensils and dishware, there is no longer a need to reorder disposable products continually. Regular maintenance and repair increases the lifetime that an item is in service and reduces the need to dispose of and replace that item.

Renting, borrowing, and sharing items that are needed only on occasion avoids the purchase and eventual disposal costs of that item. Leasing products that become outdated

TABLE 6A.2 Websites Pertaining to Source Reduction and Reuse of MSW

Organization (web site)	Description
Reuse	
Reuse Development Organization (ReDO) (www.redo.org)	Organization promoting reuse of surplus and discarded materials
Internet Resale Directory (www.secondhand.com)	Database of resale businesses
Tire retread association (www.retread.org)	Information and statistics on tire retreading and reuse
Parents, Educators, and Publishers (PEP) National Directory of Computer Recycling Programs (www.microweb.com/pepsite)	National and local organizations that distribute or accept used computers
Remanufacturing	
The National Center for Remanufacturing and Resource Recovery (www.reman.rit.edu)	Technical assistance and applied research and development to the remanufacturing industry and manufacturers interested in remanufacturing and resource recovery techniques
The Remanufacturing Institute (www.remanufacturing.org)	Resources for the remanufacturing industry
Source Reduction Information	
Source Reduction Forum of the National Recycling Coalition (www.nrc-recycle.org)	Research and information on source reduction under the programs link
Reduce Waste (www.reduce.org)	Waste prevention in Minnesota
National Waste Prevention Coalition (dnr.metrokc.gov/swd/nwpc)	Information and links for waste prevention
Waste Prevention World (www.ciwmb.ca.gov/WPW)	California Integrated Waste Management Board web site for waste prevention
INFORM, Inc. (www.informinc.org)	Strategies for waste prevention and discussion of EPR
U.S. EPA (www.epa.gov/epaoswer/ non-hw/muncpl/sourcred.htm)	Source reduction and reuse web site
Institute for Local Self-Reliance (www.ilsr.org)	Technical assistance and information on environmentally sound economic development strategies
Environmental Defense (www.edf.org/issues/Recycling.html)	Information on how to prevent waste
Indiana Institute on Recycling (web.indstate.edu/recycle/caselist.html)	Source reduction case studies
Earth's 911 (www.1800cleanup.org)	Community-specific environmental and recycling information
Californians Against Waste (www.cawrecycles.org)	Waste reduction advocacy organization
Product or Process Redesign	
Center for Sustainable Design (CfSD) (www.cfsd.org.uk/)	Discussion and research about eco-design and environmental, economic, ethical and social (e3s) considerations in product and service development and design
Alliance for Environmental Innovation (www.edfpewalliance.org)	Works with business to reduce waste and pollution
The Natural Step (www.naturalstep.org)	A systems approach to the sustainable design of products and processes
Containers and Packaging	
Container Recycling Institute (www.container-recycling.org)	Information on recycling and reuse of containers
Dual System (www.gruener-punkt.de/e/index.htm)	Information on packaging waste management in Germany
Reusable Industrial Packaging Association (www.reusablepackaging.org)	Resources for reusable industrial packaging

TABLE 6A.2 Websites Pertaining to Source Reduction and Reuse of MSW *(Continued)*

Organization (web site)	Description
Composting	
The Compost Resource Page (www.oldgrowth.org/compost)	Information on composting
Cornell Composting (www.cfe.cornell.edu/compost)	Information on science of composting
Worm Digest (www.wormdigest.org)	Information of vermicomposting
Master Composter (www.mastercomposter.com)	Information for home composting, links to state programs
Material Exchanges	
Southern Waste Information eXchange, Inc. (www.wastexchange.org)	Database of North American Material Exchange Programs
Jobs Through Recycling (www.epa.gov/jtr/comm/exchange.htm)	Database of national and state waste exchanges
Recycler's World (www.recycle.net/)	Available and wanted material listings, sorted by material
Community Waste Prevention Examples	
Davis, California (www.city.davis.ca.us/city/pworks/gguide)	Guide to recycling and waste prevention
Brockville, Ontario, Canada (www.brockville, reuses.com/)	Local directory for trading items
San Francisco, California (www.sfrecycle.org/)	Tips for waste reduction

Note: Web site addresses can change; if the link specified is not available, try a keyword search. Listing of a web site does not constitute an endorsement.

quickly has the advantage of keeping up with current technological innovation and encourages manufacturers to produce higher-quality and easily serviceable products. For example, because computers are quickly outdated, many businesses and industries are choosing to lease computer systems. Through leasing, companies are able to keep up with current technology without having to worry about eventual disposal problems, and manufacturers are encouraged to design products for end-of-life management.

New technologies that encourage paperless communication such as electronic mail and news permit the transfer of information in a more efficient form. By transferring information electronically, it is possible to use paper only when a hard copy is desired or necessary.

Industries save money by reducing product packaging, minimizing waste associated with manufacturing processes, or using scrap materials in the manufacturing process. The costs associated with delivery and marketing products are also reduced when the weight and volume of packaging used are reduced. Minimizing waste and toxicity of manufacturing processes results in a more efficient use of materials and reduces material purchasing and disposal costs.

Material exchanges divert waste products from one industry to raw materials for a different industry. Internet-based material exchanges on the national, state, and local level allow people to post ads for materials that they want as well as materials that they do not want. Garage sales also promote the local exchange of items, creating revenue from items that are no longer needed and keeping those items out of solid waste management systems.

Environmental

Many environmental benefits are associated with waste and toxicity reduction, primarily the reduced need for natural resources, less energy and pollution from avoided processing/reprocessing of materials, and a reduction in the amount of material sent to landfills and waste combustion facilities (U.S. EPA, 1995a).

Greenhouse gases, such as NO_x, CO_2, and CH_4, are released when energy is expended to mine raw materials, transport and process those materials, manufacture products, transport those products, and finally collect and dispose of the residual waste after the product's useful life has ended. Greenhouse gas emissions are also increased when trees are cut down to make paper, when waste decomposes in landfills, and when waste is combusted (U.S. EPA 1998a). Source reduction of municipal solid waste (MSW) is recognized as having a significant potential to reduce greenhouse gas emissions in the 1993 U.S. Climate Change Action Plan (CCAP). The exact impact of the uncontrolled release of such large quantities of greenhouse gases is not certain. However, it is likely that activities associated with MSW contribute to global warming.

Activities such as deposit and refund systems for beverage containers have been shown to reduce litter and increase the recovery rate for these materials to more than 80 percent in most places. Some environmental effects of waste management decisions are not clear. The use of washable products such as plates, cups, utensils, and towels instead of disposable alternatives may increase water use. Increased water use may adversely impact water supply as well as wastewater treatment processes by increasing the organic and suspended solid material in wastewater. The repair and reuse of older, less efficient appliances and electronic equipment may require more energy to operate them. While new technology may have the benefit of energy efficiency, it may also have the adverse effect of displacing the older items and adding to the waste management burden.

Because of the complexities associated with predicting a product or materials impact on the environment, a measurement known as life-cycle assessment (LCA) can be used (U.S. EPA, 1993). The process of LCA is used to assess a product or material's overall environmental footprint on the earth by considering the effects of the following processes:

- Choice of and extraction of raw materials
- Transport and processing of those materials
- Manufacture of products from those materials
- Use of those products
- Fate at end of life

Applying LCA to solid waste management systems can make it possible to consider the overall impacts that solid waste management decisions have on environmental systems, instead of considering only an individual process.

Waste Composition

As consumption habits change, the quantity and composition of solid waste generated will also change. Processes such as switching to a packaging material that is lighter or more efficient, or choosing to use packaging that can be accommodated by the existing recycling infrastructure, will also affect the characteristics of waste generated.

The removal of constituents from the solid waste stream may influence the management options associated with that waste stream. The recycling and reuse of plastics will reduce the amount of plastics in a waste stream and, for example, change the energy value of that waste stream. Food waste composting and food waste grinders will reduce the amount of food waste sent to disposal. Substantial reduction in the generation of solid wastes will reduce the amount of material that requires management, creating more capacity in the waste management system.

Source reduction in the packaging industry often consists of reducing the amount of material used and/or substitution of materials. The process of reducing the amount of a particular material per unit of product is known as *lightweighting*. For example, the weight of an aluminum beverage can has been reduced 52 percent in the last 20 years (Aluminum Association, 2000). When plastic is substituted for glass in a packaging process because it has a desirable characteristic (less weight), the amount of plastic in the waste stream will increase at the expense of the glass packaging for which it was substituted. As noted in Table 6A.1, the EPA has defined the increase in generation of a product as source expansion.

Hare (1997) estimated that the source reduction in four Canadian provinces accounted for the diversion of about 75 percent of beverage packaging waste between 1972 and 1995 (three times that of recycling diversion). The assessment is based on the reduction of material used per volume of beverage sold. Hare attributes the source reduction-diversion to technologically improved containers, introduced through competitive industry actions, and the shift in consumer purchasing.

To account for the effects of material substitution, the EPA considers products that serve a similar purpose together in functional product groupings. For example, beverage containers might consist of aluminum cans and PET bottles. If PET is then used to package some beverages that were previously packaged with aluminum, it will appear that there is less aluminum in the waste stream. To measure the change in waste generation, the cumulative effect of beverage packaging must be taken into consideration. Grouping items together in this way allows for comparison and quantification of source reduction efforts.

6A.3   INVOLVEMENT BY GOVERNMENT

A variety of programs and policies can be used to encourage or require participation in waste reduction. Federal, state, and local governments have the ability to implement measures that will reduce the amount of waste generated, including:

* Restrictions on packaging and products
* Establishing procurement guidelines
* Bans on the disposal of certain materials and products
* Legislation requiring manufacturers to meet certain packaging and product guidelines
* Taxes proportional to material use and waste fraction of a product
* Outreach and education programs
* Information clearinghouses
* Requiring waste audits and the development of source reduction plans

National

The federal government supports source reduction practices by providing technical and financial assistance programs, making policies, conducting studies, and distributing information. Currently, all source reduction policies and programs administered on the federal level are voluntary.

Programs.   The EPA administers several programs to promote voluntary participation in source reduction efforts. Several of the programs are discussed in the following, and others are presented in Table 6A.3.

The EPA has been tracking business, industry, and institutional waste prevention with the Wastewise program, a partnership between the EPA and various organizations interested in reducing costs associated with waste. An annual report is published recognizing those groups that have made significant progress in terms of waste prevention. As of 1999, over 900 organizations in the Wastewise program were actively participating in waste prevention.

In many countries, packaging has been of great concern because of its abundance in MSW

TABLE 6A.3   U.S. EPA-Supported Programs That Encourage Source Reduction

Program	Description
Extended Product Responsibility (EPR)	Voluntary program that challenges multiple players in the product chain to reduce the life-cycle environmental impacts of products.
Full Cost Accounting (FCA)	A systematic approach for identifying, summing, and reporting the actual costs of solid waste management. It takes into account past and future outlays, overhead (oversight and support services) costs, and operating costs.
Pay-As-You-Throw (PAYT)	EPA provides technical and outreach assistance to encourage communities to implement pay-as-you-throw systems for managing solid waste.
Wastewise	Voluntary partnerships between the EPA and U.S. businesses, state and local governments, and institutions to prevent waste, recycle, and buy and manufacture products made with recycled materials.
Demonstrations projects	EPA has funded more than 30 projects that demonstrate innovative waste reduction approaches with the potential to achieve significant reductions of greenhouse gas emissions.
Climate Change Action Plan (CCAP)	A blueprint for achieving voluntary reductions in greenhouse gas emissions from all sectors of our economy, initiatives include source reduction and recycling.
Comprehensive Procurement Guidelines (CPG)	Requires the U.S. EPA to designate products that are or can be made with *recovered materials,* and to recommend practices for buying these products. Once a product is designated, *procuring agencies* are required to purchase it with the highest recovered material content level practicable. (While the guidelines do not specifically focus on source reduction, it is probable that in the future items that reduce waste will be given preferential purchasing status.)
Design for the Environment (DfE)	A voluntary partnership-based program that works directly with companies to incorporate health and environmental considerations into the design and redesign of products, processes, and technical and management systems.

Source: Adapted from the U.S. EPA web site (www.epa.gov/osw/).

management systems. In the United States, packaging accounted for 33 percent of the total amount of waste generated and 28 percent of the total material disposed of. The concept of extended product responsibility (EPR) entails extending responsibility to various factions involved with a product or package life cycle.

Policy.   Several government actions, such as passing the Food Donation Act and the Comprehensive Procurement Guidelines (CPG), encourage or could be used to promote source reduction. It is expected that as problems associated with solid waste become more severe, the federal government will increase support of waste prevention policies.

Food waste accounted for 21.9 million tons of the municipal waste stream in 1997. A large fraction could have been diverted from landfilling or off-site composting through food donations. Food donations can help reduce the amount of food that becomes waste as well as provide food to those who need it. In 1996, Congress passed and the President signed into law the Bill Emerson Good Samaritan Food Donation Act. The bill protects businesses, organizations, and individuals that donate food in good faith from legal liability that might arise from their donation. Food donations can be used to feed hungry people, rendered into new products, or used as a livestock feed (U.S. EPA 1999b).

The CPG program is authorized by Congress under Section 6002 of the Resource Conservation and Recovery Act (RCRA) and Executive Order 13101. The CPG program requires the EPA to designate products that are or can be made with recovered materials, and to recommend practices for buying these products. Once a product is designated, procuring agencies are required to purchase it with the highest recovered material content level practicable. While these products were not specifically chosen for their ability to prevent waste, some activities such as the use of retread tires and remanufactured toner cartridges are recognized as contributing to source reduction and reuse of solid waste.

Studies and Information Dissemination.   Several studies have been conducted by the EPA attempting to quantify and justify source reduction efforts. The EPA also provides publications for training and program implementation manuals to assist states, communities, institutions, business, industry, and consumers participate in waste prevention.

In 1999, the U.S. EPA (1999c) published the *National Source Reduction Characterization Report,* the first attempt to quantify source reduction activity in the United States. The study endeavors to monitor increases and decreases in components of the waste stream as a function of historical records and consumer spending. A summary of the findings presented in the source reduction report is presented in Table 6A.4. Studies such as these are integral to measuring the contribution of source reduction efforts and progress toward goals.

The EPA has also published several other reports on source reduction activity. *Municipal Solid Waste Source Reduction: A Snapshot of State Initiatives* (U.S. EPA 1998b) is a summary of the initiatives taking place around the country. Another EPA published report, *Greenhouse Gas Emissions from Management of Selected Materials in Municipal Solid Waste* (U.S. EPA 1998a) describes how solid waste management decisions affect climate change. Source reduction is noted for increasing carbon sequestration in forests, avoiding emissions from material extraction and processing, and not contributing to emissions associated with waste management processes.

Since 1986, the EPA has published the *Characterization of Municipal Solid Waste in the United States* report to estimate the generation, recovery, and disposal of MSW in the United States. Information is provided in the report that can help guide solid waste management decisions. Because the report is updated annually, it is possible to observe the historical trends for various materials.

State

A study by the EPA (1998b) found that 47 U.S. states currently participate in one or more source reduction programs. State source reduction activities include planning (setting goals, mandates, quantification), in-house programs (reuse programs, procurement guidelines), res-

TABLE 6A.4 U.S. Source Reduction of MSW Based on Consumer Spending and Change in Waste Generation Rate in 1996

Waste stream	Source reduction (thousands of tons)
Durable goods (appliances, furniture, tires)	2,179
Nondurable goods (newspapers, clothing, etc.)	3,571
Containers and packaging (bottles and boxes)	4,002
Other MSW (yard wastes and food scraps)	13,534
Total source reduction	23,286

Source: Adapted from U.S. EPA (1988).

idential programs (consumer purchasing education and backyard composting and grasscycling programs), and programs that support local governments (grants, technical assistance).

Assistance.   State assistance programs generally consist of supporting local governments and businesses with financial and technical support to increase source reduction activity. Other programs provide training and workshops to educate program managers about source reduction and recycling strategies.

The state of Minnesota conducts workshops to educate school administrators and personnel on source reduction in the school environment. Massachusetts trains composting coordinators to educate homeowners on home composting and works with farmers to help compost organics generated on the farm. Some states also offer grant programs to fund source reduction initiatives. Maine supports a Master Composter program to provide training and certification of volunteers. The state of Maine awards grants to local governments to purchase composting bins for residents.

In 1999, the state of Vermont awarded $50,000 in grants to waste prevention projects. Funded projects included a demonstration project to reuse scrap wood, building a reuse center at a recycling facility, source reduction education in schools, creation of an office supply collection and reuse center, publishing of a guide on where to buy and donate used items, and a household hazardous waste education program at daycare centers (Biocycle, 1999b).

Bans and Restrictions.   Materials and items that are compostable, recyclable, repairable, or large in quantity and toxicity can be banned or restricted to keep them out of waste disposal systems. Placing bans on materials can encourage consumers and establishments to participate in source reduction activities because of the problems associated with restricted disposal. When a material or product is banned, manufacturers are also pressured to provide items that can substitute for the banned materials.

Massachusetts has banned television and computer monitors from landfills because of the lead contained in cathode ray tubes (CRTs) found in televisions and computer monitors, and the projected increase in disposal of these items due to replacement by digital screens. The sale of thermometers containing mercury and the landfilling of any mercury-containing component of MSW has been banned in Minnesota. Illinois has banned the landfilling of the toxic components of appliances (including freon and chlorofluorocarbon refrigerants). Twenty-two states have banned the landfilling of yard wastes (Glenn, 1999). The state of Minnesota is considering banning all unprocessed MSW from landfills after the year 2008 (State of Minnesota, 2000) to counteract issues such as:

- Anticipated increase in waste generation
- Pollution from landfilling of waste
- Cost of landfilling and the lack of support for the local economy
- Lower property values near landfills

Other materials that have been banned from landfills include leaves, grass clippings, whitegoods, yard waste, lead-acid batteries, tires, office and computer paper, newsprint, corrugated cardboard, paperboard, glass, plastic, aluminum, and steel containers. Material bans are most effective when a consumer education program or community outreach program is implemented, and infrastructure for alternative collection of the materials is available. It is also important to note that restricting disposal of a material may not result in source reduction. The ramifications of a material ban on commerce laws and waste management should be considered in advance.

Deposit and Refund Systems.   The principle of the deposit and refund system is that, at the time of purchase, the consumer pays a fee supplemental to the cost of the product. This fee is refunded when the package or product is returned to the manufacturer or authorized collection center. Deposit and refund systems have been implemented in 10 states and in several cities to increase recycling and to reduce litter of aluminum, glass, and plastic beverage packaging. Recovery rates for various materials have generally exceeded 80 percent. The unclaimed redemption deposits are used for program administration, support of environmental programs, or retained by the distributor.

While deposit and refund programs have been used mostly to increase recovery rates, some programs give a preference to refillable containers. The beverage industry initially used deposit and refund systems to ensure the return of glass containers for refilling. Currently only two states, Michigan and Oregon, support a pricing scheme that favors refilling. In Michigan, a 10-cent fee is placed on nonrefillable containers and a 5-cent fee is applied to refillable containers. In Oregon, a 5-cent deposit is placed on nonrefillable and a 2-cent fee for refillable.

Exchange, Donation, and Sale.   Exchange, donation, and sale of unwanted items and materials will not only prevent their disposal, but may also avoid the purchase and subsequent disposal of new items and materials. Many items such as computers, appliances, and vehicles can be donated to schools and charitable organizations. Community and personal garage sales also promote the extended life of products by transferring an unwanted item to another individual who has a use for that item.

Material exchanges are programs by which organizations can buy, sell, or donate unwanted or excess material that would otherwise end up in a disposal system. In 1999, the California Material Exchange program (CalMAX) diverted 713.5 tons of material from disposal. CalMAX is a free service sponsored by the California Integrated Waste Management Board (CIWMB), designed to assist local jurisdictions and businesses with an effective method of diverting discards, surplus, and excess materials previously sent to the landfill.

The state of Delaware has a reuse industry that employs 4500 people, with an annual payroll of $97 million; a web site (www.state.de.us/dedo/publications/reuse/) offers contacts for used building materials, office furniture repair, computer donation opportunities, second-hand clothing stores, and appliance repair (BioCycle, 2000).

Mandates.   State-imposed mandates include restrictions on products and packaging entering and being manufactured within the state, and the submission of source reduction plans by local governments, business, and industry (U.S. EPA, 1998b). For example:

- Oregon requires unit-based pricing systems, and that all rigid plastic containers be reusable 5 times and contain recycled material or meet 10 percent source reduction in 5 years.
- Wisconsin and Iowa require unit-based pricing for communities that do not reach a 25 percent recycling goal.
- Pennsylvania businesses that generate MSW must prepare a source reduction plan, reporting what types of waste they generate and identifying strategies for source reduction.
- California requires local governments to have source reduction components in their solid waste management plans and have mandated 25 percent diversion by the year 1995 and 50 percent by the year 2000, through recycling, source reduction, and composting.

Taxes.   Taxes that are applied to excessive packaged items, disposable or single-use items, or products that contain hazardous compounds, encourage source reduction. The objective of the tax can be to influence consumer purchasing decision, fund disposal of the product, or persuade production and manufacturing to adopt source reduction measures that avoid undesirable characteristics.

In some states, taxes are applied to tires, white goods, and batteries. For example, North Carolina applies a tax to tires and white goods to fund, in part, the cost of waste management for these items. While these taxes are generally used to fund recovery and management operations, it would be possible to direct taxes at reducing the toxicity and amount of waste generated.

Local

Local governments play an important role in source reduction because they work directly with residents, businesses, and institutions. On the local level, government can require businesses and institutions to conduct waste audits, prepare and submit source reduction plans, provide consumer education and outreach programs, fund programs that increase source reduction, and implement fee-based waste disposal systems.

Assistance.   Local government can provide funding for projects that encourage source reduction. Examples of funded projects include free or subsidized compost bins, consumer and student education programs, composting training programs, and funding for establishment of reuse and repair industries.

Funding waste prevention projects can be an effective way for local governments to reduce waste generation. The San Francisco Recycling Project (SFRP) awards grants to increase source reduction and recycling. Examples of funded projects include increasing pallet repair and reuse, recovering food, establishing composting and vermiculture programs, recovering and reusing furniture, and conducting workshops of textile reuse.

The city of Davis, California, operates a program whereby residents can obtain a free compost bin. Interested residents are first given a composting booklet to review. After gaining an understanding of the composting process, a short test is administered. The test consists of questions directed at reducing nuisance conditions (for example, how to avoid odors and rodents) and also fundamental questions about the composting process (such as moisture and carbon-nitrogen ratios). In addition, the city has a compost demonstration site located near the community gardening area. The city contracts with a private company for solid waste management. A guide is published annually and distributed to residents detailing the materials acceptable for recycling as well as the collection of these materials. The manual also contains information on source reduction activities and household hazardous waste collection.

Consumer and Student Education.   Programs can be implemented to educate consumers about local laws governing waste disposal practices, backyard and worm composting, grasscycling, and green shopping strategies.

Because children are the future residents and consumers, it is critical to make source reduction part of their education. Some programs have been set up in schools to educate children about waste, include conducting waste audits and backyard composting and vermicomposting demonstration projects. Seven states support source reduction through educational programs in schools (U.S. EPA, 1998b). A program funded in part by the city of Eureka, California, was designed to educate fourth graders about preventing waste through smart purchasing, vermiculture, and reusing waste materials. The program, sponsored by CIWMB, was taught by high school students who performed plays and took students to grocery stores to look at different types of packaging and discuss purchasing decisions.

A program in San Francisco was funded by local governments and the CIWMB to educate shoppers about preventing waste. The campaign to educate consumers about topics such as bringing their own bag, purchasing in bulk, not purchasing excessively packaged items, and purchasing reusable products has been publicized on radio and television stations and in newspaper ads and articles. The program has had great success, achieving a 19.4 percent increase in well-packaged products and a 36 percent decrease in overpackaged products. In 1997, a consumer survey found that, because of the survey, 30 percent purchased products with recycled packaging, 23 percent brought their own bag, and 19 percent bought products in bulk.

Reuse/Repair Industries.   Reuse and repair industries represent a way to keep products and materials from disposal through refurbishing and redistribution. Unlike recycling, the items and materials generally require little or no processing and augment the purchase of new products or materials, reducing pollution and waste generation. Examples of reuse include recovery of computers and supplies for schools and recovery of used, out-of-date, excess building materials for low-income housing projects.

Because a significant percentage of MSW is potentially reusable, some businesses such as the Loading Dock in Baltimore, Maryland, are able to keep 7000 tons of recovered building materials from landfill disposal. The Materials for the Arts in New York City and Urban Ore in Berkeley, California, are also successfully recovering, reusing, or reselling materials (ReDO, 2000).

According to ReDO, the Reuse Development Organization, there are more than 6100 reuse centers around the country, composed of:

- Thrift stores and charitable drop-off centers
- Efforts supplying charities, low-income people, food banks, and schools with reusable equipment and materials
- "Drop and swap" stations at landfills
- Used equipment stores and salvage yards
- Local and regional material exchanges

Remanufacturing and refurbishing are also forms of repair and reuse in which components from used or broken products are used to construct new products. Commonly refurbished or remanufactured products include pallets, toner cartridges, appliances, engines, and single-use cameras.

Unit Pricing for Waste Reduction.   In a conventional municipal waste collection system, bags or bins of waste are placed on the curb and picked up, usually once a week. When households pay for the collection service out of local taxes, the price at the margin is zero; a family that fills four bins with garbage each week pays no more than an elderly couple that fills one. From an economic perspective, however, the marginal cost of waste disposal is not zero. The more waste people throw away, the more collectors are needed, and the higher the cost of landfill tipping fees. If waste disposal is free, people will throw away too much. To reduce the generation of waste, people need an incentive to throw away less.

Systems that encourage people to reduce waste generation and increase recycling are known as *unit pricing, variable rates,* or *pay-as-you-throw* systems. There are four basic types of unit pricing systems:

- *Can systems.* Customers choose the number of waste containers they will set out for collection. Each can's size represents a different gallon or weight limit. Disposal fees are based on the number of cans used.
- *Bag systems.* The waste a consumer wants collected must be put in a bag with a special color or logo. The disposal fee must be prepaid when the customer buys the bag at a local store or some other designated location. Purchase of the bag guarantees collection, but the more bags are needed, the more the customer pays to buy them. An alternative is tags and stickers, which, once purchased and placed on a container or bag, guarantees collection and disposal.
- *Two-tier.* A combination of traditional funding from property taxes or monthly fee combined with a user fee. In a two-tier system a customer pays a flat fee for waste removal through a tax or monthly bill. This fee usually provides for collection of one can or one bag. Collection of any additional waste is charged through a bag or sticker system.
- *Weight-based systems.* Collection charges in such a system are assessed in accordance with the number of pounds of waste that is put out for collection. Weight-based systems are fairer than volume systems because the volume of waste generated in a volume system can be compressed to fit into a given bin or bag. However, the garbage has to be weighed by the collector with scales on the truck, and the technology of weight-based systems is expensive and subject to mechanical failure.

Weight-based systems have been tried in Hampton, Virginia, for about a year and in Milwaukee, Wisconsin, from July 1993 to October 1995. It was found that it took longer to collect the waste, and there were problems with the accuracy of the scale. The scales must meet national standards for accuracy or they cannot be used to charge for the garbage collection. Because of these problems, both experimental programs were terminated.

Unit pricing provides customers with economic incentive to reduce the amount they discard and provides a link between the amount of waste they set on the curb and the garbage collection bill. Consequently, studies of municipalities with variable-rate waste disposal programs indicate that waste reductions ranging from 25 to 45 percent can be achieved. Furthermore, increased participation in recycling and yard waste programs has been observed.

A study by the EPA found that charging variable rates for residential collection can reduce cost and improve service. To be successful, however, the municipality must find the right mix of prices and options for the particular location and circumstances of the system. To determine which waste collection system is most economical, Leith (1996) recommends that municipalities use full cost accounting (FCA). Such a system helps communities to identify and assess the total costs associated with solid waste services. With this information, they can select the best method, shape cost-cutting services, and foster better decision making and long-term planning.

FCA differs from cash-flow accounting typically used by governments because, along with indirect (overhead) costs such as administration and legal services, FCA incorporates also past and future expenses using depreciation and amortization. Several states mandate full cost accounting, and many communities are implementing it voluntarily. Houston, for example, used FCA to identify the program elements in the total solid waste disposal. It highlighted collection costs, specifically labor and worker's compensation, as the largest line item in the program budget. Armed with this information, the city took measures to decrease these costs through the use of automated collection trucks saving between $5 and 7 million annually. Although the challenges of scarce resources, incomplete records, and a lack of standardized methodology can be daunting, FCA can help communities reduce their waste management costs. More information on full cost accounting is available from the U.S. EPA in Washington, D.C. A free copy of the EPA's *Full Cost Accounting for Municipal Solid Waste Management: A Handbook* (U.S. EPA, 1997) can be obtained by calling 1-800-553-7672.

The EPA has also provided a program for balancing costs and revenues for strong unit pricing programs. The following seven steps are proposed to balance costs and revenues: (1) estimate the demand for services, (2) plan for services to be offered by the community, (3) estimate the costs of the services, (4) develop a rate structure, (5) calculate the resulting revenues, (6) compare program costs against anticipated revenues, and (7) revise the rate structure until the services are at a price residents can support. For a free copy of the EPA's booklet *Pay-as-You-Throw: Lessons Learned About Unit Pricing,* call the RCRA hotline at 1-800-424-9346.

An EPA-funded study at Duke University on the reduction in waste generation from the implementation of pay-as-you-throw (PAYT) system found that on average waste reduction in the PAYT communities ranged from 14 to 27 percent. However, implementing successful paid programs requires planning. They should be designed with cost savings in mind but provide convenient access to a variety of recycling opportunities. These opportunities must be available so residents can respond to price signals. Last but not least, the program must be accepted by the community. Some communities charge twice as much for the second can of garbage and offer essentially free recycling services (Starkey, 1996).

In one mandatory pay-by-the-bag program instituted in Carlisle, Pennsylvania, in 1990, the 18,000 residents did not receive a bill for the private waste services. Instead, residents financed the service by purchasing specially marked 30-gallon blue waste bags. These waste bags, which sold for $2.10 each, could be purchased in several locations. In addition, weekly recycling of different materials and a spring cleanup were offered at no additional charge.

In one of the most studied unit pricing programs, residents in Perkasie, Pennsylvania, were offered two bag sizes for purchase: a large, 40-pound-capacity bag for $2.25 and a smaller, 20-pound bag for $1.25. The prices reportedly covered the collection and disposal of municipal solid waste and the recycling costs. It has been reported that the borough has decreased the amount of MSW sent to landfills by 41 percent (Slovin, 1995).

In July 1999 (as reported in BioCycle, 1999c), Forest, one of six Ohio communities that received a grant to adopt pay-as-you-throw garbage and recycling systems, parlayed its grant into a 44 percent waste reduction rate and nearly doubled the amount of recyclables recovered. Crews took only half the time to collect waste as they did before the change. Also, in the North Central Ohio Solid Waste Management District, which is developing a self-sustaining PAYT program for its municipalities, Bellefontaine claimed that participation increased from 50 to 97 percent.

Waste Audits and Source Reduction Plans.   Requiring businesses and institutions to conduct waste audits and prepare source reduction plans is the first step to waste prevention. An analysis of the waste source and composition reveals the nature of the waste and where waste prevention initiatives should be directed.

Waste audits can consist of evaluating solid waste hauling records (if they exist), observations of the procedures and activities that result in waste generation, and waste characterization studies. Solid waste hauling records are useful for determining the bulk amount of waste generated and the associated specific costs of waste management. It is important to perform observation and characterization studies over a representative period of time to ensure an accurate examination is made.

Yard Waste Programs.   Grass clippings and other yard wastes make up about 14 percent of MSW (U.S. EPA 1999a). Thus, keeping yard wastes out of solid waste management systems can have a significant impact on the amount of waste that requires management. For example, in 1998 the city of Markham, Ontario (population 172,000), banned grass clippings from the landfill and ended curbside collection of the clippings, saving an estimated $665,000/year in avoided collection costs (Canterbury 1998).

On-site composting is the management of organic materials (usually yard wastes and food scraps) at or near the source of generation. Many communities are utilizing backyard composting programs in conjunction with landfill bans to reduce the amount of material that requires management. Grasscycling is the practice of leaving grass clippings on the lawn after cutting. Because grass clippings make up a relatively large fraction of MSW, grasscycling has the potential to substantially reduce the amount of waste being sent to landfills.

Composting and grasscycling are supported through workshops and training programs, educational brochures, demonstration sites, and by providing subsidized compost bins to residents. As of 1998, 14 states were active in promoting composting and grasscycling programs. The U.S. EPA (1999c) estimated that due to source reduction efforts, over 13 million tons of yard wastes and food scraps never entered the waste stream, and subsequently did not require external management. Composting and grasscycling are encouraged through community and state outreach programs, landfill bans on yard waste, and PAYT programs.

International

Because of health concerns related to the toxic component of MSW and declining landfill space, other countries are adopting more stringent source reduction measures. European and Asian countries are adopting a "producer take-back" system to reduce solid waste. In the United States, producer and manufacturer source reduction initiatives are still voluntary. However, in some other countries a similar concept known as extended producer responsibility, which shifts the costs of solid waste management to the manufacturers of the material, has been successful in minimizing packaging waste. Generally, a fee is imposed on a product and/or its packaging to fund its life in solid waste management systems. For example, a tax proportional to the weight, volume, and type of material used to package an item will encourage manufacturers to pursue the least costly alternative. Because the tax system is structured in a way that favors environmentally favorable products, the most economically efficient alternative will also be resource-efficient. The revenues from the tax can be used to fund collection and recycling of the residual waste.

Most Canadian provinces have deposit-return systems set up to reduce beverage container waste. A large percentage of beer produced in Canada is packaged in refillable containers that carry a refundable deposit. To promote reusable containers, Denmark banned the sale of beverages in metal cans and in nonrefillable glass and plastic containers (BioCycle, 1999c).

6A.4   *DEVELOPING A SOURCE REDUCTION PLAN*

Planning is central to developing effective source reduction programs. Before source reduction planners start developing specific source reduction initiatives for their communities, it is extremely important that they know what they are trying to reduce, how much reduction they want to achieve, and how they will measure their results. Municipal solid waste plans need to include an explicitly stated source reduction policy, clearly defined goals, and meaningful measurement strategies. Without these measures, it is not possible to evaluate the effectiveness of such programs.

Implementing a source reduction program also involves developing an infrastructure to support it. Specifically, an effective program requires independent leadership, authority, appropriate staffing, and an adequate budget.

Source Reduction Policy

The first step in planning for source reduction is a clear statement of policy, including a definition of terms that clarifies what source reduction means so that it can be differentiated from other waste management options, such as recycling. In other words, instead of a policy of diversion from landfills, which leaves ambiguity as to whether the strategy should be source reduction, recycling, or (in some cases) incineration, a clear policy would state explicitly that the goal is source reduction, include a definition of that term, and then specify goals and measurement methodology.

Setting Source Reduction Goals and Establishing Measurement Methodologies

The next steps in source reduction planning are setting goals and establishing measurement methodologies. Goals and measurement systems are important for effective source reduction programs because they help communities establish program priorities, track and evaluate progress, and recognize accomplishments and target areas for further efforts.

As of 1997, only five states had specific source reduction goals: Connecticut, Maine, Massachusetts, Minnesota, and Rhode Island. An additional 18 states had waste reduction goals that do not differentiate between the amount designated for recycling and the amount designated for source reduction (U.S. EPA, 1998b).

To set goals most effectively and establish measurement methodologies, communities need to take the following four steps.

1. Establish an overall source reduction goal that is separate from the recycling goal with specification of:

   - The baseline year
   - Target year
   - Type of reduction to be measured (from the current total waste generation levels, from current per capita generation levels, or from the projected increase)

2. Determine separate goals desired for:

   - Generating sectors (residential, commercial, and institutional)
   - Materials (paper, glass, plastics, organics, etc.)
   - Products (Styrofoam cups, glass bottles, tires, cardboard boxes, newspapers, etc.)

3. Select unit of measurement:

   - Weight
   - Volume
   - Weight and volume (preferable, if possible)

4. Selected measurement methodology:

   - Waste audits
   - Sampling (including weighing-in places such as transfer stations)
   - Surveys
   - Purchases (tracking sales)

Information Needs for Measuring Source Reduction.   Good data collection is vital for measuring source reduction, because communities need to know which sources are generating which types of waste materials and how much they are generating. Thus, at a minimum, communities need to collect data on:

- Amount of residential waste
- Amount of commercial waste
- Residential population
- Total employment
- Projections of population change
- An index of economic activity

The Importance of Waste Composition.   Knowledge of the composition of the waste stream is useful in setting realistic goals because it allows communities to prioritize source reduction efforts. Materials can be targeted for source reduction if they constitute a major proportion of the waste stream, are easy to reduce, or are major contributors to pollution during disposal. Because the waste stream varies from community to community, in-depth information about waste stream requires a waste audit—an actual sampling of waste generated to determine its composition by material, product, and generating sector.

Yard waste, for example, is a good target for source reduction. When yard waste is burned, various compounds are emitted; it represents a large component of the waste stream (approximately 14 percent, nationally); and it can readily be reduced through grasscycling and backyard composting. In suburban areas, yard wastes can constitute a much higher fraction of the solid waste stream than in densely populated urban areas, and should therefore be considered an easy target for source reduction efforts.

Administration and Budget

Departments charged with managing solid waste have traditionally been staffed by officials knowledgeable primarily about waste disposal and, more recently, about recycling. Their responsibilities have been the collection, transport, and disposal of waste, and the processing and marketing of recyclable materials. Their key concerns have been diminishing disposal capacity, siting new facilities, and controlling costs.

Implementing source reduction programs involves vastly different staff skills and concerns. It requires staff with a broader, long-term view of the use of materials in society and an understanding of how behavior can be changed to optimize the use of resources and minimize the waste generated. Staff members need diverse skills so they can work on planning, program development, technical assistance, education, outreach, legislation, data collection, program evaluation, waste audits, and enforcement. Their concerns must encompass broad issues (e.g., impacts on economic development) that go well beyond questions of how to manage waste.

Administration.   Efforts to provide independence and authority for source reduction are essential if it is to become a viable policy option. For the most effective administrative structure, source reduction would be separate from and independent of waste management functions. The head of the source reduction effort would have authority at least equal to that of the individuals in charge of recycling and disposal, and would have a commitment to minimizing the amount of materials actually entering the waste stream.

Source reduction is much broader in scope than recycling or disposal and is, in fact, *resource* management rather than *waste* or *material* management. That is, it involves decisions about what products and packages are made, how they are made, and how they are used. An effective source reduction program deals with producers, distributors, and consumers. It can thus be argued that source reduction does not belong in sanitation or solid waste departments at all, and should not be a function of waste managers. Theoretically, it might make more sense to place source reduction activities in a department of economic development. On a more practical level, however, the motivation to promote source reduction is generally the need to reduce waste, so it is likely to remain in the purview of solid waste departments.

If source reduction functions are placed in a solid waste department, they need some independence from the recycling functions because the immediate, everyday demands of recycling can tend to overwhelm the longer-term, more complex source reduction activities. Larger budgets and more personnel are required for recycling because it includes collection, processing, and marketing; the scale and urgency of these management tasks may result in eclipsing the attention given to source reduction.

Budget.   Source reduction does not require the costly collection and processing operations involved in conventional waste management options, but it is not free and it cannot be accomplished without an adequate budget. The costs of source reduction programs are in the form of an upfront investment in data collection, waste audits, legislative development, education, technical assistance, equipment, and planning.

A barrier to funding source reduction is that results may not happen immediately, so there may be no return on the investment in the budget year in which the expense is incurred. To assure continual and adequate funding, source reduction could be funded from a designated income stream. This might be a portion of the funds raised from charging residents for the amount of waste disposed or other waste collection fees, environmental taxes or fees, or possibly unreturned beverage container deposits. Source reduction could also be funded as a specific percentage of a recycling budget. For instance, if 5 percent of the recycling budget was dedicated to source reduction, a recycling budget of $10 million would allocate $500,000 to source reduction.

6A.5 STRATEGIES FOR SOURCE REDUCTION

Enormous potential exists to implement programs that will prevent the generation of waste. In 1996, the EPA estimated that 23 million tons of MSW were source-reduced in the United States, or 11 percent of the total 209.7 million tons that were generated that year (U.S. EPA 1999c). Yard wastes and food scraps accounted for the largest fraction of waste prevented (58.1 percent of the 23 million tons), followed by containers and packaging (17.2 percent), nondurable wastes (15.3 percent), and durable goods (9.4 percent).

There are many strategies available to accomplish source reduction. The EPA (1999c) has identified the following examples of source reduction activities:

- Redesigning products or packaging to reduce the quantity or toxicity of the materials used, substitution of lightweight materials, or making them reusable
- Reusing existing materials, products, or packaging; for example, refillable bottles, reusable pallets, reconditioned toner cartridges, and copying on both sides of a sheet of paper
- Reducing the amount of a product or packaging used
- Lengthening the lives of products or materials to postpone disposal, such as through regular maintenance or choosing to repair an item
- Using packaging that reduces the amount of damage or spoilage to a product
- Managing organic wastes (such as food scraps and yard trimmings) through on-site composting or other alternatives to disposal (such as leaving grass clippings on the lawn)

Residential

The residential sector contributes 55 to 65 percent of the total MSW generation (U.S. EPA, 1999a). It is estimated that an average home can reduce its waste by 30 percent through source reduction practices; see Table 6A.5. To increase residential participation in source reduction programs, it is necessary to educate consumers about how their own actions affect

TABLE 6A.5   Potential for Reduction of Typical Household MSW

Component	Typical (%)	Constituents targeted (source reduction activity)	Reduction, %
Organic:			
Food wastes	9.0	Waste from food preparation and spoilage (on-site vermiculture or composting)	50
Paper	34.0	Newspaper (electronic versions, used in on-site composting systems)	20
		Single-sided copies (use both sides of paper)	
		Bulk mail (request to be taken off mailing lists; increased cost to send bulk mail)	
		Misc. notes (e-mail)	
		Grocery bags (reusable shopping bags)	
Cardboard	6.0	Packaging (avoid purchasing excessively packaged products)	10
Plastics	7.0	Excess packaging (avoid purchasing; producer responsibility)	25
		Food and beverage containers (avoid purchasing; utilizing refillable and reusable containers)	
		Grocery bags (reusable shopping bags)	
		Appliances (lease; choose quality/repairable alternatives)	
Textiles	2.0	Unwanted clothes (donate to charity)	—
Rubber	0.5		—
Leather	0.5		—
Yard wastes	18.5	Waste from lawn and garden activities (on-site composting)	90
Wood	2.0		—
Misc. organics	—		—
Inorganics:			
Glass	8.0	Containers (avoid purchasing, on-site reuse)	10
Tin cans	6.0	Food containers (avoid purchasing)	10
Aluminum	0.5	Beverage cans (avoid purchasing)	—
Other metal	3.0		—
Dirt, ash, etc.	3.0		—
Total	100.0		31.7

Source: Adapted from Tchobanoglous et al. (1993).

solid waste and measures they can take to have a positive impact. A list of common residential source reduction activities is presented in Table 6A.6.

Resourceful Living.   Various ways exist for people to prevent the generation of household waste, including green purchasing and product use, and participation in reuse and exchange programs.

Green purchasing and product use is giving consideration to the environmental impacts of purchasing decisions and how the item is used. Choosing to purchase products in bulk or concentrated formulations, for example, reduces the generation of packaging waste. Purchasing produce grown locally reduces energy used to transport and preserve fruits and vegetables grown elsewhere.

Buying products that are more durable or have longer warranties may increase both the life span and the time until disposal. Reusable utensils and dishware, concentrated soaps, and rechargeable batteries save money through avoided purchases and waste disposal costs. Other examples of green product use include saving food containers for reuse, maintaining and repairing appliances, and reusing shopping bags.

Participating in reuse and exchange programs can include shopping at second-hand stores, visiting and hosting yard sales, purchasing salvaged or refurbished materials, or making food

TABLE 6A.6   Consumer Strategies for Source Reduction

Strategy	Examples
Avoid unnecessary packaging.	Choose items with least or no packaging. Purchase economy size, bulk, or concentrates.
Adopt practices that reduce waste toxicity.	Use alternative cleaners that do not have hazardous compounds. Use integrated pest management instead of pesticides. Choose batteries with reduced mercury. Use digital thermometers instead of mercury. Use household hazardous waste collection. Use nontoxic inks, dyes, and paints.
Consider reusable products.	Reusable cups, dishware, and utensils. Cloth napkins and towels instead of paper. Rechargeable batteries. Refillable detergents.
Maintain and repair durable products.	Choose long-lasting and efficient appliances and electronic equipment. Follow proper maintenance schedule. Long-lasting tires. Mend and repair clothes, footwear, and bags. Long life fluorescent light bulbs.
Reuse bags, containers, and other items.	Reuse paper, plastic, and cloth bags. Reuse scrap paper and envelopes. Wash and reuse cans, jugs, and containers. Save scrap wood for projects.
Borrow, rent, or share items.	Rent or borrow party supplies, tools, appliances and electronic equipment, floor and rug cleaners, ladders, etc. Offer before discarding items such as cameras, tools. Share newspapers and magazines.
Sell or donate goods instead of throwing them out.	Donate clothes, textiles, appliances, and furniture to thrift stores and charity. Sell at garage sales.
Compost yard trimmings and food scraps.	Backyard composting. Worm composting. Xeriscaping.

Source: Adapted from U.S EPA (1996).

and product donations. Choosing to rent, lease, or borrow items that are not often needed also reduces waste generation. For example, activities such as renting tools, sharing a lawnmower, or leasing a computer all prevent the purchasing of new items, the production of new products, and the eventual disposal of unwanted possessions.

Backyard Composting, Grasscycling, and Xeriscaping.   The management of organic materials at or near the home can substantially reduce the amount of waste generated for management. Backyard or on-site composting can consist of composting with or without a compost bin. Composting can also be facilitated with worms, known as vermiculture, vermicomposting, or worm composting. Grasscycling returns grass clippings to the lawn instead of requiring additional management or disposal. Xeriscaping is a form of landscaping that reduces water use and the generation of yard-related wastes.

Backyard composting usually consists of the collection and biological transformation of food and yard wastes. A container, known as a compost bin, is used to keep materials together and keep animals out. Compost bins can be constructed from used pallets, bricks, or wire fencing. Materials containing carbon (usually brown, such as leaves or wood chips) and nitrogen (usually green, such as grass or food waste) are combined together, along with appropriate moisture, to produce a mixture conducive to biological degradation. The mass is seeded with soil organisms and turned occasionally to introduce oxygen and accelerate the composting process. The resulting humus material (known as compost) can be used as a soil amendment or mulch.

The use of worms to compost food waste is an effective way to keep food scraps out of waste disposal systems. Worms and soil organisms consume the waste and produce a material composed of worm casting and decomposed waste, known as vermicompost. Vermicompost is higher in nutrients than yard waste compost and can be used in potting soil mixtures. A training program is useful to introduce and educate residents about the composting process.

Grasscycling is accomplished by the use of a mulching lawnmower. Grass clippings are simply left on the lawn and assimilated into the soil, instead of being collected in a bag. In addition to reducing the amount of waste that needs to be disposed of, grasscycling can also reduce the amount of fertilizer required, because nutrients are recycled when the grass clippings are not removed. For grass recycling to be effective, certain steps must be taken to avoid clumping and accumulation of clippings on the lawn surface after cutting, such as the use of a mulching lawnmower, and not removing more than one inch of grass height at one time (U.S. EPA 1992). To implement a grasscycling program, it is useful to work with local retailers of lawn and garden supplies, lawn care services, and residents to provide information on solid waste policy, mulching lawnmowers, and grasscycling guidelines.

Xeriscape is the practice of landscaping with vegetation that reduces water use and maintenance requirements. Plants are usually chosen based on how well they are adapted to the environment and for their aesthetic value. Grasses that require large amounts of water and trimming are completely avoided. Xeriscaping not only reduces water needs, but also reduces the quantity of yard waste generated.

Commercial and Institutional

Commercial (stores, restaurants, hotels, and service stations) and institutional (government, schools, correctional facilities, hospitals, and libraries) sources of waste can include large quantities of paper, cardboard, food waste, plastics, and hazardous wastes. Because of the large number of people associated with these establishments, significant potential for solid waste prevention exists. Successful initiatives that have been undertaken include efficient use of office paper, on-site food waste composting, and switching to reusable supplies. Several commercial and institutional source reduction case studies are summarized in Table 6A.7.

Strategies for implementing source reduction in commercial facilities and institutions can be classified in two main categories: (1) changing procurement policies, and (2) modifying operations.

TABLE 6A.7 Typical Examples of Source Reduction in Commercial Facilities and Institutions.

Business	Method of source reduction	Revenue/savings
Memorial Hospital	Changing batteries from mercury to zinc	Last 25% longer; 342 pounds of mercury in first year
BankAmerica Corporation	Using lighter envelope in ATM machines	773,000 pounds of paper; $570,000
Park Plaza Hotel & Towers (Boston)	Refillable soap and shampoo dispenser (instead of disposable)	2 million plastic bottles
Village of Hoffman Estates	Pay-as-you-throw waste collection	30% waste cut
Seattle, Wash.	Unit pricing	3.5 cans to 1.7 cans (50%)
Itasca County, Minn.	Reusable air filters in garages	$4700 annually
New York City, N.Y.	Dry cleaners accepting used hangers; reformatting phone book	750,000 hangers from landfill 100 pages per book, 107 tons of paper
U.S. EPA	Paper waste reduction program	16% less paper waste at photocopiers
U.S. Postal Service	Promoting change-of-address program	1087 tons of bulk mail
U.S. DOE	E-mail instead of paper bulletins	154,000 sheets of paper annually; reductions in printing costs ($9000 to $624) and delivery costs ($1900 to $120)
Larry's Markets	Providing reusable bags to customers	15% less bag waste
Rosenberger's Dairies	Supplying 220 schools with refillable milk bottles; can be refilled 100 times	90,000 bottles each day

Source: Adapted from U.S. EPA (1995b) and U.S. EPA (1999d).

Procurement.   Changes in procurement policies to favor source reduction not only have the potential of reducing the amount of waste generated, but also may set an example for the private sector and encourage manufacturers to develop less wasteful products and packages that would then be available to all purchasers. While the procurement policies favoring recycled goods are becoming more common, procurement policy is rarely used to achieve source reduction objectives. Procurement guidelines could require the purchase of reusable, refillable, repairable, more durable, and less toxic items. Procurement policy could also require minimal and reusable packaging that would not only reduce waste, but may also reduce the cost of purchasing, mailing, and disposal. Purchasing strategies for source reduction are presented in Table 6A.8.

Life-cycle costing is helpful in comparing the costs of durable and reusable products with the costs of disposable items, because it assesses the annual cost of products over their useful life. Durable products generally cost more initially, but the annual cost over their lifetime may be lower than that of disposable products. For example, the formula for lawnmowers could include fuel use and durability.

There is an abundance of opportunities for reducing waste through procurement policies. Some additional options are:

- Setting a price preference for reusable, refillable, and durable equipment that reduces waste, such as double-sided (duplexing) copy machines
- Requiring companies that ship goods to package them in reusable shipping containers and/or to take back the packaging; for example, furniture that can be delivered in reusable shipping blankets
- Negotiating for longer and more comprehensive warranties and service contracts when purchasing durable goods
- Leasing equipment instead of buying it to provide manufacturers with an incentive to keep it in good repair
- Purchasing items that can reduce paper use, such as double-sided photocopy machines, laser printers, and equipment and computer software that permit faxing from a computer to reduce printouts

TABLE 6A.8 Purchasing Strategies for Source Reduction

Strategy	Examples
Reduce product use.	• Double-sided copies. • Electronic mail and news.
Rent or lease products or equipment or contract for services.	• Rent tools that are needed only occasionally. • Lease technological products (computers, copiers). • Contract services (cleaning, waste removal, printing).
Purchase remanufactured, rebuilt, or refurbished products.	• Specify rebuilt parts and machinery. • Remanufactured laser toner cartridges. • Refurbished furniture.
Purchase more durable products.	• Long- and extended-life products (tires, lightbulbs). • Extended warranties.
Purchase products containing nonhazardous materials.	• Green cleaning products. • Water-based paints and inks. • Reduce battery use.
Purchase products that are returnable, reusable, or refillable.	• Refillable cups. • Reusable utensils and dishware in cafeterias.
Purchase products in bulk.	• Bulk food and concentrates. • Supplies in bulk.
Purchase products with less packaging or reuse packaging.	• Purchase concentrates. • Avoid purchasing products with secondary packaging.
Share or reuse resources.	• Share computers. • Reference library.

Source: Adapted from National Recycling Coalition (1999).

Operations.   Operations could also be changed to promote source reduction. For example, offices with lawns and campuses can compost yard waste on-site and leave grass clippings on lawns. Employees can be educated to reduce paper use and reliance on disposable products and to reuse materials that might otherwise be discarded.

Office paper comprises a large fraction of the institutional waste stream, and organizations have a relatively high degree of control over its use and disposal, making paper an excellent candidate for source reduction. Some paper reductions can be achieved solely by increasing double-sided (duplex) photocopying. Even greater reductions can be made by also reducing the number of copies made and increasing the intensity of use. *A document that is double-spaced and single-sided uses four times as much paper as a document that is single-spaced and double-sided.* Some additional strategies for reducing paper and waste include:

- Using e-mail instead of paper for communications
- Eliminating fax cover sheets
- Editing and careful proofreading on the computer before printing
- Storing files on computer disks and printing only when necessary
- Loading laser printer paper trays with paper used on one side for drafts
- Reducing mailings by targeting audiences as narrowly as possible
- Using scrap pieces of paper for short memos

Switching to reusable plates, glasses, and utensils can prevent waste from food service facilities. People can be encouraged to bring their own dishware and utensils by providing a discount or, preferably, an additional cost for use of disposable supplies. Separate collection of organic waste in conjunction with an on-site composting program is also an effective source reduction activity. On-site composting of organic waste has been implemented in correctional facilities, primary and secondary schools, universities, resorts and hotels, camps and conference centers, hospitals, restaurants, and military installations (Goldstein et al., 1998).

The disposal of computers is a concern because of the growing number of obsolete computers and the metals associated with them. Because computers are outdated quickly, usually in 2 to 3 years, many organizations are choosing to lease computers instead of purchasing. Leasing is a way of extending manufacturer responsibility to maintain and take back computer systems. Leasing computers also encourages manufacturers to produce computers that can be easily upgraded for refurbishing and reuse. Other options that exist for used computers include donation to schools and nonprofit organizations, resale, exchange, or recycling. The Wisconsin Department of Natural Resources (State of Wisconsin, 1999) has published a guide for managing used computers that is available from their web site (www.dnr.state.wi.us).

Industrial

Industry not only has the ability to prevent waste during manufacturing processes (through procurement and material use), but also the unique capacity to manipulate the packaging and/or product being manufactured before production, distribution, marketing, and sale. In many cases waste prevention activities have resulted in reduced manufacturing costs from more efficient use of resources (see Table 6A.9).

The EPA (1995b) has provided the following list of items to increase industrial source reduction:

- Recover plant materials such as solvents, metal, paper, oil, and cooling water
- Increase production efficiency to reduce the generation of scrap material
- Limit production to what is required
- Reuse and repair used pallets
- Reuse and refill packaging containers, such as bags and drums
- Return packaging materials for reuse and/or reuse packing material
- Redesign products to prevent waste associated with packaging and manufacturing
- Use materials from a materials exchange program in place of virgin materials

An analysis of waste generation associated with a particular industry may reveal how redesign of a product and manufacturing processes can prevent waste—for example, observing which manufacturing processes generate waste; considering how the wastes could be reduced, reused, or avoided; and implementing a waste prevention solution. Product redesign for source

TABLE 6A.9 Typical Examples of Industrial Source Reduction

Business	Method of source reduction	Revenue/savings
Dupont	Reuse of 25% of packaging materials	$3 million
Herman Miller, Inc.	Changing packaging design and switching to returnable packaging	270 tons of cardboard, 8 tons polystyrene, and $422,000
Johnson & Johnson	Reduced packaging waste since 1988	$2.8 million
Sprint	Printing two-sided telephone bills	450 tons paper annually
Quebecor Printing	Repairing and reusing broken shipping pallets	$14,000, repair costs 20% of new pallets
Asbury Park Press	Switching to cloth rags (from a laundry service) instead of disposable	$38,000 annually

Source: Adapted from U.S. EPA (1995) and U.S. EPA (1999d).

reduction may include reducing the weight of a product through material lightweighting or substitution, producing the product in concentrated or bulk form, marketing the product with reduced or no packaging, and producing a more durable, reusable, or repairable product.

Construction and Demolition

Construction and demolition activities contribute 136 million tons, or 2.8 lb per person per day, to waste disposed of in landfills in the United States (U.S. EPA, 1998c). New construction debris can be reduced by building only as necessary, using materials with reduced or no toxicity, and choosing building materials with reduced packaging. Waste can also be prevented by choosing to use refurbished or reusable products and incorporating materials that have an extended life span into the project.

Demolition waste can be reduced through deconstruction efforts. Deconstruction is the careful dismantling of structures before or instead of demolition to maximize the recovery of materials. Typically, electrical circuits and plumbing fixtures are recovered for reuse, metals and lumber are reused or recycled, wood flooring is remilled, and doors and windows are refinished for use in new construction (U.S. EPA 1998c). Salvage companies bid on buildings slated for demolition, materials are then salvaged before demolition, and materials are usually stored in warehouses for eventual resale. Salvaging materials keeps them out of the landfill, prevents additional waste generation and creates revenue through reuse and resale of the materials (BioCycle 1999b). The EPA reported that in one apartment building deconstruction project, 76 percent of the materials by weight were diverted to reuse or recycling. Other concepts related to waste prevention in the construction and demolition field include conducting concurrent deconstruction and new construction projects to increase material reuse, and designing and constructing buildings with future disassembly in mind (Goldstein, 1999).

The city of San Jose, California, has proposed a deposit system for construction and demolition debris. The city would collect money when issuing a building permit and provide a refund after the materials have been delivered to a certified recycling facility. The deposit would be based on type and quantity of material used, and the refund would be based on diversion rate of the recovery operation. (BioCycle, 2000)

Special Events

Special events, such as festivals, fairs, and sporting events, represent good opportunities to both implement source reduction plans and educate participants and the public about waste prevention. For source reduction initiatives to be successful at large events, it is important to manipulate the system upstream to avoid relying on the public to make the effort to (or even realize that they can) prevent waste. Activities such as not making disposable products an option ensures that they will be reduced in the waste stream.

The Whole Earth Festival at the University of California–Davis takes place in May each year, attracting around 30,000 people (Leverenz and Van Horn, 1999). The event organizers have attempted various measures to prevent and reduce the generation of wastes associated with the festival. Strategies to prevent and reduce waste have included:

- Use of biodegradable utensils and can liners
- Separate collection of compost
- Promotion of foods that do not require utensils
- Serving foods such as pizza with a napkin instead of a plate
- Educational booths to inform people about composting
- Signs with sayings such as "bring your own fork"

- Use of durable items (plates, utensile, etc.)
- Reward program for food vendors utilizing innovative waste prevention programs
- Not allowing the use of materials that would require disposal

A waste audit conducted by volunteers during the festival found that over 50 percent of the waste stream was compostable, 30 percent was recyclable, and the remaining 20 percent was composed of disposable products brought from outside. The residual wastes were composed of diapers, batteries, locally nonrecyclable plastics, heavily waxed cups, and other materials.

The collection system was a three-bin collection system labeled *compost, glass,* and *everything else.* The words *waste* and *trash* were purposely left out of the bin labeling to encourage thought and active participation. It was observed that many people were confused by the collection of compost or disregarded the collection system, resulting in contamination between the bins. Posting volunteers at the collection stations to answer questions was identified as a possible education strategy.

Contaminants were removed from the compost collection bin and cocomposted with manure from a dairy farm on campus. The resulting compost was used as a soil amendment on a university farm. Students tour the facility to learn about the composting process and waste management.

A similar event, known as the Festival for the Eno, takes place annually in Durham, North Carolina. Event organizers found that with a large volunteer base and separate collection of materials they were able to divert 88 percent of materials from landfilling. It was also observed that many people disregarded signs posted at collection centers to guide material discarding.

REFERENCES

Aluminum Association (2002) "The Aluminum Can" Online, Internet, January 20, 2002. URL = <http://www.aluminum.org/Content/NavigationMenu/The_Industry/Can/TheAluminumCan.htm>

BioCycle (1999a) "Salvage Opportunities in Deconstruction," *BioCycle.* vol. 40, no. 1, p. 14.

BioCycle (1999b) "Waterbury, Vermont: Waste Prevention Projects Receive State Grants," *BioCycle,* vol. 41, no. 7, pp. 24–25.

BioCycle (1999c) "Pay As You Throw Pays Dividends In Ohio," *BioCycle,* vol. 40, no. 7, pp. 43–44.

BioCycle (2000) "San Jose California, Deposit Would Insure C&D Debris Recycling," *BioCycle,* vol. 41, no. 2, p. 27.

CalMAX (2000) Communication with Deborah Orrill, California Materials Exchange, California Integrated Waste Management Board, Sacramento, CA.

Canterbury, J. (1998) "How to Succeed with Pay as You Throw," *BioCycle,* vol. 39, no. 12, pp. 30–35.

Clarke, M. J., A. D. Read, and P. S. Phillips (1999) "Integrated Waste Management Planning and Decision Making in New York City." *Resources, Conservation and Recycling.* vol. 26, no. 2, pp. 125–141.

Glenn, J. (1999) "The State of Garbage in America: 11th Annual Nationwide Survey," *BioCycle.* vol. 40, no. 4, pp. 60–71.

Goldstein, G. (1999) "Waste Not, Want Not," *Architecture,* vol. 88, no. 3, p. 131.

Goldstein N., J. Glenn, and K. Gray (1998) "National Overview of Food Residuals Composting," *BioCycle,* vol. 39, no. 8, pp. 50–60.

Hare, M. J. (1997) "The Case of the Missing R: The Importance of Hidden Source Reduction Diversion in the Canadian Soft Drink Industry during the 1972 to 1995 Assessment Period," *Journal of Solid Waste Technology and Management.* vol. 24, no. 4, pp. 196–206.

Leith, A. (1996) "Accounting Method Reinvents Waste Programs," *World Wastes.* vol. 39. no. 2.

Leverenz, H. and M. Van Horn (1999) "Minimizing Festival Trash," *BioCycle,* vol. 40, no. 9, pp. 45–47.

National Recycling Coalition (1999) "Purchasing Strategies to Prevent Waste and Save Money," Source Reduction Forum of the National Recycling Coalition, Inc., Alexandria, VA.

ReDO (2000) E-mail communication with Julie Rhodes, Reuse Development Organization, Indianapolis, IN.

Salter, C. (1999) "Festival Seeks Independence from the Landfill," *BioCycle,* vol. 40, no. 9, pp. 48–49.

Slovin, J. (1995) "Communities Take a Closer Look at Unit Pricing," *World Wastes,* vol. 38, no. 4, pp. 14–20.

State of Minnesota (1996) *Waste Prevention: Source Reduction Now, How to Implement a Source Reduction Program in Your Organization,* State of Minnesota, Office of Environmental Assistance, St. Paul, MN.

State of Minnesota (2000) *Solid Waste Policy Report: Waste management in Minnesota, A Transition to the 21st Century,* State of Minnesota, Office of Environmental Assistance, St. Paul, MN.

State of Wisconsin (1999) *Managing Used Computers: A Guide For Business and Institutions,* Pub WA-420 99, State of Wisconsin, Department of Natural Resources, Madison, WI.

Sutin, P. (November 16, 1995) "Officials are Pondering New Proposal for Methods of Collection of Garbage," *St. Louis Post Dispatch,* St. Louis, MO.

Tchobanoglous, G., H. Theisen, and S. Vigil (1993) *Integrated Solid Waste Management: Engineering Principles and Management Issues,* McGraw-Hill, Inc., Boston, MA.

U.S. EPA (1989) *The Solid Waste Dilemma: An Agenda for Action,* EPA/530-SW-89-019, U.S. Environmental Protection Agency, Washington, DC.

U.S. EPA (1992) *Environmental Fact Sheet: Recycling Grass Clippings,* EPA/530-R-92-012, U.S. Environmental Protection Agency, Washington, DC.

U.S. EPA (1993) *Life Cycle Assessment: Inventory Guidelines and Principles,* EPA/600-R-92-245, U.S. Environmental Protection Agency, Washington, DC.

U.S. EPA (1995a) *Decision Maker's Guide to Solid Waste Management,* Volume II, EPA/530-R-95-023, U.S. Environmental Protection Agency, Washington, DC.

U.S. EPA (1995b) *Spotlight on Waste Prevention: EPA's Program to Reduce Solid Waste at the Source,* EPA/530-K-95-002, U.S. Environmental Protection Agency, Washington, DC.

U.S. EPA (1996) *The Consumer's Handbook for Reducing Solid Waste,* EPA/530-K-96-003, U.S. Environmental Protection Agency, Washington, DC.

U.S. EPA (1997) *Full Cost Accounting for Municipal Solid Waste Management: A Handbook,* EPA/530-K-95-041, U.S. Environmental Protection Agency, Washington, DC.

U.S. EPA (1998a) *Greenhouse Gas Emissions From Management of Selected Materials in Municipal Solid Waste,* EPA/530-R-92-013, U.S. Environmental Protection Agency, Washington, DC.

U.S. EPA (1998b) *Municipal Solid Waste Source Reduction: A Snapshot of State Initiatives,* EPA/530-R-98-017, U.S. Environmental Protection Agency, Washington, DC.

U.S. EPA (1998c) *Characterization of Building-Related Construction and Demolition Debris in the United States,* EPA/530-R-98-010, U.S. Environmental Protection Agency, Washington, DC.

U.S. EPA (1999a) *Characterization of Municipal Solid Wastes in the United States: 1998 Update,* EPA/530-R-99-021, U.S. Environmental Protection Agency, Washington, DC.

U.S. EPA (1999b) *Waste Not, Want Not: Feeding the Hungry and Reducing Solid Waste Through Food Recovery,* EPA/530-R-99-040, U.S. Environmental Protection Agency, Washington, DC.

U.S. EPA (1999c) *National Source Reduction Characterization,* EPA/530-R-99-034, U.S. Environmental Protection Agency, Washington, DC.

U.S. EPA (1999d) *Wastewise Fifth-Year Progress Report,* EPA/530-R-99-035, U.S. Environmental Protection Agency, Washington, DC.

CHAPTER 6

SOURCE REDUCTION: QUANTITY AND TOXICITY
Part 6B.   Toxicity Reduction

Ken Geiser

The problems caused by municipal solid waste typically involve two factors: volume and toxicity. Since 1960, the volume of municipal solid waste has grown from 87 million pounds per year to a record 209 million pounds in 1997 (U.S. EPA, 1998a). The toxicity of solid waste is more difficult to measure. Toxic materials have always appeared in household wastes, but since mid-century, as synthetic materials began to replace many traditional materials, the proportion of synthetically derived toxic materials in waste has increased appreciably. The toxic constituents in solid waste include heavy metals, particularly lead, cadmium, nickel, and mercury; chlorinated hydrocarbons, such as perchloroethylene, trichloroethylene, and methylene chloride; aromatic compounds, such as naphthalene and toluene; pesticides and other biocides; and used motor oil.

Some of these toxic materials enter municipal solid waste streams because they are waste products from domestic or commercial processes. Waste oil from automobile service stations is such an example. Some toxic materials are toxic products discarded once a portion of the product has been used. Waste paints are a good example. Most of the toxic materials, however, appear in solid waste as constituents of commercial products whose useful life is over. For example, over 4 billion dry cell batteries are sold each year in the United States. These include batteries containing mercury or mercuric oxide, magnesium, zinc, silver oxide, nickel and cadmium, and lithium. Many of these metals are not dangerous in the battery itself, but a dry cell battery has a useful life of somewhere between a few hours to several months, after which it is discarded. When a battery is disposed of in a landfill, it eventually deteriorates. During this deterioration, the metals can be released to the ground and groundwater. When a battery is incinerated, some of the metals fall out in the bottom ash and some are released in incinerator gases. Incinerator filters will collect a portion of the metals as fly ash. Both bottom ash and fly ash must be disposed of in landfills where again the ground and groundwater may become contaminated.

6B.1   THE TOXICITY OF TRASH

Batteries are only one of the conventional constituents of municipal trash that lead to its toxicity. Table 6B.1 lists a number of common toxic materials found in municipal solid waste, their sources in products, and their known health effects.

How Toxic Is Trash?

The toxicity of trash can be estimated using two basic approaches: sampling or modeling. *Sampling* involves drawing samples of various waste streams, sorting each waste stream into

TABLE 6B.1   Common Toxic Materials in Municipal Solid Waste

Substance	Sources	Health effects
Cadmium	Batteries, inks, paints	Carcinogen, ecotoxin, reproductive effects
Lead	Batteries, varnishes, sealants, hair dyes	Neurotoxin, reproductive effects
Mercury	Batteries, paints, fluorescent lamps	Ecotoxin, neurotoxin, reproductive effects
Methylene chloride	Paint, paint strippers, adhesives, pesticides	Carcinogen
Methyl ethyl ketone	Paint thinner, adhesives, cleaners, waxes	Neurotoxin, reproductive effects
Perchloroethylene	Rug cleaners, spot removers, fabrics	Carcinogen, ecotoxin, reproductive effects
Phenol	Art supplies, adhesives	Ecotoxin, developmental effects
Toluene	Paint, nail polish, art supplies, adhesives	Ecotoxin, mutagen, reproductive effects
Vinyl Chloride	Plastics, apparel	Carcinogen, mutagen, reproductive effects

its specific components, and weighing the components. The most systematic effort to gather empirical data on municipal solid waste has involved the sampling of landfills in several communities across the country. This study found that household maintenance products made up the largest percentage by weight of the household hazardous substances. This was followed by batteries, cosmetics, cleaners, and automobile and yard maintenance products (Wilson and Rathje, 1988). While such studies provide some data on the hazardous constituents of household trash, they do not provide evidence on the toxic materials generated by nonhousehold sources.

Municipal solid waste, or trash, is made up of the discards—from households as well as commercial and industrial settings—that are otherwise not classified as hazardous waste. Businesses that generate less than 100 kilograms of hazardous waste per month are allowed to deposit the waste into the municipal solid waste stream. It is estimated that there are some 450,000 of these so-called very small generators. They include conventional retailers, bakers, beauty shops, dentists, dry cleaners, photography labs, printers, restaurants, schools, and vehicle maintenance shops. Some of the hazardous waste generated by very small generators is flushed into the municipal sewer system and some of it is discarded as solid waste. Altogether, it is estimated that very small generators produce roughly 197,000 tons of hazardous waste each year, an unknown portion of which is released as municipal solid waste (Abt Associates, 1985).

The *modeling* approach requires estimating the material flows through each waste stream. These material flows are calculated from materials production data and adjusted for imports and exports, materials recovery, energy conversions, and losses during production or use. The remaining volumes are then assumed to enter the solid waste stream.

Simple materials flow analyses have been done on some toxic metals (lead, cadmium, and mercury). During the 1980s, the federal Bureau of Mines conducted materials flow studies that revealed that most cadmium was used in coatings, plating, and batteries; most lead was used in storage batteries; and most mercury was used in electrical equipment. Today, over 85 percent of lead consumed each year in the United States is used in auto batteries and 54 percent of cadmium is used in household (dry cell) batteries (U.S. Bureau of Mines, 1985; U.S.G.S., 1999).

Each approach requires identifying the toxic constituents. This is limited by the available research. Toxic chemicals are those that scientific studies have shown to cause serious health effects. Today, some 70,000 chemicals are in common use, and many are toxic. Yet the U.S. Environmental Protection Agency (U.S. EPA) estimates that only 7 percent of the largest-volume chemicals used in industrial production have been screened with the basic toxicity tests. Of the 17,000 chemicals used in food, cosmetics, pesticides, and drugs, the National Research Council has found that less than 30 percent have been fully tested. This lack of information means that results from both the modeling and the sampling approaches are limited by substantial uncertainties (U.S. EPA, 1998b; NRC, 1984).

Is Trash Toxicity a Problem?

When products containing toxic chemicals are disposed of, the toxic constituents enter municipal landfills and incinerators. From these disposal facilities the toxic chemicals are dispersed into the environment as underground leachate, waste water effluents, air emissions, or hazardous waste. Once released to the environment, toxic chemicals may threaten ecological systems, wildlife, or public health.

The toxic constituents of trash in landfills have a long history of public concern. During the 1890s, the American Public Health Association completed a decade-long study of municipal refuse documenting public health threats from municipal landfills in over 150 cities. While early concerns focused on potentially infectious pathogens in municipal waste, the toxic characteristics of dump site leachate and runoff were well identified as a source of river and stream contamination. The leaking and leaching of halogenated hydrocarbons and heavy metals proved to be a major source of contamination of drinking water sources during the 1960s and 1970s (Melosi, 1981).

Toxic materials in solid waste streams destined for municipal incinerators also pose serious concerns. Incineration breaks down the paper, plastics, fibers, and containers of the municipal waste stream and liberates the heavy metals contained in consumer products. Toxic metals such as lead, cadmium, arsenic, mercury, selenium, and beryllium, among others, remain in the postincineration ash, where they are in a more concentrated form than in the raw waste stream. Organic pollutants of concern in incinerator air emissions include hydrogen chloride, dioxins, and furans. Incineration tends to volatilize some metals that then condense onto small fly ash particulates. Other metals such as mercury are easily converted to gaseous states. Still other metals may react with the organics to form complex compounds such as metal chlorides. Once released from incinerators on particulates or as gases, the metals are easily mobilized and readily available for ecological uptake. Because the metals are released from conventional matrices and easily dispersed by air or water currents, there is an increased potential for direct (inhalation) or indirect (food chain contamination) human exposure (Florini et al., 1990).

Many of the metals released from solid waste treatment facilities are neurotoxins; others, such as lead, cadmium, arsenic, and beryllium, are human carcinogens; some, such as lead and mercury, are recognized human reproductive toxins; and some others, such as mercury, copper, and zinc, are acutely toxic to aquatic life. One chemical of particular concern is chlorine because it can be involved in the formation of hydrogen chloride, dioxins, and other chlorinated organics during incineration. Recent international estimates suggest that municipal waste incinerators account for 69 percent of the dioxins in the global environment (U.N. Environment Program, 1999). Dioxins are among the most toxic compounds known to science. Chlorine occurs in many products, including solvents, biocides, bleaches, disinfectants, paper, and plastics. Wastepaper and plastics appear to be the major source of chlorine in municipal solid waste.

Where municipalities have introduced recycling programs, the toxic nature of the trash has continued to remain a problem. Workers in recycling centers may be exposed to the toxic, constituents of the materials they separate for recycling. Mismanaged recycling centers can pollute soil and groundwater with the toxic materials in the stored products. Finally, municipalities that recycle toxic products may incur liabilities for the future handling of the materials that they collect, process, and send on for recycling.

Reducing the Toxicity of Trash

Reducing the toxicity of solid waste requires policies that are well targeted, efficient, and cost-effective. Because the toxicity of most trash is directly linked to the toxicity of consumer products, some of the most effective policies for reducing the risks of trash involve the redesign of products and the processes that produce them.

There are three broad policy approaches to reducing the toxic constituents of solid waste. The first involves improving waste management practices to reduce, primarily through recycling, the amount of toxic waste that is ultimately disposed. This is the most common approach. Its immediate effects are countered by its longer-term limits. The second focuses on changing the material constituents of the products that are used in domestic and commercial activities. While this approach has more long-term impacts, the immediate prospects are less promising. The third seeks to change the processes of industrial production to reduce toxic inputs. Again, this is a long-term, but potentially a highly effective approach. Table 6B.2 presents these approaches.

6B.2 WASTE MANAGEMENT POLICY

The most immediate approach for reducing the toxicity of solid waste involves improving municipal waste management programs. Once products containing toxic materials have been mixed into conventional municipal refuse, the costs and problems of safely managing the waste dramatically increase. Therefore, most programs that seek to manage the toxic constituents of solid waste begin by separating the materials or by keeping materials separate from the first point of disposal.

Toxic Waste Disposal Bans

Prohibiting the disposal of wastes containing specified toxic chemicals is a very direct approach to detoxifying the waste stream. Forty-seven states have adopted bans on the disposal of some discarded products. Forty-three of these state ban lead-acid vehicle batteries. Others ban used oil, tires, and major appliances. Only Massachusetts, South Dakota, and Wisconsin ban recyclable paper from disposal facilities. Massachusetts has banned the disposal of cathode ray terminals (computer and television screens) because of their high lead content. Table 6B.3 identifies the states that have adopted product bans and indicates the products covered. By diverting toxic materials from disposal facilities, these disposal bans may reduce the amount of toxic materials ending up in landfills or incinerators, but there is little information available on their effects on the quantity of toxic materials in the solid waste stream.

Toxic Waste Collection

With or without targeted waste prohibitions, products containing toxic materials can be diverted from the municipal waste stream by separate waste collection programs. There are two types of programs: those directed at specific products such as batteries or tires, and those directed at specific waste generators such as households or commercial offices.

The product-specific approach is well illustrated by battery collection programs. Basically there are two types of batteries—household (dry cell) batteries and automobile (lead-acid) batteries—and each provides a different program approach. Lead-acid batteries are handled

TABLE 6B.2 Policy Strategies for Reducing the Toxicity of Trash

Strategy	Feasibility	Effectiveness
Waste management	Immediate	Modest
Product management	Intermediate	Significant
Production management	Longer-term	Significant

TABLE 6B.3   Product Disposal Bans by State

State	Vehicle batteries	Tires	Motor oil	White goods	Others
Arkansas	X	X			
Arizona	X	X			
Connecticut					X[a]
Florida	X	X	X	X	X[b]
Goergia	X				
Hawaii	X				
Idaho	X	X			
Illinois	X	X		X[c]	
Iowa	X	X	X		X[d]
Kansas		X			
Louisiana	X	X		X	
Maine	X				
Maryland		X			
Massachusetts	X	X		X	X[e]
Michigan	X				
Minnesota	X	X	X	X	X[f]
Mississippi	X				
Missouri	X	X	X	X	
New Hampshire	X				
New Jersey	X				
New York	X				
North Carolina	X	X	X	X	
North Dakota	X		X	X	
Ohio	X	X			
Oregon	X	X	X	X	
Pennsylvania	X				X[g]
Rhode Island	X				
South Carolina	X	X	X	X	
South Dakota	X	X			
Tennessee	X	X			
Texas	X	X	X		
Utah	X				
Vermont	X	X	X	X	X[h]
Virginia	X				
Washington	X		X		
West Virginia	X	X			
Wisconsin	X	X	X	X	X[i]

[a]Mercury oxide batteries, [b]Demolition debris, [c]White goods containing CFC gases, mercury switches, and PCBs, [d]Non-degradable grocery bags and carbonate beverage containers, [e]Glass and metal containers, [f]Nickel-cadmium rechargeable batteries, [g]Discarded vehicles, [h]Various dry cell batteries, [i]Metal, glass, and plastic containers, and recyclable paper.

Source: Adapted from Jim Glenn, "The State of Garbage in America," *Biocycle,* May 1992, p. 33.

by a private market dependent on the price of reprocessed lead. An average automotive battery weighs 36 pounds, of which about one-half the weight is lead. Lead-acid batteries were classified under RCRA as a hazardous waste in 1985, and today between 93 and 98 percent of the lead available from lead-acid batteries is recovered for reclamation by an increasingly comprehensive secondary lead industry. These batteries are processed at some 32 active secondary lead smelters in the United States, and these smelters rely on used batteries for some 70 percent of their lead supply (U.S.G.S., 1998; Breniman et. al., 1994).

Over the past 10 years, governments have been increasingly active in encouraging household battery collection. There are household battery collection programs in operation in the

United States, Japan, and at least 11 European countries. Most local programs in Japan and Europe rely on voluntary collection programs at special government recycling centers.

Dry cell household batteries contain a host of metals that can be usefully recycled. These include nickel, cadmium, mercury, silver, lead, lithium, and zinc. Nearly 50 percent of the cadmium and 88 percent of the mercury consumed in the United States goes into dry cell batteries, which have traditionally ended up in the municipal solid waste stream. Historically, batteries have accounted for nearly three-quarters of the mercury in municipal trash. While an individual battery may contain only a small amount of these metals, nearly four billion dry cell batteries are sold in the United States each year; in aggregate, this adds up to a large amount of these metals. Recycling of dry cell household batteries is less well developed than the recycling of wet cell batteries and is hampered by the limited number of processing facilities. Currently there are only three U.S. facilities capable of recycling household batteries, so the large percentage of batteries recovered through collection programs is either shipped offshore for reprocessing or sent to hazardous waste landfills. Still, nearly 500 tons of cadmium are recovered from nickel-cadmium batteries each year (U.S. EPA, 1989; U.S. EPA, 1992).

The generator-specific approach can be illustrated by household hazardous collection days. Each year, Americans generate 1.6 million tons of hazardous household wastes that range from solvents, paints, antifreeze, and used motor oil to pesticides and explosives. Such household hazardous waste is exempt from some federal hazardous waste regulations, but still must be handled by a licensed hazardous waste treatment operator. Typical programs set up by municipalities may include permanent collection centers, special collection days, or local businesses designated as drop-off sites. In 1997, the EPA estimated that there were over 3000 such programs in operation across the country.

Toxic Waste Recycling

Collecting toxic materials before they enter the municipal waste stream is an important prerequisite to toxicity reduction, but if those materials are not recycled back into products the overall toxicity of the solid waste stream is not reduced. For instance, most household batteries collected by local recycling programs in the United States are sent to commercial processing facilities. While mercuric and silver oxide batteries are processed to recover mercury and silver for remarketing, lithium is treated to make it less reactive and then sent to a landfill.

Fluorescent light tubes form another waste that is ripe for recycling. Discarded fluorescent lamps generate the second-largest source of mercury in the waste stream. Approximately 550 million mercury-containing lamps are sold in the United States each year; of these, 95 percent are fluorescent tubes. The EPA has added spent fluorescent lamps to the list of universal wastes, requiring that the lamps be recycled or treated as hazardous wastes. While over 50 million fluorescent lamps are collected each year in Europe, programs to collect and process fluorescent bulbs in the United States remain limited. The largest program is in California where three firms reclaim 600,000 lamps each month. While the metal and glass can usually be sold for reuse, the mercury is often sent to a landfill. Because the current price of mercury is low, efforts to capture and recycle the mercury requires a high processing cost.

Household thermostats are another common source of mercury. Several of the largest manufacturers of household thermostats formed the Thermostat Recycling Corporation in 1999 to collect thermostats, remove the mercury, and reprocess it for recycling. During its first year, the operation collected and processed some 500 pounds of mercury (Erdheim, 2000).

Used oil is another candidate for recycling. Approximately 1.2 billion gallons of used vehicle or lubricant oils are generated in the United States each year. About 360 million gallons are generated by home oil changes, most of which is disposed of in the trash. Yet, only 100 million gallons per year are rerefined into reusable oils. This low level of reprocessing is primarily due to the low cost of virgin oil and the environmental problems of reprocessing. Contaminants that appear in used oil, particularly the additives that have been added to virgin oil since the 1970s, produce a hazardous sludge that is expensive to treat. The wastewater from the distilling operations is contaminated with hydrocarbons as well. Thus, roughly two-thirds of all used oil is recycled by burning it as a fuel.

Much effort has been put into programs for recycling plastics. In 1995, 38 billion pounds of plastics were disposed of and, while plastic makes up only 9 percent by weight of the municipal waste steam, it represents 20 percent of the waste stream by volume (U.S. EPA, 1997). Typical plastic products contain a host of toxic materials. The resins themselves may have toxic effects. For instance, the combustion of the chlorinated polymer, polyvinyl chloride (PVC), has been linked to the formation of dioxins in incinerators. The additives are also often toxic. The phthalates and lead used as plasticizers have well-identified toxic effects. Common antioxidants include phenolics and thioesters. Colorants include titanium dioxide, lead chromate, chromium oxide, and cadmium, selenium, and mercury compounds. Alumina trihydrate and halogenated compounds are used as flame retardants. Heat and light stabilizers such as organotin mercaptide, methyl and butyl tins, and cadmium/zinc and barium/cadmium are frequently added. Finally, there are toxic compounds in the surface printing and treatment. Metallic inks and dyes are often used for decoration and labels (Wolfe and Feldman, 1991).

Most plastic recycling programs do not account for these various toxic constituents. Little research has been done on chemical exposure from recycled plastics. The federal Food and Drug Administration has been quite restrictive in allowing recycled plastic material to be used in food containers due to uncertainties about contaminants and the difficulties of sterilizing plastics. The high cost of transporting postconsumer plastics and the absence of a reprocessing infrastructure have limited the recycling of plastics. Today, just over 5 percent of plastics in the waste stream are recycled.

Because the material structure of plastics degrades during reprocessing, recycled plastic typically goes into low-grade products like carpet fibers, fiberfill for pillows and jackets, industrial paints, and nonstructural lumber products. While these second uses for the recycled plastics prevent the plastics from disposal at that moment, many of these second-use items are disposed of eventually. Thus, recycling generally delays, but does not eliminate, the possibility of environmental release of the toxic constituents in plastics.

Reducing the toxicity of municipal waste by banning, collecting, diverting, and recycling toxic waste materials can be readily implemented today and is an increasingly common practice. Such processes provide a fine opportunity to educate consumers about toxic materials and can be a significant community-building activity. Yet, the recycling of toxic materials in the waste stream is fraught with technical and economic limitations. It is further limited by the low level of solid waste recyling nationally; somewhere between 27 and 31 percent of municipal solid waste is currently recycled (U.S. EPA, 1997, Glenn, 1999). Thus, even if recycling were to become a significant approach to reducing the toxicity of trash, it still could not account for a substantial reduction without a significant change in waste management practices.

6B.3   PRODUCT MANAGEMENT POLICY

A second general approach to detoxifying trash focuses on the toxic materials contained in products. Instead of focusing on better management of the toxic materials in waste, this approach seeks to reduce the toxicity of the waste stream by reducing the toxic constituents of the products thrown out as waste. While the focus on improved waste management may have more immediate effects on the toxic materials entering disposal facilities, the focus on products offers more long-term efficiencies, because less toxic products mean a reduced need for highly selective waste management techniques. By focusing policy attention on products, the emphasis is shifted to an earlier point in the life cycle of a toxic material.

Life-Cycle Analysis

Trash is a product of the linear process that supplies households and commercial establishments with consumable goods. In considering the environmental effects of trash, it is important to consider the entire life cycle of a product from synthesis or manufacture through distribution and use to waste and disposal. There are environmental and human health effects associated with each stage of the life cycle. Effective environmental protection requires that improvements in the environmental performance at one stage not worsen the effects at another stage.

This broadened perspective on the role of a product in the environment has been incorporated in a new technique called *life-cycle analysis,* (Curran, 1996). Ideally, a life-cycle analysis is composed of an inventory of resource inputs and waste outputs for each stage, as well as an assessment of risks associated with each of these inputs and outputs. Such life-cycle analyses have been used in solid waste management for comparing plastic to paper packaging and disposable to reusable diapers. While the methodology is limited by the product focus, the large amounts of data requirements, and the necessity to set boundaries, the concept opens up a broad awareness of the environmental impacts of products before they become waste.

Product Bans

Governments may use their authority to prohibit the production, trade, or use of specific products or activities as a means of reducing toxicity in trash. Some states have tried to target toxic materials directly by focusing on bans at the point of use or sale. California, Oregon, Minnesota, New York, New Jersey, Vermont, and Connecticut have passed legislation banning the use of mercury in dry cell batteries. The New Jersey law, passed in 1992, sets a standard for mercury in dry cell batteries and prohibits the sale of batteries unable to meet the standard. In 1996, Congress passed the Mercury-Containing and Rechargeable Battery Management Act, which prohibits the sale of mercuric oxide button-cell batteries and alkaline-manganese and zinc-carbon batteries that contain intentionally introduced mercury.

The Great Lakes Binational Toxics Strategy, a joint program of the United States and Canada committed to the virtual elimination of persistent toxic substances in the Great Lakes, has set a goal of 50 percent reduction in the deliberate use of mercury; to meet this goal, the United States has pledged to achieve a 50 percent reduction in the release of mercury from human sources by 2006. The EPA has established its own national EPA Action Plan for Mercury through its Persistent, Bioaccumulative Toxics Initiative. In addition, all of the New England states, Florida, Indiana, Kansas, New Jersey, New York, Michigan, Minnesota, and Wisconsin have established state mercury-reduction strategies. The Massachusetts strategy, for instance, is committed to the virtual elimination of the use and release of mercury from human activities, with an interim goal of a 75 percent reduction in mercury emissions by 2010 (State of Massachusetts, 2000).

Packaging Policies

Packaging materials account for one-third of municipal solid waste, by weight, in the United States (U.S. EPA, 1998a). The packaging industry is the largest user of plastic, accounting for over one-third of the annual plastic resin consumption. Concern over the volume of packaging in landfills and the environmental hazards of incinerating plastics has led some local governments to try to ban the use of plastic packaging.

In 1988 Suffolk County, New York, passed a highly controversial packaging ban. The law, which was scheduled to take effect in the summer of 1989, banned polystyrene foam in food packaging, including produce and meat trays, grocery bags, and fast-food "clamshells," but it was challenged in court by a legal suit filed by several plastics trade groups. Minneapolis and

St. Paul have passed ordinances that would permit the prohibition of nonrecyclable plastic food packaging. Maine, Minnesota, and Rhode Island have passed legislation banning the use of polystyrene foam food packaging made with ozone-depleting chlorofluorocarbons.

The most prominent packaging conversion involved the use of public pressure to force the substitution of the plastic food packaging used by the McDonald's chain of fast-food restaurants. Concern over the toxic emissions from the incineration of plastic packaging led the Citizens' Clearinghouse on Hazardous Waste to launch a three-year national campaign to pressure McDonald's to convert to paper packaging. That campaign set the conditions for the Environmental Defense Fund and McDonald's in 1990 to negotiate a well-publicized phase-out of plastic packaging and the substitution of paper (Dennison, et al., 1990).

Product Labeling

Like product and packaging use bans, product labeling seeks to reduce the use of toxic materials by changing consumer patterns. Product labels that reveal the toxic constituents of products are likely to affect the purchasing decisions of those consumers who read labels and do comparative shopping where there are alternative products available. More significantly, product labeling may affect the material selection decisions of those manufacturers who fear that product labeling will affect consumer decisions.

In 1986, California citizens passed a ballot initiative, Proposition 65, that required warnings on product labels to be posted on products containing chemicals that can cause cancer or adverse reproductive effects. To avoid such labeling, several firms reformulated their products to remove the toxic chemicals of concern: The Gillette Corporation removed trichloroethylene from its Liquid Paper typewriter correction fluid; Dow Chemical reformulated K2r spot-lifter to eliminate perchloroethylene; and Pet, Inc., accelerated the elimination of lead from its food cans (Fishbein and Gold, 1992). Since the 1980s, many national and private product labeling programs have been established to provide information on environmental compatibility. There are national programs in Germany, Canada, Japan, and the Scandinavian countries. There are two certifying and labeling programs run as private operations in the United States. Each of these programs provides the right to use a special eco-label when a product is found to meet a set of environmental criteria that may include recyclability, stratospheric ozone impact, toxicity, energy input, and pollution. While these labels may have some notable impacts on toxicity, the high degree of generality involved in each label means that for most products these eco-labels are of limited value.

Targeted Product Procurement

Various state and local governments have pioneered targeted product procurement policies that tend to avoid products with toxic constituents. For instance, in developing its municipal program, the city of Santa Monica, CA, has published a detailed list of environmental specifications and has screened over 200 cleaning products. The state of Minnesota has established a scorecard system for evaluating multiple product attributes and has screened over 400 products in some 33 categories.

The EPA has produced lists of so-called environmentally preferred products as guides for government procurement departments. These programs, which are typically voluntary, have listed hundreds of products and targeted agencies that range in purchasing power from local school districts to the federal Department of Defense. A 1993 Executive Order of the President directed federal agencies to give preference to the purchase of products and services demonstrating the least burdens to the environment. Quotas on recycled pulp in purchased paper products is a particularly well-accepted requirement in some state and federal programs, although efforts to stipulate against paper products bleached with chlorine have met with substantial industry opposition (U.S. Office of the President, 1993).

Extended Producer Responsibility

If recycling programs were designed more consciously to close the loop on material flows, such that materials never return to the environment as waste, then toxic materials could be recycled without creating a toxic waste stream. Germany began such a program in 1991 with a bold new law called the Ordinance on Avoidance of Packaging Waste, which required that manufacturers and distributors of products must be responsible for the reclamation and processing of postconsumer packaging wastes. Under the ordinance, the government sets mandatory targets for recycling and allows industry to set fees on packaging materials. Industry responded by establishing the Duales System Deutschland, a consortium of over 600 firms that collects, processes, and recycles any of the members' packaging wastes, all of which are identified by a green dot on the packaging. Today, Germany requires recycling for 75 percent of glass containers, 70 percent of tin cans, 60 percent of aluminum packaging, 60 percent of paper and cardboard, and 60 percent of composites. Over 75 percent of all packaging carries the green dot. The result has been a 13 percent decrease in packaging in Germany between 1992 and 1997, compared to a 15 percent increase in the United States for the same period (Thorpe, 1999).

The German take-back system has encouraged a host of different programs in Europe loosely referred to as "extended producer responsibility," all basically designed to require that product producers carry responsibility for their products throughout their life cycle, or at least at the point of disposal. The Netherlands uses agreements with product manufacturers called covenants to encourage "integrated chain management" that creates a set of product responsibilities all along the life cycle of a product. The Swedish Eco-Cycle Act of 1994 also sets out producer responsibility plans for a wide range of consumer products including automobiles, electrical appliances, batteries, packaging, and tires. More recently, a European Union directive has been proposed that would require manufacturers of electronic products to carry responsibility for the collection and recycling of all used electronic products, including computers (Lifset, 1993).

In the United States, the most advanced producer responsibility programs involve state-mandated beverage container collection programs and battery take-back and recycling. Minnesota and New Jersey led with laws requiring battery producers to carry the financial burden of recovery and recycling of rechargeable nickel-cadmium batteries. In 1995, manufacturers of nickel-cadmium batteries launched a voluntary national take-back and recycling program with the establishment of the nonprofit Rechargeable Battery Recycling Corporation (RBRC). Today, the RBRC involves 285 U.S. and Canadian companies (80 percent of the rechargeable battery market), who pay a license fee to participate, and some 26,000 retail stores and recycling centers willing to accept used battery charged products and send them to a central facility in Pennsylvania for recycling. In addition, some firms such as IBM, DuPont, and Castrol have set up pilot programs for testing product take-back schemes. Behind these various initiatives and proposals is a desire to close the loop on consumer products in the hopes of both reducing their contribution to waste streams and encouraging manufacturers to design products more easy to recycle and less likely to contain toxic materials that are costly to handle in reprocessing (Davis et al., 1997).

Product Substitutes

During the past decade there has been a growing awareness among consumers about the environmental effects of the products they use and dispose. Retailers across the country have found that consumers will respond to literature, educational materials, and warnings about products. Educational campaigns in schools, as well as manuals and guides for environmentally conscious shopping, have raised further the selective capacity of consumers to choose environmentally sound products.

Bans, labeling, and educational programs have all been used to target products containing toxic chemicals. The measurable results have been limited, due in part to the relatively small number of initiatives, but also because many of the actual product changes that have resulted from the more indirect sensitivities of the market in which there are no simple causal connections.

A focus on changes in products to reduce the toxicity of solid waste yields a more fundamental approach than a focus solely on better waste management. A product management approach provides significant opportunities to educate consumers and to raise awareness about toxic chemical exposure. Eliminating toxic product use in the community will clearly reduce the volume of toxic materials disposed of, but there is only so much that a local community can do to change the product mix. Ultimately, it will take changes in the production systems to fully relieve products of their toxic constituents.

6B.4   PRODUCTION MANAGEMENT POLICY

Detoxifying industrial production systems provides a third approach to reducing the toxicity of solid waste. Products from industrial or agricultural production systems. are likely to be toxic where those systems are dependent on toxic chemicals.

Clean Production

In Europe this approach is called *clean or cleaner production.* Clean production implies more than better waste management or pollution control. The essence of clean production is to fundamentally change industrial production processes in order to manufacture in a more environmentally sound manner.

In 1989, the United Nations Industry and Environment Office established the Cleaner Production Program to promote environmentally sound production. The program defined the concept of clean production for processes to mean ". . . conserving raw materials and energy, eliminating toxic raw materials, and reducing the quantity and toxicity of all emissions and wastes before they leave the process." (U. N. Environment Program, 1994, p. 1.) Both the Netherlands and Denmark have set up special government-funded clean technologies programs. There have been important initiatives in the United States as well, particularly at the state level and among leading firms. Much of this is referred to as pollution prevention and is guided by evidence that preventing pollution can both reduce industrial operating costs and improve environmental performance. In 1990, Congress passed the Pollution Prevention Act to encourage the EPA to promote pollution prevention.

Design for the Environment

One of the most effective points in a product's life cycle for considering the use of toxic substances is during the initial design period. At the time that new products are undergoing concept development and materials specification, attention to alternative, nontoxic materials can reduce the costs and environmental impacts of the product at points further into its life cycle.

The idea of incorporating environmental criteria into the initial product design phase has been called "design for the environment." This term, first coined by the American Electronics Association and heavily incorporated by AT&T in its product development research centers, has been adopted as one of the EPA's most innovative product-focused programs (Graedel and Allenby, 1996; Fiksel, 1996). At the design stage, products can be developed that can more easily be recycled, that last longer, that can more easily be repaired, that contain no toxic material, or that require no toxic material during manufacture. For instance, plastic products or containers can be limited to one type of plastic to improve recyclability. Durable goods can be designed for take-back, disassembly, and reuse of components. Electronic equipment can be designed as a set of components that can easily be repaired by removing and replacing malfunctioning elements.

The reduction in heavy metals in printing inks provides a good illustration. Traditional printing inks contain various metals, including cadmium and lead. Consumer pressure, concern about occupational exposures, and efforts to reduce hazardous waste has led newspapers to switch to less toxic inks. In the mid-1970s, the American Newspaper Publishers Association prohibited the use of lead in inks approved by the Association and established a logo to identify environmentally acceptable inks. Starting in 1987 with only six newspapers, colored soy inks had become common in over half of the nation's 9000 newspapers by 1994. Low-toxicity inks have also become common in product packaging. For instance, Procter and Gamble has eliminated the use of all metal-based inks for printing on packaging (Glaser and Gajewski, 1994; U.S. OTA, 1989).

Unfortunately, little attention is given to the toxicity of products as waste when firms design new products. Although manufacturers have incentives to reduce the costs and liabilities of the hazardous wastes they generate, they have little incentive to consider the disposal costs of the products they make. Some of this may be changing due to waste take-back laws in Europe. For instance, the European Union has been moving forward on a directive to require automobile manufacturers to take responsibility for vehicles at the end of their useful life. In anticipation, several auto manufacturers, including BMW and Volkswagen, have developed automobile designs that enhance disassembly. Both firms have built pilot plants based on the take-back and reuse principle. BMW has developed a prototype automobile that is 100 percent recyclable (U.S. OTA, 1992, p. 59).

Toxics Use Reduction

Since 1989, several states have passed laws that promote programs designed to reduce the use of toxic chemicals in production processes. Toxics use reduction is a form of pollution prevention that focuses on reducing or eliminating toxic chemicals in industrial production as a means of reducing the toxicity of industrial waste streams. Most of these laws require or encourage firms to prepare plans demonstrating how they would reduce the use of toxic chemicals or the generation of toxic wastes. Typically, these toxics use reduction programs encourage firms to adopt one or several of a set of techniques, including substitution of the chemical inputs, changes in production equipment or processes, redesigning products to reduce toxic chemical use, improvements in production operations and maintenance, and installing closed-loop recycling systems. While these laws were enacted to reduce the generation of hazardous waste and the chemical risks of industrial production, several of the techniques can reduce the toxicity of the products as well. (Rossi et al., 1991).

For instance, a firm may redesign a product to reduce the requirement for a known toxic constituent; a firm may change the chemicals used to manufacture the product, thus reducing the residual toxic chemicals that may remain in or on a product; or a firm may change the production of a product to reduce the generation of waste toxic scrap, small amounts of which may have been disposed of as municipal solid waste.

The Polaroid Corporation has eliminated the use of mercury and reduced the use of cadmium in the batteries used in its film cassettes by developing a carbon-zinc cell with a zinc anode designed by the Rayovac Corporation. This project was initiated in anticipation of new regulations such as those in Switzerland that now require labeling and set limits on allowable concentrations of metals in batteries.

Integrated Pest Management

Yard, home, and agricultural activities that rely on toxic biocides contribute to the toxic constituents of solid waste. Each year thousands of pounds of chemical pesticides, herbicides, rodenticides, termiticides, fungicides, and fertilizers are sold to domestic customers. While much of this is used on lawns, gardens, basements, garages, and backyard orchards, some of it is also sent off as solid waste. Out-of-date product, unused portions of opened containers, and residuals in the bottom of "empty" containers may be set out as trash. In addition, some portion of used pesticides may be discarded as solid waste on grass clippings and other yard wastes.

A more significant contributor of toxic agricultural products in municipal solid waste may be small farms, nurseries, and agricultural product transporting firms that meet the RCRA definition of very small generators of hazardous waste. There is little research on the contribution of toxic agricultural products to solid waste.

Today, there are a host of new pest management practices that can reduce or eliminate the use of toxic chemical products. While there are specific, safer products that can simply replace the more toxic products, in general the preferred approach is to change the processes of yard or farm management. Knowing when and how to intervene in order to control pests is as important as the range of substances used. This new approach is often called "integrated pest management" because, like clean production, it requires a rethinking of the production system itself (Gipps, 1987).

In the yard, integrated pest management relies on natural controls (pathogens, parasites, predators, and repellents), improved yard management (increased sanitation, cultivation, aeration, and manual grooming), and selection of pest-resistant plantings. In buildings, integrated pest management means natural controls (pathogens; predators such as cats, and repellents), improved household management (increased sanitation), and architectural remedies (barriers and dampness prevention).

Sweden, Denmark, and the Netherlands have each adopted comprehensive national pesticide reduction programs over the past 10 years. Each of these programs establishes national goals for the reduction in use of various categories of active pesticide ingredients (up to 50 percent reduction over 5 years) and then employs a combination of regulation, education, financial incentives, and research to assist pesticide users to move to more integrated forms of pest management (Hurst, 1992).

While integrated pest management is not without some reliance on toxic chemical use, the general thrust is to minimize that use. Reducing the use of toxic pest controls will further reduce the toxic materials disposed of as waste from homes and farms.

6B.5   A SUSTAINABLE ECONOMY

Reducing the toxicity of materials in the municipal waste stream is the most fundamental way to reduce the health and environmental risks associated with trash management. Wherever communities seek a more sustainable future, there will need to be a focus on managing solid waste. That waste stream may prove to be a rich material resource for recycling and reuse, but the toxic constituents will surely inhibit the best of efforts. Although simple collection and recycling programs may have the effect of reducing some of the materials destined for landfills and incinerators in the short run, a focus on products and production processes in the longer term is likely to prove to be the most effective.

For now, there is a wide array of government policy options that could promote reduction in the toxicity of trash. Among the most important are the development of a comprehensive database on toxic materials in the municipal solid waste stream and the enlargement of research programs on environmentally sound materials, processes, and products.

Waste management programs could encourage better separation of waste products containing toxic materials from the solid waste stream, ban highly toxic products from landfills and incinerators, and promote manufacturer take-back of products containing highly toxic constituents. Better product management could be enhanced by product labeling schemes that inform consumers of toxic constituents, public education and media campaigns on the use and avoidance of products containing highly toxic materials, and using government procurement programs to encourage the purchase of environmentally preferred products. Production management programs could encourage reduction in the use of toxic chemicals in manufacturing and processing, and the development of more environmentally benign production materials.

None of these options alone will suffice. There is only so much that governments can do. Reducing toxicity in trash will take significant effort on the part of governments, businesses, and consumers. Although improvements in waste management and recycling can reduce the hazards of toxic materials in waste streams, the most effective programs will require changes in the production and consumption practices that precede the generation of waste. Progress in reducing toxic materials in solid waste requires a preventive and precautionary approach that ensures that the wastes of the future will be more environmentally compatible and less threatening to human and ecological health.

REFERENCES

Abt Associates (1985) *National Small Quantity Hazardous Waste Generator Survey: Final Report,* report prepared for the U.S. Environmental Protection Agency, Office of Solid Waste, Cambridge, MA.

Breniman, G. R., S. D. Cooper, W. H. Hallenbeck, and J. M. Lyznicki (1994) "Recycling Automotive and Household Batteries," in F. Krieth, ed., *Handbook of Solid Waste Management,:* McGraw-Hill, New York.

Curran, M. A., ed. (1996) *Environmental Life Cycle Assessment,* McGraw-Hill, New York.

Davis G. A., C. A. Witt, and J. N. Barkenbus (1997) "Extended Product Responsibility: A Tool for a Sustainable Economy", *Environment,* 39:7, pp. 10–15, 36–37.

Dennison, R., et. al. (1990), "Good Things Come in Smaller Packages: The Technical and Economic Arguments in Support of McDonald's Decision to Phase Out Polystyrene Foam Packaging," Environmental Defense Fund, Washington, DC.

Erdheim, R. (August, 2000) Unpublished Memo, Thermostat Recycling Corporation, Rosselyn, VA.

Fiksel, J., ed. (1996) *Design for the Environment: Creating Eco-Efficient Products,* McGraw-Hill, New York.

Fishbein, B, and C. Geld (1992) *Making Less Garbage: A Planning Guide for Communities,* INFORM, New York.

Florini, K., R. Dennison, and J. Ruston (1990) "An Environmental Perspective on Solid Waste Management," in F. Krieth, ed., *Integrated Solid Waste Management: Options for Legislative Action,* pp. 179–181, Genium, Schenectady, NY.

Gipps, T. (1987) *Breaking the Pesticide Habit: Alternatives to 12 Hazardous Pesticides,* International Alliance for Sustainable Agriculture, Minneapolis, MN.

Glaser, L., and G. Gajewski (1994) "Agricultural Products for Industry: The Situation and Outlook," *International Journal of Environmentally Conscious Manufacturing,* vol. 3, no. 2, pp. 63–66.

Glenn, J. (1999) "The State of Garbage in America," *BioCycle,* vol. 40, no. 4, pp. 60–71.

Goldstein, N., and J. Glenn, (1997) "The State of Garbage in America," *BioCycle,* vol. 38, no. 5, pp. 71–75.

Graedel, T. E., and B. R. Allenby (1996) *Design for the Environment,* Prentice-Hall, Upper Saddle River, NJ.

Hurst, P. (1992) *Pesticide Reduction Programmes in Denmark, the Netherlands and Sweden,* World Wildlife Fund International, Gland, Switzerland.

Lifset, R. (1993) "Take it Back: Extended Producer Responsibility as a Form of Incentive-Based Environmental Policy," *The Journal of Resource Management and Technology,* 21, p. 166.

Lillenthal, N., M. Ascione, and A. Flint (1991) *Tackling Toxics in Everyday Products,* INFORM, New York.

Melosi, M. V. (1981) *Garbage in the Cities: Refuse Reform and the Environment,* 1880–1890, University of Texas Press, Austin TX.

NRC (1984) *Toxicity Testing: Strategies to Determine Needs and Priorities,* U.S. National Research Council, Washington, DC.

P.L. 89-272, Sec 202(6).

P.L. 89-272, Sec 205.

P.L. 89-272, Sec 206.

P.L. 89-272, Sec 206(2).

P.L. 94-580.

P.L. 94-580, Sec 2.

P.L. 98-616.

Rossi, M., M. Ellenbecker, and K. Geiser (1991). "Techniques in Toxics Use Reduction," *New Solutions,* 2:2, Fall, pp. 25–32.

State of Massachusetts (June, 2000) *Massachusetts Zero Mercury Strategy,* State of Massachusetts, Executive Office of Environmental Affairs, Boston, MA.

Thorpe, B. (1999) *Citizen's Guide to Clean Production,* Lowell Center for Sustainable Development, University of Massachusetts, Lowell, MA.

U.N. Environment Program (1994) Cleaner Production Program, *Government Strategies and Policies for Cleaner Production,* U.N. Environment Program, Paris.

U.N. Environment Program (1999) *Dioxin and Furan Inventories: National and Regional Emissions of PCDD/PCDF,* U.N. Environment Program, Geneva, Switzerland.

U.S. Bureau of Mines (1985) *Mineral Facts and Problems,* 1985 ed., Bulletin 675, U.S. Bureau of Mines, Washington, D.C.

U.S. EPA (1989) *Characterization of Products Containing Lead and Cadmium in Municipal Solid Waste in the United States, 1970–2000,* U.S. Environmental Protection Agency, Washington, DC.

U.S. EPA (1992) *Characterization of Products Containing Mercury in Municipal Solid Waste in the United States,* U.S. Environmental Protection Agency, Washington, DC.

U.S. EPA (1997) *Characterization of Municipal Solid Waste in the United States, 1996 Update,* U.S. Environmental Protection Agency, Washington, DC.

U.S. EPA (1998a), *Characterization of Municipal Solid Waste in the United States, 1997 Update,* U.S. Environmental Protection Agency, Washington, DC.

U.S. EPA (1998b) *Availability Study: What Do We Really Know About the Safety of High Production Volume Chemicals?,* U.S. Environmental Protection Agency, Washington, DC.

U.S.G.S. (1999) *Minerals Yearbook: Metals and Minerals,* 1997, U.S. Geological Survey U.S. Government Printing Office, Washington, DC.

U.S. Office of the President (October 20, 1993) Executive Order 12873, "Federal Acquisition, Recycling and Waste Prevention", Washington, DC.

U.S. OTA (1989) *Facing America's Trash: What Next for Municipal Solid Waste,* U.S. Congress, Office of Technology Assessment, Washington, DC.

U.S. OTA (1992) *Green Products by Design: Choices for a Cleaner Environment,* U.S. Congress, Office of Technology Assessment, Washington, DC.

Wilson, D., and W. Rathje (1988) "Quantities and Composition of Household Hazardous Wastes: Report on a Multi-Community, Multi-Disciplinary Project," paper presented at the Third National Conference on Household Hazardous Waste Management, Boston, MA.

Wolf, N., and E. Feldman (1991) *Plastics: America's Packaging Dilemma,* Island Press, Covelo, CA.

CHAPTER 7
COLLECTION OF SOLID WASTE*

Hilary Theisen

Collection of commingled (unseparated) and separated (recyclables) solid waste is a critical part of any solid waste management program. As used here, collection starts with the containers holding materials that a generator has designated as no longer useful (solid waste and recyclables) and ends with the transportation of solid wastes or recyclables to a location for processing (e.g., a materials recovery facility), transfer, or disposal. Solid waste collection involves both the provision of a service and the selection of appropriate technologies. The service aspect is set through an agreement between waste generators and the waste collector or collection agency, and the waste collection contractor or agency selects the technology to be used for collection. The purpose of this chapter is to identify the various combinations of service and technology that are now used for the collection of wastes. Six specific topics to be addressed include:

1. The logistics of solid waste management
2. The types of waste collection services
3. The types of collection systems, equipment, and personnel requirements
4. The collection routes
5. The management of collection systems
6. The collection system economics

7.1   THE LOGISTICS OF SOLID WASTE COLLECTION

The management of collection is most difficult and complex in an urban environment because the generation of residential and commercial-industrial solid waste and recyclables takes place in every home, every apartment building, and every commercial and industrial facility, as well as in the streets, parks, and even vacant areas. As the patterns of waste generation become more diffuse and the total quantity of waste increases, the logistics of collection become more complex. Managers of collection systems must recognize and deal with the concerns of a population paying bills for services that reflect the high cost of fuel and labor. Of the total amount of money spent on solid waste management (collection, transport, processing, recycling, and disposal), approximately 50 to 70 percent is spent on the collection activity. Because such a large fraction of the total cost is associated with the collection operation, a small percentage improvement in the collection operation can affect a significant savings in the overall system cost.

---

* Adapted from Tchobanoglous et al. (1993).

7.2   TYPES OF WASTE COLLECTION SERVICES

The term *collection* includes not only the collection of solid wastes from the various sources, but also the hauling of these wastes to the location where the contents of the collection vehicles are emptied and the unloading of the collection vehicle (Tchobanoglous et al., 1993). While the activities associated with hauling and unloading are similar for most collection systems, the gathering or picking up of waste will vary with the characteristics of the facilities, activities, or locations where wastes are generated and the ways and means used for on-site storage (at the point of generation) of accumulated wastes between collections. The principal types of collection services that are now used for

- Commingled (unseparated) wastes
- Source-separated wastes

are summarized in Table 7.1. Because most service providers identify their service accounts according to the characteristics of the waste generator, it is convenient to develop groups of waste generators for presentation and analysis of data. The groups presented are representative of current waste management practice and are not intended to be all-inclusive of all waste generators.

Collection of Commingled (Unseparated) Wastes

The collection of wastes from low-rise detached dwellings, from medium-rise apartments, from high-rise apartments, and from commercial-industrial facilities is considered in the fol-

TABLE 7.1   Typical Collection Services for Commingled and Source-Separated Solid Waste*

Preparation method for waste collected	Type of service
Commingled wastes	Single collection service of large container for commingled household and yard waste
	Separate collection service for (1) commingled household waste and (2) containerized yard waste
	Separate collection service for (1) commingled household waste and (2) noncontainerized yard waste
Source-separated and commingled waste	Single collection service for a single container with source-separated waste placed in plastic bag along with commingled household and yard wastes
	Separate collection service for (1) source-separated waste placed in a plastic bag and commingled household waste in same container and (2) noncontainerized yard wastes
	Single collection service for source-separated and commingled household and yard wastes using a two-compartment container
	Separate collection service for (1) source-separated and commingled household wastes using a two-compartment container and (2) containerized or noncontainerized yard waste
	Separate collection service for (1) source-separated waste and (2) containerized commingled household and yard wastes
	Separate collection service for (1) source-separated waste, (2) commingled household waste, and (3) containerized yard wastes
	Separate collection service for (1) source-separated waste, (2) commingled household waste, and (3) noncontainerized yard wastes

* The method of waste preparation for collection is often selected for convenience and efficiency of collection services and subsequent materials processing activities.

lowing discussion. The collection of wastes separated at the source is considered following this discussion.

From Low-Rise Detached Dwellings.   The five most common types of residential collection services used for low-rise detached dwellings include (1) curb, (2) alley, (3) setout-setback, and (4) setout, and (5) backyard carry as summarized in Table 7.2. In some communities, more than one type of service may be offered. Where curb service is used, the homeowner is responsible for placing the containers to be emptied at the curb on collection day and for returning the empty containers to their storage location until the next collection event. Where alleys are part of the basic layout of a city or a given residential area, alley storage of containers used for solid waste is common. In setout-setback service, containers are set out from the homeowner's

TABLE 7.2 Comparison of Residential MSW Collection Services for Both Commingled and Source-Separated Solid Waste According to the Placement of Containers for Collection

Considerations	Type of service					
	Curb	Curb (mechanized)	Alley	Setout-setback	Setout	Backyard carry
Requires homeowner cooperation						
To move full containers	Yes	Yes	Optional	No	No	No
To move empty containers	Yes	Yes	Optional	Yes	Yes	No
Requires scheduled service for homeowner cooperation	Yes	Yes	No	No	Yes	No
Aesthetics						
Spillage and litter problem	High	Moderate	High	Low	High	Low
Containers visible	Yes	Yes	No	No	Yes	No
Attractive to scavengers	Yes	Yes	Yes	No	No	No
Prone to upsets	Yes	Yes	Yes	No	Yes	No
Number of persons in crew						
Typical	1	1	1	3	3	3
Range	1–3	1–2	1–3	3–7	1–5	3–5
Crew time	Low	Low	Low	High	Medium	Medium
Collector injury rate due to lifting and carrying	Low	Low	Low	High	Medium	High
Trespassing complaints	Low	Low	Low	High	High	High
Special considerations			Requires alleys and vehicles that can maneuver in them, less prone to block traffic, high vehicle and container depreciation rate			Requires wheeled caddy to roll filled containers or the use of burlap carry cloths or hand-carry bin, works best with driveway
Cost due to crew size and time requirements	Low	Low	Low	High	Medium	Medium

property and set back after being emptied by additional crews that work in conjunction with the collection crew responsible for loading the collection vehicle. Setout service is essentially the same as setout-setback service, except that the homeowner is responsible for returning the containers to their storage location. The characteristics of these services are compared in Table 7.2.

Manual methods commonly used for the collection of residential wastes include the following:

- The direct lifting and carrying of loaded containers to the collection vehicle for emptying
- The rolling of loaded containers on their rims to the collection vehicle for emptying
- The rolling of loaded containers equipped with wheels to the collection vehicle for mechanically assisted emptying
- The use of small lifts for rolling loaded containers to the collection vehicle.

In the past, large containers (referred to as *tote containers*) or drop cloths (often called *tarps*) into which wastes from small containers were emptied before being carried and/or rolled to the collection vehicle were commonly used.

For manual curb collection, where collection vehicles with low loading heights are used, wastes are transferred directly from the containers in which they are stored or carried to the collection vehicle by the collection crew (see Fig. 7.1). In other cases, collection vehicles are equipped with auxiliary containers into which the wastes are emptied. The auxiliary containers are emptied into the collection vehicle by mechanical means. Still another variant involves the use of small satellite vehicles. Wastes are emptied into a large container carried by a satellite vehicle. When loaded, the satellite vehicle is driven to the collection vehicle, where the container is emptied into the truck by mechanical means.

Where mechanized self-loading collection vehicles are used, the container used for the on-site storage of waste must be brought to the curb or other suitable location for collection. As noted previously (see also Table 7.2), the containers may be brought to the curb and returned to their normal location by the homeowner or by collection agency personnel. Typically, large containers are used in conjunction with mechanized collection vehicles (see Fig. 7.2).

From Low and Medium-Rise Apartments.   Curbside collection service is common for most low- and medium-rise apartments. Typically, the maintenance staff is responsible for transporting the containers to the street for curbside collection by manual or mechanical means. In many communities, the collector is responsible for transporting containers from a storage location to the collection vehicle (see Fig. 7.3). Where large containers are used, the contents of the containers are emptied mechanically using collection vehicles equipped with unloading mechanisms.

From High-Rise Apartments.   In high-rise apartment buildings (higher than seven stories), the most common methods of handling commingled wastes involve one or more of the following:

- Wastes are picked up by building maintenance personnel from the various floors and taken to the basement or service area.
- Wastes are taken to the basement or service area by tenants.
- Wastes, usually bagged, are placed by the tenants in a waste chute system, which is used for the collection of commingled waste at a centralized service location (see Fig. 7.4a).

In some of the more recent apartment building developments, especially in Europe, underground pneumatic transport systems have been used in conjunction with the individual apartment chutes (see Fig. 7.4b). The underground pneumatic systems are used to transport the wastes from the chute discharge points to centralized processing facilities. Both air pressure and vacuum transport systems have been used in this application.

The collection service for large apartments depends on the type of containers and processing equipment that is used. Typically, large containers with and without compaction are

(a)

(b)

FIGURE 7.1   Collection of wastes from containers placed at curb by homeowner (*a*) with a side-loading vehicle equipped with a right-hand standup drive mechanism, and (*b*) with rear-loading collection vehicle. The rear-loaded type of collection vehicle is commonly used with two- and three-person crews for the collection of residential wastes in many parts of the United States.

FIGURE 7.2 Typical example of mechanized collection vehicle with mechanical articulated pickup mechanism used for the collection of domestic source separated and comingled waste placed in a dual compartment container (see insert) (courtesy Heil Environmental Industries, Ltd.). The large containers equipped with wheels are brought to the curb by the homeowner. In some locations, helpers are used to bring the loaded containers to the curb, and the homeowner is responsible for returning the container to its storage location.

FIGURE 7.3 Emptying containers used for both commingled and source-separated wastes at an apartment complex. In the situation shown in the photo, the collector is responsible for bringing the loaded containers to the collection vehicle to be unloaded.

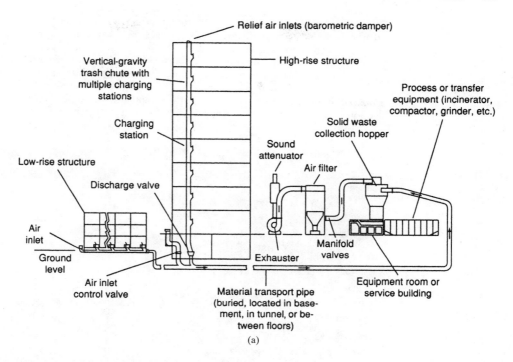

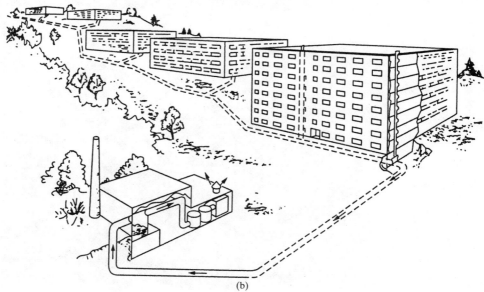

FIGURE 7.4 Schematic of a trash chute system for the collection of wastes from high-rise apartments. (*a*) For an individual apartment, the chute system will normally terminate in the basement. (*b*) In a large apartment complex, composed of a number of buildings, the wastes from the individual apartment building are transported using an underground pneumatic system to a centralized processing facility.

used. Depending on the size and type of container used, the contents of the containers may be emptied mechanically using collection vehicles equipped with unloading mechanisms (see Fig. 7.5), or the loaded containers are hauled to an off-site location (e.g., a materials recovery facility) where the contents are unloaded. Building maintenance personnel are normally responsible for the handling or processing of the wastes accumulated in the service areas.

From Commercial-Industrial Facilities.   Both manual and mechanical collection are used for the collection of wastes from commercial facilities. Because many large cities have extreme traf-

(a)

(b)

FIGURE 7.5   Mechanically loading collection vehicles. (*a*) Front-loading vehicle equipped with internal compactor. (*b*) Collection vehicle, with collection mechanism mounted on truck chassis, that can be used to collect two types of waste at the same time (e.g., a mixture of source-separated recyclable wastes and commingled waste). Such collection vehicles are used to collect wastes from large containers used in apartment complexes and commercial establishments.

fic congestion during the day, solid wastes from commercial establishments are collected in the late evening and early morning hours. Where manual collection is used during the evening hours, wastes from commercial establishments are put into plastic bags, cardboard boxes, and other disposable containers, which are placed on the curb for collection. Waste collection is usually accomplished with a three- or, in some cases, four-person crew, consisting of a driver and two or three collectors who load the wastes from the curbside into the collection vehicle. In most evening collection operations, the driver remains with the collection vehicle for reasons of safety.

Where traffic congestion is not a major problem and space for storing containers is available, the collection service provided to commercial-industrial facilities is centered around the use of large movable containers (see Fig. 7.6) and containers that can be coupled to large sta-

(a)

(b)

FIGURE 7.6 Typical examples of large containers used for the collection of wastes from commercial establishments (a) at rear of large department store in a shopping mall, and (b) in downtown location.

tionary compactors (see Fig. 7.7). As with the containers used at high-rise apartments, depending on the size and type of container used, the contents of the containers may be emptied mechanically, or the loaded containers are hauled to an off-site location where the contents are unloaded. Mechanized collection is also accomplished during the evening hours with a driver and helper.

Collection of Wastes Separated at the Source

Typically, wastes separated at the source are separated for recovery and reuse (recycled). The three principal methods now used for the collection of recyclable materials from residential sources include:

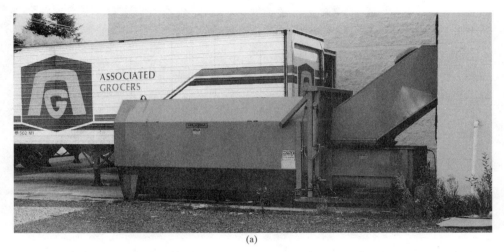

(a)

(b)

FIGURE 7.7 Typical examples of a stationary compactor used in conjunction with a large container for the collection of wastes from commercial establishments.

1. Curbside collection using conventional and specially designed collection vehicles
2. Incidental curbside collection by charitable organizations
3. Delivery by residents to drop-off and buyback centers

One, two, or more vehicles may be used where waste is separated at the source (see Table 7.1 for types of service). The collection of wastes separated at the source from different locations is considered in the following discussion.

From Low-Rise Detached Dwellings.   In a curbside system, source-separated recyclables are collected separately from commingled waste at the curbside (see Fig. 7.8), in the alley, or at commercial facilities. Because residents and businesses do not have to transport the recyclables any further than the curb, participation in curbside programs is typically much higher than for drop-off programs. Curbside programs vary greatly from community to community.

(a)   (b)

FIGURE 7.8   Typical examples of source-separated materials placed at the curbside for collection. (*a*) From a residential area. (*b*) Recyclable wastes along with commingled waste from commercial establishments placed on sidewalk in New York City to be collected in the evening or early morning hours.

Some programs require residents to separate several different materials (e.g., newspaper, plastic, glass, and metals) that are stored in their own containers and collected separately. Other programs use only one container to store commingled recyclables or two containers, one for paper and the other for "heavy" recyclables (e.g., glass, aluminum, and tin cans). Clearly, the method used to collect source-separated wastes will impact directly the layout and design of separation and processing facilities (see Table 7.1).

The two principal types of collection vehicles used for the collection of separated wastes are

1. Standard collection vehicles
2. Various specialized collection vehicles, including closed-body recycling trucks, recycling trailers, modified flatbed trucks, open-bin recycling trucks, and compartmentalized trailers.

Some of the most commonly used vehicles for the collection of separated wastes are shown in Fig. 7.9. The characteristics of these specialized vehicles are reviewed in Table 7.3.

Collection of Noncontainerized Residential Yard Waste

One option for the collection of noncontainerized yard waste involves the use of a specialized piece of equipment known as the *claw* and a modified collection vehicle. The claw clamps around piles of yard wastes left in the street by the homeowners (see Fig. 7.10*a*). The collected wastes are then emptied into a specially equipped compactor-type collection vehicle with a

(a)   (b)

(c)   (d)

FIGURE 7.9 Typical collection vehicles used for the collection of source-separated waste: (*a*) stand-up right-hand-drive, side-loaded collection vehicle with three separate compartments, using low collection troughs that are emptied mechanically. (*b*) Stand-up right-hand-drive open-top side-loaded collection vehicle. (*c*) Stand-up right-hand-drive side-loaded collection vehicle with three low-loading height compartments. (*d*) Stand-up-drive collection vehicle with mobile containers. When the containers are filled, they are emptied with a forklift.

wide receiving hopper (see Fig. 7.10*b*). The collected yard wastes are typically taken to a compost facility (see Chap. 12). The streets are swept after the yard wastes have been collected. Separate collection of yard wastes is typically done to meet a mandatory diversion goal. It should be noted that the claw has also been used to collect yard waste from apartment complexes and commercial facilities.

From Low- and Medium-Rise Apartments. The two principal methods now used for the collection of source-separated materials from low- and medium-rise apartments include:

1. Curbside collection using conventional and specially designed mechanized collection vehicles
2. Collection from designated storage areas with mechanized collection vehicles

In many low- and medium-rise apartments, large waste storage containers for recyclable materials are located outdoors in special enclosures (see Fig. 7.11). In some apartment buildings, the waste storage containers are located in the basement. Residents carry their waste and recyclable materials to the storage area and deposit them in the appropriate containers. Typi-

TABLE 7.3   Characteristics of Vehicles Used for the Collection of Wastes Separated at the Source

Item	Comment
Standard packer trucks	Packer trucks used for waste collection can also be used for collection of recyclables. Many communities use packer trucks in their recycling programs. Rear-loading packers have been used for newspaper, cardboard, and magazines with trailers attached to them for cans and glass. Front-end loaders have been used to service large containers containing newspaper recovered from apartment buildings. Some cities use side- and rear-loading packer trucks to pick up newspaper one week and glass and cans the following week. When collecting glass and cans, the compacting mechanism is not used because glass is highly abrasive and would damage the packer plate. Also, by not compacting, the majority of the glass remains unbroken and is, therefore, easier to sort into different colors at the processing site.
Closed-body recycling truck	This truck consists of an enclosed steel body installed on a lowered truck chassis, and a low-entry walk-in cab with dual left- and right-hand driving controls (allowing one-person operation). Adjustable hinged dividers on the body can be used to create from two to four compartments for different materials. One or both sides are opened for manual loading. Removable aluminum side panels are used to contain the load as the level of material rises. The overall capacity of the truck can range from 27 to 31 yd^3, although operational capacity when manually loading is 20 to 25 yd^3. The truck is equipped with a front-mounted telescopic hoist and rear body hinge for dumping. Each compartment is discharged separately by opening the rear door, unlocking the appropriate divider, and tipping the body.
Mobile container system	The mobile container system is essentially a steel frame with sets of hydraulic forks that can be used to transport large bins. The number of bins on a trailer range from three to six, and have a low-pull or gooseneck (fifth-wheel) style. To load the trailer, the forklifts are lowered to the ground and the bins are wheeled over them so that the forks slide into channels on the underside of the bins. The bins are then hydraulically raised and secured to the trailer frame. An empty set of bins can be left to replace the full ones. A pickup truck is used to pull the trailer.
Modified flatbed truck	Some curbside programs use a standard flatbed truck with a hydraulic dumping box mounted on the truck bed. The box is usually divided into three or four compartments and has a standard capacity of approximately 15 yd^3.
Open-bin recycling truck	The open-bin recycling truck is a specially designed vehicle with two or three open-top, self-dumping bins. Source-separated wastes are emptied into low-mounted troughs, which are emptied mechanically into the open bins. The front bins are typically 6 to 8 yd^3 and can be specified to unload right or left. The back bin, which dumps to the rear, has a capacity of 10 to 12 yd^3. The cab can be designed for right-hand stand-up drive to allow the loading function to be performed by the driver.

Source: Adapted from Tchobanoglous et al. (1993).

cally, the containers used for recyclables are located next to or near the containers used for commingled waste. Large containers are emptied mechanically using collection vehicles equipped with unloading mechanisms (see Fig. 7.3). In some cases, the apartment maintenance staff is responsible for moving the containers to the collection point.

From High-Rise Apartments.   In high-rise apartment buildings, the most common methods of handling commingled and source-separated wastes involve one or more of the following:

• Recyclable and commingled wastes are picked up by building maintenance personnel from the various floors and taken to the basement or service area and placed in separate containers.

| (a) | (b) |

FIGURE 7.10   Collection of noncontainerized yard waste placed in the street by the homeowner. (*a*) View of claw device mounted on a wheeled tractor used to pick up yard wastes. (*b*) Modified-compaction-type collection vehicle used in conjunction with claw. The collected yard wastes are hauled to a processing facility to be composted.

- Recyclable and commingled wastes are taken to the basement or service area by tenants and placed in separate containers.
- Recyclable wastes are taken to the basement or service area by tenants or building maintenance personnel and placed in separate containers, and, where available, other commingled waste is placed by the tenants in specially designed waste chutes, as described previously for commingled collection.

As with commingled waste, source-separated wastes are collected in large containers, which are emptied mechanically.

From Commercial Facilities.   Source-separated materials from commercial establishments are usually collected by private haulers. In many cases, the haulers have contracts with the facility for the separated material. The wastes to be recycled are stored in separate containers. In some cities, cardboard is often bundled and left at curbside, where it is collected separately. In large commercial facilities, baling equipment is often used for paper and cardboard, and can crushers are used for aluminum cans. Commingled municipal solid waste (MSW), generated in addition to the separated materials, is most commonly collected by private haulers or by city crews, if the city provides collection services.

7.3   TYPES OF COLLECTION SYSTEMS, EQUIPMENT, AND PERSONNEL REQUIREMENTS

Over the past 20 years, a wide variety of collection systems and equipment have been used for the collection of solid wastes. The two principal types of collection systems now used and the corresponding personnel requirements for these systems are described in this section. When considering collection technology, the basic components are surface streets and roadways, over-the-road trucks, and sturdy containers for storage. There have not been dramatic changes to these components since motor-driven vehicles replaced horse-drawn carts (Merrill, 1998). Technology changes will make the truck and labor more efficient, but the basic collection truck will be used for many more years.

(a)

(b)

FIGURE 7.11 Typical examples of containers and enclosures used for solid waste storage at apartment complexes. The small containers for recyclable materials are emptied with a collection vehicle of the type shown in Figs. 7.3 and 7.9a. The large containers for commingled nonrecyclable materials are emptied with a collection vehicle of the type shown in Fig. 7.5.

TABLE 7.4   Systems for the Collection of Solid Waste

Schematic of operational sequence	System description
(a) Hauled container system (conventional mode) 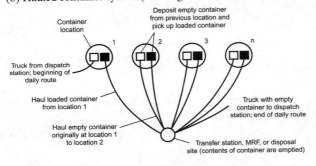	Containers used for the storage of wastes are hauled to an MRF, transfer station, or disposal site, emptied, and returned to their original location.
(b) Hauled container system (exchange container mode)	Containers used for the storage of wastes are hauled to an MRF, transfer station, or disposal site, emptied, and returned to a different location in the exchange mode of operation. The exchange mode works best when the containers are of a similar size. In the exchange mode, the driver must begin the collection route with an empty container on the vehicle to be deposited at the first collection site.
(c) Stationary container system	Containers used for the storage of wastes remain at the point of generation, except when they are moved to the curb or other location to be emptied. The collection vehicle is driven from pickup location to pickup location until it is loaded fully.

Types of Collection Systems

Solid waste collection systems may be classified from several points of view, such as the mode of operation, the equipment used, and the types of waste collected. In this Handbook, collection systems have been classified into two categories, according to their mode of operation:

1. Hauled container systems
2. Stationary container systems

The individual systems included in each category lend themselves to the same method of engineering and economic analysis. The principal operational features of these two systems are delineated in Table 7.4 and a discussion follows.

Equipment and Personnel Requirements for Hauled Container System (HCS).   These are collection systems in which the containers used for the storage of wastes are hauled to a materials recovery facility (MRF), transfer station, or disposal site, emptied, and returned to either their original location or some other location. Hauled container systems are ideally suited for the removal of wastes from sources where the rate of generation is high because relatively large containers are used (see Table 7.5). The use of large containers eliminates handling time

TABLE 7.5   Typical Data on the Container Types and Capacities Available for Use with Various Collection Systems

Collection system/vehicle	Container type	Typical range of container capacities*
Hauled container system		
Hoist truck	Used with stationary compactor	6–12 yd³
Tilt-frame	Open top, also called debris boxes or roll-off	12–50 yd³
	Used with stationary compactor	15–40 yd³
	Equipped with self-contained compaction mechanism	20–40 yd³
Truck-tractor	Open-top trash trailers	15–40 yd³
	Enclosed trailer-mounted containers equipped with self-contained compaction mechanism	30–40 yd³
Stationary container systems (compacting type)		
Compactor, mechanically loaded	Open top and closed top with side-loading	1–10 yd³
Compactor, mechanically loaded	Special containers used for the collection of residential wastes from individual residences	90–120 gal
Compactor, mechanically loaded with divided hopper	Special split cart containers used for the collection of recyclables and other nonrecyclable commingled waste	90–120 gal
Compactor trailer with mechanical lift assembly on semi-tractor	Special split cart containers used for the collection of recyclables and other nonrecyclable commingled waste	90–120 gal
Compactor, manually loaded	Small plastic or galvanized metal containers, disposable paper and plastic bags	20–55 gal
Stationary container systems (noncompacting type)		
Collection vehicle with series of manually loaded side-dump containers	All type of containers used for the temporary storage of recyclable materials	32 gal
Collection vehicle with semiautomatic manually loaded side troughs	All type of containers used for the temporary storage of recyclable materials	32 gal
Collection vehicle with semiautomatic manually loaded side troughs capable of unloading wheeled containers	All type of containers used for the temporary storage of recyclable materials plus wheeled containers	60–120 gal
Collection vehicle with mechanical lift assembly	Special containers used for the collection of source separated wastes from individual residences	60–120 gal

* *Note:* yd³ × 0.7646 = m³; gal × 0.003785 = m³.

as well as the unsightly accumulations and unsanitary conditions associated with the use of numerous smaller containers. Another advantage of HCSs is their flexibility: Containers of many different sizes and shapes are available for the collection of all types of wastes.

Because containers used in this system usually must be filled manually, the use of very large containers often leads to low-volume utilization unless loading aids, such as platforms and ramps, are provided. In this context, container utilization is defined as the fraction of the total container volume actually filled with wastes. While HCSs have the advantage of requiring only one truck and driver to accomplish the collection cycle, each container that is picked up requires a round trip to an MRF, transfer station, or disposal site. Therefore, container size and utilization are of great economic importance. Further, when highly compressible wastes are to be collected and hauled over considerable distances, the economic advantages of compaction are obvious.

There are three main types of HCSs:

1. Hoist truck
2. Tilt-frame container
3. Trash trailer

Typical data on the collection vehicles used with these systems are reported in Table 7.6.

TABLE 7.6   Typical Data on Vehicles Used for the Collection of Solid Waste

Collection vehicle			Typical overall collection vehicle dimensions				
Type	Available capacity range,[a,b] yd³	Number of axles	With indicated capacity,[c] yd³	Width, in	Height, in	Length,[d] in	Unloading method
Hauled container systems							
Hoist truck	6–12	2	10	94	80–100	110–150	Gravity, bottom opening
Tilt frame	12–50	3	30	96	80–90	220–300	Gravity, inclined tipping
Truck-tractor trash trailer	15–40	3	40	96	90–150	220–450	Gravity, inclined tipping
Stationary container system							
Compactor (mechanically loaded)							
Front-loading	20–45	3	30	96	140–150	240–290	Hydraulic ejector panel
Side-loading	10–36	3	30	96	132–150	220–260	Hydraulic ejector panel
Rear-loading	10–30	2 or 3[e]	20	96	125–135	210–230	Hydraulic ejector panel
Compactor (manually loaded)							
Side-loading	10–37	3	37	96	132–150	240–300	Hydraulic ejector panel
Rear-loading	10–30	2 or 3[e]	20	96	125–135	210–230	Hydraulic ejector panel
Noncompactor (mechanically loaded)	10–32	2 or 3		96			Gravity, inclined tipping
Noncompactor (manually loaded)	10–32	2 or 3		96			Gravity, inclined tipping

[a] See Table 7.5.
[b] *Note:* yd³ × 0.7646 = m³; in × 0.0254 = m.
[c] Capacity of truck body for hauled container or stationary container system.
[d] From front of the truck to rear of container or truck body.
[e] Drop axle now used by some operators when routes periodically require increased capacity.

FIGURE 7.12   Hoist-truck mechanism mounted on truck frame. Photo was taken in the 1960s at a naval installation. Although the truck style has changed, the hoist mechanism is essentially unchanged and is still used today.

Hoist-Truck Systems.   In the past, hoist trucks were used widely at military installations (see Fig. 7.12). With the advent of self-loading collection vehicles, however, this system appears to be applicable in only a limited number of cases, the most important of which follows:

- For the collection of wastes by a collector who has a small operation and collects from only a few pickup points at which a considerable amount of wastes are generated. Generally, for such operations the purchase of newer and more efficient collection equipment cannot be justified economically.

- For the collection of bulky items and industrial rubbish not suitable for collection with compaction vehicles.

Tilt-Frame Container Systems.   Systems that use tilt-frame-loaded vehicles (see Fig. 7.13) and large containers, often called *drop boxes* or *roll-off containers,* are ideally suited for the collection of all types of solid waste and rubbish from locations where the generation rate warrants the use of large containers. As noted in Table 7.5, various types of large containers are available for use with tilt-frame collection vehicles. Open-top containers are used routinely at warehouses and construction sites. Large containers used in conjunction with stationary compactors are common at apartment complexes, commercial services, and transfer stations. Because of the large volume that can be hauled, the use of the tilt-frame HCS has become widespread, especially among private collectors servicing commercial accounts.

Trash Trailer Systems.   The application of trash trailers is similar to that for tilt-frame container systems. Trash trailers are better for the collection of especially heavy rubbish, such as sand, timber, and metal scrap, and often are used for the collection of demolition wastes at construction sites (see Fig. 7.14).

Personnel Requirements for the Hauled Container System.   In most HCSs, a single collector-driver is used. The collector-driver is responsible for driving the vehicle, loading full con-

FIGURE 7.13   Truck with tilt-frame loading mechanism used with large debris boxes, sometimes called *roll-off containers*. Contents of debris box is being unloaded at a combined recycling and transfer facility.

FIGURE 7.14   Contents of trash trailer used for demolition wastes being unloaded at landfill.

tainers onto the collection vehicle, emptying the contents of the containers at the disposal site (or transfer point), and redepositing (unloading) the empty containers. In some cases, for safety reasons, both a driver and helper are used. The helper usually is responsible for attaching and detaching any chains or cables used in loading and unloading containers on and off the collection vehicle; the driver is responsible for the operation of the vehicle. A driver and helper should always be used where hazardous wastes are to be handled.

Equipment and Personnel Requirements for Stationary Container Systems (SCSs).   In the SCSs, the containers used for the storage of wastes remain at the point of generation, except when they are moved to the curb or other location to be emptied. The operational sequence for the SCSs is illustrated in Table 7.4. Stationary container systems may be used for the collection of all types of wastes. The systems vary according to the type and quality of wastes to be handled, as well as the number of generation points. There are two main types:

1. Systems in which mechanically loaded collection vehicles are used (see Figs. 7.2 and 7.5)
2. Systems in which manually loaded collection vehicles are used (see Figs. 7.1 and 7.9).

Because of the economic advantages involved, almost all of the collection vehicles now used are equipped with internal compaction mechanisms, especially where long-haul distances are involved. Operational data on the collection vehicles used in this system are reported in Table 7.7. To optimize the payload, many newer collection vehicles contain onboard scales, including load cells on the arms of mechanical lifting devices (to weigh individual containers) and/or load cells on the truck chassis (to weigh the loaded material).

Systems with Mechanically Loaded Collection Vehicles.   Container size and utilization are not as critical in SCSs using collection vehicles equipped with a compaction mechanism as they are in hoist-truck systems. Trips to the disposal site, transfer station, or processing station are made after the contents of a number of containers have been collected and compacted, and/or the collection vehicle is full. For this reason, the utilization of the driver in terms of the quantities of wastes hauled is considerably greater for these systems than for HCSs.

TABLE 7.7   Typical Operational Data for Computing Equipment and Labor Requirements for Various Collection Systems

Collection data			Time required to pick up loaded container and to deposit empty container, h/trip	Time required to empty contents of loaded container, h/container	At-site time, h/trip
Vehicle	Loading method	Compaction ratio, r			
Hauled container system					
Hoist truck	Mechanical		0.067	—	0.053
Tilt-frame	Mechanical		0.40	—	0.127
Tilt-frame	Mechanical	2.0–4.0	0.40	—	0.133
Stationary container system					
Compactor	Mechanical	2.0–2.5	—	0.050	0.10
Compactor	Manual	2.0–2.5	—	—	0.10
Stationary container system					
Noncompactor	Mechanical	—	—	—	0.10*
Noncompactor	Manual	—	—	—	0.10*

* Actual unloading time will depend on the number of compartments.

A variety of container sizes are available for use with these systems (see Table 7.5). They vary from relatively small sizes (1 yd³) to sizes comparable with those handled with a hoist truck. The use of smaller containers offers greater flexibility in terms of shape, ease of loading, and special features available. By using small, easier-to-load containers, utilization of containers can be increased considerably. These systems can also be used for the collection of residential wastes where one large container can be substituted for a number of small containers.

Because truck bodies are difficult to maintain and because of the weight involved, these systems are not ideally suited for the collection of heavy industrial wastes and bulk rubbish, such as that produced at construction and demolition sites. Locations where high volumes of rubbish are produced are also difficult to service because of the space requirements for the large number of containers.

Systems with Manually Loaded Collection Vehicles.   The major application of manual loading methods is in the collection of residential source-separated and commingled wastes and litter (see Fig. 7.1). Manual loading is used in residential areas where the quantity picked up at each location is small and the loading time is short. In addition, manual methods are used for residential collection because many individual pickup points are inaccessible to mechanized mechanically loaded collection vehicles.

Special attention must be given to the design of the collection vehicle intended for use with a single collector. At present, it appears that a side-loaded compactor, such as the one shown in Fig. 7.1a, equipped with stand-up right-hand drive, is best suited for curb and alley collection.

Personnel Requirements for Stationary Container Systems.   The personnel requirements for the SCS will vary, depending on whether the collection vehicle is loaded mechanically or manually. Typically, system selection is a function of worker fatigue (with the potential for injury) and capital and maintenance cost of the collection vehicle. Manually loaded collection vehicle systems cause the most fatigue while mechanically loaded vehicle systems are the highest cost.

Labor requirements for mechanically loaded SCSs are essentially the same as for HCSs. Where a helper is used, the driver often assists the helper in bringing loaded containers mounted on rollers to the collection vehicle and returning the empty containers. Occasionally, a driver and two helpers are used where the containers to be emptied must be rolled (transferred) to the collection vehicle from inaccessible locations, such as in congested downtown commercial areas.

In SCSs where the collection vehicle is loaded manually, the number of collectors varies from one to three, in most cases, depending on the type of service and the collection equipment. Typically, a single collector-driver is used for curb and alley service, and a multiperson crew is used for backyard carry service (see Table 7.2). In satellite-vehicle collection systems, one collector-driver is used for the main collection vehicle and one collector-driver is used with each satellite collection vehicle. While the satellite vehicles are being loaded, the collector-driver of the main vehicle picks up wastes from curb locations along the route. While the aforementioned crew sizes are representative of current practices, there are many exceptions. In many cities, multiperson crews are used for curb service as well as for backyard carry service.

7.4   COLLECTION ROUTES

In either private or public operations, it is important to set labor and equipment requirements for each type of service. Once the equipment and labor requirements have been determined, collection routes must be laid out so that both the collectors and equipment are used effectively. In general, the layout of collection routes involves a series of trials (Quon et al., 1965; Shuster and Schur, 1974; Truitt et al., 1970). There is no universal set of rules that can be applied to all situations. Thus, collection vehicle routing remains today a heuristic (common-sense) process.

Heuristic Guidelines for Laying out Collection Routes

Some heuristic guidelines that should be taken into consideration when laying out routes are as follows (Shuster and Schur, 1974):

- Existing policies and regulations related to such items as the point of collection and frequency of collection must be identified.
- Existing systems, such as crew size and vehicle types, must be coordinated.
- Wherever possible, routes should be laid out so that they begin and end near arterial streets, using topographical and physical barriers as route boundaries.
- In hilly areas, routes should start at the top of the grade and proceed downhill as the vehicle becomes loaded.
- Routes should be laid out so that the last container to be collected on the route is located nearest to the disposal site.
- Wastes generated at traffic-congested locations should be collected as early in the day as possible.
- Sources at which extremely large quantities of wastes are generated should be serviced during the first part of the day.
- Scattered pickup points where small quantities of solid waste are generated that receive the same collection frequency should, if possible, be serviced during one trip or on the same day.

Electronic data processing equipment and its appropriate software are being used to assist in the planning and evaluation of collection routes (Ernsdorff, 1999; SWANA, 1997a).

Layout of Collection Routes

The four general steps involved in establishing collection routes include:

1. Preparation of location maps showing pertinent data and information concerning the waste generation sources
2. Data analysis and, as required, preparation of information summary tables
3. Preliminary layout of routes
4. Evaluation of the preliminary routes and the development of balanced routes by successive trials

In many large cities and counties, some form of a *geographic information system* (GIS) is now used to identify each customer's location. In addition, a variety of other complimentary programs have been coupled to the GIS to both optimize the collection process and to improve the service provided. As noted previously, many of the newer collection vehicles contain onboard scales to optimize the payload.

It should be noted that the balanced routes prepared in the office are then given to the collector-drivers who implement them in the field. Based on the field experience of the collector-driver, each route is modified to account for specific local conditions, and information on the new route is entered into a database. In large municipalities, route supervisors are responsible for the preparation of collection routes. In most cases, the routes are based on the operating experience of the route supervisor, gained over a period of years working in the same section of the city. Some of the issues involved in laying out collection routes are illustrated in Fig. 7.15.

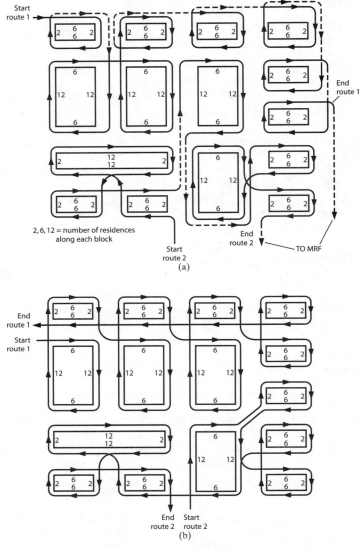

FIGURE 7.15 The effectiveness of the collection routes can be assessed by the amount of route overlap. (*a*) Route layout with overlap shown by the dotted lines. (*b*) Route layout without overlap.

Schedules

A master schedule for each collection route should be prepared for use by the engineering department and the transportation dispatcher. A schedule for each route, which includes the location and order of each pickup point to be serviced, should be prepared for the driver. Using the GIS database and the scheduling information, real-time monitoring of collection operations is now possible. In addition, each truck driver should maintain a route book. The driver uses the route book to check the location and status of accounts. It is also a convenient place in which to record any problems with the accounts. The information contained in the route book is useful in modifying the collection routes.

7.5 MANAGEMENT OF COLLECTION SYSTEMS

Solid waste collection systems are managed in various ways to provide generators with services. Management can be by private or public entities. The objective of management is to preserve public health and to maintain cost-effective service. Public health concerns have guided management decisions for more than 100 years, but only since the early 1970s has the public accepted higher costs for collection services that include waste separation and materials recovery (Kasperson, 2000; U.S. EPA, 1989; Warren, 1999).

Private-Entity Operations

Nationwide, private entities have a significant role in providing collection services. The advantages of private operations include unrestricted access to capital for equipment purchase, flexible use of workers on collection routes, and competition in setting system costs for the service. Private operations are available to generators either as service contracts directly between the generator and the entity, or indirectly through a public agency that authorizes the service under a franchise or service agreement with the entity.

Public-Entity Operation

Public entities also have a significant role in providing collection services. The advantages of public operations include control of the waste management system for public health considerations and public access to data regarding system costs (Ernsdorff, 1999). In some communities, both private and public operations are used. Such a management system is good for keeping a competitive business environment between the entities.

7.6 COLLECTION SYSTEM ECONOMICS

In the twenty-first century, the economics of collection include the costs of storage containers placed at the point of waste generation, the cost of providing collection service, and the costs of transfer stations that process the materials for recovery, consolidation, and movement to disposal sites (APWA, 1987). System managers measure cost-effectiveness of collection as one component within the cost of the integrated waste management system.

Labor Requirements

Labor efficiency in collection is measured by the productivity of each person on collection routes. Important parameters for measuring collection route productivity are crew size, service time, travel time, and time at the discharge site. Crew size is a function of the type of service provided to generators. In large cities where there is vehicle congestion and intense competition for available space on streets, crews are as large as four persons per vehicle, and collection is manual. In other communities with less vehicle congestion and where space is available for convenient placing and moving of waste storage containers, the crew is one person, and the collection vehicle is loaded mechanically.

Collection system managers evaluate equipment and labor combinations in selecting the best service for generators. In most applications, the most cost-effective system is one that employs a one-person crew with a mechanically loaded collection vehicle. The use of mechanically loaded collection vehicles is also favored because of a reduction in the number of job-related injuries and worker fatigue. Typical values for labor requirements for curbside collection using a one-person crew are given in Table 7.8.

TABLE 7.8 Typical Labor Requirements for Curbside Collection with Manually and Mechanically Loaded Collection Vehicle Using a One-Person Crew

Average number of containers and/or boxes per pickup location	Time, min/location*	
	Manual pickup	Mechanical pickup
1 (60–90 gal)		0.5–0.6
1 or 2	0.5–0.6	
3 or 4	0.6–0.9	
Unlimited service[†]	1.0–1.2	

* The values given are for a typical residential area with lot sizes varying from ¼ to ⅓ ac.
[†] Not all residents take advantage of unlimited service each collection day.

Collection Costs

The cost of collection includes equipment capital and maintenance costs and a significant labor cost. Private and public entities are evaluating numerous combinations of collection frequency, equipment, and labor in an attempt to develop the most cost-effective system (SWANA, 1997b–d). Typical costs for collection vehicles for commingled waste are in the range from $100,000 to $140,000, based on an Engineering News Record Construction Cost Index (ENRCCI) value of 6500. Compartmentalized collection vehicles for the collection of source-separated waste are in the range from about $120,000 to $140,000, again depending on the specific features. Typical collection costs, based on an ENRCCI value of 6500, are summarized in Table 7.9. The collection costs given in Table 7.9 will vary with the type of service provided, type of collection vehicle employed, local labor rate, travel times, and the characteristics of the community. Depending on the quantity of recyclable materials, the collection costs for source-separated and commingled waste as compared with the collection of commingled waste only will be higher by a factor of about $30 to $50/ton based on the total tonnage collected.

TABLE 7.9 Typical Costs for the Collection of Commingled and Separated Residential Wastes*

Type of waste collected	Collection cost, $/ton[†]	
	Manual collection with a one-person crew	Mechanized collection with a one-person crew
Commingled	60–80	50–70
Commingled waste remaining after recyclable materials have been removed	80–100	70–90
Source-separated	100–140	100–140

* Costs are based on an Engineering News Record Construction Cost Index (ENRCCI) of 6500, the current value for January 2002.
[†] Costs will vary with type of service, type of collection vehicle, labor rate, and the characteristics of the collection area.

REFERENCES

APWA (1987) *Solid Waste Collection and Disposal,* 3rd ed., Public Administration Service, American Public Works Association, Chicago.

BCUA (1988) *Bergen County Apartment Recycling Manual,* Bergen County Utilities Authority, Little Ferry, NJ.

Ernsdorff, S. M. (1999) "Using Spreadsheet Models for Estimating Collection Costs for Residential and Commercial Customers," *MSW Management,* vol. 9, no. 7, pp. 54–59.

Kaspersen, J. (2000) "Automation Options 2000," *MSW Management,* vol. 10, no. 5, pp. 86–88.

Merrill, L. (1998) "Collecting for the Millennium," *MSW Management,* vol. 8, no. 7, pp. 40–45.

Quon, J. E., A. Charnes, and S. J. Wenson (1965) "Simulation and Analysis of a Refuse Collection System," *Proceedings ASCE, Journal of the Sanitary Engineering Division,* vol. 91, no. SA5.

Shuster, K. A., and D. A. Schur (1974) *Heuristic Routing for Solid Waste Collection Vehicles,* U.S. Environmental Protection Agency, Publication SW-113, Washington, DC.

SWANA (1997a) *Getting More for Less: Cost Cutting Collection Strategies, Part 1—Case Study on Improved Routing,* City of Charlotte, North Carolina, Solid Waste Association of North America, Silver Springs, MD.

SWANA (1997b) *Getting More for Less: Cost Cutting Collection Strategies, Part 2—Changes in Collection Frequency,* City of Mesa, Arizona, Solid Waste Association of North America, Silver Springs, MD.

SWANA (1997c) *Getting More for Less: Cost Cutting Collection Strategies, Part 3—Automated Collection,* Rochester, New York, Solid Waste Association of North America, Silver Springs, MD.

SWANA (1997d) *Getting More for Less: Cost Cutting Collection Strategies, Part 4—Dual Collection,* City of Loveland, Colorado, Solid Waste Association of North America, Silver Springs, MD.

Tchobanoglous, G., H. Theisen, and S. A. Vigil (1993) *Integrated Solid Waste Management, Engineering Principles and Management Issues,* McGraw-Hill, New York.

Truitt, M. M., J. C. Liebman, and C. W. Kruse (1970) *Mathematical Modeling of Solid Waste Collection Policies,* vols. 1 and 2, U.S. Department of Health, Education, and Welfare, Public Health Service, Publication 2030, Washington, DC.

U.S. EPA (1989) *Decision-Maker's Guide to Solid Waste Management,* EPA/530-SW89-072, U.S. Environmental Protection Agency, Washington, DC.

Warren, G. (1999) "Collection and Transfer Trends," *MSW Management,* vol. 9, no. 7, pp. 30–33.

CHAPTER 8

RECYCLING

Harold Leverenz
George Tchobanoglous
David B. Spencer

During the 1980s, recycling took on much greater significance than just providing an alternative method for treatment of our solid waste. Recycling became an American philosophy, a public mandate. Source reduction and recycling became the only popularly accepted methods for dealing with the management of America's solid waste. However, to place a large part of the responsibility of solid waste management for a community on recycling alone puts an undue burden on recycling and could damage a strong, sound recycling initiative if it results in excessive cost or excessive contamination of high-value products. Just as the sanitary landfill came to be viewed as a disposal panacea at midcentury, only to be later discredited, the euphoria over recycling will need to be tempered with a strong record of tangible results. Topics discussed in this chapter include: (1) an overview of recycling, (2) the recovery of recyclable materials from solid waste, (3) development and implementation of materials recovery facilities, (4) equipment for processing of recyclable materials, (5) environmental and health impacts of recycling, and (6) recycling economics.

8.1 OVERVIEW OF RECYCLING

In the United States, the recycling rate for municipal solid waste (MSW) is about 22 percent, not including composting, based on estimates by Franklin Associates (U.S. EPA, 2001). However, the goal of 50 percent diversion by 2000 is close to being achieved in California. One significant trend is the emergence of a greater number of mandatory and voluntary programs for the source separation of recyclable materials. These so-called curbside programs require the participation of residents to separate recyclable materials into one or more fractions for collection. In 1989, 1042 curbside programs existed in 35 states. There has been considerable growth since that time, with the implementation of ambitious programs in New York, Florida, California, Ohio, and other states. By 2000, the number of curbside programs had grown to more than 4000.

Quantities and Composition of Recyclables

Between 1960 and 1990, U.S. MSW production rose from 2.7 to 4.5 lb per day per capita. For the period from 1990 to 2000, MSW generation increased from 205 million to more than 230 million tons per year (see Table 8.1). The estimates given in Table 8.1 include residential, commercial, and institutional solid waste. While per capita daily generation rates may be leveling off, population increases will continue to increase overall volumes of waste produced into the future. Municipal solid waste generation is estimated to be increasing at rates of 2 to 3 percent per year.

Growth in the quantities of waste generated is not the only problem contributing to the present problems associated with solid waste management. The composition and complexity

TABLE 8.1   Generation, Recovery, Composting, and Discards of Municipal Solid Waste

| Year | Generation, 10^6 tons | Percent of generation | | | |
		Recycling	Composting*	Total recovery	Discards
1960	88.1	6.4	N/A	6.4	93.6
1970	121.1	6.6	N/A	6.6	93.3
1980	151.6	9.6	N/A	9.6	90.4
1990	205.2	14.1	2.0	16.2	83.8
1994	214.4	19.7	4.0	23.6	76.4
1995	211.4	21.4	4.5	26.0	74.0
1997	219.1	21.6	5.5	27.1	72.9
1998	223	21.7	5.9	27.6	72.4
1999	229.9	22.1	5.7	27.8	72.2

* Composting of yard trimmings and food wastes does not include mixed MSW composting or backyard composting.

Source: U.S. EPA (2001).

of materials in the current waste stream may be more of a problem than the volume or weight produced. Recycling must deal with not only the vast quantity of bottles, cans, and containers present in the affluent U.S. society waste stream, but also the considerable complexity of these highly engineered products.

In 2000, 100 billion lb of plastics were produced in the United States. Of the 75 billion lb consumed domestically, 33 percent was used for packaging. Although these plastics represent only about 10 percent or less by weight of the waste stream, they represent about 20 percent of the solid waste stream by volume. These packages are a complex composite of many materials, making recycling in the twenty-first century a highly complex discipline, especially when the purity of the finished product is a critical limitation to market demand for recyclables.

Recycling Programs

There are many ways to implement a recycling program. The program can be either voluntary or mandatory. The materials to be recycled can include paper (newspaper, cardboard, mixed paper, etc.), glass (amber, green, and/or flint), cans (aluminum, ferrous, bimetal), and plastics (PET, HDPE, PS, PVC, PP, LDPE, etc.), as well as other items.

Recycling program alternatives include the following:

- Return of bottle bill containers or use of reverse vending machines
- Drop boxes, drop-off centers, or buyback centers for recyclables
- Curbside collection of homeowner-separated materials
- Curbside separation of homeowner-commingled recyclables
- Materials recovery facilities (MRFs) for the separation of commingled recyclables (collected at curbside, collected in drop boxes, or collected in special blue bags) using various levels of mechanization for waste processing
- Mechanically assisted hand separation of recyclables from raw waste (front-end processing or mixed-waste processing)
- Fully automated separation of recyclables from raw waste

The collection process itself can occur in many different ways. Materials can be collected at curbside in a multicompartment recycling truck either with or without compaction of the various segregated materials, or they can be collected commingled in either a dump truck or a packer truck (see Chap. 7).

Material separation can be done by the homeowner, by the collector at curbside, or by workers at a central processing plant. When items are commingled, they must be separated at a processing facility before they can be delivered to end markets. For source-separated collection programs, materials are divided by the homeowner and placed at the curbside for collection. Source-separated materials also require processing; however, the processing systems used are different from the systems used for commingled materials, as discussed in Sec. 8.3.

Typical source separation programs collect paper, glass, mixed plastics and metals, and the remaining household waste. The more materials in the program, the greater the collection and processing problems encountered. As the number of materials collected separately increases, the systems required to deal with these materials seem to grow geometrically.

When considering all of the different types of materials that can be included in a recycling program, the various methods for segregation, and the various means and methods of collection, as well as the various types of processing and separation systems that are available, the combinations and permutations seem endless. Specific expertise is required to evaluate the optimum method for a given community, based upon its population, geographic location, and proximity to markets.

8.2 RECOVERY OF RECYCLABLE MATERIALS FROM SOLID WASTE

There are three main methods that can be used to recover recyclable materials from MSW:

1. Collection of source-separated recyclable materials by either the generator or the collector, with and without subsequent processing

2. Commingled recyclables collection with processing at centralized materials recovery facilities (MRFs)

3. Mixed MSW collection with processing for recovery of the recyclable materials from the waste stream at mixed-waste processing or front-end processing facilities

Collection of Source-Separated Materials

The separation of recyclable materials into individual components, either by the generator or at curbside by the collector, is known as *source separation*. The setout of source-separated recyclables at curbside is depicted in Fig. 8.1a. The separated materials can be collected individually in single-compartment trucks, or more commonly, they are collected at the same time in a specially designed multicompartment recycling vehicle (see Fig. 7.9). The segregated components are then transported to a consolidation site for further processing and subsequent shipment to markets (see Sec. 8.3).

Usually, in the case of small communities, there is no further processing at the consolidation site. Processes such as can flattening, glass bottle crushing, and paper baling are performed by local scrap and paper dealers or recyclers who prepare the materials as necessary for final markets. In larger communities, each component may be further processed at the consolidation site and/or directly marketed to an end user when the materials meet buyers' specifications. Drop-off centers, buyback centers, and bottle-bill return stations are variations of the source separation approach.

FIGURE 8.1   Waste materials set out for curbside collection: (*a*) source-separated recyclable materials are placed in three separate containers (one for paper, one for glass, and one for cans and plastics), cardboard is bundled for collection with recyclable materials, residual nonrecyclable wastes are placed in separate containers, and yard wastes are placed in the street for collection with specialized collection equipment (see Fig. 7.10 in Chap. 7); (*b*) waste collection system employing three separate large containers [one for nonrecyclable materials (container on left), one for commingled recyclable materials (container in center), and one for yard wastes (container on right)]; (*c*) commingled recyclable materials received at a materials recovery facility for sorting; and (*d*) commingled mixed wastes in a single large (90 gal) container.

Collection of Commingled Recyclable Materials

Recyclable materials set out at curbside for commingled collection are shown in Fig. 8.1*a* and 8.1*b*. Here the generator only needs to separate recyclable materials from nonrecyclables. Newspapers are often kept separate from the rest of the commingled recyclables to prevent contamination and to improve collection vehicle efficiency.

The recyclable materials are transported to an MRF (see Sec. 8.3) where they are segregated into each recyclable component (glass, metal cans, plastic bottles, etc). Processing operations at MRFs can vary from facilities with relatively low mechanization, depending primarily on the manual sorting of waste materials, to highly mechanized automated sorting processes.

A variation of the commingled collection approach to recycling is the use of blue bags in a mixed-waste collection program. The color blue was chosen because it is distinctly different from the typical black or green trash bag, and studies have shown that the blue bag can be easily identified in a mixture of trash bags. Commingled recyclables are placed in the blue bags by the generators. The blue bags are taken along with trash bags to a central processing plant where the blue bags are hand-separated from the trash and sent to a commingled-recyclables processing facility for materials recovery. The bags can be filled with paper, commingled metals, plastic, and/or glass, depending on the design of the program. The objective of this type of program is to take advantage of the reduced collection costs of mixed-waste collection while still implementing an MRF that processes only the mixed recyclables, not the entire solid waste stream.

Collection of Mixed MSW

In the third approach to recycling, there is no segregation of recyclables from other waste materials. Mixed wastes (including recyclables) are set out at curbside (see Fig. 8.1c), as would be done for landfilling or incineration. One collection vehicle is required for collection of the mixed waste—normally, the familiar packer truck. A vehicle with a mechanical pickup mechanism for the collection of commingled recyclables is shown in Fig. 7.2. The mixed waste is then transported to a central processing facility, which employs a high degree of mechanization, including separation equipment such as shredders, trommels, magnets, and air classifiers to recover the recyclables. Mixed-waste processing of recyclables is also known as *front-end processing* or *refuse-derived fuel* (RDF) processing of MSW.

Comparison of Collection Methods for Recycling Materials

The first of the three approaches to recycling, source separation, requires a high degree of homeowner involvement, and has high collection costs but low processing costs. The second approach, commingled collection, requires an intermediate amount of generator effort, an intermediate amount of added collection cost, and processing costs that lie somewhere between those for source separation and mixed-waste collection and processing. The third approach, mixed-waste collection, requires no extra effort by the generator and results in no incremental collection costs, but it is accompanied by high processing costs plus some risk regarding technology, operating costs, and market economics due to uncertain capital and operations costs and potentially low recovery efficiency and material purity.

The quantity and quality of recyclable materials separated, collected, processed, and recycled can depend largely on which of the aforementioned approaches a community selects. Each method can affect attitudes regarding the mental and physical work required by a recycling program, and thus the extent of generator participation. In addition, each method has a different capital and operating cost, requiring varying levels of community financial commitment. Finally, each produces materials of differing composition or quality and thus can affect the amount of residue that is generated and the markets for products produced. Resident participation can also be affected by legislative actions, such as mandatory recycling. For example, motivation to recycle can be impacted by the requirement to participate in recycling to receive trash collection services, or by the levying of fines and penalties.

Defining Recyclable Materials

Determining the quantity of recyclables generated, by whatever method of separation, first requires a determination of what is to be considered a recyclable and how recycling performance is measured. Recycling programs can be compared if data are standardized, but in general, they are not. Consistent, standard, and meaningful measurement terminology is needed if communities are to effectively plan and assess their recycling programs. It would be wise to state at the outset of a program what will be counted as recyclables. Scenarios for the quantification of recyclables may include the following:

- All of the materials collected at curbside
- Those materials actually sold to market
- All recyclables collected and processed at an MRF
- Only those recyclables that are sold to market after separation and processing, with the residues that are generated at the MRF subtracted from the total

Measures of Recycling Performance

Although difficulties remain in quantitatively measuring the performance of recycling programs on a consistent, standard basis, the following useful performance criteria have been defined:

1. Capture rate
2. Participation rate
3. Recycling rate
4. Diversion rate

Other terms related to material recovery are defined in Sec. 8.3.

Capture Rate.   The term *capture rate* (also referred to as the *source recovery factor*) denotes the weight percent of an eligible material in the total solid waste stream actually separated out for recycling. Capture rate applies to a single material, not recyclables in general. This measure of performance is of greatest importance in measuring the success of a separation and collection program. Thus, for example, a capture rate for aluminum would be used to describe how much aluminum is captured by the community's curbside program versus how much is captured through the bottle-bill program.

Participation Rate.   The term *participation rate* denotes the percent of households (or businesses) that regularly set out recyclables. For example, in a particular community, on a monthly basis, 75 percent of the citizens participate in the curbside program. Participation may be different on a weekly basis than on a monthly basis, as fewer residents may participate weekly. The participation rate does not indicate the quantities of materials recycled or what materials were recycled. The participation rate term may actually provide misleading information regarding the success or failure of a recycling program, but does provide some useful measure of the extent of household involvement in the community's recycling program.

Recycling Rate.   The term *recycling rate* is sometimes used to denote the quantity of recyclables collected per household per unit of time (e.g., 35 lb/residence · month). The recycling rate normally addresses what was collected without regard to whether the material was actually sold or what amount of contamination was present in the recyclables. The term *recycling rate* is sometimes quoted as a percentage of the total quantity of waste generated in the community.

Diversion Rate.   Another performance factor in gauging the success of a recycling program is the *diversion rate,* which represents the weight of total solid waste that is not landfilled (or not incinerated). Thus, if the objective of the program is to minimize the weight of solid waste (including processing residues and incinerator ash) sent to landfill through a combination of strategies (such as source reduction and recycling), the ultimate performance measure is the net diversion rate. Again, diversion is often reported as a percent rather than weight or volume. It may be more useful to determine the net volumetric diversion rate, as it is a better measure than weight to estimate the savings in landfill life achieved by the integrated program. Landfills fill up long before they get too heavy.

Factors Affecting Material Recovery

It is widely believed that the easier it is for the public to participate in a recycling program, the higher will be the diversion rate or recycling rate, all other things being equal. A program that accomplishes recycling without any changes to the residents' disposal patterns, such as a mixed-waste processing system (which automatically separates recyclables from mixed trash), will achieve 100 percent participation, by definition. The diversion rate will be a function of the design of the program and the efficiency of waste processing.

The degree of source separation required can be expected to have a direct impact on the participation rate and capture rate. A complex recycling program in which many items are recycled and in which the resident separates each material, removes every label, and washes every container before setting it out to the curb fully segregated is more difficult for the resident than a program in which materials are set out commingled, unwashed, and with labels, in a single container.

Generally, a well-designed program for the collection of recyclables from homes will:

- Provide weekly collection
- Distribute household material storage container(s)
- Pick up recyclables on the same day as other wastes are collected
- Promote the program vigorously

The extent of resident participation is dependent on many factors beyond the complexity level and effort required in a source separation program. Public education is a substantial factor, as are the demographic characteristics of the community, including income, education, and location (suburban versus urban). When taking part in a separation program, households have to learn some new, and unlearn some old, behavior patterns. Learning new behavior involves the expenditure of time, mental and physical effort, and sometimes money. These patterns are reinforced when education, moral satisfaction, and ease of program implementation are encouraged. However, generally speaking, as a program proceeds and recycling becomes more of a habit, the perceived effort to accomplish recycling diminishes. In the course of time, less mental effort is needed to separate the domestic waste, and households become more positive regarding the costs and benefits of recycling.

Same-day collection of both recyclables and residual solid waste has been found to significantly improve the level of household participation. Moreover, the collection of yard debris, preferably on the same day as the other collections, is often required to achieve maximized recovery rates. To summarize the issue, with a high-cost system such as mixed-waste processing, participation is high because everyone participates, incremental collection cost is negligible, and recovery efficiency is high. However, the potential for product contamination is higher than for source-separated recycling; mixed-waste processing is expensive (often requiring tens of millions of dollars in one-time capital cost and millions per year in operating costs); and automated front-end processing programs do not focus the efforts of the public on a continuing basis in dealing with the solid wastes they generate.

On the other hand, the *good feeling* of recycling is lost when there is little or no household involvement. In a source separation program, although collection at curbside is more expensive and the percentage of participation may be lower, the ability to recover materials that will meet product quality specifications can be much higher. The resident will receive satisfaction from involvement in the program, and the diversion rate for the individual recyclables can still be high.

Mandatory versus Voluntary

A discussion of the impact of state recycling legislation is provided in Chap. 3. Clearly, if a community is unable to achieve its recycling goals through a voluntary recycling program, the alternative remains to make recycling mandatory. While mandatory programs should, in theory, increase the participation level of households, there is no evidence to indicate that a well-implemented, well-communicated voluntary program cannot achieve the same levels of participation and the same waste diversion as a mandated program.

Unit Pricing–Based Systems

The concept of *unit pricing,* also referred to as *pay-as-you-throw* or *variable rate,* is that residents pay a fee proportional to their waste generation. By assessing a fee on material put out for waste collection, residents can be encouraged to increase their participation and source separation factors. More details of unit pricing of MSW are available in Sec. 6.3.

Bottle-Bill Legislation and Recycling

Before the first deposit laws were enacted in 1970, there was virtually no recycling of aluminum cans or plastic bottles; glass bottles were recycled at just 1 percent. Bottle-bill programs were implemented, not so much for their impact on materials recycling but rather because of the very positive impact they have on litter control. However, such legislation has proved to be a part of many successful recycling programs.

As discussed in Chap. 3, 10 states and Columbia, Missouri, currently (in 2000) have mandatory bottle deposit legislation. In each case, the consumer pays a deposit on each beverage container purchased and receives that amount as a refund when the container is returned for recycling or refilling. These bottle deposit bills primarily affect soft-drink beverage containers; in some cases, however, as in the state of Maine, other items such as wine and juice containers are included in the deposit system.

The arguments against bottle bills put forth by the beverage industry over the last 20 years have included loss of jobs, higher cost to the consumer, and the concern that a bottle bill would compete with other recycling alternatives. Clearly, the stores that have to comply with the storage and handling requirements associated with a bottle bill are concerned because valuable store space is consumed. There is also the potential for insects and rodents, which can be attracted from storage of bottles that have not been cleaned or washed out. Moreover, maintaining the bottle return system requires significant clerical effort.

When a community is operating under a state bottle bill, there is always some concern regarding the impact on the amount of material that will be recovered under a curbside recycling program, as compared with other communities or states that do not have a bottle bill. There is substantial evidence that demonstrates the value of deposit laws working in tandem with curbside recycling collection programs. The presence of a bottle bill is expected to increase recycling levels of beverage containers and reduce overall solid waste management costs.

Drop-Off and Buyback Programs

Drop-off and buyback centers are centralized locations where a specified class of waste generators (typically residential generators) may voluntarily bring certain recyclable materials (see Fig. 8.2). One of the largest advantages of drop-off centers is that they are inexpensive to implement. A drop-off center can be as simple as several small-capacity containers that temporarily store the materials for regular pickup and transportation to market or central consolidation facility, or it can consist of drop-off at the central consolidation facility itself.

Because programs of this nature are voluntary, participation can often be poor. However, there can be notable exceptions, especially where curbside waste collection is not performed and citizens must take their trash to a disposal facility where they may also drop off their recyclables. Moreover, participation is enhanced by public education and by ordinances that increase the difficulty to otherwise dispose of recyclable materials. Economic incentives could also be a factor. For example, a variation of the drop-off center is the buyback center, where

(a)   (b)

(c)

(d)   (e)

FIGURE 8.2 Typical examples of drop-off and buyback centers for the recovery of recyclable materials: (*a*) igloo-type in a residential area, (*b*) modified igloo-type in city center, (*c*) in rural area, and (*d*) and (*e*) buyback centers located near supermarkets.   *(From Tchobanoglous et al., 1993.)*

the generators are financially compensated for materials. Both buyback centers and drop-off centers seldom capture more than 10 percent of the waste stream.

The physical layout of a drop-off center varies by the volume and number of recyclable materials processed, site characteristics, and level of supervision. A conventional drop-off center would be centrally located within a service area and provide bins or compartmentalized containers for waste generators to deposit recyclable materials. To ensure material quality and public safety as well as to prevent scavenging, many drop-off centers have controlled access, limited hours of operation, and are monitored by attendants. Once a sufficient quantity of a material has been collected, it can be shipped to end users or intermediaries in the container in which it was collected or alternatively transferred to a larger container or truck. Correct sizing and type of containers are key design features to address, along with traffic access and security.

The smallest drop-off center might be a neighborhood kiosklike or igloo container, unattended and conveniently located to maximize its use. This method of recycling is particularly applicable to heavily rural areas. However convenient these unattended containers are, they must be inspected frequently to determine if they are full, present an unsightly litter problem, or have contaminated contents. Drop-off centers are also vulnerable to odors, vectors, and vandalism, aside from incurring transportation and handling costs.

Drop-off containers should be located in areas that have a high volume of traffic and are familiar to the local populace such as schools, shopping centers, and fire stations. The presence of drop-off containers in such locations makes it easier for citizens to recycle when they are shopping, picking up their children, or running errands. Participation can be improved through advertising, special events, and local mailings to citizens informing them of the location of drop-off boxes and advising them how and what to recycle.

Drop-off centers have low capital and operating costs, little or no technical risk, limited changes in waste generator behavior (provided they are conveniently located), and are flexible to changes in waste composition or participation rates as well as the targeted recyclable materials. However, drop-off recovery suffers from lower participation rates because residents are required to store materials and physically bring them to a remote location, and the products can often be low-quality if there is no supervision. In fact, some products may be unmarketable, owing to the high degree of contamination that can occur.

8.3 DEVELOPMENT AND IMPLEMENTATION OF MATERIALS RECOVERY FACILITIES

As noted in the introduction to this chapter, the further separation and processing of wastes that have been separated at the source and the separation of commingled wastes usually occurs at MRFs or at large, integrated *materials recovery/transfer facilities* (MR/TFs). The successful development and implementation of an MRF or MR/TF requires that proper attention be paid in the planning and design phases to both technical engineering and environmental considerations. Both of these factors are considered in the following discussion. Individual elements that comprise an MRF are discussed in Sec. 8.4. Additional details on the development and implementation of MRFs are cited in Tchobanoglous et al. (1993) and the U.S. EPA (1991).

Technical Considerations in the Planning and Design of MRFs

The technical planning and design of MRFs involves three basic steps:

1. Feasibility analysis
2. Preliminary design
3. Final design

These planning and design steps are common to all major public works projects, such as landfills or wastewater treatment plants. In some cases, the feasibility analysis has already been accomplished as part of the integrated waste management planning process. A brief discussion of these topics follows. Some specific topics, including the functions of an MRF, development of MRF process flow diagrams, materials balances and loading rates, and examples of different types of MRFs, are considered in greater detail to illustrate the issues involved.

Feasibility Analysis.   The purpose of the feasibility analysis is to decide whether the MRF should be built. The feasibility study should provide the decision makers with clear recommendations on the technical and economic merits of the planned MRF. A typical feasibility analysis may contain sections or chapters dealing with the following topics:

- Functions of the MRF
- Conceptual design
- Siting
- Economics
- Ownership and operation
- Procurement

Issues associated with each of these factors are considered in Table 8.2.

Preliminary Design.   Preliminary design considerations for an MRF include development of the following:

- Process flow diagrams
- Prediction of materials recovery rates
- Development of materials mass balances and loading rates for the unit operations (conveyors, screens, shredders, etc.), which make up the MRF
- Selection of processing equipment
- Facility layout and design
- Staffing needs
- Environmental issues
- Health and safety issues

The cost estimate developed in the feasibility study is usually refined in the preliminary design report, using actual price quotes from vendors.

Final Design.   Final design includes the following:

- Preparation of final plans and specifications that will be used for construction
- Preparation of environmental documents
- Preparation of detailed cost estimates
- Preparation of the procurement documents

Functions of an MRF and Materials to Be Recovered

The functions of an MRF will depend directly on the following:

- The role that the MRF is to serve in the waste management system
- The types of material to be recovered
- The form in which the materials to be recovered will be delivered to the MRF
- The containerization and storage of processed materials for the buyer

For example, the function, equipment, and facilities required for the separation of source-separated waste will differ significantly from those required for the separation of recyclables from commingled MSW. The functions of an MRF may also change as a function of size.

TABLE 8.2   Important Technical Considerations in the Planning and Design of MRFs

Step 1—Feasibility analysis	
Function of MRF	The coordination of the MRF with the integrated waste management plan for the community. A clear explanation of the role and function of the MRF in achieving landfill waste diversion and recycling goals is a key element.
Conceptual design, including types of wastes to be sorted	What type of MRF should be built, which materials will be processed now and in the future, and what should be the design capacity of the MRF. Plan views and renderings of what the final MRF might look like are often prepared.
Siting	While it has been possible to build and operate MRFs in close proximity to both residential and industrial developments, extreme care must be taken in their operation if they are to be environmentally and aesthetically acceptable. Ideally, to minimize the impact of the operation of MRFs, they should be sited in more remote locations where adequate buffer zones surrounding the facility can be maintained. In many communities, MRFs are located at the landfill site.
Economic analysis	Preliminary capital and operating costs are delineated. Estimates of revenues available to finance the MRF (sales of recyclables, avoided tipping fees, subsidies) are evaluated. A sensitivity analysis must be performed to assess the effects of fluctuating prices for recyclables and the impacts of changes in the composition of the waste.
Ownership and operation	Typical ownership and operation options include public ownership, private ownership, or public ownership with contract operation.
Procurement	What approach is to be used in the design and construction of the MRF? Several options exist, including: (1) the traditional architect-engineer and contractor process; (2) the turnkey contracting process in which design and construction are performed by a single firm; and (3) a full service contract in which a single contractor designs, builds, and operates the MRF.
Step 2—Preliminary design	
Process flow diagrams	One or more process flow diagrams are developed to define how recyclable materials are to be recovered from the MSW (e.g., source separation or separation from commingled MSW). Important factors that must be considered in the development of process flow diagrams include: (1) characteristics of the waste materials to be processed, (2) specifications for recovered materials now and in the future, and (3) the available types of equipment and facilities.
Materials recovery rates	Prediction of the materials flow to the MRF is necessary to estimate the effectiveness or performance of the recycling program. The performance of a recycling program, the overall component recovery rate, is generally reported as a materials recovery rate or recycling rate, which is the product of three factors: (1) participation factor, (2) composition factor, and (3) source recovery factor. Component capture rates for the recyclable materials most commonly collected in source separation recycling programs are reported in Table 8.7. Composition factors are measured in waste composition studies. Typical component recovery rates are given in Table 8.8.
Materials balances and loading rates	One of the most critical elements in the design and selection of equipment for MRFs is the preparation of a materials balance analysis to determine the quantities of materials that can be recovered and the appropriate loading rates for the unit operations and processes used in the MRF.
Selection of processing equipment	Factors that should be considered in evaluating processing equipment are summarized in Table 8.30.
Facility layout and design	The overall MRF layout includes: (1) sizing of the unloading areas for commingled MSW and source-separated materials, (2) sizing of presorting areas where oversized or undesirable materials are removed, (3) placement of conveyor lines, screens, magnets, shredders, and other unit operations, (4) sizing of storage and outloading areas for recovered materials, and (5) sizing and design of parking areas and traffic flow patterns in and out of the MRF. Many of these layout steps are also common to the layout and design of transfer stations.

TABLE 8.2 Important Technical Considerations in the Planning and Design of MRFs (*Continued*)

Step 2—Preliminary design	
Staffing	Depends on type of MRF (i.e., degree of mechanization). Staffing is discussed in Sec. 8.6.
Economic analysis	Refine preliminary cost estimate prepared in feasibility study.
Environmental issues	Important environmental issues are summarized in Sec. 8.5.
Health and safety issues	Important health and safety issues are summarized in Table 8.31.

Step 3—Final design	
Preparation of final plans and specifications	Plans and specifications will be used for bid estimates and construction.
Preparation of environmental documents	Preparation of the necessary environmental documents (e.g., Environmental Impact Report).
Preparation of detailed cost estimate	A detailed engineers cost estimate is made based on materials takeoffs and vendor quotes. The cost estimate will be used for the evaluation of contractor bids, if the traditional procurement process is used.
Preparation of procurement documents	Use of a bidding process to obtain supplies, equipment, and services related to the construction, operation, and maintenance of the facility.

A typical classification of MRFs according to size and the degree of mechanization is presented in Table 8.3. In general, as reported in Table 8.3, small MRFs are usually not very mechanically intensive because of the capital investment cost and the ongoing operation and maintenance costs. While the terms *small, intermediate,* and *large* are used in Table 8.3, the terms *low, intermediate,* and *high-tech* have been used to classify the same sizes. The latter terms are not favored, because in recent MRF designs, a small MRF can also have a high degree of technical sophistication.

MRFs for Source-Separated Materials.   The types of materials that are typically processed at MRFs for source-separated wastes are summarized in col. 1 of Table 8.4. The functions that must be carried out at one or more types of MRFs to process the source-separated materials are identified in col. 2 of Table 8.4. The particular combination of materials to be separated will depend on the nature of the source separation program the community has adopted. For example, as given in the following list, a typical source separation program might involve the use of three separate containers for recyclable materials in conjunction with one or more additional containers for other wastes; yard wastes will be collected separately. The materials would be separated as follows:

- Recycle container 1: mixed paper (bundled cardboard collected separately)
- Recycle container 2: glass
- Recycle container 3: mixed plastics (HDPE and PET), aluminum cans, and tins cans
- Mixed (commingled) residual MSW (collected separately)
- Yard wastes (collected separately)

For the preceding mix of source-separated materials, three or four separate processing activities will be required at an MRF to separate and/or to process the individual components. The processing of the yard wastes would normally be done at a separate facility or at a large regional facility.

TABLE 8.3 Typical Types of MRFs, Capacity Ranges, and Major Functions and System Components Based on the Degree of Mechanization

Type of system	Capacity, ton/d	Major system components
Materials recovery		
Low	5 to 20	Processing of source-separated materials only; enclosed building, concrete floors, elevated hand sorting conveyor, baler (optional), storage for separated and prepared materials for 1 month, support facilities for the workers
Intermediate	20 to 100	Processing of source-separated commingled materials and mixed paper; enclosed building, concrete floors, elevated hand sorting conveyor, conveyors, baler, storage for separated and baled materials for 2 weeks, support facilities for the workers, buyback center
High	>100	Processing of commingled materials or MSW; same facilities as the intermediate system plus mechanical bag breakers, magnets, shredders, screens, and storage for baled materials for up to 2 months
Composting		
Low-end system	5 to 20	Source-separated yard waste feedstock only; grinding equipment, cleared level ground with equipment to form and turn windrows, screening equipment (optional)
High-end system	>20	Feedstock derived from source-separated yard waste or from the processing of commingled wastes. Facilities include enclosed building with concrete floors, in-vessel composting reactors; enclosed building for curing of compost product, equipment for bagging and marketing compost product

MRFs for Commingled MSW.   In the case of commingled MSW, the materials to be separated and function and equipment requirements for the MRF will depend directly on the role the MRF is to serve in the waste management system (see Table 8.5). An MRF can be used to separate and process source-separated materials, as well as separate materials from commingled MSW, to meet mandated diversion goals. Another common use of an MRF for commingled MSW is to remove contaminants from the waste and to prepare the waste for subsequent uses (e.g., a fuel for combustion facilities or a feedstock for composting). Another MRF function might be to recover high-value items and to process the residual waste for the production of compost to be used as intermediate landfill cover (see Chap. 14). Clearly, an endless number of variations of an MRF are possible. The types of materials and/or contaminants removed and the associated activities carried out at the different types of MRFs previously identified are summarized in cols. 2 and 3, respectively, of Table 8.5.

Development of MRF Process Flow Diagrams

Once a decision has been made on how and what recyclable materials are to be recovered from MSW (e.g., source separation or separation from commingled MSW), MRF process flow diagrams must be developed for the separation of the desired materials and for processing the materials, subject to predetermined specifications. A *process flow diagram* for an MRF is defined as the assemblage of unit operations, facilities, and manual operations to achieve a specified waste separation goal or goals. The following factors must be considered in the development of process flow diagrams:

TABLE 8.4 Typical Examples of the Materials, Functions, and Equipment and Facility Requirements of MRFs Used for the Processing of Source-Separated Materials

Materials	Function/operation	Equipment and facility requirements
Mixed paper and cardboard/1	Manual separation of high-value paper and cardboard or contaminants from commingled paper types. Baling of separated materials for shipping. Storage of baled materials.	Front-end loader, conveyors, baler, forklift
Mixed paper and cardboard/2	Manual separation of cardboard and mixed paper. Baling of separated materials for shipping. Storage of baled materials.	Front-end loader, conveyors, open sorting station, baler, forklift
Mixed paper and cardboard/3	Manual separation of old newspaper, old corrugated cardboard, and mixed paper from commingled mixture. Baling of separated materials for shipping. Storage of baled materials.	Front-end loader, conveyors, enclosed elevated sorting station, baler, forklift
PETE and HDPE plastics	Manual separation of PETE and HDPE from commingled plastics. Baling of separated materials for shipping. Storage of baled materials.	Receiving hopper, elevated sorting conveyor, storage bins, baler, forklift
Mixed plastics	Manual separation of PETE, HDPE, and other plastics from commingled mixed plastics. Baling of separated materials for shipping. Storage of baled materials.	Receiving hopper, elevated sorting conveyor, storage bins, baler, forklift
Mixed plastics and glass	Manual separation of PETE, HDPE, and glass by color from commingled mixture. Baling of separated materials for shipping. Storage of baled materials.	Receiving hopper, elevated sorting conveyor, glass crusher, storage bins, baler, forklift
Mixed glass (with sorting)	Manual separation of clear, green, and amber glass. Storage of separated materials.	Receiving hopper, elevated sorting conveyor, glass crusher, storage bins, forklift
Mixed glass (without sorting)	Storage of separated mixed glass.	Storage bunker for mixed glass, glass crusher, storage bins, forklift
Aluminum and tin cans	Magnetic separation of tin cans from commingled mixture of aluminum and tin cans. Baling of separated materials for shipping. Storage of baled materials.	Receiving hopper, conveyor, overhead suspended magnet, magnet pulley, storage containers, baler or can crusher and pneumatic transport system, forklift
Plastic, aluminum cans, tin cans, and glass	Manual or pneumatic separation of PETE, HDPE, and other plastics. Manual separation of glass by color, if separated. Magnetic separation of tin cans from commingled mixture of aluminum and tin cans. Magnetic separation may occur before or after the separation of plastic. Baling of plastic (typically two types), aluminum cans, and tin cans, and crushing of glass and shipping. Storage of baled and crushed materials.	Receiving hopper, conveyor, elevated picking conveyor, magnet pulley, overhead suspended magnet, glass crusher, storage containers, baler or can crusher, and forklift
Yard wastes/1	Manual separation of plastic bags and other contaminants from commingled yard wastes, grinding of clean yard waste, size separation of waste that has been ground up, storage of oversized waste for shipment to biomass facility, and composting of the undersized material.	Front-end loader, tub grinder, conveyors, trommel or disc screen, storage containers, compost-turning machine
Yard wastes/2	Manual separation of plastic bags and other contaminants from commingled yard wastes followed by grinding and size separation to produce landscape mulch. Storage of mulch and composting of undersized materials.	Front-end loader, tub grinder, conveyors, trommel or disc screen, storage containers, compost-turning machine
Yard wastes/3	Grinding of yard waste to produce a biomass fuel. Storage of ground material.	Front-end loader, tub grinder, conveyors, storage containers or transport trailers

TABLE 8.5 Typical Examples of the Functions, Materials Recovered or Contaminants Removed, and Activities Associated with MRFs Used for the Processing of Commingled MSW

Function of MRF	Materials recovered or contaminants removed	Activities
Recovery of recyclable materials to meet mandated first-stage diversion goals	Bulky items, cardboard, paper, plastics (PETE, HDPE, and other mixed plastic), glass (clear and mixed), aluminum cans, tin cans, other ferrous materials	Manual separation of bulky items, cardboard, plastics, glass by color, aluminum cans, and large ferrous items. Magnetic separation of tins cans and other ferrous materials not removed manually. Baling of separated materials for shipping. Storage of baled materials.
Recovery of recyclable materials and the further processing of source-separated materials to meet second-stage diversion goals	Bulky items, cardboard, paper, plastics (PETE, HDPE, and other mixed plastic), glass (clear and mixed), aluminum cans, tin cans, other ferrous materials. Additional separation of source-separated materials, including paper, cardboard, plastic (PETE, HDPE, other), glass (clear and mixed), aluminum cans, tin cans	Manual separation of bulky items, cardboard, plastics, glass by color, aluminum cans, and large ferrous items. Magnetic separation of tin cans and other ferrous materials not removed manually. Baling of separated materials for shipping. Storage of baled materials.
Preparation of MSW for use as a fuel for combustion	Bulky items, cardboard (depending on market value), glass (clear and mixed), aluminum cans, tin cans, other ferrous materials	Manual separation of bulky items, cardboard, and large ferrous items. Mechanical separation of glass, aluminum cans. Magnetic separation of tin cans and other ferrous materials not removed manually. Fuel preparation. Storage of fuel feedstock. Baling of cardboard for shipping. Storage of baled materials.
Preparation of MSW for use as a feedstock for composting	Bulky items, cardboard (depending on market value), plastics (PETE, HDPE, and other mixed plastic), glass (clear and mixed), aluminum cans, tin cans, other ferrous materials	Manual separation of bulky items, cardboard, plastics, glass by color, aluminum cans, and large ferrous items. Magnetic separation of tin cans and other ferrous materials not removed manually. Baling of separated materials for shipping. Storage of baled materials. Storage of compost feedstock.
Selective recovery of recyclable materials	Bulky items, office paper, old telephone books, aluminum cans, PETE and HDPE, and ferrous materials. Other materials depending on local markets	Manual separation of bulky items, cardboard. Manual separation of selected materials depending on market demands. Baling facilities, can crushers, and other equipment depending on the materials to be separated.

8.16

- Identification of the characteristics of the waste materials to be processed
- Consideration of the specifications for recovered materials now and in the future
- Available types of equipment and facilities

For example, specific waste materials cannot be separated effectively from commingled MSW unless bulky items (e.g., lumber, white goods, and large pieces of cardboard) are first removed and the plastic bags in which waste materials are placed are opened and the contents exposed. The specifications for the recovered material will affect the degree of separation to which the waste material is subjected.

The principal methods and types of processing equipment used for the separation of recyclable materials are reported in Table 8.6. The methods and types of processing equipment identified in Table 8.6 are considered in detail in Sec. 8.4. Some typical examples of process flow diagrams utilizing the methods and process equipment identified in Table 8.6 are illustrated subsequently in Figs. 8.14 through 8.17 and 8.19 through 8.21. It is also important to note that, depending on the form in which the material to be recycled is collected for recycling, a number of process flow diagrams may be employed in a single MRF.

TABLE 8.6   Typical Methods and Equipment Used for the Processing and the Recovery of Individual Waste Components from MSW*

Processing options	Description
Manual sorting	Unit operation in which personnel physically remove items from the waste stream. Typical examples include: (1) removal of bulky items that would interfere with other processes, and (2) sorting material off an elevated conveyor into large bins located below the conveyor.
Size reduction	Unit operation used for the reduction of both commingled MSW and recovered materials. Typical applications include: (1) hammermills for shredding commingled MSW; (2) shear shredders for use with commingled MSW and recycled materials such as aluminum, tires, and plastics; and (3) tub grinders used to process yard wastes.
Size separation	Unit operation in which materials are separated by size and shape characteristics, most commonly by the use of screens. Several types of screens are in common use including: (1) reciprocating screens for sizing shredded yard wastes, (2) trommel screens used for preparing commingled MSW prior to shredding, and (3) disc screens used for removing glass from shredded MSW.
Magnetic field separation	Unit operations in which ferrous (magnetic) materials are separated from nonmagnetic materials. A typical application is the separation of ferrous from nonferrous materials (e.g., tin from aluminum cans).
Densification (compaction)	Densification and compaction are unit operations that are used to increase the density of recovered materials to reduce transportation costs and simplify storage. Typical applications include: (1) the use of baling for cardboard, paper, plastics, and aluminum cans; and (2) the use of cubing and pelletizing for the production of densified RDF.
Materials handling	Unit operations used for the transport and storage of MSW and recovered materials. Typical applications include: (1) conveyors for the transport of MSW and recovered materials, (2) storage bins for recovered materials, and (3) rolling stock such as fork lifts, front-end loaders, and various types of trucks for the movement of MSW and recovered materials.
Automated sorting	Unit operation in which materials are separated by material characteristics. Typical examples include: (1) optical sorting of glass by color, (2) x-ray detection of PVC, and (3) infrared sorting of mixed resins.

* Additional details on the material presented in this table are given in Sec. 8.4 of this chapter.

Source: Adapted from Tchobanoglous et al. (1993).

Materials Recovery Rates

An MRF is a major component of an integrated waste management system. An MRF developed to process source-separated waste can be thought of as an extension of the recycling program. To predict the materials flow to the MRF, it is necessary to estimate the effectiveness or performance of the recycling program. The percentage of the various recyclable components that will be recovered with the waste collection system, the component capture factor at source, can be estimated as the product of three factors, as given by Eq. (8.1).

Component capture rate at the point of collection
$$= \text{[participation factor] [composition factor] [source recovery factor]} \quad (8.1)$$

where: participation factor = fraction of the public that participates in a recycling program
composition factor = fraction of waste component in total waste
source recovery factor = fraction of material recovered at the source

The overall component recovery factor after processing at an MRF, as defined in Eq. (8.2), can be estimated by accounting for the recovery efficiency of the separation processes at the facility.

Component recovery factor at the MRF
$$= \text{[participation factor] [composition factor]}$$
$$\times \text{[source recovery factor] [MRF recovery factor]} \quad (8.2)$$

where MRF recovery factor = fraction of material recovered at the MRF. Other terms are as defined previously.

Participation factors will vary with the type of recycling program and long-term education. In communities where recycling has been well established, participation rates as high as 80 percent have been achieved. Composition factors are measured in waste composition studies. Typical data are given in Chap. 5. Typical source component recovery factors for the recyclable materials collected in source separation recycling programs are reported in Table 8.7. Source recovery factors for MRFs will depend on the type of MRF (source-separated or commingled waste). For commingled wastes, the participation factor and the source recovery factor equal 1. Typical component recovery values for MRFs are given in Table 8.8.

TABLE 8.7   Recovery Factors for Source-Separated Recycled Materials at the Point of Collection

Material	Percent recovery	
	Range	Typical
Mixed paper	40 to 60	50
Cardboard	25 to 40	30
HDPE	70 to 90	80
PET	70 to 90	80
Mixed plastics	30 to 70	50
Glass	50 to 80	65
Tin cans	70 to 85	80
Aluminum cans	85 to 95	90

Source: Adapted in part from Tchobanoglous et al. (1993).

TABLE 8.8   Recovery Factors for Source-Separated Recycled Materials at an MRF

Material	Percent recovery			
	Manual sorting of source-separated waste		Machine sorting of commingled MSW	
	Range	Typical	Range	Typical
Mixed paper	60–95	90		
Cardboard	60–95	90		
HDPE	80–95	90		
PET	80–95	90		
Mixed plastics	80–98	90		
Glass (overall)	80–98	90	50–90	80
Amber glass				
Clear glass				
Green glass				
Tin cans	80–95	90	65–95	85
Aluminum cans	85–95	90	60–90	75
Light fraction*			80–95	
Heavy fraction†			90–98	

* Varying amounts of the light fraction will be retained with the heavy fraction.

† Varying amounts of the heavy fraction will be carried over with the light fraction.

Source: Adapted from Tchobanoglous et al. (1993).

Materials Balances and Loading Rates

One of the most critical elements in the design and selection of equipment for MRFs is the preparation of a materials balance analysis to determine the quantities of materials that can be recovered and the appropriate loading rates for the unit operations and processes used in the MRF. The steps involved in the preparation of a materials balance analysis and in determining the required process loading rates are as follows.

Step 1.   The first step in performing a materials balance analysis is to define the system boundary. The system boundary can be drawn around the entire MRF or around an individual unit operation (e.g., manual separation) within the MRF. In some cases, it is appropriate to draw the boundary around the community and to account for waste diversions that may occur prior to the MSW being delivered to the MRF. An aluminum can buyback program is an example of such a diversion.

Step 2.   The second step is to identify all of the waste or material flows that enter or leave the system boundary (e.g., an MRF or a unit processing operation) and the amount of material stored within the system boundary. Typically, for an MRF these can include MSW; source-separated materials; processed materials; nonrecyclable wastes to be landfilled; crushed glass; and baled paper, cardboard, plastics, and aluminum and tin cans.

Step 3.   The third step involves the application of the materials balance concept to the processes occurring within the system boundary. The materials mass balance can be formulated as follows:

1. General word statement:

   Rate of accumulation of material within the system boundary

        = [rate of flow of material into the system boundary]

        − [rate of flow of material out of the system boundary]

        + [rate of generation of waste material within the system boundary]   (8.3)

2. Simplified word statement:

$$\text{Accumulation} = \text{inflow} - \text{outflow} + \text{generation} \tag{8.4}$$

3. Symbolic representation (refer to Fig. 8.3):

$$\frac{dM}{dt} = \sum M_{\text{in}} - \sum M_{\text{out}} + r_w \tag{8.5}$$

where dM/dt = rate of change of the weight of material stored (accumulated) within the study unit, lb/d

$\sum M_{\text{in}}$ = sum of all of the material flowing into study unit, lb/d

$\sum M_{\text{out}}$ = sum of all of the material flowing out of study unit, lb/d

r_w = rate of waste generation, lb/d

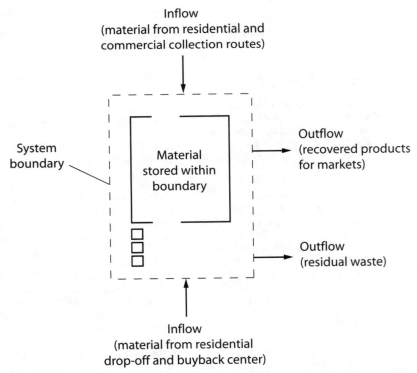

FIGURE 8.3   Definition sketch for materials balance analysis used to determine quantities of material that can be recovered and process loading rates.

In some biological transformation processes (e.g., composting), the weight of organic matter will be reduced and, therefore, the term r_w will be negative. In writing the mass balance equation, the rate term should always be written as a positive term. The correct sign for the term will be added when the appropriate rate expression is substituted for r_w. The analytical procedures used for the solution of mass balance equations usually are governed by the mathematical form of the final expression. Computationally, a materials balance is most easily accomplished by the use of a spreadsheet program. If no material is stored and no waste generation (or loss) is involved, Eq. (8.5) reduces to the following simple equation:

$$\sum M_{in} = \sum M_{out} \qquad (8.6)$$

Step 4.   The fourth and final step is to develop materials loading rates for the individual operations and processing steps in the MRF using the data from the materials balance analysis. Generally, MSW or source-separated materials delivered to the MRF are expressed in terms of tons per day (ton/d). Unit operations such as conveyors or screens must be specified in terms of tons per hour (ton/h), so the tons-per-day rate must be converted into tons per hour, taking into account the effective working day. The hourly loading (or processing) rate is given by the following expression:

$$\text{Loading rate} = \frac{(\text{ton/d})}{(\text{processing h/d})} \qquad (8.7)$$

Typically, separation processes at MRFs with manual sorting will be operational for 6 h/d where one nominal 8-h shift per day is used. Mechanized MRFs are sometimes designed for 16 h/d effective operation to maximize the utilization of expensive equipment. To allow for scheduled and unscheduled equipment downtimes, some designers suggest that the base loading rate of the facility should be increased by about 10 to 15 percent.

Layout and Design of MRFs

The layout and design of the physical facilities that make up the processing facilities will depend on the types and amounts of materials to be processed. Important factors that must be considered in the layout and design of such systems include:

- Consideration of the methods and means by which the wastes will be delivered to the facility
- Estimation of materials delivery rates
- Definition of the materials loading rates
- Development of materials flow and handling patterns within the MRF facility
- Development of performance criteria for the selection of equipment and facilities
- Careful consideration of space requirements for maintenance and repair

Because there are so many combinations in which the separation processes can be grouped, it is extremely important to view as many operating facilities as possible before settling on a final design. Some typical layouts are detailed in the following discussion of different types of MRFs.

Typical MRF for Source-Separated Wastes

In the following discussion, a typical MRF for source-separated material, owned and operated by Davis Waste Removal, Davis, CA, is described in detail to provide a more comprehensive view of the design, layout, and operation of such a facility.

FIGURE 8.4 Collection vehicle used for the collection of residential source-separated materials. The first compartment behind the cab is for mixed plastics and cans, the middle compartment is for glass, and the last compartment is for mixed paper. The bubble behind the last compartment is for the collection of flattened and bundled cardboard.

MRF Characteristics. The materials to be processed include:

- Mixed paper
- Glass
- Mixed plastic and aluminum and tin cans
- Cardboard from residential and commercial sources.

The collection vehicle used for the collection of the separated wastes from residential sources is shown in Fig. 8.4. A buyback center for recyclable materials and an oil collection facility (see Fig. 8.5) where community residents can bring used motor oil and oil filters are also part of the MRF. In addition, containers for recyclable materials, shown in Fig. 8.6, are available on a 24-h basis for residents to bring in waste materials for recycling.

The following quantities of source-separated material are received at the MRF during each weekday:

Material	Amount, ton/d
Paper	20.0
Glass	2.0
Mixed recyclables	5.5
Cardboard	6.5
Total	35.0

FIGURE 8.5 Drop-off facility for the recovery of used oil (note containers for the collection of used oil filters, and secondary containment berm in case of accidental spill).

FIGURE 8.6 Example of drop-off center for the 24-h recovery of recyclables brought by community residents. Drop-off bins are located at entrance to MRF (see Figure 8.8).

MRF Process Flow Diagrams Process and Layout.   The process flow diagrams for the MRF are illustrated in Fig. 8.7. As shown, there are four separate processing and handling activities associated with

1. Glass
2. Mixed waste comprising plastic bottles (HDPE and PET), tin cans, and aluminum cans
3. Mixed paper
4. Cardboard

The process flow diagrams will be described in detail in the discussion of the MRF operation.

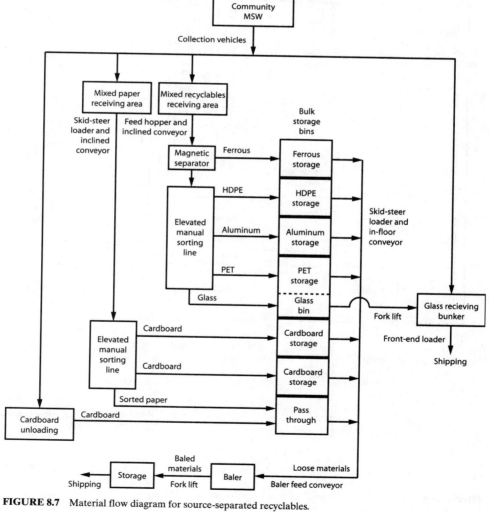

FIGURE 8.7   Material flow diagram for source-separated recyclables.

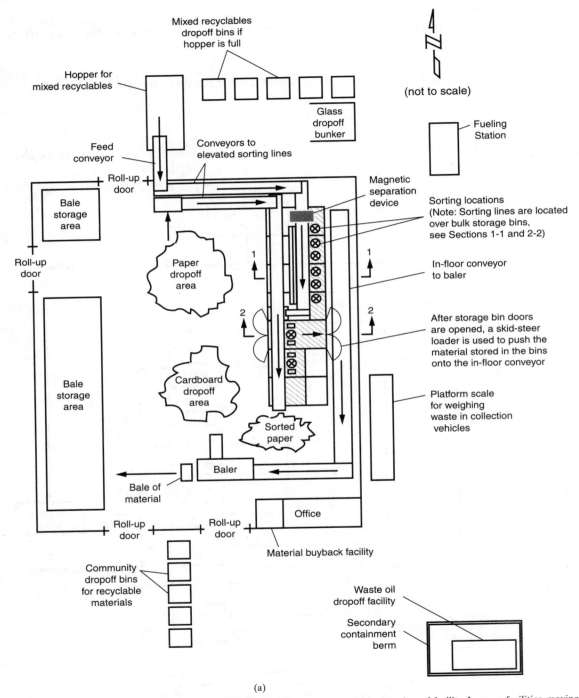

(a)

FIGURE 8.8 Layout of MRF for processing of source-separated materials: (*a*) plan view of facility. In some facilities, moving floors are used to transfer the stored material to the cross conveyor leading to the baler.

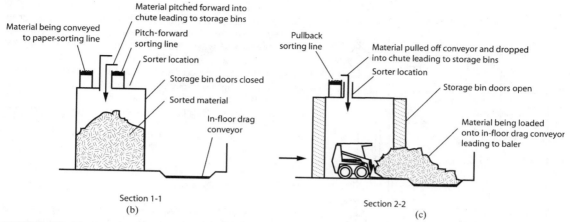

FIGURE 8.8 (*Continued*) Layout of MRF for processing of source-separated materials: (*b*) section through mixed-paper conveyor and mixed-recyclables sorting line, and (*c*) section though mixed-paper sorting line (with storage bin screen doors in open position for emptying contents onto in-floor conveyor). In some facilities, moving floors are used to transfer the stored material to the cross conveyor leading to the baler.

The layout of the MRF is shown in Fig. 8.8. The key features on the outside of the MRF are

- Containers for recyclable materials, located in the front of the MRF
- Oil recycling center locater at the southeast corner
- Platform scale located on the east side of the MRF
- Glass storage bunker located behind the northeast corner of the building
- Temporary storage containers located directly behind the MRF
- Mixed-waste receiving hopper, also located behind the MRF building

Vehicle Flow Diagram.   The process flow diagram for the vehicles used for the collection of source-separated waste on delivering waste to the MRF is shown in Figs. 8.9 and 8.10. Immediately after entering the MRF grounds, the collection vehicle is driven to the electronic platform scale located on the east side of the MRF, where it is weighed to determine the amount of waste entering the facility (see Fig. 8.11*a*). Because the tare weight on each vehicle is known, the determination of the waste weight is direct. The next stop for the collection is at the glass bunker storage area, located at the northeast end of the MRF, where glass is unloaded (see Fig. 8.11*b*). Glass is unloaded into the outside bunker for storage and subsequent shipping. After unloading the glass, the collection vehicle is driven to the outside mixed-waste storage hopper where mixed waste comprising aluminum cans, tin cans, and plastic containers (HDPE and PET) is unloaded (see Fig. 8.11*c*). In the event the receiving hopper is full, the mixed waste is unloaded into temporary storage bins (see Fig. 8.11*d*). Once the glass and mixed recyclables have been unloaded, the vehicle is driven into the MRF, and paper is unloaded near an in-floor conveyor system, from the rear compartment of the collection vehicle (see Fig. 8.11*e*). Once the paper is unloaded, the collection vehicle is driven forward where the cardboard is unloaded from the specially designed bubble compartment located at the back of the collection vehicle (see Fig. 8.11*f*). The last step is to either park the collection vehicle in the corporation yard, located adjacent to the MRF, or to proceed to another collection route.

Operation of the MRF.   Glass discharged to the outside storage bunker is currently not sorted because it has insufficient economic value. Glass is recycled to meet the mandatory diversion requirements (50 percent by 2000). The mixed wastes are transported from the out-

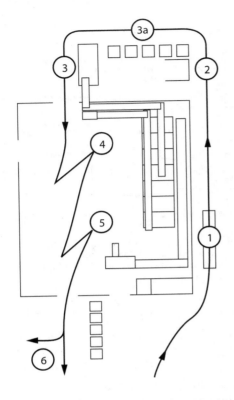

FIGURE 8.9 Collection vehicle route while dropping off recovered source-separated materials, highlighting the six basic unloading procedures: (1) collection vehicle containing collected material for recycling is weighed; (2) glass is unloaded into outside storage bunker; (3) mixed waste is unloaded into feed hopper, or into bins (3a) if feed hopper is full; (4) paper is unloaded onto tipping floor; (5) cardboard is unloaded onto tipping floor; and (6) collection vehicle continues to new route or to corporation yard.

side hopper into the MRF by an elevated conveyor (see Fig. 8.12a). Inside the MRF, the mixed wastes are transferred from the outside conveyor to another elevated conveyor that leads to the mixed-waste sorting line located above a series of storage bins (see Fig. 8.12b). A pulley magnet located at the top of the inclined conveyor is used to separate ferrous materials as the mixed waste is discharged to the sorting belt. The mixed-waste sorting conveyor is shown in Fig. 8.12c. The sorters use a positive pitch-forward technique to separate

- Any ferrous metal not removed by the pulley magnet
- HDPE plastic
- Aluminum cans
- PET plastic

Any glass that may have been placed by mistake with the mixed waste is left on the belt where it falls into a chute at the end of the belt. The positively sorted materials are pitched forward into chutes (see Fig. 8.12c) that lead to the storage bins located below the sorting platform. In addition, all of the sorters also remove other contaminants from the moving conveyor. The contaminants are placed in barrels located next to sorters. Periodically, when the storage bins are full, the wire-mesh doors are opened on either side of the storage bin, and the waste falls onto the in-floor conveyor, which leads to the baler (see Figs. 8.10 and 8.12d). Waste that falls on the floor is pushed with a bobcat onto the conveyor through the storage bin. The baled material is stored inside on the west side of the MRF.

The source-separated paper that was unloaded onto the MRF floor is pushed with a bobcat onto the inclined conveyor leading to the paper sorting conveyor (see Fig. 8.13a). A large rotating drum with protruding tines, located at the top of the conveyor, is used to keep the paper loading onto the paper sorting belt at a more-or-less uniform rate. Waste paper is sorted to remove contaminants using a pullback method of sorting. Cardboard and contaminants such as plastic film are removed. Cardboard removed from the paper is pulled back and dropped into a chute (see Fig. 8.13b) leading to two storage bins located beneath the sorting line. The negatively sorted paper drops off the end of the conveyor onto the MRF floor. Periodically, the sorted paper is pushed through the pass-through storage bin onto the in-floor conveyor leading to the baler (see Figs. 8.10 and 8.13c).

Source-separated cardboard deposited on the MRF floor is also pushed through the pass-through storage bin to be baled. Similarly, the cardboard removed from the paper is also baled by opening the wire-mesh doors and pushing the accumulated material onto the conveyor leading to the baler. The baled paper and cardboard material is also stored within the MRF (see Fig. 8.13d) to avoid the UV deterioration, which occurs if these materials are stored outside.

Buyback Center. The MRF also serves as a buyback center for aluminum cans, plastic, glass, and newsprint. Operationally, recyclable materials brought in by residents are unloaded and weighed, and the person bringing in the material is paid based on the weight of the material. Prices paid for recyclable materials as of January 1, 2002, are listed in Table 8.9. In some buyback centers, electronic scales are used, and the person bringing in the recyclable materials is given a printout along with being paid for the returned materials.

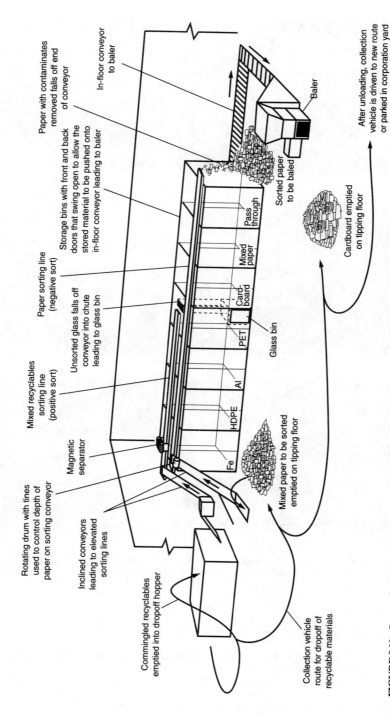

FIGURE 8.10 Perspective view of vehicle path and material flow in a typical MRF for source-separated materials.

Rotating drum with tines used to control depth of paper on sorting conveyor

Inclined conveyors leading to elevated sorting lines

Magnetic separator

Mixed recyclables sorting line (positive sort)

Unsorted glass falls off conveyor into chute leading to glass bin

Paper sorting line (negative sort)

Storage bins with front and back doors that swing open to allow the stored material to be pushed onto in-floor conveyor leading to baler

Paper with contaminates removed falls off end of conveyor

In-floor conveyor to baler

Fe

HDPE

Al

PET

Card-board

Mixed paper

Pass through

Glass bin

Sorted paper to be baled

Baler

Mixed paper to be sorted emptied on tipping floor

Cardboard emptied on tipping floor

After unloading, collection vehicle is driven to new route or parked in corporation yard

Commingled recyclables emptied into dropoff hopper

Collection vehicle route for dropoff of recyclable materials

FIGURE 8.11 Procedures associated with vehicle unloading after collection of source-separated materials: (*a*) weighing of vehicle contents, (*b*) glass unloading, (*c*) mixed waste (plastics and cans) unloading into feed hopper, (*d*) mixed waste unloading into overflow bins if hopper is full, (*e*) unloading of mixed paper, and (*f*) unloading of cardboard.

Automated Separation for Source-Separated Wastes

While not common, some MRFs utilize automated processes for the separation of source-separated wastes. Automated systems utilize a combination of sensors and computer processors to differentiate between materials in a specified feed stream. As shown in Fig. 8.14, processes that utilize sensors generally require a high degree of preprocessing and monitoring to achieve effective material separation (i.e., low product contamination). Two systems of

FIGURE 8.12 Steps in processing of source-separated wastes: (*a*) inclined conveyors used to transport recyclable materials to elevated sorting lines, (*b*) elevated sorting lines are located over bulk material storage bins, (*c*) pitch-forward chutes leading to bulk material storage bins, and (*d*) recovered material being conveyed from bulk storage to baler.

automation, *binary* and *array,* are shown in Fig. 8.14*a, b,* respectively. The binary system is used to separate the process feed stream into two components; the array system utilizes multiple sensors to separate the material in the feed stream into more components.

Automated systems are expensive and require skilled mechanics to perform maintenance activities. While the use of automated separation is somewhat limited, in some cases it can be incorporated into existing MRFs. Efficient operation of automated systems requires long operating times to compensate for the capital expenditure, adequate preprocessing, effective equipment maintenance, and personnel for quality control purposes.

Typical MRFs for Commingled Wastes

The separation of waste components from commingled wastes and their processing is a necessary operation in the recovery of materials for direct reuse and recycling and for the production of a feedstock that can be used for the recovery of energy and/or the production of compost. The purpose of this section is to illustrate how the unit operations and facilities identified in Table 8.6 can be grouped together to achieve the separation of materials from commingled MSW. An MRF designed to process commingled construction and demolition wastes is also described.

FIGURE 8.13   Steps in processing source-separated paper: (*a*) mixed paper being pushed onto inclined conveyor using skid steer loader, (*b*) elevated pull-back sorting belt and chutes leading to bulk material storage bins located below, (*c*) baler with paper bales being produced, and (*d*) paper bales being stored within the facility.

MRF for Recovery of Materials from Commingled MSW.   Recognizing that meeting mandated waste diversion goals with source separation programs alone will be difficult, many communities have developed plans for MRFs that can be used to both separate materials from commingled MSW and to process materials from source separation programs. A typical process flow diagram for an MRF employing manual and mechanical separation of materials from commingled MSW and manual separation of source-separated wastes is illustrated in Fig. 8.15. The layout of an MRF for the processing of commingled waste is shown in Fig. 8.16. Commingled MSW from residential and other sources is discharged in the receiving area. Recyclable, reusable, and oversized materials (e.g., cardboard, lumber, white goods, and broken furniture) are removed in the first-stage presorting operation before the commingled waste is loaded onto an inclined conveyor. Source-separated materials in see-through plastic bags would also be removed from the commingled MSW. Additional cardboard and large items are handpicked from the conveyor at the second-stage presorting station as the waste material is transported to the bag-breaking station. The next step involves breaking open the plastic bags, which can be accomplished either manually or mechanically (see discussion in following section).

The next step in the process involves the first stage of manual separation of specific waste materials. Materials typically removed include paper, cardboard, all types of plastic, glass, and

TABLE 8.9 Buyback Prices for Recycled Materials at an MRF as of January 1, 2002

Material	Fully segregated, $/lb	Commingled, $/lb
Aluminum	0.760	0.74
Glass	0.051	0.040
Bimetal	0.16	Segregated only
Plastic		
PET	0.41	0.35
HDPE	0.25	0.15
PVC	0.26	Segregated only
LDPE	0.88	Segregated only
PP	0.52	Segregated only
PS	1.74	Segregated only
Other	0.16	Segregated only

metals. In some operations, separate types of plastic are separated simultaneously. Mixed plastics are usually separated by type in a secondary separation process. Material remaining on the conveyor is discharged into a trommel (or disc screen) for size separation. The oversized material is sorted manually a second time (second-stage sorting). Commingled source-separated materials (collected separately from residential and commercial sources) and the source-separated materials contained in see-through bags (removed from the commingled MSW in the first-stage presorting operation) are also sorted using the second-stage sorting line. Source-separated mixed paper and cardboard would be processed separately using a process flow diagram and recovery system such as those given in Figs. 8.17 and 8.18, respectively. It should be noted that the first- and second-stage sorting activities would normally be carried out in an air-conditioned facility. Depending on the extent of the first- and second-stage sorting operations, the undersized material from the trommel and the material remaining after the second-stage sorting operation is either hauled away for disposal in a landfill or processed further and combusted or used to produce compost to be used as intermediate landfill cover. As shown in Fig. 8.15, further processing of the residual materials usually involves shredding and magnetic separation. The results of a detailed materials balance analysis for the MRF, described previously, are summarized in bold type on Fig. 8.15.

The following excerpt from a text published in 1921 (Hering and Greeley, 1921), provides a historical perspective on current materials separation activities at MRFs.

The most developed case of sorting refuse in Europe is at Puchheim, a suburb of Munich, where the refuse from a population of more than 600,000 is picked over and finally disposed of. First, the finer materials and dust are sifted out on a moving and vibrating belt, and the bulky salable articles are picked out. In the adjoining room, about 40 women stand on each side of the belt, each one picking out a designated material and throwing it into a designated wire basket. The substances thus removed are chiefly: Paper, white and green glass, rags, leather, bones, tinned cans, iron, brass, copper, tin, etc. The bones are treated with benzine, and, on the premises, are converted into grease, glue, bone meal, or charcoal. Garbage is cleaned, sterilized, and fed to hogs in an adjoining building. Paper is freed from dust, pressed into bales, and utilized for the manufacture of pasteboard. Wood is burned under the boilers. Bottles are cleaned, disinfected, and sold. Tinned cans are sold as iron. No one enters the works until after donning working clothes, nor leaves them until after a good wash or bath. The working rooms are washed twice a day with dilute carbolic acid. It is reported by De Fodor that this very effective sorting contains the germ of faulty economics, in the fact that the total revenue hardly covers three-quarters of the necessary expenditure.

With the exception that many modern sorting facilities are located in air-conditioned facilities, the similarities are striking. The economic issue remains the same today, but environmental costs were not generally considered in the 1920s.

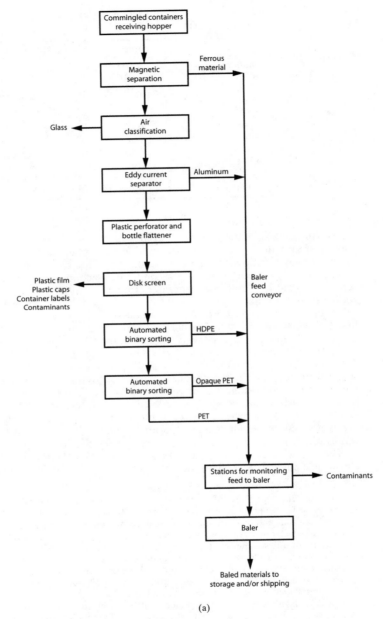

(a)

FIGURE 8.14 Materials flow diagrams for separation processes that include automated sorting systems: (*a*) binary sorting.

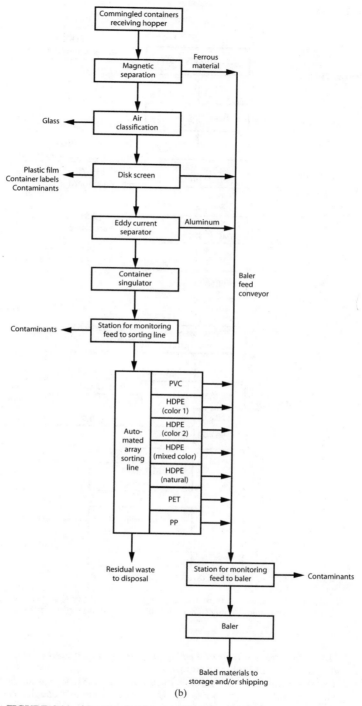

FIGURE 8.14 (*Continued*) Materials flow diagrams for separation processes that include automated sorting systems: (*b*) array sorting.

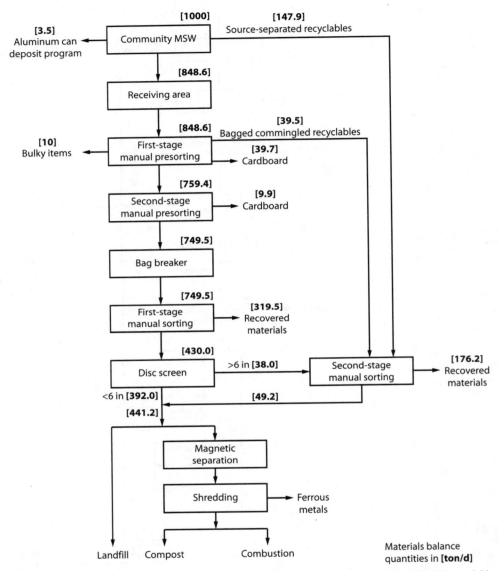

FIGURE 8.15 Materials flow diagram for an MRF processing source-separated and commingled recyclable materials. (Note: boldface numbers represent typical material recovery rates.)

MRFs for Preparation of Feedstock from Commingled MSW.

The separation of commingled MSW in a highly mechanized system is illustrated in Fig. 8.19. As shown in both process flow diagrams given in Fig. 8.19, the commingled MSW is first discharged in the receiving area where lumber, white goods, and oversized items are usually removed manually before the material is loaded onto the first conveyor. In Fig. 8.19a, the commingled MSW is shredded as the first step in the process. Air classification is then used to recover the mainly organic fraction of the MSW. The corresponding mass balance quantities for the MRF shown in Fig. 8.19a

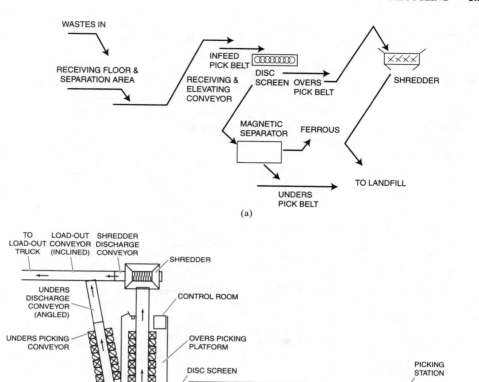

FIGURE 8.16 Typical example of the development and layout of a MRF for the processing of source-separated and commingled recyclables: (*a*) basic process flow diagram, and (*b*) layout of physical facilities.   (*Courtesy Brown and Caldwell Engineers.*)

are summarized in Fig. 8.20. In Fig. 8.19*b*, a trommel is used to achieve a better separation of the organic fraction of the MSW and to remove small contaminants more effectively. The flow diagrams shown in Fig. 8.19 represent two of the many different approaches that have been, and continue to be, used for the mechanical separation of waste components from commingled MSW for the production of a feedstock for the production of energy. The mainly organic fraction of MSW remaining after processing is known as fluff *refuse-derived fuel,* commonly known as RDF. In some operations, the mainly organic fraction is used to produce a densified refuse-derived fuel known as d-RDF.

Flow diagrams similar to those shown in Fig. 8.19 have also been used for the preprocessing of MSW for the production of compost. Unfortunately, shredding the commingled MSW before metal objects and other contaminants have been removed has resulted in the production of poor-quality compost with respect to contaminants. Because of the serious problems associated with the production of poor-quality compost, many communities have developed MRF process flow diagrams, similar to the one given in Fig. 8.15 for the production of feed-

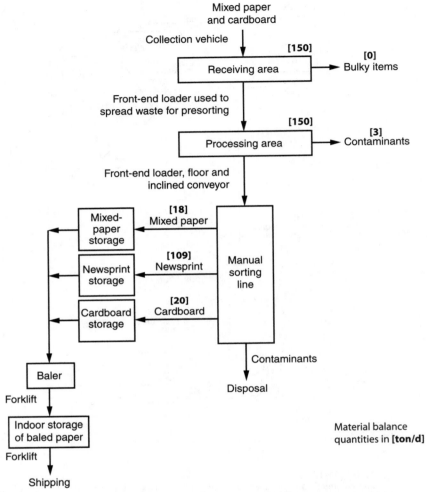

FIGURE 8.17 Materials flow diagram for a mixed paper and cardboard sorting system (see also Fig. 8.18). (Note: boldface numbers represent typical material recovery rates.)

stock for composting. The sorting conveyor and disk or trommel screen are used to remove plastics, glass, aluminum and tin cans, and other contaminants before the waste is shredded to reduce the particle size for composting.

MRF for Commingled Construction and Demolition Wastes.   A typical process flow diagram for processing commingled construction and demolition wastes is shown in Fig. 8.21*a*. The commingled wastes are brought to the site and dumped in an open area, spread out, and all of the wood and metal is removed manually (see Fig. 8.21*b*). The wood is taken to a large wood grinder where it is converted to wood chips. After the wood and metal have been removed, the waste is picked up with a front-end loader and discharged onto a two-stage vibrating screen (see Fig. 8.21*c*). The first screen is used to eliminate large pieces of concrete, roots, and similar materials. The second screen, located immediately below the first screen, is

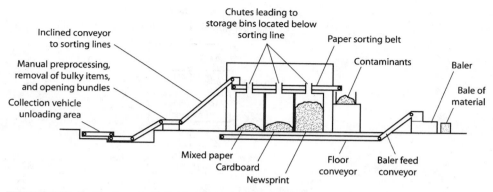

FIGURE 8.18 Profile view of a facility for separation of mixed paper and cardboard.

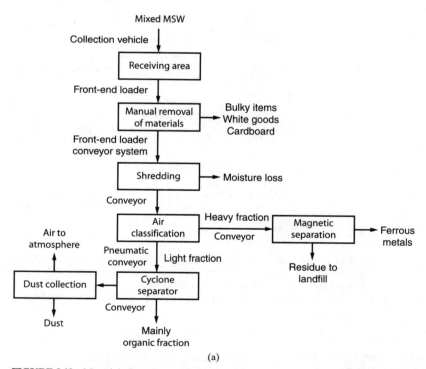

(a)

FIGURE 8.19 Materials flow diagram for mixed MSW processing for the recovery of waste components: (*a*) conventional flow diagram with shredder.

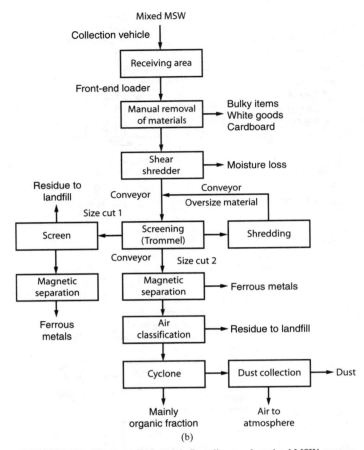

FIGURE 8.19 (*Continued*) Materials flow diagram for mixed MSW processing for the recovery of waste components: (*b*) with shear shredder and trommel screen.

used to remove finer pieces of broken concrete and other smaller-sized contaminants. The fine material passing through the two screens is then conveyed to a second vibrating screen, where additional fine contaminants are removed. The final product is stockpiled for sale as fill material. The material removed by the screens is stockpiled and eventually hauled to the landfill for disposal.

8.4 UNIT OPERATIONS AND EQUIPMENT FOR PROCESSING OF RECYCLABLES

To meet mandated waste diversion goals, various unit operations and equipment are utilized to separate waste components into relatively homogeneous groups. To make recycling operations cost-effective, the costs of buying, operating, and maintaining recycling equipment, operations, and facilities must be balanced with the value of the materials recovered. Specifications and

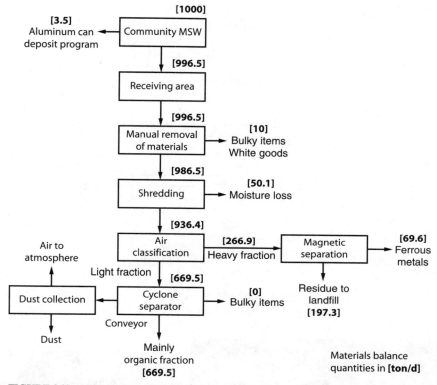

FIGURE 8.20   Materials flow diagram from Fig. 8.19a, with corresponding mass balance results for material loading and recovery shown in boldface type.

descriptions for the equipment and unit operations identified in the process flow diagrams (presented in Sec. 8.3) are provided in the following discussion.

The principal unit operations and equipment employed in processing materials at MRFs, as reported in Table 8.10, include the following:

- Manual sorting facilities
- Equipment and facilities for materials transport
- Equipment for size reduction
- Equipment for component separation
- Equipment for densification
- Weighing facilities
- Movable equipment
- Storage facilities

For the most part, the unit operations and equipment used at most MRFs are similar, whether the materials to be processed or separated are obtained from source separation, commingled collection, or mixed-waste collection programs. The purpose of this section is as follows:

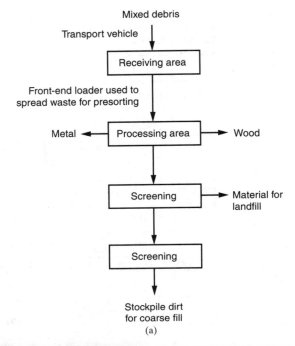

Mixed debris

Transport vehicle

Receiving area

Front-end loader used to
spread waste for presorting

Metal ← Processing area → Wood

Screening → Material for
landfill

Screening

Stockpile dirt
for coarse fill

(a)

(b)   (c)

FIGURE 8.21   Views of a MRF for construction and demolition wastes: (*a*) typical materials flow diagram, (*b*) waste spread on ground where wood is removed manually, and (*c*) waste being screened to produce usable product.   [*Figures (b) and (c) from Tchobanoglous et al., 1993.*]

- To introduce the basic unit operations and equipment used at MRFs
- To provide some specifications for some of the equipment available
- To present guidelines regarding the selection and operation of these unit operations and associated equipment

A directory and buyers' guide for manufacturers and distributors of recycling and waste management equipment is published annually by *Waste News* (www.wastenews.com). Information

TABLE 8.10   Unit Operations for the Processing of MSW

Item	Function/material processed	Preprocessing
Manual separation	Separation of waste components/all types of waste	
Materials transport		
Conveyors	Movement of waste materials from one location to another/all types of waste	Removal of large bulky items, removal of stringy materials
Wheeled equipment	Movement of waste materials from one location to another/all types of waste typically in containers	
Size reduction		
Hammermills	Size reduction/all types of wastes	Removal of large bulky items, removal of contaminants
Flail mills	Size reduction, also used as bag breaker/all types of wastes	Removal of large bulky items, removal of contaminants
Shear shredder	Size reduction, also used as a bag breaker/all types of wastes	Removal of large bulky items, removal of contaminants
Glass crushers	Size reduction/all types of glass	Removal of all nonglass materials
Wood grinders	Size reduction/yard trimmings/all types of wood wastes	Removal of large bulky items, removal of contaminants
Separation processes		
Screening	Separation of over- and undersized material; trommel also used as bag breaker/all types of waste	Removal of large bulky items, large pieces of cardboard
Cyclone separator	Separation of light combustible materials from airstream/prepared waste	Material is removed from airstream containing light combustible materials
Air classifiers	Separation of light combustible materials from airstream	Removal of large bulky items, large pieces of cardboard, shredding of waste
Eddy current	Separation of nonferrous conductors (aluminum) from nonconductors (wood, plastic)	Removal of ferrous materials, bulky items, and overburden
Magnetic separation	Separation of ferrous metal from commingled wastes	Removal of large bulky items, large pieces of cardboard, shredding of waste
Sensors	Separation of plastic by resin type, color, etc.; also separation of glass by color, contaminants	Separated feed stream (e.g., mixed plastic containers)
Wet separation	Separation of glass and aluminum	Removal of large bulky items
Densification		
Balers	Compaction into bales/paper, cardboard, plastics, textiles, aluminum	Balers are used to bale separated components
Bale binding systems	Holds baled material together	Material compressed into bales; binding system may be automatic or manual
Can crushers	Compaction and flattening/aluminum and tin cans	Removal of large bulky items
Plastic perforators	Tines punch holes in plastic containers for improved baler performance	Separated feed stream
Weighing facilities		
Platform scales	Operational records	
Small scales	Operational records	

on the processing of recyclables can also be obtained by contacting Clean Washington Center (www.cwc.org). For additional details on recycling systems and recovery of specific materials, see Tchobanoglous et al. (1993) and Lund (2001).

Manual Sorting Facilities

While automated sorting systems become more sophisticated, the ability of humans to recognize and separate materials is unique. Most facilities depend, in some capacity, on laborers to hand-separate recyclable materials (see Fig. 8.22). As described in Table 8.11, several manual sorting techniques are commonly used for the removal of items from a moving conveyor, including the *pitch-forward* and the *pullback* methods. Manual separation may also occur on the tipping floor for removing cardboard or other materials, or for the removal of oversized or hazardous materials to protect downstream operations.

The most common system for the manual sorting of material is the elevated sorting belt with storage bins located beneath (see Figs. 8.10 and 8.18). Materials to be sorted are transported to the elevated sorting conveyor with an inclined conveyor. In this system, sorters are located at stations on one or both sides of the elevated, moving conveyor (see Table 8.12). Sorting stations should be placed so that sorters are not interfering with other sorters. In other

(a) (b) (c) (d)

FIGURE 8.22 Typical examples of the manual separation of recyclable materials at two turn-of-the-century MRFs and at two modern-day MRFs: (*a*) and (*b*), men sorting commingled materials, circa 1905 (from Parsons, 1906; also note absence of plastic material in waste materials that are being sorted); (*c*) sorting commingled mixed recyclable materials; and (*d*) sorting mixed glass.

TABLE 8.11   Manual Sorting Techniques and Facilities Employed at MRFs

Sorting method	Description	Item to be removed
Presorting	Materials to be sorted are examined for items that may interfere or damage downstream equipment or processes. Items are removed before downstream processing.	Bulky items (e.g., cardboard) Hazardous materials
Pitch-forward	Materials to be removed are picked off a moving conveyor and tossed or pitched forward into a collection hopper or bin.	Items that can be tossed forward accurately (e.g., aluminum, rigid plastics, ferrous metals, glass)
Pullback	Materials to be removed are picked off a moving conveyor and pulled toward the sorter and deposited into a collection hopper or bin.	Items that are not amenable to tossing because of aerodynamic properties (e.g., paper, plastic film)
Elevated platforms	Sorting conveyors are located above bulk storage bins. Manual (and automated) sorters remove selected materials from the mixed materials and deposit them into the bins.	All items to be recovered or removed from the feed stream
Positive sort	Materials to be recovered are removed from a mixed-waste stream.	Depends on characteristics of waste stream
Negative sort	Contaminants are removed from the material to be recovered.	Depends on characteristics of waste stream

words, if sorting occurs only on one side of the conveyor, sufficient spacing should be provided; if sorting occurs on both sides of a conveyor belt, the stations should be staggered.

Typically, each sorter is responsible for a specific material, and controls are in place so that the conveyor speed can be adjusted or stopped. In some cases, reducing the speed of the conveyor may result in a higher-quality product and not significantly reduce overall processing rates. Other factors that affect the efficiency of sorting operations include the skill and training of the sorter, the presentation of the materials to be sorted (e.g., burden depth, time-varying distribution of materials), and worker fatigue. Fatigue can be reduced by adjustment of environmental variables such as temperature control, lighting, and ventilation. Additional worker safety and MRF staffing requirements are discussed in Secs. 8.5 and 8.6, respectively, of this chapter.

TABLE 8.12   Characteristics of Manual Sorting Systems

Parameter	Unit	Value	
		Range	Typical
Belt speed	ft/s	0–1.6	0.5
Belt width			
Stations on one side	in	30–42	36
Stations on both sides	in	48–72	60
Belt height	in	36–42	42
Material depth	in	0–6	4

TABLE 8.13   Manual Sorting Rates for Commingled Materials from a Moving Belt

| | Tons/sorter · h | | Recovery | |
Materials	Range	Typical	efficiency, %	Remarks
Commingled MSW				
Residential and commercial	0.3–4	2.5	60–95	Recovery efficiency reduced at higher sorting rates
Commercial	0.4–6	3.0	70–95	
Source-separated materials				
Mixed paper	0.5–4	2.5	60–95	
Paper and cardboard	0.5–3	1.5	60–95	Two products
Mixed plastics	0.1–0.4	0.2	80–95	PETE and HDPE
Mixed glass and plastic	0.2–0.6	0.5	70–95	Two products
Glass	0.2–0.8	0.4	80–95	Clear, emerald, amber
Plastics, glass, aluminum	0.1–0.5	0.3	80–95	Four products

Source: Adapted from Tchobanoglous et al. (1993).

Materials are removed from the mixed-waste stream moving on the conveyor and pitched forward or pulled back and deposited into collection chutes. The collection chutes lead to bulk storage bins for the separated materials. Waste containers may also be present for the storage of contaminants. A summary of typical sorting rates and efficiencies for various materials is presented in Table 8.13.

Equipment and Facilities for Material Transport

Transport of waste materials between the various separation and processing operations requires a reliable and effective conveyor system. Conveyor systems include the horizontal and inclined belt, drag and auger style, pneumatic, vibration, and bucket elevators. Several conveyor types are shown in Fig. 8.23. Drive systems for conveyors include friction, chain, or vibratory motion. Various conveyor types are described in Table 8.14.

Inclined belt conveyors (see Fig. 8.24) are frequently used to transport waste from mixed-waste hoppers or the tipping floor to elevated materials sorting lines. Materials are then removed from the conveyor by manual or mechanical methods and deposited into their respective bins located below the conveyor. For manual separation processes, conveyors should be sized to enable sorters to comfortably reach items on the belt. Wide chain-driven belt conveyors with crossbars are often used for transporting material from the sorting bins, tipping floor, or other bulk storage, to a baler for compaction.

Several factors should be considered when selecting a conveyor system. Conveyors for solid waste are often subjected to extreme operating conditions, including particulate material, liquids, heavy loading, and loose string and wire. Some conveyors are trough shaped or have skirt boards to keep materials from falling from the belt during transport. In addition, the speed and capacity (see Fig. 8.25) of the conveyor will need to be coordinated with the processing requirements of subsequent processes. The *Conveyor Equipment Manufacturers Association Handbook* (CEMA, 2000) and the ASME/ANSI Standards can be used to guide for more detailed design specifications for conveyors.

Equipment for Size Reduction

Several types of size reduction equipment are utilized to rip, cut, tear, and pulverize commingled recyclables, liberating materials that are bound together so they can be separated from each other in downstream unit operations. Size reduction is also utilized to densify materials before shipping to reduce storage, handling, and transportation costs. The two classes of size

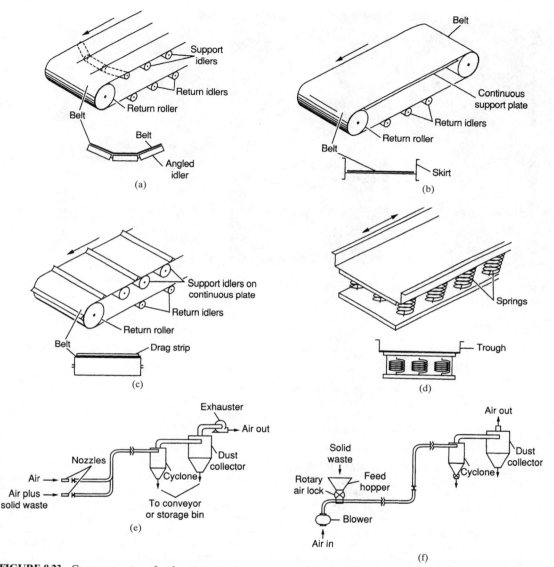

FIGURE 8.23 Conveyor systems for the transport of materials at MRFs: (*a*) trough-type belt conveyor; (*b*) belt conveyor with belt supported by continuous flat support plate; (*c*) belt conveyor with crossbars; (*d*) vibratory-type conveyor; (*e*) vacuum-type pneumatic conveyor; and (*f*) positive-pressure-type vacuum conveyor.

reduction equipment are distinguished as high-speed impact and high-torque shear action. Common size reduction equipment is summarized in Table 8.15.

High-Speed Impact Equipment. Hammermills, flail mills, and rotary grinders are used for processing various types of materials. These units operate at a relatively high rate of speed. Hammers or flails attached to a central rotating shaft collide with waste materials as the materials are fed into the chamber. High-speed impact size reduction systems and the size distribu-

(a)   (b)   (c)

FIGURE 8.24   Inclined belt conveyors used to transport recyclable materials: (*a*) mixed paper to elevated sorting platform, (*b*) HDPE plastics to baler, and (*c*) PET plastics to baler.

tion for common waste materials after processing with a hammermill are shown in Fig. 8.26. Some units may use components such as breaker plates, cutting bars, and grates to refine material processing action.

High-Torque Shear Equipment.   Shearing and shredding equipment are characterized by counterrotating blades that shear material (see Fig. 8.27). Size distribution characteristics of commingled waste after shear shredding are summarized in Table 8.16. These units generally operate at a lower rotation speed but with much higher torque and are usually driven by hydraulic power.

Selection of Size Reduction Equipment.   The selection of size reduction equipment is dependent primarily on the characteristics of the feed stream and the process needs of the size-reduced materials. Pulverizers and crushers are preferred where friable materials are being processed and size reduction can be accomplished by impact alone. Flail mills are used when materials require only coarse shredding and reduction to a specific particle size is not a factor. Hammermills are normally employed where coarse size reduction via cutting is required, and a wide-ranging particle size can be accepted with a controlled maximum particle size passing through the grate. Vertical-shaft hammermills are typically applied in place of a horizontal hammermill when the maximum particle size passing through the grate is not critical.

Pulverizers, flail mills, and hammermills tend to be noisy and are likely to generate dust during operations. They are also susceptible to explosions due to the presence of flammable materials and pressurized containers (e.g., aerosol cans, gas canisters, propane cylinders). Shear shredders are selected where the potential for explosions is high and the need to minimize dust generation is important. Because of its lower operating rotation speed, abrasion on the cutters of the shear shredder is less, and the horsepower requirement is typically lower than for a hammermill. Knife mills are employed where fine particle-sized discharge is required and tight control over size distribution is important. Care must be taken to ensure that difficult-to-shred items (metals, rocks, etc.) are removed before entering a knife mill to prevent damage to the knives.

TABLE 8.14 Typical Conveyor Systems Used in MRFs

Conveyor type	Description
Belt	Belts can travel over flat plates or idlers (rollers).
Friction	An endless rubber or synthetic belt is stretched between two pulleys. The pulley is connected to a drive mechanism and friction between the belt and pulley causes the belt to move.
Chain-driven	Chain-driven belts can be made of various materials, including hinged steel segments and steel-belted rubber. The belt is attached to a chain and is driven with a sprocket mechanism.
Horizontal	For waste-sorting (picking) lines in which mixed-waste materials to be sorted move with the conveyor while sorters standing over the moving material remove items to be segregated. Also to move material between different unit processes.
Inclined	Used to move waste material to elevated positions for various waste management processes, including feed chutes for horizontal balers and elevated sorting lines. Inclined conveyors often have crossbars to keep material from sliding down the conveyor.
Apron	Chain-driven belt composed of interlocking steel sections with upward-pointing side panels to keep material from falling off conveyor.
Trough	The edges of the conveyor belt are angled upward to keep material from falling over the edge of the conveyor.
Drag	Spaced scrapers are attached to an endless chain. As the chain and scrapers are dragged along a channel, loose materials are conveyed to a discharge point.
Vibrating	A spring- or hinge-mounted bed is oscillated in an eccentric motion, causing the material upon the bed to be transported in one direction.
Screw	A rotating auger in a narrow channel pushes loose materials forward.
Pneumatic	The transmission of materials using air; may use positive pressure or vacuum to convey material.
Bucket elevators	Buckets are mounted on chains or a belt. The bucket is loaded at the feed area and travels to the discharge point, empties its contents, and returns to obtain another load.

Equipment for Component Separation

Materials that are to be recycled need to meet certain purity criteria, depending on the requirements of the end use. Commingled waste, for example, contains many components that need to be removed from the bulk waste stream if they are to be reclaimed. The techniques used in sorting processes classify materials based on a property of the material or a characteristic of the item. The technologies discussed in this section include those that separate materials according to the following criteria:

- Size
- Magnetism
- Density
- Electrical conductivity
- Color

Size Separation.   The basis of size separation processes is that the dimensions of the items found in various waste streams can be used to differentiate between materials. The most common type of size separation process is screening, in which items are given the opportunity to

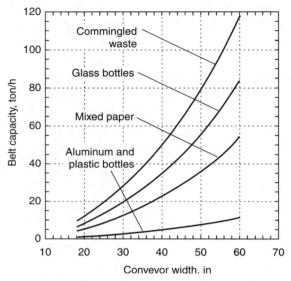

FIGURE 8.25 Typical performance curves of horizontal belt conveyor capacity for a belt traveling at 1.6 ft/s transporting various MSW components. (Note: belt capacity depends on belt style, speed, and degree of inclination.)

pass through (underflow) a certain size opening. The efficiency of a screen can be evaluated in terms of the percentage recovery of the material to be separated from the feed stream by the following expression:

$$\text{Recovery (\%)} = \frac{U \cdot w_u}{F \cdot w_f \cdot 100} \qquad (8.8)$$

where U = weight of material passing through screen (underflow)
F = weight of material fed to screen
w_u = weight fraction of material of desired size in underflow
w_f = weight fraction of material of desired size in feed

Screens operate at the highest efficiency when the materials to be separated are either much larger or much smaller than the screen opening. That is to say, when materials to be separated are of divergent sizes, the screening efficiency is high. Specifically, when a spherical particle of diameter d impinges on a square hole with side length (or round hole with diameter) of a, where $a > d$, the probability that it will pass through the hole is expressed as follows:

$$p = 2Q\left(1 - \frac{d}{a}\right) \qquad (8.9)$$

The quantity Q is the ratio of the area of the openings to the total screen surface area. As the ratio of d/a approaches 1, the probability of material falling through the screen approaches 0 (i.e., as the material to be screened approaches the size of the screen opening, the screening efficiency becomes very low and approaches 0). Meanwhile, very large materials flow on top of the screen as if it were a solid surface, and very fine materials fall through the openings of the screen with a very high efficiency or probability of passage.

TABLE 8.15 Grinding and Shredding Equipment for Size Reduction of MSW

Process	Description
High-speed impact	
Horizontal-shaft hammermill	Material is fed through a feed hopper to a hammer circle. The hammers, which are attached to a rotor or shaft, impact the infeed material, breaking it into smaller pieces. Below the hammer circle is a series of cast grates, the material remains inside the hammermill and is crushed or torn between the hammers and the grates, until its size is sufficiently reduced to pass through the grates, where it is discharged onto a belt or vibrating pan conveyor below. Some units allow the rotation of the hammers to be reversed, requiring less maintenance.
Vertical-shaft hammermill	The infeed material is placed in a chute that feeds into a breaker plate and hammer area. As the material is hammered, it works its way down a cone-shaped passage. The distance between the hammers and the breaker plate constantly decreases, thus continuously working to reduce particle size.
Vertical-shaft ring grinder	Similar in action to the hammermill, however, a gear-type device grinds the infeed material in place of the hammers. This grinding action is particularly good in densifying materials such as metal cans and tends to produce a nuggetized metal product of high density.
Flail mill	A flail mill is somewhat like a hammermill, but without grates. Material is fed into the top of the single- and double-shaft mills through a feed chute. The flails that are attached to a rotating shaft function as knives. Paper is torn and ripped, cans pass through the mill relatively unaffected, while glass is pulverized into very fine sizes. Because the flail mill does not have grates, it is not a good device for controlling particle size, especially the particle size of rags and other similar material that is hard to shred.
Pulverizer (glass crusher)	A pulverizer is much like a flail mill but utilizes a breaker plate and hammers rather than knives. These machines have impact bars and impact plates that assist in the pulverization of glass and other friable materials. As these materials fall into the mill, glass is struck by the hammers and thrown against the impact blocks where it is again smashed into smaller pieces.
High-torque shear	
Knife shredder (granulator)	A granulator or knife shredder employs very sharp, long knives for cutting materials such as rags and plastic bottles into small pieces for later separation. The knives are attached to a rotor and are positioned horizontally across the entire width of the shredder. As the shredder rotor rotates, the knives pass by an impact or cutting block at high speeds. Material caught between the impact block and the knife is cut. Hard materials such as glass and metal should not be fed into this type of equipment, as they would damage the knives. Granulators are typically used in plastics processing operations to reduce the particle size of bottles and increase density.
Rotary shear shredder	The rotary shear shredder is essentially a continuous rotary shear or scissor. Material is fed into counterrotating shafts with closely spaced cutters. This type of shredder tends to cut feed material into strips, which are the same dimension as the cutter width or spacing. The cutters are not circular, but rather oblong. Material passes down through openings that form between the tops of opposing cutters from opposite shafts. Hooks are positioned on each cutter to grab material that enters the mill and pull it into the shear where it is cut. Unlike the hammermills described earlier, the shear shredder operates at very slow speeds and does not pulverize glass or significantly reduce the size of cans, many of which, depending upon the size of the cutters, actually pass through the machine unaffected for a machine having 4-in cutters.

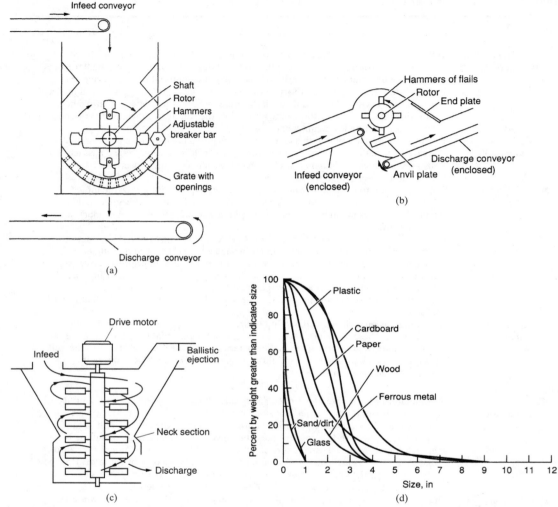

FIGURE 8.26 Types of high-speed impacting equipment used for size reduction of solid waste: (*a*) horizontal-shaft hammermill, (*b*) vertical-shaft hammermill with ballistic ejection, (*c*) horizontal-shaft flail mill, and (*d*) size distribution of various waste components after processing in a hammermill.   (*From Tchobanoglous et al., 1993.*)

The most common types of screens in recycling applications—the vibrating screen (see Fig 8.28), the rotary drum screen or trommel (see Fig. 8.29), and the disc screen (see Fig. 8.30)—are described in Table 8.17. Typical operating characteristics for a trommel screen are shown in Table 8.18. Factors to be considered in selection of screening equipment include the following:

- Particle size, particle size distribution, bulk density, moisture content, particle shape, and potential for the material to stick together or entangle
- Screen design characteristics, including materials of construction, size of screen openings, shape of screen openings, total surface screening area, rotational speed for rotary drum

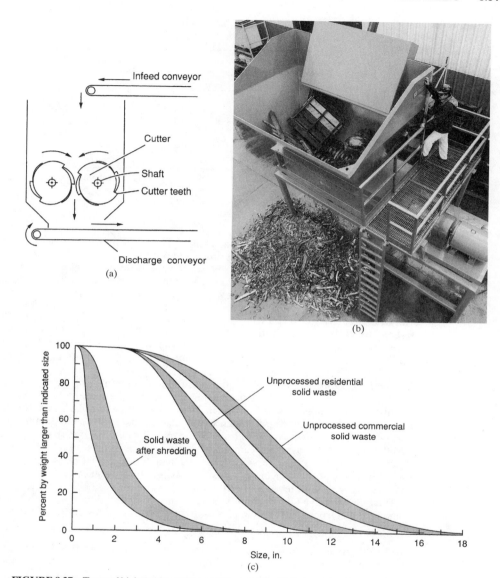

FIGURE 8.27 Types of high-torque shearing equipment used for size-reduction of solid waste: (*a*) shear shredder, (*b*) view of shear shredder used to reduce size of wood waste, and (*c*) typical distribution of solid waste before and after processing in a shear shredder. (*From Tchobanoglous et al., 1993.*)

screens and oscillation rate for vibrating screens, length and width for vibrating screens, or length and diameter for rotary screens

- Separation efficiency and overall effectiveness
- Operational characteristics (e.g., energy requirements, routine maintenance, simplicity of operation, reliability, noise and vibration, potential for plugging)

TABLE 8.16   Size Distribution Characteristics of Shear Shredders*

Screen size, in	Percent retained		
	Avg	Min	Max
+4	19.3	9.7	32.9
+2	30.7	23.0	35.0
+1	19.2	14.9	23.9
+½	13.2	8.9	17.4
+¼	9.3	6.2	12.0
Pan	8.3	5.1	10.3

* Based on test results from 11 samples of unseparated MSW.

Source: Adapted from Tchobanoglous et al. (1993).

Thus, the selection of screens for a given application requires considerable attention to the characteristics of the materials being processed. Vibrating screens offer inexpensive sizing for free-flowing granular materials (e.g., glass) that do not tend to blind the screen or become entrapped with other materials. Trommel screens are better suited for large-particle applications where blinding is anticipated or where materials become entrapped and require tumbling to free them for removal by the screen. The performance of disc screens lies between that of vibratory screens and trommels, as does the cost and size of the equipment. Typically, disc screens are utilized in applications where rigid materials such as wood chips are being screened to remove the grit and dirt. Disc screens are prone to wrapping (and thus higher maintenance) when long flexible items such as wire, rope, and textiles are present in the feed stream.

Magnetic Separation.   The most common method for removing ferrous metals from commingled recyclables involves the use of magnetic separation systems. Magnets can be classified as either *electromagnets,* which use electricity to magnetize or polarize an iron core, or *permanent magnets,* which utilize permanently magnetized materials to create a magnetic field. The various types of magnet configurations, the suspended belt magnet, the magnetic head pulley, and the suspended magnetic drum, all of which have been used in recycling applications, are described in Table 8.19 and shown in Fig. 8.31. In addition, specialized solid waste magnets, also shown in Fig. 8.31, have been designed, involving multiple stages of magnetic separation to shake nonmagnetic material loose from tin cans while they are being separated.

The following five factors should be taken into consideration in the selection of either permanent or electromagnets and in choosing the type and configuration of the magnetic installation:

1. The physical relationship between the feed conveyor and the discharge conveyor and whether the magnetic product is to be conveyed in-line or at 90° to the infeed material
2. The width of the feed conveyor and the size, weight, and cost of the magnet required to effectively cover the entire conveyor width with an adequate magnetic field strength
3. The largest size of materials on the belt and/or the tendency of the feed conveyor to encounter piling and surges of flow that could affect the physical mounting arrangement and the distance of the magnet from the feed conveyor
4. The amount of contamination and the shape of contaminants, which are intermixed with the magnetic product
5. Operating requirements such as electrical consumption, space requirements, structural support requirements, conveyor speeds, conveyor widths, type of magnetic cooling systems required, magnetic strength, materials of construction, maintenance, and physical access

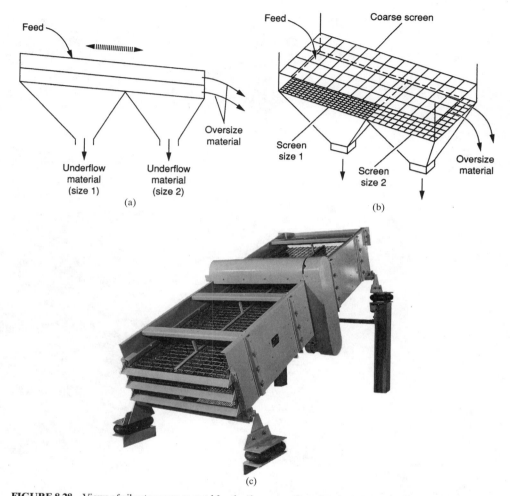

FIGURE 8.28 Views of vibratory screen used for the size separation of waste components: (*a*) profile diagram, (*b*) perspective diagram, and (*c*) typical vibratory screen (see also Figure 8.21*c*). *(From Tchobanoglous et al., 1993.)*

A head pulley magnet is typically used where low-cost separation is required to remove small amounts of magnetic particles from materials being processed. Where large quantities of highly magnetic materials are involved, permanent magnetic separators are usually employed. These can be either drum magnets or overhead suspended belts, depending on the space requirements and personal preference. For less magnetic materials, electromagnetic separators can be used. In-line magnetic separation tends to provide higher recovery efficiency for separation of magnetic materials; however, space constraints frequently dictate the need for cross-belt magnets. In considering whether to use a belt magnet versus a drum magnet, care should be taken to identify the potential for belt damage that can result from nails, wire, and other sharp objects. Drum magnets employ a metal surface, which is more resistant to damage from projectiles.

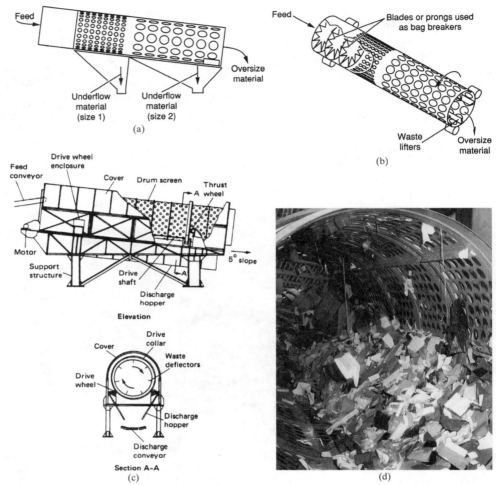

FIGURE 8.29 Views of trommel (rotary) screen used for the size separation of waste components: (*a*) profile diagram, (*b*) perspective diagram, (*c*) diagram detailing components of trommel screen, and (*d*) typical trommel screen in operation. *[Figures (a), (b), and (c) from Tchobanoglous et al., 1993; Fig. (d) from Triple/S Dynamics Systems, Inc.]*

Magnetic separators typically can accomplish a recovery efficiency of 95 to 99 percent for magnetic materials, depending on the application and burden depth of materials being processed. The contamination level of the recovered magnetic fraction varies, depending on the particle size and characteristics of the feed stream. Purities of 95 to 98 percent for the recovered magnetic fraction are considered typical. In mixed-waste processing systems, recovery efficiencies of 80 percent are typical, and the grade or purity of the magnetic product can be as low as 60 to 80 percent ferrous metal. Mixed-waste magnetic material often requires reprocessing before the material can be sold to end users.

Density Separation. Density separation can be accomplished with an airstream (air classification), water, or other dense fluid (flotation), or by light material diversion (chain curtain). Systems for density separation are discussed in Table 8.20; of these processes, air classification is the most commonly used. When mixed materials are subjected to a moving airstream, the

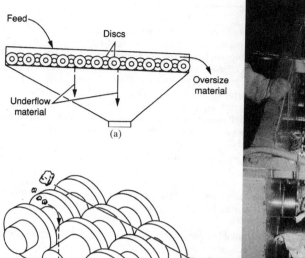

FIGURE 8.30   Views of disk screen used for the size separation of waste components: (*a*) profile diagram, (*b*) perspective diagram of disks, and (*c*) typical disk screen in operation.   *[Figures (a) and (b) from Tchobanoglous et al., 1993; Fig. (c) from Triple/S Dynamics Systems, Inc.]*

more lightweight materials will be carried away if the air current is large enough. Applications of air classifiers include separation of the following:

- Labels from granulated plastic bottles
- Lighter plastic bottles and cans from heavier glass bottles
- Paper and plastic films from bottles and cans after these materials are liberated
- Fine glass and dirt from coarse glass

Diagrams of air classification systems are presented in Fig. 8.32 and fluidizing velocities for various materials are given in Table 8.21.

   The selection of an air classifier for a particular application is dependent primarily on the separation efficiency required and the money and facility space that can be committed. The least expensive equipment is a vertical column that requires a minimum of space but is the least efficient type of air classifier. The zigzag air classifier is slightly taller than a vertical column unit and has internal veins that add to the cost. The zigzag offers higher separation efficiency than the vertical column but is sensitive to the infeed particle size. Performance ranges for typical air classifier units are presented in Table 8.22.

   The horizontal air classifier can process materials having a larger particle size than a zigzag unit; however, it requires a considerable amount of floor space. The vibroelutriator provides a higher efficiency than the horizontal air classifier because the materials are spread uniformly

TABLE 8.17 Size Separation Systems for MSW

Type of screen	Description
Vibratory deck	Vibrating screens typically have flat decks and are mounted on an incline to assist in material movement. The screens may be designed with one deck to make a bimodal product or may have multiple. A wire-mesh or solid-metal plate with holes punched into it, powered by an electric motor and drive mechanism, vibrates the material and throws it up and down on the screen so that the material impinges many times on the deck, providing numerous opportunities to pass through an opening. The gyratory motion can often be adjusted to change the throw and alter the extent of upward travel versus horizontal travel down the length of the screen.
Rotary (trommel)	Material to be separated is fed into one end of a tubular, rotating screen with a downward slope (around 5°), so that the material will flow down the screen as it is dropped and tumbled. Lifters are sometimes placed within the screen to increase the degree of lifting and dropping of material. Blades or prongs may be included on the inlet end to open bags.
Disk screen	Horizontal bars or shafts that run across the screen width are arranged perpendicular to the material flow. On each shaft are several serrated or star-shaped discs spaced evenly across the width of the screen. As the shaft turns, it carries material across the discs and bounces it into the air. The disks require periodic replacement as they wear down.

and the vibration tends to stratify the lighter materials on top for easy retrieval. The vibro-elutriator is considerably higher in cost than the horizontal air classifier. The rotary drum air classifier is the most efficient of all air classifiers and can accommodate a wide range of particle sizes. Materials are repeatedly tumbled and have multiple opportunities for separation. However, the rotary drum is the highest in cost and requires the greatest amount of floor space.

Eddy Current Separation. Separation of conductors from nonconducting material by electric field is known as the *eddy current process*. The principle of separation relies upon Faraday's law of electromagnetic induction. In essence, when a magnetic field passes through a conductor (e.g., when a conductor experiences a change in an applied magnetic field), it induces in that conductor an electric current. That electric current also has a secondary magnetic field associated with it that always opposes the primary magnetic field, as shown in Fig. 8.33. Several eddy current separation systems are described in Table 8.23.

TABLE 8.18 Operating Characteristics of a Typical Trommel Screen

Parameter	Unit	Value
Diameter	m	3.5
Screen length	m	4.0
Screen size	mm	50
Screen open area	%	53
Inclination angle (variable)	degrees	3–7
Rotational speed (variable)	rev/min	11–13

Source: Adapted from Tchobanoglous et al. (1993).

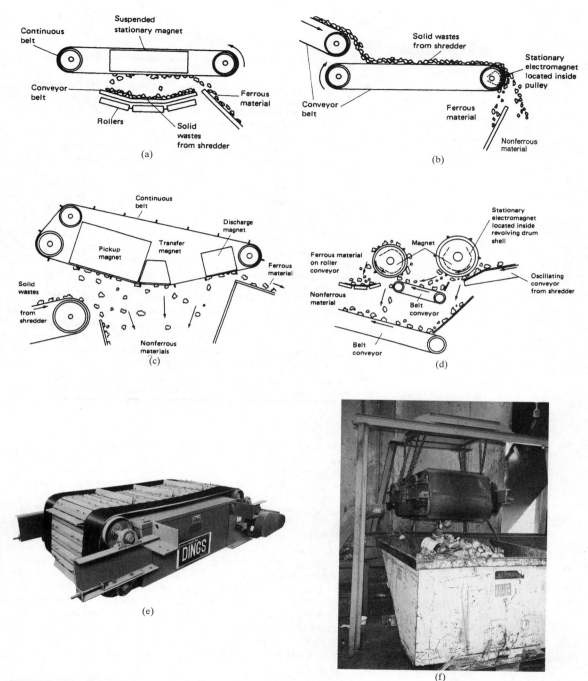

FIGURE 8.31 Magnetic separators used for the separation of ferrous materials include (*a*) suspended-belt magnet, (*b*) magnetic head pulley, (*c*) multiple-stage magnetic separation with a belt magnet to shake contamination loose from tin cans while they are being separated, (*d*) multiple-stage magnetic separation using drum magnets, (*e*) typical belt magnet, and (*f*) typical view of an operating belt-type magnetic separator.   [*Figures (a), (b), (c),(d),and (f) from Tchobanoglous et al., 1993; Fig. (e) from Dings Co., Magnetic Group.*]

TABLE 8.19   Magnetic Separation Systems for MSW

Process	Description
Suspended-belt	The magnet is located above the feed stream. As the feed stream moves under the magnet, material is lifted off the belt. As the material comes into contact with the belt moving around the magnet, it gains momentum and is directed to separate collection.
Crossbelt	Material is directed at a 90° angle from the primary feed conveyor into a bin or downstream item of processing equipment. Material overlying ferrous material (overburden) may also be picked up and held between the magnet and ferrous material, resulting in contamination.
In-line	As the material falls off the end of a fast-moving conveyor, a belt magnet exerts a force on the suspended ferrous materials, causing the ferrous material to leave the waste stream in a direction parallel to the direction of flow. (See *in-line suspended magnetic drum* later in this table.)
Head-pulley	The head-pulley magnet is installed as an integral part of the belt conveyor. As material falls off the end of the conveyor, the head-pulley magnetic forces hold magnetic material to the belt, attracting the magnetic materials and changing their trajectory as they fall off the end of the belt. Underburden material, which is entrapped under the magnetic metal, will be held onto the belt by the magnetic material above it and carried over into the magnetic product, resulting in contamination.
In-line suspended drum	Similar function as the in-line belt magnet, a rotating metal drum with an internal magnet separates ferrous metals from a suspended waste stream. Since separation takes place while fully suspended in the air, the potential for entrapment of nonmagnetic material is reduced.
Multiple and combination systems	Various processes have been developed to deal with overburden and underburden materials, including reversing polarity and transfer magnets.

If the conductor is in the shape of a ring, current flows in one direction and can easily be measured. If the conductor is a solid piece of metal, current is more difficult to observe and measure, owing to the complex current path, but the current is there just the same. Such currents, which are enclosed within more or less solid pieces of metal, are known as *eddy currents,* because they resemble eddies observed in liquids. Eddy currents give rise to a physical force that is the basis of a separation process. In general, all conductors resist changes in magnetic field strength.

The repulsive force set up by eddy currents is a function of particle size, particle geometry, and the ratio of conductivity to mass density, in addition to such factors as magnetic field strength and frequency. The opposing forces between the primary and secondary magnetic fields are utilized in eddy current separators to make a separation between materials. The principal criteria used to make the separation have to do with conductivity and mass. Aluminum has a lower conductivity than copper, but also has a lower mass. Aluminum has the lower mass-conductivity ratio and thus is more affected by an eddy current separator; that is, aluminum sees a greater trajectory than copper when subjected to an eddy current separator. The force exerted on a particle increases with the fourth power of the radius, and thus bigger particles are much more affected than small particles. Shape and thickness can also be factors in such separations.

Automated Separation.   Until recently, sorting of materials has been accomplished either through source separation practices or by manual sorting at an MRF. Manual sorting is not ideal, however, owing to the potential for injury to workers, high labor and training costs, and the susceptibility to error. Automated sorting systems have thus been developed in an attempt to replace manual labor with an automated technology. Automated systems have been devel-

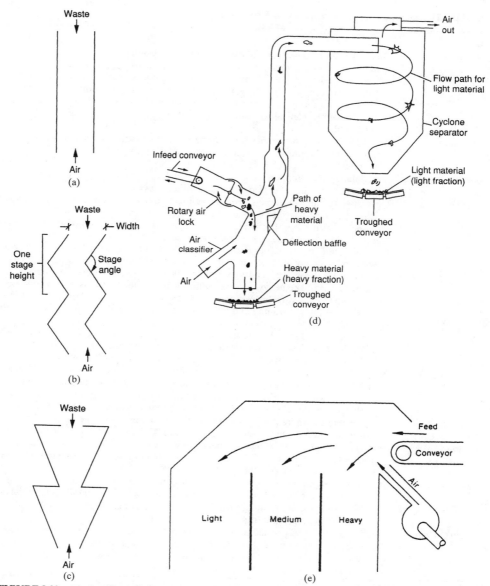

FIGURE 8.32   Air classifiers used for the separation of waste components: (*a*) vertical column air classifier, (*b*) zigzag air classifier, (*c*) stacked triangle, (*d*) typical air classification system, and (*e*) diagram of an air knife classification system.

oped for the sorting of ceramics from glass, sorting of glass by color, the separation of plastic resins, the separation of paper by grade, and the sorting of plastics by color. A description of sensors used to classify material is presented in Table 8.24.

Several commercial systems for the automated identification and separation of individual components from a mixed-waste stream have been developed. Automated detection and separation systems, as shown in Fig. 8.34, are generally composed of the following components:

TABLE 8.20 Density Separation Systems for MSW

Process	Description
Horizontal	Material is fed by conveyor and dropped into a horizontal airstream. Heavy objects that are more affected by gravitational force than by pneumatic forces of the air current drop through the air current. The lighter fraction (more air-buoyant objects) are carried by the airstream farther distances or are carried away with the airstream into a cyclone that acts to separate the light entrained matter from the airstream itself. Material that is not dropped out of the airstream by the cyclone mechanical separator is later separated from the airstream using either a bag fitter or a wet scrubber.
Air knife	An air knife is similar in concept to a horizontal air classifier. Moving mixed waste is exposed to an airstream that removes the light fraction from the bulk of the material. The light fraction is then directed to subsequent processing.
Vertical-column	Infeed material is dropped straight down through the air column inside a vertical chamber. The separation is bimodal in that material is either heavy or light (goes up or down).
Baffled-column	Developed to allow multiple stages of classification to occur, improving the quality of separation. This type of air classifier is similar to the vertical air classifier, except that material tumbles down deflectors as it drops through the air current, falling from one shelf onto the next. Examples include the zigzag and stacked V-shaped columns.
Rotary drum	Material is fed into the upper end of a slightly inclined large barrel. The barrel is oriented on its side and rotates slowly while a gentle airstream passes through it. The drum normally has lifters inside of it that lift the waste and drop it repeatedly as air is being drawn up the inclined drum and while the drum turns. The heavy material, which is more affected by gravitational forces than by pneumatic forces, travels down the incline and out the bottom of the drum while the lighter material either is entrained in the air current and flies up the drum or alternatively travels up the drum after repeated drops in the airstream.
Stoner	A mixed-waste stream falls into the center of a sloped vibratory deck screen. A uniform air current applied from below the deck fluidizes the bed and causes light material to float downslope while the heavy fraction is transported through contact with the vibratory action of the deck.
Vibroelutriator	Material to be separated is introduced at the upper end of a device somewhat resembling a horizontal inclined vibrating screen, which has a hood above it. As the material is vibrated down the screen deck, air is sucked up the hood both from the bottom discharge area and from under the screen cloth. Light material (e.g., labels and other fine particles) is lifted by the air currents and drawn up the feed hood where it is removed from the airstream by cyclones and bag filters. The heavier materials travel down the screen cloth and are discharged at the lower end.
Cyclone separator	Material from an air classification operation is fed into a cylindrical tank with a tapered bottom. The material whirls around the inside of the cylinder and gradually settles to the bottom while the air leaves from the top of the unit.
Flotation	A unit operation that uses a fluid to separate components. When immersed in the fluid, heavy materials sink to the bottom and are removed with a scraper, and light materials are skimmed from the surface. Typical examples include the separation of glass from the light fraction of MSW, wood from debris, and plastics from other organic materials.
Chain curtain	A series of chains suspended from a rotating belt placed over a moving conveyor. As the chain curtain intercepts material, light items are directed in the direction of the moving chain curtain, while heavy items pass through the chains. The material flow rate can affect the efficiency of this process.

- Feed system to meet requirements for separation (e.g., uniform layer, individual bottles, no overburden)
- Sensors for the detection of material properties
- Microprocessors for classification of the materials and to send signal to air jet
- Pulsed, compressed air jets to eject selected material out of waste stream

Automated sorting systems are expensive and require a high degree of maintenance for effective operation. However, the sensitivity of recycling equipment to contaminants and the

TABLE 8.21 Fluidizing Velocities for Air Separation of Various Solid Waste Components

Component	Velocity, ft/min Zigzag classifier with 2-in throat	Velocity, ft/min Straight 6-in-diameter pipe
Plastic wrapping (shirt bags)	Less than 400 (electrostatic)	—
Dry, shredded newspaper (25% moisture)	400–500	350
Dry, cut newspaper		
1-in rounds	500	350
3-in squares	—	350
Agglomerates of dry, shredded newspaper and cardboard	600	—
Moist, shredded newspaper (35% moisture)	750	—
Dry, shredded corrugated cardboard	700–750	450–500
Dry, cut corrugated cardboard		
1-in rounds	980	700
3-in squares	—	1000
Styrofoam, packing material	750–1000 (electrostatic)	—
Foam rubber (½-in squares)	2200	—
Ground glass, metal, and stone fragments (from automobile body trash)	2500–3000	—
Solid rubber (½-in squares)	3500	—

Source: Adapted from Tchobanoglous et al. (1993).

TABLE 8.22 Typical Ranges of Air Classifier Performance

Parameter	Typical range
Critical air/solids ratio	2.0–7.0
Input waste composition to air classifier (%)	
Ferrous metals	0.1–1.0
Nonferrous metals	0.2–1.0
14-mesh fines	15.0–30.0
Paper and plastic	55.0–80.0
Ash	10.0–35.0
Output light fraction from air classifier (%)	
Ferrous metals	2.0–20.0
Nonferrous metals	45.0–65.0
14-mesh fines	80.0–99.0
Paper and plastic	85.0–99.0
Ash	45.0–85.0
Column loading (ton/h)/m^2	5.0–40.0

Source: Adapted from Tchobanoglous et al. (1993).

increased value of higher-purity materials have made automated sorting systems cost-effective in some cases. The systems that have been developed require a preprocessed feed stream. Depending on the particular system, operational requirements can include physical properties (e.g., whole, flaked, flattened, screened), preprocessing (e.g., mixed colors, single resin), and feed style (e.g., singulated containers, specified density). The degree to which these requirements are met can have a significant impact on the process efficiency.

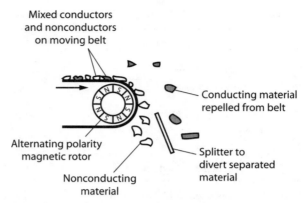

FIGURE 8.33 Schematic diagram of a rotating-drum-type eddy current separation unit.

Advances in products and product packaging have resulted in the introduction of new packaging and material combinations. As new products continue to be developed, they will also need to be evaluated for compatibility in recycling systems. In the future, waste materials and electronic systems for sensing and identifying materials will also become more sophisticated, including systems for the following:

- Advanced imaging and identification systems
- Coding and marking materials for positive identification
- Robotic separation

The feasibility of these highly mechanized processes for sorting waste materials will depend on technological advances, cooperation between the various stakeholders, and the overall economics.

Equipment for Densification (Compaction)

The need to reduce the space requirements for storage of materials is often addressed by compaction. The degree of compaction for a material is a function of the equipment operating pressure and characteristics of the material. The volume reduction and compaction ratio can be calculated using Eqs. (8.10) and (8.11), respectively.

$$\text{Volume reduction (\%)} = \left(\frac{V_i - V_f}{V_i}\right) \cdot 100 \tag{8.10}$$

where V_i = initial volume of wastes before compaction, yd^3
    V_f = final volume of wastes after compaction, yd^3

$$\text{Compaction ratio} = \left(\frac{V_i}{V_f}\right) \tag{8.11}$$

where terms are as defined in Eq. (8.10).

As summarized in Table 8.25, compaction processes can occur at various locations of solid waste management, including at the point of generation, collection, processing, and disposal. For MRFs, the most common forms of compaction are can flatteners, can densifiers, pelletizers, and balers. Design parameters for compaction equipment are summarized in Table 8.26.

TABLE 8.23   Eddy Current Separation Systems

Process	Description
Linear motor	A traveling magnetic field is generated by an electromagnet energized by an electromagnetic field. The electromagnetic forces move the metal particles, which pass over the linear induction motor, laterally across the belt, and roll off one side or the other.
Popper	Particles are passed over a rapidly changing magnetic induction coil capable of generating very high currents with large capacitor banks that are discharged intermittently. The conductors that are over the coil when it discharges are propelled off the belt.
Sliding ramp	Magnets of alternating polarity are arranged in stripes on a flat, inclined board or ramp. The materials to be separated are conveyed to the top of the ramp and slide down using gravitational force. As the conductors slide over the top of the magnets, they see an oscillating magnetic field similar to the active field that is generated in the stator of an electric motor. The conductors see a force that moves them over to one side while the nonconductors slide down the ramp unaffected by the oscillating magnetic field.
Rotating drum	The drum-type eddy current separator is much like the ramp-type separator, except that the magnets are attached to a drum that is inclined to a continuous ramp. The permanent magnets are rotated under an outer shell or under a belt. The conductors are then moved by the eddy current forces.

Can Flattener.   For small facilities, one approach to increase the density of loose metallic cans is to utilize a can flattener. These flatteners are useful only for aluminum and tin cans and should not be fed such metallic items as castings and heavy metal objects that could severely damage the equipment. Can flatteners are often used in conjunction with a blower. After being crushed, the cans are blown into a roll-off container or semitrailer for storage or transport (see Fig. 8.35a).

TABLE 8.24   Sensors Used for Detection of Material Properties

Sensor type	Description
Optical sensors	Optical sensors can be used to detect color for glass, plastic, and paper.
Image recognition	Scans an object and a microprocessor compares the object with a database.
X-ray fluorescence	Scans material surface for presence of chlorine atoms in PVC.
X-ray transmission	Chlorine atoms in PVC are detected by sending x-ray though material.
Infrared	Distinguishes between clear, translucent, and opaque materials.
Near infrared (NIR)	Measures the NIR absorbance of plastics to distinguish between resins.
Electrostatic	Electrical permittivity is used to separate different nonconductors, such as plastic from paper.
Eddy current	Materials with electrical conductivity are detected, such as ferrous and aluminum.

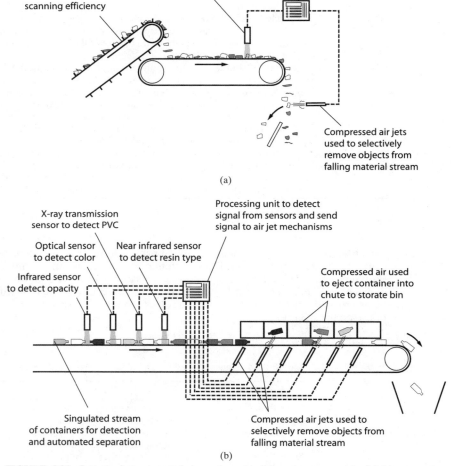

FIGURE 8.34 Systems for automated separation of solid waste components: (*a*) binary sorting device, and (*b*) an array of sensors for material identification.

Can Densifier. Small aluminum bales or biscuits can be produced with a can densifier. A densifier is a small version of the baler, the discussion of which follows. Densified aluminum biscuits weigh about 40 lb and can be stacked and shipped on pallets. Bale density can be increased by flattening or shredding the material before densification.

Pelletizer. Plastics are sometimes pelletized through compression and extrusion for storage and shipment purposes. The light fraction of MSW can also be pelletized for use as a fuel in combustion processes.

TABLE 8.25   Compaction Equipment Used for Volume Reduction

Location or operation	Type of compactor	Remarks
Solid waste generation points	Stationary/residential	
	Vertical	Vertical compaction ram; may be mechanically or hydraulically operated; usually hand-fed; wastes compacted into corrugated box containers or paper or plastic bags; used in medium- and high-rise apartments.
	Rotary	Ram mechanism used to compact wastes into paper or plastic bags on rotating platform, platform rotates as containers are filled; used in medium- and high-rise buildings.
	Bag or extruder	Compactor can be chute-fed; either vertical or horizontal rams; single or continuous multibags; single bags must be replaced and continuous bags must be tied off and replaced; used in medium- or high-rise apartments.
	Undercounter	Small compactors used in individual residences and apartment units; wastes compacted into special paper bags; after wastes are dropped through a panel door into a bag and door is closed, they are sprayed for odor control; button is pushed to activate compaction mechanism.
	Stationary/commercial	Compactor with vertical or horizontal ram; waste compressed into steel container; compressed wastes are manually tied and removed; used in low-, medium-, and high-rise apartments, commercial and industrial facilities.
Collection	Stationary/packer	Collection vehicles equipped with compaction mechanisms.
Transfer and/or processing station	Stationary/transfer trailer	Transport trailer, usually enclosed, equipped with self-contained compaction mechanisms.
	Stationary	
	Low pressure	Wastes are compacted into large containers.
	High pressure	Wastes are compacted into dense bales or other forms.
Disposal site	Movable wheeled or tracked equipment	Specially designed equipment to achieve maximum compaction of wastes in situ.
	Stationary/track-mounted	High-pressure movable stationary compactors used for volume reduction at disposal sites.

Source: Adapted from Tchobanoglous et al. (1993).

Baler.   Most recycling facilities employ at least one baler (see Fig. 8.35*b, c*). In addition to the traditional function of baling paper and corrugated cardboard, the baler can also serve to densify ferrous metals, aluminum, and plastics (see Table 8.27). Furthermore, balers can be an efficient means of reducing the volume and thus the disposal cost of residues or rejects from various recycling operations. It should be noted that when baling plastic bottles, caps should be removed to release entrapped air. Alternately, a conditioning unit can be installed ahead of the feed hopper to puncture bottles so air can be released on compression, thus eliminating the need to remove caps before baling.

Balers can be categorized into two main types: (1) vertical balers and (2) horizontal balers. A description of different baler types is presented in Table 8.28. Wire ties are normally utilized

TABLE 8.26 Typical Design Factors for Compaction Equipment

Factor	Value Unit	Value Range	Remarks
Size of loading chamber	yd³	<1–11	Fixes the maximum volume of wastes that can be placed in the unit.
Cycle time	s	20–60	The time required for the face of the compaction ram, starting in the fully retracted position, to pack wastes in the loading chamber into the receiving container and return to the starting position.
Machine volume displacement	yd³/h	30–1500	The volume of wastes that can be displaced by the ram in 1 h.
Compaction pressure	lb/in²	15–50	The pressure on the face of the ram.
Ram penetration	in	4–26	The distance that the compaction ram penetrates into the receiving container during the compaction cycle. The further the distance, the less chance there is for wastes to fall back into the charging chamber and the greater the degree of compaction that can be achieved.
Compaction ratio	Unitless	2:1–8:1	The initial volume divided by the final volume after compaction. Ratio varies significantly with waste compaction.
Physical dimensions of unit	Variable	Variable	Affects the design of service areas in new buildings and provision of service to existing facilities.

Source: Adapted from Tchobanoglous et al. (1993).

in either a manual or automatic configuration to secure the bale so that on ejection from the compression chamber, the bale does not expand or break apart. All balers have the following four features:

1. Feed hopper or area into which the recyclables are fed
2. One or more hydraulic or mechanically driven rams that compress the infeed material
3. Compression chamber where the materials are densified
4. Discharge area opening from which the completed bales are ejected

Selection of Bailing Equipment. Vertical balers are often the choice of small recycling operations because of their low purchase price. They are slower and operational costs are higher than for horizontal balers. Larger recycling operations and MRFs normally employ horizontal balers. The size and baling densities vary depending upon the size of the unit selected, as shown in Table 8.29. The price increases as the capability for higher bale density is increased.

High-density, automatic-tie horizontal balers are preferred when it is necessary to generate export bales. High-density bales of paper being moved with a forklift are shown in Fig. 8.35d. The added density is important, especially in baling plastics (which are more difficult to bale) because they require high density to hold the bale together. High density is also desirable for improving the value of corrugated and paper bales. The labor savings and improved shipping densities for materials offset the capital costs incurred.

An MRF, which has a processing line that processes only paper and corrugated cardboard, or a high-volume paper-only facility, may best utilize a two-ram baler. However, this baler can

(a)

(b)

(c)

(d)

FIGURE 8.35   Compaction of recyclable materials: (*a*) can flattener, and blower to eject flattened cans into trailer, (*b*) two-ram automatic-tie horizontal baler, (*c*) open-end, automatic-tie horizontal baler, and (*d*) bales of paper being stacked with a forklift. *(From Tchobanoglous et al., 1993.)*

only make whole bales. When changing over from one material to another, a mixed bale is produced. Typically, an open-end, horizontal baler is preferable to the two-ram baler when there are frequent changeovers from one type of material or one grade of paper to another. These in-line balers can make short bales and full bales of high density. They are often more appropriate for recycling operations, while the two-ram units are best for high-volume, paper-only processing facilities or lines.

Weighing Facilities

The weighing of materials is a necessary part of any recovery operation. Depending on the operation, facilities for weighing materials can range from a small portable scale used at buy-back centers (Fig. 8.36*a*) to large platform scales for weighing loaded collection vehicles (see Fig. 8.36*b*).

TABLE 8.27 Typical Densities of As-Received Source-Separated Materials

Material	Typical density, lb/yd³	Baled density,* lb/yd³
Paper		
Newspaper	475	950
Corrugated cardboard	350	800
High grades	300–400	
Glass—whole bottles		
Clear	500	
Green or amber	550	
Glass—crushed		
Semicrushed	1000	
1½-in mechanically crushed	1800	
¼-in Furnace ready	2700	
Aluminum Cans		
Whole	50	950
Flattened	175	
Tin plated steel cans ("tin cans")		
Whole	150	1400
Flattened	850	
Plastics		
PET, whole	34	750
PET, flattened	75	
HDPE (natural), whole	30	
HDPE (natural), flattened	65	
HDPE (colored), whole	45	
HDPE (colored), flattened	90	

* Based on bale size of 45 × 30 × 62 in.

Source: Adapted from Tchobanoglous et al. (1993).

Movable Equipment

Front-end loaders and forklifts are common at MRFs. As shown in Fig. 8.37, loaders are used to move material from the receiving location to belt conveyors and to load processed material into vehicles or containers for shipping. Forklifts are commonly used for moving bales from the baler to the storage location or onto trucks for transport. Forklifts can be equipped with a rotating mechanism for elevating and tipping specially designed metal storage bins (see Fig. 8.37d).

Storage Facilities

Materials that are to be processed or shipped require storage space. Common storage methods include metal containers and feed hoppers, large screened storage bins (such as those used with elevated separation processes), brick or concrete bunkers, and ground area for loose or baled material. The type and size of storage facilities needed will depend on the design of the processing facility and the materials to be processed. Excess capacity should be provided in case of equipment downtime or increased material flow. Long-term storage capacity may allow more flexibility in marketing processed materials.

TABLE 8.28 Balers Used for MSW

Baling system	Characteristics
Vertical	Material is repeatedly fed into the baler and compressed until the appropriate-sized bale is formed. The bale is then manually wound with metal or plastic strapping. Relatively inexpensive and low output.
Downstroke	A hydraulic ram is used to compress the material in a downward motion. Requires vertical clearance for operation.
Upstroke	Units utilize mechanical, chain-driven densification rams. The compression chamber is below grade, thus installation costs are higher and unit is not movable.
Horizontal	Horizontal balers are fed from the top through a feed chute, and the hydraulic ram is arranged in a horizontal configuration. May incorporate continuous feed and automatic tying mechanisms. Fluffers are frequently used on horizontal balers to loosen incoming newsprint before baling, thus improving the stability and integrity of the bales.
Closed-door, manual-tie	May be hand-fed or conveyor-fed into a charging hopper rather than into the baling chamber itself. Can also be fed continuously (the ram can be in a compression cycle while material is being fed into the charging hopper). Intermittent bale production due to manual tie system.
Open-end, automatic-tie	Material is continuously extruded (allowing for higher throughput rates) from the baler. Tension cylinders control bale density and movement through the system. An automatic tying mechanism wraps and ties wire around the emerging bale and is capable of producing bales of variable lengths.
Two-ram horizontal, automatic-tie	One ram is used to compress the material, which is continuously fed into the feed hopper, and the second is used to eject the bale after it has been tied off. The bale ejection ram is situated at right angles to the compression stroke ram. Produces whole bales only.

Selection of Equipment and Facilities for MRFs

Selection of the actual equipment and physical facilities that will be used in the MRF separation process flow diagram is perhaps the most challenging engineering aspect of implementing an MRF. Factors that should be considered in evaluating processing equipment are summarized in Table 8.30. Of the factors listed in Table 8.30, special attention must be given to the following:

- Proven reliability and flexibility of the available equipment and facilities
- Proven process performance efficiency
- Ease and economy of operation

Because obtaining meaningful data on these factors is, at present, difficult, it is recommended that visits be made to actual operating installations to obtain firsthand information on performance and maintenance requirements.

Because of the abrasive nature of many of the components found in solid wastes, the rate of wear on much of the processing equipment and the downtime has been greater than antic-

TABLE 8.29   Typical Design Factors for Baling Equipment

Factor	Unit	Typical value	Comment
Small baler			
Bale size	in	$42 \times 30 \times 30$	Highly variable, depends on manufacturer
Motor size	hp	10	
Operating pressure	lb/in²	100	
Bale weight	lb	500	Corrugated cardboard
Medium baler			
Bale size	in	$62 \times 45 \times 30$	Highly variable, depends on manufacturer
Motor size	hp	75	
Operating pressure	lb/in²	225	
Hourly production	ton/h	5–9	Corrugated cardboard
		15–20	Commingled MSW
Bale weight	lb	To 1350	Corrugated cardboard
		To 1000	Aluminum cans
		To 2800	Commingled MSW
Large baler			
Bale size	in	$62 \times 45 \times 30$	Highly variable, depends on manufacturer
Motor size	hp	200	
Operating pressure	lb/in²	225	
Hourly production	ton/h	12–20	Corrugated cardboard
		30–40	Commingled MSW
Bale weight	lb	To 1350	Corrugated cardboard
		To 1100	Aluminum cans
		To 2800	Commingled MSW

Source: Adapted from Tchobanoglous et al. (1993).

ipated. As a result, it has been found that more equipment failures and operational problems will occur in processing operations as compared with other waste management operations. As a result of operational problems and other equipment limitations, many system designers now recommend the installation of two or more independent process lines (also known as trains), especially where electric power is to be produced on a continuous basis. Further, because many of the firms in this developing field do not have a long history, it is recommended that the equipment selected be such that it can be repaired with standard parts and components that, if necessary, can be rebuilt or remade locally. The availability of a local distributor is also important.

8.5 ENVIRONMENTAL AND PUBLIC HEALTH AND SAFETY ISSUES

Among solid waste management alternatives, recycling, including only the collection and processing of recyclables, is believed by many to be environmentally benign. There are, however, very few technical data to support this hypothesis and the environmental impacts that are associated with transportation, separation of recyclables from MSW, and the reformulation of recyclables into new products.

(a)   (b)

FIGURE 8.36   Weighing facilities used at MRFs: (*a*) typical scale used at buyback centers and (*b*) platform scale used for determining weight of recyclables obtained from curbside collection program.

Environmental Impacts

Environmental impacts from recycling on groundwater, dust, noise, vector, odors, and the atmosphere are considered briefly in the following discussion.

Groundwater Contamination.   Groundwater resources are largely unaffected by recycling. The MRFs for curbside separation programs typically are constructed on a concrete pad that prevents seepage of any waste pollutants into the soils. Moreover, these facilities typically handle precleaned, dry, and solid components of the waste stream. Most facilities are new and therefore subject to state-of-the-art design and regulatory scrutiny with respect to surface drainage and runoff. Potential groundwater impacts of a mixed-waste processing facility would be similar to an RDF processing plant or a composting plant.

Dust Emissions.   Dust emissions from recycling programs are derived from two sources: (1) collection operations and (2) processing facilities. Dust emissions are minimal on route. Operations are usually conducted indoors where ventilation and localized dust suppression measures are taken as required for stationary sources. Mixed-waste processing results in greater emission of dust, but more sophisticated ventilation and collecting devices, such as cyclones and fabric filters, are typically used.

Significant emission of dust comes from the blowing of paper and plastics from containers, tipping operations, storage, and loading operations. This requires operators and collectors to frequently monitor the area to clean up windblown or spilled materials. If attention is not given to collecting loose paper and plastics around the area of an operating MRF, significant nuisance and litter conditions will create an unattractive facility.

Noise.   Potential noise impacts are from two sources: (1) collection vehicles and (2) machinery. Collection vehicles are equipped with conventional noise abatement devices. Vehicles typically contain loading machinery that is considerably less noisy than conventional packer

(a)

(b)

(c)

(d)

FIGURE 8.37 Movable equipment used for the handling of materials at MRFs: (*a*) front-end loader equipped with bucket and solid rubber tires; (*b*) skid steer loader with claw bucket and solid rubber tires (note rubber bumper on bottom of bucket to prevent floor damage); (*c*) forklift used internally within a MRF for moving bales, and also to load bales into containers for shipping; and (*d*) forklift equipped with rotary fork mechanism for emptying contents of smaller containers into larger containers.

trucks for MSW. Processing machinery noise is suppressed by restriction of operations to the interior of buildings.

Vector Impacts.   Potential vector impacts are minimal in front-end processing systems in general due to the enclosure of processing operations, ventilation, and pest control. The MRFs for curbside source separation programs also process a cleaner fraction of the waste, which often is prewashed by the waste generator of food and other organic residues. The putrescible waste content of the commingled source-separated recyclable stream entering an MRF can be virtually eliminated with a carefully controlled collection program.

TABLE 8.30 Factors That Should Be Considered in Evaluating Processing Equipment

Factor	Evaluation
Capabilities	What will the device or mechanism do? Will its use be an improvement over conventional practices?
Reliability	Will the equipment perform its designated functions with little attention beyond preventive maintenance? Has the effectiveness of the equipment been demonstrated in use over a reasonable period of time or merely predicted?
Service	Will servicing capabilities beyond those of the local maintenance staff be required occasionally? Are properly trained service personnel available through the manufacturer or the local distributor?
Safety of operation	Is the proposed equipment reasonably foolproof so that it may be operated by personnel with limited mechanical knowledge or abilities? Does it have adequate safeguards to discourage careless use?
Efficiency	Does the equipment perform efficiently? What is the specific energy consumption (kWh/ton) relative to other equipment with similar capacity?
Environmental effect	Does the equipment pollute or contaminate the environment?
Health hazards	Does the device, mechanism, or equipment create or amplify health hazards?
Economics	What are the economics involved? Both first and annual costs must be considered. Full operation and maintenance costs must be assessed carefully. All factors being equal, equipment produced by well-established companies and having a proven history of satisfactory operation should be given appropriate consideration.

Odor Emissions.   Odor emissions are controlled with design features for vehicles and machinery similar to those used to control noise and dust. In addition, in mixed-waste processing systems such as front-end systems, the tipping floor areas can be designed to maintain a slightly negative pressure to control odors. Again, due to the minimal putrescible waste content of commingled or source-separated recyclables entering an MRF, odor is typically not a problem.

Vehicular Emissions.   The most significant air pollution from collection operations and processing facilities, particularly curbside recycling programs that employ dedicated vehicles, is from vehicular emissions to the atmosphere. Atmospheric emissions data from MRFs processing commingled recyclables are largely unavailable, but air pollution data from vehicle exhaust are well known. The total pollution depends on the distance traveled per ton of waste collected, but it is exacerbated by frequent stops and starts, particularly in cold weather.

Other Environmental Emissions.   Another source of pollution is the energy that is necessary to operate MRF facilities. This pollution occurs at the site of energy generation where coal, nuclear reactors, or natural gas is used and waste products are created. This type of pollution can be estimated from available data.

Public Health and Safety

Materials recovery facilities are a relatively new type of industrial facility without a long history of experience in terms of public health and safety issues. Nevertheless, special attention must be devoted to these issues during the design of the process. Two principal types of public health and safety issues are involved in the design of MRFs. The first issue is related to the public health and safety of the employees of the MRF. The second issue is related to the health and safety of the general public, especially for MRFs that will also be used as drop-off and buyback centers.

TABLE 8.31 Health and Safety Issues in Design and Operation of MRFs

Component	Safety issue
Mechanical	High-speed rotating and reciprocating parts Exposed drive shafts and belts High-intensity noise Broken glass, sharp metal objects Explosive hazards
Electrical	Exposed wiring, switches, and controls Ground faults
Architectural	Ladders, stairways, and railings Vehicle routing and visibility Ergonometrics of handpick conveyor belts Lighting Ventilation and air conditioning Drainage
Operational	Housekeeping practices Safety training Safety and first-aid equipment
Hazardous materials	Hazardous wastes from households and small-quantity generators Biohazards such as human blood products and pathogenic organisms
Personal safety equipment	Puncture-proof, impermeable gloves; safety shoes, uniforms, eye protection, noise protection

Worker Issues. Materials recovery facilities are potentially dangerous work environments unless proper precautions are taken during design and operation. Some of the most important safety and health issues are summarized in Table 8.31. Because of the moving equipment and conveyors used in most MRFs, special attention must be devoted to materials flow and worker involvement at each stage of the process. Where the manual separation of waste materials from commingled MSW is used, careful attention must be given to the types of protective clothing, air-filtering headgear, and puncture-proof gloves supplied to the workers. In addition, worker fatigue is another important issue that must be addressed. Where sorting from moving belts is used, the height of the worker relative to the moving belt must be adjustable. The federal government through the Occupational Safety and Health Administration (OSHA) and state OSHA-type programs now requires the development of comprehensive health and safety programs for workers at MRFs.

Public Access Issues. Because the activities involved with the operations of MRFs are potentially dangerous, the public should be excluded from access except under careful control as during conducted tours. Convenience stations for the deposit of recyclables should be provided for public access away from the main traffic pattern.

8.6 RECYCLING ECONOMICS

In some instances, the choice of recycling method can be heavily influenced by the population that is to be served. If there are not enough recyclable materials generated by an individual village or small town, unless a large group of towns can consolidate their recyclables for processing, a central processing facility will be prohibitively expensive. Large cities or regional waste districts have more options open to them in terms of recycling operations, but that does not mean that they have a significant cost advantage over the smaller population bases. Collection costs, processing or consolidation costs, and net revenues vary widely from method to method, but their estimated net costs of recycling are relatively close.

Capital Cost

Capital costs for small, intermediate, and large MRFs (including separation, processing, and transformation systems) can range from $15,000 to $45,000 per ton of daily capacity, as summarized in Table 8.32. The lowest collection cost is associated with front-end processing; in

TABLE 8.32 Typical Capital Costs for MRF*

System	Major system components	Capital costs, $/ton capacity · d
	Materials recovery	
Low mechanical intensity[†] (5–20 ton/d)	Processing of source-separated materials only; enclosed building, concrete floors, hand picking stations and conveyor belts, storage for separated and prepared materials, support facilities for the workers	10,000–20,000
Intermediate mechanical intensity (20–100 ton/d)	Processing of source-separated commingled materials and mixed paper; enclosed building, concrete floors, elevated first-stage hand picking stations and conveyor belts, baler, storage for separated and baled materials for 2 weeks, support facilities for the workers; buyback center	12,000–25,000
High mechanical intensity[‡] (>100 ton/d)	Processing of commingled materials or MSW; same facilities as the low-end system plus mechanical bag breakers, magnets, shredders, screens, and storage for up to 3 months; also includes a second-stage picking line	20,000–40,000
	Composting	
Low-end system	Source-separated yard waste feedstock only; cleared, level ground with equipment to turn windrows	10,000–20,000
High-end system	Feedstock derived from processing of commingled wastes; enclosed building with concrete floors, MRF processing equipment, and in-vessel composting; enclosed building for curing of compost product	25,000–50,000
Waste to energy	Integrated system of a receiving pit, furnace, boiler, energy recovery unit, and air discharge cleanup	75,000–125,000

* All cost data have been adjusted to an Engineering News-Record Construction Cost index of 6500, which corresponds to the value of the index on January 2002.

† Low-mechanical-intensity MRFs contain equipment to perform basic material separation and densification functions.

‡ High-mechanical-intensity MRFs contain equipment to perform multiple functions for material separation, preparation of feedstock, and densification.

this case, the recyclables are collected with the remainder of the MSW. There are no extra trucks, extra labor, or any of the other collection costs. The cost of collecting the recyclables is the same as the cost to collect the MSW. This lower collection cost is offset by the cost of processing, as the recyclables now need to be recovered from a very complex mix of difficult-to-handle materials. A separate collection for recyclables is inevitably more expensive than collecting them with the rest of the MSW. Collection of commingled recyclables as a group for subsequent separation and processing in a centralized facility reduces collection costs to an intermediate level [see Sec. 7.6 (in Chap. 7) of this Handbook]. However, this savings is offset by the cost of building and operating an MRF.

Staffing for MRFs

Recycling programs that do not do any processing can save on the costs of centralized facilities. Such programs merely consolidate individual materials for shipment to other processors and markets, such as scrap dealers. Capital costs are limited to the cost of consolidation containers and a simple method for loading into them. However, such a simple system does not reap the same revenues as the programs that produce a high-purity product and that densify recyclables for economical shipping over long distances. On the collection side of these types of programs, the homeowner or the collector can do the separations. Paying employees to separate materials is expensive relative to the cost of having the homeowner do it. However, participation can suffer if the homeowner is asked to do too much separation or follow a complicated collection schedule. The net result is a lower amount of material recycled and lower revenues. Typical staffing requirements for MRFs of various sizes are reported in Table 8.33.

TABLE 8.33 Typical MRF Personnel Requirements

Personnel	Capacity, ton/d		
	10	100	500
Office			
Manager	1	1	1
Bookkeepers	0	1	1
Clerk	0	0–1	2–3
Janitor	0	0	1
MRF			
Foreman/operator	1	1–2	3–4
Sorters	1–2	13–25	40–80
Forklift/loader operator	0	2–3	5–6
Maintenance	0	1	4

Source: Adapted in part from U.S. EPA (1991).

In summary, when designing a recycling program, many factors need to be taken into consideration. A diverse array of equipment and processing methods are available for the separation of materials, ranging from manual sorting on conveyor belts to fully mechanical. Markets for the separated materials may be local, regional, national, or overseas; however, all markets for recovered materials fluctuate.

REFERENCES

Clean Washington Center (CWC) (2002) Seattle, WA; <www.cwc.org>.

Conveyor Equipment Manufacturers Association (CEMA) (2001) *Conveyor Equipment Manufacturers Association Handbook,* CEMA, Naples, FL; web site (www.cemanet.com).

Hering, R., and S. A. Greeley (1921) *Collection and Disposal of Municipal Refuse,* 1st ed., McGraw-Hill, New York.

Lund, H. F. (2001) *Recycling Handbook,* 2nd ed., McGraw-Hill, New York.

Parsons, H. de B. (1906) *The Disposal of Municipal Refuse,* 1st ed., John Wiley & Sons, Inc., New York.

Tchobanoglous, G., H. Theisen, and S. A. Vigil (1993) *Integrated Solid Waste Management: Engineering Principles and Management Issues,* McGraw-Hill, New York.

U.S. EPA (1991) *Material Recovery Facilities for Municipal Solid Waste Handbook,* EPA625/6-91/031, Office of Research and Development, U.S. Environmental Protection Agency, Washington, DC.

U.S. EPA (2001) *Municipal Solid Waste in the United States: 1999 Facts and Figures, Executive Summary,* EPA530/S-01/014, Office of Solid Waste, U.S. Environmental Protection Agency, Washington, DC.

Waste News (2001) *Resource 2001—A Directory and Buyers Guide for the Solid Waste and Scrap Processing/Recycling Markets,* Crain Communications, New York.

CHAPTER 9
MARKETS AND PRODUCTS FOR RECYCLED MATERIAL

Harold Leverenz
Frank Kreith

Diminishing landfill space and concern about pollution from landfills and incinerators led to a reevaluation of waste disposal practices in the 1980s. "Reduce, Reuse, and Recycle" became the challenge during the 1990s. The public embraced recycling because it felt a sense of pride in their belief that recycling is beneficial for environmental quality. To many citizens, the act of recycling consisted of placing solid waste in containers to be picked up by a waste collector and taken to a recycling center. Organizations involved in recycling soon discovered, however, that the collection of recyclable material is only the beginning of a circular process in which recyclable resources must be sold and used to make new products and buyers for the recycled products must be found (Fig. 9.1). Today, recycling faces a new and different challenge: sufficient recycling markets to financially sustain collection programs must be developed and maintained.

Despite the many advantages obtainable from developing an effective recycling process, experience has shown that several barriers hinder the development of effective recycling markets:

1. Lack of consumer awareness about recycled products.
2. Lack of consumer confidence in the quality of products made from recycled materials.
3. The social costs and benefits are not reflected in the price of products. Despite environmental advantages, recycled products are generally more expensive than their counterparts made from virgin materials.
4. The high cost of transporting recyclables from the point of collection in many cities and rural areas to centralized processing plants.
5. Uncertainty about supply-and-demand stability of recyclable products deters financial investment in facilities using recycled materials.
6. Recovery and sorting of certain recyclable materials such as plastics, oil, tires, and demolition products is difficult, and technological improvements are necessary to increase the efficiency of recovery.

For recycling to be successful, markets must be available for the materials diverted from the waste stream. Key terminology related to recycling markets is presented in Table 9.1.

9.1 SUSTAINABLE RECYCLING

In a truly sustainable recycling economy, for each truckload of recyclable commodities leaving a region, a truckload of recycled consumer goods must enter. A sustainable recycling system requires a balance of inflows and outflows. In other words, the energy and resources

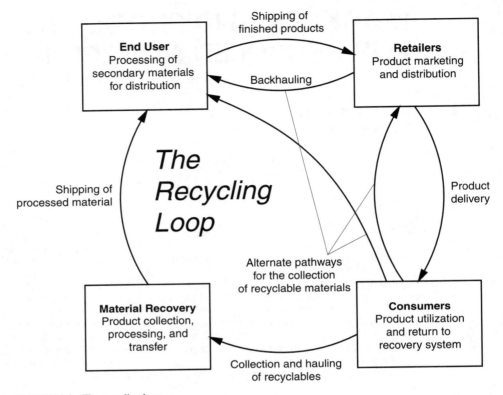

FIGURE 9.1   The recycling loop.

invested in the recovery of materials should be equal to that invested in the delivery of goods to consumers. For this balance to be approached, several steps are required.

1. *Collections.* Mechanisms must be developed for used products and recyclable materials to flow from the recovery and sorting locations to reprocessing and remanufacturing facilities. Systems such as backhauling from homes and businesses would have to be added to existing recovery methods.

2. *Preliminary processing.* An infrastructure of disassembly facilities needs to be developed. This infrastructure must be operated concurrently with the existing infrastructure that generates products and packaging.

3. *Transport.* The final transport system necessary to move recyclable materials from sorting facilities to end users for processing must be improved. Existing means of transportation such as truck, rail, or maritime transport need to be adapted to provide transportation of recyclable materials at minimal cost.

4. *End-use industries.* To sustain open- and closed-loop recycling systems, end-use, recycled-content industries capable of utilizing the output from the entire manufacturing process must be developed.

5. *Promotion and education.* Manufacturers will have to be encouraged to invest in recyclability alongside other product attributes. Furthermore, consumers will have to be educated on the need to purchase products made of recyclable material even if the cost is higher than the cost of similar products made of virgin material.

TABLE 9.1   Terminology

Term	Definition
End user	An industry that utilizes recyclable material in the manufacture of new products
Backhauling	Utilizing empty return transport (after product delivery) to return collected recyclable materials to end users.
Recycled content	
Preconsumer	Material that is scrap from an industrial process, usually known exactly where it is from and what it contains
Postconsumer	Material that has been collected from residential or commercial sources, usually from many different locations and of unknown content
Market development	Strategies used for the development and expansion of industries that utilize recyclable materials in the manufacture of new products
Recycling rate	The fraction (by weight) of material recycled to the total amount of material consumed
Primary materials	Materials used for the first time, also known as virgin materials
Secondary materials	Materials that can be used as a substitute for primary materials, including recovered and scrap materials

6. *Buy recycled content.* Manufacturers will need to increase the recycled contents of their products and market them with sufficient intensity to ensure consumer demand and support of a circular recycling process.

In a utopian total reverse logistics system, manufacturers would use recycled materials to the maximum possible extent and create a virtual superhighway for recyclable commodities. Such a system would put an end to products with one-way, dead-end trips from the manufacturer to the consumer, and then to the landfill or incinerator. It is interesting to note that other economic world powers such as Japan and Germany have already taken initial steps in this direction.

9.2   RECYCLING MARKETS

In classical markets, a company making a product must charge a price that covers the manufacturing, transportation, marketing, and other associated costs as well as a reasonable profit. Buyers interested in obtaining the commodity generate a demand for the product and then negotiate a price. Recycling markets, on the other hand, are driven by the demand for collection services, not by the supply of recovered materials resulting from the collection services. In other words, the generation of the supply is not influenced by the demand. State and local governments must find someone to accept the recovered materials as part of a solid waste management process. When the supply exceeds the demand, it is often necessary to sell the recyclable material for an arbitrarily low price or send it to a landfill. The challenge for a viable recycling process is to foster a diverse market with varied users that will create a strong and stable outlet for recyclable materials recovered from the waste stream. A study of the available markets, known as a *market analysis,* will help to identify these outlets and markets that need further development.

Market Analysis

To develop recycling markets, it is first necessary to know what markets are available for recovered materials. Such an analysis should identify the following factors:

1. *The existing and potential markets.* Recyclable materials may be sold to intermediate processors or directly to end users. Alternative uses for materials should also be investigated, such as using scrap tires to make a mulch or playground cover.

2. *Which recyclable products have potential buyers.* Potential buyers should be able to provide information such as how much they pay for materials, the material specifications including processing and purity requirements, types of contracts available (e.g., long or short term, guaranteed or variable prices), material transport expectations, seasonal market fluctuations, and the amount of the various materials acceptable.

3. *Which materials will have to be stockpiled or disposed of because no present market is available.* The analysis should take into consideration the current and projected types and amounts of materials in the waste stream.

4. *Economic feasibility of recycling a commodity.* The analysis should address which markets are already saturated and those that need further development to reach their full potential.

5. *What tools should be pursued for market development.* The tools listed in the sections that follow can be used to modify the supply and demand of recyclable materials.

Market surveys can provide valuable insight into existing recycling markets and help to direct market development efforts. Surveys that have been conducted in the past have shown that recycling activities can make a significant contribution to local economies. Recycling market surveys should include the following components:

- Geographic and demographic study area
- Waste stream analysis
- Status of existing markets for recyclables, including capacity and demand
- Projected future market for recyclables
- Market development initiatives
- Summary of conclusions and recommended actions

These studies may be useful as models for conducting market surveys:

Assessment of Markets for King County Recyclable Materials
King County Commission for Marketing Recyclable Materials
400 Yesler Way, Suite 200
Seattle, WA 98104
Phone: (206) 296-4439
www.metrokc.gov/market

Recycling Economic Information Study
Northeast Recycling Council
139 Main Street, Suite 401
Brattleboro, VT 05301
Phone: (802) 254-3636
www.nerc.org

Materials for Recycling

The quality and degree of separation of recovered material has an impact on the potential end use for that material. A list of some possible end uses is presented in Table 9.2, and the general handling for each material is discussed below. When marketing materials, it is important to be aware of the material specifications required. Material specifications have been prepared by the Institute for Scrap Recycling Industries (ISRI) and generally include:

TABLE 9.2 Examples of Recycled Materials and Products Made from Them

Material	Application
Glass	New containers, water filtration, sandblasting, various fills, asphalt and aggregate blends, creative uses, insulation
Plastics	
HDPE	Rigid containers, film, pallets, lumber
PET	Carpet, textiles, rigid bottles, clothing
Other plastics	Lumber, bags
Aluminum	Beverage containers
Tires	Fill material, mixed with asphalt, mulch groundcover
Ferrous metals	Steel products
Yard waste	Compost, mulch
Wood	Fiberboard, mulch, paper
Paper	New paper, insulation, mulch, animal bedding, wallboard, packing and fill material, molded packaging
Waste oil	Rerefined motor oil
Textiles	Yarn, paper, industrial wiping cloths
Batteries	Reclamation of silver oxide, mercuric oxide, and nickel-cadmium

1. A description of the material and its physical state
2. Materials other than the material in question, also known as outthrows
3. Materials that will make a product unusable at a specified amount, also known as prohibitive materials
4. Other processing requirements, such as degree of separation, baling, pelletizing, and shredding

Paper. Paper is recovered for sale in domestic and foreign markets. Many different grade descriptions exist, 51 standard specifications and 33 specialty grades, including news, magazines, corrugated, sorted office paper, telephone books, and kraft. The price of paper can fluctuate significantly, based on the demand by foreign and domestic markets. Paper is generally sorted by local processors, baled, and then exchanged via a broker.

Plastics. Seven grades of plastic are labeled to aid in recycling. Plastics can be separated by grade and color; the most popular grades of plastic for recycling are high-density polyethylene (HDPE) and polyethylene terephthalate (PET) because of higher market demand. Consumer products made with HDPE include milk jugs and plastic lumber, while most other containers for household products, including soda bottles, are packaged in PET. Due to plastic material properties and processing methods, there is potential for unintended human exposure to toxic compounds when using recovered plastics of unknown origin. To ensure safety, the FDA approves plastics recycling for food packaging on a case-by-case basis. Plastics are generally separated and baled, flaked, or pelletized for sale in recycling markets.

Metals. Metals separated for recovery include ferrous (iron, steel, tin) and nonferrous (aluminum, copper, brass). Steel is recovered from many sources including automobiles, white goods, cans, and structural members, and is remanufactured back into those same items. When recovering steel from appliances and automobiles, contaminants must be removed by disassembly or magnetic separation. Aluminum cans recovered from municipal recycling are baled for shipment to manufacturing facilities. End markets for steel include export markets, detinning facilities, steel mills, and foundries.

Glass.   Glass for recycling can be separated by color (clear, brown, and green) and be free of contamination. Processing typically involves crushing and cleaning the glass, producing what is known as *cullet*. Glass cullet can by used to make new containers or, if sufficient prohibitive materials exist, used as an additive in construction materials. Contaminants include ceramics, paper labels, and metal caps.

Organics.   Green waste can be collected for large-scale composting. Compost can be given or sold back to the community, used in nurseries and for landscaping, placed as a daily cover at landfills, and added to agricultural fields as a soil amendment, among other applications. Green waste for composting can be collected at curbside or at designated drop-off locations. Generally, the composting of green waste is not economical but can substantially reduce the volume of waste going to landfills, thus reducing tipping fees. Various standards for gauging compost quality are available depending on the end use, including duration of active composting (at specified temperatures and turning frequencies) and curing; nutrient and organic matter contents; lack of contamination, viable weed seeds and pathogens; and values for pH, metals, salts, and nonbiodegradable chemicals (i.e., pesticides) present in the feed stock.

Commodity Prices

A rapid expansion of curbside collection programs combined with underdeveloped capacity to use recycled materials can create an imbalanced market where supplies exceed demands. When such a situation develops, prices paid for recyclable materials will decrease, thereby increasing the net cost to governments and taxpayers to operate recycling programs. The glut of materials and low prices in the early 1990s illustrated that the success of solid waste management policies is closely linked to private sector material markets. If markets for recycled commodities do not exist, the collected recyclables may have to be incinerated or landfilled. Well-intentioned policies to encourage recycling market development can create instability if end-use opportunities are inadequate.

Marketing of recycled materials is an extremely complicated process. Prices paid by consumers of recycled materials are in constant fluctuation. They vary not only with time, but also differ substantially with geographic location. There exists as yet no on-line clearance center that provides information for prices on a geographic basis in real time. *Waste News* gives prices of major recycled materials for eight major cities, but regional price movements are constant. Consequently, recycle managers rely mostly on a networking system for successful operation.

The cost of processing recycled material varies with location. Many recycle managers believe that transportation costs can be the key to a successful recycling operation. This is particularly true for paper products, which find some of their most important markets overseas. Consequently, the demand-and-supply picture for paper products depends heavily on whether or not Asian countries buy the wastepaper.

Although the total market for recycled products is only a small fraction of the market for virgin products, the recycle market is enormous. Working in the recycle market, however, is more difficult than in conventional markets because it is in constant fluctuation. Moreover, the economic conditions overseas are crucial to its success; for example, the market for recycling paper depends on conditions in China and Japan. When Asia buys, prices go up; when Asia does not buy, prices go down.

In addition to the prices that can be obtained for recycled products from manufacturers, the cost of shipping the material to the manufacturers varies enormously and can play havoc with the market. For example, the cost of shipping depends on the availability of C containers, which can be transported both by truck and by ship. Depending upon political considerations, the availability of C containers varies. If they are not available, recycled materials need to be stored. The cost of shipping also depends on the value of the material in the container and insurance cost.

Several organizations provide information on commodity prices or facilitate material trading, including the following:

Global Recycling Network (www.grn.com)

Plastics News (www.plasticsnews.com)

Recycler's World (www.recycle.net)

Waste News (www.wastenews.com)

Identifying Markets

Recycling markets can be categorized in three ways: (1) by the activities or services performed, these activities or services encompass collectors, haulers, processors or intermediaries, and manufacturers that use the recyclable material to make new products; (2) by geographic locations, market location can be local, regional, national, or international; or (3) by the type of recyclables marketed, recyclable commodities typically include paper, metal, glass, and plastic.

Information obtained from a market analysis can be used to identify markets. Potential markets should be reliable and close enough to ensure that transportation costs are not restrictive. Only materials that have a market should be considered for recycling. The contract components listed should be clearly stated in contracts with intermediaries and end users (Powelson and Powelson, 1992).

- Name and contact info for both parties.
- Commonly recognized material specifications.
- Quantity per week/month to be provided.
- Prices paid for materials (fixed or fluctuating, and basis for price).
- Packaging and shipping requirements.
- Terms of payment.
- Length of agreement and terms of renewal.
- Rights to terminate agreement.
- Finally, the contract should be reviewed by an attorney.

Markets should be monitored for changes in the supply and demand for materials, as well as the status of processing facilities. Possible outlets for selling recyclable materials include scrap brokers, scrap processors and dealers, and end users.

Scrap Brokers.   Intermediaries who do not take possession of recyclable commodities, but do take part in facilitating material exchange, are known as brokers. Brokers offer experience on the cost of material transport and knowledge of current markets. In some cases, the use of a broker can make material markets accessible, including foreign markets.

Scrap Processors.   Scrap processors are intermediaries that process material for sale. The method by which the scrap processor acquires the recyclable materials can include collection from discrete locations such as individual residences, collection from centralized drop-off sites, or direct drop-off at the processing facility. Scrap processors may also accept recyclables at various degrees of separation, ranging from homogeneous material to commingled waste. Processing activities increase the market value of the materials and can entail cleaning, sorting into grades, and baling of the material. Scrap processors can be identified by searching in phone directories and trade organizations.

End Users.   End users are manufacturers that will use the materials to make a usable product. Marketing materials to end users may offer the advantage of leaving out costs associated with intermediate processors; however, end users can have more rigid requirements for material exchange. Generally, end users require a larger, more consistent supply of material. To increase the amount of material, it may be possible to work cooperatively with other organizations within close proximity, pooling recovered materials for marketing. In this type of joint effort, the portion of the recyclables with the poorest quality will limit the value of the material. End users may also have specific processing, handling, and shipping requirements.

Identification of end users can be facilitated by trade organizations, Internet sites, phone directories, and state environmental departments. Many organizations representing the users and processors of materials can provide information to assist with market analysis efforts. A partial list of these trade organizations along with web site addresses is presented:

Institute of Scrap Recycling Industries (www.isri.org)

Glass Packaging Institute (www.gpi.org)

American Plastics Council (www.americanplasticscouncil.org)

American Forest and Paper Association (www.afandpa.org)

Aluminum Association (www.aluminum.org)

Steel Recycling Institute (www.recycle-steel.org)

U.S. Composting Council (www.compostingcouncil.org)

9.3 MARKET DEVELOPMENT

Developing new and existing markets for recycled materials can create an economically sustainable recycling system that does not require subsidies or government intervention to operate. Markets for recycled materials can be improved by utilizing strategies that enhance the supply of recyclables or increase the demand for recycled products.

Factors that improve the supply are known as *supply-side* tools, and factors that increase the demand are known as *demand-side* tools. A summary of some factors influencing supply and demand for various recyclables is presented in Table 9.3. The supply and demand are

TABLE 9.3 Some Factors Influencing Material Processing and Value

Material	Supply	End-use demand
Aluminum	Contaminants (bottle caps, steel, lead)	More efficient to recycle than to use virgin material
Ferrous	Hazardous components	More efficient to recycle than to use virgin material
Glass	Mixed colors; contaminants (ceramics, window glass, etc.); cost of grinding	Cost of competing aggregates
Paper	Grade of paper; degree of separation; contamination with other grades (office, newsprint, magazines)	Cost of various pulps; limitations of recycling (e.g., length of fiber); capital costs to retrofit existing process to accept recovered paper
Plastic	Contaminants (caps, residual contents, medical waste); UV degradation; mixed resins and colors	Required content laws, procurement programs, FDA-approved uses
Yard waste	Contaminants, landfill bans	Quality requirements, availability of markets

inherently linked, so it is important to consider both aspects when considering recycling markets. When the supply or demand is insufficient, an unbalanced market may result and threaten the feasibility of recycling. Thus, the primary goal of recycling market development is to create a system of material flow that is stable and economically sustainable.

A report prepared by Mt. Auburn Associates for the office of Recycling Market Development in New York identified five steps that should be taken before selecting specific methods for secondary material market development.

1. *Set principles and goals.* Development of policies should be guided by clearly identified goals because each material requires a different policy approach. Manufacturers have unique uses for materials, and the market infrastructure development and the range of market structures will depend on the demand for various materials.

2. *Identify priority materials.* In addition to supply-and-demand issues, opportunities to conserve resources, minimize solid waste, and reduce toxic pollution, as well as to pursue economic development and job creations, should be included.

3. *Identify key leverage points for action.* This step should analyze how difficult it is to recover the material and what the environmental effects might be. This step should also identify whether collection and transporting recycled material would be a problem and should calculate the processing capability and capacity for the material.

4. *Select appropriate marketing tools.* Separate marketing tools may be necessary for different materials.

5. *Evaluate the program impact.* A continuing evaluation of the program should be planned at the outset because it is often possible to use this evaluation to make changes to improve the program's effectiveness after it has been initiated. There should be a mechanism for constant feedback to the responsible organization with suggestions to improve the program or policy.

Demand-side

Measures that stimulate a stable demand for goods made from recycled materials are called demand-side tools. Creating a demand for recovered material is attractive because good markets help to cover the cost of collecting recyclables, avoid disposal cost, conserve resources, and provide economic development opportunities. These measures can also create jobs in depressed areas and provide economic stimuli.

Government Procurement Programs. State and federal governments have enormous purchasing power that can be used to promote markets for recycled materials. Such programs are usually incorporated into purchasing guidelines that require "buy recycled" purchases and/or price preferences for recycled products. Some states have formed regional coalitions to increase the cost-effectiveness of recycled procurement programs. All 50 states, as well as the federal government, have laws and regulations that address the purchase of recycled products in some way. Federal agencies are required to purchase recycled products by Executive Order 13101 (1999), an update to Executive Order 12873 (1993). The new mandate, for example, increases the amount of postconsumer recycled material from 20 to 30 percent, and calls for building and improving markets for other recyclable materials.

Government procurement practices traditionally have focused on recycled paper products. Recently, however, several states have included other goods such as oil, tires, paint, and asphalt made from secondary materials. In 1995, the U.S. Environmental Protection Agency (EPA) adopted comprehensive procurement guidelines (CPGs) for federal agencies. These guidelines form the basis of the federal government's purchasing strategies to increase the use of recycled materials, including paper and nonpaper office products, construction products (pipes, carpet, and tiles), transportation products (traffic control cones and barriers), and park and recreation products (playground surfaces, running tracks, and park benches). The CPGs are updated periodically to reflect changes in product availability and to increase the recycled content requirements. Currently, 54 products have been specified by CPGs. As of January 2000, the following products had been given purchasing guidelines by the EPA:

Construction products
  Building insulation products
  Carpet
  Cement and concrete containing coal fly ash and ground or granulated blast furnace slag
  Consolidated and reprocessed latex paint
  Floor tiles
  Laminated paperboard
  Patio blocks
  Shower and restroom dividers or partitions
  Structural fiberboard
  Carpet cushion
  Flowable fill
  Railroad grade crossing surfaces
Paper and paper products
  Commercial and industrial sanitary tissue products
  Miscellaneous papers
  Newsprint
  Paperboard and packaging products
  Printing and writing papers
Park and recreation products
  Plastic fencing
  Playground surfaces
  Running tracks
  Park benches and picnic tables
  Playground equipment
  Landscaping products
  Garden and soaker hoses
  Hydraulic mulch
  Lawn and garden edging
  Yard trimmings compost
  Food waste compost
  Plastic lumber landscaping timbers and posts
Nonpaper office products
  Binders (paper, plastic-covered)
  Office recycling containers
  Office waste receptacles
  Plastic desktop accessories
  Plastic envelopes
  Plastic waste bags

Printer ribbons

Toner cartridges

Solid plastic binders

Plastic clipboards

Plastic file folders

Plastic clip portfolios

Plastic presentation folders

Transportation products

Channelizers

Delineators

Flexible delineators

Parking stops

Traffic barricades

Traffic cones

Vehicular products

Engine coolants

Rerefined lubricating oils

Retread tires

Miscellaneous

Pallets

Absorbents and adsorbents

Awards and plaques

Industrial drums

Mats

Nonroad signs, including sign supports and posts

Manual-grade strapping

The EPA also issues recycled material advisory notices (RMANs) as recommendations for recycled content in various products. The RMAN guidelines are based on currently available recycled content ranges. While the CPG and RMAN product listings are intended for use by the federal government, the recycled product guidelines can also be used as a research tool to find products made with recycled content.

Price preferences allow government agencies to purchase products made from recycled materials at a higher price than similar products made with virgin material. Such price preferences typically range between 5 and 10 percent, although some states allow preferences as high as 20 percent. Some organizations require that a certain percentage of purchases contain recycled materials, also known as a set-aside. All 50 states have enacted price preference or set-aside laws.

Education.   Consumer education about recycling is a mandatory step for the success of any integrated recycling system. All too often, people enthusiastic about recycling simply place the material at the curbside or take it to a recovery center and then assume that someone else will take care of the essential steps involved in selling, reusing, and marketing the products. Educated consumers realize that, in addition to diverting their recyclables from disposal, purchasing products made from recycled material creates demand for those products and closes the recycling loop.

"Buy Recycled" campaigns can be developed to educate consumers. Eco-labeling is a way that manufacturers can inform consumers of the environmental qualities of their products.

For eco-labeling to be successful, consumers need to be familiar with recycling terminology and have confidence that environmental claims such as "this product is made from recycled material" are true. It is also important that the percentage of recycled material in products such as packing or plastic be clearly identified. In 1992 the Federal Trade Commission (FTC) issued guidelines for the use of environmental marketing claims; changes and additions are regularly published in the federal register. Educated consumers will differentiate between labeling that identifies products as postconsumer, preconsumer, recycled, or recyclable. The more educated that consumers are, the more willing they will be to participate in purchasing recycled products. Educational measures that encourage citizens and organizations to purchase recycled products include the distribution of recycled product guides, advertising on the radio and in newspapers, and school programs.

Recently, states have developed recycled products directories to improve awareness of product availability and thereby increase demand for recycled products. Some states have promoted labeling on recycled products, including the amount of recycled and virgin material. Most important in the education program, however, is the need to accept the fact that recycled goods can be more expensive then similar products made with virgin material. Without the consumers' willingness to contribute financially, recycling programs are likely to fail in the long haul.

Recycled Content Laws and Utilization Programs.   Utilization rate programs require manufacturers to ensure that a certain percentage of their products meet a minimum recycled content. Requiring recycled content is a way of directly increasing the use of recycled material in products, although it is difficult to enforce and is a barrier to interstate commerce. Products such as plastic containers and newsprint have been targets for recycled content mandates.

To increase the recycling rate of plastics, California passed the Rigid Plastic Packaging Container (RPPC) Act (1991) that requires rigid plastic containers offered for sale in the state to meet certain performance or content requirements. The RPPC must (1) have a recycling rate of 25 percent, unless made from PET or brand-specific plastic, in which case the recycling rate is increased to 55 and 45 percent, respectively, (2) be made with a minimum 25 percent (soon to increase to 35 percent) postconsumer material, (3) be lightweighted 10 percent, or (4) be reusable or refillable 5 times. Exemptions to the law include RPPCs intended for food, medication, and other sensitive applications. In addition, California requires a minimum recycled content in trash bags, glass, and newspaper. Oregon has a similar law for RPPCs requiring minimum recycling, postconsumer content, or reusability.

Product Standards and Specifications.   Standards and specifications for recycled products build consumer confidence in the recycled products market and often allow those products to be used in more applications. Standards and specifications for a product can guarantee a certain level of performance or ensure that a product is safe for its intended use.

The plastic lumber industry continues to expand after the recent development of standards. There are many different methods and constituents used in the production of plastic lumber, including fiberglass, wood, and rubber, making comparison difficult. Currently, plastic lumber is used in a broad range of applications, including bridges, railroad ties, benches, picnic tables, and fences. The development of standards by the American Society for Testing and Materials (ASTM) is allowing the plastic lumber industry to expand into new applications. In many cases, plastic lumber is being considered as a replacement for pressure-treated lumber, which typically presents a disposal problem after its useful life. Standards have allowed the market to expand to new applications and may soon include structural members.

Financial Tools.   Financial tools can be used to give products made from recycled materials a cost advantage over products made with virgin materials. The capital investment to obtain equipment for processing the recyclable material; the ongoing costs of collection, separation, and processing of the recyclable material; and market fluctuations can make recycling financially unstable. In addition, the financial incentives that exist for the extraction of virgin materials may also hinder the ability of recycling to compete economically. Thus, financial tools can be used to generate market competition with virgin materials.

A number of tools that have been used to stimulate markets for recyclable materials include grants and loans from government agencies to manufacturers, tax policies favoring recycled material, deposit laws for beverage containers, and advanced disposal fees (ADFs). These tools can affect the demand side or supply side of recycling markets and should be considered in the development and assessment of a recycling policy.

Grants and loans can be used to fund the development of industries to utilize recovered materials. Local, state, and federal funds are often available for this type of activity but must be budgeted for accordingly. Grant and loan programs can also be used to educate consumers on the importance of correctly separating materials and purchasing products made with recovered material. Funding demonstration and pilot projects can provide a basis for decision making.

Tax policies can be used to increase the cost of products that are not recyclable or made from recycled materials (virgin material fees), increase the cost of disposal, or provide credit for investing in recycling equipment or utilizing scrap materials.

Deposit laws are most often applied to increase the recovery of beverage containers. Generally, a deposit is charged at the time the product is purchased and refunded when the empty container is returned. Initially, bottle deposits were intended to reduce the amount of roadside litter; however, bottle deposits now represent one way that communities can meet recovery mandates.

Advanced disposal fees are applied to products at the time of sale to fund their future management in solid waste management systems. Advanced disposal fees are often applied to tires and white goods.

Financial incentives can also be used to encourage the use of processes that conserve resources. For example, recycling often requires less energy and water, and produces less waste and emissions, than the mining and processing of virgin materials. The reduced environmental impact of recycling will become more important as energy and water resources become limited.

Other Resources.   In addition to the above marketing tools, various management techniques are available for market development. These techniques include export promotion, technical assistance, recycling market development zones (RMDZs), and Internet services to exchange information and material prices. Although none of these tools alone is adequate to maintain a program, they can help stabilize or support an existing program.

Export promotion can include activities such as hosting workshops to educate businesses on accessing export markets, monitoring and providing information on current export markets and trade issues, and providing financial and consulting support to organizations interested in export. Networking can be facilitated by publishing newsletters, hosting workshops, and attending conferences and trade shows.

Technical assistance programs can provide manufacturers with information such as how they could utilize recycled materials in their production process, the potential problems that may be encountered in using recycled material, and the status of material supply. Technical assistance programs can also be used to promote new or improved technologies for reprocessing materials. For example, recent advances in deinking paper have made magazine recycling more economical.

The EPA's Jobs Through Recycling (JTR) program (www.epa.gov/jtr) provides financial, technical, and networking assistance to create and improve recycling market development. Most state and local governments provide technical support, generally via the Internet or recycling hot lines. In addition, the following organizations may be able to provide assistance on market development issues:

Clean Washington Center (www.cwc.org)

Materials for the Future Foundation (www.materials4future.org)

Northeast Recycling Council (www.nerc.org)

Chelsea Center for Recycling and Economic Development (www.chelseacenter.org)

Institute for Local Self-Reliance (www.ilsr.org)

Some states delineate geographic areas as RMDZs and provide financial and technical assistance to foster recycling operations. Incentives that have been offered include low-interest loans, consistent material supply, marketing services, and reduced taxes, permitting, and licensing.

The widespread use of the Internet to transfer information on material prices, availability, and markets has made it possible to dramatically increase the trading of recyclable materials. Services to develop and advertise web sites make it possible to reach markets that would otherwise be unavailable.

Supply Side

In many communities, recycling has been mandated by the local or state government. The effect of large-scale recycling is generally an oversupply of the recyclable material. Thus, the challenge to the recycling marketer is to provide the appropriate quality and quantity of material to the various end users at a minimum cost.

Measures that improve the supply of recyclable material to recycling operations are known as supply-side tools. Improving the supply (quantity and quality) will increase the stability of recycling markets and increase investment in recycling industries. The supply of recyclable goods can be modified with activities such as disposal bans and mandatory recycling programs, removing the barriers to the transport of the material, improving the recovery technologies to correspond with the specifications of available materials, and improving sorting and processing operations to produce a more consistent supply of material. These measures can also create jobs in depressed areas and provide economic stimuli.

Disposal Bans and Disincentives.   Possibly the most effective way to increase the supply of a material is to prohibit its disposal. In some cases, disposal bans and mandatory recycling can cause an oversupply of the material along with reduced commodity prices. Thus, it is important to make sure that sufficient capacity exists to utilize the supply or that the market will be able to adjust to the excess supply.

Making it less economical to dispose of material will encourage people to consider recycling. For example, increasing tipping fees provides an economic incentive to find an alternate market for materials. Factors that should be included in disposal costs include landfill closure and monitoring costs, development of new landfills, and environmental remediation.

Technology.   New technologies to improve or modify the supply of material can be developed to meet the needs of end users. These technologies can focus on areas such as advanced methods of separating materials in a commingled waste stream, cleaning and sterilization techniques of bottles to be reused, or new processing techniques that allow a broader range of materials or material qualities to be used.

Glass and metal are commonly reused or recycled for use in applications where the material will be in contact with food. Because these materials are impermeable, there is little risk of contaminants coming in contact with food. Currently, recycled plastics are approved by the FDA on a case-by-case basis for use in food contact applications. Because of the potential for unknown contaminants to be mixed in with recycled plastic, it is not common to use recycled plastic in contact with food unless the plastic is from a known source, such as industrial scrap. Technological improvements such as coextrusion of virgin and recycled plastics may increase the use of recycled plastic without increasing the risk of contaminants migrating from the plastic to the food. As our understanding of associated risks and technologies improves for eliminating these risks, the market for recycled plastic will increase.

Recycling markets can be promoted by helping manufacturers retrofit existing processing equipment to facilitate the use of recyclable material in place of virgin material. Installing new equipment generally requires a large capital investment, but may result in reduced long-term costs. For example, the technologies used to process paper from wood pulp and recovered materials require different equipment. Improving the technology to be compatible with existing processing techniques will improve the markets for the recovered materials.

Paper-sorting machines have been developed that sort paper many times faster than can be sorted by hand. Automated, high-rate sorting of paper into various grades will increase the economics of paper sorting and result in a higher value material. Technologies that streamline sorting processes and result in a reliable, high-quality supply will make end-use markets more accessible.

Another strategy to improve product recycling is known as *design for recycling*. Design for recycling requires that products be manufactured in such a way that, after their useful life, they can be easily disassembled into various recoverable materials. The electronics industry may be able to make the greatest contribution to this field.

Logistics.   A reliable supply of both quality and quantity of recyclable material is critical to building the relationship between providers of material and end users. Improving the transport and transfer of materials can reduce the distance between the materials and the end users and create a more robust supply of recyclables.

Rail transport can increase the available markets for recyclable materials. While rail transport is more efficient than hauling by truck, rail transport may not be as reliable. Factors such as the cost of fuel, the value of the materials compared to the cost to ship, and state-to-state transport limits and requirements also need to be considered. Backhauling of recovered materials to manufacturing locations may be possible in some areas and can further reduce transport costs.

Disassembly facilities can be used to separate composite wastes into its parts. Before automobiles and appliances are recycled, the hazardous and reusable components are removed. The scrap metal is then shredded and magnetically separated into ferrous and nonferrous metals for recycling. Some auto manufacturers provide disassembly information to increase the recovery of materials from their products.

In some cases, the supply of material from a business may not be sufficient to support recycling markets, or the business may not have enough space to accumulate materials for recycling. If neighboring industries work cooperatively to implement recycling operations and storage of materials, the process will be more effective and will increase the possibility of finding markets for their recyclable material.

Waste Exchanges.   Waste exchanges are programs that are organized to facilitate the trading of recyclable or reusable materials. Benefits of waste exchange programs include: definition of strict material standards to allow trading without being present to confirm the quality, identification of potential markets that otherwise may not have been known, and stabilization of a potentially volatile financial market.

Many waste exchanges are available online, allowing individuals or organizations to post advertisements for available or wanted materials. A database of waste exchanges and their web site addresses may be obtained from the Southern Waste Information eXchange, Inc. (SWIX, www.wastexchange.org).

Material Quality.   The quality of the recyclables can be modified at the source (for example, how residents separate their recyclables), during the collection process, or during sorting. Glass, for example, has a higher market value if the colors are separated, the glass is free of contaminants (such as ceramic), and the bottles are not broken. Several strategies could be used to achieve these results: collectors could take care not to damage the bottles, residents could be educated about the glass recycling process and encouraged to cooperate, and processing systems could be arranged to separate glass into the respective colors.

To gain the highest value for a material, it should be processed to meet the requirements of the end user. In some cases this can entail meeting a certain level of purity, being processed or separated in a specified way, or being free from certain compounds or materials.

The recycling marketer must be aware of the rules for shipping. An important source of information is the Institute of Scrap Recycling Industries (ISRI) (www.isri.org/specs/) circular, which publishes the specifications and shipping rules for scrap materials every 2 years. This publication deals with the four most important recyclable materials: paper, plastics, glass, and metal. A uniform set of standards allows for more efficient exchange of recyclable materials.

Recycling marketers must also be aware that there are many grades of materials on the market. Common types of paper include mixed paper, office paper, newspaper, and magazines. In addition, the quality of the material can be further defined to meet specific needs; for example, newspaper can be sunburned, can contain magazines, or can be blank.

The result of introducing new materials to the market generally makes separation processes more challenging. While new materials may improve packaging efficiency and product shelf life, recycling systems may suffer from increased contamination and higher sorting costs.

Producer Responsibility. Producer responsibility laws require manufacturers to create an infrastructure for collection and management of their products. This type of system has been successful in Europe for the management of packaging waste. Other products, such as batteries, appliances, electronics, and hazardous materials, have also been the targets of producer responsibility laws. Internalizing the costs of waste management makes manufacturers more aware of the problems associated with their products.

Producer responsibility programs can also be used to apply fees to materials based on type and amount. This type of sliding scale encourages manufacturers to use environmentally preferable materials, to consider the life cycle of their products, and to reduce the amount of material used. The money generated from this type of system can then be used to fund collection and processing of the material.

9.4 TRADE ISSUES

The trading of scrap materials can be complicated by trade agreements between countries and restrictions between state and local governments. In addition, the value of commodities and the cost of shipping can change and needs to be monitored. The international markets may require different material specifications for some materials; thus, it is important to be familiar with the current material specifications.

The commerce clause in the U.S. Constitution has been used to establish that states may not restrict wastes originating out of state more than that of waste originating in-state. For example, if in-state or county recycling laws require separation of waste into recyclable and nonrecyclable components, out-of-state waste may also be restricted.

Two important international treaties may influence SWM programs. They are the North American Free Trade Agreement (NAFTA) and the World Trade Organization (WTO), formerly known as the General Agreement on Tariffs and Trades (GATT).

The intent of NAFTA, launched in 1994 (see NAFTA citation), is to remove most tariff and trade barriers between the United States, Canada, and Mexico. This agreement established the largest trading block in the world, and some fear that NAFTA could impede adversely on U.S. recycling efforts. For example, the Canadian virgin newsprint producers might claim that state minimum recycled contents laws are restrictions of trade and therefore illegal. Proponents of NAFTA argue that state laws and regulations can be useful if they have a justifiable objective and are only as stringent as necessary to meet their objective. In NAFTA's dispute resolution process, the agreement places the burden of proof on the complaining party. Article 905 of NAFTA is designed to provide federal, state, provincial, and municipal governments with opportunities to set their own norms of environmental conservation. NAFTA will place increasing scrutiny on recycling requirements, and at this point it is not clear what the outcome will be. Additional details on NAFTA may be found at the following web sites: www.fas.usda.gov/itp/policy/nafta/nafta.html, www.citizen.org/trade/nafta/index.cfm, www.nafta-sec-alena.org/, and www.tech.mit.edu/Bulletins/nafta.html.

The WTO, established in 1995, was designed to help eliminate tariffs and other trade restrictions among countries worldwide. One of its goals is to "harmonize" global, environmental, and food safety standards. Nations that sign on to the WTO must treat imports from other countries on an equal basis as products made domestically. Furthermore, regulations must not be used to give an advantage to recyclable products or punish a product whose manufacturing creates pollution. Opponents of the WTO fear that imported products made from harmful materials or made with environmentally irresponsible processes could not be prohibited. The full effect of the WTO is likely to be decided in the dispute resolution process that will determine if state or national environmental standards are a hindrance to international trade. A description of the WTO documentation system and instructions for using the online documents database has been published by Mesa (2001). Additional details on WTO may be found at the following web sites: www.wto.org/, usinfo.state.gov/topical/econ/wto/, and www.llrx.com/features/wto2.htm.

REFERENCES

ISRI (1998) *Guidelines for Nonferrous Scrap, Ferrous Scrap, Glass Cullet, Paper Stock, Plastic Scrap,* ISRI Scrap Specifications Circular 1998, Institute of Scrap Recycling Industries, (www.isri.org/specs/).

Mesa, J. M. (2001) "Legal and Documentary Research at WTO: The New Documents On-line Database," *Journal of International Economic Law,* vol. 4, p. 245.

NAFTA *North American Free Trade Agreement Implementation Act,* Title 19, United States Code Section 3301.

Powelson, D. R., and M. A. Powelson (1992) *The Recycler's Manual for Business, Government, and the Environmental Community,* Van Nostrand Reinhold, New York.

CHAPTER 10

HOUSEHOLD HAZARDOUS WASTES

David Nightingale
Rachel Donnette

10.1 INTRODUCTION

Household Hazardous Waste (HHW) is a subgroup of solid waste commonly found in MSW as well as in wastewater streams. As the "HHW" term implies, these special wastes originate from households. HHW is categorically and unconditionally excluded from regulation as a hazardous waste under RCRA Subtitle C, so long as it is not mixed with a Subtitle C regulated hazardous waste. Because of the broad hazardous waste exclusion, the "HHW" term is loosely defined and does not necessarily parallel the usual hazardous waste definitions. It is used slightly differently from state to state, and in some cases its exact meaning differs between local jurisdictions.

Programs that handle these wastes are typically concerned about the proper purchase, use, handling, and disposal of products that contain hazardous constituents. The products of concern are not necessarily limited to the RCRA Subtitle C, nor to the Consumer Product Safety Commission's definitions of "hazardous." Discarded fluorescent lamps and consumer electronics are two examples of wastes often viewed as HHW despite mixed results when actually tested as hazardous wastes.

As with other special solid wastes, HHW present inherent problems in safe handling. They also involve human health and environmental hazards. A unique feature of this special waste, due to its ambiguous place between the solid and hazardous waste arenas, is that regulation of facilities that manage HHW vary significantly between jurisdictions. For example, California and New York State regulate HHW under their hazardous waste rules. A few other states regulate HHW with a blended approach of hazardous and solid waste laws. Most states that have rules bearing on HHW regulate it under their solid waste laws.

An illustration of a typical listing of HHW is contained in the draft HHW definition for the revised solid waste regulations from the Washington State Dept. of Ecology (Washington State Department of Ecology, 2000).

> HHW are any household wastes which are generated from the disposal of substances identified by the department as hazardous household substances including but not limited to the following listed waste sources and types:
>
> (i) Repair and Remodeling wastes including: adhesives, glues, cements, roof coatings and sealants, caulkings and sealants, epoxy resins, solvent based paints, solvents and thinners, painter removers and strippers,
> (ii) Cleaning Agent wastes including: oven cleaners; degreasers and spot removers; toilet, drain, and septic cleaners; polishes, waxes, and strippers; deck, patio, and chimney cleaners; solvent cleaning fluids,
> (iii) Pesticide wastes including: insecticides, fungicides, rodenticides, molluscides, wood preservatives, moss retardants and chemical removers, herbicides, and fertilizers containing pesticides,

(iv) Automotive maintenance wastes including: batteries, waxes and cleaners, paints, solvents, cleaners, additives, gasoline, flushes, auto repair materials, motor oil, diesel fuel, and antifreeze,

(v) Hobby and recreation wastes including: paints, thinners, solvents, photo chemicals, pool chemicals, glues, adhesives and cements, inks, dyes, glazes, chemistry sets, pressurized gas containers, white gas, charcoal lighter fluid, and household batteries,

(vi) Other household wastes including: ammunition, asbestos, fireworks, and any other household wastes identified as moderate-risk waste in the planning area's local hazardous waste plan.

In this example, HHW is defined but left open for change at the state or local level over time. Other special wastes sometimes categorized as HHW include fluorescent lamps; freon recovered from white goods; and electronics, including computer components (CPU, monitors, keyboards), televisions, and other electronic equipment. Electronics wastes are an unusual HHW because they often contain both valuable (silver and gold) as well as very toxic (cadmium, lead, and other) heavy metals.

Many of these waste materials share the chemical characteristics that would make them subject to regulation as hazardous wastes if the household exclusion did not exist in federal law. The basic federal categories for hazardous waste determination are: *ignitability, corrosivity, reactivity,* and *toxicity.* These are delineated in Title 40 of the Code of Federal Regulations (CFR) in subsection 261.21.

Many HHW collection programs ship these wastes off-site. In addition, there is an increasing trend in the United States toward collection and shipment of hazardous waste from non-household conditionally exempt small quantity generators (CESQGs) of hazardous waste by HHW programs. In sufficient quantities, these materials must conform to the hazardous waste shipping regulations of the federal Department of Transportation (DOT). DOT relies on the Title 40 CFR definitions to determine some of the shipping requirements. Consequently it is useful to briefly describe those hazardous waste categories:

Ignitability includes liquids with a flash point, at standard temperature and pressure, less than 140°F. It also includes liquids that are ignitable through heat of friction or spontaneous chemical change, oxidizers, and ignitable compressed gas.

Corrosivity includes aqueous wastes with a pH at or below 2.0 (acids) or at or above 12.5 (bases). Corrosives can also be designated by steel corrosion testing over time or animal tests showing irreversible damage to skin tissue.

Reactivity includes unstable chemicals, violent reactions with water, formation of explosive mixtures when mixed with water, formation of toxic gases when mixed with water, detonation or explosive reaction when exposed to pressure or heat, and/or detonation or explosive decomposition or reaction at standard temperature and pressure.

Toxicity includes poisons and other toxic substances that pose a threat to human health, domestic livestock, pets, or wildlife through ingestion, inhalation, or absorption. This is most often expressed in terms of toxicological lethal dose rates of a substance and a certain species.

Solid waste professionals often make waste management program and design decisions based on recycling rate and waste characterization studies. These studies result in measures such as pounds of recyclables set out weekly per household, tons per year of solid waste requiring disposal, percent by weight of the compostable fraction of the waste stream, or other metrics. Similar studies have been performed to quantify the amounts of hazardous wastes from households disposed of in the municipal solid waste stream.

The quantities of HHW have been estimated to be from less than 0.01 percent to as much as 3.4 percent of MSW by weight. This variability is due to a number of factors, including the lack of standardization of what is considered to be HHW, normal variability in generation of HHW (which differs by income and other demographic factors), the method of weighing (with or without the container weight), and statistical variability of measuring a very small part of the waste stream with limited sampling. A commonly assumed value for HHW in the MSW stream is 1 percent by weight.

In addition to HHW present in MSW, HHW is also placed in wastewater treatment systems, including septic systems, dumped on the ground, and diverted to energy recovery and recycling or reuse. King County, Washington, estimates that approximately 35 to 40 percent of HHW is disposed of outside MSW systems. Improperly managed, HHW can negatively affect wastewater systems, local surface- and groundwater, and local air quality, and otherwise impact human health and environmental resources.

Although HHW is a relatively small proportion of the MSW stream, it represents the most toxic part of this waste stream. The hazards of HHW in integrated waste management systems as well as in other contexts are reviewed in the following text.

10.2   PROBLEMS OF HOUSEHOLD HAZARDOUS PRODUCTS

Household hazardous products (HHP) pose risks to personal and environmental health through home use and storage, transport, and disposal. Adverse health effects are most likely to be caused by pesticides, oil-based paints, solvents, adhesives, automotive products, pool chemicals, drugs, and corrosive cleaners. Adverse environmental effects are most likely to result from pesticides and fertilizers, automotive products, and solvent-containing products.

First we will look at risks of HHPs to human health and the environment during the course of home use; then the increased risk of fire; then we will look at the risks of HHPs in the normal course of municipal solid waste management; and finally, impacts to wastewater.

Health Risks

Chemicals in household products can enter the body and cause adverse health effects through ingestion, inhalation, or absorption. Examples of acute effects (felt soon after exposure) from HHP include poisoning from a toxic substance such as antifreeze; burns from an acidic product such as battery acid; or injuries from an exploding aerosol can left too close to a stove. Some products emit toxic fumes that may produce acute reactions such as headaches, fatigue, burning eyes, runny noses, and skin rashes. Examples of HHP, their typical hazardous ingredients, and health hazards are listed in Table 10.1.

Children are at a much higher risk than adults of being poisoned by accidental exposure to household chemicals. The most common products to which children are exposed, as reported to poison control centers, are cosmetics (146,661 calls in 1997); cleaning products (129,346); pesticides (45,391); arts/craft/office supplies (29,771); and antimicrobials (40,546) (Litovitz et al., 1998).

Half of all deaths from cleaning substances are attributed to suicide: In 1997, 25 poisoning deaths from cleaning substances and 17 from pesticides were reported to poison control centers in the United States (Litovitz et al., 1998).

Death also occurs from substance abuse; 48 adolescents died in 1997 from inhaling air fresheners, hydrocarbons, or fluorocarbons. "Huffing" HHP can cause heart and lung failure, paralysis of the breathing mechanisms, or accidents from intoxication. It also can permanently damage the brain, heart, lungs, kidneys, and bone marrow. Inhalant abuse is most common in 10 to 12-year-olds; according to the National Inhalant Prevention Coalition, by the eighth grade one in five young people has tried huffing (NIPC, 1998).

Chronic health effects may result from repeated, long-term exposure to highly toxic products such as automotive solvents, oil-based paints, or pesticides. Chemicals may be stored in the body's fatty tissues and accumulate over time, causing liver or kidney damage, central nervous system damage, cancer and birth defects, paralysis, sterility, and suppression of immune functions.

Pesticide exposure can occur through foods, drinking water, indoor and outdoor home use, and occupational use (including structural application and farming). Children are exposed to pesticides used on school grounds and parks as well as home yards. Toddlers, intimate with the

TABLE 10.1 Hazards of Typical Household Hazardous Products

Product type	Typical hazardous ingredients	Typical product hazards
Drain and oven cleaners	Lye, sulfuric acid	Extremely caustic, eye and skin damage, also reactive, toxic, may be flammable
Spot remover	Trichloroethane, ethylene dichloride, benzene, toluene	Skin and lung irritants, central nervous system depression, liver and kidney damage, flammable
Oil-based paint, paint remover	Petroleum distillates, methylene chloride, toluene, acetone, methanol	Eye, skin, and lung irritants; headaches, nausea, respiratory problems, muscle weakness, liver and kidney damage, flammable
Garden insecticides	Diazinon, acephate, malathion, chlorpyrifos	Skin, eye, and lung irritation; headache, dizziness, nausea, muscle cramps, coma, organ damage
Disinfectants	Chlorine, quats, pine oil, phenol	May be toxic, corrosive, or reactive
Rubber cement	Hexane, heptane, petroleum distillates	Skin and lung irritants, sensitizer, lethal in high concentrations, extremely flammable
Antifreeze	Ethylene glycol	Central nervous system depression, vomiting, drowsiness, respiratory failure, kidney damage

carpet, have high potential exposure to tracked-in pesticides as well as to lead and other automotive by-products. Infants are particularly susceptible to "toxic house dust": their rapidly developing organs are more prone to damage; they have a small fraction of the body weight of an adult and may ingest five times more dust—100 mg a day on the average (Ott and Roberts, 1998). Children may experience neurological effects as well as increased cases of asthma and allergies from pesticide exposure.

Indoor air pollution from everyday household items is now more of a threat to human health than industrial pollution, even for people in communities surrounded by factories (Ott and Roberts, 1998). Paints, dry cleaning solvents, home pesticides, air fresheners, particle board, and glues create indoor air levels of toxic substances that may be 25 times (and occasionally more than 100 times) higher than outdoor levels (U.S. EPA, 1993). These levels of indoor air pollutants are of particular concern because it is estimated that most people spend as much as 90 percent of their time indoors. A steep increase in asthma in the United States has been related to poor indoor air quality (Dickey et al., 1995).

Case studies dating back to the 1940s have documented chemical sensitivities and chronic illness due to synthetic, human-made chemicals (Randolph, 1962). More recently, estrogen mimics have been found in many common chemicals such as pesticides and common surfactants (Dickey et al., 1995). The plethora of new chemicals continually introduced into our environment has created an ever-changing exposure regimen of chemicals new to our species and to others as well.

Increasing cases of indoor air pollution, multiple chemical sensitivities, and various chronic illnesses may be due in part to the untested introduction of chemicals in our environment of our own making and distribution (Randolph, 1962). Multiple chemical sensitivity is usually initiated by acute or prolonged exposure to toxic chemicals such as formaldehyde, epoxy resins, and pesticides. Reactions typically involve the nervous system and can range from respiratory effects to seizures (McDonnell, 2000). Avoidance of the materials involved is the best way to control these hypersensitivity reactions.

Elimination of these materials from households would eliminate the related household hazardous waste generated from disposal of HHP.

Environmental Risks

Environmental risks depend on a particular product's characteristics: its solubility and mobility (chance of moving into surface water or groundwater), persistence and degradability (how long it stays hazardous), toxicity to nonhuman target species, potential for penetrating landfill liners, and potential for being broken down by sewage treatment processes (King County, 1997).

Chemicals that persist in the environment and bioaccumulate in the food chain are of particular concern for environmental quality. Heavy metals such as mercury, lead, and cadmium build up in soils, water, and animals. The U.S. Environmental Protection Agency (EPA) has called for elimination of persistent, bioaccumulative, toxic chemicals (PBTs) from use and in the environment (www.epa.gov/pbt, www.watoxics.org/).

Mercury, a potent nerve toxin, is one of the highest-priority PBTs identified by the U.S. EPA for elimination. Mercury is used in numerous items, including thermometers, blood pressure cuffs, dental fillings, and batteries. The single largest source of mercury in the solid waste stream is medical supplies, which account for about 17 tons per year. Incineration of solid waste can speed the movement of volatilized mercury into the environment. Atmospheric deposition of mercury is 3.4 times what it was 150 years ago (Moore, 1996). When people eat fish and wildlife with bioaccumulated chemicals such as mercury, they can suffer health problems; notably impaired nervous system development in developing or nursing infants.

Storm Water Runoff and HHP.   Storm water runoff is another leading cause of environmental pollution from household hazardous products. Rainfall picks up pesticides and fertilizers used in yards and antifreeze and motor oil spilled on driveways and washes them into local streams and rivers. The U.S. Geological Survey (USGS) found that pesticides commonly sold in the Puget Sound, Washington region for use on lawns and gardens contributed to the occurrence of several pesticides in urban streams (Voss et al., 1999; Bortleson and Davis, 1997). Twenty-three pesticides were detected in water from urban streams during rainstorms; the concentrations of five of these pesticides exceeded limits set to protect aquatic life. Pesticides and fertilizers also leach into groundwater and can result in pollution of nearby water bodies or drinking wells.

Some residents pour unwanted liquids such as motor oil and paint down storm drains, perhaps thinking they lead to wastewater treatment, although storm drains more often lead to streams or ditches. Drips and spills on the driveway are also a problem; antifreeze unfortunately tastes sweet, and can kill pets that lap up too much. Pets are also victims of slug bait and other yard pesticides.

Improper storage of household hazardous products can lead to accidental spills. A hazardous product spill during an earthquake, flood, fire, or hurricane adds to the dangers of these disasters by increasing the hazard of fire, explosion, and water contamination.

Fire Risks

The amount and variety of chemical products in the average home has increased substantially since the early 1960s. In the heat of a house fire, chemical products that might otherwise be relatively safe and kept separated from each other may combine and react. Cans of gasoline or kerosene or exploding aerosol cans that have been heated and contain butane or toluene may act as an accelerant, increasing the intensity and spread-rate of the fire.

An additional hazard may result from toxic compounds that are heated and released in the fire. Certain plastics as well as products that contain pesticides or poisons may exacerbate the danger to firefighters called to a house fire. Vaporized poisonous combustion products are a significant threat to firefighters. If homes in general contain less of these flammable and toxic products, due to the collection and proper management of products no longer useful to the homeowner, there is a direct reduction in the potential threat to firefighters from these home fire hazards.

Articles 79 and 80 in the Uniform Fire Code are dedicated to flammable and other hazardous materials commonly found in buildings that present special dangers to structures, occupants, and firefighters. The other major model buildings and fire codes used in North America have similar provisions. Consequently, fire departments are often interested in HHW programs. Evidence of this is seen in San Bernardino County, California. The San Bernardino County Fire Department serves the largest county in the United States and also has a Hazardous Materials Division that runs one of the oldest and most extensive HHW collection programs in the country.

HHW Toxic Loading and Fate in MSW Handling Systems

The storage, transportation, processing, and final disposal of solid wastes include incidents of personal injury, equipment damage, and toxic loading from the existence of hazardous wastes from households as well as nonhousehold sources. The total loading of HHW in MSW is small and variable as a percentage by weight. Some data suggest that natural degradation and adsorption of some common organic compounds in Subtitle D landfills is significant and may be enhanced in bioreactor landfills.

Toxic Loading in MSW.   Waste sorting studies of HHW in MSW typically select truckloads from residential collection routes. Hazardous wastes attributed to HHW from solid waste characterization studies range from 0.01 to over 3.4 percent. Weights of HHW in MSW are typically found to be in the 0.5 to 1 percent range. Additional toxics are typically found in the larger MSW stream from nonhousehold waste generators. The level of hazardous materials in the waste stream is highly variable.

Because the percentage of HHW in MSW is small, it takes a relatively extensive solid waste sorting effort to create a data set that is statistically significant and representative of HHW present in a particular community. For instance, a waste sort that is adequate to represent the paper fraction, or yard waste in MSW, would typically fall short of providing a parallel level of accuracy for HHW. Nonetheless, many communities include HHW in their broader residential MSW waste characterization efforts in recognition of the inherent problems of this most toxic part of the MSW stream. Other communities relegate the HHW fraction of MSW to categories such as "other" or "special wastes."

For simplicity we will assume 1 percent HHW in MSW. If each person was to generate 1 ton of MSW per year, there would be approximately 20 lb of HHW generated by the average citizen per annum. With an estimated U.S. population of 275 million people, as of June 2000, there would be a potential of 2.75 million tons of HHW entering the MSW stream annually (U.S. CB, 2000).

Using a household-based analysis, as of July 1, 1998, the U.S. Census Bureau estimated there were 101 million U.S. households. To generate 2.75 million tons of HHW per year each household would have to contribute over 54 pounds of HHW per year to the MSW stream. These averages will certainly vary by community and the existence of local programs to divert HHW from the MSW.

Toxics and Selected MSW Management Options.   A U.S. EPA study focused on the effects of toxic compounds in MSW on solid waste management options. While some toxic compounds are generated during the decomposition of MSW, most solvents, metals, and other synthetic toxic compounds are attributable to HHW and similar hazardous material sources. The study recognized the importance of removing HHW from MSW prior to composting to reduce metals and organic compounds. In the literature review, the study found "little data on the effect of toxics in MSW on landfill liner materials." It also concluded that:

> toxics in landfilled MSW may remain in the landfill or be released to the air through volatilization, fugitive dust, and landfill gas emissions or through ground and surface water via landfill leachate. (U.S. EPA, 1995a)

Additional research on HHW regarding MSW management was recommended.

Variations in HHW Generation.   There has been limited study into variation in the generation of HHW. A few factors that have been shown to influence the generation of HHW include the concepts that (1) lower-income households will generate more automotive product-related wastes, and (2) higher-income households with larger well-maintained properties will generate more yard chemical wastes and cleaners and less automotive-related wastes.

Groner performed a more in-depth study of residential habits that identified who was improperly disposing the most HHW. He found a more complex structure of psychological and demographic profiles in a Los Angeles County random stratified survey of nearly 1200 citizens. He discerned six groups of respondents, three of which were likely to improperly dispose of 80 to 90 percent of their HHW. These three groups included (1) middle-class suburban families that are do-it-yourselfers and avid recyclers, (2) middle-aged, middle-class male homeowners who enjoy do-it-yourself projects and are willing to do what is right for the environment; and (3) young live-at-home adults who are self-absorbed, not community oriented, and more concerned with the future than the environment (an interesting paradox) (Groner, 1997).

Another factor in the generation of HHW is that the materials brought into collection sites often average 6 to 10 years old. Clearly, citizens tend to collect and store HHW in the home before throwing it away or taking it in for proper management. The implication here is that products that may have been taken off the market many years ago may still enter the MSW as HHW if not otherwise managed.

There is an apparent connection between residential garbage container size and quantities of HHW placed in MSW. Tucson, Arizona, found that when an MSW collection system goes to a larger household refuse container there is an immediate jump in the HHW disposed of in the solid waste. Rathje has documented a significant rise in HHW, yard waste, textiles, and other recyclables, that coincided with the Tucson switch to larger mechanized collection containers in 1988 (Rathje, 2000). Citizens seem to want to completely fill the larger, one-size-fits-all waste containers. Households fill the larger residential containers with HHW and large-volume recyclables, the very items that many municipal collection programs are increasingly striving to remove from their MSW stream.

HHW Diversion Programs.   HHW may be diverted from MSW through educational programs that change the behavior of households as well as by collection of HHW at special events or permanent collection facilities. Clark County, Washington, which includes the City of Vancouver, provides an interesting case study in measuring HHW diversion (Clark County, 1999). From 1990 to 1993 the county relied on a series of special HHW collection events. Such events typically attract a few percent of the households in their early years but are more valued as an educational tool. During 1992, eight used-oil collection centers were set up in the community and a used-oil curbside collection program was begun. In 1993, two year-round permanent HHW collection facilities began operation. In 1993 and again in 1995 to 1996 Clark County performed solid waste characterization studies including the HHW faction. The HHW percentage in the total MSW waste stream showed a noticeable decline between 1993, when the HHW diversion program was institutionalized and made relatively convenient, and 1996. The results are contained in Table 10.2 (Clark County, 1999)

TABLE 10.2   Clark County HHW Percent by Weight in MSW, 1993 and 1995–1996

Residential waste source	1993 sort	1995–1996 sort	Percent HHW reduced
Residential single family	0.81%	0.52%	35%
Residential multifamily	1.62%	0.57%	65%
Residential self-haul	1.78%	1.26%	29%

Source: Clark County (1999).

Another approach to measuring HHW diversion resulting from collection programs alone is to compare per capita collection with a default baseline generation rate. By 1999 Clark County collected approximately 4.4 pounds per capita of HHW. Assuming a baseline generation rate of 20 pounds per capita per year, the Clark County HHW collection program represents a 22 percent HHW diversion rate, before accounting for any education-induced behavior changes that also result in HHW reductions. This level of toxicity reduction in the MSW stream from HHW removal means that there will be significantly less toxic loading to the landfill leachate and volatile gases from this community's solid waste disposal.

Landfill Toxicity In-Situ Treatment.   Sanin and Barlaz (1998) found that toluene is readily absorbed and biodegraded in excavated MSW from a landfill, and relatively little (less than 8.5 percent) of the toluene is volatilized. This in-situ landfill treatment of toluene, and likely other volatile organic compounds (VOCs), is enhanced by the addition of moisture analogous to leachate recirculation. In addition, this experiment demonstrated the dehalogenation of 1,2,dichloroethane (DCA) to chloroethane (CA) and finally to ethylene when exposed to MSW (Sanin and Barlaz, 1998). These findings show that some common hydrocarbons and halogenated hydrocarbons may be at least partially immobilized and biodegraded in a landfill, especially where accelerated decomposition is fostered by leachate recirculation. This finding suggests that some common household solvents that are landfilled would be likely to be naturally absorbed and degraded in the landfill environment, and the largest fraction of these materials is likely to remain in the landfill system instead of volatilizing.

The broad reduction in leachate strength from leachate recirculation is evident from the Kootenai County (Idaho) Farm Landfill, which began full-scale leachate recirculation in early 1995. The leachate treatment quality dramatically improved from the spring to fall of 1995, as is shown in Table 10.3.

These studies suggest that leachate recirculation may significantly improve the rate of detoxification of the leachate, including some common metals, for standard water quality parameters in addition to the solvents previously identified. As the leachate strength is reduced at a faster rate, the relative threat from landfill liner leakage of leachate to any underlying water resources is also more quickly diminished. It is important to note that although this in-situ landfill leachate treatment reduces the strength of the leachate, it does not remove the need for further treatment by a wastewater treatment plant. However, the toxicity reduction through hazardous materials diversion programs in combination with in-situ treatment may make the leachate less costly to treat.

TABLE 10.3   Leachate Strength Reduction by Leachate Circulation, 1995 Leachate Lagoon Sampling at Kootenai Co., Idaho, Farm Landfill

Leachate lagoon constituent	6/1/95 measurement, mg/l	9/21/95 measurement, mg/l
BOD_5 (carbonaceous)	3891	121
COD	7230	1040
Total suspended solids	898	93
NO_2 and NO_3 as N	0.41	0.15
Chloride	951	1200
Sulfate	320	88
TOC	2440	1160
Iron	46	5.1
Manganese	1.96	0.58
Zinc	1.31	0.05

Source: Adapted from Miller and Emge (1996).

Landfill Gas Quality.   A study of hazardous wastes in MSW and VOCs in landfill gas over time suggest that hazardous wastes are found in relatively large quantities from nonresidential sources and that HHW contribute the smaller fraction of VOCs to landfill gases. In 1981 the County Sanitation Districts of Los Angeles began a hazardous waste screening and exclusion/enforcement program for hazardous waste, focused on businesses as opposed to households. In 1981 the hazardous materials from nonhouseholds were characterized through unannounced searches of random loads that contained up to one-half percent hazardous materials (Huitric, 1999). These hazardous wastes were typically found in large commercial-size containers and case lots of discarded products readily identifiable when MSW was spread about 1 ft thick on a pad. This was quickly followed with upgraded employee training, full-time waste screening staff, and an enforcement program through the district attorney's office (Huitric, 2000).

Between 1981 and 1984, the hazardous waste found in the Los Angeles MSW from predominantly nonhousehold sources was reduced to less than 10 percent of the original quantities based on the waste exclusion program and the associated random sampling. By 1999 the average hazardous waste content was "less than 1 percent of the 5000 ppm initially found in 1981" for nonhousehold sources of hazardous waste (Huitric, 1999).

In parallel with this waste exclusion data were dramatically declining trends in five common VOC landfill gases from active landfills receiving this waste stream. The VOCs monitored in the Los Angeles landfill gas study were perchloroethylene, trichloroethylene, vinyl chloride, benzene, and toluene. These compounds are also found in many household and commercial as well as industrial products. The hazardous waste exclusion program focused primarily on nonhousehold quantities of hazardous wastes in MSW and reportedly diverted 99 percent of those hazardous wastes from that source. The landfill gas monitoring demonstrated that the hazardous waste exclusion program "resulted in typical VOC reduction of 80 percent over a 10-year period" again ending in 1999 (Huitric, 1999). Of course these VOCs would have originated from household as well as nonhousehold sources within the MSW being landfilled.

The Los Angeles area typically serves 1 to 2 percent of their households with HHW collection services per year (Huitric, 2000), so the diversion of HHW is certainly much less than the 99 percent diversion of the hazardous waste due to the landfill exclusion program. The typical remaining 20 percent of historic levels of VOCs in landfill gas indicates that there is a significant fraction of volatile compounds remaining in MSW despite the hazardous waste exclusion program. The hazardous waste exclusion program has effectively eliminated the vast majority of large nonhousehold sources. Possible remaining VOCs are likely from HHW in combination with similar harder-to-find small quantities of hazardous waste from nonhousehold sources, CESQGs, and residual outgasing of VOCs from MSW placed before the exclusion program became effective.

It is interesting to note that the hazardous waste from nonhousehold sources was estimated at approximately 0.5 percent by Los Angeles at the beginning of the exclusion program. This is comparable with the normal range typically used for HHW in MSW of between 0.4 to 1 percent. Based on this, diverting HHW and CESQG wastes in significant proportions from MSW could reasonably be expected to further reduce VOCs in landfill gas.

VOCs from Landfills and Off-Site Groundwater Contamination.   In 1995 it was estimated that 70 percent of 544 landfills that had contaminated adjacent aquifers in California included VOC contamination (Pickus, 1996). The VOCs were contributed to the aquifers from either or both the liquid migration of leachate as well as gas migration of VOCs. The fraction of VOCs that volatilize to gas have the potential to migrate to soils adjacent to the landfill and contaminate off-site groundwater.

Landfills with intact modern liner systems have shown off-site groundwater contamination solely from the migration of gas phase VOCs. Careful landfill gas extraction system design and operation can minimize this potential off-site contamination. Reducing the initial VOC loading of MSW from HHW and CESQG waste sources can further reduce the likelihood of any significant negative impacts from off-site landfill gas migration. Cleanup of VOC contamination of groundwater is often very expensive because cleanup levels are often set at drinking water standards (Pickus, 1996).

MSW Landfill Liners, Operations, and Leachate.   In the past it was common practice to place drums of commercial chemical wastes into MSW landfills without liners. The result was that approximately 20 percent of federal Superfund sites were former landfills. The federal Subtitle D landfill standards require removal of the larger quantities of hazardous wastes through waste screening programs; the new Subtitle D landfills are in some aspects approaching the hazardous waste landfill standards. A follow-up question can be stated as, "Are the remaining toxics from HHW and CESQG sources in MSW a significant environmental or operational threat?"

Kinman and Nutini (1988) concluded that the small amounts of HHW combined with such a large mass of solid waste and the additional protection of modern liner systems could not conceivably be problematic. Their conclusion was based on a number of bench-scale test cells of 454 kg MSW each and a few g spike of simulated HHW added to certain cells. Their data show that no significant variation between spiked and unspiked simulated leachate was measured. The mass ratio of 4000:1 was provided as the level of simulated HHW spiking. Assuming a nominal value of 1 percent HHW in MSW based on actual waste sorting data, there should have been about 4 kg of simulated HHW added instead of a few grams. A robust experimental design appears to have been lacking in this study. Consequently, the Kinman and Nutini study, although widely cited, does not provide a valid answer to the question just posed.

Most currently operating landfills include composite landfill liners. The geomembrane, typically high-density polyethylene (HDPE), of the modern liner system is capable of withstanding the typically harsh chemical and physical conditions in the landfill. A U.S. EPA study has shown that the chemical-mechanical properties of these HDPE geomembrane materials should last well over a century (Bonaparte and Othman, 1995). This time factor is significant because the strength and quantity of leachate rapidly diminishes in the first 15 years after "dry-tomb" landfill closure, and faster in a leachate recirculation landfill. In addition, landfill closure caps have been shown to dramatically reduce the quantity of leachate generated by modern landfills. The same U.S. EPA study documents these improvements and suggests the reason for lower-strength leachate is due to reductions in the hazardous chemical loading to MSW by recent environmental initiatives. Specifically cited initiatives were

> increasing the number of chemicals listed as . . . (RCRA) hazardous wastes, lowering the cut-off levels for small quantity hazardous waste generators, and implementation of household hazardous waste pick-up programs. (Bonaparte and Othman, 1995).

In modeling landfill liner performance, it is not assumed that landfills are a sealed system. However, current liner technology and installation methods may be approaching that ideal for leachate containment. These are very encouraging trends, which should result in reducing the likelihood of future leachate impacts to local ground or surface waters. Nonetheless, landfill liner installation technology and best management practices are still evolving. In addition, most landfill liners in use were installed prior to current standard methods and practices.

The effectiveness of a landfill liner and cap system to prevent environmental degradation can be significantly compromised through improper design or installation or operation. The level of leachate leakage through the liner is calculated using an assumed maximum leachate depth (leachate head), typically 2 ft or less. In practice an undersized leachate collection system design, materials failure, improper installation, foundation settlement, adverse weather, poor landfill surface water management, and leachate collection pipe or drainage layer clogging can each result in leachate heads exceeding the maximum design depth. Any liner leakage would be exacerbated with increased leachate head.

The leachate collection system clogging problem was common enough in early composite liner and leachate collection systems that in 1995 U.S. EPA established a new method to determine the long-term permeability of geotextile fabrics used in leachate collection systems. This research resulted in a new ASTM standard for testing geotextile filter fabrics, D 1987 (U.S. EPA, 1995b).

Few MSW landfills have direct measurement of the operating leachate head level. MSW landfills with leachate detection systems and secondary liner systems substantially reduce the potential environmental impact from excess leachate head accumulation and liner leakage. Most MSW landfills are not required to have leachate detection systems and instead rely on monitoring wells to detect contamination in the surrounding aquifer. The assumption with such an approach is that most liners are installed correctly and leachate collection systems will successfully capture all significant quantities of leachate during the active and postclosure life of the landfill. As the majority of modern landfills have been built in the last decade and many are still accepting MSW, it is too early to know the ultimate success of these containment systems in avoiding significant incidences of groundwater contamination.

The presence of HHW and similar materials in MSW would tend to create a leachate of higher strength. Any release of lower-strength leachate to the environment, from MSW with HHW removed, would impact any adjacent water resources less. It can be reasoned that the need for or likelihood of groundwater cleanup resulting from a leachate release can be reduced by removing the most toxic fraction of MSW, which is HHW and CESQG waste. Consequently some solid waste managers consider removing HHW and CESQG wastes as a normal cost of doing business for long-term supplemental pollution liability insurance.

Modern MSW composite-lined landfills have a relatively short operational history and are still not universal. Consequently, it is too early to conclude that the MSW landfills are an environmentally secure place for HHW and CESQG wastes.

MSW Energy Recovery Facilities Ash Quality.   The combustion of MSW involves the potential for pollutants to exit in the stack emissions and generate ash containing pollutants scrubbed from the gas. Both air emissions and ash residue pollutants can be attributed in large part to the quality of the fuel, MSW. During complete combustion the organics in MSW are consumed and the residual ash contains minerals and metals. Of most concern with respect to ash quality are the toxic metals—mercury, cadmium, and lead. These are found in ash from MSW energy-recovery facilities. With the current federal U.S. EPA standards for air pollution control equipment, essentially all of these metals remain in the ash. The major sources in MSW of these three metals were identified by U.S. EPA and are listed in Tables 10.4 through 10.6.

The accumulation of these heavy metals is greater in fly ash, from the pollution control equipment, than in bottom ash that falls to the bottom of the combustor due to gravity. Since the 1990s, battery manufacturers have significantly reduced the mercury content in household

TABLE 10.4   Sources and Tons of Mercury in the Municipal Solid Waste Stream

Product	1970	1980	1989	2000
Household battery	310.8	429.5	621.2	98.5
Electric lighting	19.1	24.3	26.7	40.9
Paint residue	30.2	26.7	18.2	0.5
Fever thermometers	12.2	25.7	16.3	16.8
Thermostats	5.3	7.0	11.2	10.3
Pigments	32.3	23.0	10.0	1.5
Dental uses	9.3	7.1	4.0	2.3
Special paper coating	0.1	1.2	1.0	0.0
Mercury light switches	0.4	0.4	0.4	1.9
Film pack batteries	2.1	2.6	0.0	0.0
Total discards	421.8	547.5	709.0	172.7

Source: U.S. EPA (1992).

TABLE 10.5 Sources and Tons of Lead in the Municipal Solid Waste Stream

Product	1970	1986	2000
Lead-acid battery	83,825	138,043	181,546
Consumer electronics	12,233	58,536	85,032
Glass and ceramics	3465	7956	8910
Plastics	1613	3577	3228
Soldered cans	24,117	2052	787
Pigments	27,020	1131	682
All others	12,567	2,537	1,701
Total discards	164,840	213,652	281,887

Source: U.S. EPA (1989a).

TABLE 10.6 Sources and Tons of Cadmium in the Municipal Solid Waste Stream

Product	1970	1986	2000
Household battery	53	930	2035
Plastics	342	502	380
Consumer electronics	571	161	67
Appliances	107	88	57
Pigments	79	70	93
Glass and ceramics	32	29	37
All others	12	8	11
Total discards	1196	1788	2684

Source: U.S. EPA (1989a).

batteries to minimal levels. The industry-funded Rechargeable Battery Recycling Corporation (RBRC) now collects nickel-cadmium batteries across North America and is instituting a similar collection program for nickel-metal-hydride, lithium-ion, and nonautomotive small sealed lead-acid batteries (RBRC, 2000).

Unless manufacturers reduce the suite of toxic substances present in their products, which eventually may be incinerated, MSW energy-recovery plants will continue to have incentives to manage HHW as well as other toxic-containing products separately from the MSW going for energy recovery. The U.S. EPA studies identified the largest source of mercury and cadmium in MSW as household batteries. These findings led Minnesota, Pennsylvania, and other states to divert these and other HHW from MSW.

MSW Energy Recovery Facilities Explosions.   In addition to the metals loading to the ash residue of energy recovery facilities, there have been a number of damaging explosions from the processing or combustion of MSW. A common incident is a small propane tank or flammable liquid container that is shredded prior to combustion or heated in the combustor, resulting in a damaging explosion. Such incidents have occurred across the country, and can result in from a few hours' downtime and insignificant damage to a few years' downtime and damaged equipment costs approaching a million dollars. These incidents have driven various state and local programs to implement HHW diversion in communities that are served by energy recovery facilities.

Solid Waste Handling and Processing Worker Hazards.   Hazardous materials in the MSW stream are a concern for workers and managers responsible for handling waste. Garbage truck collection haulers and transfer station operators have been exposed to these hazards for many years. This concern has been increasing as MSW is handled more intensely through materials recovery facilities (MRFs), composting operations, and other places where MSW comes in close proximity to workers, and the output product quality of the facility must meet contamination standards to remain marketable. Incidents that exemplify these problems include repeated events of pool chlorine reacting with liquids to form chlorine gas at a transfer station (Austin, 1997); acid sprayed under the pressure of a garbage truck compactor blade, killing a garbage collector (Van Golder, 1996); fumes of pesticides or other unidentified vapors causing acute exposures to landfill workers (Brown, 1998); and other undesirable uncontrolled reactions or exposures (Waste News, 1998).

The Occupational Safety and Health Administration (OSHA) uses standard industrial codes (SIC) to identify various business types where injuries and accidents occur. There is no SIC classification for transfer stations, MRFs, or HHW collection facilities. A statistical analysis was performed by Drudi of the various segments of the MSW system. Refuse collection and disposal activities data showed physical hazards as the top three causes, comprising two-thirds of the fatalities in this group. On the other hand, recycling facilities for paper, plastic, metal scrap, and similar operations showed a pattern of fatalities from 1992 to 1997 as shown in Table 10.7 (Drudi, 1999).

TABLE 10.7   Fatalities at Recycling Facilities, 1992–1997

Fatality event resulted from	Number of fatalities	Percent
Struck by object	42	22
Caught in equipment or objects	36	19
Fires and explosions	24	13
Struck by vehicle or equipment	19	10
Homicides	16	9
Other	50	27
Totals	187	100

Source: Drudi, 1999.

The fires and explosions category is only 13 percent but nevertheless is the third-most-common type of event resulting in fatalities. At a recycling facility, fire or explosion could be caused by many common industrial hazards, such as overheated equipment fires, equipment maintenance, flammable liquid spills, and unsafe fueling of vehicles. It is also likely that some of these events occurred in the presence of hazardous chemicals from households or small businesses mixed in the MSW and handled or processed in a way that lead to an explosion or fire as noted in the preceding sections.

MSW handling and processing is inherently a dangerous occupation. Removing the hazardous fraction of the MSW to the extent practical will reduce the risk to employees and equipment. As accident rates are reduced, insurance premiums often follow.

HHW in Wastewater

Household hazardous products are not only of concern in solid waste planning; they also enter wastewater in a variety of ways. During use and disposal, HHP are washed down the drain into municipal wastewater treatment systems or on-site sanitary systems. Local governments prohibit the discharge of hazardous substances—which may include petroleum products, antifreeze, metals, acids or alkalis, paints, degreasers, solvents, and pesticides—to storm water drains. Clearly many HHWs would exacerbate these pollutant levels if they were diverted from MSW to wastewater treatment systems.

States issue NPDES permits to regulate the discharge of significant quantities of wastewater and materials that could adversely affect the collection system, sewage treatment plants, workers, or the environment.

A nationwide study on cleaning-product disposal (NPD Group, Inc., 1995) found that 70 percent of homes disposed of either a partially full or empty cleaning product container in the past three months; 8 percent disposed of at least one unused (full) product. Renters were more likely to dispose of products (10 percent) than homeowners (7 percent). There was no difference in level of product disposal between those who have septic systems and those on municipal sewage treatment. The product most often disposed of was liquids in bottles, followed by aerosols. Two-thirds (67 percent) of those who disposed of a product left it in the container and put it in the trash, while 10 percent poured the product down the drain or toilet; 1.5 percent brought items to a HHW site, and 0.3 percent poured them outside on the ground or into a storm sewer.

The recommended disposal for HHPs varies, depending on product type and who is providing the recommendation. For example, industry and government have different disposal recommendations for cleaners. The national trade association for the cleaning product industry recommends disposing of general cleaning products, drain openers, toilet bowl cleaners, and liquid metal polishes by pouring them down the sink with running water. Local health departments and others recommend pouring some cleaning products down the drain with lots of water, but advocate saving drain and metal cleaners for HHW collection.

The reason for the stricter recommendations is that sewage treatment plants and on-site sewage systems do not treat heavy metals and pesticides.

Municipal Treatment Plants and HHW.   Secondary treatment removes 85 to 95 percent of conventional pollutants (bacteria, nutrients, solids, and oxygen-demanding substances); 77 percent of the metals; 69 percent of volatile organics; and 78 percent of extractable organics (Brook et al., 2000). This still leaves a significant amount of hazardous materials in the wastewater. Most metals stay in the sludge. Heavy metals are toxic, mobile, persistent, and tend to bioaccumulate, and therefore have very low acceptable concentrations in drinking water. Volatile solvents may evaporate from wastewater treatment plant aeration tanks and become air pollutants. In strong concentrations, solvents, acids, bases, and poisons can cause problems with wastewater treatment plant effluent, worker safety, sludge, and groundwater contamination (Breiteneicher, 1997).

The Massachusetts Water Resources Authority sampled sewers that contained only residential wastewater and found chromium, silver, zinc, pesticides, phenols, acetone, and toluene. They estimate that households make up one-third of their total flow (Breiteneicher, 1997). The Palo Alto (California) Regional Water Quality Control Plant conducted a study of mercury sources to the plant's influent (WasteWatch Center, 1997). They calculated that the plant received 23 pounds of mercury in 1997, and determined that residences accounted for 46 percent of the total.

When sanitary and storm sewers are combined, treatment plants may receive storm water, which carries waste oil and antifreeze poured down storm drains. Used oil and the metals in used antifreeze can disrupt the treatment process. Treatment plants also may receive landfill leachate, with the myriad traces of hazardous chemicals leaking from liquid items disposed in the trash. Even small amounts of pesticides can cause the treatment plant to fail toxicity tests (Breiteneicher, 1997).

Septic Systems and HHW.   Hazardous chemicals disposed of in septic systems can pass through untreated to groundwater (Kolega, 1989). This groundwater may supply private or public wells, leading to the contamination of the drinking water. Septic systems on shorelines must be particularly well maintained to avoid contaminating water, fish, and shellfish. Excess phosphates from detergents can move through well-drained soils to "fertilize" nearby water bodies, which along with nitrates from septic systems and lawn fertilizer runoff result in the overgrowth of algae. Eutrophication occurs when the excess algae use up so much dissolved oxygen that fish die off. Lakeside residents or managers may then treat the algae with an herbicide, temporarily killing the weeds but not addressing the underlying problem of excess nutrients.

High concentrations of common household cleaners, such as liquid bleach, Lysol, and Drano, can destroy bacteria in septic tanks. In one study (Gross, 1987), the products tested and the concentrations that destroyed bacteria were: liquid bleach 7 L (1.85 gal), Lysol 19 L (5.0 gal), and Drano 11.3 g (0.4 oz), tested in a 3780 L (1000 gal) septic tank. Bacterial populations recovered in several days, but during that time effluent was not treated.

Home businesses that can cause problems on septic systems are home day cares (excessive use of bleach), jewelers (metal etching acids), photographers (processing solutions), even illegal methamphetamine drug labs (solvents, fuels, acids).

National Pollutant Discharge Elimination System (NPDES).   The NPDES permit system was the original point-source, end-of-the-pipe, discharge regulatory system under the modern federal Clean Water Act. It has now been expanded to include storm water runoff pollution—that is, rainwater contaminated as it runs off from impervious surfaces or disturbed soils. This is also referred to as non-point-source water pollution because it does not come from a single discharge point, such as a pipe. Storm water pollutants have been found to exceed the effluents discharge contamination levels from some wastewater treatment plants.

Non-point-source pollution is the leading remaining cause of water pollution in the United States today (U.S. EPA, 1999a). These non-point sources include common activities such as construction, commercial chemical, and agricultural products, and waste handling, as well as home maintenance. Improper disposal of antifreeze, motor oil, solvents from painting, and other types of HHW into storm water collection systems is remarkably common in residential areas.

In 1990, Phase I of the NPDES Storm Water Program was implemented for cities with more than 100,000 population. In some of these storm water permits, HHW programs were required as a storm water pollution source control measure. In 1999, U.S. EPA promulgated Phase II of the NPDES Storm Water Program, to include smaller cities in Census Bureau "urbanized areas" not already brought into the NPDES permit system under Phase I. The NPDES authority can also designate cities outside urbanized areas to be under storm water permits. Suggested criteria for inclusion as a designated storm water NPDES permit community include storm water systems in jurisdictions with 10,000 or more population and a density of more than 1000 persons per square mile. These areas will be required to be under a storm water NPDES permit by March 10, 2003, and have all storm water management programs fully implemented by the end of the first permit term, typically five years.

The 1999 storm water rule preamble recognized that HHW collection and education programs have provided a successful means to reduce the improper discharge of household toxics into storm water systems in urban areas (U.S. EPA, 1999b). The Phase II rule will typically be implemented through general permits that require six minimum control measures. HHW collection programs can be used to satisfy at least part of two of these requirements. HHW education and collection programs fit well within the best management practices described by U.S. EPA as minimum control measures of "public education and outreach" as well as "illicit discharges" (U.S. EPA, 2000a; U.S. EPA, 2000b).

This is an opportunity to collaborate on the implementation efforts of HHW programs with local storm water program partners. Existing or planned programs run by others in the community are allowed to substitute for storm water program minimum control measures to avoid duplication and gain overall efficiencies. The U.S. EPA estimates the six minimum control measures to cost nearly $9 per year per household when fully implemented. This presents a significant opportunity to jointly fund existing or desired HHW programs.

HHW and Wellhead Protection

In addition to NPDES water quality permits, protection of water quality through the Safe Drinking Water Act's wellhead protection program also ties into proper use and management of hazardous products and HHW. Public water supplies are required to develop a wellhead protection plan to reduce the likelihood of contaminants polluting water supplies. This typically includes an evaluation of sources of water pollution, which are mapped in a 5- and 10-year time-of-travel wellhead recharge area. In these areas, restrictions on activities or types of land use are enforced through local ordinances. Improper use or disposal of household products and wastes from CESQGs are part of the pollutant sources that are examined in wellhead protection programs. This connection to HHW and CESQG programs is an opportunity to collaborate and increase the effectiveness of both programs.

10.3   HHW REGULATION AND POLICY

This section reviews the layers of government that regulate HHP—from federal and state laws to fire codes, OSHA, training programs and standards, and labeling laws.

Federal Overview

Hazardous waste generated by households is excluded from the federal hazardous waste regulations promulgated under Subtitle C of the Resource Conservation and Recovery Act. Lawmakers considered it impossible to regulate the numerous products containing hazardous chemicals in every house in the United States. This exclusion applies to wastes generated by normal household activities (such as routine house and yard maintenance) from the definition of hazardous waste. The U.S. EPA has expanded the exclusion to include household-like areas, such as bunkhouses, ranger stations, crew quarters, campgrounds, picnic grounds, and day-use recreation areas. While household hazardous waste is excluded from Subtitle C (which establishes the "cradle to grave" management system of hazardous wastes), it is regulated under Subtitle D as a solid waste (40 CFR 261.4[b]).

Subtitle D of RCRA encourages states to develop and implement solid waste management plans. These plans are intended to promote the environmentally sound management of solid waste, including household hazardous waste. The U.S. EPA provides state and local agencies with information, policies, and regulations.

State Overview

While federal law does not require households to separate household hazardous waste from trash, every state in the United States has some type of HHW management program. Rules and regulations vary from state to state. California has the most stringent rules regarding HHW; state law does not allow HHW in the solid waste stream. Many states have laws pertaining to waste oil; some also have rules and regulations on special wastes such as batteries, fluorescent lamps, and mercury. Several states (e.g., California, Vermont, Washington, Minnesota) provide funding to local or regional governments to develop plans to manage HHW.

California, Florida, Massachusetts, New Jersey, Minnesota, and Washington have the largest number of HHW programs, with over 500 collection events or permanent facilities each, including oil recycling facilities. Forty-four states have at least one permanent HHW facility.

Other Regulatory Programs

There are several common regulatory programs and constraints other than waste handling regulations that bear on HHW diversion and collection programs. The intents behind each of these constraints drives the focus of each set of requirements.

Fire Codes.   There are various fire codes in use in the United States and Canada. There has been great effort recently to reconcile the U.S. fire and building codes with each other. The intents of the fire codes are to protect buildings, occupants, and firefighters. One of the major U.S. code systems is the Uniform Code system. In the Uniform Code system, the Uniform Fire Code contains articles that directly affect the design and operation of HHW collection and storage. Specifically, Article 79, Flammable and Combustible Liquids, and Article 80, Hazardous Materials, can significantly influence the cost and design of HHW collection facilities. To design, build, and operate an HHW collection facility requires good working knowledge or a detailed study of these articles or their equivalents in the parallel fire codes. Because these codes are subject to interpretation and exception based on the local fire official's professional judgment, close and early facility design coordination is needed with those officials.

Provisions in the fire codes have typically been developed after significant loss of life or major property damage. Consequently, the requirements are continually under review and revision. In some areas they overlap with employee health and safety requirements, for instance, emergency showers and eye-wash stations. There is also some overlap with environmental regulatory features such as secondary containment of wastes and sprinkler water. In addition, it is not uncommon for the local fire official to require a hazardous materials management plan and hazardous materials inventory statement before issuing a permit for HHW facility construction or operation.

The most common regulatory threshold triggered by the fire code is for the quantity of flammable liquids in storage and the process of bulking flammable liquids from smaller to larger containers. This later regulatory threshold is typically crossed when bulking flammable solvents and oil-based painting products into a 55-gal drum to save transportation and disposal costs. These threshold limits on bulking flammable liquids (and combustible liquids) are found in Article 79 of the Uniform Fire Code and are primarily based on flash-point temperatures of the liquids.

In the Uniform Fire Code a *flammable* liquid, also called a Class I liquid, has a flash-point temperature of less than 100°F. Within this class there are three subclasses, IA, IB, and IC. IA and IB subclasses both have flash-point temperatures below 73°F. IA differs in that the flammable liquid also has a boiling point temperature less than 100°F, which means that it will generate flammable vapors very fast at room temperature when in an unsealed container. Examples of this class are many ethers, and n-pentane. Subclass IB and IC have flash-point temperatures from 73° up to but not including 100°F. Methyl ethyl keytone (MEK), acetone, gasoline, and turpentine are examples of these flammable liquids.

Class II liquids, and higher classes, are called *combustible* liquids. Class II combustible liquids include stoddard solvent, diesel, and naphtha, and have flash points from 100° up to but not including 140°F. Class IIIA combustible liquids have flash points from 140° up to but not including 200°F. Examples of Class IIIA combustible liquids include kerosene and formaldehyde. Class IIIB combustible liquids, such as used oil and malathion, have flash points at or above 200°F.

Class I flammable liquids are the most dangerous from a fire hazard perspective and have the most stringent requirements for use and storage. Inside a conventional building, without automatic sprinklers or other fire prevention upgrades, you are limited to storing or bulking containers of 10 gal or less. The proportion of Class I flammable liquids is relatively small in the HHW waste stream. Most liquid HHW with fire hazards that are commonly bulked into 55-gal drums are Class II or Class IIIA liquids. These quantities typically trigger the need for an automatic fire suppression system and other special construction and safety features in or around the flammable/combustible liquid bulking and drum storage areas.

Hazardous materials that often require special construction, storage, and handling in a building due to fire code provisions are not limited to flammable or combustible liquids. Chemicals that may invoke these fire code requirements include pyrophorics, unstable chemicals, flammable and combustible liquids, peroxides, ammunition and fireworks, strong oxidizers, and highly toxic chemicals. When these materials are present in sufficient quantities or when consolidated into larger containers, the fire and building codes may require many additional safety features. These features may include some or all of the following:

- Secondary spill and leak containment
- Fire-resistant construction components
- Additional exits
- Exit door crash bars
- Larger fire water flow capacity (a 6-in supply pipe is not unusual) or fire water storage tank with gravity flow

- Extra nearby fire hydrants
- Automatic closing fire doors between rooms
- Internal and external alarms
- Flammable gas detection and alarm
- Fire-rated windows
- Explosion-suppressing electrical fixtures and wiring
- Water or chemical automatic fire suppression systems
- Fire extinguishers
- Intrinsically safe or flammable-atmosphere-rated motors and equipment
- Mechanical ventilation
- Safety cabinets
- Automated emergency power
- Generation equipment
- Building setbacks
- Explosion blow-out (pressure relief) panels

These features can easily push the unit costs of those areas up to $150 or $250 per square foot or more. Consequently, segregating these areas from other parts of the building with fire-rated walls can limit the higher unit costs to those areas.

OSHA.   Under the federal OSHA program requirements there is a general health and safety responsibility for all employers of more than 10 employees. This is called the "general duty clause" in Public Law 91-596. It prescribes minimum requirements for the prevention and control of conditions hazardous to workers' health and applies to HHW-handling activities. There are also specific regulations that cover workers at Superfund cleanup sites, emergency responders to hazardous materials spills, and workers at hazardous waste treatment storage and disposal facilities (TSDFs). This specific training is often referred to by its regulatory section number, OSHA Title 29 CFR Sec. 1910.120.

Because HHW is categorically excluded from the federal hazardous waste law, Sec. 1910.120 does not directly apply to workers handling HHW or other exempt hazardous wastes. Nonetheless, this regulatory section contains many concepts and features that can be used as best management practices by the manager of workplaces where hazardous chemicals are present, such as HHW collection facilities. These or similar practices can also be used to implement the general duty clause of OSHA by regulatory agencies.

HHW-Specific Training Programs and Standards

Minnesota, Florida, New Hampshire, California, and Washington have developed or contracted for specialized HHW-handling health and safety training. In addition, based largely on courses developed by Washington and California, the Solid Waste Association of North America (SWANA) and the North American Hazardous Materials Management Association (NAHMMA) have developed a suite of national training courses specific to HHW operations.

Because HHW handling is a relatively recent specialty, there are few standards available regarding safety and health training and none at the federal level. As just mentioned, OSHA regulates hazardous waste spills, hazardous waste treatment storage and disposal facilities (TSDFs), and Superfund cleanup site workers, but not the exempt category of HHW and CESQG wastes. To begin filling this void, in December 1999 the American Society for Testing of Materials (ASTM) adopted a standard guide regarding health and safety training for HHW facilities and collection events, ASTM Standard Guide D 6498. This standard was developed by HHW professionals and industry representatives, including leadership by SWANA and NAHMMA members. The standard guide provides a comprehensive outline of appropriate topics and references for health and safety training of HHW collection site staff. It lists a set of topics and describes the context for those responsible for HHW operations to determine the appropriate level and frequency of staff training. The standard also provides a basis for judging the appropriateness and completeness of training specific to individual jobs (Nightingale, 2000).

Labeling Laws

Federal programs regulate consumer and commercial product labeling to protect humans and in some cases, the environment. In 1960, the Food and Drug Administration drafted the Hazardous Substances Labeling Act (later changed to the Federal Hazardous Substance Act). This law aimed to control the problem of home poisonings that occurred as a result of the proliferation of household chemicals following World War II. It stated that products defined as hazardous had to carry labels with specific cautionary statements. Later, the law was amended to have authority to ban substances found to be too hazardous to be used safely around the household, even with cautionary labeling (U.S. CPSC, 1993).

Three federal agencies determine which household products are hazards. The U.S. EPA regulates pesticides. The Food and Drug Administration regulates food, drugs, and cosmetics. Other products fall under the Consumer Product Safety Commission (CPSC). Table 10.8 summarizes the types of product each agency regulates.

TABLE 10.8   Regulating Agencies for Household Hazardous Products

Regulating agency	Products regulated
U.S. Environmental Protection Agency (U.S. EPA)	Pesticides: insecticides, herbicides, fungicides, rodenticides, disinfectants, chlorine bleach, mildew removers, wood preservatives
Food and Drug Administration (FDA)	Food, drugs (medicines), cosmetics, and personal care products
Consumer Product Safety Commission (CPSC)	Cleaners, nonchlorine bleach, wood finishes, art supplies, other household items not covered by FDA

Source: Brook et al. (2000).

The Consumer Product Safety Commission provides information on product recalls and hazards and implements the Poison Prevention Packaging Act of 1970 (U.S. CPSC, 1993). This Act requires child-resistant packaging and labeling for many household substances, including pharmaceuticals such as aspirin, as well as furniture polish, turpentine, solvents, prepackaged fuel, ethylene glycol, and other highly toxic household chemicals.

The responsibility for pesticide packaging lies within the Environmental Protection Agency under the Federal Insecticide, Fungicide, and Rodenticide Act (FIFRA). Labeling requirements are listed in 40 CFR Chap. 1 Sec. 156.10. As of this writing, pesticide labels carry the following disposal instructions: "Securely wrap original container in several layers of newspaper and discard in trash." The U.S. EPA developed these instructions in the early 1980s to be consistent with RCRA's exclusion of household solid wastes from regulation as hazardous wastes, regardless of the wastes' characteristics (U.S. EPA 2000c). The instructions conflict with state and local instructions, which often direct the disposal of household pesticides to HHW facilities rather than mixed in MSW. The label language also does not promote the U.S. EPA's pollution prevention and waste management goals of source reduction, reuse, and recycling. The U.S. EPA Office of Pesticide programs is reviewing current pesticide labeling and may modify disposal instructions to differentiate between partly filled and empty containers. The proposed label would state, "If empty, place in trash. If partly filled, call your local solid waste management agency or 1-800-CLEANUP for disposal instructions for your area. Never pour unused product down the drain or on the ground" (U.S. EPA, 2000c).

All three programs require cautionary labeling for hazardous products. The U.S. EPA's labeling requirements (for pesticides) differ from CPSC-regulated product labeling requirements. Both require the use of the signal words "Caution," "Warning," and "Danger" to indicate the hazard of the product. The FDA does not have a hierarchy of signal words, but uses the word "Caution" to indicate that some health hazard exists. Table 10.9 compares the basic definitions of the U.S. EPA and the CPSC.

TABLE 10.9   Cautionary Label Systems for Hazardous Products

U.S. EPA		CPSC	
Hazard category	Signal word	Hazard category	Signal word
Toxicity I	Danger	Extremely hazardous	Danger
Toxicity II	Warning	Hazardous	Warning or Caution
Toxicity III	Caution		
Toxicity IV	No signal word	Not hazardous	No signal word

Source: Modified from Brook et al. (2000).

Next to the signal word are phrases that identify the nature of the hazard (i.e., "Harmful or fatal if swallowed," "Flammable," "May cause skin irritation"). In 1990 the CPSC required that art materials be labeled if they pose chronic health hazards, such as cancer, birth defects, or nervous system damage. A small square "health label" emblem is put on these products. Unfortunately, consumers who notice this emblem may believe that it indicates the product is safe for use by children, while actually the symbol is meant to convey the opposite recommendation.

Pesticide labels list some environmental hazards as well as health hazards. Pesticides must also list the chemical(s) that provide the pesticidal effect and the percentage of active ingredients, but the identity of "inert" ingredients is not listed, although they may be quite toxic.

Chronic hazard warnings only appear on products manufactured after the labeling requirements went into effect. Thus many products brought to household hazardous waste collections have been on the garage shelf or under the sink for years and have no hazard labeling at all.

If a CPSC-regulated product is considered hazardous, the label must list the ingredients that contribute to the hazard. The ingredients are not necessarily listed in order of decreasing percentage. The FDA, on the other hand, requires that products it regulates list all ingredients in decreasing order of amount. Also, unlike pesticide labels, CPSC-regulated products have little or no information about environmental hazards and are not required to state proper storage and disposal methods. Many product labels use unregulated words such as "nontoxic," "biodegradable," "environmentally safe," "green," etc.

Two organizations, Green Seal and Scientific Certification Systems (SCS), evaluate product environmental performance. Their logos appear on products that have requested and met their criteria. Note that SCS does not necessarily consider the whole product—it may just rate one aspect, such as packaging.

OSHA requires that Material Safety Data Sheets (MSDS) be available to workers who are exposed to products in their workplace. Inert ingredients in pesticides and their health effects must be listed on MSDS, even if the EPA does not require them on the product label.

In recognition of the importance and confusion surrounding label information, the U.S. EPA formed a Consumer Labeling Initiative to research consumer understanding of home and garden product labels. The results include an educational campaign titled "Read the Label First!" and a redesigned pesticide label (Website www.epa.gov/opptintr/labeling/).

10.4   PRODUCT STEWARDSHIP AND SUSTAINABILITY

Product stewardship is a part of a growing environmental concern in the broader context of sustainability and sustainable development. The terms *sustainability* and *product stewardship* have different implications and definitions, depending on who is framing the discussion. In simple terms, "sustainability" relates to the concept of maintaining or changing current practices so that actions today do not threaten the quality of the environment and ecosystem upon which future generations must depend. "Product stewardship" includes the part of sustainability related to the resources consumed and wastes generated from raw material extraction and processing, production of the product, product use, and final disposition of products.

The final disposition of products has traditionally involved disposal of end-of-life products in the solid waste system with little screening or concern for what went in the dumpster excepting relatively large quantities of hazardous wastes. With the advent of HHW programs as well as the federally mandated hazardous waste screening programs, there is much more scrutiny of what becomes part of the MSW stream. Some solid waste managers are starting to look to the manufacturers of certain products or distributors to take more responsibility for products that contain chemicals of special concern and are unwanted in the MSW. Several local, state, and national approaches to product stewardship are ongoing.

Because sustainability and product stewardship focus on the resource consumption, design and process of manufacturing, and customer use as much as the final disposition of a product, it is not a traditional solid waste disposal issue. However, sustainability is a natural extension of waste reduction and pollution prevention strategies that are widespread and a fundamental part of waste management planning and program implementation. In the same way that waste water systems and wellhead protection programs now look upstream to protect their resources, solid waste managers need to look up the waste stream to protect their systems and the environment from unnecessary degradation as well as to preserve throughput capacity for growing populations. The following paragraphs note a few sample programs and policies which indicate where sustainability concepts have begun to take root.

Product Stewardship

Product stewardship, in its broadest sense, involves taking responsibility for the entire life cycle of a product. This can include all steps of a product, from resource extraction, through manufacturing and marketing, and finally to the user and disposal/reuse/recycling phase. In considering where to begin encouraging product stewardship, solid waste professionals are examining many common solid waste subsets. A primary initial focus has been on subsets of MSW that contain significant amounts of hazardous materials.

An example of a waste that is routinely found in MSW and contains significant hazardous materials is the personal computer. A typical desktop computer weighs approximately 60 lb. One analytical study determined the proportions of plastic, silica, and various metals through a chemical analysis. The metals include precious, commodity, and toxic varieties as shown in Table 10.10.

TABLE 10.10   Composition of a Desktop Personal Computer

Selected PC materials	Pounds per ton of PCs	Percent by weight
Silica	498	24.8803%
Plastics	460	22.9907%
Iron	409	20.4712%
Aluminum	283	14.1723%
Copper	139	6.9287%
Lead	126	6.2988%
Zinc	44.1	2.2046%
Tin	20.2	1.0078%
Nickel	17.0	0.8503%
Silver	0.378	0.0189%
Titanium	0.314	0.0157%
Cadmium	0.188	0.0094%
Chromium	0.126	0.0063%
Mercury	0.044	0.0022%
Gold	0.032	0.0016%
Arsenic	0.026	0.0013%
Palladium	0.006	0.0003%

Source: SVTC (2000).

The precious metals found in the computers included gold, palladium, and silver, and had estimated recyclability levels of between 95 and 99 percent. The commodity metals of aluminum, iron, tin, copper, nickel, and zinc had estimated recyclability levels of between 60 and 90 percent. The toxic metals—lead, chromium, cadmium, mercury, and arsenic—had recyclability levels estimated at between zero and 5 percent.

The relatively high levels of lead in the computer monitors and television sets containing cathode ray tubes (CRTs) have been a concern to solid waste managers. Color CRTs are more likely to fail the TCLP test for lead than monochrome CRTs. The use of brominated additives in the plastics, which provide flame retardancy, has also been a concern. The breakdown products of these flame-retardant chemicals may be persistent bioaccumulative toxic compounds.

Then, there are many businesses salvaging the precious metals from PCs and commercial computers. In both cases, solid waste managers are often looking to the private sector entrepreneurs as well as manufacturers to manage these wastes outside the normal solid waste stream. The broader product category of consumer electronics has received much attention in the United States and in Europe. Various approaches to managing these electronics and other problem wastes, such as batteries and fluorescent lamps, are being developed.

Product Return

Some manufacturers have offered product return programs for cameras, computers, copiers, and other complex and readily recyclable or recoverable products. For instance, Xerox now typically leases copiers to businesses instead of selling copiers. Used models are returned to the manufacturer, and they are designed for easy demanufacturing and also have many parts that are interchangeable between models. In this way the company is selling a service instead of hardware and has a built-in incentive to internalize resource conservation and design more sustainable products. The used machines also provide valuable performance feedback in the design of subsequent product line improvements.

Because of the complexity and specialization of managing technologically sophisticated products at the end of their useful life, manufacturers are often in the best position to develop reasonable product return programs. An example of this is Interface Carpets. Interface manufactures commercial carpets in square carpet tiles. As the carpets wear more in the high traffic areas, the worn carpet tiles are replaced with new ones as part of a carpet lease agreement. Old carpet tiles are sent back to the factory and deconstructed. They then provide the feedstock for new carpet tiles. Large investments in research and capital were required to create the technology for recycling carpet tiles. However, the company's profitability has consistently and dramatically risen as a result of the businesses' new, sustainable approach.

Disposal Bans

Based on environmental concerns, Massachusetts has banned CRTs from landfill disposal. This is a government program intended to encourage a fully functional CRT recovery system in that state. Massachusetts worked over a number of years with industry and public sector stakeholders to develop the methods and infrastructure for waste recycling before instituting a product disposal ban. Similar laws are being considered or have been instituted for other common solid wastes in many states for tires, automotive batteries, construction and demolition wastes, and wood wastes.

Disposal bans are a relatively easy and potentially effective administrative means to divert specified wastes from disposal. Disposal bans can also be short-circuited if nearby disposal facilities do not also ban the same materials.

Industry Licensing Fee System—Batteries

The Rechargeable Battery Recycling Corporation (RBRC) is a private nonprofit corporation created by the battery manufacturers in North America to collect, transport, and recycle certain types of dry cell batteries that contain heavy metals. There are over 300 member companies in the United States and Canada. Each member manufacturer is assessed a licensing fee per weight of battery produced. This fee funds the collection and processing of the batteries so that the metals are of a quality suitable for use in manufacturing new batteries.

Initially RBRC collected and managed only nickel-cadmium (Ni-Cd) batteries. In 2001 RBRC has expanded its collection program to include nickel metal hydride (Ni-MH) and lithium ion (Li-Ion) rechargeable batteries (commonly found in cellular phones, laptop computers, and other portable electronic products) and small sealed lead-acid (Pb) rechargeable batteries.

This industry initiative provides an example of a private sector group working collaboratively to develop an industrywide collection system and recycling capacity for all manufacturers. The RBRC overcame significant technological, legal, and logistical hurdles to develop the system. Congress sponsored national legislation, the "Mercury Containing and Rechargeable Battery management Act" (see Website *www.epa.gov/osw*). RBRC has been persistent in reaching out to communities and retailers to reduce batteries' contribution to the toxicity of the solid waste stream.

European Programs

European countries are ahead of the United States in some important policy areas regarding product stewardship and sustainability issues. Two examples of HHW-related issues that have been developing across Europe are the waste electrical and electronic equipment (WEEE) directive approved by the EU in June 2000, and the ongoing Responsible Care initiative and other efforts sponsored by the European Chemical Industrial Council.

The European Commission, the EU's policy-making body, approved WEEE in June 2000 (Parliament must adopt it before it is legal, which has not yet occurred as of this writing). Waste electrical and electronic equipment includes a broad range of consumer goods including white goods and brown goods such as; (European Commission, 2000):

- Large household appliances
- Medical equipment systems (with the exception of all implanted and infected products)
- Automatic dispensers
- IT & telecommunication equipment
- Electrical and electronic tools
- Monitoring and control instruments
- Small household appliances
- Consumer equipment
- Lighting equipment
- Toys

The European Community, which has been studying WEEE issues since the mid-1990s, has collected or researched the best scientific opinions available on all aspects of this broad waste category. Among other findings were that about 22 percent of the world's mercury is used in the production of electrical and electronic products. WEEE is approximately 4 percent of the current waste stream but is the fastest growing part of the waste stream, expected to double in 12 years at current growth rates of 3 to 5 percent per annum. More than 90 percent of WEEE is landfilled or incinerated. The hazardous content in this waste substream creates an environmental burden far exceeding the remaining MSW substreams.

The preamble to the WEEE directive explains the specific heavy metal and other hazardous contents, their health assessment, exposure assessments, markets, barriers to recycling, policy and legal contexts, and other issues explored and examined in forming the directive, and concludes that the directive "will contribute to the protection of human health and the environment" (European Commission, 2000). The directive's objectives assign a large part of the responsibility for waste management to producers (manufacturers) of WEEE.

The principal objectives of this Proposal are to protect soil, water and air from pollution caused by current management of WEEE, to avoid the generation of waste, which has to be disposed of and to reduce the harmfulness of WEEE. It seeks to preserve valuable resources, in particular energy. Another objective of the proposed Directive is the harmonisation of national measures on the management of WEEE.

The objectives are to be achieved by means of a wide range of measures, including measures on the separate collection of WEEE, the treatment of WEEE and the recovery of such waste.

- Producers should take the responsibility for certain phases of the waste management of their products. This financial or physical responsibility creates an economic incentive for producers to adapt the design of their products to the prerequisites of sound waste management. The financial responsibility of economic operators should also enable private households to return the equipment free of charge.
- Separate collection of WEEE has to be ensured through appropriate systems, so that users can return their electrical and electronic equipment. In order to create a common level playing field between the Member States, a "soft" collection target is provided for.
- In order to ensure improved treatment and re-use/recycling of WEEE, producers have to set up appropriate systems. Certain requirements are prescribed as a minimum standard for the treatment of WEEE. Treatment plants must be certified by the Member State. Targets are laid down for the obligation to re-use, recycle WEEE and recover energy thereof.
- To achieve high collection rates and to facilitate recovery of WEEE, users of electrical and electronic equipment must be informed about their role in this system. The proposed Directive contains

a labeling requirement for equipment which might easily end up in a dustbin. In addition, it will be necessary for producers to inform recyclers about certain aspects of the content of such equipment.

The proposed Directive on the restriction of the use of certain hazardous substances in electrical and electronic equipment will contribute to the same objectives by ensuring that substances causing major problems during the waste management phase, such as lead, mercury, cadmium, hexavalant chromium and certain brominated flame retardants are substituted. (ibid)

The WEEE directive represents a significant governmental regulatory action for the EU after careful and methodical scientific study, risk assessment, stakeholder negotiations, and social political analysis.

The European Chemical Industry Council (CEFIC) represents chemical companies in Europe. The European chemical industry produces more than 30 percent of the world's chemical production (CEFIC, 2000). "Responsible Care" is an international program designed to demonstrate the industries' commitment to continual improvement in all aspects of health, safety, and environmental performance. Other programs of industry-sponsored basic science research include risk assessment methods and interactions between chemicals, human health, and the environment. Basic research is also funded for carcinogenicity, respiratory toxicity, immunotoxicity, and allergies. The chemical industry is focusing special research now on endocrine disruptors. The European budget for basic science research is $5 million per year.

In addition, the CEFIC is promoting programs to look at ways in which manufacturing processes can be made "cleaner, safer, more efficient and less damaging to the environment" (CEFIC, 2000). This program was initiated in 1994 due to public concerns about the chemical industries' products and processes. The objectives of this program are "to introduce new technologies and to encourage innovation by forging partnerships that spread the cost and risk of research and development." The prime motivation for this program is "in response to public concern" and to demonstrate a willingness to work with authorities" (CEFIC, 2000). Clearly the European chemical industry has identified a public concern about the nature and safety of their products. It is also clear from these quotes from their 1999 annual report that CEFIC views these issues as outside problems. CEFIC appears to see these issues only as requiring potential additional costs that must be borne due to the need to invest in new, risky technology with dubious profitability scenarios. The concepts of producer responsibility and sustainability are not yet clearly articulated.

The CEFIC fundamentally supports scientific and cost-benefit based decision-making. Any other factors that ban products or restrict innovative manufacturing or the free marketplace are considered arbitrary and heavy-handed (CEFIC, 2000). This is a very common first response by business leaders and others to suggestions that there may be a better way that was "not invented here." It is also true that outsiders typically do not know what is involved in running a chemical company. Chemists and business executives are rewarded for their ability to perform their job without surprises. Keeping unnecessary complications and untried changes in doing business is viewed as a high-stakes gamble. For most business executives, producer responsibility and integrating sustainability models into business decision methods provides a set of conditions that is unfamiliar and therefore is a barrier to acceptance or serious exploration.

In the United States, the Paint and Coatings Association has had a similar reaction as the European CEFIC when state and local officials have asked about the possibility of increasing the level of involvement or commitment from this industry in product stewardship for used paints. State and local government initiatives focused more on paints as larger and larger quantities are brought into HHW collection programs. Local programs such as in Portland, Oregon, as well as various specialty paint recyclers, have developed some recycling processing capacities to recycle paints. Other local programs have chosen to discontinue acceptance of this HHW or are trying to share the responsibility for this recycling with local paint manufacturers. A significant statewide program involves Massachusetts and Benjamin Moore Paints. Benjamin Moore has agreed to take back paint of its own brand for recycling and disposal from 85 municipal collection points.

10.5   *EDUCATION AND OUTREACH*

As demand for HHW programs has spread throughout the country, the sophistication of public information and education programs has increased. The ideal emphasis for HHW education is to follow the U.S. EPA's Waste Management Hierarchy, (U.S. EPA 1989b), which, from most preferred to least preferred, is: reduction, reuse, recycling, treatment, and finally, disposal. The most preferred option, source reduction, makes economic sense as well as environmental sense. It is much less costly to avoid problems and conserve resources than to try to fix problems after the damage is done.

Many of the risks associated with household hazardous products (e.g., poisoning, nonpoint pollution) occur not only during disposal, but during the use of products. Although hazardous products carry signal words and cautionary statements, studies show that the more familiar that consumers become with a product, the less likely they are to read the label (Sulzberg and Nightingale, 1998). Frequent users pay less attention to warning labels, perceive less danger associated with the product hazards, and are more likely to engage in risk-taking behaviors with the product than will infrequent users of the same product.

Thus education is needed to reduce the amount of hazardous products being purchased, explain precautions during use, encourage proper disposal of unwanted hazardous waste, and encourage participation in HHW collection programs. To achieve source reduction, residents must change existing beliefs and behavior. The philosophy of promoting behavior change is at the heart of successful HHW education.

Education

As with HHW collection programs, HHW education programs vary in scope throughout the country. They may be conducted from the local, regional, or state level. The regional U.S. EPA has nine offices with a HHW specialist. State environmental offices (DEP, DOE, DEQ) house coordinators for HW and HHW education. The Cooperative Extension Service runs Home*A*Syst, a self-assessment checklist to help homeowners identify household risks and take action to create a safer home environment (University of Wisconsin runs their Website: http://uwex.edu/homeasyst). County and city HHW education programs may be implemented by departments of Solid Waste, Public and Environmental Health, Fire Departments, Public Works, Resource Management, Storm Water Utilities, and the like.

HHW education programs typically follow the EPA's hierarchy and emphasize waste reduction, encouraging people to reduce their use of hazardous products through product substitution. For products that do not have a less-toxic substitute, then reuse, recycling, and finally proper disposal are advocated. Programs focus both on specific waste streams (such as pesticides, paints, cleaning products, and used motor oil) and specific audiences (such as school children, rural residents, gardeners, new residents, and do-it-yourself oil changers). In all cases, a less-toxic MSW stream is one beneficial end-result.

Education Program Design

Dana Duxbury of the WasteWatch Center suggests that 10 percent or more of a local HHW budget be devoted to education and reduction (Duxbury, 1995). The ultimate goal is to change behavior—to reduce use of hazardous products and increase appropriate recycling and disposal.

HHW educators compete with—or work with—marketing and media specialists to attract the eye and attention of message-saturated consumers. Written messages, such as brochures, may seem like the easiest way to get information across, but they do not necessarily bring about changes in behavior. Similarly, media (newspaper, radio, television, bus ads, and billboards) are useful to promote awareness; however, increased awareness is only a first step toward behavior change.

The King County (Washington) Local Hazardous Waste Program HHW program uses social marketing techniques to understand its audience and design relevant campaigns (Jenkins-McLean, 1999). The seven steps for successful social marketing (Lee 1997) are to determine the . . .

1. Purpose (Why are you doing this?)
2. Audience (Who are you trying to influence?)
3. Objectives (What specifically do you want them to think or believe or do?)
4. Audience position (How do they feel about the idea?)
5. Strategy (What can you say or do that will persuade them?)
6. Media (How will you reach them?)
7. Evaluation (What happened?)

At the outset of a new project, educators following the social marketing model will conduct a baseline survey or focus groups to gain knowledge of the audience—to measure local knowledge, behavior, and attitudes—and to later be able to assess program results. They will then prioritize particular audiences, messages, and problems and design projects that bring about behavioral change. Every two to three years they will conduct follow-up surveys or use other evaluation methods to track changes in HHW awareness, behavior, levels of concern, and action. Other evaluation methods include tracking product sales, surveys during collection events, and conducting waste sorts. Evaluation is best planned in the initial stages of an event or campaign, when it can be integrated into program design (Donnette, 2000). A guide to planning your project can be found at the Website http://www.toolsofchange.com/English/firstsplit.asp.

To bring about behavior change, HHW educators provide specific steps for people to follow, and work to break down the barriers to following these steps. For example, to overcome obstacles that keep do-it-yourself oil-changers from recycling their used motor oil, Thurston County, Washington, periodically provides free drain-pan recycling containers. It also publicizes local sites that accept used oil at 45 auto stores throughout the county (Thurston County, 2000). Educators use incentives to encourage action, such as give-aways, rebates, discounts, and free expert consultation in use of less-toxic methods.

In the early years of a HHW education program, the focus is on raising awareness of collection opportunities and recognizing what is a HHW. Collection opportunities must be publicized to be effective. Publicity techniques include newspaper articles, radio announcements, flyers, newspaper inserts, banners, door-hangers, and direct mail. Cosponsors can be sought by speaking at local service clubs and organizations. Joint sponsorship means more of the community becomes invested in the HHW collection, and separating HHW is on its way to becoming a "community norm" (McKenzie-Mohr 1996; see the book on-line at www.cbsm.com/Chapters/Ch2.html).

When communities first hold collection events, publicity materials instruct citizens to differentiate between products that are hazardous and those that are not. This may be the first time that consumers realize HHW need special care. By learning what is accepted at household hazardous waste collection sites, residents learn what is improper to dispose of in the garbage, and may then think twice, perhaps reading the label more carefully, the next time they go to purchase or use this product.

This increased consumer awareness leads to an increased demand for information on less-toxic formulas and for less-toxic commercial products. Some companies have responded by formulating low-odor, solvent-free formulations of products such as drain and oven cleaners. The King County Hazardous Waste Management Program Coalition found that consumers need specific information, such as product names, to make informed decisions. They contracted with the Washington Toxics Coalition to study hundreds of cleaners, pest controls, and fertilizers for toxicity and environmental impacts, and ranked them by impact in their publications, *Buy Smart, Buy Safe* (Dickey, 1994) and *Grow Smart, Grow Safe* (WTC, 1998).

HHW programs provide information on alternatives right at collection events, through telephone, written resources, and Websites. It is important to the credibility of programs to test alternatives for effectiveness before making recommendations. Peer review boards with representatives from industry, government, and private groups have formed around the issue of ensuring good-quality information when recommending less-toxic alternatives (see the California Peer Review Project Website www.peerreview.com). The Environmental Hazards Management Institute (EHMI) produced a "Household Product Management Wheel" listing product hazards, alternatives, and recommended disposal options. It was reviewed by approximately 60 representatives of waste management, trade, and research organizations (Website www.EHMIWORLD.org).

Much of the recognition of HHW disposal opportunities comes when a community first opens a HHW collection center or holds their first event. In the following years, when "front page coverage" has faded, it is harder to keep awareness of the program high. Yet new populations keep moving in and so the basic education—what is HHW, how to reduce use, how to properly recycle and dispose—must continue. One way to provide specific advice, reminders, and prompts of desired behavior is through "point of purchase" programs. These on-the-shelf methods disseminate HHW information while consumers are making choices, by placing information on shelves or counters in paint and hardware stores, auto supply stores, and plant nurseries or grocery stores (McKenzie-Mohr 1996).

A dedicated HHW phone line is another powerful ongoing educational aide. It is a convenient, confidential way to provide information to residents unsure of how to handle HHW. It also provides a means to track and respond to interest in advertising campaigns. Seattle's Hazards Line receives over 20,000 calls per year (Local Hazardous Waste Management Program in King County, 1997) (King County, 1995). Phone lines can be answered by staff, or provide a menu of options for taped information.

Youth Education

Many organizations have well-established school programs with carefully reviewed curricula and evaluation methods (see, for example, King County Solid Waste Division, 1995, on the Web at www.metrokc.gov/hazwaste/teachers/). Lessons cover how to identify hazardous products by reading labels; how to recognize the signal words; properties of hazardous products; routes into the environment; health hazards; and alternatives to HHP.

School programs work in several ways: They help youth protect themselves from harm; they educate current and future consumers; and they can get information home to parents. A limitation is that it is unusual for a school district to adopt HHW as part of its regular curriculum. Thus programs typically reach students only when a teacher requests a guest speaker, or is able to attend a training session. A challenge is to integrate HHW issues into the essential learning requirements for science and/or social studies (consumer education).

School programs introduce students to the familiar green Mr. Yuk stickers, produced and distributed by Poison Control Centers for parents to place on hazardous items in the home. During National Poison Prevention Week in March, the U.S. Consumer Product Safety Commission (Website, www.cpsc.gov) and the American Association of Poison Prevention Centers (Website, www.aapcc.org) issue press releases on the need to prevent poisonings.

Other approaches to youth education include scouting (a HHW badge or patch), 4H projects, and public relations television ads run during children's programming.

10.6 HHW COLLECTION, TRENDS, AND INFRASTRUCTURE

Most communities with HHW programs began their activities with occasional HHW collection events and later graduate to more permanent collection systems. Duxbury has estimated that over 85 percent of all HHW collection programs are publicly operated (Duxbury, 1997). This section will briefly review the typical HHW collection methods used by these programs.

HHW Collection Events and Mobile Systems

Collection events are the typical first entry into HHW collection for local communities. This collection method is typified by one or a few times yearly event held at a convenient parking lot that is given over to the event for a few hours or days to collect HHW from local citizens. This collection method usually relies heavily on a waste contracting firm to provide technical expertise and trained laborers.

Advantages of this method often include: requires less local staff expertise, often affords high visibility locations, reduces permitting requirements, provides an indication of community interest level over time, offers relatively low administrative overhead, provides a pilot program without long-term commitments, allows easily varying locations and timing of events. Some disadvantages of collection events include: potential to frustrate customers if lines create long waits, less control over waste processing and packaging efficiency, service may not be available to customers when needed (during moving, when parents die, spring or fall cleaning), lower levels of public participation than permanent program alternatives, and limited control over variable waste handling cost and number customers served.

A variant of the collection event is mobile HHW collection. One of the earliest examples of this method was employed in Klickitat County, Washington, in the summer of 1989. This rural county contracted with a waste firm to provide two box vans, trained staff, ground covers, traffic cones, traffic control signs, and other HHW collection event supplies and equipment for a weekend. The vans, staff, supplies and equipment were scheduled to visit various small towns, a few hours each, over the period of a weekend. This brought brief HHW collection events once or twice a year to small communities spread over a county consisting of 1880 square miles and less than 17,000 people. This mobile technique is often referred to as "tailgate HHW collection."

In September 1989, King County, Washington, implemented the "Wastemobile," a mobile HHW collection system that circulates year-round to urban and rural parts of the county outside the City of Seattle. (Seattle serves its population with two permanent collection sites.) The Wastemobile visits each part of the county about twice per year and has become one of the largest HHW collection systems in the country, serving over 26,000 customer vehicles in 1999.

Permanent Fixed Facilities

After a community has held collection events for a few years it typically looks for more permanent, ongoing collection alternatives provided by fixed HHW collection facilities. Unlike many kinds of solid waste facilities, HHW collection facilities are often cited without opposition. In fact, HHW facility development has been held out in some communities to partially offset objections of other colocated solid waste handling or processing facilities, such as transfer stations and waste-to-energy facilities (Nightingale, 1995).

Advantages of fixed facilities include increased control over the acceptance, sorting, packaging, and consolidation of HHW. This typically results in higher levels of recycling and reuse of HHW received, better education opportunities for customers, lower volumes and unit disposal costs paid to hazardous waste contractors, safer operations, increased availability and convenience of HHW services to citizens, expansion of services to surrounding communities, and increased levels of participation.

It is not unusual to realize a 20 to 50 percent reduction in cost per participant over temporary collection services when the waste is handled more intensively and efficiently at a fixed facility. One of the larger volumes of HHW collected at most programs is latex paint. In Portland, Oregon, the HHW program has a separate facility dedicated to the recycling of latex paint. The paint is blended into different colors and sold to the public in 5 gallon and larger sizes. The latex recycling facility operator expects to be financially self-supporting.

Many HHW programs have reuse areas where pesticides, solvents, cleaners, and other usable products are taken to be used as intended, which saves significant processing, transportation, and disposal costs. Some wastes brought into HHW collections should not be offered for reuse. For instance, in California and Washington, HHW facilities are encountering illegal methamphetamine lab wastes in otherwise normal looking product containers. In addition, customer selections from reuse items that can be abused as illegal inhalants need to be monitored.

One of the most expensive types of HHW disposal is pesticides. Many pesticides arrive in unopened or good condition original containers and can be reused. Care must be taken to sort out and dispose of pesticides that have been banned, suspended, or cancelled by the U.S. EPA. Some programs further restrict the pesticides diverted for reuse based on toxicity levels, available less-toxic substitutes, environmental concerns, or liability concerns.

In many communities a permanent HHW fixed facility provides the anchor for an extended collection infrastructure. Often nearby communities can be served through mobile collection programs or satellite collection stations based from or serviced by a permanent fixed facility. This can provide higher levels of service to less populated or more distant areas and spread fixed capital and operating costs over a larger service area.

There were 450 established permanent HHW collection facilities in the United States and Canada by 1998 (Bickel, 1999). These facilities are not evenly distributed. States with more than 10 permanent HHW collection facilities as of 1995 included only California, Florida, Minnesota, Washington, and Kansas. States with relatively high facility-to-population ratios include Minnesota, Washington, Vermont, Alaska, Florida, and Kansas. In this group, the average population per facility in 1995 ranged between 127,500 persons per facility (for Washington with 40 facilities) to 295,000 persons per facility (for Alaska with 2 facilities) (Sulzberg and Nightingale, 1998).

Optimal service population per HHW collection facility may be in the range of 100,000 to 200,000 persons per facility. In some ways HHW fixed facilities are similar to transfer stations and drop boxes. They can be sized to fit the local communities' needs and service level demand. There are also many variations of facilities, some of which take a select few types of high volume, low-risk wastes, while others take all types and HHW and CESQG wastes.

High Volume, Low-Risk HHW Collection Centers

For customer convenience, antifreeze, batteries, used oil, and paints are often collected at special collection sites. These sites increase the total collection volume and customers served without commensurate capital and labor expenditures required at a full-service HHW collection facility. Some local and state authorities require these facilities to be staffed while open and others are unstaffed do-it-yourself collection points much like drop-off recycling centers. In Washington State, over 500 used-oil collection sites annually collect approximately 10 million pounds of used oil. This is about the same amount of HHW as the combined total from the 48 permanent HHW collection fixed facilities, more than 60 collection events, and various mobile HHW collection programs in the state.

Collection Trends

The cost of HHW collection has been decreasing on a per pound and per participant basis. In the 1980s it was not unusual to pay $100 or more per participant at HHW collection events. There is now much more competition in the hazardous waste management field and more options available to the local HHW collection facility operator or collection event manager handling their own HHW. It is now typical to find per participant costs in the $35 to $75 range in places of high competition and metropolitan settings.

As the popularity of HHW collection programs grows in a community, sponsors often want to know at what point the participation is likely to level off. Nightingale and McLain (1997) were able to answer this question for facilities that were in operation for at least six full years. Figure 10.1 shows the results of the national survey of early HHW facilities and the average annual increase experienced over time.

Permanent HHW collection programs typically increase participation from year to year but also have seasonal patterns of participation. Usually the winter months are slower than other times of the year. The yearly increasing participation and seasonal variation can be clearly seen in Fig. 10.2, from Monterey Regional Waste Management District's HHW facility in Marina, California (Griffith, 2000).

HHW Collection Infrastructure Development

Infrastructure needs for HHW collection events are minimal. All essential functions can be contracted with private hazardous waste management firms. At some point in HHW collection programs, managers tend to seek more permanent solutions to the collection of HHW. This typically includes the option of a permanent collection facility.

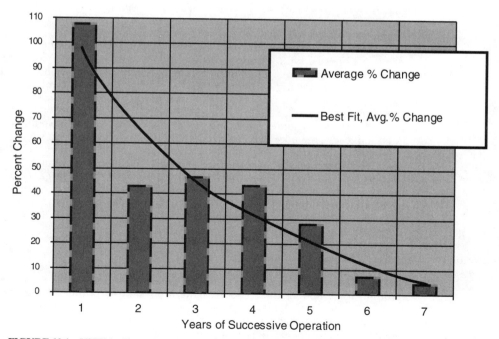

FIGURE 10.1   HHW facility average annual participation growth rate.

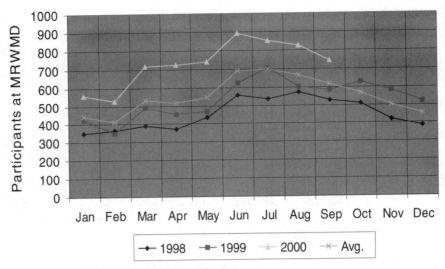

FIGURE 10.2   HHW facility seasonal participation.

A rule of thumb used for planning purposes is to develop a site which is built to serve (or can support with expansion) 10 percent of the service area households per year. Most permanent facilities begin to plateau in the 7 to 12 percent participation range in the first three to eight years of operation when considering the population within a 10-mile radius of the facility. This varies by local community and can increase with the use of satellite HHW collection facilities or drop-off centers. A determining factor in facility planning and design in a more urbanized area is the ability to store truckload quantities of HHW or access frequent smaller waste pick-ups by waste contractors with local hazardous waste facilities.

It is not unusual for a full-service metropolitan-based HHW collection facility to cost in the range of $700,000 to $1.5 million or more exclusive of land purchase, or any significant utility runs or earthwork. Permanent HHW collection facilities have been developed using existing structures or with minimal capacities and features involving significantly less capital costs. However, the unique requirements of the fire code and availability of suitable zoning areas and existing buildings often limit this option. Saving capital expenditures at these facilities often results in undersized facilities that operate inefficiently and with less safety.

In 1997, Nightingale and McLain (1997) compared the total operating cost for the three most recent years to the total capital costs from 14 HHW collection facilities in the United States and Canada. They found that, on average, the total operating cost for the three years equaled 5 times the total capital costs of the HHW collection fixed facilities. Therefore, a prudent solid waste manager will pay serious attention to how the facility site, design, features, materials process flow, and equipment may improve or hinder the operating efficiency, safety, and costs.

Design of Collection Activities and Structures

Collection events and by extension mobile collection activities can be easily set up and run according to well-understood practices of hazardous waste contractors; guidance is also found in various state and federal documents. The design practice for permanent HHW collection fixed facilities is less well developed and is by necessity unique to the site, climate, level of use, and processes anticipated at a particular location.

While it is not possible to provide exact design parameters, there are concepts and requirements that are commonly applied to HHW facility design to provide for efficient and safe handling of HHW at collection facilities. These include the following.

- Design for more space than you think you will need. It is nearly universal that HHW facilities are found to be too small by their operators within 12 months of opening.
- Design for future flexibility of waste volumes and types.
- Manage flammable/combustible liquids in separate areas in consideration of fire code requirements.
- Use hazardous materials cabinets for low volume reactive and less stable HHW.
- Provide excess capacity for spot ventilating areas where bulking/consolidation of liquids and lab-packing with dry absorbents occur. Overhead hoods take contaminants into workers' breathing zone. Spot ventilation can avoid this problem.
- Provide at least a roof over all waste handling and storage areas and heat for freeze-sensitive materials in cooler climates.
- Provide good traffic flow and access.
- Provide a clean, conditioned space for workers (heating and cooling as appropriate).
- Provide a straight-line process flow as much as possible.
- Visit and talk with other operators of HHW collection facilities.
- Provide for a waste exchange area.
- Install at least one flammable gas monitor with alarm in flammable liquids bulking area near floor level.
- Provide secondary containment with appropriate chemical-resistant coatings over concrete in most waste handling and full-drum storage areas.

Because the design and operations are interdependent and HHW is handled very intensively, it is best to develop a conceptual draft operating plan before much time is invested in the design process. It is prudent to spend time and resources ahead of the hard design process to program the facility space use, equipment options, site use, and operational needs.

REFERENCES

Abt Associates (1999) *Consumer Labeling Initiative Phase II Report,* Prepared for Office of Pollution Prevention and Toxics, U.S. Environmental Protection Agency, Washington, DC.

Austin, D. (January 10, 1997) "Chlorine Fumes From Garbage Strike Again," *The Oregonian,* Portland, OR, p. D5.

Bickel, S. (November 29, 1999) Personal Communication, Waste Watch Center, Andover, MA.

Bonaparte, R., and A. Othman (March, 1995) *Geotechnical News.*

Bortleson, G. C., and D. Davis (1997) *Pesticides in Selected Streams in the Puget Sound Basin, 1987–1995.* Washington State Department of Ecology. U.S. Geological Survey Fact Sheet 067-97, 4 p.

Breiteneicher, D. (June, 1997) "HHW in Wastewater," *Household Hazardous Waste Management News,* The Waste Watch Center, Andover, MA.

Brook, D., et al. (2000) *Training Manual for the Master Home Environmentalist Program,* American Lung Association of Washington, Seattle, WA.

Brown, B. (May 18, 1998) "Unknown Source Sickens Fresh Kills Workers," *Waste News,* p. 2.

CEFIC (2000) *CEFIC Annual Report 1999,* European Chemical Industry Council, Web Site (www.cefic.be) accessed 10/2/2000.

Clark County (December, 1999) *Clark County Comprehensive Solid Waste Management Plan,* Preliminary Draft, Clark County Department of Public Works, Vancouver, WA, pp. 11-5, 11-7.

Dickey, P. (1994) *Buy Smart, Buy Safe: A Consumer Guide to Less-Toxic Products,* Washington Toxics Coalition and the Local Hazardous Waste Management Program in King County, Seattle, WA.

Dickey, P., D. Duxbury, D. Galvin, B. Johnson, and A. Weissman (December 28, 1995) *Concerns with Household Cleaning Products—A White Paper,* Revised Draft. Metro, Seattle, WA.

Donnette, R. (2000) *Evaluating General Outreach Materials,* Thurston County Environmental Health, Olympia, WA. Web Site (www.evaltool.cjb.net/).

Drudi, D. (Summer, 1999) "Job Hazards in the Waste Industry," Compensation and Working Conditions, U.S. Bureau of Labor Statistics, pp. 19–23.

Duxbury, D. (March, 1995) "A Best HHW Program?" *Household Hazardous Waste Management News,* The Waste Watch Center, Andover, MA.

Duxbury, D. (October, 1997) "HHW Standards," *Household Hazardous Waste Management News,* The Waste Watch Center, vol. IX, no. 33, p. 1, Andover, MA.

European Commission (June 13, 2000) *Proposals for Directives of the European Parliament and of the Council on Waste Electrical and Electronic Equipment, and on the Restriction of the Use of Certain Hazardous Substances in Electrical and Electronic Equipment,* Commission of the European Communities, Brussels.

Griffith, J. (November 3, 2000) Personal Communications, Monterey Regional Waste Management District, Marina, CA.

Groner, S. (November 1997) "Honing In on Our Target to More Effectively Modify Behavior," Los Angeles Public Works; *Proceedings of 1997 Hazardous Materials Management Conference on Household, Small Business, and Universal Wastes,* San Diego, CA., Solid Waste Association of North America, publication #GR-HW 3001, p. 46.

Gross, M.A. (June, 1987) *Assessment of the Effects of Household Chemicals Upon Individual Septic Tank Performances.* Research Project Technical Completion Report, Publication No. 131. Arkansas Water Resources Research Center, Fayetteville, AR.

Huitric, R. (August 30, 1999) "Documentation of Large MSW Landfill Gas Constituent Declines From U.S. EPA AP-42 Default Values," attachment to letter from J. H. Skinner, Solid Waste Association of North America, to Laur, Thorneloe, and Huntley, U.S. EPA, Los Angeles County Sanitation Districts.

Huitric, R. (May 25, 2000) Personal Communications.

Jenkins-McLean, T. (1999) "How Can We Effect Behavior Change? Social Marketing vs. Public Education," *Household Hazardous Waste News,* Fall 1999, Local Hazardous Waste Management Program in King County, Seattle, WA.

King County Solid Waste Division (1995) *Hazards on the Homefront: A Teacher's Guide to Household Hazardous Waste,* 4th–12th grade, Local Hazardous Waste Management Program in King County, King County Solid Waste Division, Seattle, WA.

King County (January, 1997) *Local Waste Management Plan for King County,* Draft. Local Hazardous Waste Management Program in King County, King County Solid Waste Division, Seattle, WA.

Kinman, R., and Nutini (January, 1988) "Hazardous Household Waste in the Sanitary Landfill," *Proceedings of the 1988 Conference on Solid Waste Management and Materials Policy,* New York State Legislative Commission on Solid Waste Management, B 19–45.

Kolega, J. J. (1989) "Impact of Toxic Chemicals to Ground Water." in R. Seabloom, ed., *Proceedings of the 6th Northwest On-Site Wastewater Treatment Short Course,* September 18–19, 1989, pp. 247–256, University of Washington, Seattle, WA.

Lee, N. (1997) *Seven Steps for Successful Social Marketing.* Presentation to Thurston County, Washington public information staff, Social Marketing Services, Mercer Island, WA.

Litovitz, T. et al. (1998) "1997 Annual Report of the American Association of Poison Control Centers

Toxic Exposure Surveillance System," *American Journal of Emergency Medicine,* vol. 16, no. 5, pp. 443–497.

McDonnell, K. (Spring 2000) "Chemical Sensitivity Awareness," *Alternatives,* Washington Toxics Coalition, Seattle, WA.

McKenzie-Mohr, D. (1996) *Promoting a Sustainable Future: An Introduction to Community-Based Social Marketing,* St. Thomas University, Ottawa, Ontario. Canada, Web Site (www.cbsm.com).

Miller, D. E., and S. M. Emge (September, 1996) "Leachate Recirculation System Design, Operation, and Performance at the Kootenai County (Idaho) Landfill," *Proceedings of WasteCon 1996,* Solid Waste Association of North America, Portland OR.

Moore, E. (May, 1996) "Quicksilver," *Household Hazardous Waste Management News,* The Waste Watch Center, Andover, MA.

NIPC (1998) "Inhalants: The Silent Epidemic," *Synergies,* National Inhalant Prevention Coalition, Austin, TX.

Nightingale, D. E. B. (December, 1995) "Flexible Permitting of Fixed Facilities in Washington State," *Proceedings of the Hazardous Materials Management Conference of Household, Small Business, and Universal Wastes,* Philadelphia, PA, International City/County Management Association, Washington DC.

Nightingale, D. E. B., and B. McLain, (November, 1997) "Lessons from Collection Facilities Operating at Least Six Years," *Proceedings of Hazardous Materials Management Conference for Household, Small Business and Universal Wastes,* Solid Waste Association of North America and North American Hazardous Materials Management Association, San Diego, CA.

Nightingale, D. E. B. (August, 2000) "HHW and CESQG Waste Handling—"When Have I Been Fully Safety Trained?" *Integrated Solutions—Special Waste Management Focus,* Solid Waste Association of North America, p. 1.

NPD Group, Inc. (1995) *Consumer Disposal of Household Cleaning Products.* Prepared for The Soap and Detergent Association, New York.

Ott, W. R. and J. W. Roberts (February, 1998) *Everyday Exposure to Toxic Pollutants.* Scientific American. Viewed at (sciam.com/1998/0298issue/0298ott.html#links).

Pickus, W. (September, 1996) "Landfill Gas Impacts on Groundwater Contamination: A Superfund Site Case Study," *Proceedings of WasteCon 1996,* Solid Waste Association of North America, Portland OR.

Randolph, T. G. (1962) *Human Ecology and Susceptibility to the Chemical Environment,* Charles C. Thomas, Springfield IL.

Rathje, W. L. (May/June 2000) "Parkinson's Law of Garbage," *MSW Management,* vol. 10, p. 14.

RBRC (March, 2000) *Charge Up to Recycle,* Rechargeable Battery Recycling Corporation, p. 1.

Sanin, F. D., and M. A. Barlaz (October, 1998) "Natural Attenuation of Hazardous Organics during Refuse Decomposition in a Municipal Landfill," *Proceedings of WasteCon 1998 and the 36th Annual International Solid Waste Exposition,* Solid Waste Association of North America and the International Solid Waste Association, Charlotte, NC.

SVTC (2000) *Just Say No to E-Waste: Background Document on Hazards and Waste from Computers,* Silicon Valley Toxics Coalition, Web Site (svtc.org/cleancc/ecc.htm) accessed August 6, 2000, p. 4.

Soap and Detergent Association (1995) "Some Facts about Cleaning Product Disposal," Fact sheet in *Today's Cleaning: Consumer Information about Household Cleaning Products,* The Soap and Detergent Association, New York.

Sulzberg, J. D., and D. E. B. Nightingale (March, 1998) "Household Hazardous Material Management," in R. A. Meyers, ed., *Encyclopedia of Environmental Analysis and Remediation,* vol. 4, p. 2187, John Wiley & Sons Inc., New York.

Thurston County (2000) *1998–99 Progress Report,* Thurston County Local Hazardous Waste Program, Thurston County Environmental Health, Olympia, WA.

U.S. CB (June 26, 2000) U.S. Population Clock, Census Bureau, U.S. Department of Commerce, Web Site (http://www.census.gov/).

U.S. CPSC (1993) *Poison Prevention Packaging: A Text for Pharmacists and Physicians,* U.S. Consumer Product Safety Commission, Washington, DC.

U.S. EPA (1989a) *Characterization of Products Containing Lead and Cadmium in Municipal Solid Waste in the United States, 1970 to 2000—Executive Summary,* EPA/530-SW-90-042A, Office of Solid Waste, U.S. Environmental Protection Agency Washington, DC.

U.S. EPA (1989b) *The Solid Waste Dilemma: An Agenda for Action,* U.S. Environmental Protection Agency, Washington, DC.

U.S. EPA (1992) *Characterization of Products Containing Mercury in Municipal Solid Waste in the United States, 1970 to 2000—Executive Summary,* EPA/530-S-92-013, Office of Solid Waste and Emergency Response, U.S. Environmental Protection Agency, Washington, DC.

U.S. EPA (1993) *Targeting Indoor Air Pollution,* Office of Air and Radiation (6601), Document #400-R-92-012, U.S. Environmental Protection Agency Washington, DC.

U.S. EPA (1995a) *Analysis of the Potential Effects of Toxics on Municipal Solid Waste Management Options,* U.S. EPA Office of Research and Development, Risk Reduction Engineering Laboratory, EPA/600/R-95/047, U.S. Environmental Protection Agency, pp. xiv, xv, U.S. Environmental Protection Agency, Cincinnati, OH.

U.S. EPA (1995b) *Leachate Clogging Assessment of Geotextile and Soil Landfill Filters, Project Summary,* Office of Research and Development, National Risk Management Research Laboratory, EPA/600/SR-95/141, U.S. Environmental Protection Agency, Washington, DC.

U.S. EPA (1999a) *Reducing Polluted Runoff: The Storm Water Phase II Rule,* EPA/833-F-99-020, Office of Water, U.S. Environmental Protection Agency,

U.S. EPA (1999b) *National Pollutant Discharge Elimination System—Regulations for Revision of the Water Pollution Control Program Addressing Storm Water Discharges; Final Rule.* Federal Register, vol. 64, no. 235, December 8, 1999, pp. 68727, 68728.

U.S. EPA (2000a) *Storm Water Phase II Final Rule—Public Education and Outreach Minimum Control Measure,* Office of Water, Fact Sheet 2.3, EPA/833-F00-005, U.S. Environmental Protection Agency, Washington, DC.

U.S. EPA (2000b) *Storm Water Phase II Final Rule—Illicit Discharge Detection and Elimination Minimum Control Measure,* Office of Water, Fact Sheet 2.5, EPA/833-F00-007, U.S. Environmental Protection Agency, Washington, DC.

U.S. EPA (2000c) *Pesticide Registration (PR) Notice 2000-X. Notice to Manufacturers, Producers, Formulators, and Registrants of Pesticide Products,* Draft, Office of Prevention, Pesticides, and Toxic Substances, U.S. Environmental Protection Agency, Washington, DC.

U.S. EPA Region 5 (2000d) *Household Waste Management,* Agricultural and Biological Engineering, Purdue University, Lafayette, IN, Web Site (epa.gov/grtlakes/seahome/housewaste/) accessed May 2000.

Van Gelder, L. (November 13, 1996) "Trash Collector Dies After Inhaling Discarded Acid," *The New York Times,* p. B1.

Voss, F. D., et al. (April, 1999) *Pesticides Detected in Urban Streams during Rainstorms and Relations to Retail Sales of Pesticides in King County, Washington.* U.S. Geological Survey Fact Sheet 097-99.

Washington State Department of Ecology (June, 2000) *Public Review Draft of the Minimum Functional Standards for Solid Waste Handling,* Olympia, WA.

WTC (1998) *Grow Smart, Grow Safe: A Consumer Guide to Lawn and Garden Products,* Washington Toxics Coalition and the Local Hazardous Waste Management Program in King County, Seattle, WA.

Waste News (October 5, 1998) "Robbins Explosion Injures Worker," *Waste News,* p. 27, Crain Communications Inc., Akron, OH,

Waste Watch Center (October, 1997) "Mercury?" *Household Hazardous Waste Management News,* The Waste Watch Center, Andover, MA.

CHAPTER 11

OTHER SPECIAL WASTES
Part 11A   Batteries

James M. Lyznicki
Gary R. Brenniman
William H. Hallenbeck

11A.1   AUTOMOTIVE AND HOUSEHOLD BATTERIES

Introduction

Americans use considerable quantities of batteries to power a variety of household and industrial products. Batteries are used in motor and marine vehicles, electronics, watches, cameras, calculators, hearing aids, cordless telephones, power tools, and countless other portable household devices. Recently, much attention has been focused on the potential environmental and human health risks associated with the heavy metals present in batteries. Such concern has caused many municipalities to consider programs for recovering the large number of batteries discarded in municipal solid waste (MSW). Historically, residential collection and recycling efforts have been limited to used automobile batteries. In recent years, states and municipalities have begun to focus on the recovery of used household batteries. Such activities have coincided with a number of legislative and industry initiatives to reduce the toxicity of batteries and to promote their safe collection, reclamation, and disposal.

Battery Definitions and Terms

Batteries are complex electrochemical devices, composed of distinct cells, that generate electrical energy from the chemical energy of their cell components. Despite the technical distinction between them, the terms *battery* and *cell* are often used interchangeably. A battery cell consists primarily of a metallic anode (negative electrode), a metallic oxide cathode (positive electrode), and an electrolyte material that facilitates the chemical reaction between the two electrodes. Electric currents are generated as the anode corrodes in the electrolyte and initiates an ionic exchange reaction with the cathode. The electrical energy produced from this reaction is sufficient to power a variety of consumer and industrial devices.

Batteries are classified and distinguished according to their chemical components. Batteries are referred to as wet or dry cells. In wet cell batteries, the electrolyte is a liquid. In dry cell batteries, the electrolyte is contained in a paste, gel, or other solid matrix within the battery. Primary batteries contain cells in which the chemical reactions are irreversible, and they therefore cannot be recharged. This is in contrast to secondary batteries in which the chemical reactions are reversible and external energy sources can be repeatedly applied to recharge the battery cells.

Batteries are manufactured in a variety of sizes, shapes, and voltages. They are produced in rectangular, cylindrical, button, and coin shapes. In addition, many portable tools and elec-

TABLE 11A.1   Types and Uses of Batteries Found in Municipal Solid Waste

Battery type	Shapes	Uses
Wet cells 　Lead-acid	Rectangular	Cars, motorcycles, boats
Dry cells—primary 　Zinc-carbon 　Alkaline	Cylindrical, rectangular, 　button; AA, AAA, 　C, D, 9V	Flashlights, radios, 　tape recorders, toys
Mercuric oxide 　Silver oxide 　Zinc air 　Lithium	Button, cylindrical	Hearing aids, watches, 　calculators, pagers, 　camcorders, computers, 　cameras
Dry cells—secondary 　Nickel-cadmium 　Lead-acid	Cylindrical, button, 　or in battery packs	Rechargeable cordless 　products such as power 　tools, vacuum cleaners, 　shavers, phones

tronic devices utilize rechargeable batteries contained in battery packs. Refer to Table 11A.1 for a listing of the common types and general uses of batteries found in MSW.

Composition of Batteries in MSW

Batteries differ in their chemical composition, energy storage capacity, voltage output, and life span. These factors affect their overall performance, utility, and cost. Because of their different intended uses, consumer batteries are usually distinguished as automotive (i.e., lead-acid storage batteries) and household batteries.

Lead-Acid Storage Batteries (Wet Cells).   Lead-acid storage batteries are used in automobiles, motorcycles, boats, and a variety of industrial applications. They are primarily used to provide starting, lighting, and ignition for automotive products (U.S. EPA, 1989). These are wet cell batteries consisting of lead electrodes in a liquid sulfuric acid electrolyte. It is estimated that 78 to 80 million automotive and light truck batteries are sold each year in the United States (Apotheker, 1991). The average battery weighs approximately 36 lb, one-half of which is composed of the lead anode and lead dioxide cathode (U.S. EPA, 1989). It is estimated that the lead in automobile batteries accounts for approximately two-thirds of the total weight of lead in MSW (U.S. EPA, 1989). In addition to lead, each battery contains approximately 1 gal of sulfuric acid (i.e., 9 lb), almost 3 lb of polypropylene plastic casing, about 3 lb of polyvinyl chloride rubber separators, and about 3 lb of various chemical sulfates and oxides to which the lead is bound (Apotheker, 1990b). The typical useful lifetime of lead-acid storage batteries is 3 to 4 years (U.S. EPA, 1989).

Household Batteries (Dry Cells).   Americans use about 8 to 10 household batteries per person each year (Reutlinger and de Grassi, 1991). In 1992, it is estimated that almost 4 billion dry cell batteries were sold in the United States (Hurd et al., 1992). Total future sales of household batteries are expected to increase by about 6 percent each year (Hurd et al., 1992). The types and percentages of household batteries sold in the United States in 1992 are shown in Fig. 11A.1.

Dry cell batteries contain electrodes composed of a variety of potentially hazardous metals including cadmium, mercury, nickel, silver, lead, lithium, and zinc. The electrode materials and electrolytes found in household batteries are listed in Table 11A.2. In addition to elec-

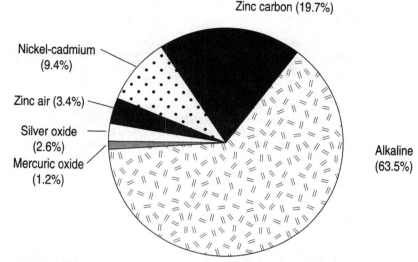

FIGURE 11A.1 1992 sales percentage of domestic batteries in the United States. (Note: Lithium batteries accounted for 0.2 percent of battery sales in 1992.) *(Source: Hurd et al., 1992)*

trodes and electrolytes, batteries also contain other materials that are added to control or contain the chemical reactions within the battery (Hurd et al., 1992; Eutrotech Inc., 1991). For example, mercury is added to the zinc anode of primary cells (e.g., alkaline, zinc-carbon) to reduce corrosion and to inhibit the buildup of potentially explosive hydrogen gas (U.S. EPA, 1992b). In addition, mercury helps to prevent the batteries from self-discharging and leaking (NYSDECED, 1992). Other components of batteries include graphite, brass, plastic, paper, cardboard, and steel.

TABLE 11A.2 Primary Chemical Components of Household Batteries in Municipal Solid Waste

Battery type	Cathode	Anode	Electrolyte
Alkaline	Manganese dioxide	Zinc	Potassium and/or sodium hydroxide
Zinc-carbon	Manganese dioxide	Zinc	Ammonium and/or zinc chloride
Mercuric oxide	Mercuric oxide	Zinc	Potassium and/or sodium hydroxide
Zinc-air	Oxygen from air	Zinc	Potassium hydroxide
Silver oxide	Silver oxide	Zinc	Potassium and/or sodium hydroxide
Lithium	Various metallic oxides	Lithium	Various organic and/or salt solutions
Nickel-cadmium (rechargeable)	Nickel oxide	Cadmium	Potassium and/or sodium hydroxide
Sealed lead-acid (rechargeable)	Lead oxide	Lead	Sulfuric acid

Sources: U.S. EPA (1989); Hurd et al. (1992).

Primary Dry Cell Batteries. When purchasing batteries, primary dry cell batteries are generally less expensive than secondary or rechargeable batteries. However, when making cost comparisons, consumers should consider that rechargeable batteries are reusable whereas primary batteries must be replaced once they are discharged. Primary dry cells accounted for almost 90 percent of U.S. battery sales in 1992 (Reutlinger and de Grassi, 1991). The majority of batteries purchased were cylindrical and rectangular varieties. Button cells represented only 5 percent of the total battery cell market in 1992 (Reutlinger and de Grassi, 1991).

ALKALINE (MANGANESE) BATTERIES. Alkaline batteries are the most common household dry cell batteries sold in the United States. It is estimated that they represent over 63 percent of the household battery market and are increasing their market share (Hurd et al., 1992). Alkaline batteries are manufactured in many sizes and shapes. Their good performance and long shelf life make them appealing for a variety of consumer uses.

Recent environmental concerns have resulted in dramatic reductions in the mercury content of alkaline batteries. For example, batteries that contained up to 1 percent mercury by weight in the mid-1980s are now being produced with mercury concentrations of 0.0001 to 0.025 percent (Hurd et al., 1992). Because of design limitations, such reductions will be more difficult to achieve for button-size batteries than for cylindrical and rectangular batteries (NYSDECED, 1992). The major alkaline battery manufacturers have established implementation dates for no-mercury-added battery designs for nonbutton cells by 1993 (NYSDECED, 1992). In addition to mercury, alkaline batteries also contain metals such as lead, cadmium, arsenic, chromium, copper, indium, iron, nickel, tin, zinc, and manganese (U.S. EPA, 1989; Hurd et al., 1992).

ZINC-CARBON BATTERIES. Zinc-carbon batteries are the second most commonly used household battery. These batteries represent about 20 percent of the household battery market; however, sales are declining (Hurd et al., 1992). These batteries are manufactured as inexpensive, general-purpose batteries as well as heavy-duty varieties. Zinc-carbon batteries have a shorter shelf life than alkaline batteries, are less powerful, and have a tendency to leak in devices once they are discharged (Eutrotech Inc., 1991; NYSDECED, 1992). Because of their anode configuration, zinc-carbon batteries require less mercury than alkaline batteries (Eutrotech Inc., 1991). Reduction and eventual elimination of added mercury to zinc-carbon batteries is anticipated in the near future (NYSDECED, 1992). In addition to mercury, zinc-carbon batteries also contain metals such as lead, cadmium, arsenic, chromium, copper, iron, manganese, nickel, zinc, and tin (U.S. EPA, 1989; Hurd et al., 1992).

SILVER OXIDE BATTERIES. Silver oxide batteries account for less than 3 percent of all household battery sales and about 5 percent of the button cell market (Reutlinger and de Grassi, 1991; Hurd et al., 1992). These batteries are manufactured in a variety of button sizes and provide a more constant voltage output than alkaline or zinc-carbon button cells (NYSDECED, 1992). Silver oxide batteries are interchangeable with mercuric oxide batteries and are increasingly being used to power hearing aids and watches (U.S. EPA, 1992b). However, silver oxide batteries are generally more expensive than mercuric oxide cells (Eutrotech Inc., 1991). Silver oxide batteries contain about 1 percent mercury by battery weight (NYSDECED, 1992).

MERCURIC OXIDE BATTERIES. Mercuric oxide batteries account for about 1 percent of annual U.S. battery sales, a percentage that is expected to decline in future years (Hurd et al., 1992). Most mercuric oxide batteries are manufactured as button cells and represent about 20 percent of that market (Reutlinger and de Grassi, 1991). Increasingly these batteries have come under scrutiny since more than one-third of their weight is mercury. Suitable alternatives to mercuric oxide cells have been developed (e.g., silver oxide, zinc air) that should reduce consumer dependence on mercuric oxide cells. Despite their decreasing use by household consumers, mercuric oxide batteries continue to be used in a variety of industrial, medical, military, and communications devices (U.S. EPA, 1992b).

ZINC AIR BATTERIES. Zinc air batteries have become increasingly popular in the United States. They represent over 3 percent of the total U.S. household battery sales (Hurd et al.,

1992) and over 60 percent of consumer button cell battery purchases (Reutlinger and de Grassi, 1991). Primary uses are in hearing aids and pagers. Zinc air batteries are advantageous since they have a longer life than silver or mercuric oxide batteries (NYSDECED, 1992). However, their use is restricted since they require ambient air to provide their oxygen cathode. Consequently, they cannot be used for tightly sealed applications such as watches (NYSDECED, 1992). These batteries contain about 1 to 2 percent mercury by weight (Hurd et al., 1992).

LITHIUM BATTERIES.   Lithium batteries represent less than 0.25 percent of the total U.S. household battery market, although their market share is expected to increase in the future (Hurd et al., 1992). They are manufactured as cylinders, buttons, or coin shapes and may also be contained in battery packs. Despite their high cost, their excellent performance characteristics make them useful in a variety of consumer electronics and computer applications as well as military and medical devices (NYSDECED, 1992). Lithium is a highly reactive material, especially when mixed with water. Consequently, safety precautions are recommended when collecting, storing, or transporting unspent batteries for disposal or reclamation (Hurd et al., 1992).

Secondary Dry Cell Batteries.   Secondary dry cells accounted for approximately 10 percent of battery sales in 1992 (Hurd et al., 1992). These rechargeable batteries are preferable to primary cells since they can be used repeatedly. However, their lower performance characteristics may be restrictive for some consumer applications (Hurd et al., 1992; NYSDECED, 1992). It is estimated that one rechargeable battery can substitute for 100 to 300 single-use batteries (Hurd et al., 1992; NYSDECED, 1992). Many rechargeable batteries are sealed within consumer products such as cordless telephones, power tools, appliances, personal computers, and other electronics. Recent attention has focused on providing easy access to the rechargeable batteries in consumer products to encourage their recycling or proper disposal. Legislative action by a number of states has resulted in mandates for manufacturers to produce products with removable rechargeable batteries by July 1, 1993 (Cohen, 1993).

NICKEL-CADMIUM BATTERIES.   Nickel-cadmium batteries (i.e., Ni-Cd or "ni-cads") are the most common of the secondary or rechargeable household batteries sold in the United States. They represent almost 10 percent of the total U.S. household battery purchases (Cohen, 1993). Future sales are expected to increase (Hurd et al., 1992). These batteries are available in sizes comparable with alkaline and zinc-carbon batteries. However, they are currently not as powerful as primary cells and tend to discharge more rapidly (NYSDECED, 1992). Nickel-cadmium batteries are a major consumer of cadmium in the United States (U.S. EPA, 1989). These batteries typically have a cadmium content ranging from 11 to 15 percent of the battery weight (Hurd et al., 1992).

SEALED LEAD-ACID BATTERIES.   In addition to automotive uses, lead-acid batteries (called sealed lead-acid batteries) are used in a variety of consumer products such as toys, video recorders, portable electronics, tools, appliances, and electric start lawn mowers. These smaller, rechargeable batteries are dry cells since the sulfuric acid electrolyte is contained on a solid separator material or in a gel (U.S. EPA, 1989). These batteries account for less than 1 percent of U.S. household battery purchases (NYSDECED, 1992).

Environmental Impacts of Batteries in MSW

The disposal of used automobile and household batteries into MSW must be assessed for its potential human and environmental health impacts. An estimated 1.7 million tons of lead-acid batteries were generated in MSW in 1990 (U.S. EPA, 1992a). This represented less than 1 percent of the total weight of MSW generated during that year. Despite a well-established recycling infrastructure, about 6 percent of lead-acid batteries were landfilled or incinerated in 1990. Household batteries accounted for about 142,000 tons of MSW in 1991 and about 146,000 tons in 1992 (Hurd et al., 1992). This represented less than 0.1 percent of the total weight of MSW generated during those years. Although the tonnage of these materials in

MSW may seem small and inconsequential, their potential toxicity must be considered in order to evaluate appropriate and safe disposal practices (Hurd et al., 1992).

The disposal of used batteries in MSW is problematic for two reasons. First of all, batteries contribute to the total quantity of potentially hazardous waste that is disposed in MSW. Second, and more important, batteries contain many potentially toxic chemicals that can have adverse environmental and human health impacts. Assessing the environmental impact of used battery disposal involves evaluating the potential for groundwater contamination due to the leaching of contaminants from MSW landfills; the emission of contaminants into the air from MSW incinerators; the presence of hazardous materials in the residual ash remaining after MSW incineration; and finally, the contamination of composted organic waste with battery components. Such assessments must continually be refined as more stringent regulations are imposed on the design of MSW landfills and incinerators and as the toxicity and types of batteries in MSW continue to change.

Batteries contain a variety of heavy metals that may become toxic contaminants in landfill leachate, incinerator emissions, incinerator ash, and compost (Hurd et al., 1992; Eutrotech Inc., 1991; NYSDECED, 1992; Arnold, 1991). Much concern has been directed to the high percentage of mercury, cadmium, and lead in MSW that is attributed to used batteries (U.S. EPA, 1989; U.S. EPA, 1992b). Other potentially toxic metals that may be present in batteries include silver, zinc, nickel, manganese, lithium, chromium, and arsenic (Hurd et al., 1992). Table 11A.3 depicts the metallic composition of the typical household batteries found in MSW.

TABLE 11A.3   Weight Percentage of Potentially Toxic Heavy Metals in Common Household Batteries

Battery type	Metal, %				
	Cadmium	Mercury	Nickel	Silver	Zinc
Alkaline	0.01	0.025–0.5			8–18
Zinc-carbon	0.03	0.01			12–20
Mercuric oxide		30–43			10–15
Silver oxide		1.0		30–35	30–35
Zinc air		2.0			35–40
Nickel-cadmium	11–15		15–25		

Source: Hurd et al. (1992).

When released, heavy metals persist in the environment. Many of them also have associated human, animal, or plant toxicities (Hurd et al., 1992; NYSDECED, 1992). Many accumulate in aquatic sediments and soil and may be metabolized by indigenous microorganisms to more toxic organic forms. Of most concern is the potential for uptake and accumulation of heavy metals or their metabolites in the food chain (Eutrotech Inc., 1991; NYSDECED, 1992). Recent attention has focused on potential human and environmental exposure risks from the metals present in used batteries, specifically lead, cadmium, and mercury (Hurd et al., 1992; Eutrotech Inc., 1991; NYSDECED, 1992; Arnold, 1991).

Lead.   Nearly 1.3 million metric tons of lead are consumed in the United States each year. Of this, over 1 million metric tons (79 percent) are used to manufacture lead-acid batteries; most of this is used to manufacture automobile batteries (U.S. EPA, 1992c). Lead-acid storage batteries comprise the largest percentage of the weight of batteries discarded in the United States. In addition, they comprise almost two-thirds of the lead in MSW. Based on 1986 figures, lead-acid batteries contributed about 65 percent of the lead in MSW (U.S. EPA, 1989). Although the total lead discards in MSW are expected to increase yearly, the percentage of lead due to lead-acid batteries is expected to remain fairly constant (U.S. EPA, 1989). Ultimately, the effectiveness of lead-acid battery recycling programs will significantly impact the weight of lead-acid batteries that are landfilled or incinerated (U.S. EPA, 1989).

Cadmium.   Based on 1986 figures, household batteries accounted for more than 50 percent of the cadmium in MSW (U.S. EPA, 1989). As nickel-cadmium batteries increase in popularity, the amount of cadmium in the waste stream is expected to increase. It is estimated that by the year 2000, dry cell batteries will account for 76 percent of the cadmium in MSW (U.S. EPA, 1989; Hurd et al., 1992).

Mercury.   Based on 1989 figures, household batteries accounted for more than 88 percent of the mercury in MSW (U.S. EPA, 1992b). Alkaline batteries accounted for the largest quantity (i.e., 59 percent) of the total weight of mercury in MSW (Hurd et al., 1992; U.S. EPA, 1992b). Although the sales of household batteries (including alkaline cells) is projected to increase, the amount of mercury in MSW is expected to decrease in the future. This is due to further reductions in the mercury content of dry cells as well as to the use of alternatives for mercuric oxide batteries. Despite reductions in their mercury content, dry cell batteries are still expected to account for about 56 percent of the mercury in MSW in the year 2000 (Hurd et al., 1992; U.S. EPA, 1992b).

Used Battery Regulations

Solid waste disposal regulations are based on Subtitles C and D of the Resource Conservation and Recovery Act of 1976 (RCRA) and are codified in Title 40 of the Code of Federal Regulations (CFR). Much of the attention for managing used batteries has centered around whether to classify them as hazardous waste. Such a designation could complicate used battery collection programs by imposing RCRA Subtitle C hazardous waste regulations on the collection, storage, transportation, reclamation, and disposal of used batteries. However, household generated waste, including used batteries, is exempt from RCRA hazardous waste rules and regulations under the "household waste" exclusion rule of 40 CFR, Part 261.4(b)(1).

Waste generated from nonhousehold sources may be declared hazardous waste if it is specifically "listed" (i.e., 40 CFR, Part 261, Subpart D) or if it exhibits one of the following "characteristics" (40 CFR, Part 261, Subpart C): ignitability, corrosivity, reactivity, or toxicity. Although waste batteries are not "listed" hazardous wastes, they may exhibit one of the four "characteristics" of hazardous waste (Hurd et al., 1992). The toxicity characteristic is of most concern to battery recyclers and is determined by a laboratory procedure called the toxicity characteristic leaching procedure (TCLP). Heavy metals such as lead, mercury, cadmium, and chromium have been detected from batteries at concentrations that exceed the TCLP concentrations. Consequently, some household batteries would be considered characteristic hazardous wastes if it were not for the U.S. EPA exemption (Hurd et al., 1992).

In addition to RCRA hazardous waste regulations, battery recyclers must adhere to a number of additional federal and state regulations. These include compliance with water and air quality standards, transportation regulations, postal laws, and applicable state hazardous waste laws. Furthermore, potential liability under the Comprehensive Environmental Response Compensation and Liability Act (CERCLA) must also be considered by battery generators and recyclers (Hurd et al., 1992).

Automobile Batteries.   In 1985, the U.S. EPA declared that lead-acid batteries were to be considered a hazardous waste (Apotheker, 1990b). Spent lead-acid batteries may exhibit the hazardous waste characteristic for toxicity (i.e., lead) as well as for corrosivity (i.e., acidic electrolyte). Regulations in 40 CFR, Part 261.6, (a)(2)(v) and Subpart G of 40 CFR, Part 266 establish a hazardous waste exemption for spent lead-acid batteries. This exemption pertains only to persons who generate, collect, store, and transport spent lead-acid batteries for reclamation but are not directly involved with battery reclamation. Lead-acid battery reclamation facilities must comply with RCRA hazardous waste regulations. If lead-acid batteries are disposed of rather than sent to a battery reclamation facility, they are considered hazardous waste and are subject to RCRA Subtitle C hazardous waste regulations.

According to the Battery Council International (BCI), by February 1993, 41 states had passed some form of lead-acid battery legislation (Battery Council International, 1993). Most states have passed laws following model legislation proposed by the BCI. In addition, states have added more specific and usually more stringent regulatory language. Examples of various state lead-acid regulations include (U.S. EPA, 1992c; Gaba and Stever, 1992):

- Prohibition on the disposal of lead-acid batteries in landfills or incinerators
- Mandated delivery of batteries to approved retailers or collection facilities
- Establishment of requirements for retailers and other collection facilities for accepting used lead-acid batteries
- Requirement of posted written notices that inform the public of lead-acid battery recycling
- Imposition of fines to enforce the regulations
- Regulation of spent lead-acid batteries as a hazardous waste
- Regulation of transporters of spent batteries
- Exclusion of lead-acid battery recycling from regulation under hazardous waste provisions
- Deposit fees on the sale of new lead-acid batteries that can be recovered upon return of the used battery
- Requirements for the state to purchase batteries with a minimum specified recycled lead content

Household Batteries.   A hazardous waste exemption for used batteries and battery cells that are returned to a manufacturer for "regeneration" is provided in 40 CFR, Part 261.6(a)(3)(ii). However, few companies in the United States are involved with the reclamation of household batteries. Additional regulations for reclaiming precious metals such as silver (i.e., from silver oxide batteries) are provided in 40 CFR, Part 261.6(a)(2)(iv).

To encourage the collection of used household batteries, the U.S. EPA has drafted and proposed a "universal waste rule" to be codified as 40 CFR, Part 273 (U.S. EPA, 1993). Ultimately, the rule is intended to encourage the proper collection, treatment, and/or recycling of "post-user" generated hazardous waste. Specific guidelines for used batteries, other than spent lead-acid (e.g., automobile batteries), are in Subpart B of the proposed rule. The rule is designed to streamline the collection process for used batteries by removing current regulatory barriers to their collection. The regulation would apply only to used batteries that exhibit an RCRA hazardous waste characteristic. It would affect their management prior to being received at a permitted hazardous waste treatment, storage, reclamation, or disposal facility.

As of March 1993, 15 states had adopted legislation regarding the management of waste household batteries (PRBA Newsletter, 1993). Laws range from required battery recycling feasibility studies and plans to mandated collection and disposal programs. A survey of existing state legislation is included in the New York State household battery report (NYSDECED, 1992). The most significant legislation regarding consumer batteries has been enacted by Connecticut, Minnesota, New Jersey, New York, and Vermont. Essentially, state legislation focuses on reducing the toxicity of dry cell batteries in MSW. It is hoped that such initiatives will create more cost-effective battery reclamation and recycling options than currently exist in the United States. Notable components of state legislation include (Hurd et al., 1992):

- Manufacturers are directly responsible for the costs of properly disposing of their products. This can include used batteries as well as products powered by rechargeable batteries that may have adverse environmental impacts.
- Mandate maximum mercury content standards for batteries sold in the state. These range from 0.0001 to 0.025 percent by weight for alkaline batteries and 25 mg for alkaline button

batteries. These standards also require manufacturers to redesign batteries to minimize or eliminate hazardous components.

- Require accessibility to the rechargeable batteries in consumer products.
- Ban the disposal of recyclable batteries with unregulated MSW.
- Impose regulations to reduce or eliminate various metals (e.g., mercury) from batteries to facilitate reclamation activities in the future.
- Require labeling on all batteries to assist consumers in separating battery types.

Important points for state legislators to consider when developing household battery management programs include (Hurd et al., 1992): (1) deciding whether to regulate household batteries as hazardous waste if they exhibit any of the RCRA hazardous waste characteristics; (2) deciding whether to enact the RCRA "household waste" exemption rule for household batteries; (3) deciding whether exemptions should be granted for batteries collected for reclamation or for other specified management options in states that choose not to allow the RCRA "household waste" exemption; or (4) deciding whether to adopt specific legislation regarding the composition, collection, transport, processing, and disposal of household batteries.

Collection of Used Batteries for Recycling and Disposal

Battery collection programs are intended to separate batteries from mixed MSW and keep them out of MSW incinerators, MSW landfills, and compost. In addition, such programs are designed to recover certain types of batteries for reclamation and recycling of their components. Used batteries are collected through community-sponsored drop-off locations, residential curbside collection programs, household hazardous waste collection centers, and retailers (e.g., automotive shops, jewelry stores). To encourage battery recovery, some manufacturers now provide prepaid mailers for the return of used rechargeable batteries (Cohen, 1993).

Battery collection programs must ensure convenient facilities for the safe collection and storage of batteries prior to shipment for reclamation or disposal. Program designs should achieve high participation and be cost-effective (NYSDECED, 1992). Collection and storage procedures will depend on applicable state regulations as they relate to the classification of household batteries as hazardous waste. Ultimately, a successful battery collection program should include ongoing public education to increase consumer participation and awareness of the types of batteries included in the program.

Automobile Batteries.   State and federal regulations have established an effective infrastructure for the collection and reclamation of lead-acid batteries. Disposal bans as well as mandatory take-back and deposit programs have created a collection system that includes high retailer and consumer participation rates (U.S. EPA, 1992c). The success of automobile lead-acid battery collection programs results from the implementation of centralized and convenient collection locations such as automotive parts stores and service centers. In addition, the collection of lead-acid batteries is supported by the secondary lead industry, which reclaims and markets the battery components.

Household Batteries.   Collection of household batteries is hampered by the lack of any centralized recovery and recycling network such as that which exists for spent lead-acid batteries. Deliberation continues over the types and components of batteries in MSW that present the greatest potential environmental and human health risks. Consequently, communities may choose: (1) not to collect any household batteries, (2) to collect only those batteries that can be reclaimed, (3) to collect only the most toxic batteries, or (4) to collect all used household batteries.

Consumers are encouraged to bring their spent household batteries to approved collection facilities for proper handling. Collected batteries should be stored in well-ventilated areas to avoid the buildup of heat as well as mercury and hydrogen gases. Facilities should also have adequate safety and fire-prevention equipment. Batteries should be stored in a dry environment and packed to minimize the potential hazards from short-circuits, leaking cells, and unspent lithium cells (Hurd et al., 1992; NYSDECED, 1992; Arnold, 1991).

Recycling Used Batteries

Used batteries cannot be recycled in the same sense that aluminum or glass containers are recycled into new containers. It is more appropriate to reserve the term "recycling" for those battery components that can be reclaimed and reused. These components include metals (e.g., lead, mercury, silver, nickel, cadmium, steel) and plastic (e.g., the battery case of automobile batteries). Some of the reclaimed materials may then be recycled into new battery components or manufactured into other products.

States have been more aggressive than the federal government in promoting used battery reclamation and recycling. Since automobile lead-acid batteries comprise such a large percentage of the lead consumed in the United States, many states and communities have established programs to collect automobile batteries for reclamation (U.S. EPA, 1992c). In contrast, programs for household batteries are not well established and essentially focus primarily on battery collection and safe disposal rather than reclamation (Hurd et al., 1992). Such programs are more concerned with reducing the potential environmental and human health effects from household batteries by disposing of them in approved hazardous waste landfills. Information and technical assistance regarding lead-acid and household battery reclamation is available from the organizations listed in Fig. 11A.2.

Battery Council International (BCI) 401 N. Michigan Ave. Chicago, Ill. 60611-4267 (312) 644-6610	Independent Battery Manufacturers Association, Inc. (IBMA) 100 Larchwood Dr. Largo, Fla. 34640 (813) 586-1408
National Electrical Manufacturers Association (NEMA) 2101 L Street, N.W. Ste. 300 Washington, D.C. 20037-1581 (202) 457-8400	Portable Rechargeable Battery Association (PRBA) 1000 Parkwood Circle Ste. 430 Atlanta, Ga. 30339 (404) 612-8826

FIGURE 11A.2   Sources of information on automobile and household battery recycling.

Automobile Batteries.   Federal and state regulations have raised the costs of battery recycling and have greatly consolidated the lead recycling industry (Apotheker, 1991). Recovery of lead-acid batteries for reclamation has varied from 60 percent to well over 90 percent (U.S. EPA, 1992a). In 1990, the U.S. EPA estimated that 1.6 million tons (i.e., 94 percent) of the lead-acid batteries in MSW were recovered for recycling. The recycling rates calculated by the BCI for 1990 and 1991 were 98 and 97 percent, respectively (Smith, Bucklin and Associates, 1993). Recovery rates have improved as a result of regulations that ban the landfilling and incineration of lead-acid batteries. Historically, lead-acid battery recycling rates have reflected the market conditions for lead. The goal of the U.S. EPA and state governments is to maintain a high recycling rate despite market fluctuations in lead prices or reductions in processing capacity.

Lead-acid batteries are recycled to reclaim the lead, sulfuric acid, and polypropylene plastic housing. Batteries are processed by secondary smelters who rely on used batteries for more than 70 percent of their lead supply (Apotheker, 1991). There are 22 active secondary processors in the United States (Table 11A.4) (Battery Council International, undated). Most of these are independently owned and operated.

TABLE 11A.4 Secondary Lead Smelters in the United States

Secondary lead smelter	Location
ALCO Metals	Los Angeles, Calif.
The Doe Run Co.	Boss, Mo.
East Penn Manufacturing Co.	Lyon Station, Pa.
Exide Corp.	Reading, Pa. (General Battery Corp.); Muncie, Ind.; Dallas, Tex. (Dixie Metals Corp.)
GNB Incorporated	Columbus, Ga.; Frisco, Tex.; Los Angeles, Calif.
General Smelting & Refining	Cottage Grove, Tenn.
Gopher Smelting & Refining	Minneapolis, Minn.
Gulf Coast Lead	Tampa, Fla.
Interstate Lead Co.	Leeds, Ala.
Refined Metals Corp.	Beech Grove, Ind.; Memphis, Tenn.
Ross Metals	Rossville, Tenn.
RSR Corp.	Middletown, N.Y.; Indianapolis, Ind.; Los Angeles, Calif.
Sanders Lead Co.	Troy, Ala.
Schuylkill Metals	Baton Rouge, La.; Cannon Hollow, Mo.

Source: Battery Council International (undated).

At the smelter, the batteries are crushed and then processed to recover the battery components. The sulfuric acid can be reclaimed and used in fertilizer or neutralized for disposal (Apotheker, 1990a). The plastic battery case can be recycled into new cases or other recycled plastic products. All lead-containing components are loaded into reverberatory furnaces in which the lead is melted and extracted (Apotheker, 1990a). The furnace residue is further processed in blast furnaces to recover more of the lead. The slag that remains still contains lead and must be tested prior to disposal to determine its hazardous waste characteristics. Alternative lead smelting technologies are now available that significantly reduce the amount of potentially hazardous slag generated during the smelting process (Apotheker, 1991; Apotheker, 1990a).

Household Batteries.   Recycling programs for household batteries are not widespread and have been hampered by the limited number of processing facilities available. Such programs may even be misleading, since only certain types of batteries are reclaimed in the United States. As mentioned previously, the largest percentage of household batteries sold in the United States are alkaline and zinc-carbon batteries. However, there is no facility in the U.S. that reclaims these batteries. Instead, they are either shipped overseas for reclamation or disposed in domestic hazardous waste landfills. The mercury present in these batteries complicates the reclamation of other battery components (Hurd et al., 1992). Expensive mercury recovery systems are required before the zinc, steel, brass, manganese, and carbon can be recovered safely from alkaline and zinc-carbon batteries. Two foreign companies, Sumitomo Heavy Industries (Japan) and Recymet (Switzerland), have developed processes for reclaiming alkaline and carbon-zinc batteries (Hurd et al., 1992).

Currently, only three U.S. facilities reclaim household batteries. The Mercury Refining Company (MERECO) in New York recovers mercury from mercuric oxide and silver from silver oxide batteries. INMETCO in Pennsylvania recovers nickel and steel from nickel-cadmium batteries. After the batteries are processed, the residue is sent to another U.S. firm to recover cadmium. The Bethlehem Apparatus Company, also in Pennsylvania, recovers mercury from mercuric oxide batteries (Arnold, 1991). Detailed descriptions of U.S. as well as foreign battery reclamation processes have been published previously (Hurd et al., 1992; Arnold, 1991). Refer to Table 11A.5 for a listing of U.S. firms that accept waste household batteries for disposal or reclamation (Arnold, 1991; Adams and Amos, 1993). It is recommended that battery reclamation, disposal, and storage facilities be contacted directly regarding their

TABLE 11A.5   U.S. Companies Accepting Waste Household Batteries for Disposal or Reclamation

Company and location	Types of batteries processed
BDT 4255 Research Parkway Clarence, N.Y. 14031 (716) 634-6794	Accepts alkaline and lithium batteries for disposal as hazardous waste. Lithium batteries are neutralized prior to disposal.
Bethlehem Apparatus Co. 890 Front St., P.O. Box Y Hellertown, Pa. 18055 (215) 838-7034	Accepts mercuric oxide batteries for reclamation of mercury.
Environmental Pacific Corp. P.O. Box 2116 Lake Oswego, Oreg. 97055 (503) 226-7331	Accepts all batteries and provides hazardous waste storage facilities.
F. W. Hempel & Co., Inc. 1370 Avenue of the Americas New York, N.Y. 10019 (212) 586-8055	Accepts nickel-cadmium batteries only for shipment to France for processing.
INMETCO P.O. Box 720, Rte. 488 Ellwood City, Pa. 16117 (412) 758-5515	Accepts nickel-cadmium batteries only for reclamation of nickel. Residual sent to Zinc Corporation of America for cadmium reclamation.
Kinbursky Brother Supply 1314 N. Lemon St. Anaheim, Calif. 92801 (714) 738-8516	Accepts lead-acid and nickel-cadmium batteries. Lead plate and nickel sent to a smelter; cadmium sent to France.
Mercury Refining Co., Inc. (MERECO) 790 Watervliet-Shaker Rd. Latham, N.Y. 12110 (518) 785-1703, (800) 833-3505	Accepts all household batteries. Mercuric oxide and silver oxide batteries refined on-site; other cells marketed for disposal or reclamation.
NIFE Industrial Blvd. P.O. Box 7366 Greenville, N.C. 27835 (919) 830-1600	Accepts nickel-cadmium batteries only. Batteries sent to parent smelting company in Sweden for cadmium reclamation.
Quicksilver Products, Inc. 200 Valley Drive, Ste. 1 Brisbane, Calif. 94005 (415) 468-2000	Accepts mercuric oxide batteries only.
Universal Metals and Ores Mt. Vernon, N.Y. (914) 664-0200	Accepts nickel-cadmium batteries only. Batteries are marketed overseas for metals reclamation.

Sources: Arnold (1991); Adams and Amos (1993).

specific guidelines and restrictions for used batteries. In addition, it is essential to carefully evaluate the business practices of these facilities to minimize future RCRA and CERCLA liability.

Summary

- The recovery of used automobile batteries for reclamation continues to be successful as a result of federal and state legislation and cooperative efforts between the battery manufacturers, secondary lead smelters, retail stores, automotive shops, and consumers. Ultimately, these efforts can significantly reduce the amount and potential toxicity of lead in MSW.

- The framework for lead-acid battery collection and reclamation programs can potentially be applied to programs for household batteries. However, consensus legislation and an infrastructure for the collection and reclamation of household batteries is not well established in the United States.

- Debate continues regarding the effectiveness and feasibility of household battery collection and recycling programs. Currently, most collected batteries are disposed of in hazardous waste landfills and are not reclaimed, much less recycled. Currently, reclamation of used household batteries is not feasible in the United States and is hindered by the presence of potentially toxic metals (e.g., mercury) in primary cells.

- Source reduction initiatives must continue at both the industrial and municipal levels. Emphasis should be placed on the redesign of batteries to reduce potential toxicity. In addition, battery manufacturers should develop cost-effective reclamation technologies for their products. Source reduction also involves the removal of batteries from mixed MSW prior to composting, burning in MSW incinerators, and disposal in MSW landfills.

- Source reduction can also be promoted by encouraging consumers to use more rechargeable batteries in household products. This would reduce the number of alkaline and zinc-carbon batteries in MSW and also reduce the amount of mercury (and other potentially toxic metals) in MSW. Since nickel-cadmium batteries can be reclaimed in the United States, manufacturers and retailers should more actively promote programs for their collection. The effectiveness of such programs should be monitored to ensure that any increase in nickel-cadmium battery sales is not reflected by an increase in the amount of cadmium in MSW.

- If the toxicity of household batteries in MSW decreases, future assessments should be performed to consider the impact of their disposal in modern, well-designed, and regulated MSW landfills rather than in hazardous waste landfills. Ultimately, the development of cost-effective technologies for the reclamation of used household batteries is needed to ensure that recycling, rather than disposal, becomes the preferred waste management alternative.

REFERENCES

Adams, A., and C. Amos, Jr. (1993) Batteries, in H. F. Lund (ed.), *The McGraw-Hill Recycling Handbook,* McGraw-Hill, New York.

Apotheker, S. (1990a) Batteries Power Secondary Lead Smelter Growth, *Resource Recycling,* vol. 9, pp. 46–47.

Apotheker, S. (1990b) Does Battery Recycling Need a Jump? *Resource Recycling,* vol. 9, pp. 21–23.

Apotheker, S. (1991) Get the Lead Out, *Resource Recycling,* vol. 10, pp. 58–63.

Arnold, K. (1991) *Household Battery Recycling and Disposal Study,* Minnesota Pollution Control Agency, St. Paul, MN.

Battery Council International (undated) Personal communication, Chicago, IL.

Battery Council International (1993) *List of States without Lead-Acid Recycling Laws,* memorandum from Saskia Mooney of Weinberg, Bergeson, and Neuman, Washington, DC.

Cohen, S. (1993) Recycling Nickel-Cadmium Batteries, *Resource Recycling,* vol. 12, pp. 47–54.

Eutrotech Inc. (1991) *Used Batteries and the Environment: A Study on the Feasibility of Their Recovery,* EPS 4/CE/1, Environment Canada, Montreal, Canada.

Gaba J., and D. Stever (1992) *Law of Solid Waste, Pollution Prevention and Recycling,* Clark, Boardman, and Callaghan, Deerfield, IL.

Hurd, D., D. Muchnick, M. Schedler, and T. Mele (1992) *Feasibility Study for the Implementation of Consumer Dry Cell Battery Recycling as an Alternative to Disposal,* Recoverable Resources/Boro Bronx 2000, Inc., New York.

NYSDECED (1992) *Report on Dry Cell Batteries in New York State,* New York State Departments of Environmental Conservation and Economic Development.

PRBA Newsletter (1993) *The Recharger,* Portable Rechargeable Battery Association, Atlanta, GA.

Reutlinger, N., and D. de Grassi (1991) Household Battery Recycling: Numerous Obstacles, Few Solutions, *Resource Recycling,* vol. 10, pp. 24–29.

Smith, Bucklin and Associates (1993) *Battery Council International 1991 National Recycling Rate Study,* Smith, Bucklin and Associates, Inc., Chicago, IL.

U.S. EPA (1989) *Characterization of Products Containing Lead and Cadmium in Municipal Solid Waste in the United States, 1970 to 2000,* Office of Solid Waste, EPA/530-SW-89-015A, U.S. Environmental Protection Agency, Washington, DC.

U.S. EPA (1992a) *Characterization of Municipal Solid Waste in the United States: 1992 Update,* Office of Solid Waste and Emergency Response, EPA/530-R-92-019, U.S. Environmental Protection Agency, Washington, DC.

U.S. EPA (1992b) *Characterization of Products Containing Mercury in Municipal Solid Waste in the United States, 1970 to 2000,* Office of Solid Waste and Emergency Response, EPA/530-R-92-013, U.S. Environmental Protection Agency, Washington, DC.

U.S. EPA (1992c) *States' Efforts to Promote Lead-Acid Battery Recycling,* Office of Solid Waste and Emergency Response, EPA/530-SW-91-029, U.S. Environmental Protection Agency, Washington, DC.

U.S. EPA (1993) *Modification of the Hazardous Waste Recycling Regulatory Program: Proposed Rule,* Federal Register, vol. 58, pp. 8101–8133, Washington, DC.

CHAPTER 11

OTHER SPECIAL WASTES
Part 11B   Used Oil

Stephen D. Casper
William H. Hallenbeck
Gary R. Brenniman

11B.1   USED OIL

Introduction

Used oil is a problem waste because its generation is ubiquitous and it can contain hazardous liquid wastes and other contaminants. When used lubricating oil qualifies as a hazardous waste, its disposal becomes complicated and costly. Even when properly handled, rerefined oil carries a misperception of low quality with respect to new lubricating oil and often must sell at a lower price than new oil.

The most recent survey of used oil (1988) and its ultimate disposition indicates that of the 1.351 billion gal of used oil (automotive and industrial) generated annually, 901 million gal were reused in some manner such as fuel, secondary industrial use, rerefined lube oil, or road oil (Temple, Barker & Sloane, Inc., 1989a; McHugh, 1991). Over 400 million gal were disposed of through dumping on the ground or in water, landfilling, or non-energy-recovery incineration. Though inappropriate, the legality of these disposal methods depends on the disposal path, the generator, the type of oil, and the specific state regulations.

Burning as a fuel and rerefining are the two major methods for recycling used oil. By volume, the waste oil fuel industry consumes 58 percent of available used oil while rerefining consumes only 2 percent. Used oil has an excellent heating value (13,000 to 19,000 Btu/lb) and can help meet the growing national energy demand. However, because of the constituents present in used oils, air emission controls may be necessary when burning. A better secondary use of lube oils is its rerefining back into a usable base stock oil. In automotive engines, lube oil becomes dirty and the additives break down, but the base stock oil (roughly 80 percent by volume of marketed product) does not break down, allowing for its redistillation. The rerefining process concentrates contaminant metals into a "bottoms" residue, which is reused in asphalt production. Producers of rerefined crankcase oil have shown their product to meet or exceed the engine lubrication testing requirements for "new" oil by the American Petroleum Institute. For these reasons, the rerefining of used oils into a usable product is a preferred energy conservation and pollution prevention alternative.

A lack of regulatory control of used oil disposal and lack of an infrastructure for the collection and processing of used oil from the public has resulted in improper disposal of used oils. To minimize improper disposal, the management system for used oil must be modified:

- To divert the improper disposal of used oil in refuse and the environment and get it into the used-oil management system of collectors, handlers, and processors

11.15

- To correctly manage used oil after entering the system by including standards that address accountability
- To increase the flow of used oil into rerefineries

This section discusses oil consumption, hazardous contaminants in new and used oil, legislation and regulation of used oil, methods to increase used oil collection, the rerefining process, and recovery of oil and scrap material from oil filters.

Oil Consumption

Methods of Estimating Oil Usage.   There are three methods for estimating oil usage (Hegberg et al., 1991). In the first method, states annually report fuel consumption because of the motor vehicle fuel sales tax. One can base oil consumption estimates on current motor vehicle fuel consumption. The automotive industry recommends changing engine oil after every 3000 miles driven. This translates into approximately 4 or 5 qt of oil generated per 150 gal of gasoline purchased. Second, since motor vehicles use more oil than any other application, motor vehicle registrations may be used to estimate engine-related oil consumption. Third, estimating state oil consumption using population data is a straightforward and commonly used method; although population data may not exactly reflect oil users.

Of all engine oils produced, over 80 percent is multigrade crankcase oil. Engine oils are distributed largely through the retail and commercial sector, which accounts for 88 percent of the sales of engine lubricants (NPRA, 1990). Retail distribution occurs through service stations and other retail outlets such as automotive parts stores or chain stores. Commercial distribution includes sales to commercial truck fleets, governments, railroads, commercial marine, airlines, or industrial plants.

Estimates of Used-Oil Generation Factors.   A significant portion of engine oils are nonrecoverable owing to combustion in the engine, residual left in engine components (e.g., oil filters), and inadvertent spilling. Generation factors are defined as the volume of waste oil generated compared with the volume of oil initially purchased. Industrial and engine oil generation factors range from 10 to 80 percent, depending on the application. Engine oil generation factors are higher because of the frequency with which crankcase oil is changed. Various studies have put the engine oil generation rate at about 60 percent of annual purchases (Temple, Barker & Sloane, Inc., 1989b; Franklin Associates, Ltd., 1985, 1987).

Disposal of Used Oils.   Used oil is improperly disposed of by:

- Direct disposal into the environment by dumping
- Collection with municipal solid waste with subsequent disposal in a landfill or incinerator
- Burning on site
- Disposal to a liquid effluent treatment system
- On-site secondary use such as a machinery lubricant or dust suppressant

The group responsible for the majority of improper disposal is known as do-it-yourselfers (DIY), or people who change the oil on their own vehicles. Only 5 percent of the DIY oil generated in 1988 was channeled into the used-oil management system by collection at gas stations, quick-lube oil change stations, repair shops, or municipal recyclers (Temple, Barker & Sloane, Inc., 1989a). Fifty-one percent of non-DIY used engine oil managed on site was improperly disposed of through dumping. Much of this was from off-road construction and mining sources. This indicates the need to examine regulations and collection methods for used engine oils.

Hazardous Contaminants

Virgin Lube Oil Characteristics. New and used lubricating oils can contain hazardous constituents such as metals, chlorinated compounds, and polynuclear aromatic hydrocarbons (PAHs). *Finished lubricating oil* is the term used for oil available in the marketplace. It contains a combination of base stock lubricant and various additives. Limited data have shown that virgin base stock oils contain metals such as barium, cadmium, lead, and zinc on the order of 0 to 1 ppm. Lesser amounts of chromium (0 to 0.05 ppm) and benzo(a)pyrene (at <1 ppm), a carcinogenic PAH, have also been identified (Franklin Associates, Ltd., 1985).

Additive compounds enhance the effectiveness of lubricating oil and greatly influence its composition. They comprise 10 to 30 percent by volume of finished engine oil products. A typical formulation for gasoline engine oil is shown in Table 11B.1. Additives inhibit metal corrosion and oxidation of oil, and act as detergents, dispersants, and antiwear compounds. They contain hazardous constituents such as magnesium, zinc, lead, and organics, and they also increase the concentrations of sulfur, chlorine, and nitrogen in lube oil.

TABLE 11B.1 Typical Formulation of Gasoline Engine Oils

Ingredient	Percent by volume
Base oil (solvent 150 neutral)	86
Detergent inhibitor (ZPDD-zinc dialkyl dithiophosphate)	1
Detergent (barium and calcium sulfonates)	4
Multifunctional additive (polymethyl-methacrylates)	4
Viscosity improver (polyisobutylene)	5

Source: Weinstein (1974).

Used-Oil Contaminants. During service, lubricating oils become contaminated with metal particles from engine wear, gasoline from incomplete combustion, and rust, dirt, soot, lead compounds, and water vapor from engine blowby (i.e., material that leaks from the engine combustion chamber into the crankcase). Oil additives can oxidize during combustion, forming corrosive acids. Table 11B.2 shows the results of used automotive oil analyses when the samples were collected directly from the generator. The table contains 19 constituents, 17 of which are part of the U.S. EPA's list of hazardous constituents listed in Title 40, Code of Federal Regulations, Part 261, Appendix VII (40 CFR 261). Other samples taken from collectors, processors, and refiners that were identified by the respective source as used engine oil showed lower levels of some metals and higher levels of chlorinated solvents, indicating the mixing of different oil types.

A sampling program of curbside-collected used crankcase oil was conducted in Oregon in 1986. The state has over 100 curbside used-oil programs, affecting a population of 2 million. More than 400 individual samples were used to form 20 composite samples for testing. For the samples, average total halogen level was 357 ppm; lead, 662 ppm; arsenic, 0.21 ppm; cadmium, 1.2 ppm; and chromium, 3 ppm (Spendelow, 1989). The lead content exceeds the U.S. EPA lead limit of 100 ppm defining off-specification used oil for fuel. This figure should decrease over time owing to the federal phase-down of allowable lead additive levels in leaded gasoline. Overall, the study concluded that used oil collected from households was generally not contaminated with household hazardous material or other inappropriate wastes.

Minimum Testing of Used Oil. When testing used oil through the services of a laboratory or as part of a waste disposal firm's services, minimum analysis should address metals, halogenated solvents, aromatics, and polychlorinated biphenyls (PCBs). A disposal service will often retain a sample of used oil in order to identify the source. A good practice for the used-oil generator is to split the sample collected and retain half in order to serve as the check

TABLE 11B.2   Concentration of Potentially Hazardous Constituents in Used Automotive Oil Samples Taken Directly from Generators

Constituent	Number tested	Samples above detection limit, %	Mean concentration,* ppm	Median concentration, ppm	Concentration range, ppm		Regulatory limit for used-oil fuel, ppm
					Low	High	
Metals							
Arsenic	24	8	9.9	5	<5	14	5
Barium	113	95	209.5	94	0.78	3,906	
Cadmium	64	93	1.7	1	<0.2	10	2
Chromium	99	97	10.8	8	0.5	50	10
Lead	40	97	2,573.7	1,470	5	21,700	100
Zinc	116	100	982.3	1,000	4.4	3,000	
Chlorinated solvents							
1,1,1-Trichloroethane	22	18	401.3	6	<1	1,000	
Trichloroethylene	22	9	2.5	5	<1	16	
Tetrachloroethylene	22	36	180.1	9	<2	660	
Total chlorine	36	100	1,200.0	800	<100	4,700	4,000
Other organics							
Benzene	22	45	589.0	9	1	3,600	
Toluene	22	86	1,010.7	190	1	6,500	
Xylene	22	90	2,005.2	490	2	14,000	
Benzo(a)pyrene	21	100	9.7	10	1.3	17	
PCBs	22	5	39	—	—	—	

* Calculated for detected concentrations only.
Source: Franklin Associates, Ltd. (1985).

against the sample taken by a hauler or processor. An initial test that is typically performed on loads of oil at the generator site is a total organic chlorine test. Such a test is only an initial evaluation of whether waste oil may contain chlorinated hazardous waste greater than the 1000-ppm threshold level (see later).

Legislation and Regulations Surrounding Used Oil

Federal Regulation of Used Oil.   Congress and the U.S. EPA have been attempting to deal with used-oil regulation since 1976 when the Resource Conservation and Recovery Act (RCRA) was legislated. Recent years have seen a growing debate over whether used oil should be classified as a hazardous waste. Table 11B.3 gives limits for specification used-oil fuel. Listing used oil as a hazardous waste, some argue, would discourage the recovery and recycling of used oil and is contrary to the intent of regulation. Essentially, used oil that is not burned for energy recovery is not currently regulated unless it has been mixed with a listed hazardous waste or exhibits a hazardous waste characteristic of ignitability, reactivity, corrosivity, or toxicity.

At the federal level, household waste is exempt from regulation as a hazardous waste whether it contains hazardous waste or not (40 CFR 261.4). Household waste means any material derived from any type of dwelling, and includes waste that has been collected, stored, transported, treated, disposed, recovered, or reused.

The hazardous waste industry and petroleum rerefiners support the listing of used oil as hazardous. They would like to see full RCRA Subtitle C (hazardous waste) regulations for all recycling facilities and transporters, because many of the large companies involved are already permitted for handling RCRA Subtitle C waste and, based on technical grounds, used

TABLE 11B.3   Limits for Specification Used-Oil Fuel*

Contaminant and property	Limit
Arsenic	≤5 ppm
Cadmium	≤2 ppm
Chromium	≤10 ppm
Lead	≤100 ppm
Total halogens†	≤4000 ppm
Flash point	≥100°F

* The specification does not apply to used-oil fuel mixed with a hazardous waste other than that from a small-quantity generator.

† Used oil containing more than 1000 ppm total halogens is presumed to be a hazardous waste. Such used oil is subject to 40 CFR, Part 279, Subpart G, unless rebutted by demonstrating that the used oil does not contain hazardous waste.

oil can qualify as a hazardous waste. Service stations and other such entities would be exempt from such regulations. Rerefined oil products and their users would also be exempt.

Oil processors and others, including the American Petroleum Institute (API), are opposed to such a listing. Since processors essentially filter and remove the water from oil for reuse as fuel, the cost to operate is less than a rerefinery. This gives the processor a cost-competitive edge over rerefining. A hazardous listing could force many processors out of business. Additionally, any hazardous listing would require users of such fuel to become a permitted hazardous waste incineration facility.

Current Regulatory Status.   The U.S. EPA has attempted to resolve the aforementioned debate by issuing its final ruling on used oil on September 10, 1992 (U.S. EPA, 1992). They decided not to list used oil as hazardous waste because existing hazardous waste regulations based on hazardous characteristics (e.g., toxic, corrosive, reactive, and ignitable) adequately address the disposal of oil exhibiting hazardous properties. There are also sufficient existing federal and state regulations to control the disposal of nonhazardous used oils.

The U.S. EPA simultaneously promulgated used-oil handling standards for generators, transporters, processors, rerefiners, burners, and marketers. These standards apply to DIY-generated oil only after it has been collected and aggregated through public or private collection services (e.g., municipal collections, service stations, etc.). These standards are located in 40 CFR, Part 279.

The generator regulations apply to facilities that produce more than 25 gal of oil per month (not farmers or DIY). Generators must:

- Maintain storage tanks and containers.
- Label storage tanks "Used Oil."
- Clean up any leaks or spills.
- Engage a used-oil transporter possessing an EPA identification number.

Service station owners who comply with the preceding may accept oil from DIY without liability for subsequent handling mishaps.

There are about 300 used oil processors and rerefiners in the United States. They must follow these management standards:

- Obtain an EPA identification number.
- Maintain storage tanks.

- Handle or store oil only in areas with impervious floors and secondary containment.
- Plan oil testing for halogen content.
- Keep records.
- Safely manage processing residue.
- Plan for proper facility closure.

Transporters or collectors deliver oil from one site to another for recycling. Transfer areas (e.g., loading docks, parking areas) must comply with transporter storage requirements, if oil is held for more than 24 h in route to its final destination. Generators who transport less than 55 gal of their own oil are exempt. Transporters must comply with the same requirements as processors (where applicable) and:

- Limit storage at transfer facilities to 35 days.
- Test waste in storage tanks that are out of service for hazardous characteristics and, if wastes are hazardous, close them according to existing hazardous waste management requirements.

Used-oil burners must comply with the same storage requirement as transporters. There is no significant new regulation of used-oil marketers.

Used Oil Burned for Energy Recovery.   Federal regulation addresses and controls two of the primary mismanagement activities of used oil: (1) burning contaminated used oils in nonindustrial boilers, and (2) mixing hazardous wastes into used oils. Any oil that meets the specification levels shown in Table 11B.3 is subject only to analysis and recordkeeping requirements. Oil that does not meet the requirements of Table 11B.3 is "off-specification used-oil fuel." Used oil containing more than 1000 ppm total halogens is presumed a hazardous waste because it was mixed with a halogenated hazardous waste (40 CFR, Part 279, Subpart G).

Off-specification used oil is subject to the used-oil burning regulations (40 CFR, Part 279, Subpart G). It can be burned in industrial applications such as cement kilns, blast furnaces, manufacturing plant boilers, and utility boilers, but not in nonindustrial boilers such as those located in office buildings, schools, and hospitals. Off-specification used oil may also be burned in space heaters with less than 500,000 Btu/h capacity provided only DIY oil is burned and the heater is vented to the atmosphere. Those who treat off-specification used oil by processing, blending, or other treatment to meet the specification shown in Table 11B.3 must document that the used oil meets the specification. Off-specification used oil incurs additional restrictions and requirements on marketers and burners.

Liability Concerns.   Pollution from a leak, spill, or improper disposal of used oil is the primary liability concern that owners and operators of used-oil collection and processing facilities should consider. The Comprehensive Environmental Response, Compensation, and Liability Act (CERCLA), passed in 1980 and amended in 1986, allows the courts and government to hold those parties that created dangerous conditions at a hazardous waste disposal site financially responsible for the required cleanup. Under CERCLA's strict liability standard, it is not a defense that the generator exercised "due care" in arranging for disposal through another entity or that the disposal facility complied with all contemporary environmental and safety requirements (Nolan et al., 1990).

The only specific exclusion from the definition of a *hazardous substance* in CERCLA is for *petroleum,* which Congress defined to encompass crude oil or any fraction thereof (including used oil) that is not otherwise specifically listed or designated as a hazardous substance. A follow-up definition of *petroleum* by the U.S. EPA provides guidance on whether used-oil products may be required to meet hazardous waste regulations: (1) *Petroleum* must be interpreted to include all hazardous substances, such as benzene, which are indigenous to petroleum substances. Inclusion of hazardous substances that are found naturally in crude oil and its fractions is necessary for the petroleum exclusion to have any meaning. (2) *Petroleum* must be interpreted to also include hazardous substances that are normally mixed with or added to crude oil or crude oil fractions during the refining process. (3) *Petroleum* does not include hazardous substances that are added to petroleum or increase in concentration solely as a result of contamination of petroleum during use (Nolan et al., 1990). This means that a hazardous substance added to petroleum during use (e.g., as a result of contamination) is not part of the petroleum and cannot be excluded from the requirements of CERCLA.

Regulatory Methods to Increase Used-Oil Collection

Methods for increasing the collection of used oils should be aimed at impacting do-it-yourself behavior. A number of different methods are available at the municipal, state, and federal levels to increase used-oil recovery (McHugh, 1991):

- Impose regulations on the generators of used oil.
- Provide a deposit-refund system on the purchase and return of new and used oil.
- Provide a tax on engine-related purchases to subsidize used-oil collection and recycling.
- Require sellers of lube oil to maintain collection facilities.
- Develop a state-supported infrastructure of public and private collection centers.
- Rely on public education and labeling as a method of impacting end use.
- Use government procurement policies to stimulate the market for rerefined used oil.
- Require lube oil producers to reuse a certain amount of used oil (either through rerefining or as a fuel).

Generator Regulation.   Imposing regulations on the generator of used oils can significantly change generator behavior. Such regulations must be limited in their scope because much problem oil is a result of DIY, and environmental compliance monitoring and enforcement of such regulations at the household level is not practical. Semienforceable regulations, such as the banning of used-oil disposal in refuse, may have the most impact on DIY generators.

Point-of-Sale Collection.   Requiring retail sellers of lube oil to accept used oil and maintain a collection facility can have both positive and negative impacts. Such a requirement may discourage nonautomotive retailers from selling crankcase oil at all. On the other hand, automotive franchises may be well suited to coordinate a uniform policy for collection, handling, and disposal. Valvoline, through its subsidiary EcoGard, has established a retail used-oil collection service in which the store owner receives training materials, consumer education materials and signage, a wheeled collection tank (which is used behind the counter by a store employee), generator documentation material, and a collection service.

Government-Supported Used-Oil Infrastructure.   Project ROSE (Recycled Oil Saves Energy) is a successful, state-supported program in Alabama that works with private-sector businesses such as retail stores and service stations to develop public- and private-sector collection of used oil. The state's portion of the program, which is funded by the Department of Economic and Community Affairs, has four main objectives:

1. Educate citizens about the energy and environmental benefits of used-oil recycling.
2. Create a statewide awareness of the implications of improper disposal of used oil.
3. Organize and promote used-oil collection centers for every county in the state.
4. Document energy savings for the state.

Project ROSE works with retailers, government agencies, and public-service groups, and provides citizens with up-to-date information concerning used-oil recycling in their area. The program also maintains a service for identifying used-oil haulers and processors. Thanks to Project ROSE, the state of Alabama has used-oil collection in 45 of 67 counties with over 200 collection centers.

Germany uses taxes levied on auto or lube oil purchases to subsidize a used-oil recycling infrastructure and to stimulate development of the used-oil market. Automobiles are taxed based on engine size and revenues are deposited into a central fund. The recyclers are then reimbursed by the government based on the difference between the cost of rerefining or reprocessing and the market price for the oil. (McHugh, 1991) Such a system taxes the consumer to create a relatively new industry; however, this may inhibit competition within the used-oil recycling business.

Mandatory Recycling for Lube Oil Producers. Mandatory recycling requires lube oil producers to recycle a percentage of their annual oil production or support the recycling infrastructure through the purchase of recycling credits. Such a regulation would encourage development of the used-oil recycling industry (particularly rerefining), provide a market demand for used oil (by viewing it as a resource rather than a waste product), and minimize cost impact because of private-sector competition. Development of a "credit system" allows lube oil producers to support recycling activities if they opt not to physically enter into the used-oil collection and recycling business. Recycling credits are also meant to reduce the incentives for illegal behavior on the part of the generator by creating the market for collection.

Education and Collection of Used Engine Oil

Participation in Collection Programs. Education and the availability of collection programs are the most important elements in minimizing used-oil mismanagement. DIY oil changers tend to be the primary group mismanaging used oil. A statewide survey in Minnesota showed that 58 percent of the population are DIY, and only 37 percent of these disposed of their oil in a responsible manner (Shull et al., 1987). (See Table 11B.4.) Any used-oil management program should address these issues. When compared with the rural population, the urban DIY population recycled at a substantially higher rate.

TABLE 11B.4 Used-Oil Disposal Practices of Do-It-Yourselfers in Minnesota

Disposition	Statewide, %	Urban, %	Rural, %
Recycled	37	54	14
Taken to a service station or store	22	31	11
Taken to a recycling center	15	23	3
Thrown away	24	19	30
In the trash	17	14	20
Taken to a landfill	1	0	1
Dumped into the sewer	1	1	0
Dumped on the ground	6	4	9
On-site reuse and disposal	39	27	56
Road dust control	15	15	14
Reused	10	6	16
Used as fuel	1	1	1
Weed killer	2	0	4
Burned	7	3	14
Kept	4	2	6

Source: Shull et al. (1987).

Most of the DIY population would make the effort necessary to recycle if it were marginally convenient or if minor compensation were involved. Shull also showed that convenience in collection is an important factor in recycling DIY oil, particularly in urban areas. Rural residents are less likely to recycle regardless of the options open to them.

Education and Promotion.   An educational campaign to promote proper management of used oil should focus on three groups: (1) current DIY, (2) young people in school, and (3) the general public.

In educating consumers about used oil, there should be three goals: (1) educate about the problems raised by mismanagement of used oil; (2) encourage more responsible used-oil management; and (3) inform DIY exactly how to recycle oil in their locality. When presenting the problems caused by mismanagement, it is important to note that used oil is a valuable resource.

To have a lasting impact in the community, it is necessary to educate young people who will soon be driving. Impressing upon young people that used oil can be rerefined back into a usable product or can be reused in the crude oil production process will show that used oil has value. High schools and driver's education programs are natural places to present short courses on the benefits of used-oil recycling and how to change oil properly.

A number of promotional methods can be used to promote a used-oil program: program kickoff day, a used-oil recycling hotline, handouts, brochures, posters, mailings, windshield service stickers, mailing inserts in utility bills, editorials in newspapers, and broadcast public service announcements. Periodic used-oil collection programs should be ideally held during the spring and fall because these are the times when people clean house and may dispose of accumulated used oil.

The U.S. EPA has developed a number of documents on promoting the proper management of used oil. They present clear, simple ways to initiate oil recycling programs and include sample brochures, press releases, signs, and letters to encourage participation. These documents are available from the U.S. EPA, Office of Solid Waste, 401 M Street, S.W., Washington, D.C. 20460; telephone: (800) 424-9346.

Used-Oil Collection in Rural Areas.   It is estimated that an average farmer in Illinois buys 50 gal of motor oil and 40 gal of hydraulic fluid per year (Peterson, 1991). A survey in Minnesota identified that 80 percent of the farmers either burn, use for dust suppression, or lubricate machinery with some of the used oil generated (Shull et al., 1987). Nonetheless, the farmers' comments indicated that they thought the best way to handle the used oil would be to have someone collect it.

Used-Oil Collection Days.   Used-oil collection in farming areas is a particular problem because waste oil haulers have to travel long distances between stops and the load collected per stop is minor in relation to the truck tank size. The result is a large collection fee to the farmer. An alternate oil collection method for sparsely populated areas is to organize used-oil collection days. This provides a service to the generator, cuts down on the cost of collection, consolidates the collection locations, and eliminates the need for filing a permanent storage permit application. A collection of once or twice per year may be adequate in rural areas because storage space is usually not a great concern for the generator.

The organizer of the collection day typically makes one or more tanks available at a farm cooperative or arranges for a waste oil hauler's truck to be on-site. Using a tank instead of a waste oil hauler truck for collection allows the hauler to perform the normal route for the day and simply make an additional stop at the collection day site. In some cases haulers do not charge for collection because of the oil value and the ease with which it is collected, and in other cases the sponsoring cooperative has paid the waste oil hauling charges so that no cost is passed along to its customers.

To track any potential problems with collected used oil, a small sample of oil should be collected from each generator and a simple form signed by the generator stating, ". . . the oil is free of contamination such as water, gasoline, antifreeze, solvents, and farm chemicals . . . if said product is contaminated, the generator may be held liable for a disposal fee." The cooperatives additionally provide "Waste Oil Only" stickers for its customers to label drums and tanks at home.

Promotional methods for the collection days have included direct mail, notices included in monthly statements, and spots in local newspapers and radio stations.

Two collection days per year in agricultural areas, one in early spring (March) and another in late summer (end of August), may be sufficient. Early spring collection handles the waste oil from the fall and winter, before spring planting, and late summer handles the spring and summer oil when farm machinery is used most.

Employer Collection.   Another method for managing used oil in rural areas is to locate a collection station at places of employment. Amana Refrigeration in Amana, Iowa, has opened up a recycling center for employees. Collection includes used oil as well as newspapers and clear high-density polyethylene (HDPE) bottles. The recycling center consists of a skid-mounted shed divided into three sections that can be transported with a forklift. Amana provides free oil change recycling tubs with disposable plastic liners or employees simply use 1-gal milk jugs and put waste oil on a shelf in the recycling center for disposal. Janitors empty the center daily by putting used oil in 55-gal drums. Once a drum is filled with used oil, a total halogen check is performed on the material, and then it is mixed in a large tank with waste process oil from the company's manufacturing process. The waste oil is then picked up by a waste oil hauler at a collection cost of about 10 cents per gallon. Since some of Amana's employees are also farmers, the company has supplied these workers with 55-gal drums to return to the company when full.

Used-Oil Collection in Urban Areas.
There are a number of methods for collecting used oil in urban areas. Examples include curbside collection, drop-off at recycling stations or in conjunction with local business, drop-off at dedicated used-oil collection depots, point-of-purchase collection, door-to-door pickup by appointment on designated days, used-oil drop-off collection days, or as part of household hazardous waste collection days.

Curbside Collection of Used Oil.   Curbside programs are by far the most successful recycling programs because they make it convenient for the public to participate. An earlier study indicated that 70 to 75 percent of people would save their used engine oil for recycling if it were collected at home (U.S. DOE, 1981). Nationwide, it is estimated there are 170 used-oil curbside collection programs, of which 43 are in California. (Arner, 1991). Curbside collection of used oil requires the separation of oil in a sealed container by the generator. As with source separation of any recyclable material, curbside collection promotes attitude change and behavior modification.

Curbside collection of used oil is fairly simple. The most popular method has been to attach a collection tank to the side of a refuse collection or recycling vehicle. The drain funnel for the tank is sized to hold common collection containers such as 1-gal milk jugs or 1-qt oil bottles. Another method has been to fasten a collection rack to the side of a refuse truck or install an additional compartment on a recycling truck. This allows the operator to collect entire containers of used oil and then empty them at a central facility. Whether waste oil is emptied on a route or at a central facility, it is imperative that operators be trained to watch for non–waste oil products being disposed of. It is also important to locate the tank or rack in a location that would not cause a problem or spill on the ground if the collection truck is involved in an accident.

The town of Florence, Alabama, participates in Project ROSE, the statewide oil collection program mentioned earlier. The town retrofitted its two recycling vehicles with 75- and 150-gal collection tanks. The tanks are located underneath the recycling truck near the rear with the collection funnel piped to the side of the truck. To keep costs low, the town's sanitation division performed the retrofit. The coordinator for the project, R. Holst of the Northwest Alabama Council of Local Governments, indicates collection amounts are fairly steady at a rate of 100 gal per week for the 5000-household collection area. As expected, initially large amounts were collected that were previously saved up, and there is seasonal variation, with large amounts in the fall and spring. To help resolve this, residents are asked to put out no more than 2 gal per week.

The used-oil management program in Florence is noteworthy because it represents a coordination of municipal government and private business to their mutual benefit. The project coordinator arranged for a local franchised quick-lube center, Express Oil Change, to accept the used oil collected in the curbside program. Express Oil is not paid to accept the oil but does receive a nominal price from the waste oil hauler. In exchange, the municipality has put Express Oil signs on the recycling truck tanks. Also, an opening day kickoff press conference (with resulting front-page coverage) for the curbside program was held at the Express Oil station where drop-off occurs.

Drop-Off Recycling Stations.   Establishing a local drop-off station for used oil is one of the simplest methods for collecting used oil. The basic features include tank- or barrel-type collection above ground with a raised curb, a roof and side walls to prevent water entry, and a fence for security protection. The cost of constructing the Rockford, Illinois, used-oil collection station was roughly $1500 for the shed and concrete base; $1100 for the two 150-gal tanks, complete with valving, sight glass, drain pan, and piping; and $300 for signage.

Preassembled, igloo-shaped collection stations made of fiberglass have been used in Europe for many years, and are gaining popularity in the United States. These should be placed in locations under occasional observation (e.g., in front of a service station).

Point-of-Sale Collection.   The solid waste management board of Snohomish County, Washington, has jointly coordinated the countywide point-of-sale collection of used crankcase oil with a retail automotive parts chain (Wolfin, 1991). They decided that since most oil is purchased at automotive parts stores, such a location was appropriate for used-oil collection. The county has a population of 480,000 and contains urban as well as rural areas.

A local automotive retail subsidiary expressed interest in participating in such a program as a site sponsor. The county placed outdoor collection tanks at each outlet store as well as solid waste handling facilities (a total of 16 tanks). Responsibilities of the retailer were to:

- Provide a site for placing the collection tank (a spot in adjacent parking lots).
- Obtain approval from the actual property owner.
- Lock up the collection tank at night.
- Clean up minor spills with oil-absorbent or cleanup equipment.
- Monitor oil levels in the tanks.
- Call for pickup if off schedule.
- Maintain a log.

The county was responsible for:

- Providing the used-oil collection tanks and curbing around the tanks.
- Establishing contracts.
- Countywide education and promotion of the project.
- Cleanup of major spills (e.g., knocking over a collection tank).
- The material collected.

The county established three contract agreements:

1. With the site sponsor parent company, outlining the responsibilities of the county and the sponsor.
2. With the waste oil hauler.
3. With a hazardous waste hauler for handling "hot" loads.

A key component of the program was a ruling by the state environmental agency indicating that, since the collection sites are for consumer use, any hazardous waste collected qualifies as household hazardous waste. This ruling exempted the oil from RCRA hazardous waste requirements and relieved the county of generator responsibility.

The county now consistently collects 6000 to 7000 gal per month at its 16 collection sites; less than 2 percent of total volume collected has been contaminated. A key recommendation by the county for keeping hazardous waste disposal in the tanks to a minimum is to ensure that alternative disposal means are available for other household hazardous wastes.

The capital cost for the collection tanks, curbing, and site setup was approximately $45,000; annual operating cost is $12,000; and the cost for removing contaminant disposal in 1990 was $6000. Based on a full year of collection, total cost to the county is 28 to 35 cents per gallon of used oil (Wolfin, 1991).

Rerefining Used Oil

Rerefining used oil is analogous to recycling an aluminum beverage can (i.e., remanufacturing a waste commodity into a new product of the same type). The rerefining process is up to 98 percent efficient in converting used engine and industrial oils into high-quality lubricants for identical applications. One rerefining process consists of four steps: (1) dehydration, (2) defueling, (3) extraction and distillation, and (4) hydrotreating. When waste oil arrives at a rerefining plant, it is first tested for contamination and then bulked and mixed in a storage tank to achieve a uniform feedstock. The first processing step, dehydration, is needed because waste oil coming into the plant can contain 12 to 15 percent water by volume. Oil is piped to the dehydration tank and heated to 135°C under atmospheric pressure. This boils off any water and some lighter petroleum fractions. The wastewater produced is treated on site (Safety-Kleen, 1993).

From the dehydrator, the oil is fed to the defueling system and the temperature is raised to 230°C under a vacuum of 100 mmHg. This process removes more light fuel and lube oils, which are then condensed and used as a fuel on site.

In the next process step, the oil is completely vaporized at 400°C under a vacuum of 3 mmHg. It is then condensed into three separate oil fractions and pumped to holding tanks. Of the material not collected, the lightest fraction is marketed as an industrial fuel, and the heavy fraction, or "bottom," is marketed as an asphalt extender. This product contains the additives, polymers, wear metals, contaminants, and oxidized materials removed in the distillation process.

In the final rerefining step, each of the three distilled oils is fed into a reactor at high pressure and temperature with hydrogen and a catalyst. This process removes sulfur, nitrogen, chlorine, oxygenated compounds, heavy metals, and other impurities. The hydrogen is then removed and light distillates stripped.

The end product of the rerefining process is base oil, which can be used to formulate new engine, gear, and hydraulic oils. API has certified many rerefined oil products, using the same standards applied to virgin oil products, API SG/CD for 10W30 crankcase oil, as an example. Figure 11B.1 lists known U.S. and Canadian oil rerefiners (Hegberg, 1991; Wolfe, 1992; Arner, 1992).

Recovery of Used Oil and Scrap Material from Oil Filters

Oil filters and their contents are one component of the waste oil stream that is nearly always disposed of in landfills. A study at the University of Northern Iowa (Konefes and Olson, 1991) has evaluated the recovery of used oil from filters as well as the filter material and scrap metal. The evaluation was divided into three phases: (1) methods to reduce residual oil in waste oil filters at the point of generation by simple draining; (2) hydraulic compaction of waste oil filters to extract additional quantities of residual oil; and (3) recyclability potential of the resulting used oil, filter media, and scrap metal.

Oil Filter Recovery Results

Reducing Residual Oil by Draining.   Two independent studies found that only about half of the oil contained in a filter can be removed by simple gravity draining (Konefes and Olson,

Breslube-Safety Kleen
P.O. Box 130
Breslau, Ont. N0B 1M0
519-648-2291

Breslube-Safety Kleen
7001 W. 62nd St.
Chicago, Ill. 60638
312-229-1500

Cibro Petroleum Products
Bronx, N.Y.
718-824-5000

Consolidated Recycling
8 Commerce Dr.
P.O. Box 55
Troy, Ind. 47588
606-264-7304

Demenno/Kerdoon
2000 N. Alameda St.
Compton, Calif. 90222
213-537-7100

Ecoguard, Inc.
Promax Division
301 E. Main St.
P.O. Box 14047
Lexington, Ky. 40512
606-264-7389

Evergreen Oil
5000 Birch St.
Ste. 500
Newport Beach, Calif. 92660
714-757-7770

International Recovery Corp.
Miami Springs, Fla.
305-884-2001

Lyondell Petrochemical Co.
12000 Lawndale Ave.
P.O. Box 2451
Houston, Tex. 77252
713-652-7200

Mid-America Distillations
P.O. Box 2880
Hot Springs, Ark. 71914
501-767-7776

Mohawk Lubricants
130 Forester St.
N. Vancouver, B.C. V7112M9
604-929-1285

Motor Oils Refining Co.
7601 W. 47th St.
McCook, Ill. 60525
708-442-6000

Shannon Environmental Services
Toronto, Ont.
416-466-2133

FIGURE 11B.1 Used-oil rerefiners and marketers. *(Sources: Hegberg, 1991; Wolfe, 1992; Arner, 1992.)*

1991; MTAP, 1991). Using this method of oil recovery alone would certainly not be practical, especially in a service station setting, owing to the low recovery and time constraints.

Oil Recovery Through Mechanical Compaction. Compacting oil filters with a hydraulic press removes about 88 percent of the oil contained in a used filter (Konefes and Olson, 1991). The remaining 12 percent cannot be recovered because it is absorbed into the filter media or remains as a residue inside the filter. Compaction also reduces the volume of the filter by 73 percent. Figure 11B.2 lists North American manufacturers of oil press filters (Hegberg, 1991).

Recyclability of Used Oil, Filter Media, and Scrap Metal. Oil recovered from crushed filters is subject to the same regulations discussed earlier. The crushing could significantly add to the quantity of used oil collected. A maintenance shop that performs 50 oil changes daily could recover an additional 2 to 3 gal per day of used oil that would otherwise be disposed of with the filters.

The compacted filters from this study were processed through a scrap-metal shredder, which resulted in the separation of canister metal from the filter media. The recovered metal was essentially "oil-free" and acceptable to the existing scrap-metal smelting market.

Air Boy Sales & Mfg. Co. P.O. Box 2649 Santa Rosa, Calif. 95405 800-221-8333	Morris Enterprises 2393 Teller Rd. Newbury Park, Calif. 91320 800-833-3409
Danco Development Corp. 10832 Normandale Blvd. Bloomington, Minn. 55437 612-888-3255	Sun Fire Mfg. Corp. 126 Bonnie Crescent Elmira, Ont. N3B 3G2 519-669-1514
Graham Resources, Inc. 220 S. Edwards St. P.O. Box 15 Pierz, Minn. 56364 800-228-0901	United Marketing International P.O. Box 989 Everett, Wash. 98206 800-848-8228

FIGURE 11B.2   Used-oil filter press manufacturers.   *(Source: Hegberg, 1991.)*

Laboratory analysis of the filter media was inconclusive in determining if the media should be considered a hazardous waste. The U.S. EPA has stated that the TCLP test is not appropriate for oil- and solvent-based waste.

Regulatory Classification of Waste Oil Filters.   The U.S. EPA in 1990 issued a regulatory interpretation regarding the crushing of waste oil filters and subsequent reclamation of contents (U.S. EPA, 1990). Such an interpretation serves as a legal interpretation of U.S. EPA regulations. It basically indicates that if crushed or drained filters are recycled, it is not necessary to determine the hazardous waste status of used filters because of exemption due to recycling of the scrap metal. However, the filter must be drained to the point of having no free-flowing liquid, or crushed. The U.S. EPA recommends that the generator or recycling facility do both. The act of crushing filters is not regulated provided that the oil is collected for recycling. The interpretation makes no specific mention of the oil contained within a filter or the filter media. The standards mentioned earlier for used oil are assumed to apply to the oil contents.

Summary

- The greatest environmental threat from used oil comes from individuals who change the oil on their own vehicles. Many of these people use improper methods to dispose of their oil (e.g., dumping on land or water) that may or may not be legal.
- Used-oil generators should retain a sample of each load of oil taken by a waste hauler, to resolve any questions about possible contamination.
- Currently, the U.S. EPA does not regulate used oil as a hazardous waste. However, there are management standards for generators, transporters, processors, rerefiners, burners, and marketers.
- Many different oil collection policies have been tried, ranging from mandated recycling to public drop-off sites. Collection schemes with public- and private-sector cooperation have proved very effective.
- Oil collection procedures must be tailored to the community. Urban and rural collection systems will be quite different.
- Rerefined motor oil is subject to the same testing and performance standards as virgin oil.
- Mechanical compaction can recover significant amounts of oil from used oil filters.

Other Information Sources

American Petroleum Institute
1220 L Street, N.W.
Washington, D.C. 20005
(202) 682-8000

Association of Petroleum Re-refiners
P.O. Box 605, Ellicott Station
Buffalo, N.Y. 14205
(716) 855-2757, FAX 716-855-0339

Center for Earth Resources Management
5528 Hempstead Way
Springfield, Va. 22151
(703) 941-4452

Community Coalition for Oil Recycling
P.O. Box 141255
Dallas, Tex. 75214
(214) 821-3000

Hazardous Waste Treatment Council
1440 New York Ave., N.W.
Washington, D.C. 20005
(202) 783-0870

National Institute of Governmental Purchasing
115 Hillwood Avenue
Falls Church, Va. 22046
(703) 533-7715

National Oil Recyclers Association
277 Broadway Avenue
Cleveland, Ohio 44115
(216) 791-7316

National Recycling Coalition
1101 30th Street, N.W.
Washington, D.C. 20007
(202) 625-6406

Natural Resources Defense Council
40 West 20th St.
New York, N.Y. 10011
(212) 727-2700

Service Station Dealers Association of America, Inc.
499 S. Capitol St., S.W.
Suite 1130
Washington, D.C. 20003-4013
(202) 479-0196

Sierra Club
408 C Street, N.E.
Washington, D.C. 20002
(202) 547-1141

Society of Automotive Engineers
400 Commonwealth Drive
Warrendale, Pa. 15096
(412) 776-4841

U.S. EPA, Office of Solid Waste
401 M Street, S.W.
Washington, D.C. 20460

RCRA Hotline: 800-424-9346
Specific information on used-oil rule:
Ms. Rajani D. Joglekar [(202) 260-3516] or
Ms. Eydie Pines [(202) 260-3509]

Waste Oil Heating Manufacturers Assoc.
c/o Patton, Boggs, & Blow
2550 M Street, N.W.
Washington, D.C. 20037
(202) 457-6420

REFERENCES

Arner, R. (1991) "Curbside Programs: Success or Failure?" Presented at the 6th International Conference on Used Oil Recovery and Reuse, San Francisco, CA.

Arner, R. (1992) *Used Oil Recycling Markets and Best Management Practices in the United States,* Northern Virginia Planning District Commission, Annandale, VA.

Hegberg, B., W. Hallenbeck, and G. Brenniman (1991) *Used Oil Management in Illinois,* OTT-10, School of Public Health, Office of Technology Transfer, University of Illinois, Chicago, IL.

Franklin Associates, Ltd. (1985) *Background Report: [Michigan] Used Motor Oil Development Study,* Michigan Department of Natural Resources, Lansing MI.

Franklin Associates, Ltd. (1987) *Composition and Management of Used Oil Generated in the United States,* EPA/530-SW-013, Washington, DC.

Konefes, J., and J. Olson (1991) Motor Vehicle Oil Filter Recycling Demonstration Project, Iowa Waste Reduction Center, University of Northern Iowa, Cedar Falls, IA.

McHugh, R. (1991) "Incentives for Recycling: A World Review," Presented at the 6th International Conference on Used Oil Recovery and Reuse, San Francisco, CA.

MTAP (1991) *Management Options for Motor Vehicle Oil Filters* (draft), Minnesota Technical Assistance Program, Minneapolis, MN.

Nolan, J., C. Harris, and P. Cavanaugh (1990) *Used Oil: Disposal Options, Management Practices and Potential Liability,* 3d ed., Government Institutes, Inc., Rockville, MD.

NPRA (1989) *1989 Report on U.S. Lubricating Oil Sales,* National Petroleum Refiners Association, Washington, DC.

Peterson, G. (1991) Personal communication, Fuels and Lubricants Marketing Division, Growmark, Inc., Bloomington, IL.

Safety-Kleen (1993) *Safety-Kleen Re-Refined Base Oils/Products,* company publication, Chicago, IL.

Shull, H., M. Barnes, V. Leak, J. Powers, M. Rouse, and T. Van Hale (1987) *Feasibility Study on Long-Term Management Options for Used Oil in Minnesota,* Minnesota Waste Management Board, Crystal, MN.

Spendelow, P. (1989) "Analysis of Used Oil Collected through Curbside Recycling," Presented at the 5th Conference on Used Oil Recovery and Reuse, Baltimore, MD.

Temple, Barker & Sloane, Inc. (1989a) "Generation and Flow of Used Oil in the United States in 1988," Presented at the 5th Conference on Used Oil Recovery and Reuse, Baltimore, MD.

Temple, Barker & Sloane, Inc. (1989b) Memorandum to U.S. EPA, Flow Estimate Documentation, Washington, DC.

U.S. DOE (1981) Market Facts, *Analysis of Potential Used Oil Recovery from Individuals, Final Report,* DOE-AC19-79BC10053, U.S. Department of Energy, Washington, DC.

U.S. EPA (1990) Regulatory Determination on Used Oil Filters, memorandum, S. Lowrance, Dir. Office of Solid Waste, to R. Duprey, Dir. Hazardous Waste Management Div., Region VIII, October 30, 1990, Washington, DC.

U.S. EPA (1992) *Hazardous Waste Management System; Identification and Listing of Hazardous Waste; Recycled Used Oil Management Standards,* Federal Register, 57 FR 41566, September 10, 1992, Washington, DC (final rule, 1992).

Weinstein, K. (1974) *Waste Oil Recycling and Disposal,* EPA-670/2-74-052, Washington, DC.

Wolfin, J. (1991) "The Snohomish County Used Oil Collection and Recycling Project," Presented at the 6th International Conference on Used Oil Recovery and Reuse, San Francisco, CA.

CHAPTER 11

OTHER SPECIAL WASTES
Part 11C   Scrap Tires

John K. Bell

11C.1   BACKGROUND

Based on 2001 data, scrap tires represented nearly 5.7 million tons, or about 1.8 percent, of the total solid waste stream generated annually in the United States. In terms of quantity, this percentage translates to nearly 281 million waste tires (RMA, 2002a). These in turn are part of the estimated 1.4 billion scrap tires that are generated worldwide. Markets consumed approximately 218 million scrap tires, whole or shredded, from this annual waste stream. Fifty-three percent of these were used as tire-derived fuel (TDF), 19 percent as ground and stamped rubber products, 18 percent as civil engineering applications, 7 percent as exports, and 3 percent as miscellaneous exports (RMA, 2002a). The remainder of this waste stream, roughly 6 million tires, went to stockpiles, landfill disposal, single-material tire "monofills," or was disposed of illegally in some manner.

The term *scrap tire* generally refers to an inflatable rubber tubular covering encircling the wheel of a vehicle (automobile, truck, bus, or aircraft) that has been thrown away because it is no longer suitable for its original intended use due to wear, damage, or defect. This definition can be further refined to relate to specific tire processes or to be incorporated into various jurisdictional regulations. It is often used interchangeably with the term *waste tire,* though a waste tire can also refer to a tire that cannot be reused.

Scrap tires are composed of natural and manufactured synthetic rubbers, along with various additives, as shown in the following list:

Typical Materials of Which Tires Are Composed (RMA, 2002b)

Synthetic rubber

Natural rubber

Sulfur and sulfur compounds

Silica

Phenolic resin

Oil (aromatic, naphthenic, paraffinic, etc.)

Fabric (polyester, nylon, etc.)

Petroleum waxes

Pigments (zinc oxide, titanium dioxide, etc.)

Carbon black

Fatty acids

Inert materials

Steel wire

Additives are used to aid the vulcanization process, enhance tire life, act as coloring agents, and increase tensile strength and resistance to abrasion.

It is interesting to note that the typical composition of an average 100-lb scrap truck tire, by weight, is 27 percent natural rubber and 14 percent synthetic rubber, whereas the average 20-lb scrap passenger car tire is 14 percent natural rubber and 27 percent synthetic rubber. All other major components, by weight, are basically the same for both truck and passenger tires (RMA, 2002b). Table 11C.1 gives a typical passenger car tire's composition by percentage (CIWMB, 1996).

TABLE 11C.1   Typical Passenger Tire Composition

Material	Percentage
Styrene butadiene	46.78
Carbon black	45.49
Aromatic oil	1.74
Zinc oxide	1.40
Stearic acid	0.94
Antioxidant 6C	1.40
Wax	0.23
Sulfur	1.17
Accelerator CZ	0.75

Source: CIWMB (1996).

Besides the tires used by passenger vehicles and trucks, there is another type of tire that is manufactured for off-road use by farm, mining, and construction industries. The chemical composition of these tires varies by manufacturer and function.

Scrap tires from autos and trucks can come from many sources, including:

- Tire retailers
- Car dealers
- Auto equipment and auto parts stores
- Tire wholesalers
- Tire retread and repair shops
- Cab companies
- Rental car companies
- Fleet owners, including the government
- Auto salvage yards
- Scrap tire stockpile cleanups

Dealing with nearly 281 million new scrap tires annually, as well as existing stockpiles, presents a unique challenge for all the governmental and private entities involved. Scrap tire management efforts will continue to be influenced and driven by changes in the scrap tire markets and by changes in government regulations and financial incentives.

11C.2   SOURCE REDUCTION AND REUSE

Source reduction is a waste management technique that can help reduce the quantity of scrap tires generated over a given time period by extending the useful life of new and existing tires. Source reduction, in turn, reduces the demand for new tires.

A sure way to increase tire life is to get the maximum wear from tires that are in use. This can be accomplished through proper tire maintenance that includes proper wheel alignment, maintaining shock absorbers, and proper tire inflation. Lower highway speeds also help by reducing tire flex and temperature increase.

Tire longevity can also be enhanced through the use of preservatives. These help dissipate heat caused by road friction and maintain proper air pressure by preventing porosity and leaky valve stems.

Designing increased wearability and longevity into tires is another facet of source reduction as is decreasing miles driven.

Reusing existing, partially worn tires parallels source reduction by reducing the need for new tires by extending the life of existing ones. One example is selling used tires with legally remaining tread.

Another example is retreading, which involves removing the worn outside tread layer of a used tire and adding a new tread. Retreading extends the usable life of a tire and, in addition, saves more than 400 million gal of oil each year in North America (U.S. EPA, 1998) by using approximately 7 gal of oil per tire as compared with 22 gal to produce a new tire (U.S. EPA, 1999). Generally, retread tires can be used in the same manner as new tires and have safely been used on all types of vehicles.

11C.3 DISPOSAL OF WASTE TIRES

Tires present unique and challenging disposal problems because of their size, shape, and physical and chemical properties.

Landfilling of whole tires consumes a large volume of landfill space because the tires are relatively incompressible and 75 percent of the space a tire occupies is empty (Clark et al., 1993). Tires can also migrate, or "float," upward to the landfill surface where they can breach the landfill cover. As a further complication, tires can harbor vectors and are by design resistant to breakdown by mechanical or thermal means as well as by biological degradation.

Burying whole tires in municipal solid waste landfills avoids processing costs but does nothing to mitigate the disposal problems associated with whole tires. Therefore, whether by regulation or choice, the shredding or splitting of tires is becoming increasingly common as a part of the disposal process (Clark et al., 1993). Some other forms of tire reduction that have been considered use ultra-high-pressure water (Frenzel, 1993) or a cryogenic process using liquid nitrogen to produce crumb material (NASA, 1997).

Tire reduction can effectively eliminate the problems that are associated with whole tires. The main disadvantage of tire reduction is that it is an energy-intensive extra step that can add appreciably to disposal costs.

In a variation of landfilling, shredded tires can be buried in special single-waste landfills called *monofills*. Monofills allow easy recovery of tire shreds for potential use at a later date. Tire shreds in monofills that are located above the water table and have a low-permeability cover layer will have minimal contact with surface and groundwater. However, if substantial contact with water does occur, test results have shown that the concentrations of hazardous constituents detected in tire rubber samples did not exceed the concentration values necessary to be defined as hazardous waste (CIWMB, 1996). In addition to a potential for leaching, catastrophic internal heating has been identified in shredded tire monofills with a thickness of greater than 8 m (STMC, 1998).

Disposing of waste tires in stockpiles poses a number of environmental and public health and safety hazards. Rainwater can accumulate in tires within a stockpile, thus providing an ideal breeding environment for large populations of potential disease vectors (mosquitoes). In the southern United States, two exotic mosquito species predominate in tires: (1) *Aedes aegypti* and (2) *Aedes albopictus*), which are known to be the principle vectors of yellow fever and dengue. In temperate regions of North America, *Aedes triseriatus* and *Aedes atropalpus* have been known to be competent vectors of eastern equine encephalitis (EEE) and

LaCrosse encephalitis (LACV). The *Aedes triseriatus* also transmits dog heartworm. *Aedes albopictus,* the Asian tiger mosquito, was accidentally transported from Japan in the mid-1980s in shipments of scrap tires. It is also a competent vector for EEE and LACV and has been found in many states (RIDEM, 2000).

Another hazard from scrap tire stockpiles is their tendency to catch fire. Open-tire fires produce many unhealthful products of incomplete combustion that are released directly into the atmosphere. Tire fires are variable, and exact emissions and concentrations cannot be predicted because many factors come into play (e.g., amount of fuel, fire temperatures, meteorologic conditions, area topography). The airborne emissions of tire fires have been shown to be more toxic (i.e., mutagenic) than those of a controlled combustion source, regardless of fuel. Tire fire emissions include "criteria" pollutants (pollutants for which emissions standards have been set) such as particulates, carbon monoxide, sulfur oxides, oxides of nitrogen, and volatile organic compounds. They also include "noncriteria" hazardous air pollutants (HAPs), such as polynuclear aromatic hydrocarbons (PAHs), dioxins, furans, hydrogen chloride, benzene, and polychlorinated biphenyls (PCBs). Metals present can include arsenic, cadmium, nickel, zinc, mercury, chromium, and vanadium (U.S. EPA, 1997).

Tire fire emissions should not be inhaled or permitted to contact the skin. In the short term, health effects can range from mild irritation to acute exposure symptoms in firefighters and nearby residents. In the long term, high PAH exposure can relate to an association with increased mortality from lung cancer (CIWMB, 1997).

Tire fires also produce pyrolytic oil (a free-flowing, oily tar) that can contaminate soil, ground, and surface water. The pyrolytic oil consists of naphthalenes, anthracene, benzenes, thiazoles, amines, ethyl benzene, toluene, and other hydrocarbons. Metals such as cadmium, chromium, nickel, and zinc are also present (U.S. EPA, 1997).

The ash from tire fires also typically contains heavy metals, including zinc. This ash can contaminate soil and surface water.

Illegal dumping and stockpiling of scrap tires constitute one of the most serious nonresidential waste threats. Illegal sites pose the dual health threats from disease and fire that have already been discussed in detail. In particular, scrap tire fires are difficult to deal with. Once tire fires start, whether from lightning, grass fires, forest fires, arson, or unknown causes, they are extremely difficult to extinguish. Large piles are often left to burn themselves out and they can burn for days, releasing heavy black smoke that can be seen for miles and releasing pyrolytic oils to surface and groundwater while contaminating areas of the soil with ash. No federal laws or regulations specifically govern scrap tires, so each state must deal with illegal dumping on its own.

11C.4 ALTERNATIVES TO DISPOSAL

Tire-Derived Fuel

As a fuel, tires are equivalent to coal and, as such, are an excellent energy resource (CIWMB, 1992). (See Table 11C.2.) When scrap tires are shredded and processed into chips that can be used as fuel in a boiler or other combustion unit, they are generally referred to as *tire-derived fuel* (TDF). The types of facilities where tires are used as fuel include power plants, tire manufacturing plants, cement kilns, pulp and paper plants, and small-package steam generators (Clark et al., 1993). Coal cogeneration plants also use TDF as a supplement with the coal.

With the exception of cement kilns and some power plants that, by design, burn or have burned whole waste tires, most facilities that burn tires for fuel use TDF. When TDF is used, there are always cost trade-offs on the amount of radial steel and bead wire to leave in or take out. TDF that is not wire-free will have a decrease in fuel value of roughly 10 to 15 percent and an ash content typically of 14 to 18 percent, as compared with approximately 3 to 5 percent with wire removed (RMA, 2002c). Scrap tire fly ash contains almost 51.5 percent zinc and more than 32 percent carbon. Bottom ash, by comparison, contains nearly 96 percent iron

TABLE 11C.2 Comparison of fuel characteristics
of various materials

Fuel type	Energy content (Btu/lb)
Coal (anthracite)	12,000–14,000
Coal (bituminous)	11,000–13,000
Tires	12,000–16,000
Mixed MSW	3,500–5,500
No. 6 Fuel oil	18,000–18,500
Typical RDF	5,200–7,300
Newspaper	7,975

Source: CIWMB (1996).

(RMA, 2002b). Fly ash, if not marketable for its zinc content, can be combined with bottom ash and placed in environmentally safe landfills.

Facilities that burn TDF should have air pollution control equipment in place and are required to comply with the emission control requirements in effect in their respective areas, thereby providing for the protection of air quality. All facilities within the United States must ensure that federal New Source Performance Standards (NSPS) and National Emission Standards for Hazardous Air Pollutants (NESHAP) are met when applicable. The actual quantity of emissions produced by burning tires as a supplemental fuel, as well as the resulting relative emissions, compared with operating the facility without supplemental fuel, can only be determined by emissions testing.

The cement manufacturing process in particular, is able to use a variety of fuels. Some cement kilns supplement their standard fuel, like natural gas or coal, with TDF or whole tires. By using tires, cement kilns reduce their emissions of criteria air pollutants like oxides of nitrogen and sulfur, while the steel belts in the tires provide a source of iron for the cement manufacturing process (CIWMB, 1992). Thus, unlike other facilities that use tires for fuel, even the ash becomes part of the chemistry of the cement. However, tires cannot exceed 30 percent of the kiln fuel without adversely altering the chemistry of the cement's curing process (DiChristina, 1994).

Besides direct burning, waste tires can undergo *pyrolysis* (also known as *gasification*) and be broken down, in the absence of oxygen, into three recoverable fractions or products (CIWMB, 1996) (see Table 11C.3): (1) char, (2) oil, and (3) gas. Pyrolytic char is a fine particulate composed of carbon black, ash, clay fillers, sulfur, zinc oxide, calcium, and magnesium carbonates and silicates. Pyrolytic oils consist of heavy oils, light oils, benzene, and toluene. Pyrolytic gases are typically composed of paraffins and olefins with carbon numbers from 1 to 5 (Frenzel, 1993). Carbon black, a major raw material that is reclaimed, is of mixed grade but can be used for printing inks and pigments. The oil from the process can be used as low-grade fuel.

TABLE 11C.3 Products of Waste Tire Pyrolysis

Product	Composition	Properties
Gas	Hydrocarbon mixture, low sulfur content	Calorific value = 500–1,200 Btu/ft^3
Oil	Contains <1% sulfur	Calorific value = 18,000 Btu/lb^3
Char (solid)	Contains 2–3% sulfur and approximately 4–5% zinc	Calorific value = 12,000–14,000 Btu/lb^3

Source: CIWMB (1996).

Recycling Options

Based on 1998 data, approximately 27 percent of the annual scrap tire waste stream is used in alternate applications other than TDF and exports. The recycling applications for these scrap tires are governed by tire particle size (Frenzel, 1993). (See Table 11C.4.)

TABLE 11C.4   Alternative Applications for Waste Tires Based on Particle Size

Particle size	Applications
Whole tire	Artificial reefs and breakwaters Playground equipment Erosion control Highway crash barriers
Split or punched tire	Gaskets, seals, washers, shims, and insulators Floor mats, belts, and shoe soles Dock bumpers Muffler hangers
Shredded tire	Lightweight road construction material Playground gravel substitutes Alternative daily cover at landfills and leachate drainage material Sludge composting
Ground rubber	Rubber and plastic products (e.g., molded floor mats, mud guards, carpet padding, and plastic adhesives) Rubber railroad crossings Stadium playing surfaces and running tracks Friction brake material Injection-molded products and extruded goods Additives for asphalt pavements

Source: Frenzel (1993).

The civil engineering markets uses scrap tires and scrap tire–derived material in a wide range of structural and nonstructural applications. Baled tires have been used as retaining walls, berms, and fences. Shredded tires have been used for highway embankments and subgrade insulation, slope stabilization, levee slurry walls, landfill leachate collection systems, and for alternative cover at landfills.

The U.S. EPA estimates that from 1996 to 1998, scrap tire material in civil engineering applications, such as fill and drainage aggregate, increased 100 percent. Experts cite the development of American Society for Testing and Materials (ASTM) specifications as a key to expanding the civil engineering as well as other tire markets (U.S. EPA, 1999).

REFERENCES

CIWMB (1992) *Tires as a Fuel Supplement: Feasibility Study—Report to the Legislature,* California Integrated Waste Management Board, Sacramento, CA.

CIWMB (1996) *Effects of Waste Tires, Waste Tire Facilities, and Waste Tire Projects on the Environment,* publication no. 432-96-029, California Integrated Waste Management Board, Sacramento, CA.

CIWMB (1997) *Evaluation of Employee Health Risk from Open Tire Burning,* LEA advisory no. 46, California Integrated Waste Management Board, Sacramento, CA.

Clark, C., K. Meardon, and D. Russel (1993) Scrap Tire Technology and Markets, *Pollution Technology Review,* no. 211, U.S. Environmental Protection Agency, Durham, NC.

DiChristina, M. (1994) Mired in Tires, *Popular Science,* vol. 345, no. 4.

Frenzel, L. (1993) *Feasibility Study—Ultra-High-Pressure Water Jetting for the Processing of Waste Tires,* California Integrated Waste Management Board, Contract no. TR-92-0043-03, Sacramento, CA.

NASA (1997) Commercial Benefits—Spinoffs, *Tire Recycling,* informational material, National Aeronautics and Space Administration.

RIDEM (2000) *Mosquitoes, Disease, and Scrap Tires,* Office of Mosquito Abatement Coordination, Division of Agriculture, Rhode Island Department of Environmental Management.

RMA (2002a) *Scrap Tires—Facts and Figures—2001,* <www.rma.org/scraptires/facts_figures.html> Rubber Manufacturers Association, Washington, DC.

RMA (2002b) *Scrap Tires—Scrap Tire Characteristics,* n.d., <www.rma.org/scraptires/characteristics.html> Rubber Manufacturers Association, Washington, DC.

RMA (2002c) *Scrap Tires—Markets,* n.d., <www.rma.org/scraptires/markets.html> Rubber Manufacturers Association, Washington, DC.

STMC (1998) *Design Guidelines to Minimize Internal Heating of Tire Shred Fills,* Scrap Tire Management Council.

U.S. EPA (1997) *Air Emissions from Scrap Tire Combustion,* EPA-600R97115, U.S. Environmental Protection Agency, Washington, DC.

U.S. EPA (1998) *Retread Tires,* informational material, U.S. Environmental Protection Agency, Washington, DC.

U.S. EPA (1999) *Commodities: Rubber,* informational material, U.S. Environmental Protection Agency, Washington, DC.

CHAPTER 11

OTHER SPECIAL WASTES
Part 11D   Construction and Demolition Debris

George Tchobanoglous

Construction and demolition (C&D) debris results from the construction, renovation, and demolition of structures including buildings of all types (both residential and nonresidential), road repaving projects, bridge repair, and the cleanup associated with natural and human-made disasters. Components of C&D debris typically include concrete, asphalt, wood, metals, gypsum wallboard, and roofing. Typically, C&D debris comprises about 40 to 50 percent rubbish (concrete, asphalt, bricks, blocks, and dirt), 20 to 30 percent wood and related products (pallets, stumps, branches, forming and framing lumber, treated lumber, and shingles), and 20 to 30 percent miscellaneous wastes (painted or contaminated lumber, metals, tar-based products, plaster, glass, white goods, asbestos and other insulation materials, and plumbing, heating, and electrical parts). Land clearing debris, such as stumps, rocks, and dirt, is also included in some state definitions of C&D debris.

Over the past 10 years, significant strides have been made in the recycling of construction and demolition debris. It is anticipated that significantly greater amounts will be recycled in the future as a result of higher tipping fees, mandatory landfill diversion legislation, and the success of entrepreneurs in processing both source-separated and mixed wastes. Many landfills already use rubble for road building and daily cover, which may be considered as diversion by regulators. For those municipalities where C&D debris is presently combined with household wastes, recycling programs afford an excellent opportunity to meet diversion goals and extend landfill life. Because so much information is available on the Internet, the following is only meant to serve as an introduction to the management of C&D debris. Topics to be discussed include (1) sources, characteristics, and quantities of C&D debris; (2) regulations governing C&D materials and debris; (3) management opportunities for C&D debris; (4) specifications for recovered C&D debris; and (5) the management of debris from natural and humanmade disasters.

11D.1   SOURCES, CHARACTERISTICS, AND QUANTITIES OF C&D DEBRIS

The sources and characteristics and quantities of C&D debris are considered briefly in the following subsections. It is important to note that both the sources and the composition of C&D debris vary widely with the season of the year as well as with the strength of the economy.

Sources and Characteristics of C&D Wastes

Typical sources and the corresponding types of C&D debris are reported in Table 11D.1. As noted in Chap. 5, many nonhazardous wastes, such as municipal sludges, combustion ash, nonhazardous industrial process wastes, C&D debris, and automobile bodies, now often landfilled along with MSW, are not included in the definition of municipal solid waste. There is, however, some confusion concerning the classification of C&D debris from residential construction. Under normal circumstances, C&D debris is not included in the definition of municipal solid waste. The distribution of C&D debris from building and residential construction is given in Table 11D.2. The wide variation in the values reported in Table 11D.2 reflect the variability associated with different types of construction, construction practices, and geographic location (i.e., Midwest versus west or east coast).

TABLE 11D.1   Typical Sources and Characteristics of Construction and Demolition Debris

Source	Characteristics
Building construction (reusable materials)	Clean bricks, concrete blocks, concrete or stone facades, tiles, ceramics, roofing tiles, undamaged windows, roofing and metal/vinyl siding, wooden cabinets, counters, flooring, staircases/trim, plumbing/electrical fittings, carpeting, clean insulation, and wooden beams/facades.
Building construction and demolition (recyclable materials)	Broken bricks, concrete blocks, concrete or stone facades, ceramics, and roofing tiles, damaged or broken window glass, fixtures, wooden beams, trim, trees, metal siding, roofing material, and scrap aluminum door and window frames.
Building construction (reusable materials)	Mixed waste not suitable for separation, materials that cannot be reused or recycled, asphalt shingles, linoleum flooring, hazardous wastes including asbestos. Wood wastes consist of framing and form lumber, treated wood, plywood and particleboard, and wood contaminated by paint, asbestos, or insulation.
Demolition of physical facilities including concrete structures	Concrete (without metal reinforcing), concrete (with metal reinforcing), fill material (earth, gravel, sand), ferrous metals (beams, wall studs, piping) brick, stone, wood products, electrical and plumbing fixtures, electrical wiring and mixed rubble, and miscellaneous wastes.
Excavation/leveling	Earth, earth-contaminated wood, sand, stones, and mixed materials found during excavation.
Heavy construction	Mixed waste including wood products, roofing materials, wallboard, insulation materials, ferrous and nonferrous metals (wall studs, piping, wiring, ductwork), and carpeting.
Humanmade disasters (acts of sabotage or terrorism)	Mixed waste not suitable for separation, materials that cannot be reused or recycled, asphalt shingles, linoleum flooring, hazardous wastes including asbestos. Concrete (with and without metal reinforcing), fill material (earth, gravel, sand), miscellaneous wastes, plus materials from the demolition of buildings as previously discussed.
Natural disasters (hurricanes, tornadoes, earthquakes)	Mixed waste not suitable for separation, materials that cannot be reused or recycled, trees, asphalt shingles, linoleum flooring, and hazardous wastes including asbestos. Concrete (without metal reinforcing), concrete (with metal reinforcing), fill material (earth, gravel, sand), miscellaneous, plus materials from the demolition of buildings as previously discussed.
Road construction	Asphalt, concrete (without metal reinforcing), concrete (with and without metal reinforcing), fill material (earth, gravel, sand), and miscellaneous (separated metal reinforcing, metal signs, signposts, guard rails, culverts).
Site clearing	Timber, underbrush, earth, concrete, steel, rubble, and other waste materials (paper, plastic, brick, organics).

Source: Adapted in part from SWANA (1993), U.S. EPA (1996), and Franklin Associates (1998).

TABLE 11D.2   Distribution of Construction Debris from Building and Residential Construction

Type	Percent of total	
	Range*	Typical
Lumber (untreated and treated)	20–35	28
Roofing materials (e.g., asphalt shingles)	15–30	20
Drywall	12–20	14
Rubble (mixed waste including concrete block, slump stone, and rocks)	10–25	18
Metals (ferrous and nonferrous)	4–10	8
Mixed paper/cardboard, plastic	2–12	5
Soil	2–10	5
Green wastes (i.e., trees, brush, grass trimmings)	1–5	2

* Range reflects the variability in the construction activity during different seasons and as the result of regional differences and the strength of the economy.
Source: Adapted in part from U.S. EPA (1996) and U.S. EPA (1998).

Quantities of Construction and Demolition Wastes

In 1996, it was estimated that 136 million tons of building-related C&D debris was generated in the United States, or about 2.8 lb/capita · d (U.S. EPA, 1996). Of the C&D debris generated, 43 percent is from residential sources and 57 percent is from nonresidential sources (see Table 11D.2). Further, building demolition accounted for 48 percent of the total, 44 percent is from renovations, and 9 percent is from new construction. The distribution of C&D debris from building and residential construction and demolition is given in Table 11D.3.

TABLE 11D.3   Distribution of Construction and Demolition Debris from Buildings

Type*	Percent of total	
	Range†	Typical*
Nonresidential demolition	30–40	33
Residential demolition	20–30	23
Nonresidential renovation	15–25	21
Residential demolition	10–20	15
Residential new construction	3–10	5
Nonresidential new construction	2–10	3

* Adapted in part from U.S. EPA (1996) and U.S. EPA (1998).
† Range reflects the variability in the construction activity during different seasons and under different economic conditions.

Unfortunately, similar detailed estimates are not available for debris from the construction of nonresidential and commercial facilities and the demolition of physical facilities including concrete structures, steel bridges, road beds, and site and land clearing. Based on a review of multiple sources, it is estimated that the nonbuilding C&D debris is on the order of 2.0 lb/capita · d. Using a population of 280 million, the corresponding quantity of non-building-related debris in the United States is about 100 million tons per year.

11D.2 REGULATIONS GOVERNING C&D MATERIALS AND DEBRIS

At the present time (2002), C&D debris is not classified as a Resource Conservation and Recovery Act (RCRA) hazardous waste or as a RCRA municipal solid waste. As a result, C&D landfills are not subject to federal design and operational criteria. If, however, C&D debris is disposed of in municipal solid waste landfills (MSWLFs) or landfills that accept conditionally exempt small quantity generator (CESQG) wastes, the landfills must meet all of the federal regulations set forth in RCRA, Subtitle D (Part 258 for MSWLFs and Part 257, Subpart B for CESQG) (U.S. EPA Region 9, 2002).

State programs for C&D debris landfills vary widely. In 11 states C&D landfills must meet MSWLF regulations, 24 states have adopted separate regulations for C&D landfills, 8 states have adopted separate regulations for onsite and off-site C&D landfills, and in 7 states onsite C&D landfills are exempt from regulation (Franklin Associates, 1998).

11D.3 MANAGEMENT OF C&D DEBRIS

There are four options for the management of C&D debris: (1) source reduction, (2) reuse, (3) recycling, and (4) landfilling. In recent times, the primary focus of C&D debris management programs throughout the country is on the first three: source reduction, reuse, and recycling. It is estimated that there are more than 3500 facilities for the recovery and recycling of C&D debris currently (2002) operating in the United States (Brichner, 1997).

Source Reduction

Source reduction involves reducing the amount of material used through more careful estimating to eliminate waste. Increasing costs for the disposal of C&D debris by landfilling, as a result of new regulations, continues to provide a stimulus for reducing the quantity of waste.

Reuse

In any construction or demolition project, a wide variety of reusable and unused items will be found, including lumber of different sizes (typically, two-by-fours), plywood, asphalt shingles, insulation, paint, heating ducts, and piping. In addition, other wastes such as broken concrete block and bricks can be used in a number of applications (e.g., as fill material). Short sections of drywall can be saved for other uses or where a small section is required. It is not the purpose here to list the many ways in which materials can be reused, but rather to note that a paradigm shift is needed to bring about the reuse of materials. Fortunately, almost every state has developed information on reuse opportunities. Further, most states have set up exchange programs for reusable and recyclable materials.

Recycling

Reuse and recycling opportunities for C&D debris depend on the markets for the individual materials comprising the wastes and the ability to process the commingled waste or separate the individual materials. The principal materials that are now recovered from C&D debris for recycling include concrete (see Fig. 11D.1), wood, asphalt shingles, drywall, metals, and soil. In 1996, it was estimated that from 20 to 30 percent of the C&D debris was recycled (U.S. EPA, 1996).

FIGURE 11D.1   Concrete demolition debris brought to a recycling center to be processed for reuse/recycling.

Typical processing methods and reuse/recycling opportunities are summarized in Table 11D.4. The processing methods used for C&D debris are relatively simple, involving, for example,: (1) manual separation for a variety of items including wood (see Fig. 11D.2), concrete blocks, brick, metals, etc.; (2) crushing, grinding, pulverizing, and screening for concrete; (3) grinding and/or pulverizing for asphalt; (4) grinding and/or pulverizing and screening for wood (see Fig. 11D.3); (5) magnetic separation; and (6) multistage screening for soil (see Fig. 11D.4). The equipment used for processing C&D debris is discussed in greater detail in Chap. 8.

Landfilling

Much of the C&D debris generated in the United States now ends up in separate C&D landfills. At the present time (2002), it is estimated that 35 to 45 percent of the building-related C&D debris is disposed of in C&D landfills. An additional 20 to 40 percent of the building-related C&D debris is disposed of in MSWLFs (IFC Inc., 1995; Franklin Associates, 1998). The exact disposition of the non-building-related C&D debris is not well known at the present time. It is, however, estimated that about half of the total quantity is placed in on- and off-site C&D landfills and/or in MSWLFs. Because C&D debris is not classified as an RCRA waste, most C&D landfills are not required to provide the same level of protection as MSWLFs. Because of fewer constraints, tipping fees are usually lower for C&D landfills.

Although, for the most part, C&D debris is inert, some problems have developed when C&D debris is mixed with residential wastes. For example, drywall landfilled with household wastes is not entirely inert or benign when moistened, and anaerobic decomposition can result in the production of hydrogen sulfide gas. For this reason, many cities require that old drywall be bagged or boxed separately for disposal. It should be noted that the organic matter in C&D debris is inert because of the relatively low moisture content (e.g., the moisture content of dry wood is about 5 to 6 percent). However, when the moisture content increases to about 14 to 16

TABLE 11D.4   Typical Processing and Recycling Opportunities for C&D Debris

Material	Processing	Reuse/recycle opportunity
Wood	Manually sorted, shredded in a tub or other commercial wood grinder, and passed through a classifier or trommel, where the oversize pieces are separated. Ferrous metals are removed magnetically, and the fines (undersized materials, which are often sold for mulch or soil amendments) are separated by screening.	Shredded and/or chipped wood can be used for landscape mulch, for animal bedding, in compost, as boiler fuel source, in engineered building products, or as intermediate cover material in landfills.
Concrete	Concrete chunks are crushed and ferrous materials such as bolts or reinforcing bar are removed; the resultant aggregate is screened to sizes suitable for various applications.	Crushed and graded concrete can be used as aggregate for new construction applications. Aggregate must meet standard specifications, such as those of the ASTM. Crushed concrete can also be used as road subbase material. Larger material can be used for riprap on roads/lagoons.
Asphalt shingles	Shingles are shredded and reduced in size with a hammermill, and nails and other ferrous materials are removed magnetically. The final material is screened.	Crushed asphalt (1) can be added to the production of new roofing materials, (2) can be used as hot and cold mix asphalt for paving, and (3) can be used in pothole repair.
Drywall (gypsum)	Gypsum interior is pulverized; paper backing is repulped.	Pulverized gypsum (1) can be mixed with virgin gypsum and remanufactured and (2) can be used as an animal bedding, as a soil amendment (as substitute for lime for lawns), and as a cat litter. Repulped paper can be used to make new wallboard paper backing.
Metals	Separated by category (i.e., ferrous, aluminum, copper, and brass).	Recycled by metals processors.
Soil	Screened to remove wood, metal, rocks, and other miscellaneous materials.	Screened soil can be used for landscaping, agricultural and residential fill, and landfill cover.

Source: Adapted in part from SWANA (1993), Tchobanoglous et al. (1993), and U.S. EPA (1998).

percent, biological activity will also begin. In time, if moist, most of the organic matter in C&D debris can be degraded biologically. In some cases, the biological degradation of the organic matter in C&D debris has resulted in the formation of leachate that may contain constituents that can contaminate local groundwater. Because of the production of leachate, a number of states now require C&D landfills to meet MSWLF regulations.

11D.4   SPECIFICATIONS FOR RECOVERED C&D DEBRIS

There are no industrywide specifications for C&D debris. Specifications are negotiated individually with buyers of the separated materials. Wood and asphalt shingles will be used as examples. The specifications for recovered wood vary, depending on the markets available to wood

FIGURE 11D.2 Typical wood waste separated for processing. The wood waste in the foreground is from a cabinet shop and the larger pile is from a construction site. The tub grinder used to process the wood waste is shown in the background. The processed waste is sold as a boiler biomass fuel.

FIGURE 11D.3 Tub grinder used to process wood, trees, brush, and other green waste. Processed wood is sold as a boiler biomass fuel, and green waste is typically used for the production of compost.

FIGURE 11D.4   Two-stage vibrating screen for the first-stage processing of excavated soil containing broken concrete, bricks, and tree roots. The screened soil from the first-stage screening is conveyed to a second, finer, set of screens for further processing. The final screen product is stockpiled in the background for reuse.

processors. Processors accept a variety of wood wastes, depending on the supply available and the end market. Plants producing boiler fuel prefer clean construction and demolition waste, pallets and containers, and clean brush and tree trimmings; some will accept small, clean stumps. Processors do not want pressure-treated wood, telephone poles or railroad ties (which are treated with tar or creosote), plywood, leaves, grass clippings, large tree trunks, or dirty stumps, because these materials affect boiler performance and may cause air pollution violations.

Asphalt shingles contain up to 30 percent asphalt, and several asphalt pavement manufacturers use shredded postconsumer shingles as a portion of their mix for both road base and paving. The limiting factor for the recycling of asphalt shingles is meeting the specifications for paving and roofing materials. In the future, as more C&D debris is recovered for recycling, materials specifications will become ever more important.

11D.5   MANAGEMENT OF DEBRIS FROM NATURAL AND HUMANMADE DISASTERS

The management of debris from both natural and humanmade disasters is a serious issue that must be considered in developing solid waste management plans for C&D debris. The waste from natural and humanmade disasters is typically classified as C&D debris.

Natural Disasters

Natural disasters (floods, windstorms, hurricanes, tornadoes) can cause large quantities of solid waste (debris), especially when the disaster occurs in an urban environment. Collection and storage of the wastes can greatly exceed the number and capacity of collection trucks used in normal operations. One example is the debris left by Hurricane Andrew when it passed through Miami/Dade County. About 40 million tons of construction- and demolition-type debris had to be collected, stored, and processed. Most debris was collected using a zone designation for the affected areas (multiple contractors were awarded management contracts in each zone) so that the streets could be cleaned first to allow contractors to move materials to processing sites. After debris removal from streets, the second stage of collection was the loading and movement of materials from the demolition of damaged homes and businesses.

The processing sites were large enough to serve as the staging area for dispatching collection trucks to debris collection routes. The initial collection of debris from streets was done on an emergency basis, but the second-stage collection of demolition wastes had to be coordinated with the timing of insurance company investigations and certification of structures for type of damage. All collection routing and contracts for collection, processing, and disposal of hurricane debris were managed through an emergency response center made up of local, state, and federal officials.

The landfills in the Miami area did not have the capacity to hold 40 million tons of wastes, so the organic debris was burned at the processing sites. Large burn pits were dug at the processing sites, and air supply units were installed at each pit to add combustion air above the surface of the burning organics. Engine-driven compressors provided air, and the organic fuel was added to the pit and mixed in the pit by excavators with a bucket on an extended articulated arm.

Humanmade Disasters

The tragic events of September 11, 2001, resulting in the destruction of the twin towers of the World Trade Center in New York City, created more than 1.6 million tons of debris (see Fig. 11D.5). Because of the need to sift through all of the debris for personal belongings, the debris

FIGURE 11D.5   Cleanup activities at the World Trade Center disaster area in New York (Photo by James Tourtellotte).

from the twin towers was hauled to the Fresh Kills landfill, which was closed officially on July 4, 2001. At the landfill site, located on Staten Island, the debris is unloaded, and the painstaking job of sifting through all of the debris is carried out. Debris that has been sorted is then processed for recycling and/or disposal. Concrete is taken to the concrete crushing plant located on the northwest edge of the Fresh Kills site. The concrete crushing plant is part of the construction and demolition recycling facility, which serves the tri-county state area. Metal is processed and hauled to metal recycling facilities. The remaining debris is landfilled.

REFERENCES

Brickner, R. H. (1997) "Overview of C&D Debris Recycling Plants," *C&D Debris Recycling,* January/February.

Franklin Associates (1998) *Characterization of Building-Related Construction and Demolition Debris in the United States,* EPA/530-R-98-010, U.S. Environmental Protection Agency, Washington, DC.

IFC Inc. (1995) *Construction and Demolition Waste Landfills,* May 18, 1995 Draft Report, Prepared for the U.S. Environmental Protection Agency, Washington, DC.

SWANA (1993) *Construction Waste and Demolition Debris Recycling—A Primer,* Publication GR-REC 300, Solid Waste Association of North America, Silver Springs, MD.

Tchobanoglous, G., H. Theisen, and S. A. Vigil (1993) *Integrated Solid Waste Management: Engineering Principles and Management Issues,* McGraw-Hill, New York.

U.S. EPA (1996) *Characterization of Municipal Solid Waste in the United States,* PB/96-152 160, U.S. Environmental Protection Agency, Washington, DC.

U.S. EPA (1998) *Characterization of Construction and Demolition Debris in the United States,* EPA/530-R-96-010, U.S. Environmental Protection Agency, Washington, DC.

U.S. EPA Region 9 (2002) "Construction and Demolition (C&D) Debris," U.S. Environmental Protection Agency Region 9, Web Site www.epa.gov/region09/waste/solid/debris.htm#1.

CHAPTER 11

OTHER SPECIAL WASTES
Part 11E   Computer and Other Electronic Solid Waste

Gary R. Brenniman
William H. Hallenbeck

11E.1   INTRODUCTION

Most consumers are unaware of the toxic materials in the products on which they rely for word processing, data management, and access to the Internet, as well as for electronic games. In general, computer equipment is assembled from more than 1000 materials (e.g., chlorinated and brominated substances, metals, biologically active materials, acids, plastics and plastic additives). The health impacts of the materials in the products often are not known. The production of semiconductors, printed circuit boards, disk drives, and monitors uses particularly hazardous chemicals. Cancer clusters have been reported in workers involved in chip manufacturing. In addition, computer recyclers may have high levels of dangerous chemicals in their blood.

Computers have a short life span because hardware and software companies constantly generate new programs that fuel the demand for more speed and memory. Today, it is frequently cheaper and more convenient to buy a new computer to accommodate the newer generations of technology than it is to upgrade the old.

The following statistics indicate why the United States has a big computer solid waste problem:

- Computer solid waste is growing at an escalating rate, and consumers do not know what to do with it. It has been estimated that over three-quarters of all computers ever bought in the United States are currently stored in attics, basements, and closets (MCTC, 1996). If everyone disposed of these, the United States would face a huge waste problem all at once.

- A recent U.S. study found that over 315 million computers will become obsolete by the year 2004. Reliable numbers were not available for the number of computers manufactured between 1980 and 1992 (National Safety Council, 1999). Many of these obsolete computers will end up in landfills, incinerators, or hazardous waste exports.

- Americans buy more computers than people in any other nation. Currently, more than 50 percent of U.S. households own a computer (National Safety Council, 1999).

- The lifespan of computers is decreasing. In 1997 the average life span of a computer tower was between 4 and 6 years and computer monitors between 6 and 7 years. This will soon fall to 2 years before 2005 (U.S. EPA, 1998).

- By the year 2005, one computer will become obsolete for every new one put on the market (National Safety Council, 1999).

- By the end of 1999, 24 million computers in the United States became obsolete. Only about 14 percent, or 3.4 million, of these will be recycled or donated. The rest will be dumped, incinerated, shipped as waste exports, or put into temporary storage in the home (National Safety Council, 1999).

- In 1998 only 6 percent of older computers were recycled.

The European Union is developing a policy that will require producers to take back their old products. This legislation, which includes take-back requirements and toxic material phaseouts, also encourages cleaner product design and less waste generation. To date no such initiative has occurred in North America. In fact, the U.S. trade representative, at the request of the American electronics trade associations, is currently lobbying against this European Union initiative.

11E.2 HAZARDOUS COMPONENTS IN COMPUTERS AND ELECTRONIC WASTE

In Table 11E.1, a breakdown of materials in a 60-lb desktop personal computer is presented. Most of these materials are hazardous and the recycling efficiencies are poor. The following presents a brief summary of risks to some humans due to the toxics found in computers.

Lead

Lead can damage the central and peripheral nervous systems, blood system, and kidneys in humans (SVTC, 2000). Effects on the endocrine system have also been observed, and its serious negative effects on children's brain development has been well documented. Lead accumulates in the environment and has toxic effects on plants, animals, and microorganisms. Consumer electronics constitute 40 percent of lead found in landfills. Between 1997 and 2004, over 315 million computers will become obsolete in the United States. This adds up to about 1.2 billion lb of lead. The main concern regarding the presence of lead in landfills is the potential for the lead to leach and contaminate drinking water supplies.

Cadmium

Cadmium compounds are classified as toxic with a possible risk of irreversible effects on human health. It affects the kidneys and is adsorbed through respiration and ingestion. Due to its long half-life (30 years) in the body, cadmium can easily be accumulated in amounts that cause symptoms of poisoning. It also can accumulate in the environment.

Between 1997 and 2004, more than 315 million computers will become obsolete. This represents almost 2 million lb of cadmium that could end up in landfills.

Mercury

When inorganic mercury enters water, it is transformed to methylated mercury in the bottom sediments (SVTC, 2000). Methylated mercury easily accumulates in living organisms and concentrates through the food chain, particularly via fish. Methylated mercury causes damage to the brain.

It is estimated that 22 percent of the annual world consumption of mercury is used in electrical and electronic equipment. It is basically used in thermostats, sensors, relays and switches (e.g., on printed circuit boards and in measuring equipment), and discharge lamps. Furthermore, it is used in medical equipment, data transmission, telecommunications, and mobile phones. Mercury is also used in batteries, switches and housing, and printed wiring boards. Although the amount is small for any single component, 315 million obsolete computers by the year 2004 represent more than 400,000 lb of mercury that could end up in landfills.

TABLE 11E.1 Materials Used in a Desktop Computer and
the Efficiency of Current Recycling*

Name	Content (% of total weight)	Weight of computer material (lbs)	Percent recycling efficiency (current recyclability)
Plastics[†]	22.9907	13.8	20
Lead	6.2988	3.8	5
Aluminum	14.1723	8.5	80
Germanium	0.0016	<0.1	0
Gallium	0.0013	<0.1	0
Iron	20.4712	12.3	80
Tin	1.0078	0.6	70
Copper	6.9287	4.2	90
Barium	0.0315	<0.1	0
Nickel	0.8503	0.51	80
Zinc	2.2046	1.32	60
Tantalum	0.0157	<0.1	0
Indium	0.0016	<0.1	60
Vanadium	0.0002	<0.1	0
Terbium	0	0	0
Beryllium	0.0157	<0.1	0
Gold	0.0016	<0.1	99
Europium	0.0002	<0.1	0
Titanium	0.0157	<0.1	0
Ruthenium	0.0016	<0.1	80
Cobalt	0.0157	<0.1	85
Palladium	0.0003	<0.1	95
Manganese	0.0315	<0.1	0
Silver	0.0189	<0.1	98
Antimony	0.0094	<0.1	0
Bismuth	0.0063	<0.1	0
Chromium	0.0063	<0.1	0
Cadmium	0.0094	<0.1	0
Selenium	0.0016	0.00096	70
Niobium	0.0002	<0.1	0
Yttrium	0.0002	<0.1	0
Mercury	0.0022	<0.1	0
Arsenic	0.0013	<0.1	0
Silica	24.8803	15	0

* Composition of a typical desk personal computer weighing about 60 lbs.
[†] Plastics contain polybrominated flame-retardants, and hundreds of additives and stabilizers *not* listed separately.
   Source: MCTC (1996).

Hexavalent Chromium (Chromium VI)

Chromium VI can easily pass through membranes of cells and is easily absorbed, producing various toxic effects within cells (SVTC, 2000). It causes strong allergic reactions even in small concentrations (e.g., asthmatic bronchitis). Chromium VI may also cause DNA damage. In addition, hexavalent chromium compounds are toxic for the environment. It is well documented that contaminated wastes can leach from landfills. Incineration results in the generation of fly ash from which chromium is leachable. Wastes containing chromium should not be incinerated. The 315 million computers destined to become obsolete between 1997 and 2004 will contain about 1.2 million lb of hexavalent chromium.

Plastics

Based on the estimate that more than 315 million computers will become obsolete between 1997 and 2004 and that plastics make up 13.8 lb per computer on average, there will be more than 4 billion lb of plastic present in this computer waste (SVTC, 2000). An analysis commissioned by the Microelectronics and Computer Technology Corporation (MCC) estimated that total electronics plastic scrap amounted to more than 1 billion lb per year (500,000 tons per year). This same study estimated that the largest volume of plastics used in electronics manufacturing (at 26 percent) was polyvinyl chloride (PVC), which creates more environmental and health hazards than most other types of plastic. Although many computer companies have recently reduced or phased out the use of PVC, there is still a huge volume of PVC contained in the computer scrap that continues to grow at a rate of up to 250 million lb per year.

The use of PVC in computers has been used mainly in cabling and computer housing, although most computer moldings are now being made of ABS plastic. PVC cabling is used for its fire-retardant properties, but there are concerns that if it starts to burn, fumes from PVC cabling can be a major contributor to fatalities, and hence there are pressures to switch to alternatives for safety reasons. Such alternatives are low-density polyethylene and thermoplastic olefins.

Polyvinyl chloride is a difficult plastic to recycle, and it contaminates other plastics in the recycling process. Of more importance, however, the production and burning of PVC products generate dioxins and furans. This plastic, commonly used in packaging and household products, is a major cause of dioxin formation in open burning and garbage incinerators. Hospitals are now beginning to phase out the use of PVC products, such as disposable gloves and IV bags, because of the dangers of incinerating these products.

Many local authorities in Europe have PVC-free policies for municipal buildings, pipes, wallpaper, flooring, windows, and packaging. Recent concerns about the use of softeners in PVC plastic toys leaching out into children's mouths have led to further restrictions on PVC.

Brominated Flame-Retardants

Brominated flame-retardants [polybrominated biphenyls (PBBs) and polybrominated diphenylethers (PBDEs)] are used in electronic products to reduce flammability (SVTC, 2000). In computers, they are used mainly in four applications: (1) in printed circuit boards, (2) in components such as connectors, (3) in plastic covers, and (4) in cables. They are also used in plastic covers of TV sets and in domestic kitchen appliances.

Polybrominated diphenylethers may act as endocrine disrupters. The levels of PBDEs in human breast milk are doubling every five years. Other studies have shown PBDEs, like many halogenated organics, reduce levels of the hormone thyroxin in exposed animals and have been shown to cross the blood-brain barrier in the developing fetus. Thyroxin is an essential hormone needed to regulate the normal development of all animal species, including humans.

Researchers in the United States found that exposure to polybrominated biphenyls may cause an increased risk of cancer of the digestive and lymph systems. The study looked at cancer incidence in individuals who were exposed to PBBs after a 1973 food incident in Michigan. About 1 ton of PBB fire-retardant was added to cattle feed in error and contamination spread through animals to humans. Some 9 million people were affected. A study published in 1998 found that the group with the highest exposure was 23 times more likely to develop digestive cancers, including stomach, pancreas, and liver cancers. Preliminary results also found a 49-fold increase in lymph cancers.

The presence of PBBs in Arctic seal samples indicates a wide geographical distribution. The principal known routes of PBBs from point sources into the aquatic environment are PBB plant areas and waste dumps. Polybrominated biphenyls are almost insoluble in water and are primarily found in sediments of polluted lakes and rivers. Polybrominated biphenyls have been found to be 200 times more soluble in landfill leachate than in distilled water, which may result in a wider distribution in the environment. Once they have been released into the environment, they can reach the food chain, where they are concentrated. Polybrominated biphenyls have been detected in fish from several regions. Ingestion of fish is a source of PBB transfer to mammals and birds. Neither uptake nor degradation of PBBs by plants has been recorded. In contrast, PBBs are easily absorbed by animals.

In May 1998, Sweden's National Chemicals Inspectorate called for a ban on PBBs and PBDEs while urging the government to work for a European-wide ban and for controls on the international trading of these chemicals.

As a consequence, PBBs should no longer be used commercially. Between 1997 and 2004, over 315 million computers will become obsolete and contain over 350 million lb of PBBs and PBDEs in monitors. This is an underestimate because it does not take into account the amount present in the computer tower or printed wiring boards.

11E.3   DISPOSING OF COMPUTERS IS HAZARDOUS

In addition to the recent evidence of worker exposure to flame-retardants, the environmental risks posed by landfilling and burning are also significant. In particular, when computer waste is incinerated or landfilled, it poses contamination problems due to air emissions and leachate to water sources.

The Hazards of Incinerating Computer Waste

The stream of waste from electronic and electrical equipment contributes significantly to the heavy metals and halogenated substances contained in the municipal waste stream (SVTC, 2000). Because of the variety of different substances found together in electroscrap, incineration is particularly dangerous. For instance, copper is a catalyst for dioxin formation when flame-retardants are incinerated. This is of particular concern as the incineration of brominated flame-retardants at a low temperature (between 600 and 800°C) may lead to the generation of extremely toxic polybrominated dibenzodioxins (PBDDs) and polybrominated dibenzofurans (PBDFs).

The burning of electronic waste results in high concentrations of metals in slag, fly ash, flue gas, and the filter cake. Municipal incineration is the largest point source of dioxins in the U.S. and Canadian environments and among the largest point sources of metal contamination of the atmosphere.

Electronic waste is burned in cement kilns as an alternative fuel. Also, smelting presents the same dangers as incineration. Indeed, there is growing concern that the Noranda Smelter in Quebec, where much of the North American electroscrap is sent, is producing dioxins due to the presence of PVC plastics in the scrap.

The Hazards of Landfilling Computer Waste

It has become common knowledge that all landfills leak (SVTC, 2000). Even the best state-of-the-art landfills are not completely leakproof, and a certain amount of chemical and metal leaching will occur. The situation is far worse for older or less stringently regulated landfills.

Mercury will leach when certain electronic devices, such as circuit breakers, are destroyed. The same is true for PCBs from condensors. When brominated flame-retardant plastics are landfilled, PBDEs may leach into the soil and groundwater. It has been found that significant amounts of lead ions are dissolved from broken lead containing glass (e.g., when the cone glaze of cathode-ray tubes mixes with acid waters in landfills). In addition, the vaporization of metallic mercury and dimethyl mercury is also of concern in the operation of a landfill.

The Hazards of Recycling Computer Waste

Recycling of hazardous products has little environmental benefit because it simply moves the hazards into secondary products that eventually will have to be disposed of via landfills or incineration. The list of toxic components in computers includes the following:

- Computer circuit boards containing heavy metals (e.g., lead and cadmium)
- Computer batteries containing cadmium
- Cathode ray tubes containing lead oxide and barium
- Brominated flame-retardants used on printed circuit boards, cables, and plastic casing
- Polyvinyl chloride– (PVC-) coated copper cables and plastic computer casings that release dioxins and furans when burned
- Mercury switches
- Mercury in flat screens
- Polychlorinated biphenyls (PCBs) present in older capacitors and transformers

The presence of polybrominated flame-retardants in plastic makes recycling dangerous and difficult. It has been shown that PBDEs form the toxic PBDFs and PBDDs during the extruding process, which is part of the plastic recycling process. As a consequence, the German chemical industry stopped the production of these chemicals in 1986. In addition, high concentrations of PBDEs have been found in the blood of workers in recycling plants. A recent Swedish study found that when computers, fax machines, or other electronic equipment are recycled, dust containing toxic flame-retardants is spread in the air. Workers at dismantling facilities had 70 times the level of one form of flame-retardant than was found in hospital workers. Because of their presence in the air, clerks working full-time at computer screens also had levels of flame-retardants in their blood that were slightly higher than for hospital workers. Humans may directly absorb PBDEs when they are emitted from electronic circuit boards and plastic computer and TV cabinets.

Due to the halogenated substances found in plastics, both dioxins and furans may be generated as a consequence of recycling the metal content of electronic waste. Because of the risk of generating dioxins and furans, recyclers usually do not recycle flame-retardant plastics. However, due to the lack of proper identification of plastics containing flame-retardants, most recyclers do not process any plastic from this electronic waste.

It is difficult to find data on the amount of computer scrap leaving the United States for countries such as Taiwan and China. This is because of past bad publicity and the fact that generators will sell scrap to recyclers and not bother to find out the final destination and fate of their end-of-life product. The export of scrap is profitable because the labor costs are cheap and regulations are lax compared with U.S. law. A pilot program that collected electronic scrap in San Jose, California, estimated that it was 10 times cheaper to ship cathode ray tube (CRT) monitors to China than it was to recycle them in the United States. The overwhelming majority of the world's hazardous waste is generated by industrialized market economies. Exporting this waste to less developed countries has been one way in which the industrialized world has avoided having to deal with the problem of expensive disposal and close public scrutiny at home. In 1989, the world community established the Basel Convention on the Transboundary Movement of Hazardous Waste for Final Disposal to stop industrialized nations of the Organization for Economic Cooperation and Development (OECD) from dumping their waste on less developed countries. The United States, however, has declined to sign the Convention. In 1994, over 60 OECD countries participating in the Basel Convention agreed to an immediate ban on exports of hazardous waste destined for final disposal in non-OECD countries. It was clear, however, that this was not enough to stop the export of waste for recycling. Seventy-seven non-OECD countries and China pushed for a ban on the shipping of waste for recycling. As a result, the Basel Ban was adopted, promising an end to the export of hazardous waste from rich OECD countries to poor non-OECD countries for recovery operations by December 31, 1997.

The United States has declined to participate in the Basel Ban. In fact, the United States has lobbied governments in Asia to establish bilateral trade agreements to continue dumping their hazardous waste after the Basel Ban came into effect on January 1, 1998. The amount of computer scrap exported from the United States will continue to grow as product obsolescence increases.

11E.4   EXTENDED PRODUCER RESPONSIBILITY AND ELECTRONIC TOXIN PHASEOUTS

Europe has taken the lead in reducing electronic waste from electronic products through *extended producer responsibility* (EPR) (i.e., making the producers responsible for taking back their products). The aim of EPR is to encourage producers to prevent pollution and reduce resource and energy use in each stage of the product life cycle through changes in product design and process technology. Producers will bear a degree of responsibility for all the environmental impacts of their products. This includes impacts arising from the choice of materials and from the manufacturing process, as well as impacts resulting from the use and disposal of products. However, product take-back must go hand-in-hand with mandatory legislation to phase out electronic toxins. Extended producer responsibility focuses on the responsibility that producers assume for their products at the end of their useful life (post-consumer stage). The model example of EPR is product take-back where a producer takes back a product at the end of its useful life either directly or through a third party. Other terms used are *take-back, product liability,* or *life-cycle product responsibility.*

The European Union has drafted legislation on waste from electrical and electronic equipment based on the concept of EPR. The objective of this legislation is to require manufacturers to improve the design of their products in order to avoid the generation of waste and to facilitate the recovery and disposal of electronic scrap. This can be achieved through the phaseout of hazardous materials, as well as the development of efficient systems of collection, reuse, and recycling. The ultimate aim is to close the loop of the product life cycle so that producers, who are in charge of designing the product, get their products back and assume full responsibility for end-of-life-cycle costs. Ensuring this feedback to the producer and making them financially responsible for end-of-life waste management should create an incentive to design products with less hazardous and more recyclable materials. This change in the market economics, the internalization of costs that are currently passed off to the general public, should encourage the design of products for repair, upgrade, reuse, dismantling, and safer recycling.

What the European Union Has Proposed as a Solution for Electronic Scrap

- This legislation will phase out the use of mercury, cadmium, hexavalent chromium, and two classes of brominated flame-retardants in electronic and electrical goods by the year 2004.
- It puts full financial responsibility on producers to set up collection, recycling, and disposal systems.
- Between 70 and 90 percent by weight of all collected equipment must be recycled or reused. In the case of computers and monitors, 70 percent recycling must be met.
- Recycling does not include incineration with energy recovery.

- For disposal, incineration with energy recovery is allowed for the 10 to 30 percent of waste remaining. However, components containing the following substances must be removed from any end-of-life equipment that is destined for landfill, incineration, or recovery: lead, mercury, hexavalent chromium, cadmium, PCBs, halogenated flame-retardants, radioactive substances, asbestos, and beryllium.

- Member states shall encourage producers to integrate an increasing quantity of recycled material in new products. Originally, the European Union stipulated that by 2004, new equipment must contain at least 5 percent of recycled plastic content, but this provision was recently dropped because of intense industry lobbying. This is a major weakening of the legislation because, on the one hand, it encourages recycling, but on the other hand, it does not stipulate recycled content in new products. Instead, the revised legislation encourages member states to set recycled content in their procurement policies.

- Producers must design equipment that includes labels for recyclers that identify plastic types and location of all hazardous substances.

- Member states must collect annual information from producers about quantities and weight of equipment put on the market and on market saturation in respective product sectors. This information will be transmitted to the European Union Commission by 2004 and every three years after that date.

- Producers can undertake the recycling operation in another country, but this should not lead to shipments of electronic waste to non–European Union countries that have no, or lower, standards. Accordingly, producers shall deliver this waste only to those establishments that comply with the certain treatment and recycling requirements, and producers shall verify compliance through adequate certifications.

It is envisaged that the extra costs of waste management will be reflected in a 1 to 3 percent higher retail price on some items. However, the European Union believes this is likely to diminish as economies of scale and innovation bring down the costs of separately collecting and treating this waste. Also, the issue of who should pay is at the heart of EPR, because it is actually a mechanism to implement the "polluter pays" principle. Consumers who buy the product should pay the full price of that product's waste management rather than the general taxpayer who may never purchase that particular product. Companies that learn how to produce products that are less hazardous, easier, and less costly to recycle will develop a competitive advantage because their recycling costs will be lower.

What Has Been the Response of Industry, Member States, and the U.S. Government?

Some industry representatives support the objectives of this legislation. However, many object to mandatory phaseouts of the most toxic materials, although most agree in principle with the need to minimize their use. Industry also objects to the financial responsibility for collection of this waste from private households, but it accepts a certain involvement in the recycling stage of their products.

The 15 member states of the European Union in general welcome the legislation. No country favors a voluntary approach, and there is general agreement about involving producers in the waste management phase of electrical and electronic equipment. Some countries favor the involvement of municipalities in the collection of this waste, but maintain that the responsibility for treatment, recovery, and disposal should be assigned to producers.

The U.S. trade representative, the U.S. Mission in Brussels, and U.S. trade associations [e.g., the American Electronics Association (AEA) and the Electronics Industry Alliance (EIA)] have expressed strong disagreements with the European Union initiative. In a September 9, 1998, letter from the AEA, the EIA, and other trade groups, several high-tech trade associations sought assistance from the U.S. State Department and the U.S. trade representative to derail the proposal. They reiterated that the prohibition on the use of certain materials "that are essential to the functionality, safety, and reliability of electronic and electrical products will impede the development of new technologies and products, increase costs, and restrict global trade in these products." The AEA also lobbied against the 5 percent mandatory recycled content in new products, and the financial responsibility of producers for collection and treatment.

In a January 11, 1999, position paper that cited "Trade Concerns," the U.S. Mission in Brussels has stated that the directive may constitute "unnecessary barriers to trade, particularly the ban on certain materials, burdensome take-back requirements for end-of-life equipment, and mandated design standards." They further state that substitutes may be as problematic or more problematic than the materials they are replacing, and that exemptions for certain uses could lead to uncertainty and confusion in the marketplace.

In response to the lobbying position of the trade organizations and the U.S. government's apparent support, a coalition of public advocacy groups, organized by the International Campaign for Responsible Technology based in Silicon Valley, California, petitioned the European Union not to cave in to U.S. lobbying. At a meeting of the President's Council on Sustainable Development in April 1999, they issued a press release supported by hundreds of nongovernmental organizations (NGOs) from around the world, asserting that the United States had no right to interfere in other countries' environmental protection.

In response to this NGO position, the AEA wrote to Vice President Gore, defending their position. They reiterated that they shared the goal of waste minimization and increased recycling, but the material bans and design requirements went ". . . far beyond the establishment of environmental standards applicable to waste of electrical and electronic equipment, and will hamper global trade of high-tech products, impede technological innovation, and fail to benefit the environment."

In August 1999, a legal opinion prepared for the AEA and EIA asserted again that the European Union legislation would violate several international trade rules and would be an invitation to further trade disputes. A previous assessment by the same law firm that the 5 percent recycled plastic content in new products posed a serious barrier to trade was successful in getting the European Union to drop this recommendation in the latest draft legislation. It now seems that the toxic material phaseouts are the next main focus of the U.S. high-tech lobbyists.

The European Union has always maintained that the legislation does not impose a barrier to trade and that European legal experts had studied the draft thoroughly. The European Union also maintains that the phaseouts only apply when technically feasible and safer substitutes already exist.

11E.5 CAN A CLEAN COMPUTER BE DESIGNED?

Many companies have shown that they can design cleaner products. Industry is making some progress to design cleaner products, but they need to move beyond pilot projects and ensure that all products are upgradable and nontoxic. Some examples of a clean computer design are as follows:

- Hewlett-Packard has developed a safe cleaning method for chips using carbon dioxide cleaning as a substitute for hazardous solvents.

- Printed circuit boards can be redesigned to use a different base material that is self-extinguishing. This should eliminate the need for flame-retardants.

- Matsushita is accelerating efforts to eliminate toxic substances and develop more environmentally benign materials (e.g., lead-free solder, nonhalogenated lead wires, and nonhalogenated plastics). Matsushita also developed the first-ever lead-free solder for flow soldering applications and in Japan has recently started their first totally recyclable television sets. Sony Corporation has developed a lead-free solder alloy that is usable in conventional soldering equipment. There is a range of lead-free solders now available. Obviously, substitutes need to be tested for safety.

- Pressures to eliminate halogenated flame-retardants and to design products for recycling have led to the use of metal shielding in computer housings.
- In 1998, IBM introduced the first computer that uses 100 percent recycled resin (PC/ABS) in all major plastic parts for a total of 3.5 lb of resin per product.
- Researchers at Delft University in the Netherlands are investigating the design of a wind-up laptop, similar to the wind-up radio that plays one hour for every 20 seconds of hand-winding.
- Toshiba is working on a modular upgradable and customizable computer to cut down on the amount of product obsolescence. They are also developing a cartridge that can be rewritten without exchanging parts or modules, allowing the customer to upgrade at low cost.

11E.6   WHAT CAN YOU DO AS A COMPUTER OWNER?

- Write to or phone your computer manufacturer asking them to take back your old computer free of charge to you, or just bring it back to them and tell them that you want them to take it back.
- Sign the letter to the European Union, urging them to stand strong against aggressive lobbying efforts by the high-tech industry and the U.S. trade representative.
- Ask the manufacturer to phase out hazardous materials in your computer.
- Write to the U.S. State Department, telling them that you do not support their lobby against European take-back plans. Europe should be able to protect its own environment. Tell them that you want to see take-back legislation here, too.
- Contact your local or state government representatives, and explain to them why you are concerned. Ask them to get involved in developing solutions. They can ban the landfilling and incineration of electronic waste; they can help to promote computer reuse and recycling infrastructure; they can support EPR for computer manufacturers.

11E.7   CONTACTS AND RESOURCES FOR DEALING WITH COMPUTER WASTE

British Columbia, Canada
Society Promoting Environmental Conservation
Contact: Helen Spiegelman
Vancouver, B.C. V651J3
Tel: 604-731-8464
E-mail: helens@axionet.com
The province of British Columbia has some of the best product take-back programs in Canada.

Environmental Data Services
ENDS Environment Daily
E-mail: envdaily@ends.co.uk
Fax: +44 171 415 0106
Also on the web at www.ends.co.uk/envdaily

Grassroots Recycling Network
Contact: Bill Sheehan
Network Coordinator
P.O. Box 49283
Athens, GA 30604-9283
Tel: 706-613-7121
Fax: 706-613-7123
E-mail: zerowaste@grrn.org
Also on the web at www.grrn.org

Minnesota
Contact: Garth Hickle
Policy Analyst
Minnesota Office of Environmental Assistance
St. Paul, MN 55155-4100
Tel: 651-215-0271
E-mail: garth.hickle@moea.state.mn.us

National Recycling Coalition
Contact: Dawn Amore
1727 King Street, Suite 105
Alexandria, VA 22314-2720
Tel: 703-683-9025, ext. 205
Fax: 703-683-9026
E-mail: dawna@nrc-recycle.org
The National Recycling Coalition is exploring voluntary product take-back programs with some industries.

Product Stewardship Advisor
Cutter Information Corporation
37 Broadway
Arlington, MA 02474
Tel: 1-800-964-5125 or 781-641-5125 outside North America
Fax: 1-800-888-1816 or 781-648-1950 outside North America
PSA specializes in product take-back around the world. See their web site at www.cutter.com.

University of Tennessee—Knoxville
Center for Clean Products and Clean Technology
Contact: Gary A. Davis, Director
University of Tennessee
Suite 311, Conference Center Building
Knoxville, TN 37996
Tel: 423-974-1835
Fax: 423-974-1838
The Center conducts research and publishes comprehensive news and reports on producer responsibility.

U.S. EPA Office of Solid Waste Management
Contact: Clare Lindsay
Project Director for Extended Product Responsibility
Tel: 703-308-7266
E-mail: LINDSAY.CLARE@epamail.epa.gov

Wisconsin
Contact: John Reindl, Recycling Manager
Dane County, WI 53713
Tel: 608-267-8815
Fax: 608-267-1533
E-mail: reindl@co.dane.wi.us

REFERENCES

MCTC (1996) *Electronics Industry Environmental Roadmap,* Microelectronics and Computer Technology Corporation, Austin, TX.

National Safety Council (1999) *Electronic Product Recovery and Recycling Baseline Report,* National Safety Council, Washington, DC.

SVTC (2000) *Just Say No to E-Waste: Background Document on Hazards and Waste from Computers,* Silicon Valley Toxics Coalition, www.igc.org/svtc/cleancc/eccc.htm.

CHAPTER 12
COMPOSTING OF MUNICIPAL SOLID WASTES

Luis F. Diaz
George M. Savage
Clarence G. Golueke

Composting is one element of an integrated solid waste management strategy that can be applied to mixed municipal solid waste (MSW) or to separately collected leaves, yard wastes, and food wastes. The four basic functions of composting are (1) *preparation*, (2) *decomposition*, (3) *postprocessing*, and (4) *marketing*. This chapter treats the last three functions. Preparation or preprocessing is described in Chap. 8. MSW composting results in a volume reduction of up to 50 percent and consumes about 50 percent of the organic mass on a dry weight basis, by releasing mainly CO_2 and water. Composting breaks down easily degradable plant and animal tissue but does not produce appreciable changes in difficult-to-degrade organics (wood, leather, polymers) or in inorganics (dirt, glass, ceramics, and metals). A typical composting process flow diagram is shown in Fig. 12.1. The most important preprocessing steps are (1) receiving, (2) removal of contaminants and recyclable materials, (3) size reduction, and (4) possibly some adjustment of the waste properties (e.g., carbon-to-nitrogen ratio). Three basic systems used for the decomposition steps are (1) static windrows (piles), (2) turned windrows, and (3) in-vessel composting.

Yard waste composting is a relatively simple open air process. The first step is to "chip" the yard waste to reduce the particle size and promote the breakdown of organic matter. It is then set out in long piles or windrows that are periodically turned over to expose all of the material to air. Alternatively, the piles can be placed on a porous pad that is connected to a blower to supply air. The processing of MSW to make a commercially valuable compost, however, is a complex process. MSW composting begins with separating the biodegradable organic materials from the rest of the waste and then shredding or grinding the organics (the remaining MSW is usually landfilled). In some cases, the organics are initially composted inside a vessel that provides mechanical agitation and forced aeration; in other cases, composting takes place entirely in the open. Enclosed composting can help to control odors through better control of aeration and temperature, but composting in a vessel is generally followed by additional open air composting.

The number of U.S. composting programs has increased steadily from about 700 in 1988 to more than 3800 in 1998. However, some of these programs compost only leaves on a seasonal basis. For composting mixed MSW, in 1998 the United States had about 20 operating plants, with a total combined design capacity of about 800 tons per day. About half a dozen composting facilities are in the planning stages. In those municipalities or states having recycling goals in excess

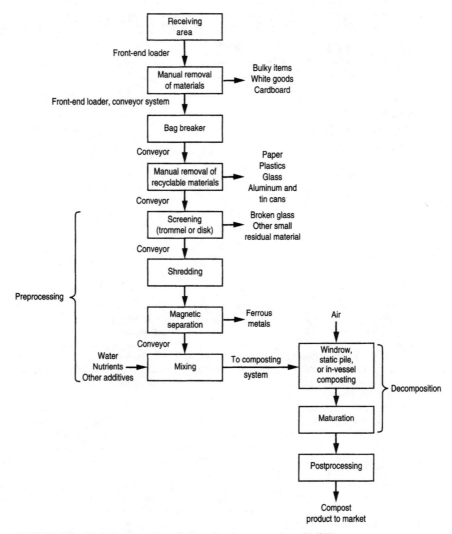

FIGURE 12.1   Typical process flow diagram for the composting of MSW.

of 15 percent, composting can be, and usually is, an important facet of an integrated waste management program. Theoretically, composting could be used to process the 10 to 30 percent of MSW in the waste stream (see Chap. 3) that is yard waste. This quantity can be increased if some of the paper fraction (e.g., soiled paper), as well as food preparation residues, are included in the feedstock to the composting facility. Because the level of technology and mechanization for composting varies widely, the costs of composting systems also vary as shown in this chapter. But for an order-of-magnitude estimate, the empirical relations shown in Fig. 12.2 can be used to estimate the capital cost in 1991 dollars as a function of capacity in tons per day.

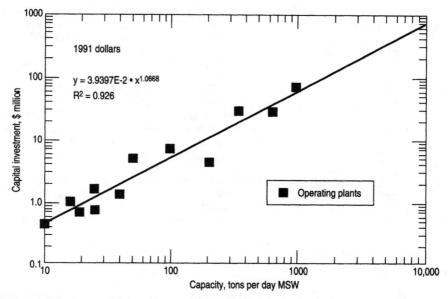

FIGURE 12.2   Capital investment for composting plants as a function of capacity (excluding cost associated with collection, e.g., trucks). *(From SRI International, 1992.)*

12.1   PRINCIPLES

Definition

A definition of composting as applied to MSW management is as follows: "Composting is the biological decomposition of the biodegradable organic fraction of MSW under controlled conditions to a state sufficiently stable for nuisance-free storage and handling and for safe use in land applications (Golueke et al., 1955; Golueke, 1972; Diaz et al., 1993)." Definitive (i.e., distinguishing) terms in the definition are "biological decomposition," "biodegradable organic fraction of MSW," and "sufficiently stable."

The specification "biological decomposition" confines composting to the treatment and disposal of the biologically originated organic fraction of MSW. The specification, "under controlled conditions," distinguishes composting from the simple decomposition that takes place in open dumps, landfills, and feedlots. The specification, "sufficiently stable," is a prerequisite for nuisance-free storage and handling and for safe use in land applications.

Biology

The organisms that are actively involved in composting can be classified into six broad groups. Named in order of decreasing abundance, the groups are: (1) bacteria, (2) actinomycetes, (3) fungi, (4) protozoa, (5) worms, and (6) some larvae. The bacteria include a wide spectrum of classes, families, genera, and species. For example, pseudomonads have been isolated and classified down to the genus level. Although actinomycetes are bacteria, they are named separately because of their particular role in the curing stage of the process (Golueke et al., 1955).

Two genera of actinomycetes have been isolated and identified, *Actinomyces and Streptomyces* (Golueke et al., 1955). The fungi rival the bacteria in terms of number and importance in the later stages of the process. The worms include nematodes and some earthworms (species of annelids). The larvae are of various types of flies.

Attempts to identify a hierarchy of microbes down to the species level on the basis of number and activity in the compost process have met with little success, because of the inevitable local differences in the gamut of environmental and operational situations. An even greater uncertainty arises from the limitations of analytical procedures and techniques presently available. However, the following very broad generalizations have proven to be adequate for routine composting, particularly of MSW. In terms of number and activity, the predominant organisms are bacteria and fungi and, to a far lesser extent, protozoa. However, some higher organisms such as earthworms and various larvae may appear in the later stages of the composting process.

Of great practical and economic importance is the fact that the presence of all of these organisms is a characteristic of all wastes—particularly of yard waste and MSW. Hence, as has been confirmed by carefully conducted scientific research, the use of inoculums (including enzymes, growth factors, etc.) not only would be unnecessary, it would also be an economic handicap.

Classification

The compost process can be classified in terms of distinguishing cultural condition and in terms of technology. This section deals with classification in terms of cultural conditions (i.e., aerobic vs. anaerobic and mesophylic vs. thermophylic). Classification on the basis of technology is reserved for Sec. 12.2.

Aerobic vs. Anaerobic. Originally, composting was classified into aerobic vs. anaerobic,* and many arguments were offered in favor of one or the other. However, in time, the aerobic approach became the usual one, and anaerobic composting fell into disfavor. In fact, a tendency has developed in recent years to define composting as "aerobic decomposition," thereby invalidating the terms "anaerobic composting." Nevertheless, many practitioners, especially those well versed in composting, are not following the trend. Regardless of one's views on the matter, maintenance of completely aerobic conditions in a composting mass would be exceedingly difficult and certainly impractically expensive. In recognition of the foul odors associated with anaerobiosis, a more realistic approach is to design the composting system such that aerobiosis is promoted and anaerobiosis is minimized as much as is feasible.

Mesophylic vs. Thermophylic. In modern compost practice, the question of relative advantages of an all-thermophylic vs. an all-mesophylic process is moot because, with few exceptions, modern composting incorporates the rise and fall of temperature levels that normally occur unless positive measures are taken to circumvent the process. Mesophilic is the temperature range from about 5 to 45°C. Thermophilic is the temperature range from about 45 to 75°C.

Compost Phases

Composting characteristically is an ecological succession of microbial populations almost invariably present in wastes. The succession begins with the establishment of composting conditions. "Resident" (indigenous) microbes capable of utilizing nutrients in the raw waste

---

* Aerobic processes are those carried out in the presence of oxygen. Anaerobic processes are those carried out in the absence of oxygen.

immediately begin to proliferate. Owing to the activity of this group, conditions in the composting mass become favorable for other indigenous populations to proliferate. Plotting the effect of the succession of total bacterial content of the mass would result in a curve, the shape of which would roughly mirror those of the normal microbial growth curves and of the rise and fall of temperature during composting (see Fig. 12.3). Judging from the curve, composting proceeds in three stages, namely (1) an initial lag period ("lag phase"), and (2) a period of exponential growth and accompanying intensification of activity ("active phase") that (3) eventually tapers into one of final decline, which continues until ambient levels are reached ("curing phase" or "maturation phase"). In practice, this progression of phases is manifested by a rise and fall of temperature in the composting mass. A plot of the temperature rise and fall would result in a curve, the shape of which would be roughly identical with that of the growth curve.

The course of the process in all its aspects and the characteristics of its product are all determined by the environmental factors to which the process is exposed, by the operational parameters being followed, and by the technology employed. An abrupt deviation during any of the phases (e.g., sharp drop in temperature) betokens a malfunction. The phase resumes upon the elimination of the malfunction.

Lag Phase. The lag phase begins as soon as composting conditions are established. It is a period of adaptation of the microbes characteristically present in the waste.

Microbes begin to proliferate, by using sugars, starches, simple celluloses, and amino acids present in the raw waste. Breakdown of waste to release nutrients begins. Because of the accelerating activity, temperature begins to rise in the mass. Pseudomonads have been routinely identified as being among the more numerous types of bacteria. Protozoa and fungi, if present, are not discernible. The lag period is very brief when highly putrescible materials and/or herbaceous yard wastes are involved. It is somewhat longer with mixed MSW and woody yard waste, and is very protracted with dry leaves and resistant wastes such as dry hay, straw, rice hulls, and sawdust.

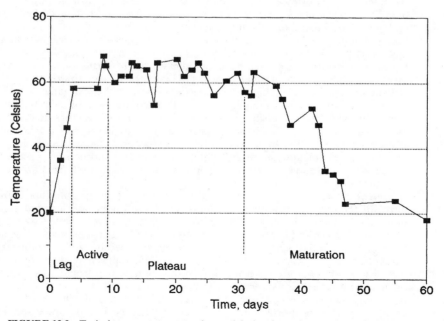

FIGURE 12.3   Typical temperature curve observed during the various compost phases.

Active Phase.   The transition from lag phase to active phase is marked by an exponential increase in microbial numbers and a corresponding intensification of microbial activity. This activity is manifested by a precipitous and uninterrupted rise in the temperature of the composting mass. The rise continues until the concentration of easily decomposable waste remains great enough to support the microbial expansion and intense activity. Unless countermeasures are taken, the temperature may peak at 70°C or higher.

The activity remains at peak level until the supply of readily available nutrients and easily decomposed materials begins to dwindle. In a plot of the temperature curve, this period of peak activity is indicated by a flattening of the curve (i.e., by a plateau). This "plateau" phase may be as brief as a few days or, if the concentration of resistant material is high, as long as a few weeks.

The duration of the entire active stage (exponential plus plateau) varies with substrate and with environmental and operational conditions. Thus, it may be as brief as five or six days or as long as two to five weeks. It should be pointed out that a sudden drop in temperature during the active stage is an indication of some malfunction that requires immediate attention (e.g., insufficiency of oxygen supply, excess moisture). Temperature drop due to turning is of brief duration.

Maturation or Curing Phase.   Eventually, the supply of easily decomposable material is depleted, and the maturation stage begins. In the maturation phase, the proportion of material that is resistant steadily rises and microbial proliferation correspondingly declines. Temperature begins an inexorable decline, which persists until ambient temperature is reached. The time involved in maturation is a function of substrate and environmental and operational conditions (i.e., as brief as a few weeks to as long as a year or two).

Environmental Factors and Parameters

Nutrients and Substrate.   In composting, *substrate* and *nutrient supply* are synonymous because the substrate is the source of nutrients. In the composting of yard waste and MSW, the biologically originated organic fraction of the wastes is the substrate. The specification "biologically originated" eliminates synthetic organic wastes. The exclusion of synthetic organics has a very practical significance because it eliminates many types of plastics. Wastes of biological origin differ from synthetic organic wastes in terms of molecular structure and arrangement. Examples of organic wastes of biological origin are wood, paper, and plant and crop debris. Plastics and vehicle tires are examples of synthetic organic materials.

There are exceptions to the biological origin requirement. In fact, the number of exceptions is growing because of strides being made in microbial genetics, gene manipulation, and molecular engineering. This is particularly true with toxic organics and chemical pesticides. However, much remains to be done before the exceptions become generalities.

Although the ideal waste would contain all necessary nutrients, in practice it may be necessary at times to add a chemical nutrient to remedy a nutrient deficiency.

Chemical Elements.   The major nutrient elements ("macronutrients") are carbon (C), nitrogen (N), phosphorus (P), and potassium (K). Among the nutrient elements used in minute amounts ("micronutrients" or "trace elements") are cobalt (Co), manganese (Mn), magnesium (Mg), and copper (Cu). Calcium (Ca) falls between macro and micronutrients. Carbon is oxidized (respired) to produce energy and metabolized to synthesize cellular constituents. Nitrogen is an important constituent of protoplasm, proteins, and amino acids. An organism can neither grow nor multiply in the absence of nitrogen in a form that is accessible to it. Although microbes continue to be active without having a nitrogen source, the activity rapidly dwindles as cells age and die. The principal use of calcium is as a buffer (resists change in pH). Phosphorus is involved in energy storage and to some extent in the synthesis of protoplasm.

Availability of Nutrients.   An aspect of nutrition is that the mere presence of a nutrient element in a substrate does not suffice. To be utilized, the element must be in a form that can be assimilated by the organism. In short, the element must be "available" to the organism. This applies even to sugars and starches, most of which are readily decomposed.

Availability to a microbe is a function of the organism's enzymatic makeup. Thus, members of certain groups of microbes have an enzymatic complex that permits them to attack, degrade, and utilize organic matter present in a raw waste. Groups that lack the needed complex can utilize as a nutrient source only the decomposition products (intermediates) produced by enzymatically endowed organisms.

The restriction regarding availability is particularly significant. The significance is in the fact that it makes the composting of a waste the result of the activities of a dynamic succession of groups of microorganisms. In this succession, groups "prepare" the way for their successor. In this context, succession does not necessarily imply that groups "come and go" in series. On the contrary, some or even most groups may persist. However, some persistent groups may become less prominent in the ongoing activity.

Certain organic substances are not readily decomposed even by microbes that have the required enzymatic complex. Such resistant materials are broken down slowly despite the maintenance of conditions at optimum levels. Examples are lignin and chitin. Lignin is the principal constituent of wood, whereas chitin is a major constituent of feathers and shellfish exoskeletons. Although the cellulose C in wood, straw, and pith is readily available to many fungi, it is resistant to most microbes.

Nitrogen is readily available when it is in the proteinaceous, peptide, or amino acid forms. On the other hand, because of the resistance of the chitin and lignin molecules to microbial attack, the small amount of nitrogen in them is released too slowly for practical composting.

Carbon-to-Nitrogen Ratio.   The available carbon to available nitrogen ratio (C/N) is the most important of the nutritional factors, inasmuch as experience shows that most organic wastes contain the other nutrients in the required amounts and ratios for composting. The ideal ratio is about 20 to 25 parts of available carbon to 1 of available nitrogen. The nitrogen content and C/N of several wastes are given in Table 12.1.

TABLE 12.1   Percent N, C/N, and Moisture of Selected Materials

Material	% N (dry wt.)	C/N (weight to weight)	% moisture (wet wt.)
Corncob	0.4–0.8	56–123	9–18
Cornstalks	0.6–0.8	60–73	12
Fruit wastes	0.9–2.6	20–49	62–88
Rice hulls	0.0–0.4	113–1120	7–12
Vegetable wastes	2.5–4.0	11–13	*
Poultry litter (broiler)	1.6–3.9	12–15	22–46
Cattle manure	1.5–4.2	11–30	67–87
Horse manure	1.4–2.3	22–50	59–79
Garbage (food wastes)	1.9–2.9	14–16	69
Paper (from domestic refuse)	0.2–0.25	127–178	18–20
Refuse	0.6–1.3	34–80	
Sewage sludge	2.0–6.9	5–16	72–84
Grass clippings	2.0–6.0	9–25	
Leaves	0.5–1.3	40–80	
Shrub trimmings	1.0	53	15
Tree trimmings	3.1	16	70
Sawdust	0.06–0.8	200–750	19–65

* Not reported.

Source: From Rynk (1992).

A C/N higher than 20/1 or 30/1 can slow the compost process. A C/N that is too low (less than 15/1 to 20/1) leads to loss of nitrogen as ammonium N. The addition of a nitrogenous waste can lower an unfavorably high C/N, whereas the addition of a carbonaceous waste can raise an undesirably low C/N. Examples of nitrogenous wastes are grass clippings, green vegetation, food wastes, sewage sludge, and commercial chemical fertilizers. Examples of carbonaceous wastes are hay, dry leaves, paper, and chopped twigs.

The requirement that the carbon be in an *available* form minimizes or even eliminates wood and woody materials as a carbon source in the composting of sewage sludge. Thus, sawdust, wood chips, or woody shavings used to bulk sewage sludge should not be regarded as a carbon source. Although the carbon in paper, dry leaves, and chopped twigs is relatively slowly available, the materials can serve as carbon sources only to a limited extent. Furthermore, because the carbon in the latter materials is only slowly available, their use can raise the permissible upper C/N to as high as 35 to 40/1.

Animal manures, sewage sludge, and commercial (agricultural) chemical fertilizers are adequate sources of nitrogen and any other element that may be needed. For example, experience indicates that the unfavorably high C/N ratio of the organic fraction of refuse can be advantageously lowered through the addition of digested sludge (Diaz et al., 1977).

Particle Size. Theoretically, the smaller the particle size, the more rapid the rate of microbial attack. In practical composting, however, there is a minimum size below which it is exceedingly difficult to maintain an adequate porosity in a composting mass. This size is the "minimum particle size" of the waste material. In composting, the practical "optimum" is a function of the physical nature of the waste material. With a rigid or not readily compacted material such as fibrous waste, twigs, prunings, and corn stover, the suitable size is from ½ in (13 mm) to about 2 in (50 mm). The particle size of the greater part of a fresh green plant mass such as vegetable wastes, fruits, and lawn clippings should be no less than 2 in (50 mm). On the other hand, depending upon their overall decomposability, their maximum particle size can be as large as 6 in (0.15 m) or even larger.

Oxygen. Oxygen availability is a prime environmental factor in composting, inasmuch as composting is an aerobic process. Oxygen is a key element in the respiratory and metabolic activities of microbes. Interruption in the availability leads to a shunt metabolism, the products of which are reduced intermediates, which characteristically are malodorous. The microbes involved in the composting process obtain their oxygen from the air with which they come in contact (i.e., the air that impinges upon them). Consequently, the oxygen content of this air must be continually replenished or the air itself must be continually replaced. The interstitial oxygen content in a windrow can be estimated by use of an oxygen probe inserted into the windrow. The oxygen content of the airstream into and out of a static windrow (forced aeration) and in-vessel systems can be directly measured. For convenience, the amount of oxygen required by the microbes is termed "oxygen demand."

Attempts to establish a universally applicable numerical rate of oxygen uptake for use as a design parameter have been unsuccessful. The underlying reason for the lack of success is the variability of key factors that influence oxygen demand. Among such factors are temperature, moisture content, size of bacterial population, and availability of nutrients. Therefore, determination of the amount of aeration that would meet a specific demand adds another level of complexity, because the capacity and performance of the aeration equipment and the physical nature of the composting mass must be taken into account. The straightforward methods (procedures) used for determining oxygen demand in wastewater treatment (e.g., COD, BOD) are poorly or not at all applicable to composting.

The variability of oxygen demand is demonstrated by the diversity of results reported in the literature. One of the earlier reports described a study in which air was passed at a known rate through composting material enclosed in a drum and the oxygen content of the influent and effluent airstream were measured (Schulz, 1960, 1964). Oxygen uptake rose from 1 mg/g of volatile matter at 30°C to 5 mg/g at 63°C. In a later study, Chrometska (1968) observed oxygen requirements that ranged from 9 mm^3/g·h for ripe compost to 284 mm^3/g·h for raw substrate. "Fresh" compost (seven days old) required 176 mm^3/g·h. Lossin (1971) reports average chemical oxygen demands that range from almost 900 mg/g on the first day of composting to about 325 mg/g on the 24th day. In a review of the decomposition of cellulose and refuse, Regan and Jeris (1970) observed that oxygen uptake was lowest (1.0 mg oxygen/g volatile matter per hour) when the temperature of the mass was 30°C and the moisture content was 45 percent. The highest uptake (13.6 mg/g volatile matter per hour) occurred when the temperature was 45°C and the moisture content 56 percent.

With respect to the design of airflow through an in-vessel reactor and to a lesser extent through a static pile, the indicated procedure would be to estimate the carbon content and to determine the amount of oxygen consumed in oxidizing the carbon. The flaw in such an approach is that it would result in an overdesign, because normally only a fraction of the carbon is available to the microbial population. Nevertheless, some overdesign is advisable because of the impossibility of aerating a mass such that all microorganisms simultaneously have access to sufficient oxygen. Perhaps the airflow requirement estimated by Schulze (562 to 623.4 m^3/tonne volatile matter per day) could be of some use. However, his estimates are based upon the use of his particular equipment and on a laboratory-scale experiment.

Moisture Content

Maximum. Theoretically, the optimum moisture content of the wastes is one that approaches saturation, provided that the material can be sufficiently aerated to meet the oxygen demand. Although meeting the demand is technologically feasible, it also is economically unfeasible. Hence, the term *permissible maximum* is introduced. It is the moisture content above which oxygen availability becomes inadequate and anaerobiosis ensues. The maximum permissible moisture content usually is also the optimum content.

Because the air entrapped in interstices between particles is the primary source of oxygen for the microbial population, interstitial ("pore") volume is a decisive factor (i.e., the more numerous the pores the greater the interstitial volume). Hence, porosity is a key consideration. The relation to moisture stems from the fact that the greater the fraction of the pore volume occupied by water, the less is the volume available for air and hence for oxygen.

Interstitial Volume, also known as porosity, is determined by: (1) the size of individual particles, (2) the configuration of the particles, and (3) the extent to which individual particles maintain their respective configuration. Maintenance of configuration depends upon the structural strength of the individual particle (i.e., resistance to flattening). Because these characteristics vary, maximum permissible moisture content varies from substrate to substrate. Thus, the maximum permissible moisture content is higher with wastes in which straw and wood chips rather than paper are the bulking materials. Several wastes and their respective permissible moisture contents are listed in Table 12.2.

TABLE 12.2 Maximum Permissible Moisture Contents

Type of waste	Moisture content, % of total weight
Theoretical	100
Straw	70–85
Rice hulls	70–85
Wood (sawdust, small chips)	80–90
Manure with bedding	60–65
Wet wastes (vegetable trimmings, lawn clippings, kitchen wastes, etc.)	50–55
MSW (refuse)	55–60

With in-vessel systems, the microbial oxygen demand is more or less met by direct exposure to air brought about by agitating the composting mass. Reliance upon interstitial air is correspondingly reduced. Although the reliance upon interstitial air may be reduced, it is never eliminated because agitation is neither complete nor uninterrupted. Therefore, moisture content continues to be a decisive factor. Moreover, excessive moisture adversely affects materials handling characteristics.

Minimum.   Moisture inadequacy is a common operational factor because the combination of relatively high temperatures and intense aeration is conducive to evaporation. Minimum moisture content becomes a consideration at moisture levels lower than those at which oxygen availability is a limiting factor. At such levels, microbial biological requirement for water becomes the determinant.

The penalty for moisture shortage is inhibition of microbial activity. Because almost all biological activity ceases at moisture contents lower than about 12 percent, the more closely the moisture content of a composting mass approaches that level, the less is the intensity of the microbial activity. The consensus is that efficient composting requires that the moisture content of the composting mass be maintained at or above 45 to 50 percent.

The effect of moisture insufficiency is illustrated by its effect on oxygen demand. In a report, Lossin states that moisture content is a determinant of oxygen needs. For example, fresh compost having a moisture content of 45 percent required 263 mm^3/g·h, whereas at a moisture content of 60 percent the demand was 306 mm^{3}g·h (Lossin, 1971). This increase in demand indicates that moisture insufficiency had inhibited bacterial activity. It would have taken another determination at a higher moisture content to determine the level at which moisture would no longer be limiting.

pH Level.   The optimum pH range for most bacteria is between 6.0 and 7.5, whereas the optimum for fungi is 5.5 to 8.0. Precipitation of essential nutrients out of solution rather than inhibition due to pH per se establishes the upper pH limit for many fungi.

In practice, little should be done to adjust the pH level of the composting mass. Owing to the activity of acid-forming bacteria, the pH level generally begins to drop during the initial stages of the compost process. These bacteria break down complex carbonaceous materials (polysaccharides and cellulose) to organic acid intermediates. Some acid formation may also occur in localized anaerobic zones. Some may be due to the accumulation of intermediates formed by shunt metabolisms. Shunt metabolism may be triggered by an abundance of carbonaceous substrate and/or perhaps by interfering environmental conditions. Whatever the cause, the early pH drop in composting MSW may be to 4.5 or 5.0. The drop could well be lower with other wastes.

Organic acid synthesis is paralleled by the development of a microbial population for which the acids serve as a substrate. The consequence is a rise in pH level to as high as 8.0 to 9.0. The mass becomes alkaline in reaction.

Buffer against the initial pH drop through the addition of lime is unnecessary. Moreover, it promotes a loss of nitrogen. The loss can be particularly serious during the active stage of the compost process. For example, in research conducted at the University of California at Berkeley in the 1950s, nitrogen loss always was greater from piles to which lime [$Ca(OH)_2$] had been added to raise the pH (Golueke et al., 1955).

Despite the potential promotion of nitrogen loss, the addition of lime might be beneficial in cases in which the raw waste is rich in sugars or other readily decomposed carbohydrates (e.g., fruit and cannery waste). Acid formation in such wastes is more extensive than in MSW and yard waste. For example, it was found in studies on the composting of fruit waste bulked with sawdust, rice hulls, or composted refuse that the three to four days' lag in temperature rise characteristic of unbuffered fruit waste could be eliminated by adding lime (NCA, 1964). However, nitrogen loss also was greater.

Occasionally, the addition of lime may lessen offensive odors because of the effect of pH. Lime addition also improves the handling characteristics of some wastes.

Temperature.   In the consideration of temperature as an environmental factor, the interest is in the effect of temperature on the well-being and activities of the microbial population, rather than in the effect of microbial well-being and activity on temperature level. In short, environmentally oriented interest is on the effect of temperature on microbial well-being and activity; whereas operationally oriented interest is on the effect of microbes on temperature.

As an Environmental Factor.   In the section on classification, brief mention was made of the relation between temperature and rate of composting. The question is not so much one of the effect of temperature within either the mesophylic or the thermophylic ranges as it is the relative advantages of one range over the other, (i.e., mesophylic vs. thermophylic). Thus, each group of mesophyles has an optimum range specific to it in the mesophylic range. Similarly, each group of thermophyles has its specific optimum level in the thermophylic range. The result is that because of the diverse population and variation in temperature in the composting mass, chances are that in any given instant in time, temperature will be optimum for some group. Conversely, chances of its being optimum for all groups at any single instant in time would be nil. For example, the optimum for the mesophyle *Pseudomonas delphinium* is 25°C; whereas, for *Clostridium acetobutylicum,* another mesophyle, it is 37°C. The high degree of activity is indicative of a satisfactory temperature for most of the microbes.

A straight-line relationship exists in terms of increase in process efficiency and speed and rise in temperature because of the overlapping of optimum temperatures at levels lower than 30°C. The slope of a curve showing efficiency or speed of the process as a function of temperature would flatten somewhat between 35 and 55°C, perhaps with some decline between 50 and 55°C. The existence of an activity plateau at the transition from the mesophylic range to the thermophylic range is due not only to the involvement of many types of organisms but also to adaptation of organisms or enrichment for organisms adapted to a given range. As the temperature rises above 55°C, efficiency and speed drop and are negligible at temperatures above 70°C. At temperatures higher than 65°C, spore formers rapidly enter the spore stage and, as such, are dormant. Most nonspore formers die off.

Mesophylic vs. Thermophylic Composting.   In the 1950s, there was some debate on the relative merits of mesophylic vs. thermophylic composting regarding nature, extent, and rate of decomposition (Wiley, 1957). The opinion in favor of thermophylic composting largely rested on the fact that experimental evidence shows that up to a certain point, chemical and enzymatic reactions are accelerated by each increment in temperature. For enzymes, the acceleration continues up to the point above which they are inactivated.

Experience shows that the upper limit for most thermophyles involved in composting is between 55 and 60°C and, accordingly, the process is adversely affected if temperatures rise above this range (Regan, et al., 1970; Finstein, 1992). In practice, the question is moot, not only because of the costs involved in establishing and maintaining a thermophylic environment, but also because the heat generated in a reasonably large or insulated mass inevitably brings the internal temperature to thermophylic levels unless something is drastically amiss with the operation. In fact, measures should be taken to avoid the inhibitory range (Finstein, 1992). Therefore, if the question of mesophylic vs. thermophylic has any significance, it would mainly concern in-vessel systems because with them, heat is dissipated by the continued agitation and ventilation of the mass in the compost unit. Of course, temperature can rise in the units if the mass of composting material is sufficiently large and the degree of agitation of the content is kept below a critical level.

Operation and Performance Parameters

Commonly used operational parameters include these eight: oxygen uptake, temperature, moisture content, pH, odor, color, destruction of volatile matter, and stability. With respect to the first four, the distinction between their status as environmental factor and that as operational parameter is very difficult to define because the two overlap in that operational parameters evolve from environmental factors.

Oxygen Uptake.   Oxygen uptake is a very useful parameter, because it is a direct manifestation of oxygen consumption by the microbial population and, hence, of microbial activity. Microbes use oxygen to obtain the energy to carry on their activities.

A very effective means of monitoring for adequacy of oxygen supply is by way of the olfactory sense, namely, detection of odors. The emanation of putrefactive odors from a composting mass is a positive indication of anaerobiosis. The intensity of the odors is an indication of the extent of anaerobiosis. Attempts to measure odoriferous constituents (e.g., H_2S) have been only indifferently successful. Because of their anaerobic origin, the malodors soon decrease after aeration is intensified. Although reliance upon the detection of objectionable odors may seem to be rather primitive, nevertheless it is a useful supplement in routine monitoring. It does have the disadvantage of being an "after-the-fact" indicator. Therefore, in operations in which an oxygen probe can be used or the oxygen of input and output airstreams can be measured, a direct monitoring of oxygen is advisable.

An important operational consideration is that although the input airstream may be sufficiently great to meet the theoretical microbial oxygen demand and the discharge airstream may contain some oxygen, localized anaerobic zones may be present. The zones may be due to inadequate mixing or to short-circuiting of air through the mass. In practice, the complete prevention or elimination of these zones would be economically, if not technologically, unfeasible. Fortunately, the complete elimination is not essential for a nuisance-free operation, provided the number and size of the zones does not become excessively large.

According to Diaz et al. (1982), four generalizations can be made, despite the many uncertainties mentioned or implied in the preceding paragraphs. The generalizations are:

1. An oxygen pressure greater than 14 percent of the total indicates that not more than one-third of the oxygen in the air has been consumed.
2. The optimum oxygen level is 14 to 17 percent.
3. Aerobic composting supposedly ceases if the oxygen concentration drops to 10 percent.
4. If CO_2 concentration in the exhaust gas is used as a parameter for oxygen concentration, then the CO_2 in the exhaust gas should be between 3 and 6 percent by volume.

Temperature.   Temperature is a very useful parameter because it is a direct indicator of microbial activity. However, in the application of temperature as an operational parameter, it must be remembered that in a practical operation, the desired temperature range should include thermophylic temperatures. The reasons are: (1) some of the organisms involved in the process have their optimum level in the thermophylic range; (2) weed seeds and most microbes of pathogen significance cannot survive exposure to thermophylic temperatures; and (3) unless definite countermeasures are taken, thermophylic levels will be reached during the active stage.

In general, any abrupt and unexplained deviation from the normal course of temperature rise and fall is an indication of an environmental or operational deficiency that requires attention. An exception to this general rule is the need to prevent the temperature from exceeding 55 to 60°C (i.e., reaching a level that is inhibitory to most microbes). Probably the most effective remedial measure is ventilation.

Moisture.   The numerical value of the operational parameter, moisture, is the maximum permissible moisture content. As stated earlier, this value varies from substrate to substrate. Table 12.2 lists several maximum permissible moisture contents. The relatively low value for MSW reflects the high paper content of the waste.

Regardless of substrate, the lowest permissible moisture content for efficient composting is about 45 percent. An unfavorably low moisture content is a common problem in compost practice, because conditions in a composting mass are conducive to evaporation (i.e., water loss). Unless this water is replaced, moisture is likely to become limiting.

pH. Unless the substrate is unusually acidic, which rarely is the case with MSW, pH level has little value as an operational parameter. If the pH level is lower than 4.5, some buffering may be indicated (e.g., adding lime). Liming may also be indicated for certain cannery wastes.

Odor. Odor as an operational parameter received some attention in the discussion of aeration. Attempts to develop a quantitative standard for odor, based on hydrogen sulfide concentration, have met with little if any success, because the olfactory nerve senses H_2S concentrations lower than the detection level of H_2S analytical tests. In waste treatment practice, all odors are regarded as being objectionable to the public.

Color. Although the color of the composting mass progressively darkens, it is a crude parameter and at best is roughly qualitative and highly subjective.

Destruction of Volatile Solids. Inasmuch as composting is a decomposition process, it is characterized by some destruction of volatile solids. Complete destruction is neither desirable nor necessary because the value of the compost product, particularly as a soil conditioner, is mostly due to its volatile (i.e., organic) solids content. Hence, rate rather than extent of destruction would be the useful parameter. The problem is in the establishment of a standard rate. Rates vary with several important factors. The best indicator that is presently available is to the effect that volatile matter is being destroyed.

Stability. "Stability" is a broad term that may refer to chemical and physical stability and/or to biological stability. As applied in composting, the composting mass is judged "stable" when it has reached a state of decomposition at which it can be stored without giving rise to health or nuisance problems. This excludes the temporary stability due to dehydration or other condition that inhibits microbial activity. Despite many claims to the contrary, a satisfactory quantitative method for determining degree of stability has yet to be developed, at least one that can be used as a "universally" applicable standard.

The search for a method of determining stability that can be sufficiently standardized is almost as old as the compost practice. The list of proposed methods is correspondingly lengthy. It includes final drop in temperature (Golucke, et al., 1955), degree of self-heating capacity (Niese, 1963), amount of decomposable and resistant organic matter in the material (Rolle et al., 1964), rise in the redox potential (Moller, 1968), oxygen uptake (Schulze, 1960), growth response of the fungus *Chaetolnium gracillis* (Obrist, 1965), and the starch test (Lossin, 1970). Of this array of tests, the final drop in temperature is the most reliable, because it is a direct consequence of the entire microbial activity, as well as of the intensity of the activity. The weakness of temperature decline as a parameter is its time element. Because the decline represents a trend, it involves a succession of readings taken over a period of days. The other tests lack the necessary universality. For example, a redox potential that characterizes stability under one set of compost conditions does not necessarily do so under another set. With certain tests, lack of universality is aggravated by the difficulty of conducting them (e.g., the *Chaetomium* test).

Phytoxicity frequently is regarded as being an indication of stability, although it is true that in the early stages of maturation, composting material often contains a substance that is inhibitory to plants (phytoxic), and which almost invariably disappears as maturation progresses. However, the disappearance does not always coincide with the attainment of the required degree of stability.

12.2 *TECHNOLOGY*

Facility Site Selection Considerations

Buffer Zone. Measures taken with waste treatment facilities regarding site selection and preparation to protect air, water, and soil resources must also be taken with a compost facility. In addition to the usual topographical, hydrological, economic, political, and sociological considerations involved in the selection process, provision of an adequate "buffer zone" between the facility and residential areas is particularly essential to the continued survival of a compost facility. Of the factors responsible for the need of a buffer zone (e.g., vehicular traffic, noise), odors rank among the highest. Moreover, because of the likelihood of offensive odors, the size of the buffer zone must be substantial. Nevertheless, the actual dimensions of the zone required for a particular facility depend upon the magnitude of the operation, the nature of the waste, the type of compost system employed, and the degree of enclosure and control of emissions. Obviously, the buffer zone indicated for a few tons per day of yard waste compost operation is far smaller than that for a full-scale MSW, manure, or sewage sludge compost facility.

Odors. In a compost facility, offensive odors usually have two main origins: (1) the raw wastes that serve as substrate; and (2) operational shortcomings and mishaps. (Raw yard wastes have very little, if any, odor unless they include high concentrations of grass or food waste.) Although operational odors can be kept at a minimum, those due to odoriferous raw wastes are inevitable. However, their intensity can be substantially reduced by proper storage and prompt processing.

The impact of the odor problem can be considerably reduced by enclosing the raw waste receiving and storage areas and the active and early maturation stages of the composting process. The ventilation of the housing structure can be designed such that air is exhausted through an air scrubber (Public Works, 1992) or an odor filter. Unfortunately, doing so is costly and is not failproof.

Compost Systems

The rationale underlying compost system design is twofold: (1) provide optimum conditions for composting in an environmentally and economically acceptable manner; and (2) determine the type and size of the compost system and other aspects of technology by the type, volume, nature of waste, and the size of the available buffer zone.

Classification of Compost Systems. Compost systems fall into two very broad groups, (1) *windrow* and (2) *in-vessel.* Reflecting their mechanisms of aeration, windrows may be the turned type, forced aeration (static pile) type, or a combination of turned and forced aeration. A typical windrow is presented in Fig. 12.4. Windrows may be sheltered (i.e., contained within a structure) or they may be unsheltered. Shelters must be provided with a ventilation system such that emissions can be satisfactorily conditioned. Also, in winter, consideration must be given to the control of condensation of moisture released by actively composting material. Reactors in in-vessel systems have one of the following configurations: *horizontal drum,* which is rotated slowly and which may be compartmentalized; *vertical silo;* and an *open tank* equipped with a stirring or an agitation device. All designs of in-vessel systems have provisions for forced aeration. Because of economic constraints, the usual procedure with in-vessel systems is to use the reactor for the lag and active phases and to rely upon windrowing for the maturation phase. A photograph of an in-vessel system is given in Fig. 12.5.

Aeration Mechanisms. The provision of satisfactory aeration is an essential feature of almost all existing compost systems. Aeration mechanisms involved in providing atmospheric oxygen fall into three broad groups, namely, *agitation, forced aeration,* and *turning.* A particu-

FIGURE 12.4 Photograph of a typical windrow composting operation. *(Courtesy of CalRecovery, Inc.)*

FIGURE 12.5 Photograph of an in-vessel composting system. *(Courtesy of CalRecovery, Inc.)*

lar system may incorporate one or a combination of mechanisms. Agitation is accomplished by tumbling, stirring, and/or the act of mixing the composting mass. In forced aeration, air is either pushed or pulled through the composting mass. Most in-vessel systems rely on a combination of the three mechanisms. In windrow composting, milling and stacking the raw waste accomplishes the initial aeration. As stated in the section on moisture content, most of the microbial oxygen requirement in windrowed material is met by the air entrapped in the windrows (i.e., interstitial air). Interstitial air is renewed by turning the windrow or by forcing air through the windrow, (i.e., by ventilating the pile). Very little oxygen comes by way of diffusion of ambient air into the outer layer of the windrow.

Windrow Systems

Site Preparation.   The preparation of concern in this discussion is that of the working area (i.e., the area in which the windrows are constructed and the associated maintenance equipment is maneuvered). For convenience, the working site is also referred to as the *compost pad* in this discussion. Not included is the preparation of the facility site as a whole (access roads, grading, construction of structures, and provision of utilities).

Pad Specifications.   Pad specifications cover a wide spectrum and are influenced by the size of the operation, the nature of the wastes to be composted, and the dictates of circumstances specific to it (e.g., proximity to residential areas, land use, financial capacity). Among the most applicable specifications are availability of essential utilities, surface and construction geared to all-weather accessibility and use regardless of whether the pad is sheltered or is exposed to the elements, appropriate aerial dimensions, prevention of water intrusion, collection and disposal (treatment) of runoff from pad, and leachate collection and disposal.

Utilities.   Access should be available to water and electricity. Water occasionally must be added to the composting mass to keep the moisture content from dropping to inhibitory levels. Water also should be at hand for fire and dust control. Although access to power is not as important as access to water, power has many useful applications (e.g., powering blowers and illumination).

Pad Surface and Construction.   All working areas should be paved and be ready for use regardless of weather. The pad should be sufficiently rugged to support the combined weight of the composting mass and associated materials handling equipment, as well as the maneuvering of the latter. The required degree of conformity with the specifications (i.e., flexibility of application) regarding surface and construction is closely related to the stage of the compost process. Pad conformity with specifications is most necessary for the lag, active, and early maturation stages, decreases as maturation progresses, and is least necessary during storage.

Calculation of Total Area of Windrow Pad.   A variety of factors combine to determine the dimensions of the area requirement. Among them are total volume of material to be accommodated during all stages of the compost process, i.e., from the construction of the windrows through disposal of the stored product, the configuration of the windrows, space required for the associated materials handling equipment and the maneuvering thereof, and the aeration system (forced or turning).

The following is a summary of the steps involved in calculating the pad area. For convenience, the steps are arranged in four main groups: (A) total volume of feedstock to be composted, (B) area occupied solely by windrows, (C) maneuvering area, and (D) total pad area.

A.   Total Volume for Feedstock.

Total volume of feedstock (ft^3 or m^3)

$$= \frac{[\text{retention time (days)} \times \text{rate of feedstock delivery (lb/day or kg/day)}]}{\text{bulk density (lb/ft}^3 \text{ or kg/m}^3)} \quad (12.1)$$

B.   Area Occupied Solely by Windrows.

Step 1. Determine the volume of each windrow.

Volume (ft^3 or m^3) = cross-sectional area (ft^2 or m^2) × length of windrow (ft or m)   (12.2)

Cross-sectional area is a function of cross-sectional configuration. Figure 12.6 illustrates four types of cross-sectional configurations. The cross-sectional area of a pile having a square or rectangular cross section (Fig. 12.6a) is given by Eq. (12.3).

Cross-sectional area (ft^2 or m^2) = base (ft or m) × height (ft or m)   (12.3)

On the other hand, the cross-sectional area of the configuration in Fig. 12.6b is defined by Eq. (12.4).

Cross-sectional area (ft^2 or m^2) = $\pi/4$ × b (ft or m) × h (ft or m)   (12.4)

The cross-sectional area of the configuration in Fig. 12.6c is given in Eq. (12.5).

Cross-sectional area (ft^2 or m^2) = ½ (a + b) (ft or m) × h (ft or m)   (12.5)

The area of the configuration in Fig. 12.6d is defined by Eq. (12.6).

Cross-sectional area (ft^2 or m^2) = ½b (ft or m) × h (ft or m)   (12.6)

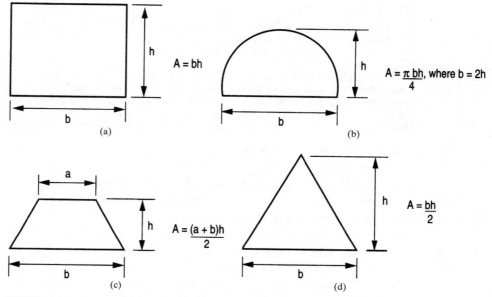

FIGURE 12.6   Areas of various potential cross sections for compost piles.

Step 2. Determine the number of windrows:

$$\text{Number of windrows} = \frac{\text{total volume of feedstock (ft}^3 \text{ or m}^3)}{\text{volume per windrow (ft}^3 \text{ or m}^3 \text{ per windrow)}} \qquad (12.7)$$

Step 3. Determine the area solely occupied by windrows:

Total windrow area (ft^2 or m^2)

$$= \text{number of windrows} \times \text{area per windrow (ft}^2 \text{ or m}^2 \text{ per windrow)} \quad (12.8)$$

C.   Maneuvering Area.

Maneuvering area is the space required to maneuver the turning and other equipment. Two such spaces must be provided for each windrow (one on each side of a windrow). The area of each space is given by Eq. (12.9).

$$\text{Area of space (ft}^2 \text{ or m}^2) = \text{windrow/length (ft or m)} \times \text{width of space (ft or m)} \quad (12.9)$$

The width of a space depends upon the type of turning machine. The following widths are approximate estimates: If turned with a bucket loader, the width may be as little as 4 ft (1.22 m). If a self-propelled turner is used, the width may be from 3 to 5 ft (0.9 to 1.5 m). If a tractor-assisted turner is used (two passes), a 6- to 8-ft (1.8 to 2.4 m) space is indicated. The space between two individually aerated piles is about 20 ft (6.1 m).

D. Total Area of Pad.

The total pad area is the sum of the area required for the windrows plus that needed for maneuvering the material (e.g., constructing windrows, turning the composting mass, water trucks, force aeration equipment, etc.).

It is emphasized that the calculations do not allow for the shrinking of the piles that will occur due to destruction of volatile matter and loss of moisture. Because of the variation in percentage loss due to differences in nature of feedstock and compost system applied, it is impractical to apply a single shrinkage value to all situations. Therefore, the value calculated for the total area will be the maximum value (i.e., the maximum requirement).

1. Volume of material to be composted = 24 m^3/day
2. Composting period (detention time) = 50 days
3. Total volume of material on pad = 50 days × 24 m^3/day = 1200 m^3
4. Dimensions of windrow: length = 50 m, height = 3 m, and width = 4 m
5. Volume of windrow: V = ⅔ × (4 × 3) × 50 m^3
6. Number of windrows = total volume of material/volume of windrow 1200/400 = 3
7. Distance between windrows = 4 m
8. Space around perimeter of composting area = 3 m
9. Length of composting area = windrow length and perimeter space = 50 m + 2(3) = 56 m
10. Width of composting area: width of windrows + distances between windrows + perimeter space = (4 × 3) + (2 × 4) + (2 × 3) = 12 + 8 + 6 = 26 m
11. Area required = length × width = 56 × 26 = 1456 m^2

These calculations do not include required setbacks or buffer zones.

Windrow Construction.   A windrow is constructed by stacking the prepared feedstock in the form of an elongated pile. The procedure involved in stacking the material is influenced by the volume and nature of the feedstock, the design and capacity of the available materials handling equipment, and the physical layout of the windrow pad. If more than one feedstock is involved (e.g., cocomposting sewage sludge and MSW, or yard waste and food waste), or an additive is to be employed, the incorporation would take place at this time. If cocomposting or additives are not involved, the windrows are set up directly after preprocessing is completed. If cocomposting is involved, one approach is to build up the windrow by alternating layers of one of the feedstocks with layers of the other feedstock or doses of the additive. The first and subsequent turning accomplish the necessary mixing of the components. If turning is not the method of aeration, necessary mixing is done immediately prior to constructing the windrow.

Conventional materials handling equipment such as a bulldozer or a bucket loader can be used for windrow construction. An alternative approach involves the use of a conveyor belt as follows: directly after having been preprocessed, the feedstock is transferred to the windrow pad by way of a conveyor belt, the discharge end of which has been adjusted to the height intended for the completed windrow.

Windrow Dimensions.   Three key factors enter into the determination of windrow dimensions, namely, (1) aeration requirements, (2) efficient utilization of land area, and (3) the structural strength and size of the feedstock particles. Structural strength, in turn, is a key factor in the maintenance of the interstitial integrity needed to ensure a sufficient oxygen supply.

All dimensions could be expanded during winter to enhance self-insulation. In windy regions, the dimensions can also be expanded so as to minimize moisture loss through evaporation.

Height. Interstitial integrity and height of the windrow are closely related, because the higher the pile, the greater is the compressive weight on the particles. Hence, the greater the structural strength, the higher is the permissible height. With the organic fraction of municipal refuse, the height is on the order of 5 to 6 ft (1.5 to 1.8 m). Depending upon the size of the shrubbery trimmings fraction, it can be slightly higher with yard waste. In practice, the actual height is determined by the type of equipment used to aerate the composting mass. It generally is lower than the maximum permissible height.

Width. Other than its determination of the ratio of surface area exposed to inward diffusion of air, width has little effect on aeration. However, the amount of inward air diffusion usually is negligible. In summary, width is dictated by convenience. If other factors do not intervene, a width of about 8 to 9 ft (2.4 to 2.7 m) is suitable.

Windrow Geometry. Windrow geometry should be geared to climatic conditions and efficient use of pad area. However, in practice, the determinant is the type of windrow construction and turning equipment, principally the latter. For example, as was indicated in the section on calculation of pad dimensions, cross-sectional configuration exerts a significant impact on the ratio of windrow volume to area—and, hence, on efficient use of land area. However, in regions in which rains are frequent or heavy and the windrows are not sheltered, the cross-sectional configuration should be conical in order to shed water. On the other hand, a flattened top (square or rectangular configuration) is appropriate where rainfall is not a problem. With such a configuration, heat loss is less and windrow volume per unit pad area is greatest.

Turned Windrow Aeration. As stated earlier, windrows can be aerated by turning, by forced aeration, or by a combination of the two. A long record of successful experience has demonstrated the efficacy of turning.

Turning is accomplished by tearing down and then reconstructing the windrow. The windrow can be reconstructed either in its original position, or immediately adjacent, or somewhat removed from its prior position. Tearing down and reconstructing the windrow exposes the composting material to the ambient air and replenishes the interstitial oxygen supply. The resulting mixing renews microbial access to nutrients and disperses metabolic intermediates. The cooling effect of turning can be used for lowering a pile temperature that has reached inhibitory levels.

To avoid defeating its purpose, turning should be done in a manner that does not compact the composting mass.

Although ideally the turning should be such that the outer layers of the original pile become the inner layers of the reconstituted pile, limitations of turning equipment make it unfeasible to do so. However, the benefits that accrue from the reversal of positions can be gained by increasing the frequency and number of turnings. Benefits to be gained are twofold: (1) it promotes uniform decomposition, and (2) it subjects all material to an eventual exposure of all material to the high temperatures characteristic of the interior of an actively composting windrow. The temperatures are high enough to be lethal to disease-causing organisms and to most weed seeds.

Frequency of Turning. Because required frequency of turning exerts a strong influence on design and size of the equipment used in accomplishing turning, a few words on frequency are appropriate. Ideally, frequency should be a function of rate of oxygen uptake by the active microbial population. For example, judging from past experience, (Golueke et al., 1955; Golueke, 1972; Diaz et al., 1993), turning every third day is sufficient to meet the oxygen uptake in actively composting MSW. Of course, this assumes that the MSW is neither waterlogged nor compacted. Incidentally, waterlogged conditions and compaction can be remedied by increasing the frequency of turning.

It should be noted that a turning frequency of three times per week may not be sufficient to kill off all pathogens. (Cooper et al., 1974). Undoubtedly, the incomplete die-off is a consequence of only about 40 to 50 percent of a typical windrow being exposed at a given interval to lethal temperatures. Each subsequent turning brings about a recontamination of "sterilized" material. One solution is to resort to attrition by increasing the frequency such that intervals between turnings become too brief for appreciable regrowth.

Equipment.   The simplest but also the least satisfactory method of turning involves the use of a bulldozer for tearing down and reforming a windrow. With such an approach, mixing and aeration are minimal and the material is compacted instead of being fluffed. The situation is somewhat improved when a bucket loader is used. However, owing to materials handling limitations, both approaches become increasingly inefficient when the volume involved exceeds a few tons per day. Nevertheless, it may be that economic circumstances render the use of a more complex turner unfeasible. Such a situation may not be unusual, which perhaps explains why the use of a bulldozer or bucket loader for turning continues to be a fairly widespread practice. If a bucket loader is used, it should be operated such that the bucket contents are discharged in a cascading manner, rather than dropped as a single mass.

Among the first of the automatic turners was one used in the mushroom industry in the 1950s. In the succeeding years, other mechanical turners began to appear in increasing numbers and design variations. Consequently, several types of turners are now available. A good idea of the diversity may be gained from the lists in Rynk (1992) and by consulting the advertisements in publications such as *BioCycle*. The several types of turners presently on the market fit one or the other of three general groups divided on the basis of the design of the turner mechanism. They are the *auger turner,* the *elevating face conveyor,* and the *rotary drum* with flails. Some types of turners are designed to be towed and others are self-propelled. As is to be expected, the self-propelled types are more expensive than the towed types. An advantage of the towed type is the fact that the tow vehicle (tractor) can be used for other purposes between turnings. In addition to convenience, the self-propelled type requires much less space for maneuvering and therefore the windrows can be closer to each other (Fig. 12.7). The turning capacity of the machines ranges from about 800 tons per h (727 tonnes/h) with the smaller models, to as much as 3000 tons per h (2727 tonnes/h) with the larger, self-propelled versions. Similarly, the dimensions and configuration of the windrows vary with type of machine, e.g., 9 to 15 ft (4.6 m) wide and 4 to 10 ft (1.2 to 3.0 m) high.

To allow for an increased frequency dictated by emergency situations (e.g., excessive moisture) and to ensure hygienic safety, the equipment capacity should be sufficient to permit daily turning.

Forced Aeration (Static Pile).   The substitution of forced aeration for turning as an effective means of aeration has long intrigued compost practitioners (Wylie, 1957; Senn, 1974). A major factor, if not the decisive factor, in favor of forced aeration is the fact that it is less expensive than turning. Supposedly, a prime saving would be the elimination of the need for expensive turning equipment. However, in practice, this saving may not always materialize because some turning inevitably is necessary for the satisfactory remedying of localized problem zones, for ensuring uniformity of decomposition, and for the adequate destruction of pathogens.

Windrow Construction.   The construction of a windrow for forced aeration begins with the installation of a loop of perforated pipe on the compost pad. The perforations are evenly spaced in a long row slightly off-center at the top of the pipe. The pipe diameter is 4 to 5 in (12.2 to 12.7 cm). The loop is oriented longitudinally and is centered under what is to be the ridge of the windrow. Short circuiting of air is avoided by not extending the piping the full length of the windrow. The perforated pipe is connected to a blower by way of a nonperforated pipe. After the pipe is in place, it is covered with a layer of bulking material or finished compost that extends over the area to be covered by the windrow. This base layer (bed) is intended to serve as a means of facilitating the movement and uniform distri-

FIGURE 12.7 Self-propelled windrow-straddling composting machine. A rotary drum, not seen within the machine, picks up the material, throws it over the drum, and redeposits it into a windrow ready for subsequent turning.

bution of air during composting. Additionally, the bed absorbs excess moisture and thereby minimizes seepage from the windrow. The compost feedstock is then stacked upon the piping and bed of bulking material to form a windrow which has the configuration diagrammed in Fig. 12.8. The finished windrow is of indeterminate length, about 13 ft (3.9 m) wide, and 8 ft (2.4 m) high.

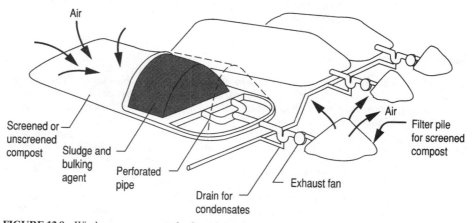

FIGURE 12.8 Windrow arrangement for forced aeration.

The completed windrow is entirely covered with a 12- to 18-in (30.5- to 47.7-cm) layer of wood chips, finished compost, or similar material. The covering absorbs objectionable odors. It also results in the occurrence of high temperatures throughout the composting mass, and in that way leads to a more complete pathogen kill as well as uniform decomposition.

Process Management.   Experience indicates that intermittent forcing of air into the windrows serves to maintain aerobiosis at an adequate level. This was confirmed by results obtained in a study that involved a 50-ft (15.2-m) windrow that contained about 73 tons (66.4 tonnes) of sludge. In the study, it was found that a timing sequence that forced air into the windrow at 16 m³/h for 5- to 10-min intervals was fully adequate. This particular rate was based upon a need of about 4 L/s/tonne of dry sludge solids. It should be noted that these numbers are only indicative. For a given situation, the required rate of air input should be determined experimentally, inasmuch as it will depend upon a number of variable factors (Epstein et al., 1976; Willson et al., 1980).

An innovation introduced during the past decade calls for tying airflow rate and timing with temperature control. The underlying rationale is to use windrow ventilation as a means of cooling the interior of the windrow (Finstein, 1992). The temperature control approach attempts to maintain optimum windrow temperatures (e.g., 130 to 140°F or 54.4 to 60°C). Because temperature directly indicates the status of the process, electronic temperature sensors, such as thermocouples or thermistors, provide a means to control airflow as well as monitor the temperature. An electronic signal from the sensor causes a control circuit to switch the blowers on or off when the windrow temperature reaches set limits. Similarly, blowers are shut off when the temperature drops below a set level. Another innovation is to use an electronic oxygen-sensing device to activate the blowers when the oxygen level drops below a predetermined level (e.g., 5 percent). From the standpoint of process management, temperature control is better aeration strategy, because it prevents the attainment of inhibitory temperature levels. However, it involves greater airflow rates, larger blowers, and more expensive and sophisticated temperature-based control systems than do timer sequenced systems.

Direction of Airflow.   The direction of the airflow through the windrow may or may not be reversed during the course of the process. A common arrangement is to initially pull air through the windrow (suction) and pass the discharged gaseous emissions through an emission conditioning filter (e.g., odor control). The filter may consist of fully composted material, organically rich soil, or other materials. The rationale is that the suction arrangement facilitates the control of gaseous emissions during the initial lag and active phases (i.e., phases during which gaseous emissions are particularly troublesome). Airflow direction is reversed during the maturing and curing phases, inasmuch as objectionable emission characteristics are likely to be at an acceptable level.

In-Vessel Systems

Currently, there are several in-vessel systems on the market. The primary objective of the design is to provide the best environmental conditions, particularly aeration, temperature, and moisture. Nearly all in-vessel systems use forced aeration in combination with stirring, tumbling, or both.

Past and current experience indicates that in-vessel composting does not guarantee a nuisance-free, specifically odor-free, operation. All recent forced closures of composting facilities were occasioned, in part, by complaints about odors. With very few exceptions, modern "in-vessel" compost systems are in reality combinations of in-vessel and windrow composting in which the vessel (reactor) is reserved for the active stage of the composting process and the windrowing is reserved for curing and maturation. Relatively high capital and operation and maintenance costs of most in-vessel composting systems, together with the long residence times required to achieve stabilization, make such a hybridization of reactor and windrow mandatory. Most of the problems leading to closure of the facilities can be traced to the discharge of the composting mass from the reactor before the completion of the active stage and the failure to make the necessary compensation in the windrowing phase.

The four basic configurations of reactor are the *vertical silo*, the *horizontal silo*, the *horizontal drum* (usually rotating), and the *horizontally oriented open tank* (rectangular or circular). The method of aerating the composting mass varies with type of reactor. Modes of aeration are forced and agitation. Agitation may be accomplished by stirring or tumbling the composting mass.

Representative Systems

1. Plug-flow vertical reactor.   Pertinent features are illustrated in Fig. 12.9. Experience with the use of the plug-flow vertical reactor has revealed difficulty in adequately aerating the contents throughout the column.

Another version of forced aeration stiffing involves the use of three completely enclosed vessels. A special feature of this version is a rotating screw device installed at the bottom of the vessel for discharging the compost. One of the three tanks serves as a storage container for carbonaceous material intended for use as a bulking agent and for correcting the C/N ratio. The composting process takes place in the second and third vessels, the *bioreactor* and the *cure reactor*. Air is fed continuously into the bottom of the bioreactor, with positive control maintained by pulling air off of the top. Composted material from the bioreactor is transferred into the cure reactor, in which further stabilization takes place. Air is fed continuously into the cure reactor to maintain aerobic conditions and to remove moisture due to evapora-

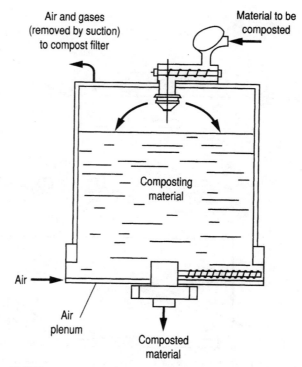

FIGURE 12.9 Schematic diagram of plug-flow vertical reactor.

tive cooling. Since the retention period in the bioreactor is 14 days, the normal daily operating sequence begins with the bioreactor outfeed discharging approximately one-fourteenth of the contents into a conveyor. The conveyor transports the material to the top of the cure reactor. At the same time, the outfeed device in the cure reactor is started and final compost product is discharged. Retention time in the cure reactor is on the order of 20 days. Problems frequently encountered in the operation of the units are: (1) a tendency of the material to "bridge" over the discharge screw, (2) failure of the rotating screw, and (3) excessive condensation of the upper layer of the bioreactor. The system originally was designed for sewage sludge and manure composting. Most experience up to now has been with sewage sludge. The extent of the experience with MSW has been limited.

2. Rotating horizontal drum.   One of the earliest in-vessel systems to utilize the tumbling mode of aeration is the rotating horizontal drum. In most versions, the major piece of equipment is a long, slightly inclined drum at least 9 ft (2.7 m) in diameter that is rotated at about 2 r/min. According to promotional literature, the retention periods in the drum may range from 1 to 6 days. However, the degree of stability acquired in such a time is not sufficient. Consequently, it is necessary to windrow the partially composted material for periods of 1 to 3 months in order to produce a properly matured (stabilized) product. The windrows should receive some aeration during the maturation period. MSW should be size reduced and sorted before it is introduced into the drum. Liquid or dewatered sewage sludge may be added to the MSW.

The relatively high capital, operational, and maintenance costs involved would seriously detract from the use, if not the economic feasibility, of a drum system for yard waste composting.

A schematic diagram indicating a design of a composting facility using a rotating horizontal drum is shown in Fig. 12.10.

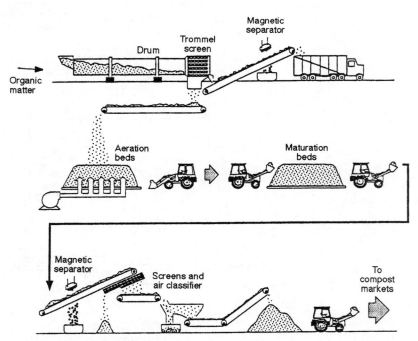

FIGURE 12.10   Schematic diagram of a composting facility using a rotating horizontal drum.

3. Open, horizontal, rectangular tank.   An in-vessel system that has much in its favor also is based upon a combination of forced aeration and tumbling like that shown in Figs. 12.5 and 12.11. It involves the use of a long, horizontal bin. In the operation of the system, properly prepared waste is placed in the bin. Tumbling is accomplished by way of a traveling endless belt, and air is forced through the perforated plates that make up the bottom of the bin and into the composting mass in the bin. The belt is passed through the composting material periodically. After a 6- to 12-day retention in the bin, the material is windrowed over a 1- to 2-month period. This system is well suited for MSW, sewage sludge, mixtures of MSW and sludge, and properly bulked high-moisture food (cannery) wastes. Times involved (in bin plus required windrowing) range from 1 to 2 months. Currently there are several variations of the bin system. The open, horizontal rectangular tank is one of the most successful in-vessel units. This is probably related to the combination of aeration and adequate detention times. Gaseous emissions from the system must, however, be properly collected and treated.

4. Vertical, mixed reactor.   A system that is based on a combination of forced aeration and stiffing involves the use of a cylindrical tank. The tank is equipped with a set of augers supported by a bridge attached to a central pivoting structure. The bridge, with its set of hollow augers, is slowly rotated. The augers are turned as the arm rotates. The hollow augers are perforated at their edges. Air is forced through the perforations and into the composting material. Retention time varies. If it is less than three weeks or so, the discharged material must be windrowed until stability is reached. A schematic diagram of the reactor is shown in Fig. 12.12. In addition, the diagram in Fig. 12.13 describes the position of a vertical mixed reactor in an overall resource recovery operation. The system is that of the Delaware Reclamation Project in which several inorganic materials, as well as RDF, are separated mechanically from mixed MSW. The process leaves a highly organic fraction, which is mixed with sludge and introduced into the reactor. As of this writing, the composting portion of the facility had been closed because of odor complaints.

5. Plug-flow, horizontal tank.   A diagram of this system is presented in Fig. 12.14. As the figure shows, the material to be composted is introduced into a rectangular tank. The material is forced into the tank by means of a hydraulic ram. After a certain detention time, the material exits the unit. As shown in Fig. 12.14, the material is aerated while in the tank. Problems that may be encountered with this reactor are due to inadequate aeration, mixing, and moisture control throughout the composting mass. In this case, the inadequacy is a result of the compaction exerted by the ram used to move the mass through the reactor.

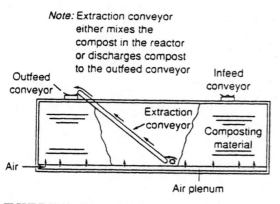

FIGURE 12.11 Schematic diagram of open, horizontal, rectangular tank.

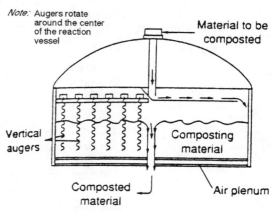

FIGURE 12.12 Schematic diagram of a vertical mixed reactor.

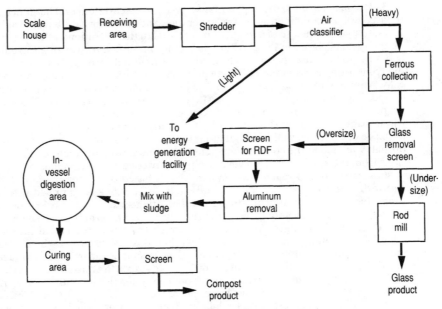

FIGURE 12.13 Schematic diagram of the Delaware Reclamation Project.

Anaerobic Processes

As stated earlier, anaerobic composting was once ranked with aerobic composting by some authorities on composting. Because of a record of environmental problems (e.g., foul odors, strong leachates), anaerobic composting fell into disfavor. However, anaerobic composting still receives attention occasionally.

It is true that the difference between high-solids anaerobic digestion and anaerobic composting is rather insignificant and is primarily a matter of objective. If the objective is recovery of energy through biogasification, the accepted classification is as anaerobic digestion or biogasification. If the objective is simply stabilization of a solid waste (e.g., refuse), the process is called anaerobic composting. However, it must be admitted that the distinction between the two objectives entails considerable differences in approach. With biogasification, process

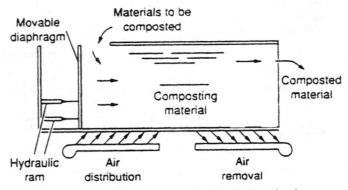

FIGURE 12.14 Schematic diagram of a plug-flow horizontal tank.

parameters are aimed at methane production; whereas in anaerobic composting, methane production is incidental and parameters center on stabilization.

In essence, the principles discussed in the preceding sections are directed to aerobic composting. To ensure anaerobiosis, the technology would involve the use of enclosed reactors. The main difference would be the absence of aeration. In fact, the presence of oxygen would inhibit the process. Finally, in order to produce an acceptable compost product, it would be necessary to follow the anaerobic phase by an aerobic phase.

Equipment and System Vendors

The number of enterprises dealing with composting has expanded substantially during the last few years. A partial listing of equipment and system vendors is given in Appendix 12.A.

12.3 ECONOMICS

Introduction

Economics is a key element in all phases of a compost undertaking. Thus, it is one of the considerations that enter into the decision to select composting as a waste management and treatment option. Economics enters into all subsequent decisions involved in the implementation of the compost option. Accordingly, it influences the selection of a particular compost system and of the associated equipment. Most important, it acts as a constraint on the continuity of the operation, in that an operation can survive only as long as its economic condition permits. This fact underlines the need to have a financial base sufficiently large to meet anticipated and unanticipated problems.

As would be true regarding the economics of all waste management alternatives, the status of the economics of composting is measured by the extent to which costs are balanced by returns. Costs include those related to the facility (capital, operation, and maintenance) and those involved in the disposal and/or marketing of the product. Returns include: (1) the extensive list of benefits associated with composting and its product, (2) avoided costs, (3) income from sale of the product, and (4) collected tipping fees.

The item of interest in making an economic analysis of composting and decisions regarding the choice of the composting option is the net cost. Net cost can be expressed arithmetically as,

$$\text{Net cost} = \text{gross cost} - (\text{income} + \text{avoided costs} + \text{other benefits}) \qquad (12.10)$$

The principal element of uncertainty in the use of the equation is the assigning of a monetary value to the benefits.

The cost of composting is discussed under three main headings, namely, general composting costs, mixed MSW composting, and yard waste composting. In keeping with its heading, the first subsection deals with composting in general (i.e., cost of the compost process as applied to all wastes). Its main emphasis is on the problems attending the interpretation and utilization of the data reported in the literature. The second and third subsections concentrate on costs peculiar to municipal solid waste composting and yard waste composting, respectively.

The economic data for composting presented in the first printing of the Handbook remains the best available data. These data were the result of several comprehensive analyses of composting in the United States. A lack of more recent comprehensive data in the literature is primarily a twofold result of lack of funding for such analyses by federal and state governments, and the maturation of the technology. The magnitude of the economic data that appear in this section remains accurate, despite the passage of time. While capital and operating costs may have increased by up to 30 percent since 1990 if only taking inflation into account, the actual increase in costs has usually been less, having been moderated by (1) increases in processing efficiency and product quality (and, therefore, increased revenues from sale of compost), and (2) substantial competitive pressure from lower-cost waste management alternatives (e.g., landfilling).

General Composting Costs

Gross Costs. Costs cited in the literature cover a wide range of values. This situation is a reflection of the bearing exerted on cost by technology, climate, topography, demography, and availability of skilled and unskilled personnel. Nevertheless, accumulated data that are presently available permit the making of reasonably realistic projections regarding the economics of existing and of proposed operations.

The range of reported costs is illustrated by data cited in two references. One reference (Cal-Recovery Systems, 1989) presents estimated capital and operation and maintenance (O&M) costs for two 400 TPD* MSW compost facilities, one of which is a windrow facility and the other, an in-vessel MSW facility. The other reference (Eastern Research Group and CalRecovery Systems, Inc., 1991) cites data on capital and O&M costs for eight compost facilities in the United States at the time the report was written. Designed capacities ranged from 10 to 800 TPD. In 1988 dollars, the capital costs reported for the eight facilities ranged from about $30,000 to $75,000 per TPD of installed capacity. Estimates for the 400 TPD facilities ranged from $49,000 to about $156,000 TPD of installed capacity. Annual O&M costs ranged from about $150,000 to $6,000,000. (CalRecovery Systems, Inc., 1989; Eastern Research Group and CalRecovery Systems, Inc., 1991). A summary of reported costs for some MSW composting facilities is presented in Table 12.3.

---

* All costs are given in TPD (U.S. tons per day). Multiply TPD by 0.907 to obtain tonnes (metric ton) per day.

TABLE 12.3 Reported Costs for Some MSW Composting Facilities

Facility	Year opened	System	Capacity, TPD	Capital cost Total, $	Capital cost $/TPD	O&M costs[a] Annual, $	O&M costs[a] $/ton	Tipping fee, $/ton	Reference
Lake of the Woods County, Minnesota	1989	TW	10	500,000	50,000	150,000	60	Zero[b]	45
Fillmore County, Minnesota	1987	A-SP	15–20	1,310,000	75,000	NA		40	46
Swift County, Minnesota	1990	A-SP	30	1,400,000	46,667	266,000	51	69	47
Portage, Wisconsin[c]	1986	Drum-TW	30	850,000[d]	28,300[d]	NA		Zero[b]	45
St. Cloud, Minnesota	1988	Drum-TW	100	NA	NA	NA		50	45
Portland, Oregon	e	Drum-A-SP	600	20,000,000	33,300	5,000,000	27	42	48
Pembroke Pines, Florida	f	A-SP	667	48,500,000	72,700	NA		NA	45
Dade County, Florida	f	TW	800	25,000,000	31,250	NA		24[g]	49

All quantities are given in U.S. tons. Multiply U.S. tons by 0.907 to obtain tonnes (metric ton).
TW = turned windrow
A-SP = aerated static pile
NA = not available
[a] Projected.
[b] Funded through taxes.
[c] Cocomposting facility.
[d] Built with used equipment.
[e] Closed as of mid-1993.
[f] Closed.
[g] Tipping fee mandated by county.

Assigning realistic costs to yard waste composting is difficult because of the paucity of reported economic information. The few published costs for operation and maintenance range from approximately as little as $4 to as much as $56 per ton of yard waste. Tipping fees for yard waste composting facilities on the West Coast are typically in the range of $15 to $30 per ton of yard waste.

Factors Underlying Uncertainty of Reported Costs.   The factors that account for the wide range of the reported costs are important contributors to the uncertainty in assessing compost economics. Therefore, they should serve as constraints upon the use of the reported information in decision making. For example, variations in size of the facility, type and nature of the technology employed, and failure to report costs in constant dollars are responsible for the wide range of reported capital and O&M costs.

An important contributor to the uncertainty is the difficulty of making accurate economic comparisons relative to operating facilities and compost technologies. The difficulty can be traced to the exceedingly broad collection of factors that not only are unique to each facility but which also determine its economics. The collection includes everything from climate, labor, and equipment, to cost accounting practices. Regarding cost accounting practice, in some cases, costs associated with a compost project are not segregated from existing solid waste operations. Cost accounting practice as applied to a yard waste compost operation may take the form of cost sharing the use and costs of items such as land, labor, and equipment with other ongoing operations. In this case, costs attributed to composting reflect estimated incremental costs. Yet another complication is the absence of consistent and precise definitions of operations and maintenance costs. The many obstacles to arriving at an accurate prediction of costs can be countered by defining proposed project requirements and performing detailed cost analyses.

Personnel and Training.   The extent of the personnel and training expenditure depends upon the size of the staff. Staffing requirements for composting operations vary with facility size and with the relative allocation of capital equipment vs. labor. The number of personnel range from part-time employment for small, seasonal leaf composting operations to approximately 30 full-time employees for large MSW compost operations. Labor requirements for manual removal of recyclables and inerts vary roughly in proportion to plant throughput. Mechanical separation reduces the need for sorters.

Personnel training requirements and costs usually do not vary markedly as facility size increases, although more people are involved with the larger facilities. However, the range of skills required is similar to that of some smaller facilities.

Impact of Technology on Costs.   In this section, the emphasis is on the impact exerted on costs by the type of technology employed, by system residence time, by equipment redundancy, and by plant utilization. The section closes with a few words on the impact of feedstock characteristics.

In windrow composting, the composting mass often is sheltered in a structure during the active and early curing stages. This practice is prevalent in regions characterized by seasonal changes, especially in winter, and in regions subjected to heavy rainfall.

Most in-vessel systems reserve the reactor for the active stage of the compost process and rely upon windrowing for the curing and maturation stages. The rationale is twofold, namely, to maintain conditions at optimum levels during the active stage, thereby accelerating the rate of microbial activity and correspondingly shortening the active phase. The economic gain in shortening the active stage is the reduction of residence time in the reactor and, hence, the increase of its processing capacity (i.e., reduction in size of the reactor).

The approach to hastening the active stage has been to increase the sophistication of the reactor. However, acceleration made possible by sophistication of the reactor ultimately is limited by the genetic makeup of the microbial population. Obviously, the more expensive the reactor, the greater the required degree of process acceleration. This requirement often is underestimated, which in turn leads to disastrous consequences. The trench type of reactor is a good compromise with respect to both economy and processing efficiency.

It is difficult to make a comparison between the impact of windrow technology and in-vessel technology. At a minimum, it is assumed in making such a comparison that the products of the two are competitive with each other. All other process variables being the same, particularly the active stage, the capital cost of structures and equipment for the active phase of mixed MSW composting would be higher for the various in-vessel systems than for a conventional windrow system. Theoretically, however, this disadvantage is balanced by lower operating costs. In-vessel proponents claim that their systems have the potential for lowering operating costs by virtue of increased automation. Many in-vessel process configurations also require less land area per unit of throughput than is required for windrow composting. An important point is that regardless of the sophistication of the system, prudence demands that a generous buffer zone be provided.

The present consensus is that because of the factors just mentioned, high labor and land costs tend to favor the selection of labor- and land-efficient in-vessel systems. Conversely, low labor and land costs tend to favor the selection of windrow composting.

Feedstock.   Feedstock differences can also have a significant effect on the cost of composting MSW, in particular if undesirable materials are removed by source separation. Virtually any MSW feedstock rich in paper and yard waste would require some shredding in order to produce a marketable compost in a reasonable amount of time. However, if glass, metal, and plastics were to be removed from the waste stream via source separation, some of the cost of preprocessing could be avoided, by eliminating the labor and equipment required for processing the recyclables. Eliminating glass, plastic, and metal from the compost plant's feedstock could reduce the unit cost of composting by 10 to 20 percent. The effect is relatively small, due to the elimination of the offsetting revenue from recyclables. The level of plant utilization also has a significant influence on the unit cost of composting. Throughput per unit of capital cost increases markedly as facility utilization moves from 40 h per week to near-continuous operation. In anticipation of growth in the waste stream, because of noise and traffic requirements, or for other reasons, some facilities are designed for 8 h per day, 5 days per week. A related plant utilization issue is the optimum level of equipment redundancy. In general, a process facility functions at or near its lowest unit cost when the process operates continuously, except for interruptions for necessary repairs and maintenance.

Income.   Among the available sources of income are tipping fees, taxation, state grants, sale of collected recyclables, and sale of compost product. In some respects, taxation and state grants could be interpreted as monetary expressions of benefits.

The collection of recyclables is a part of preprocessing in most MSW compost facilities. The range of reported prices received for MSW compost is from "no charge" (i.e., zero) to $5/yd^3 ($6.6/m^3) at the site. Unsold compost can be put to a variety of beneficial uses such as weed abatement, improvement of marginal land, or landfill cover.

Avoided Costs.   A commonly used measure of costs avoided through the use of composting is the cost of disposing of the waste by landfilling. On that basis, the avoided costs are the equivalent of landfill tipping fees, plus collection and hauling costs. Obviously, the degree of the equivalence is the extent to which the tipping fees in force reflect the true costs of the landfill, including amortization of future costs for the development of MSW landfills.

Inasmuch as composting MSW inevitably leaves some residue that must be disposed of by landfilling, the cost of landfilling this residue must be taken into consideration.

Mixed MSW Composting

In this section, the subject matter becomes more specific by way of an analysis of the cost of composting mixed MSW.

In addition to technology and size, a number of ancillary factors determine system costs. The nature of the ancillary factors and degree of their influence ultimately depend upon type of the selected system and its technology, and upon size. Among the ancillary factors specific to the selected system are site preparation costs, land costs, climate, dust and odor-level stipulations, restrictions on waste receiving and operating hours, and the amount of process redundancy included. The diversity of these factors and their variability make a simple estimate of costs extremely difficult.

The authors of an EPA manual (Eastern Research Group and CalRecovery, Inc., 1991) approached the cost projection problem by considering two facilities that rely upon windrow systems but are widely divergent as to size in that one is a 100 ton/d (91 tonne/d) facility, whereas the other is a 1000 ton/d (909 tonne/d) facility. The smaller facility involves a system in which forced aeration is the method of aeration. Turning through the use of mechanical turners is the method of aeration in the larger facility. Both have preprocessing systems. In both cases, the active stage takes place inside a building. Odor control is more expensive in the smaller facility because the degree of control is greater owing to location of the site. Moreover, in both cases, it is assumed that all compost can be sold at a net revenue of $2.50/ton ($2.27/tonne) of compost. It should be noted that revenue from sales is minor in comparison with the cost of production.

The EPA manual (Eastern Research Group and CalRecovery, Inc., 1991) presents detailed cost analyses of the two systems by way of three tables. Table 12.4 is a summary of the data presented in these three tables. As the data in Table 12.4 show, the net daily unit cost with the smaller facility is $63/ton ($57.3/tonne), and for the larger facility it is $48/ton ($43.6/tonne). Additional information is presented in Appendix 12.B.

Yard Waste Composting

Introduction.   Basing estimates of the costs of a planned yard waste facility on those of existing operations or those reported in the literature is an exceedingly difficult task. The reasons are similar to those encountered with MSW undertakings, namely, the wide range of cited costs that reflect variations in facility size, design, and ancillary conditions. Thus, collectively, existing yard waste compost facilities include very simple facilities, some fairly complex and fully equipped facilities, and a full gamut of intermediate facilities.

The nature and breadth of the gap between a fully equipped facility and a simple, minimally equipped facility can be indicated by calling attention to their similarities and differences regarding key construction and equipment features. Typically, the entire site of a fully equipped facility is graded, and the receiving, processing, and about a fourth of the composting areas are paved. A typical, fully equipped facility has a shredder, a mechanical turner, a front-end loader, screens, and necessary conveyors. "Minimal" refers to the lowest level at which environmental and public health requirements can be met. The minimal construction requirements are a graded site, and paved and fenced-in processing and active compost stage areas. The minimal equipment requirements are a shredder and a turning device (e.g., front-

TABLE 12.4   Summary of Unit Costs for Two Facilities (1990 $)

Cost item (TPD of capacity)	100 TPD (91 MT/D)[†] facility	1000 TPD (909 MT/D) facility
Capital ($)	67,300	55,300
O&M* ($/ton, $/Mton)	45 (41)	35 (31.8)
Net unit cost ($/ton, $/Mton)	63 (57.3)	48 (43.6)

* Based on 312 operating days per year.
[†] U.S. ton × 0.907 = 1 tonne.

end loader). A shredder provides a very essential function, namely, size reduction of woody yard waste (brush, branches, etc.). Its final product is inferior in quality to that from the fully equipped facility, because the product has not been screened.

Capital Costs.   An excellent detailed analysis of the likely investment requirements for two widely divergent hypothetical yard waste compost facilities is presented by the Eastern Research Group (ERG) and CalRecovery (1991). One of the two facilities is highly capital-intensive and has the characteristics of the fully equipped facility. The second facility is much less, perhaps even minimally, capital-intensive. It has the characteristics of the "minimally" equipped facility previously described. The daily throughput capacity of each of the two facilities is 70,000 yd^3 (53,200 m^3) of input waste, from which each produces 10,000 yd^3 (7600 m^3) of compost. Each facility has a total area requirement of 12 acres (4.9 ha).

A summary of the estimated initial investment costs of the two facilities, as listed in the reference are presented in Table 12.5. (The reference treats land cost as an operational expense rather than as a capital cost item, because it assumes that the land is leased.) Although relative, the dollar values listed in the table are indicative of the magnitude of the monetary outlay involved in implementing a decision to establish a yard waste compost facility, and a measure of its economic feasibility. Additional information is given in Appendix 12.B.

Site-Specific Factors.   The extrapolation of many reported costs from one facility to another is constrained by the site specificity of the costs. The cost of bringing utilities to a facility (i.e., utility hookup) is an example. The hookup cost is largely a function of the location of the nearest service and the distance between it and the facility. Construction costs serve as another example, inasmuch as they are dependent upon conditions unique to the site. Thus, differences between sites with respect to soil conditions can introduce a variability factor of 200 percent for cost of grading and paving.

The equipment costs shown in the example presented previously are typical costs for adequately sized machines. Actual equipment costs can be determined after particular pieces have been chosen in the design stage of the operation.

TABLE 12.5   Estimated Costs of Yard Waste (1990$)*

Item	Fully equipped	Minimally equipped
Initial investment costs		
Construction	392,500	116,750
Engineering	84,700	48,600
Utility hookup	40,000	40,000
Equipment	405,000	260,000
Total investment costs	922,200	465,350
Net annual costs		
Annual costs		
Amortized investment	150,100	75,730
Annual O&M	244,450	256,000
Total annual cost	394,550	331,730
Annual revenues		
Sale of compost	90,000	70,000
Net annual cost	304,550	261,730
Net unit costs ($ per ton yard waste)		
Unit cost	32	28

* Throughput 70,000 yd^3 per day (53,200 m^3 per day); composted product, 10,000 yd^3 per day (7600 m^3 per day). Area of each, 12 acres (4.9 ha).

Operation and Maintenance Expenses.   Among the many items to be considered in arriving at an estimate of the O&M cost are insurance, fuel, labor, lease on land, trailer rental, water, and power. Judging from the analysis reported by the Eastern Research Group (ERG) and CalRecovery (1991) and summarized in Table 12.5, the estimated annual O&M cost of the fully equipped hypothetical facility is on the order of $244,450; and of the minimally equipped facility, $256,000. The higher costs of the minimally equipped facility are due to higher estimated fuel and labor costs. However, the facility's maintenance, insurance, and power costs are somewhat lower.

Income and Avoided Cost.   The estimated income listed in Table 12.5 is based upon a 60 percent product yield and the existence of a market for the entire yield at a net return equal to $9/yd^3 ($11.8/m^3) of product from the fully equipped facility and $7/yd^3 ($9.2/m^3) from the minimally equipped facility.

Public versus Private Ownership

Much can be said for and against public ownership and operation. A compromise that is fairly common is public ownership and private operation. Three considerations rank high in the decision regarding ownership and operation. They are *project control, risk allocation,* and *project costs.*

Project Control.   A municipality can exercise a large measure of control over a privately operated large facility by way of a series of contracts negotiated prior to the actual construction of the facility. A relatively small privately-operated facility can be controlled through a contract which specifies areas subject to municipal control. However, implementation of major operational changes according to a timetable desired by the municipality is more easily done if the facility is municipally rather than privately operated.

Risk Assumption and Allocation.   Major risks that can adversely affect the cost of operations are increases in costs associated with system reliability, management, labor productivity, inflation, landfill, and insurance. For a privately-operated facility, the allocation of these risks is a subject of contract negotiations. It is up to the municipality to develop contracts to reduce risks that can best be borne by the private sector.

Project Costs.   Public or private operation should matter little with respect to project costs unless a municipality has surplus labor and/or equipment that could be effectively utilized through implementation of the project. The issue could become important if labor, fuel, utilities, or material costs are substantially different or can be more easily met by one of the two.

The need for the private owner or operator to show a profit can tip project cost in favor of public ownership or operation. Determination of which type of operation will be most cost-effective must be made on a case-by-case basis.

12.4 MARKETING PRINCIPLES AND METHODS

The principles involved in marketing compost are basically the same as those in marketing any commodity—whether it be a feedstuff, chemical fertilizer, or compost. Application of these principles takes the form of the following sequence of steps: (1) determine the possible uses of the product and its application; (2) identify potential users (market analysis); (3) make the potential user aware of the product characteristics and its utility, as well as the benefits from using the product; (4) persuade the potential customer to procure and use the product; and (5) establish a satisfactory distribution program. In connection with step 3, the producer and/or vendor should be able to assure the customer that the product will be unfailingly available and that its specifications remain constant. Each of these steps is discussed separately in this section.

In this section, the terms "compost product" and "compost" are used synonymously. Unless otherwise indicated, the primary emphasis is on MSW compost, although the information is readily applicable to other types of compost.

Uses of the Compost Product and Its Application

The characteristics and utility of compost were discussed in detail earlier in this chapter. In summary, the primary use of compost is as a soil amendment. A distant secondary use is as a source of fertilizer elements, especially of nitrogen, phosphorus, and potassium (NPK). Applications include landfill cover, as well as the entire gamut of agricultural activities.

Market Analysis

The objective of a market analysis is to arrive at an estimate of the full size of the potential market. It is best begun with a survey designed to obtain the information on which a realistic estimate of the full size of the market potential can be made. Preferably, the survey is conducted by way of interviews or questionnaires, or by a combination of the two, to identify the needs of prospective customers and to establish the dimensions of their potential demands. An important fact to keep in mind when designing or conducting an interview is that successful selling presupposes knowledge of the customer's values and motivations.

Parties to be interviewed are representative members of major agricultural sectors, of government agencies, of the public, of the landscapers, of soil vendors and distributors, of nurseries, and of any organization that may have a use for the product. If the feasibility of conducting personal interviews becomes a limiting factor, they can be supplemented by mailed questionnaires. Questions posed in the interviews and questionnaires should be relevant to the targeted market.

Acquainting the Customer

"Acquainting the customer" is used in the sense of ways and means of imparting a knowledge of the utility of compost to all sectors of the actual and potential market. Imparting this knowledge involves the identification and description of compost characteristics and benefits associated with its use. It includes explaining and illustrating ways of obtaining the benefits and utility. In short, the objective is to "educate the market."

The need for education is emphasized by the fact that, particularly with MSW, marketing compost is seriously encumbered by inertia and bias. The encumbrance is largely due to potential users being unaware of the true worth of compost. Obviously, the best means of removing or minimizing the obstacle is to instill in potential users an awareness of the real worth of MSW compost. This can be done through a program of education and salesmanship. The task is made easier by the fact that the product does indeed have value and genuine utility.

Public education can be accomplished through presentations in the media. The presentations may deal with the advantages and disadvantages of compost utilization; with methods of producing compost; and with information on obtaining compost and on how compost can and should be used. The presentations should be backed by carefully orchestrated demonstrations.

An important aspect of the process is the dispelling of troubling doubts. The removal of doubts should be accompanied by explanations regarding the best utilization of compost for particular applications.

The line of demarcation between education and advertising is not clear-cut. Probably, education becomes advertising when a product of a particular producer is promoted. Given the current situation, the logical course would be to precede the advertising phase by a carefully planned and conducted program of educating the largest group of prospective customers, namely, farmers from the field crop, row crop, and orchard sectors.

The participation of governmental and educational professionals who specialize in advising and guiding farmers on a local level is very helpful. Farmers tend to take the advice of such specialists seriously because of their close association with them. Moreover, the specialists have a better understanding of the problems that beset farmers locally. The specialists could furnish practical advice and assistance in designing, implementing, and publicizing demonstrations. The demonstrations could range in magnitude from small plots involving two or three types of plants to a large undertaking comparable with one conducted in Johnson City, Tennessee [see U.S. EPA (1975)].

Product Sales and Salesmanship

Having established a receptive climate through education and demonstration, the next step is to narrow the focus to particular potential users (i.e., to advance from sectors to the individuals in the sectors).

Importance.   Disposal of the waste depends upon persuading potential users to acquire and use the product. Failure to accomplish this step leaves the compost producer with the responsibility of satisfactorily disposing of the product. Failure might even reduce the value of composting as a feasible disposal option. In short, failure to find appropriate uses would defeat one of the principal objectives of MSW composting.

The position of composted MSW in the hierarchy of uses is immaterial, provided that the waste generator and compost producer are divested of the disposal responsibility. This assumes that the compost is used in an environmentally sound manner, from use as landfill cover to use as a soil amendment in food crop production.

Compost marketing specialists stress the importance of analyzing the needs and the potential of each sector, including delivery requirements, storage capabilities, and pricing policies (Snyser, 1982). It is necessary for the seller to understand the needs and requirements of the targeted market.

Methods.   The first step in persuading a customer is to emphasize the benefits that the user will gain from using the compost product in preference to a competitive product. This can be done by calling attention to the benefits described in the section on uses of compost. Some benefits that are deemed especially important by potential compost users are once again mentioned at this time. It should be pointed out that the cash value of a crop strongly influences the purchaser's decision of whether or not to buy and use compost. If the cash value of a crop is high, a potential user is more likely to purchase and use compost. The willingness to purchase declines with decline in dollar value of the increase in crop yield resulting from compost use.

Particularly important to most potential users are the compost properties that facilitate and improve crop production through remedying soil deficiencies and improving soil characteristics. For example, because it is predominantly organic and is an excellent medium for soil bacteria, compost improves the tilth of soil, thereby enhancing the soil's productivity. Moreover, tilth is a key consideration in soil management. Another benefit that should be stressed is that compost incorporated into the soil lowers fertilizer expenditures by lessening nutrient loss through leaching. For example, as much as 30 percent of applied chemical nitrogen may be dissolved and leached to the groundwater. In addition to minimizing nutrient loss through leaching, compost also increases the ability of plants to utilize nutrients efficiently. Increase in efficiency results in higher yields, and likelihood of greater financial return. One property of compost will be of particular interest to customers in arid regions or in any region where water is in limited supply either seasonally or year-round. The property is the high water-retention capacity of compost that is second only to that of peat. (Peat is more expensive than compost, and its plant and soil bacteria nutrient content is much less than that of compost.)

Particularly attractive, not only to agriculturalists but also to soil conservationists, is the reduction in loss of topsoil through wind and water erosion that compost use achieves. Loss of topsoil reduces crop yield and increases cost of soil management because the lower strata in the soil profile are more difficult to till.

In terms of percentage of the soil amendment market, steer manure offers the greatest competition at present. However, composted and noncomposted steer and other animal manures are handicapped by having a higher sodium (Na) and chloride (Cl) content than do MSW and yard waste composts. The higher Na and Cl contents are due to the presence of urine in mammalian manures. (Sewage sludge is an exception because, being of human origin, it also includes urine.)

Advertising. The means of communicating the message presented in the preceding paragraphs is through conventional advertising practices. The efficacy of conventional advertising has been amply demonstrated. With regard to compost, the task of advertising is considerably lightened by the public educational campaign that should both precede and accompany it.

Not to be ignored is word-of-mouth advertising. This form of advertising can be initiated and facilitated by carefully planned demonstrations, of which the "first user" is the most venerable. Moreover, the approach not only is venerable, it is effective. The demonstration involves persuading a representative potential user to try the product. Persuasion is strongly facilitated by providing the compost either free or at a very low price. It is imperative that the participant be carefully guided and supervised during the demonstration. A successful demonstration will attract favorable attention on the part of neighbors and onlookers. The demonstration may be expanded to include several neighborhoods and participants.

A key requisite for a successful advertising campaign is the ability to convincingly assure targeted customers that (1) the compost will be unfailing available; (2) there will be no large deviations in quality other than improvement; (3) no unwelcome deviations in product characteristics and specifications will take place; and (4) the price always will be "right."

Market Continuity. Four factors also are applicable to market continuity: product quality, availability, constant specifications, and pricing.

Product Quality. Most users of compost base their evaluation of compost quality as a soil amendment and organic fertilizer on certain characteristics of the product. Among the characteristics of importance are NPK (nitrogen, phosphorus, potassium), moisture content, extent of contamination, odors, and particle size. The concentration of NPK should be high enough to justify the application of compost. The higher the NPK, the less is the amount of product at which the crop's NPK requirements can be met. Moisture content is a factor because of its influence on ease of handling and on cost of transport. Handling is more difficult at high moisture contents. Furthermore, water adds to the weight of the product and increases the cost of transport. Pathogens and toxic and nontoxic contaminants adversely affect product quality and act as constraints on use of the product. Foul odors, even in very low intensity, adversely affect quality perception. Particle size relates to visual quality, ease of handling, and applicability.

Grading is generally recommended as a means of coping with the many variations in products with respect to visual and nutritive quality. Grading assures the most effective utilization of the product. A facility may produce only one type of compost. On the other hand, the output may be separated into different products on the basis of quality. Effective use is ensured by matching type of application with appropriate quality of compost. For example, a relatively low grade of compost would be adequate for the reclamation of excavations and denuded forests. On the other hand, a high grade of compost is required in row crop production or use by homeowners.

As previously stated, aside from the grading demanded by the market, serious efforts to establish a formal system of grading did not occur until the late 1980s. However, guidelines have been and continue to be proposed by federal and state agencies. Guidelines for grading generally are based on constraints on application of the product. A sample of grades proposed by the authors for the State of Washington is presented in Table 12.6. (CalRecovery Inc., 1990). In this grading system, there are no constraints on the use of grade 1 composts. They can be used on food chain and row crops and for all other uses. Grade 2 composts cannot be used in food chain or row crop production but can be used in orchards, viticulture, landscaping, etc. The third and lowest grade can be used only for all other applications. Assignment to a grade usually is based on toxic substance concentration (e.g., Cd, Pb, PCBs), number of viable pathogens, weed seed concentration, contaminant content (e.g., plastic, glass), plant nutrient concentration, degree of maturity, and major physical characteristics (e.g., particle size distribution, moisture content).

Maintaining product consistency is an overriding requirement for market continuity. The customer demands product consistency because efficient utilization, particularly in crop production, depends upon the use of a soil amendment of known composition and physical characteristics. Variation in consistency lessens the usefulness of the product, resulting in a loss of customer confidence and interest. Therefore, it is extremely important that the compost meet a fixed set of specifications.

Unfailingly Available. As far as market continuity is concerned, availability implies that production must not only be sufficiently large to permit adequate introduction of the product into the market, but it also must be consistently available in the future. The market could not long survive sporadic availability.

Price. Unlike conventional marketing, the pricing policy for MSW compost, and to a lesser degree for yard waste compost, is not to make a monetary profit from the operation but rather to defray as much of the cost as is possible. The main objective for composting is usually treatment and disposal. If the selling price is too high, potential users will turn to less expensive competing products. Compost must compete with other organic products for a share of the organic fertilizer market. If the spread between the price of compost and a competing product is too wide, the user buys the less expensive product despite being aware of the benefits to be gained from use of compost. Inasmuch as the primary reason for composting MSW and yard waste is waste disposal, unsold compost must be disposed of either by landfilling or by incineration.

An upper limit on the selling price for compost also is determined by the potential user's ability to pay. If the greater share of the compost market is the agricultural sector, the limit

TABLE 12.6 Marketability Standards

	Unit	Grade A	Grade B
Bulk density	lb/yd^3 (kg/m^3)	600–800 (356–475)	400–1000 (238–594)
CEC*	meq/100 g	>100	>100
Foreign matter	Maximum %	2	5
Moisture content	%	40–60	30–70
Odor		Earthy	Minimal
Organic matter	Minimum %	50	40
pH		5.5–6.5	5–8
Size distribution	Nominal, in	<1/2	<7/8
Water-holding capacity	Minimum %	150	100
C/N ratio	Maximum	15	20
Nitrogen	Minimum %	1	0.5
Conductivity (soluble salts)	mmhos/cm	<2	<3
Seed germination	Minimum %	95	90
Viable weed seeds		None	None

* CEC = cationic exchange capacity, expressed in milliequivalents (meq) exchangeable cations per 100 g of dry soil.
Source: From CalRecovery, Inc. (1990).

would be relatively low because the profit margin characteristic of agricultural enterprises is small. The average farmer can afford to make a relatively small expenditure for fertilizer, chemical or organic. Compost produced specifically for landscaping and cultivation of ornamentals should be priced according to the buyer's ability and willingness to pay, as well as by the price of competitive products.

At the time of this writing, the prices for composts produced from MSW, yard waste, and sludge range from zero (i.e., is given away free) to about $5/yd^3 ($6.6/m^3).

Public vs. Private Marketing.   Generally, public vs. private marketing of the compost is a question only when the public entity (community, district, etc.) has sole ownership of the MSW, yard waste, or sewage sludge compost. Moreover, the question applies only at the wholesale (bulk sales) level; because the consensus is that, with few exceptions, private enterprise is better qualified at the retail level. Therefore, the discussion that follows is not concerned with retail selling.

If a facility is privately owned and operated, it is to be expected that the entrepreneur owns the product and, hence, is responsible for marketing or disposing of it—unless the contract with the community states differently. If a community owns a facility, but by way of a contract has the facility operated by a private party, ownership of the product is specified in the contract. It follows that if a public entity owns and operates a facility, the entity owns the product and is responsible for marketing or otherwise disposing of it. The entity has two choices: (1) it can sell the entire compost output to a single entrepreneur, who thereupon is responsible for the disposition of the product; or (2) the entity can do the marketing.

If the entity opts to do the marketing, its success will depend upon meeting certain requirements. First and foremost, the entity must be prepared to unreservedly do everything that is needed. Selling requires the full-time input of highly qualified, knowledgeable, and dedicated professional staff. Such a staff can give the task the necessary effort and attention. The difficulty is that most entities either cannot afford such a staff or are unwilling to make the necessary expenditure. Short of these specifications, the selling will be less than adequate.

Distribution

The significance of distribution is readily apparent because it is the link between the production facility and users. Methods range from free transport in bulk form to end users, to bagging and distribution through existing channels established for other soil amendments. Despite its significance and a long record of compost production and use, progress in the distribution of compost has been very limited. Distribution channels are still not well-defined.

The rate established for shipping secondary (recovered) materials is a major element in the economy of a resource recovery operation. In general, regulated freight rates are based on cost and value of service. In turn, several factors have an impact on the rate structure established by freight carriers (motor, railroad, ship, barge). These factors are too numerous and varied to be adequately discussed in this chapter. However, of particular importance is the fact that the combination of relatively low monetary value of MSW compost and its low bulk density exacerbates the cost of long-distance transport, and consequently sharply limits the distance at which haul is economically feasible.

Motor Freight.   Motor freight is the primary mode of transport of compost in the United States. Bulk transport typically occurs using large open-top trailers. Material is loaded into the trailers using a conveyor system or a front-end loader. Freight rates vary locally. However, the curve plotted in Fig. 12.15 (CalRecovery Systems, Inc., 1989) illustrates a typical relation between motor freight rates and distance of the haul, assuming a full load in the range of 20 tons.

Railroad Freight.   Rail haul of compost is generally not practiced because for rail freight to be cost-effective, very large quantities must be transported, and rail access must exist. The uti-

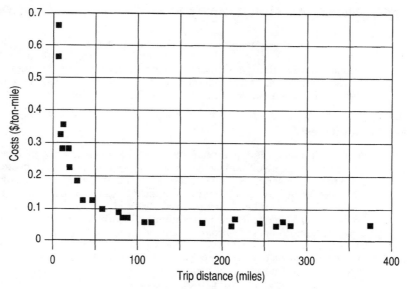

FIGURE 12.15   Intrastate motor carrier rates for all bagged composts.

lization of rail transport requires special loading and unloading systems, and access of the composting facility to the rail line.

Ship Transport.   Rates are highly variable and are dependent upon the individual shipping lines. In the near term, transporting compost by ship likely will be negligible in most locations. The exception might be transport between coastal ports, or between ports in the Great Lakes states.

Barge Transport.   Compost transport via barge could be practical in certain regions of the United States, particularly those that have access to the Ohio and Mississippi Rivers and a few canals designed for barge traffic.

Cost estimates made by the authors of this chapter on the basis of information gained in an informal survey (CalRecovery Systems, Inc., 1989) can serve as guidelines for estimating barging costs. In general, barging rates apparently are independent of the type of commodity being transported. On the other hand, weight of the load being transported and the crew time involved are primary factors.

The following considerations upon which the authors based their estimates can serve as a model for estimating barging service costs per ton of compost (dollar values are for 1989). Barges (i.e., barge train and tug) move at about 5.6 nautical miles/h. The barges are "bareboat" charter at $2300 per day. The party who charters the barge is responsible for insurance and maintenance costs, which usually amount to about $0.25 to $0.35/ton ($0.23 to $0.32/tonne). Based on 1989 labor and fuel costs, the daily rate for a 1000-hp tug for the operation would be about $7200, assuming a single rate for a 30-day operation is arrived at by combining straight time and weekend overtime. The estimated costs for one tug and four barges moving 60,000 tons per month (54,500 tonnes per month) of material is about $8.10/ton ($7.4/tonne). One tug plus three barges (15 trips per month) could haul only 24,000 tons per month (21,800 tonnes per month), in which case the estimated cost would be $13/ton or $11.8/tonne.

12.5 *ENVIRONMENTAL, PUBLIC, AND INDUSTRIAL HEALTH CONSIDERATIONS*

Major potential negative impacts of a compost operation would be lowering the quality of water and air resources, and compromising the public health and well-being. It should be emphasized that these are potential negative impacts, and that they become *actual* impacts *only* when an inadequate technology is used, a normally adequate technology is improperly applied, or preventive or corrective measures are not taken.

Water and Air Resources

Water Resource.   The quality of the water resource can only be adversely affected through contamination with either leachate from raw, composting, or composted refuse or with runoff from the compost operation. Leachate is formed only when the moisture content of the material is higher than the optimum for composting. Aside from maintaining the moisture content of the material at or below the optimum level, chances of uncontrolled increases in moisture content from rain or snow can be minimized by protecting the material from the elements. As a precautionary measure, provision should be made to keep leachate from reaching ground and/or surface water resources by conducting all phases of the operation on an impermeable surface. The surface should be equipped to collect all leachate for treatment or discharge into a public sewer. Runoff can be avoided by selecting a site where it would not be likely to occur. If this is not possible, runoff can be prevented from entering the operation site by constructing ditches to divert the runoff around the site. Runoff from the site can be intercepted and channeled to a treatment facility (e.g., conventional stabilization lagoon). It is important that leachate does not reach a body of water, since leachate from raw waste is similar to raw sewage sludge in terms of pollutant concentration. Although the compost process sharply reduces the pollutant concentration, leachate from a properly matured compost mass would still reduce the quality of ground and surface water (Diaz, 1977; Cooper et al., 1974).

Air Resource.   Biological and nonbiological agents from a compost operation most likely would enter the environment by way of dust particles and aerosols generated during the various stages of the operation and subsequently discharged into the air. Some of the microbes transported in this manner could pose a health hazard to a susceptible individual who by chance ingested the dust particle or aerosol. Aerosols, in particular, are vehicles for a wide variety of microorganisms. These microorganisms may occur as single entities, as clumps of organisms, or by adhering to dust particles. Two types of infections may be acquired from such contaminants, namely, (1) those that are limited to the respiratory tract, and (2) those that may affect another part of the body. Both are taken in by way of the respiratory tract. The existence of a hazard from the spores of *Aspergillus fumigatus* is yet to be demonstrated. The infectivity of the spores is low. Consequently, any danger posed by it would be significant to only an unusually susceptible individual. Nevertheless, prudence indicates that an open-air compost plant should not be sited in close proximity to human habitation (NSF, 1978; Golueke, 1982).

Dust suppression at all stages of a compost operation can be accomplished through the use of conventional dust control measures. Some of these measures include the use of mist sprayers in the working area, and the installation of air collection and particulate control devices such as cyclones and fabric filters.

Odors.   Although the generation of objectionable odors lowers the quality of the air resource in terms of human well-being, it does not become a health hazard until the odors become particularly foul. Some odors are inescapable (e.g., those from raw wastes). Odor control during preprocessing can be accomplished by enclosing all of the operations in a building; conditioning the feed; and treating exhaust gases through absorption, adsorption, or oxidation methods. Foul odors are generated during the composting stages, principally through improper management of the composting process (e.g., failure to maintain aerobic conditions). The use of an in-vessel system does not always ensure odor-free operation.

In the absence of proper management, all materials can become sources of foul odors until they are adequately matured. However, conventional techniques are available for treating foul odors. Control and containment are effective approaches to preventing the development of odor problems or contending with those that may escape prevention. Means of control involve trapping the odors through ventilation or containment of the composting process. Exhaust air can be treated by passing it through a chemical scrubbing system or by way of biofiltration. Difficulties with biofiltration are its large filter areas and its relatively sophisticated management requirements (Miller et al., 1988; Hay et al., 1988).

Vectors

Fly and rodent attraction is almost inevitable because of the nature of organic residues and the long time interval between reception of the raw material and the stage of the compost process in which conditions become lethal to flies and intolerable to rodents.

Most likely, flies and rodents would not constitute a serious problem in a yard waste compost operation. However, food wastes or sewage sludge cocomposted with the yard wastes could serve as strong attractants for flies and rodents. Although a fly and rodent problem could be almost completely eliminated by enclosing the entire facility, it can be considerably eliminated by using certain measures. For example, an important mitigating measure would be careful "housekeeping" throughout all stages of the operation. Storage of raw wastes should be as brief as possible. Preprocessing, particularly size reduction, wreaks a substantial destruction on fly eggs and fly larvae, and lowers the value of the refuse as a feedstuff for rodents. Migration of fly larvae that survive the preprocessing and the early compost stages can be prevented by the use of a paved surface.

Industrial Health and Safety

Detailed studies reporting the results of environmental monitoring of MSW composting facilities are very limited. The same is true for the more common materials recovery facilities, which process some or all of the materials that would serve as feedstock for an MSW composting facility, and which use some of the same types of equipment.

The greatest potential for accidents in a composting facility is in the preprocessing stage. Chances of injuries from accidents are greatest in this stage because of the extensive exposure of the workers to machinery. Standard measures for minimizing such hazards are well developed and easily available.

The greatest hazard to the health of the workers comes from the dust particles that are suspended in the air in a compost plant. Hazards associated with dust are greatest in the preprocessing phase and become much less in the subsequent stages of the compost process. In addition to the biological burden generated by the dust particles and aerosols, there is a fibrous fraction of dust that may have a health significance. Biological agents that have been identified in waste processing facilities (of which a composting plant is a subcategory) include fecal coliforms, streptococci, *Aspergillus fumigatus,* and certain cell wall components of bacteria and fungi, namely endotoxin and glucan. Based on past reported results and more recent research (Gladding, 1998), airborne concentrations of these agents are generally greater in mixed waste processing facilities than in those facilities processing source-separated wastes. Among a variety of types of waste processing facilities, measured concentrations of microorganisms generally have been reported in the range of 10^3 to 10^7 colony-forming units (cfu). The range reflects the types of wastes processed, operating conditions, and methods of control. The types of airborne agents and the magnitude of their airborne concentrations are sufficient to potentially represent a risk to susceptible workers, depending on conditions. Health problems associated with airborne dust can be substantially controlled by the use of particulate control systems, face masks, and protective clothing, and the installation of adequate sanitation facilities.

The highest levels of noise that would occur in a preprocessing plant would be from about 95 dBA to about 105 dBA (slow response). These levels can be generated by a shredder or a front-end loader. With the present state of the processing technology, some type of ear protection is needed for exposures longer than about 2 h.

Fires

There are primarily two causes of fires at composting facilities: (1) sparks produced from welding or flame-cutting equipment used to maintain preprocessing equipment; and (2) spontaneous combustion of stored, processed organic materials. Spontaneous combustion is the self-ignition of organic material. Proper welding and flame-cutting practices minimize the ignition of organic materials with hot slag. The variables associated with spontaneous combustion are numerous and difficult to control without proper management and monitoring.

To control instances of spontaneous combustion, piles of stored, processed organic materials must be constructed based on the duration of storage and on the characteristics of the stored material. The characteristics include particle size, composition, and moisture content. Piles of organic material with moisture contents in the range of 20 to 45 percent are susceptible to spontaneous combustion (Rynk, 2000). The susceptibility increases with the volume of the pile, since temperature buildup to combustion levels is a function of: (1) the volume of the material; and (2) the potential cooling effect from heat loss and water loss (via evaporation), which together are governed by the surface area of the pile. Consequently, control of occasions of spontaneous combustion entails proper sizing and maintenance of storage piles.

Proper fire protection controls include: (1) equipment such as fire extinguishers and fire hoses, (2) supply of water and sand or dirt, and (3) regular and frequent monitoring of the storage piles.

Constraints on Use of the Compost

Constraints on the use of the compost with respect to the health and safety of humans arise from the harmful substances that may be in the compost. Examples of such harmful substances are heavy metals, toxic organic compounds (including PCBs), glass shards, and pathogenic organisms. The sources of harmful substances obviously are the wastes used as feedstock for the process. Concentrations of harmful substances usually are lower in the organic fraction of municipal solid wastes than in sewage sludge, as shown by the data in Table 12.7. When considering the data, it should be kept in mind that the concentrations vary widely from operation to operation [e.g., cadmium ranges from 0 to 1100 µg/g dry sludge (Sharma, 1980)], because of a

TABLE 12.7 Composition and Characteristics of Sludge and of the Light Fraction of Air Classified Refuse

Item	Units	Air classified light fraction		Sludge cake	Refuse-sludge mixture
		As received	As analyzed		
Carbon (C) (total organic)	%	15.9	16.8	15.6	15.8–18.0
Nitrogen (N)	mg/kg dry	7080	7500	41,000	11,000–13,000
(total Kjeldahl)	%	0.7	0.75	4.1	1.1–1.3
Zinc (Zn)	mg/kg dry	226.6	240.0	2000	680–840
Cadmium (Cd)	mg/kg dry	1.0	1.1	93.0	3.7–21.0
Lead (Pb)	mg/kg dry	29.3	31.0	1000	47–110
Nickel (Ni)	mg/kg dry	6.1	6.5	150.0	10–35
Copper (Cu)	mg/kg dry	22.7	24.0	8900	

Source: From Golueke et al. (1980).

variety of site-specific differences (e.g., mostly residential vs. mostly industrial generators). Reliable information on concentrations of toxic substances and pathogens in composted yard debris is extremely scarce. The few data that are available indicate that the concentrations of heavy metals in compost from yard wastes are relatively low (<0.1 ppm to about 10 ppm for Hg, Cd, Cu, and Ni; and from 50 ppm to about 200 ppm for Pb and Zn). Similarly, concentrations of pesticides, PCBs, and pathogens are quite low (CalRecovery Systems, Inc., 1988; Miller et al., 1992).

Average concentrations of pesticides in composts produced from yard wastes in Illinois were found to be between 1 and about 10 ppm Carbaryl, Atrazine, and 2,4,5-T. All other pesticides analyzed were found at levels between about 0.01 and 1 ppm (Miller et al., 1992). Therefore, compost products, especially those from MSW and sewage sludge, should be routinely analyzed as a precautionary measure.

The harmful effect on humans and animals may be exerted directly by eating food crops grown on soil that has been amended with compost. The effect could be exerted indirectly through the consumption of meat and other products involving animals fed on such food crops. The effects are due to persistence of the inorganic contaminants and survival of certain pathogens through the food chain.

Cadmium can be used to demonstrate how a heavy metal passes through the food chain. A certain fraction of the cadmium in a compost incorporated into soil is assimilated by plants grown on that soil. The amount of cadmium assimilated by the plant depends upon a number of factors, such as the availability of the metal, plant species, and the particular part of the plant. Availability depends upon the concentration of the metal in the soil, the pH of the soil, concentration of organic matter in the soil, ion exchange capacity of the soil, and several other factors. Generally, availability decreases as the soil pH changes from acidic to alkaline. As a rule, leafy vegetables assimilate more than cereal crops. In cereal crops, concentration is greater in the root and leafy portions than in the grain. If those plants are eaten by humans, a fraction of the cadmium in the plants is assimilated in the tissues of the persons who eat the plants. If the plants are consumed by animals, some of the cadmium is assimilated by the animals and remains in their meat and in products (e.g., eggs) produced by them. This cadmium awaits assimilation by humans who consume the meat and the products. The distribution of cadmium in the soil, plant, and animal as it passes through the food chain, and the contribution of sewage sludge are described in detail in Refs. Sharma (1980), Chaney (1982), and Mennear (1978). The incidence of other metals and chemicals in the food chain is summarized in Golueke (1982).

Restraints due to the presence of pathogens in compost range from negligible to substantial, depending upon the waste composted and the conditions under which it was composted. Such constraints can be eliminated by rendering the product free of pathogens through pasteurization. Pasteurization can be accomplished by way of composting or through the application of an external source of heat. Except through contamination by contact (e.g., adhering compost particles), direct transfer of pathogenic organisms between members of the food chain even without disinfection is either nonexistent or very minor.

Types and concentrations of pathogens that might be in the product prior to pasteurization depend upon the feedstock. Yard wastes are not likely to contain human pathogens because human body wastes are not involved. However, they may contain organisms pathogenic to pets or plants. Sewage sludge, on the other hand, has a wide range of human pathogens (Golueke, 1982). MSW may contain some human pathogens because of contamination by body wastes. Because the indicators are of pet rather than human origin (Diaz et al., 1977; Cooper et al., 1974), the extent, if any, of such contamination cannot be measured by concentration of "indicator organisms" (the concentration rivals that in sewage sludge). Improperly composted food wastes could contain zoonotic organisms (trichina, ascaris, taenia) by way of meat scraps. In summary, composted yard waste is not likely to contain human pathogens, whereas inadequately composted sewage sludge and food waste could.

Health constraints are receiving legal backing in the form of "Classifications" proposed or actually promulgated by the U.S. EPA and various state regulatory bodies, although generally yard waste composts have been and are being less tightly regulated. Tables 12.8 and 12.9 are examples of such classifications based on heavy metal content. It should be noted that these tables are only examples because at present classification development is rapidly changing.

TABLE 12.8   Classification for the State of Minnesota (Maximum Allowable Limits)

Metal	Class I, ppm dry wt.	Class II
Cadmium	10	
Chromium	1000	
Copper	500	All composts that do not meet class I
Lead	500	standards are placed in class II.
Mercury	5	Therefore, there are only two classes
Nickel	100	
Zinc	1000	
PCB	1	

Source: From CalRecovery Systems, Inc. (1988).

Among the legal constraints other than those that are health-oriented is an important one pertaining to labeling. It prohibits labeling a compost product as a "fertilizer" when the product's NPK (nitrogen, phosphorus, potassium) concentration is less than a total of 6 percent (the required total may vary from one state to another). Permitted are labels of "soil amendment," "soil conditioner," or simply "compost." The NPK of a compost product depends on the NPK of the wastes from which the compost is produced. Because of the wide variations between products in terms of nutrient content, it would be misleading to list particular concentrations as being "typical." Other conventional designations are named and described in Verdonck et al. (1987) and Anon. (1988). Convincing arguments and a plea for the setting of labeling requirements and regulations on potting soil are given in Pittenger (1986).

TABLE 12.9   Regulations for the States of New York and Massachusetts (Maximum Allowable Limits)

State	Class I food-chain crops	Class II non-food-chain crops
New York (sludge, MSW compost):*		
Mercury	10	10
Cadmium	10	25
Nickel	200	200
Lead	250	1000
Chromium	1000	1000
Copper	1000	1000
Zinc	2500	2500
PCB	1	10
Particle size	<10 mm	<25 mm
Massachusetts (sludge, MSW, yard waste compost):†		
Cadmium	2	25
Mercury	10	10
Molybdenum	10	10
Nickel	200	200
Lead	300	1000
Boron	300	300
Chromium	1000	1000
Copper	1000	1000
Zinc	2500	2500
PCB	2	10

* *Source:*   From New York State (1988).
† *Source:*   From Trubiano et al. (1987).

12.6 CASE STUDY

The City of San Jose, California (population 909,000), employs a residential yard waste processing program as a key component of its solid waste management system. Approximately 130,000 tons of source-separated yard wastes were collected in 1999 from single- and multi-family dwellings and from municipal facilities, such as parks.

The yard waste is taken to two privately-operated sites for processing and marketing. The two processing systems are designed to produce a variety of products. About 70 percent of the products is in the form of compost. The compost is marketed in bulk form to primarily landscapers and agricultural users. The remaining types of products are landscaping mulch and boiler fuel. Both operators utilize size reduction, aerobic composting, and screening to produce compost from the yard waste.

Size reduction is used to prepare the yard waste for windrow composting. Mobile, mechanical turners are used to mix and aerate the waste. After biological stabilization, the waste is mechanically screened to recover the compost product as the undersize fraction. The oversize fraction, primarily wood and chips, is returned to the composting process or marketed as mulch or boiler fuel.

The processing cost to the City of San Jose is about $25/ton.

12.7 CONCLUSIONS

The application of composting to the management of municipal solid wastes (MSW) in the United States has undergone extraordinary changes during the last 45 years.

Changes since 1988 are reflected by the data presented in Table 12.10. The data clearly show the substantial increase in the number of compost facilities in the United States—particularly those used for treating yard waste. In many instances, the increase has been a direct response to regulatory constraints (e.g., bans on the disposal of yard wastes in landfills). In other cases, growth has been primarily due to a desire to apply an appropriate, environmentally benign technology to the recycling of organic wastes.

The expansion of composting practice was not entirely advantageous, however. The problem was that the rapidity and magnitude of the expansion were such that the waste management industry could not adequately meet the substantial demand. The compost bonanza attracted a wide diversity of industries, equipment and system vendors, financial institutions, and a sizable number of companies and individuals to promote and develop compost programs. This development brought to light another problem, namely, the absence of a matching "infrastructure" to satisfy program demands. Here, we use "infrastructure" in the sense of collection of human resources, equipment, markets, material specifications, guidelines, and other factors. The large demand coupled with the inadequacy of the infrastructure has led over the years to several costly and painful errors.

Among the major causes of failed programs are (1) a tendency to oversimplify the compost process, (2) underestimation of the complexity of large-scale compost facilities, and (3) insufficient understanding of mechanical and biological processes.

TABLE 12.10 Change in the Number of Composting Facilities in the United States (1988–1998)

Year	MSW	Yard waste
1988	5	650
1993	16	1500
1998	20	3800

A fourth cause that often is cited is the apparent frequency of incidents of generation of malodors. The desire to minimize malodor generation has been made the reason for the modification of some facilities and for the closure of others. When evaluating this problem, it should be remembered that there is no such a thing as an "odorless" waste treatment facility. Of itself, the delivery of feedstock entails the generation of odors that differ from customary background odors. Odors due to handling and delivering the feedstock will be present even though the entire compost process is conducted properly. Furthermore, malodors are symptoms of a variety of problems, most of which have been identified in this chapter. Among these problems are (1) an inordinately long storage time (on the order of 1 to 2 months) of the raw feedstock (e.g., yard wastes); (2) inadequate mechanical processing (i.e., insufficient size reduction); (3) unrealistically short detention times during the composting process (a few days rather that weeks); (4) abbreviated maturation time; and (5) shortage of dedicated land area. Any one of these situations can result in the generation of malodors.

The majority of these problems can be prevented and even completely avoided by complying with the basic principles of feedstock preparation and composting outlined in this chapter. Furthermore, it is emphasized that prudence and experience dictate that the entire process of system development (i.e., identification of need, procurement, system selection, and monitoring) must not be delegated to individuals who are novices in the waste management business. The seriousness of this admonition has been learned the hard way by several hapless entities.

Currently, the future of composting in waste management appears to be favorable, especially in the United States. Yard waste composting is practiced in many locations in the United States. Initially, in the late 1980s and early to mid-1990s, the proliferation of yard waste composting programs and the need for large processing capacity generally caused technical, economic, and environmental problems (i.e., there were "growing pains"). Over the past five to eight years, many of the earlier problems have been resolved or mitigated. Perhaps the largest remaining issue is odor and odor control from yard waste composting facilities. While odors can be controlled to a certain degree, they cannot be eliminated, as pointed out earlier in this chapter.

As of this writing, MSW composting has fallen out of favor in the United States due to past substantial failures and to the successful performance of yard waste composting programs.

The composting of food waste is being shown some interest in the United States. Primarily, the reason is that food waste composes a sufficient percentage of the remaining solid waste stream in some locations and, therefore, represents one of the few remaining areas to utilize for diversion of waste from landfills or incineration. Currently, organized food waste composting appears better suited as a waste management method for commercial generators than for the residential sector, due to the greater quantities of food waste produced per generator type, such as food service businesses. The exception, of course, is backyard or home composting applied to the residential sector, which has been promoted and implemented by municipalities to various degrees around the United States.

REFERENCES

Anon. (August, 1988) "Mulch: 1988 Garden Hero," *Sunset,* pp. 56–59.

BioCycle, The JG Press, Inc., Emmaus, PA.

Biocycle (1988) "Agripost News Construction of Compost Facility," *BioCycle,* vol. 29, no. 10, pp. 13–14.

CalRecovery, Inc. (1988) *Swift County Solid Waste Compost/Recycling Project,* prepared for Swift County Auditor, Benson, MN, March 1988, updated November 1988.

CalRecovery, Inc. (1990) *Compost Classification/Quality Standards for the State of Washington,* prepared for the State of Washington, Department of Ecology.

CalRecovery Systems, Inc. (1988) *Portland Area Compost Products Market Study,* prepared for The Metropolitan Service District, Portland, OR.

CalRecovery Systems, Inc. (1989) *Composting Technologies, Costs, Programs, and Markets,* prepared for the Congress of the United States, Office of Technology Assessment, Washington, DC.

Chaney, R. L. (1982) "The Establishment of Guidelines and Monitoring System for Disposal of Sewage Sludge to Land," *Proceedings, International Symposium on Land Application of Sewage Sludge,* Tokyo.

Chrometska, P. (August, 1968) "Determination of the Oxygen Requirements of Maturing Composts," *Information Bulletin 33,* International Research Group on Refuse Disposal, Union of International Associations (UIA), Brussels.

Cooper, R. C., S. A. Klein, C. J. Leong, J. L. Potter, and C. G. Golueke (1974) *Effect of Disposable Diapers on the Composition of Leachate from a Landfill,* Final Report, SERL Report 74-3, Sanitary Engineering Research Laboratory, University of California, Berkeley.

Craig, N. (1988) "Solid Waste Composting Underway in Minnesota," *BioCycle,* vol. 29, no. 7, pp. 30–33.

Diaz, L. F., et al. (1977) *Public Health Aspects of Composting Combined Refuse and Sludge and of the Leachates Therefrom,* University of California, Berkeley, CA.

Diaz, L. F., G. M. Savage, and C. G. Golueke (1982) "Final Processing," Chap. 11 in *Resource Recovery from Municipal Solid Wastes,* CRC Press, Boca Raton, FL.

Diaz, L. F., G. M. Savage, L. L. Eggerth, and C. G. Golueke (1993) *Composting and Recycling Municipal Solid Waste,* Lewis Publishers, Inc., Ann Arbor, MI.

Eastern Research Group (ERG) and CalRecovery, Inc. (1991) *Municipal Solid Waste Composting Programs: A Guidebook for Officials,* prepared for the U.S. Environmental Protection Agency, Cincinnati, OH.

Epstein, E., G. B. Wilson, W. D. Burge, D. C. Mullen, and N. K. Enkiri (1976) "A Forced Aeration System for Composting Wastewater Sludge," *Journal of the Water Pollution Control Federation,* vol. 38, no. 4, p. 688.

Finstein, M. (1992) "Composting in the Context of Municipal Solid Waste Management," Chap. 14 in *Environmental Microbiology,* pp. 355–374, Wiley-Liss, Inc., New York.

Gladding, T. (1988) "Investigating Health and Safety Hazards at MRFs," *Resource Recycling,* vol. 17, no. 10, pp. 32–37.

Goldstein, N. (1989) "Solid Waste Composting in the U.S.," *BioCycle,* vol. 30, no. 11, pp. 32–37.

Golueke, C. G., and P. H. McGauhey (1955) "Reclamation of Municipal Refuse by Composting," *Technical Bulletin 9,* Sanitary Engineering Research Laboratory, University of California, Berkeley.

Golueke, C. G. (1972) *Composting: A Study of the Process and Its Principles,* Rodale Press, Inc., Emmaus, PA.

Golueke, C. G., D. Lafrenz, B. Chaser, and L. F. Diaz (1980) *Benefits and Problems of Refuse Sludge Composting,* prepared for the National Science Foundation (PFR-791707), Washington, DC.

Golueke, C. G. (October, 1982) "Epidemiological Aspects of Sewage Sludge Handling and Management," *Proceedings, International Symposium on Land Application of Sewage Sludge.*

Gorham, D., and R. Zier (1988) "Portland Moves to Solid Waste Composting," *BioCycle,* vol. 29, no. 7, pp. 34–37.

Hay, J., J. H. Alin, S. Chang, R. Caballero, and H. Kellogg (1988) "Alternative Bulking Agent for Sludge Composting, Part II," *BioCycle,* vol. 29, no. 10, pp. 46–49.

Lossin, R. D. (1970) "Compost Studies," *Compost Science,* vol. 11, no. 6.

Lossin, R. D. (1971a) "Compost Studies, Part II," *Compost Science,* vol. 12, no. 1, pp. 12–13.

Lossin, R. D. (1971b) "Compost Studies. Part III. Measurements of Chemical Oxygen Demand of Compost," *Compost Science,* vol. 12, no. 2, pp. 31–32.

Mennear, J. H. (ed.) (1978) *Cadmium Toxicity,* 224 pp., Marcel Dekker, New York.

Miller, F. C., and B. J. Macauley (1988) "Odors Arising from Mushroom Composting: A Review," *Australian Journal of Experimental Agriculture,* vol. 28, no. 19, pp. 553–560.

Miller, T. L., R. R. Swager, S. G. Wood, and A. D. Atkins (December, 1992) *Sampling Compost Sites for Metals and Pesticides, Resource Recycling.*

Moller, F. (1968) "Oxidation-Reduction Potential and Hygienic State of Compost from Urban Refuse," *Information Bulletin 32,* International Research Group on Refuse Disposal, Union of International Associations (UIA), Brussels.

NCA (1964) "Composting Fruit and Vegetable Refuse. Part II. Investigation of Composting as a Means for Disposing of Fruit Waste Solids," Progress Report, National Canners Association Research Foundation, Washington, DC.

New York State (1988) *Solid Waste Management Facilities,* Revised 6 NYCRR, Part 360, New York State, Department of Environmental Conservation.

Niese, G. (1963) "Experiments to Determine the Degree of Decomposition of Refuse by Itself Heating Capability," *Information Bulletin 17,* International Research Group on Refuse Disposal, Union of International Associations (UIA), Brussels.

NSF (1978) "Aspergillus-Implications on the Health and Legal Implications of Sewage Sludge Composting," *Workshop Proceedings,* vol. 1, prepared by Energy Resources, Inc., for Division Problems-Focused Research Applications, National Science Foundation, Washington, DC.

Obrist, W. (1965) "Enzymatic Activity and Degradation of Matter in Refuse Digestion: Suggested New Method for Microbiological Determination of Degree of Digestion," *Information Bulletin 24,* International Research Group on Refuse Disposal.

Pittenger, "Potting Soil Label Information Is Inadequate," *California Agriculture,* vol. 40, pp. 6–8.

Public Works (1992) "Odor Control at Honolulu," *Water Environment Research,* March–April, 1992, in "Environmental Waste Control Digest," *Public Works,* vol. 132, no. 9, pp. 132–133.

Regan, R. W., and J. S. Jeris (1970) "A Review of the Decomposition of Cellulose and Refuse," *Compost Science,* vol. 11, no. 17.

Rolle, G., and E. Organic (1964) "A New Method for Determining Decomposable and Resistant Organic Matter in Refuse and Refuse Compost," *Information Bulletin 21,* International Research Group on Refuse Disposal.

Rynk, R. (ed.) (1992) *On-Farm Composting Handbook,* Northeastern Regional Agricultural Service, 152 Riley-Robb Hall, Cooperative Extension, Ithaca, NY.

Rynk, R. (2000) "Fires at Composting Facilities: Causes and Conditions, Part I," *BioCycle,* vol. 41, no. 1, pp. 54–58.

Schulze, K. F. (1960) "Rate of Oxygen Consumption and Respiratory Rate Quotients during the Aerobic Composting of Synthetic Garbage," *Compost Science,* vol. 1, no. 36.

Schulze, K. F. (1964) "Relationship between Moisture Content and Activity of Finished Compost," *Compost Science,* vol. 2, no. 32.

Senn, C. L. (1974) "Role of Composting in Waste Utilization," *Compost Science,* vol. 15, no. 4, pp. 24–28.

Sharma, R. P. (1980) "Plant-Animal Distribution of Cadmium in the Environment," Chap. 14 in Nriagu, J. O. (ed.), *Cadmium in the Environment. Part 1, Ecological Cycling,* Wiley, New York, 682 pp.

Snyser, S. (1982) "Taking the Sludge to Market," *BioCycle,* vol. 23, no. 1, pp. 21–24.

SRI International (1992) *Data Summary of Municipal Solid Waste Management Alternatives*, Stanford Research Institute International, Palo Alto, CA.

Trubiano, R. P., M. Thayer, S. J. Kruger, and F. Senske (1987) "Boston Project Tests Methods and Markets," *BioCycle,* vol. 28, no. 7, p. 25.

U.S. EPA (1975) *Composting at Johnson City: Final Report on Joint USEPA-TVA Composting Project, Vols. I, II,* EPA/530/SW-3 I s.2, U.S. Environmental Protection Agency, Cincinnati, OH.

Verdonck, O., M. DeBrodt, and R. Gabriel (1987) "Compost as a Growing Medium for Horticultural Plants," in *Compost: Production, Quality, and Use,* pp. 399–414, Elsevier Applied Science, New York.

Wiley, J. S. (1957) "Progress Report on High-Rate Composting Studies," *Engineering Bulletin, Proceedings of the 12th Industrial Waste Conference,* Series 94.

Willson, G. B., J. F. Parr, E. Epstein, P. B. Marsh, R. L. Chaney, D. Colacicco, W. D. Burge, L. J. Sikora, C. F. Tester, and S. Hornick (1980) *Manual for Composting Sewage Sludge by the Beltsville Aeration Pile Method,* EPA 600/S-80-022, Office of Research and Development, U.S. EPA, Cincinnati, OH.

Wylie, J. S. (1957) "Progress Report on High-Rate Composting Studies," *Engineering Bulletin, Proceedings of the 12th Industrial Waste Conference,* Series 94, pp. 590–595.

APPENDIX12A: PARTIAL LISTING OF VENDORS OF EQUIPMENT AND SYSTEMS FOR COMPOSTING MSW AND OTHER ORGANIC WASTES

Biosolids, MSW, and Yard Waste Systems

Ag-Bag International, Inc.
2320 S.E. Ag-Bag Lane
Warrenton, OR 97146
503-861-1644
800-334-7432
503-861-2527 (fax)
compost@agbag.com
www.agbag.com

Becker Underwood, Inc.
801 Dayton Avenue
Ames, IA 50010
515-232-5907
800-232-5907
515-232-5961 (fax)
request@bucolor.com
www.bucolor.com

Bio Gro Systems
1110 Benfield Boulevard, Suite B
Millersville, MD 21108
410-729-1440
410-729-0854 (fax)
kmihm@wm.com
www.bio-gro.com

CBI Walker, Inc.
1501 N. Division Street
Plainfield, IL 60544
815-439-4000
815-439-4010 (fax)

Double T Equipment, Ltd.
P.O. Box 3637
Airdrie, AB T4B 2B8
CANADA
403-948-5618
800-661-9195
403-948-4780
solutions@double-t.com

Earthgro, Inc.
Route 207
Lebanon, CT 06249
860-642-7591
800-736-7645
860-642-7912 (fax)
www.scottsco.com

The Fairfield Engineering Company
240 Boone Avenue
P.O. Box 354
Marion, OH 43302
740-387-3327
740-387-4869 (fax)
sales@fairfieldengineering.com
www.fairfieldengineering.com

Farmer Automatic
P.O. Box 39
Register, GA 30452
912-681-2763
912-681-1096 (fax)
farmer@g-net.net

Green Mountain Technologies
P.O. Box 560
Whitingham, VT 05361
802-368-7291
800-610-7291
802-368-7313 (fax)
gmt@sover.net
www.gmt-organic.com

GSI Environment, Inc.
855 Pepin Street
Scherbrooke, PQ J1L 2P8
CANADA
819-829-2818
819-829-2717 (fax)
sherbroo@serrener.ca

Hallco Manufacturing Company, Inc.
6605 Ammunition Road
P.O. Box 505
Tillamook, OR 97141
503-842-8886
800-542-5526
503-842-8499 (fax)
info@hallco-mfg.com
www.hallco-mfg.com

Keith Manufacturing Company
401 N.W. Adler
P.O. Box 1
Madras, OR 97741
541-475-3802
800-547-6161
541-475-2169 (fax)
sales@keithwalkingfloor.com
www.keithwalkingfloor.com

Kolberg-Pioneer, Inc.
700 W. 21st Street
P.O. Box 20
Yankton, SD 57078
605-665-8771
800-542-9311
605-665-8858 (fax)
mail@kolbergpioneer.com
www.kolbergpioneer.com

LH Resource Management
RR1
Walton, ON N0K 1Z0
CANADA
519-887-9378
519-887-9011 (fax)

Longwood Manufacturing Corporation
816 E. Baltimore Park
Kennett Square, PA 19348
610-444-4200
610-444-9552 (fax)

NaturTech Composting Systems, Inc.
44 28th Avenue N, Suite J
St. Cloud, MN 56303
320-253-6255
320-253-4976 (fax)
naturtech@composter.com
www.composter.com

Rexius Forest Bay Products
750 Chambers Street
Eugene, OR 97402
541-342-1835
541-343-4802 (fax)

SGE Environment
1 Cours Ferdinand de Lesseps
Rueil-Malmaison 92851
FRANCE
33-147164764
33-147164074 (fax)
sge-env@calva.net

U.S. Filter
2650 Tallevast Road
Sarasota, FL 34243
941-355-2971
800-345-3982
941-351-4756 (fax)

U.S. Filter/CPC IPS Composting
  Systems
441 Main Street
P.O. Box 36
Sturbridge, MA 01566
508-347-7344
508-347-7049 (fax)
www.usfilter.com

WPF Corporation
P.O. Box 381
Bellevue, OH 44811
419-483-7752
419-483-6150 (fax)

Chippers

American Recycling Equipment
3141 Bordentown Avenue
Parlin, NJ 08859
732-525-1104
732-525-0488 (fax)

Arctic, Inc.
465-507 Wilson Avenue
Newark, NJ 07105
973-589-1670
888-589-2888
973-589-4236 (fax)

Bandit Industries, Inc.
6750 Millbrook Road
Remus, MI 49340
517-561-2270
800-952-0178
517-561-2273 (fax)
brushbandit@eclipsetel.com
www.banditchippers.com

Karl Kuemmerling, Inc.
129 Edgewater Avenue NW
Massillon, OH 44646
330-477-3457
330-477-8528 (fax)

Montgomery Industries International
P.O. Box 3687
Jacksonville, FL 32206
904-355-5671
904-355-0401

Morbark Sales Corporation
8507 S. Winn Road
P.O. Box 1000
Winn, MI 48896
517-866-2381
800-233-6065
517-866-2280 (fax)
morbark@worldnet.att.net
www.morbark.com

Nelmore Company, Inc.
44 Rivulet Street
North Uxbridge, MA 01538
508-278-5584
508-278-6801 (fax)
www.nell.com

Nordfab Systems
P.O. Box 429
Thomasville, NC 27361
336-889-5599
800-222-4436
336-884-0017 (fax)
tomballus@vortexgrinder.com

Packer Industries, Inc.
5800 Riverview Road
Mableton, GA 30126
404-505-0522
800-818-2899
404-505-1450 (fax)
packerind@aol.com
www.packer2000.com

Peterson Pacific Corporation
29408 Airport Road
Eugene, OR 97402
541-689-6520
541-689-0804 (fax)
sales@petersonpacific.com
www.petersonpacific.com

Rayco Manufacturing Company
4255 Lincoln Way E.
Wooster, OH 44691
330-264-8699
800-392-2686
330-264-3697 (fax)
www.raycomfg.com

Rotochopper
N591 County Road P1
Coon Valley, WI 54623
608-452-3651
608-452-3031 (fax)
rotochpr@mwt.net
www.rotochopper.com

Royer Industries, Inc.
341 King Street
Myerstown, PA 17067
717-866-2357
717-866-4710 (fax)
royer@royerind.com
www.royerind.com

Simplicity Engineering/Gruendler Crusher
212 S. Oak Street
Durand, MI 48429
517-288-3121
800-248-3821
517-288-4113 (fax)
www.simplicityengineering.com

Strong Manufacturing Company
498 Eight Mile Road
Remus, MI 49340
517-561-2280
517-561-2530 (fax)

Universal Refiner Corporation
217 W. Pioneer
P.O. Box 151
Montesano, WA 98563
360-249-4415
800-277-8068
360-249-4773 (fax)

VC Marketing, Inc.
119 Dorsa Avenue
Livingston, NJ 07039
973-992-8514
973-992-4219 (fax)

Vermeer Manufacturing Company
2411 Vermeer Road
P.O. Box 200
Pella, IA 50219
515-628-3141
888-VERMEER
515-621-7734 (fax)
www.vermeer.com

Wendt Corporation
2080 Military Road
Tonawanda, NY 14150
716-873-2211
800-936-WENDTCO
716-873-9309 (fax)
sales@wendtcorp.com
www.wendtcorp.com

West Salem Machinery Company
665 Murlark Avenue N.W.
P.O. Box 5288
Salem, OR 97304
503-364-2213
800-722-3530
503-364-1398 (fax)
sales@westsalem.com
www.westsalem.com

Williams Patent Crusher & Pulverizer
Company, Inc.
2701 N. Broadway
St. Louis, MO 63102
314-621-3348
314-436-2639 (fax)
info@williamscrusher.com
www.williamscrusher.com

Wood/Chuck Chipper Corporation
P.O. Drawer 400
Shelby, NC 28150
704-482-4356
800-269-5188
704-482-7349 (fax)

Dewatering Systems

Ashbrook Corporation
11600 E. Hardy
Houston, TX 77093
281-449-0322
800-362-9041
281-449-1324 (fax)

Atlantic Screen & Manufacturing, Inc.
118 Broadkill Road
Milton, DE 19968
302-684-3197
302-684-0643 (fax)
atlantic@ce.net

Atlas-Stord, Inc.
309 S. Regional Road
Greensboro, NC 27409
336-668-7728
336-668-0537 (fax)
atlas-stord@atlas-stord.com
www.atlas-stord.com

Bio Gro Systems
(see "Biosolids, MSW, and Yard Waste
Systems")

Brown Bear Corporation
602 Avenue of Industry
P.O. Box 29
Coming, IA 50841
515-322-4220
515-322-3527 (fax)
brnbear@mddc.com
www.brownbearcorp.com

Carrier Vibrating Equipment, Inc.
P.O. Box 37070
Louisville, KY 40233
502-969-3171
502-969-3172 (fax)
cve@carriervibrating.com
www.carriervibrating.com

HUWS Corporation
RR 1
Palgrave, ON L0N 1P0
CANADA
519-942-1008
519-942-1060 (fax)
huws@headwaters.com

Hydropress Environmental Services, Inc.
59 Dwight Street
Hatfield, MA 01038
413-247-9656
800-292-2325
413-247-9401 (fax)

K-F Environmental Technologies, Inc.
210 W. Parkway, Unit 5
P.O. Box 277
Pompton Plains, NJ 07444
973-616-0700
973-616-9504 (fax)

Komline-Sanderson Engineering Corporation
12 Holland Avenue
Peapack, NJ 07977
908-234-1008
800-225-5457
908-234-9487 (fax)
www.wateronline.com/storefronts/
komline.html

Marathon Equipment Company
Highway 9 S.
P.O. Box 1798
Vernon, AL 35592
205-695-9105
800-269-7237
205-695-8813 (fax)
sales@marathon-equipment.com
www.marathon-equipment.com

Mobile Dredging & Pumping Company
3100 Bethel Road
Chester, PA 19013
610-497-9500
800-635-9689
610-497-9708 (fax)

Oliver Manufacturing Company, Inc.
P.O. Box 512
Rocky Ford, CO 81067
719-254-7814
888-254-7813
719-254-6371 (fax)
oliver@ria.net

Rayfo, Inc.
15629 Clayton Avenue
Rosemount, MN 55068
612-437-4441
612-437-2272 (fax)

Roediger Pittsburgh, Inc.
3812 Route 8
Allison Park, PA 15101
412-487-6010
412-487-6005 (fax)
www.roediger.com

Sebright Products
127 N. Water Street
Hopkins, MI 49328
616-793-7183
800-253-0532
616-793-4022 (fax)

Somat Company
555 Fox Chase
Coatesville, PA 19320
610-384-7000
800-23-SOMAT
610-380-8500 (fax)
www.somatcorp.com

SP Industries, Inc.
2982 Jefferson Road
Hopkins, MI 49328
616-793-3232
800-592-5959
616-793-7451 (fax)
sburk@sp-industries.com
www.sp-industries.com

Magnetic Separation

Countec Recycling Systems, Inc.
1901 NW 92nd Court, Suite B
Clive, IA 50325
515-457-3131
515-457-3137 (fax)

Dings Company Magnetic Group
4740 W. Electric Avenue
Milwaukee, WI 53219
414-672-7830
414-672-5354 (fax)

Douglas Manufacturing Company, Inc.
300 Industrial Park Drive
Pell City, AL 35125
205-884-1200
205-884-1207 (fax)
sales@douglasmanufacturing.com
www.douglasmanufacturing.com

Getz Recycle, Inc.
2800 W. Lincoln Street
P.O. Box 6249
Phoenix, AZ 85005
602-278-7600
888-234-7660
602-272-5668 (fax)
rgetz@uswest.net
www.getzrecycle.com

Global Equipment Marketing, Inc.
P.O. Box 810483
Boca Raton, FL 33481
561-750-8662
561-750-9507 (fax)
info@globalmagnetics.com
www.globalmagnetics.com

Hamos USA
P.O. Box 2089
Skyland, NC 28776
828-684-4910
828-654-0957 (fax)
hamosusa@mindspring.com

Industrial Magnetics, Inc.
1240 M-75 S.
P.O. Box 80
Boyne City, MI 49712
616-582-3100
800-662-4638
616-582-2704 (fax)
imi@magnetics.com
www.magnetics.com

Magnetic Products, Inc.
683 Town Center Drive
P.O. Box 529
Highland, MI 48357
248-887-5600
800-544-5930
248-887-6100 (fax)
info@mpimagnet.com
www.mpimagnet.com

Newell Industries, Inc.
P.O. Box 10629
San Antonio, TX 78210
210-227-9090
800-933-9090
210-227-7038 (fax)

Ohio Magnetics, Inc.
5400 Dunham Road
Maple Heights, OH 44137
216-662-8484
800-486-8446
216-662-2911

REM
6512 N. Napa Street
Spokane, WA 99217
509-487-6966
800-745-4736
509-483-5259 (fax)

Steinert, Inc.
14 Summerwinds Drive
Lakewood, NJ 08701
732-920-6838
732-920-6922 (fax)
steinertmg@aol.com

Odor Control Systems

A-Odormaster
292 Alpha Drive
RIDC Industrial Park
Pittsburgh, PA 15238
412-252-7000
800-556-0111
412-252-1005 (fax)
stopodors@surcopt.com
www.surcopt.com

AAF International
215 Central Avenue
P.O. Box 35690
Louisville, KY 40232
502-637-0011
888-AAF-2003
502-637-0321 (fax)
www.aafintl.com

Adwest Technologies, Inc.
151 Trapping Brook Road
Wellsville, NY 14895
716-593-1405
716-593-6614 (fax)
adwestny@eznet.net
www.adwestusa.com

Aireactor, Inc.
(see "Turning and Mixing Equipment")

Air-Scent International
290 Alpha Drive
RIDC Industrial Park
Pittsburgh, PA 15238
412-252-2000
800-247-0770
800-351-7701 (fax)
www.airscent.com

Ecolo Odor Control Systems, Inc.
1222 Fewster Drive, Unit 9
Mississauga, ON L4W 1A1
CANADA
905-625-8664
800-NO-SMELL
905-625-8892 (fax)
info@ecolo.com
www.ecolo.com

Epoleon Corporation
19160 S. Van Ness Avenue
Torrance, CA 90501
310-782-0190
800-376-5366
310-782-0191 (fax)
info@epoleon.com
www.epoleon.com

The Fogmaster Corporation
1051 S.W. 30th Avenue
Deerfield Beach, FL 33442
954-481-9975
954-480-8563 (fax)
info@fogmaster.com
www.fogmaster.com

Howe-Baker Engineers
3102 E. 5th Street
P.O. Box 956
Tyler, TX 75710
903-597-0311
903-581-6178 (fax)

Kuma Corporation
19114 Halcon Crest Court
Grass Valley, CA 95949
530-268-7070
530-268-7080 (fax)
sales@kumacorp.com
www.kumacorp.com

LKB/Emtrol
One Aerial Way
Syosset, NY 11791
516-938-0600
516-931-6344

Met-Pro Corporation
1550 Industrial Drive
Owosso, MI 48867
517-725-8184
517-725-8188 (fax)
dualldiv@shianet.org
www.met-pro.com/duall.html

NCM Marketing
P.O. Box 108
Reeders, PA 18352
717-620-1856
717-874-7610 (fax)

Nature Plus, Inc.
555 Lordship Boulevard
Stratford, CT 06615
203-380-0316
203-380-0358 (fax)
info@nature-plus.com
www.nature-plus.com

NuTech Environmental Corporation
5350 N. Washington Street
Denver, CO 80216
303-295-3702
800-321-8824
303-295-6145 (fax)
nutech@sni.net

Odor Management, Inc.
Suite 200
18-4 E. Dundee Road
Barrington, IL 60010
847-304-9111
800-NO2-ODOR
847-304-9977 (fax)
ecosorb@goldengate.net
www.odormanagement.com

Piian Corporation, Inc.
1243 S. Gene Autry Trail
Palm Springs, CA 92264
760-778-4366
760-778-4368 (fax)
gregincal@msn.com

SciCorp Systems, Inc.
Building 3, Suite 203B
247 Burton Avenue
Barrie, ON L4N 5W4
CANADA
705-733-2626
705-733-2618 (fax)
scicorp@ibm.net
www.scicorpbiologic.com

The Spencer Turbine Company
600 Day Hill Road
Windsor, CT 06095
860-688-8361
800-232-4321
860-688-0098 (fax)
marketing@spencer-air.com
www.spencerturbine.com

U.S. Filter
(see "Biosolids, MSW, and Yard Waste
  Systems")

Wheatec, Inc.
2 S. 076 Orchard Road
Wheaton, IL 60187
630-682-3024
800-745-ODOR
630-682-5337 (fax)
wheatec@aol.com

Zep Manufacturing Company
3008 Olympic Industrial Boulevard
Atlanta, GA 30301
404-355-3120
877-IBUYZEP
404-350-0255 (fax)
www.zepmfg.com

Screens

Action Equipment Company, Inc.
P.O. Box 3100
Newberg, OR 97132
503-537-1111
503-537-1117 (fax)
actioneqco@aol.com

Aggregates Equipment, Inc.
9 Horseshoe Road
P.O. Box 39
Leola, PA 17540
717-656-2131
717-656-6686

Amadas Industries
1100 Holland Road
Suffolk, VA 23434
757-539-0231
757-934-3264 (fax)
amadas@amadas.com
www.amadas.com

American Pulverizer Company
5540 W. Park Avenue
St. Louis, MO 63110
314-781-6100
314-781-9209 (fax)
american@ampulverizer.com
www.ampulverizer.com

Banner Environmental & Recycling
 Equipment
N. 117 W18200 Fulton Drive
Germantown, WI 53022
262-253-2900
262-253-2919 (fax)
bob_k@bannerweld.com
www.bannerweld.com

Broer Services, Ltd.
702 Talbot Street W.
Aylmer, ON N5H 2V1
CANADA
519-773-9261
519-773-2150 (fax)
broersvc@kanservu.ca

Bulk Handling Systems, Inc.
1040 Arrowsmith Street
Eugene, OR 97402
541-485-0999
541-485-6341 (fax)
bhsequip@rio.com
www.bulkhandlingsystems.com

CBT Wear Parts, Inc.
13658 Hilltop Valley Road
Richland Center, WI 53581
608-538-3290
888-228-3625
608-538-3289 (fax)
cbtwear@mwt.net

Central Manufacturing, Inc.
P.O. Box 1900
Peoria, IL 61656
309-387-6591
309-387-6941 (fax)
centralmfg@flink.com

Continental Biomass Industries, Inc.
22 Whittier Street
Newton, NH 03858
603-382-0556
603-382-0557 (fax)
info@cbi-inc.com
www.cbi-inc.com

Erin Screens, Inc.
41 Evergreen Drive
Portland, ME 04103
207-878-3661
800-789-3746
207-878-3674 (fax)
escreens@erinscreens.com

Extec of USA, Inc.
P.O. Box 355
Essington, PA 19029
610-521-1448
800-447-2733
610-521-1863 (fax)
info@extecscreens.com
www.extecscreens.com

Fecon, Inc.
10350 Evendale Drive
Cincinnati, OH 45241
513-956-5700
800-528-3113
513-956-5701 (fax)
fecon@fuse.net
www.fecon.com

Fuel Harvesters Equipment, Inc.
P.O. Box 7908
Midland, TX 79708
915-694-9988
800-622-7111
915-694-9985 (fax)
earthsaver@planetwide.com
www.earthsaver-fhe.com

Green Mountain Technologies
(see "Biosolids, MSW, and Yard Waste Systems")

Heil Engineering Corporation
205 Bishops Way, Suite 201
Brookfield, WI 53005
414-789-5530
414-789-5508 (fax)
heilco@execpc.com
www.execpc.com/~heilco

Kinergy Corporation
7310 Grade Lane
Louisville, KY 40219
502-366-5685
502-366-3701 (fax)
kinergy@kinergy.com
www.kinergy.com

Knight Manufacturing Corporation
1501 W. Seventh Avenue
P.O. Box 167
Brodhead, WI 53520
608-897-2131
608-897-2561 (fax)
kmc@knightmfg.com
www.knightmfg.com

Lubo USA
78 Halloween Boulevard
Stamford, CT 06902
203-967-1140
203-967-1199 (fax)
lubousa@aol.com

Machinex Industries, Inc.
2121 Olivier Street
Plessisville, PQ G6L 3G9
CANADA
819-362-3281
819-362-2280 (fax)
sales@machinex.ca
www.machinex.ca

Magnatech Engineering, Inc.
P.O. Box 52
St. Charles, MO 63302
636-949-0096
888-949-0096
636-723-7879 (fax)
graveman73@aol.com
www.magnatech.org

McCloskey Brothers Manufacturing
403 Frankcom Street
Ajax, ON L1S 1R4
CANADA
905-683-4915
800-561-6216
905-683-9566 (fax)

Mill Power, Inc.
3141 S.W. High Desert Drive
Prineville, OR 97754
541-447-1100
541-447-1101 (fax)

Morbark Sales Corporation
(see "Chippers")

Multitek, Inc.
700 Main Street
P.O. Box 170
Prentice, WI 54556
715-428-2000
800-243-5438
715-428-2700 (fax)
multitek@win.bright.net
www.multitekinc.com

Nordberg-Read Corporation
25 Wareham Street
P.O. Box 1298
Middleboro, MA 02346
508-946-1200
800-992-0145
508-946-0721 (fax)

Nordfab Systems
(see "Chippers")

Ohio Central Steel Company
(see The Screen Machine)

Peterson Pacific Corporation
(see "Chippers")

Powerscreen of America
11901 Wesport Road
Louisville, KY 40245
502-326-9300
502-326-9305 (fax)
mail@powerscreen.co.uk
www.powerscreen.co.uk

Ptarmigan Machinery Company
5027 Broadway, Suite B
San Antonio, TX 78209
210-930-2757
800-648-0637
210-930-2758 (fax)

Rader Resource Recovery, Inc.
P.O. Box 181048
Memphis, TN 38181
901-795-7722
901-795-4077 (fax)
www.beloit.com

Rawson Manufacturing, Inc.
99 Canal Street
Putnam, CT 06260
860-928-0844
860-928-0366 (fax)
rawman@neca.net

ReTech
341 King Street
Myerstown, PA 17067
717-866-2357
800-876-6635
717-866-4710 (fax)
retech@re-tech.com
www.re-tech.com

Royer Industries, Inc.
(see "Chippers")

Satellite Screens
P.O. Box 366
DeWitt, IA 52742
319-659-3799
800-922-2493
319-659-8387 (fax)

The Screen Machine
7001 Americana Parkway
Reynoldsburg, OH 43068
614-866-0112
800-837-3344
614-866-1181 (fax)
email@screenmach.com
www.screenmach.com

Screen USA
1772 Corn Road
Smyrna, GA 30080
770-433-2440
770-433-2669 (fax)

Simplicity Engineering/Gruendler Crusher
(see "Chippers")

Triple "E" Company
4704 W. Mt. Vernon Road
Cedar Falls, IA 50613
319-266-4723
319-268-0394 (fax)

Triple/S Dynamics, Inc.
1031 S. Haskell Avenue
P.O. Box 151027
Dallas, TX 75223
214-828-8600
800-527-2116
214-828-8688 (fax)
www.sssdynamics.com

Vesco Engineering & Sales
P.O. Box 2007
New Hyde Park, NY 11040
516-746-5139
516-747-6911 (fax)

West Salem Machinery Company
(see "Chippers")

Wildcat Manufacturing Company, Inc.
P.O. Box 1100
Freeman, SD 57029
605-925-4512
800-627-3954
605-925-7536 (fax)
wildcat@sd.cybernex.net

Williams Patent Crusher & Pulverizer
Company, Inc.
(see "Chippers")

Shredders

Advanced Manufacturing
5780 I-10 Industrial Parkway
Theodore, AL 36582
334-653-6888
800-329-6888
334-653-6617 (fax)
info@hydraulicdesign.com
hydraulicdesign.com

Allegheny Paper Shredders Corporation
Old William Penn Highway E.
P.O. Box 80
Delmont, PA 15626
724-468-4300
800-245-2497
724-468-5919 (fax)
solutions@alleghenyshredders.com
www.alleghenyshredders.com

Ameri-Shred Corporation
P.O. Box 46130
Monroeville, PA 15146
412-798-7322
800-634-8981
412-798-7329 (fax)
info@ameri-shred.com
www.ameri-shred.com

American Pulverizer Company
(see "Screens")

Ball & Jewell
44 Rivulet Street
P.O. Box 328
North Uxbridge, MA 01538
508-278-9930
508-278-6452 (fax)

Blower Application Company, Inc.
N114 W19125 Clinton Drive
Germantown, WI 53022
800-959-0880
414-255-3446 (fax)
bac@bloapco.com
www.bloapco.com

Bridgestone/Firestone Off Road Tire
    Company
565 Marriott Drive, Suite 600
Nashville, TN 37214
615-231-5734
800-905-2367
615-231-5799 (fax)
www.bfor.com

Broer Services, Ltd.
(see "Screens")

Buffalo Hammer Mill Corporation
222 Chicago Street
Buffalo, NY 14204
716-855-1202
716-855-1204 (fax)
www.hammermills.com

CB Manufacturing & Sales Company
4455 Infirmary Road
P.O. Box 37
West Carrollton, OH 45449
937-866-5986
800-543-6860
937-866-6844 (fax)

CBT Wear Parts, Inc.
(see "Screens")

CMI Corporation
I-40 at Morgan Road
P.O. Box 1985
Oklahoma City, OK 73101
405-787-6020
405-491-2471 (fax)
cmicorp@cmicorp.com
www.cmicorp.com

Columbus McKinnon Company
1920 Whitfield Avenue
Sarasota, FL 34243
941-755-2621
800-848-1071
941-753-2308 (fax)
www.cmshredders.com

Conair
317 Meadow Street
Chicopee, MA 01013
413-789-1990
800-999-5677
413-786-3658 (fax)

Concept Products Corporation
Paoli Corporate Center
16 Industrial Boulevard, Suite 110
Paoli, PA 19301
610-722-0830
610-647-7210 (fax)
sales@conceptproducts.com
www.conceptproducts.com

Continental Biomass Industries, Inc.
(see "Screens")

Cumberland Engineering Division
100 Roddy Avenue
South Attleboro, MA 02703
508-399-6400
508-399-6653 (fax)

CW Manufacturing, Inc.
14 Commerce Drive
Sabetha, KS 66534
785-284-3454
800-743-3491
785-284-3601 (fax)
hogzilla@jbntelco.com
www.hogzilla.com

DuraTech Industries International, Inc.
3780 Highway 281 SE
P.O. Box 1940
Jamestown, ND 58401
701-252-4601
701-252-0502 (fax)
indsales@dura-ind.com
www.dura-ind.com

Ecco Business Systems, Inc.
55 W. 39th Street, Suite 11N
New York, NY 10018
212-921-4545
800-682-3226
212-921-2198 (fax)

Endura-MAX, Inc.
3490 U.S. 23 N.
P.O. Box 205
Alpena, MI 49707
517-356-1593
800-356-1593
517-358-7065 (fax)
emi-info@endura-max.com
www.endura-max.com

Entoleter
251 Welton Street
Hamden, CT 06517
203-787-3575
800-729-3575
203-787-1492 (fax)

Extec of USA, Inc.
(see "Screens")

Fecon, Inc.
(see "Screens")

Foremost Machine Builders
P.O. Box 644
Fairfield, NJ 07006
973-227-0700
973-227-7307 (fax)

Foresight, Inc./Svedala Arbra
7300 Pyle Road
Bethesda, MD 20817
301-229-0090
301-320-5971 (fax)
yaacovn@aol.com

Franklin Miller, Inc.
60 Okner Parkway
Livingston, NJ 07039
973-535-9200
800-932-0599
973-535-6269 (fax)
info@franklinmiller.com
www.franklinmiller.com

Garbalizer Machinery Corporation
Newhouse Office Building
20 Exchange Place, Suite 507
Salt Lake City, UT 84111
801-359-7583
801-363-1701 (fax)
garb.oil@burgoyne.com

Gensco America, Inc.
5307 Dividend Drive
Decatur, GA 30035
770-808-8711
800-268-6797
770-808-8739 (fax)
info@genscoequip.com
www.genscoequip.com

GrindStar, Inc.
1900 County Road I
Wrenshall, MN 55797
218-384-3066
218-384-3087 (fax)

Hazemag USA, Inc.
Mt. Braddock Road
P.O. Box 1064
Uniontown, PA 15401
724-439-3512
800-441-9144
724-439-3514 (fax)
hazemag@hhs.net

Heil Engineering Corporation
(see "Screens")

Industrial Paper Shredders, Inc.
707 S. Ellsworth Avenue
P.O. Box 180
Salem, OH 44460
330-332-0024
888-637-4733
330-332-4535 (fax)
info@industrialshredders.com
www.industrialshredders.com

Innovative Distributors & Manufacturing,
  LLC
17335 S.W. Johnson
Beaverton, OR 97006
503-591-9532
503-591-9502 (fax)
idm@spiritone.com

Jackson & Church
P.O. Box 169
Augres, MI 48703
517-876-6365
517-876-6640 (fax)

Jeffrey Company
398 Willis Road
Woodruff, SC 29388
864-476-7523
800-615-9296
864-476-7510 (fax)
www.jeffreycompany.com

Komar Industries, Inc.
4425 Marketing Place
Groveport, OH 43125
614-836-2366
614-836-9870 (fax)
komarindustries@worldnet.att.net
www.komarindustries.com

MacKissic, Inc.
P.O. Box 111
Parker Ford, PA 19457
610-495-7181
800-348-1117
610-495-5951 (fax)

Magnatech Engineering, Inc.
(see "Screens")

Marathon Equipment Company
(see "Dewatering Systems")

Maren Engineering Corporation
111 W. Taft Drive
P.O. Box 278
South Holland, IL 60473
708-333-6250
708-333-7507 (fax)
sales@marenengineering.com
www.marenengineering.com

McDonald Services, Inc.
1734 University Commercial Place
Charlotte, NC 28213
704-597-0590
800-468-3454
704-597-7415 (fax)
mrsjrm@aol.com
www.msibalers.com

Miller Manufacturing
2032 Divanian Drive
P.O. Box 336
Turlock, CA 95381
209-632-3846
209-632-1369 (fax)

Montgomery Industries International
(see "Chippers")

Morbark Sales Corporation
(see "Chippers")

Nelmor Company, Inc.
(see "Chippers")

Newell Industries, Inc.
(see "Magnetic Separation")

Nordfab Systems
(see "Chippers")

Pacific Shredder Technologies, Inc.
1335 N.W. Northrup Street
Portland, OR 97209
503-223-4980
800-417-4733
503-224-5052 (fax)

Packer Industries, Inc.
(see "Chippers")

Peterson Pacific Corporation
(see "Chippers")

Polymer Systems, Inc.
63 Fuller Way
Berlin, CT 06037
860-828-0541
860-829-1313 (fax)
polymer.systems@snet.net

Priefert Manufacturing Company
P.O. Box 1540
Mt. Pleasant, TX 75456
903-572-1741
800-527-8616
903-572-2798 (fax)
sales@priefert.com
www.priefert.com

Process Control Corporation
6875 Mimms Drive
Atlanta, GA 30340
770-449-8810
770-449-5445 (fax)

Prodeva, Inc.
100 Jerry Drive
Jackson Center, OH 45334
937-596-6713
800-999-3271
937-596-5145 (fax)
www.prodeva.com

Rapid Granulator, Inc.
P.O. Box 5887
Rockford, IL 61125
815-399-4605
800-272-7431
815-399-0419 (fax)

Rawlings Manufacturing, Inc.
P.O. Box 4485
Missoula, MT 59801
406-728-6182
406-728-7957 (fax)
rhog@bigsky.net

Rotochopper
(see "Chippers")

Royer Industries, Inc.
(see "Chippers")

Saturn Shredders
201 E. Shady Grove Road
Grand Prairie, TX 75050
972-790-7800
972-790-8733 (fax)
size-reduction@mac-corp.com
www.saturn-shredders.com

Shred-Tech, Ltd.
295 Pinebush Road
Cambridge, ON N1T 1B2
CANADA
519-621-3560
800-465-3214
519-621-0688 (fax)
shred@shred-tech.com
www.shred-tech.com

Shred-Vac Systems
15501 Little Valley Road
Grass Valley, CA 95949
530-477-7240
530-477-7488 (fax)
shred-vac@iname.com
www.shredvac.com

Simplicity Engineering/Gruendler Crusher
(see "Chippers")

SSI Shredding Systems, Inc.
9760 S.W. Freeman Drive
Wilsonville, OR 97070
503-682-3633
503-682-1704 (fax)
info@ssiworld.com
www.ssiworld.com

Sundance
P.O. Box 2437
Greeley, CO 80632
970-339-9322
970-339-5856 (fax)

Svedala Industries, Inc.
800 First Avenue N.W.
Cedar Rapids, IA 52405
319-365-0441
800-995-9149
319-369-5440 (fax)

Tire Resource Systems, Inc.
4444 S. York Street
Sioux City, IA 51106
712-255-5701
800-755-8473
712-255-9239 (fax)
tirecut@pionet.net
www.vitalsite.com/recycle/tires

Tri-C Manufacturing
3100 W. Capital Avenue
West Sacramento, CA 95662
916-371-0800
916-371-3591 (fax)

Triple/S Dynamics, Inc.
(see "Screens")

Tryco/Untha International
Harryland Road
Route 6, Box 105A
P.O. Box 1277
Decatur, IL 62525
217-864-4541
217-864-6397 (fax)
tryco@midwest.net
www.tryco.com

Universal Engineering Corporation
800 First Avenue N.W.
Cedar Rapids, IA 52405
319-365-0441
319-369-5440 (fax)

Universal Refiner Corporation
(see "Chippers")

Van Dyk Baler Corporation
78 Halloween Boulevard
Stamford, CT 06902
203-967-1100
203-967-1199 (fax)
info@vandykbaler.com
www.bollegraaf.com

Vermeer Manufacturing Company
(see "Chippers")

Warren & Baerg Manufacturing, Inc.
39950 Road 108
Dinuba, CA 93618
559-591-6790
800-344-2131
559-591-5728 (fax)
sales@warrenbaerg.com
www.warrenbaerg.com

Waste Reduction Systems
8482 Old Kings Road N.
Jacksonville, FL 32219
904-766-0882
904-766-0724 (fax)

Wendt Corporation
(see "Chippers")

West Salem Machinery Company
(see "Chippers")

W.H.O. Manufacturing Company, Inc.
P.O. Box 1153
Lamar, CO 81052
719-336-7433
800-772-0301
719-336-7052 (fax)
who@who-mfg.com
www.who-mfg.com

Williams Patent Crusher & Pulverizer
  Company, Inc.
(see "Chippers")

Spreaders and Applicators

Amadas Industries
(see "Screens")

Gehl Company
143 Water Street
West Bend, WI 53095
262-334-9461
262-338-7517 (fax)
www.gehl.com

Knight Manufacturing Corporation
(see "Screens")

Thermometers and Monitoring

Hanna Instruments, Inc.
584 Park East Drive
Woonsocket, RI 02895
401-765-7500
800-426-6287
401-765-7575 (fax)
sales@hannainst.com
www.hannainst.com

Marcom Industries, Inc.
948 Highland Avenue
Greensburg, PA 15601
724-832-0140
800-338-1572
724-832-8185 (fax)
compost@marcom-ind.com
www.marcom-ind.com

Omega Engineering, Inc.
One Omega Drive
P.O. Box 4047
Stamford, CT 06907
203-359-1660
800-848-4286
203-359-7700 (fax)
sales@omega.com
www.omega.com

Reotemp Instrument Corporation
11568 Sorrento Valley Road, Suite 10
San Diego, CA 92121
619-481-7737
800-648-7737
619-481-7415 (fax)
reotem@reotemp.com
www.reotemp.com

Tel-Tru Manufacturing Company
408 St. Paul Street
Rochester, NY 14605
716-232-1440
800-232-5335
716-232-3857 (fax)
info@teltru.com
www.teltru.com

Tub Grinders

Bandit Industries, Inc.
(see "Chippers")

Becker Underwood, Inc.
(see "Biosolids, MSW, and Yard Waste Systems")

Burrows Enterprises, Inc.
2024 E. 8th Street
Greeley, CO 80631
970-353-3769
800-724-5498
970-353-0839 (fax)
rotogrind@ctos.com
www.rotogrind.com

CBT Wear Parts, Inc.
(see "Screens")

CMI Corporation
(see "Shredders")

CW Manufacturing, Inc.
(see "Shredders")

Diamond Z Manufacturing
11299 Bass Lane
Caldwell, ID 83605
208-585-2929
800-949-2383
208-585-2112 (fax)
diamondz@micron.net
www.diamondz.com

DuraTech Industries International, Inc.
(see "Shredders")

Eurohansa, Inc.
P.O. Box 6416
High Point, NC 27262
336-885-1010
336-885-1011 (fax)
eurohansa@aol.com

Fecon, Inc.
(see "Screens")

Franklin Miller, Inc.
(see "Shredders")

Fuel Harvesters Equipment, Inc.
(see "Screens")

Grinder Equipment Technology
P.O. Box 700
Ontario, OR 97914
541-889-2558
888-412-8060
541-881-1302 (fax)
gspath@uswest.net

Jeffrey Company
(see "Shredders")

Jones Manufacturing Company
1486 12th Road
P.O. Box 38
Beemer, NE 68716
402-528-3861
402-528-3239 (fax)

Kolberg-Pioneer, Inc.
(see "Biosolids, MSW, and Yard Waste Systems")

Morbark Sales Corporation
(see "Chippers")

Nordfab Systems
(see "Chippers")

Packer Industries, Inc.
(see "Chippers")

Peterson Pacific Corporation
(see "Chippers")

Re-Tech
(see "Screens")

Royer Industries, Inc.
(see "Chippers")

Sundance
(see "Shredders")

Svedala Industries, Inc.
(see "Shredders")

The Toro Company
8111 Lyndale Avenue S.
Minneapolis, MN 55420
612-888-8385
800-525-6841
612-887-7211 (fax)
www.toro.com

Universal Refiner Corporation
(see "Chippers")

U.S. Manufacturing Company
104 N. Main Street
New Providence, IA 50206
515-497-5260
800-800-1812
515-497-5224 (fax)
usm@adiis.net

Vermeer Manufacturing Company
(see "Chippers")

W.H.O. Manufacturing Company, Inc.
(see "Shredders")

Turning and Mixing Equipment

Aireactor, Inc.
(see "Odor Control Systems")

Allied Construction Product, Inc.
(see "Compost Turning and Mixing Equipment")

American Recycling Equipment
(see "Chippers")

Arctic, Inc.
(see "Chippers")

Athey Products Corporation
1839 S. Main Street
Wake Forest, NC 27587
919-556-5171
919-556-0122 (fax)
sales@athey.com
www.athey.com

Brown Bear Corporation
(see "Dewatering Systems")

CBT Wear Parts, Inc.
(see "Screens")

DETCON
5039 Industrial Road
Farmingdale, NJ 07727
732-938-2211
732-938-9674 (fax)

Double T Equipment, Ltd.
(see "Biosolids, MSW, and Yard Waste Systems")

DuraTech Industries International, Inc.
(see "Shredders")

Fecon, Inc.
(see "Screens")

Frontier Industrial Corporation
192-B Young Street
Woodburn, OR 97071
503-982-2907
503-982-5449 (fax)
frontier@gervais.com

Fuel Harvesters Equipment, Inc.
(see "Screens")

Green Mountain Technologies
(see "Biosolids, MSW, and Yard Waste Systems")

Knight Manufacturing Corporation
(see "Screens")

Littleford Day, Inc.
7451 Empire Drive
P.O. Box 128
Florence, KY 41022
606-525-7600
606-525-1446 (fax)
sales@littleford.com
www.littleford.com

Lubo USA
(see "Screens")

McLanahan Corporation
200 Wall Street
P.O. Box 229
Hollidaysburg, PA 16648
814-695-9807
814-695-6684 (fax)

Met-Pro Corporation
(see "Odor Control Systems")

Midwest Bio-Systems, Inc.
28933-35 E Street
Tampico, IL 61283
815-438-7200
800-335-8501
815-438-7028 (fax)
treo@compuserve.com

ODB
5118 Glen Alden Drive
Richmond, VA 23231
804-226-4433
800-446-9823
804-226-6914 (fax)
info@theodbco.com
www.theodbco.com

Pike Agri-Lab Supplies, Inc.
RR 2, Box 710
Strong, ME 04983
207-684-5131
207-684-5133 (fax)
pike@inetme.com
www.maine.com/tse/pals

Processall, Inc.
10596 Springfield Pike
Cincinnati, OH 45215
513-771-2266
513-771-6767 (fax)
info@processall.com
www.processall.com

Resource Recovery Systems of Nebraska,
   Inc./KW Composters
511 Pawnee Drive
Sterling, CO 80751
970-522-0663
970-522-3387 (fax)
rrskw@kci.net
www.rrskw.com

Re-Tech
(see "Screens")

Scarab Manufacturing
HCR 1, Box 205
White Deer, TX 79097
806-883-7621
806-883-6804 (fax)
scarab@arn.net
www.scarabmfg.com

Waste Reduction Systems
(see also "Shredders")

Wastech Equipment
8302 Dunwoody Place, Suite 130
Atlanta, GA 30350
770-594-0922
888-9-WASTECH
770-594-1214 (fax)

Scat Engineering
503 Gay Street
P.O. Box 266
Delhi, IA 52223
319-922-2981
800-843-7228
319-922-2700 (fax)
info@scat.com
www.scat.com

Sludge Systems International
5039 Industrial Road
Farmingdale, NJ 07727
732-938-2211
732-938-9674 (fax)

Wildcat Manufacturing Company, Inc.
(see "Screens")

APPENDIX12B: COSTS FOR COMPOSTING MSW AND YARD WASTES

TABLE 12.11 Projected Capital Construction Cost Estimate for Two Hypothetical MSW Composting Facilities (1990 $)

Category	100 TPD aerated windrow	1000 TPD turned windrow
Site preparation (mobilization, earthwork, paving, utilities, connections, landscape, fencing, etc.)	$ 400,000	$ 5,000,000
Building, structures, foundations (receiving floor, equipment areas, office building, scale, scale house, etc.)	3,000,000	26,000,000
In-plant mobile equipment	450,000	2,500,000
Composting equipment (conveyors, shredders, screens, preprocessing, postprocessing, etc.)	1,300,000	11,000,000
Miscellaneous (supplies, office, furnishing, insurance, etc.)	100,000	800,000
Engineering, permits, construction management	470,000	4,500,000
Haul and package equipment	Not included	Not included
Land purchase	650,000	2,500,000
Working capital (start-up)	360,000	3,000,000
Total	$6,730,000	$55,300,000
$ per TPD of capacity	67,300	55,300

Quantities given in TPD (U.S. tons per day). Multiply TPD by 0.907 to obtain tonnes per day.

TABLE 12.12 Projected Annual Operation and Maintenance Cost Estimate for Two Hypothetical MSW Composting Facilities (1990 $)

Category	100 TPD aerated windrow	1000 TPD turned windrow
Labor (O&M personnel, scale operations, supervisory and office personnel including fringes)	$ 580,000	$ 4,500,000
Maintenance and materials	300,000	2,500,000
Utilities (water, sewer, electric)	90,000	900,000
Administration and insurance	150,000	750,000
Regulatory compliance	40,000	100,000
Miscellaneous (contract services)	55,000	350,000
Haul and residue disposal*	175,000	1,747,000
Total	$1,390,000	$10,847,000
Unit O&M cost, $ per ton MSW[†]	$45	$35

Quantities given in TPD (U.S. tons per day). Multiply TPD by 0.907 to obtain tonnes per day.
* 11.2 percent of MSW input at $50 per ton.
[†] Based on 312 operating days per year.

TABLE 12.13   Cost Summary for Hypothetical MSW Composting Facilities (1990 $)

Cost item	100 TPD aerated windrow	1000 TPD turned windrow
Net annual costs:		
A. Amortized investment cost*	$ 811,000	$ 6,608,000
B. Annual operating and maintenance cost	1,390,000	10,847,000
Total annual cost	2,201,000	17,455,000
C. Recyclables revenue	214,000	2,140,000
Compost revenue	31,200	312,000
Total annual revenue	245,200	2,452,000
D. Net annual cost	1,955,800	15,003,000
E. Avoided annual cost[†]	1,248,000	12,480,000
F. Annual net, minus avoided cost	$ 707,800	$ 2,523,000
Net unit costs[‡],[§] ($ per ton)		
A. Amortized investment cost*	$26	$21
B. Operating and maintenance cost	45	35
Total unit cost	71	56
C. Revenue	8	8
D. Net unit cost	63	48
E. Avoided cost	40	40
F. Unit cost minus avoided cost	$23	$8

Quantities given in TPD (U.S. tons per day). Multiply TPD by 0.907 to obtain tonnes per day.
  * Interest rate = 10 percent per year. Building and most stationary equipment are assumed to have a life of 20 years. Mobile equipment is assumed to have a life of 10 years.
  [†] Avoided cost = $40 per ton of MSW.
  [‡] Unit costs are given in dollars per ton of MSW feedstock. Based on 312 operating days per year.
  [§] Round-off affects numerical values presented.

TABLE 12.14   Variations between the Yard Waste Compost Plant Designs Considered for the Economic Analysis (1990 $)

	Option A	Option B
Construction:		
Grading	Entire site	Entire site
Paving	Processing area, receiving area, and ¼ of composting area	Processing area
Fencing	Entire site	Processing area
Equipment:		
Grinder	1	1
Compost turner	1	0
Front-end loader	1	1
Screen	1	0
Conveyors	4	2

Quantities given in TPD (U.S. tons per day). Multiply TPD by 0.907 to obtain tonnes per day.

TABLE 12.15 Estimated Initial Investment Costs for Options A and B in Table 10.14 for Composting Yard Wastes (1990 $)

	Option A	Option B
Construction:		
Grading site	$ 50,000	$ 50,000
Paving site	300,000	46,000
Road in	5,000	5,000
Fencing	30,000	8,250
Water distribution system	7,500	7,500
Engineering:		
Design	72,700	39,100
Construction supervision	5,000	2,500
Training of site operators	7,000	7,000
Utility hookups:		
Water	20,000	20,000
Power	20,000	20,000
Equipment:		
Compost turner	125,000	0
Front-end loader	75,000	125,000
Grinder	100,000	100,000
Screen	50,000	0
Conveyors	40,000	20,000
Miscellaneous equipment	15,000	15,000
Total investment costs	$922,200	$465,350

Quantities given in TPD (U.S. tons per day). Multiply TPD by 0.907 to obtain tonnes per day.

TABLE 12.16 Estimated Annual Operating and Maintenance Costs for Options A and B in Table 10.14 for Composting Yard Wastes (1990 $)

	Option A	Option B
Maintenance	$ 20,250	$ 13,000
Insurance	8,100	5,200
Fuel	16,800	21,900
Labor	70,000	87,600
Lease on land ($800 per acre per month)	115,200	115,200
Trailer rental	2,100	2,100
Water	2,000	2,000
Power	10,000	9,000
Total operating and maintenance costs	$244,450	$256,000

Quantities given in TPD (U.S. tons per day). Multiply TPD by 0.907 to obtain tonnes per day.

TABLE 12.17 Cost Summary for Hypothetical Yard Waste Composting Facilities in Table 10.14 (1990 $)

	Option A	Option B
Net annual costs		
Annual costs:		
Amortized investment cost*	150,100	75,730
Annual operating and maintenance costs	244,450	256,000
Total annual cost ($)	394,550	331,730
Annual revenues:		
Sale of compost[†]	90,000	70,000
Total revenues	90,000	70,000
Net annual cost	304,550	261,730
Annual avoided cost[‡]	350,000	350,000
Annual net minus avoided cost	(54,450)	(88,270)
Net unit costs ($ per ton of yard waste)[§]		
Amortized investment cost	16	8
Operating and maintenance cost	26	27
Total unit cost	42	35
Revenue	10	7
Net unit cost	32	28
Avoided cost	37	37
Net unit cost minus avoided cost	(5)	(9)

Quantities given in TPD (U.S. tons per day). Multiply TPD by 0.907 to obtain tonnes per day.
* Based on a 10 percent discount rate, 10-year term, and no salvage value.
[†] 10,000 yd^3 at $9/yd^3 for option A and $7/yd^3 for option B.
[‡] Based on $5/yd^3 of yard waste.
[§] Based on 70,000 yd^3 of annual yard waste at an average bulk density of 270 lb/yd^3.

CHAPTER 13

WASTE-TO-ENERGY COMBUSTION: Introduction

Frank Kreith

Waste-to-energy combustion is an important technology for municipal solid waste management. But its growth has recently slowed while communities wrestle with issues that range from flow control to impact on recycling to cost effectiveness, and to political acceptability. Nevertheless, waste-to-energy combustion can be an important factor in an overall fully integrated solid waste management strategy. The traditional term *incineration* has acquired a bad connotation in the mind of the public due to the poor operation of some waste combustors in the past. Therefore, the term *waste-to-energy combustion* is now widely used in its place. The term *incineration,* as used in this chapter, refers to the modern practice of incineration of waste that cannot be recycled economically. The technology offers great opportunities for reducing the volume of waste to be landfilled, as well as for generating heat and power. Raw solid waste has a heating value between 4000 and 7000 Btu/lbm compared to coal, which releases about 10,000 Btu/lbm. Hence, a large amount of heat can be released by burning municipal waste, and that heat can be used to generate electric power. It has been estimated that waste-to-energy facilities could supply as much as 2 percent of the electrical power needed in this country. But, more important, incineration reduces the volume of waste dramatically, up to tenfold. Thus, incineration can be attractive for large metropolitan areas where landfills are a long distance from the population center.

The major constraints on waste-to-energy combustion facilities are their cost, the level of sophistication needed to operate them safely, and the fact that the American public lacks confidence in their safety. The public is concerned about stack emissions of dioxins and the toxicity of ash residues. This concern exists, despite the assurance of experts that incineration in a modern plant with proper air pollution control equipment does not pose any dangers to health and environment. A panel of experts at the 1990 U.S. Conference of Mayors, evaluating the health and environmental impact of waste incineration, concluded that "inclusive of ash residue management, properly designed, operated and maintained incinerators equipped with state-of-the-art pollution devices can be used [to burn solid waste] in a manner that maintains associated risks below levels set by regulatory bodies for the protection of human health."

Incineration has been used widely in Europe and Japan without any adverse health impacts. Switzerland, a country with high environmental standards, incinerates about 75 percent and Japan more than 50 percent of their solid waste, according to a survey by the Integrated Waste Services Association in the spring of 1993. Sweden incinerates 60 percent and composts up to 25 percent. But waste-to-energy combustion is only slowly gaining public acceptance in the United States. But as more information on this technology becomes available, political support for siting new facilities is likely to increase and pave the way for full integration of combustion in waste management schemes. In any successful integrated waste management system in the United States designed for the twenty-first century, waste-to-energy combustion is bound to perform an important role.

This chapter presents technical information needed in designing and siting a modern waste-to-energy combustion facility. The first part of the chapter has been written by Calvin R. Brunner, a consulting engineer who has previously published *The Handbook of Incineration Systems*. Part A of this chapter has been adapted from this large and extensive work to reflect the current technology and cost of municipal solid waste systems. The chapter also deals with the incineration of medical wastes, which is becoming increasingly important in many metropolitan areas.

Significant contributions have been made in Parts B and C of this chapter by Floyd Hasselriis. Part B covers the disposal of ash residues from waste-to-energy incineration. In addition to addressing concerns about ash toxicity, it also presents novel ways of converting ash to useful products, thereby completely recycling the solid waste. Part C deals with the problem of mitigating emissions from the stack by means of various control devices. The data in this section show that the emissions of dioxins and furans can be reduced to levels that will not pose a health hazard, as the panel of experts for the U.S. mayors concluded. Results also show that, on a comparative basis, power can be generated from municipal waste with less pollution than from a coal-fired power plant.

There are over 150 waste-to-energy plants in operation today and more are either planned or under construction. The technology has proven reliable and the information presented in this chapter should assist in integrating waste-to-energy combustion in future municipal solid waste management systems.

CHAPTER 13
WASTE-TO-ENERGY COMBUSTION
Part 13A
Incineration Technologies

Calvin R. Brunner

13A.1 INCINERATION*

One of the most effective means of dealing with many wastes, to reduce their harmful potential and often to convert them to an energy form, is incineration. In comparing incineration (the destruction of a waste material by the application of heat) to other disposal options such as land burial, the advantages of incineration are:

- The volume and weight of the waste are reduced to a fraction of their original size.
- Waste reduction is immediate; it does not require long-term residence in a landfill or holding pond.
- Waste can be incinerated on-site, without having to be carted to a distant area.
- Air discharges can be effectively controlled for minimal impact on the atmospheric environment.
- The ash residue is usually nonputrescible, or sterile (see Part 13B).
- Technology exists to completely destroy even the most hazardous of materials in a complete and effective manner.
- Incineration requires a relatively small disposal area, compared to the land area required for conventional landfill disposal.
- By using heat-recovery techniques the cost of operation can often be reduced or offset through the use or sale of energy.

  Incineration will not solve all waste problems. Some disadvantages include:

- The capital cost is high.
- Skilled operators are required.
- Not all materials are incinerable (e.g., construction and demolition wastes).
- Supplemental fuel is required to initiate and at times to maintain the incineration process.

---

* Adapted from *Handbook of Incineration Systems*, published by McGraw-Hill.

Incinerable Waste

The Incinerator Institute of America was a national organization attempting to quantify and standardize incinerator design parameters. It went out of business over 20 years ago; however, a number of its standards are still in use. One such standard, given in Table 13A.1, is used by manufacturers of small and packaged incinerators in rating their equipment. The classifications in the table represent incinerable wastes, wastes which are combustible and are viable candidates for incineration.

Incinerability can be defined more specifically by consideration of the following factors:

Waste moisture content. The greater the moisture content, the more fuel is required to destroy the waste. An aqueous waste with a moisture content greater than 95 percent or a sludge waste with less than 15 percent solids content would be considered poor candidates for incineration.

Heating value. Incineration is a thermal destruction process where the waste is degraded to nonputrescible form by the application and maintenance of a source of heat. With no significant heating value, incineration would not be a practical disposal method. Generally, a waste with a heating value less than 1000 Btu/lb as received, such as concrete blocks or stone, is not applicable for incineration. There are instances, however, where an essentially inert material has a relatively small content (or coating) of combustibles and incineration would be a viable option even with a small heating value. Two such cases are incineration of empty drums with a residual coating of organic material on their inner surfaces and incineration of grit from wastewater treatment plants. The grit adsorbs grease from within the wastewater flow which results in a slight heating value to the grit material, normally less than 500 Btu/lb.

Inorganic salts. Wastes rich in inorganic, alkaline salts are troublesome to dispose of in a conventional incineration system. A significant fraction of the salt can become airborne. It will collect on furnace surfaces, creating a slag, or cake, which severely reduces the ability of an incinerator to function properly.

High sulfur or halogen content. The presence of chlorides or sulfides in a waste will normally result in the generation of acid-forming compounds in the offgas. The cost of protecting equipment from acid attack must be balanced against the cost of alternative disposal methods for the waste in question.

Radioactive waste. Incinerators have been developed specifically for the destruction of radioactive waste materials. Unless designed specifically for radioactive waste disposal, however, an incinerator should not be used for the firing of a radioactive waste.

Load Estimating

The quantity of solid waste generated in the United States, industrial and municipal, is approximately 300 million tons/year. Of this figure approximately 2000 lb of household refuse is produced per year per capita.

The estimation of incinerator loading, where the waste quantity is not known, usually requires a survey of the area in question including a study of past records, demographic trends, etc. Table 13A.2 can be used as a guide in determining the solid waste generated from various sources.

Table 13A.3 lists the average weight of various solid wastes, and Table 13A.4 lists per capita waste generation in the United States.

Another major waste, sewage sludge, can be estimated to be generated at the rate of 0.2 lb/day of sludge solids per capita.

TABLE 13A.1 Classification of Wastes to Be Incinerated

Type	Description	Principal components	Approximate composition, % by weight	Moisture content, %	Incombustible solids, %	Refuse as fired, Btu/lb	Btu of auxiliary fuel per lb of waste to be included in combustion calculations	Recommended min Btu/h burner input per lb waste
0	Trash	Highly combustible waste, paper, wood, cardboard cartons, including up to 10% treated papers, plastic or rubber scraps; commercial and industrial sources	Trash 100%	10	5	8500	0	0
1	Rubbish	Combustible waste, paper, cartons, rags, wood scraps, combustible floor sweepings; domestic, commercial, and industrial sources	Rubbish 80% Garbage 20%	25	10	6500	0	0
2	Refuse	Rubbish and garbage; residential sources	Rubbish 50% Garbage 50%	50	7	4300	0	1500
3	Garbage	Animal and vegetable wastes, restaurants, hotels, markets; institutional, commercial, and club sources	Garbage 65% Rubbish 35%	70	5	2500	1500	3000
4	Animal solids and organic wastes	Carcasses, organs, solid organic wastes; hospital, laboratory, abattoirs, animal pounds, and similar sources	100% Animal and human tissue	85	5	1000	3000	8000 (5000 Primary) (3000 Secondary)
5	Gaseous, liquid or semi-liquid wastes	Industrial process wastes	Variable	Dependent on predominant components	Variable according to wastes survey	Variable according to wastes survey	Variable according to wastes survey	Variable according to wastes survey
6	Semi-solid and solid wastes	Combustibles requiring hearth, retort, or grate burning equipment	Variable	Dependent on predominant components	Variable according to wastes survey	Variable according to wastes survey	Variable according to wastes survey	Variable according to wastes survey

[a] The above figures on moisture content, ash, and Btu as fired have been determined by analysis of many samples. They are recommended for use in computing heat release, burning rate, velocity, and other details of incinerator designs. Any design based on these calculations can accommodate minor variations.

Source: Incinerator Institute of America (1972).

TABLE 13A.2   Incinerator Capacity Chart

Classification	Building types	Quantities of waste produced
Industrial buildings	Factories Warehouses	Survey must be made 2 lb/(100 ft^2 · day)
Commercial buildings	Office buildings Department stores Shopping centers Supermarkets Restaurants Drugstores Banks	1 lb/(100 ft^2 · day) 4 lb/(100 ft^2 · day) Study of plans or survey required 9 lb/(100 ft^2 · day) 2 lb per meal per day 5 lb/(100 ft^2 · day) Study of plans or survey required
Residential	Private homes Apartment buildings	5 lb basic & 1 lb per bedroom 4 lb per sleeping room per day
Schools	Grade schools High schools Universities	10 lb per room & ½ lb per pupil per day 8 lb per room & ½ lb per pupil per day Survey required
Institutions	Hospitals Nurses' or interns' homes Homes for aged Rest homes	15 lb per bed per day 3 lb per person per day 3 lb per person per day 3 lb per person per day
Hotels, etc.	Hotels—1st class Hotels—Medium class Motels Trailer camps	3 lb per room and 2 lb per meal per day 1½ lb per room & 1 lb per meal per day 2 lb per room per day 6 to 10 lb per trailer per day
Miscellaneous	Veterinary hospitals Industrial plants Municipalities	Study of plans or survey required

Do not estimate more than 7-h operation per shift of industrial installations.
Do not estimate more than 6-h operation per day for commercial buildings, institutions, and hotels.
Do not estimate more than 4-h operation per day for schools.
Do not estimate more than 3-h operation per day for apartment buildings.
Whenever possible an actual survey of the amount and nature of refuse to be burned should be carefully taken. The data herein are of value in estimating capacity of the incinerator where no survey is possible and also to double-check against an actual survey.

Source: Incinerator Institute of America (1972).

TABLE 13A.3   Average Weight of Solid Waste

Type	lb/ft^3
Type 0 waste	8 to 10
Type 1 waste	8 to 10
Type 2 waste	15 to 20
Type 3 waste	30 to 35
Type 4 waste	45 to 55
Garbage (70% H$_2$O)	40 to 45
Magazines and packaged paper	35 to 50
Loose paper	5 to 7
Scrap wood and sawdust	12 to 15
Wood shavings	6 to 8
Wood sawdust	10 to 12

Source: Incinerator Institute of America (1972).

TABLE 13A.4   Average Solid Waste Collected (lb per person per day)

Solid wastes	Urban	Rural	National
Household	1.26	0.72	1.14
Commercial	0.46	0.11	0.38
Combined	2.63	2.60	2.63
Industrial	0.65	0.37	0.59
Demolition, construction	0.23	0.02	0.18
Street and alley	0.11	0.03	0.09
Miscellaneous	0.38	0.08	0.31
Totals	5.72	3.93	5.32

Source: Black and Klee (1968).

TABLE 13A.5 Typical Moisture Content of Municipal Solid Waste (MSW) Components

Component	Moisture, percent	
	Range	Typical
Food wastes	50–80	70
Paper	4–10	6
Cardboard	4–8	5
Plastics	1–4	2
Textiles	6–15	10
Rubber	1–4	2
Leather	8–12	10
Garden trimmings	30–80	60
Wood	15–40	20
Glass	1–4	2
Tin cans	2–4	3
Nonferrous metals	2–4	2
Ferrous metals	2–6	3
Dirt, ashes, brick, etc.	6–12	8
Municipal solid waste	15–40	20

Source: Brunner and Schwarz (1983).

TABLE 13A.6 Typical Heating Value of MSW Components

Component	Energy, Btu/lb	
	Range	Typical
Food wastes	1500–3000	2000
Paper	5000–8000	7200
Cardboard	6000–7500	7000
Plastics	12000–16000	14000
Textiles	6500–8000	7500
Rubber	9000–12000	10000
Leather	6500–8500	7500
Garden trimmings	1000–8000	2800
Wood	7500–8500	8000
Glass	50–100	60
Tin cans	100–500	300
Nonferrous metals	—	—
Ferrous metals	100–500	300
Dirt, ashes, brick, etc.	1000–5000	3000
Municipal solid wastes	4000–6500	4500

Source: Brunner and Schwarz (1983).

Estimating Solid Waste Quality

While a general figure for waste generation can be obtained as noted in the previous sections, a more accurate means of determining the quality of a solid waste stream is by use of Table 13A.5 and/or Table 13A.6. By a visual inspection of the waste, a percentage of each waste component as listed in these tables can be established. By multiplying the moisture percentage or heating value or density of each of these components by the indicated moisture, heating value, or density, a more accurate figure for the total waste quality can be estimated. (A more detailed analysis of heating value of wastes is included in this chapter.)

As an example, to estimate the heating value of a particular municipal solid waste, with the waste components as listed below, using the heating value listed in Table 13A.6, the total waste heating value is calculated as follows:

Component	Solid wastes, %	Inherent energy, Btu/lb	Total energy contribution, Btu/lb
Food wastes	15	2,000	300
Paper	40	7,200	2880
Cardboard	5	7,000	350
Plastics	5	14,000	700
Wood	15	8,000	1200
Glass	10	60	6
Tin cans	10	300	30
Total	100		5466

Total energy content is therefore 5466 Btu/lb.

Solid Waste Incineration

Solid waste incinerators are usually categorized according to the nature of the material which they are designed to burn (i.e., refuse or industrial waste). However, more than one waste type can often be burned in a given unit.

Incinerators for destruction of solid waste are the most difficult class of incinerators to design and operate, primarily because of the nature of the waste material. Solid waste can vary widely in composition and physical characteristics, making the effects of feed rates and parameters of combustion very difficult to predict. Solid waste incinerators most often burn wastes over a range of low and high heat values (i.e., from wet garbage with an as-received heat value as low as 2500 Btu/lb, to plastic wastes, over 19,000 Btu/lb). Materials handling, firing, and residue removal equipment are more critical, cumbersome, expensive, and difficult to control with these than with other types of incinerators.

Types of Solid Waste Incinerators

Waste incineration includes the following techniques:

1. Open burning
2. Single-chamber incinerators
3. Tepee burners
4. Open-pit incinerators
5. Multiple-chamber incinerators
6. Controlled air incinerators
7. Central-station disposal
8. Rotary kiln incinerators

Open Burning.   Open burning is the oldest technique for incineration of wastes. Basically it consists of placing or piling waste materials on the ground and burning them without the aid of specialty combustion equipment.

This type of system is found in most parts of the United States. It results in excessive smoking and high particulate emission, and it presents a fire hazard.

Open burning has been utilized to dispose of high-energy explosives such as dynamite or TNT. For proper incineration, the waste is placed on a refractory pad which is in turn placed over gravel, in a cleared location, remote from populated areas.

Single-Chamber Incinerators.   Single-chamber incinerators will, in general, not meet the air pollution emission standards that have been developed over the past 10 to 15 years. A typical single-chamber incinerator is shown in Fig. 13A.1. Solid waste is placed on the grate and fired. These incinerators have also been manufactured in top-loading (flue loading) configuration for apartment house waste disposal, firing waste in 55-gal drums or wire baskets or in a concrete or refractory-lined structure with a cast-iron grate, etc. This equipment may or may not have a firing system to ignite the waste. As with open burning, smoking and excessive air pollution emissions can occur.

Attempts have been made to control emissions to reasonable levels by the addition of an afterburner. Normally a temperature of 1400°F is required, at a retention time of 0.5 s, and the afterburner is used to obtain these combustion parameters in the exiting off-gas.

A *jug incinerator* is another type of single-chamber unit. A typical jug incinerator is shown in Fig. 13A.2. This is a specialty incinerator used for the destruction of cotton waste and other waste agricultural products. It is a brick-lined vertical cylindrical or conical structure. Waste is fed through the top section of the incinerator and falls to its floor, which may or may not be provided with grates. Waste is pneumatically conveyed to the incinerator charging system, and the transfer air is the only combustion air supplied to the incinerator. Afterburners are provided in the stack to control air emissions, although many such incinerators discharge from their conical top, without provision of a stack or afterburning equipment.

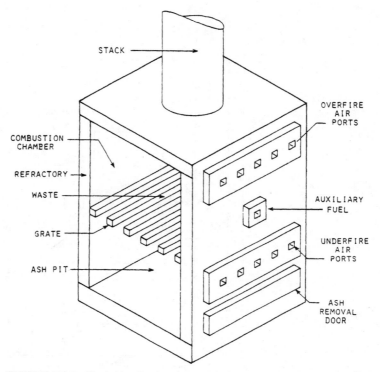

FIGURE 13A.1   Single-chamber incinerator. *(Source: U.S. EPA, 1980.)*

Open-Pit Incinerators.   Open-pit incinerators have been developed for controlled incineration of explosive wastes, wastes which would create an explosion hazard or high heat release in a conventional, enclosed incinerator. They are constructed as shown in Fig. 13A.3 with an open top and a number of closely spaced nozzles blowing air from the open top down into the incinerator chamber. Air is blown at high velocity, creating a rolling action (i.e., a high degree of turbulence). Burning rates within the incinerator provide temperature in excess of 2000°F with low smoke and relatively low particulate emissions discharges.

Incinerators of this type may be built either above or below ground. They are constructed with refractory walls and floor or as earthen trenches. The width of an open-pit incinerator is normally on the order of 8 ft, with a depth of approximately 10 ft. The length varies from 8 to 16 ft.

Overfire air nozzles are 2 to 3 in in diameter, located above one edge of the pit. They fire down at an angle of 25° to 35° from the horizontal. The incinerator is normally charged from a top-loading ramp on the edge opposite the air nozzles. Some units have a mesh placed on their top to contain larger particles of fly ash. Residue cleanout doors are often provided on aboveground incinerators.

For a waste with a heating value of 5000 Btu/lb note the following typical parameters of design for open-pit incinerators:

- Heat release of 3.4 MBtu/h per foot of length.
- Provision of 100 to 300 percent excess air.
- Overfire air of 850 standard ft^3/min per foot of pit length at 11 in water column (WC). *Note:* standard (st) conditions are 1 atm pressure and 60°F temperature.

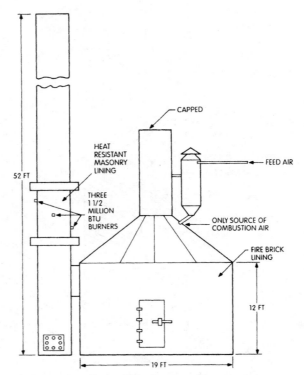

FIGURE 13A.2   Modified jug incinerator. *(Source: U.S. EPA, 1980.)*

Particulate emissions are normally below 0.25 gr/dry st ft³, corrected to 12 percent CO_2, which is unacceptable with regard to current air pollution control standards. (Most current statutes limit air pollution emissions from burning refuse to 0.08 gr/dry st ft³ corrected to 12 percent CO_2.) Other than combustion control by control of overfire air, there is no mechanism practicable for control of exhaust emissions. This incinerator, while effective in destruction of some waste, cannot normally be used without relaxation of local air pollution emission requirements.

Multiple-Chamber Incinerators.   In an attempt to provide complete burnout of combustion products and decrease the airborne particulate loading in the exiting flue gas, multiple-chamber incinerators have been developed. A first, or primary, chamber is used for combustion of solid waste. The secondary chamber provides the residence time, and supplementary fuel, for combustion of the unburned gaseous products and airborne combustible solids (soot) discharged from the primary chamber. There are two basic types of multiple-chamber incinerators: the retort and the in-line systems.

Retort Incinerator.   This unit is a compact cubic-type incinerator with multiple internal baffles. The baffles are positioned to guide the combustion gases through 90° turns in both lateral (horizontal) and vertical directions. At each turn ash drops out of the flue gas flow. The primary chamber has elevated grates for discharge of waste and an ash pit for collection of ash residual. A cutaway view of a typical retort-type incinerator is shown in Fig. 13A.4. Figure 13A.5 gives dimensional data for typical retort units.

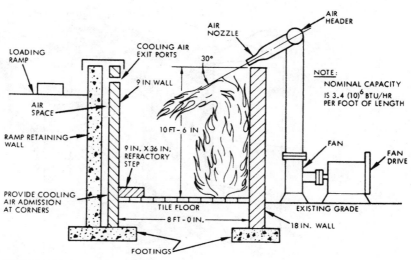

FIGURE 13A.3 Open-pit incinerator.

Overfire air and underfire air are provided above and below the primary chamber grate. This air is normally supplied by forced-air fans at a controlled rate. Flue gas exits the primary chamber through an opening, termed a *flame port,* which discharges to the secondary chamber or to a smaller *mixing chamber* immediately before the secondary chamber. The flame port is actually an opening atop the bridge wall, separating the primary from the secondary chamber.

Air ports are provided in the secondary combustion chamber and, when present, in the mixing chamber. Supplemental fuel is provided in the secondary and primary chambers. Depending on the nature of waste charged, the fuel supply in the primary chamber may be unnecessary after start-up (i.e., after bringing the chamber temperature to a level high enough for the waste to ignite and sustain its own combustion). The secondary chamber normally requires a continuous supplemental fuel supply.

As the flue gas enters and exits the secondary combustion chamber, larger airborne particles settle out of the gas stream. Temperatures in the secondary chamber are high enough (in the range of 1400°F for refuse and other carbonaceous waste) to destroy unburned airborne particles. This equipment therefore has relatively low particulate emissions and in many cases can meet an emission standard of 0.08 gr/dry st ft^3 corrected to 12 percent CO_2, without additional air pollution control equipment.

In-Line Incinerator.   This is a larger unit than the retort incinerator. Flow of combustion gases is straight through the incinerator, axially, with abrupt changes in the direction of flow only in the vertical direction, as shown in Fig. 13A.6, an in-line incinerator using natural gas as supplemental fuel. Waste is charged on the grate, which can be stationary or moving. A moving grate lends itself to continuous burning whereas stationary grates, as with the retort incinerator, are used for batch or semicontinuous operation. In-line incinerators are often provided with automatic ash removal equipment or ash discharge conveyers which also contribute to continuous operation of the incinerator.

As with the retort type, changes in the flow path and flow restrictions in an in-line incinerator provide settling out of larger airborne particles and increase turbulence for more effective burning. Supplemental fuel burners in the primary chamber ignite the waste whereas secondary-chamber supplemental fuel burners provide heat to maintain complete combustion of the burnable components of the exhaust gas.

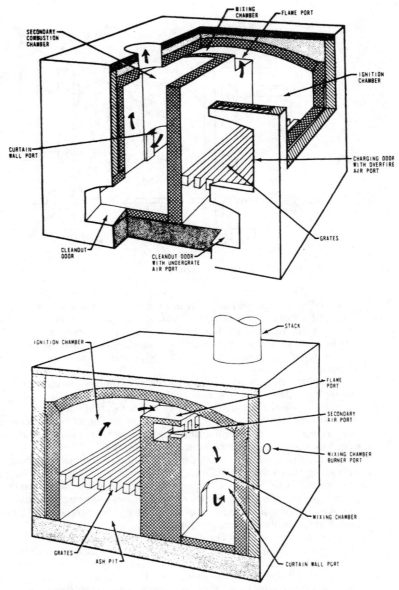

FIGURE 13A.4 Cutaway of a retort multiple-chamber incinerator. *(Source: Danielson, 1973.)*

The retort incinerator is used in the range of 20 lb/h to approximately 750 lb/h. In-line incinerators are normally provided in the range of 500 to 2000 lb/h and greater with automatic charging and/or ash removal equipment not usually provided for units smaller than 1000 lb/h in capacity. Figure 13A.6 shows an in-line incinerator with moving grates for charging and ash disposal. Figure 13A.7 is another type of in-line incinerator utilizing manual charging, i.e., fixed grates. Typical in-line unit dimensions are shown in Fig. 13A.8.

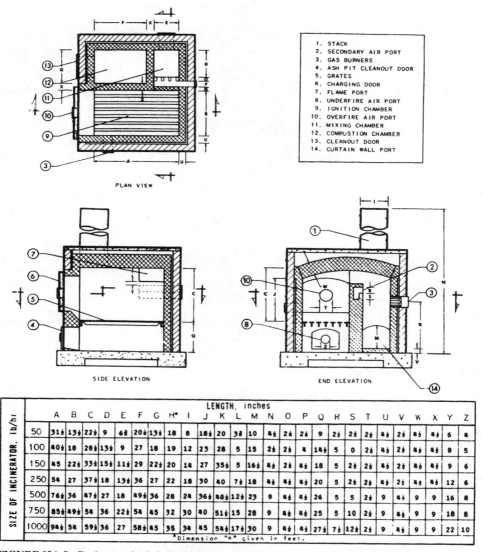

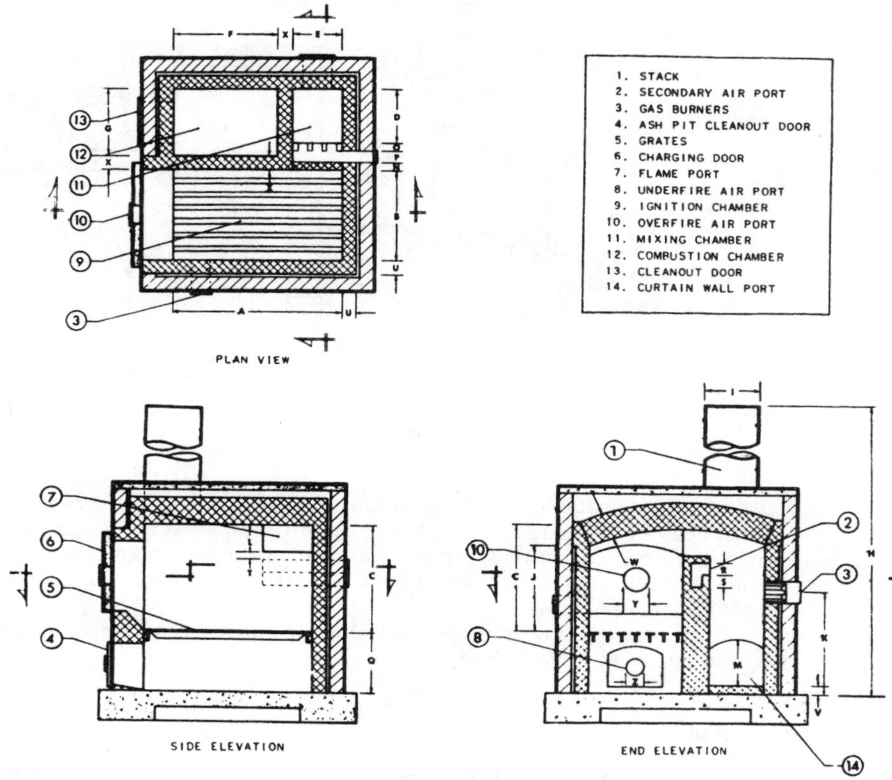
1. STACK
2. SECONDARY AIR PORT
3. GAS BURNERS
4. ASH PIT CLEANOUT DOOR
5. GRATES
6. CHARGING DOOR
7. FLAME PORT
8. UNDERFIRE AIR PORT
9. IGNITION CHAMBER
10. OVERFIRE AIR PORT
11. MIXING CHAMBER
12. COMBUSTION CHAMBER
13. CLEANOUT DOOR
14. CURTAIN WALL PORT
PLAN VIEW
SIDE ELEVATION
END ELEVATION

SIZE OF INCINERATOR, lb/hr	LENGTH, inches: A	B	C	D	E	F	G	H*	I	J	K	L	M	N	O	P	Q	R	S	T	U	V	W	X	Y	Z
50	31½	13½	22½	9	6¾	20½	13½	18	8	18½	20	3¾	10	4½	2½	2½	9	2½	2½	2½	4½	2½	4½	4½	6	4
100	40½	18	28½	13½	9	27	18	19	12	23	28	5	15	2½	2½	4	14½	5	0	2½	4½	2½	4½	4½	8	5
150	45	22½	33½	15½	11½	29	22½	20	14	27	35½	5	16½	4½	2½	4½	18	5	2½	2½	4½	2½	4½	4½	9	6
250	54	27	37½	18	13½	36	27	22	18	30	40	7½	18	4½	4½	4½	20	5	2½	2½	4½	2½	4½	4½	12	6
500	76½	36	47½	27	18	49½	36	28	24	36½	48½	12½	23	9	4½	4½	26	5	5	2½	9	4½	9	9	16	8
750	85½	49½	54	36	22½	54	45	32	30	40	51½	15	28	9	4½	4½	25	5	10	2½	9	4½	9	9	18	8
1000	94½	54	59½	36	27	58½	45	35	34	45	54½	17½	30	9	4½	4½	27½	7½	12½	2½	9	4½	9	9	22	10

*Dimension "H" given in feet.

FIGURE 13A.5 Design standards for multiple-chamber retort incinerators. *(Source: Danielson, 1973.)*

Combustion air requirements are the same for either of these incinerators: approximately 300 percent excess air. Approximately half the required air enters as leakage through the charging port and other areas of the incinerator. Of the remaining air requirement, 70 percent should be provided in the primary combustion chamber as overfire air, 10 percent as underfire air, and 20 percent in the mixing chamber or in the secondary combustion chamber.

Multiple-chamber incinerators will produce significantly lower emissions than single-chamber incinerators, as illustrated in Table 13A.7. Water curtains across the path of the flue gases exiting the secondary combustion chamber will decrease emissions even further. Multiple-chamber incinerators have been designed for specialty wastes. Typical are pathological waste incinerators such as that shown in Fig. 13A.9. Table 13A.8 lists chemical composition and com-

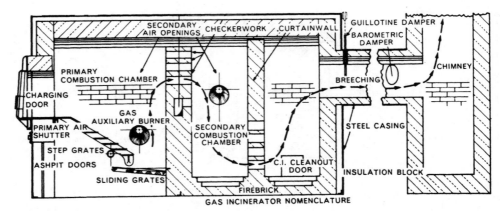

FIGURE 13A.6   Schematic diagram of a gas incinerator. *(Source: Incinerator Committee, Industrial & Commercial Gas Section, American Gas Association, New York.)*

bustion data for pathological waste. Design factors and gas velocities for pathological waste incinerators are listed in Tables 13A.9 and 13A.10 respectively. Air emissions from pathological incinerators based on two test runs with and two runs without afterburner firing are listed in Tables 13A.11 and 13A.12 respectively. Note the significant decrease in emissions with the afterburner in operation.

A crematory retort is shown in Fig. 13A.10. Its operating parameters for typical crematory waste are listed in Table 13A.13.

Rotary Kiln Technology

The rotary kiln incinerator is the most universal of thermal waste disposal systems. It can be used for the disposal of a wide variety of solid and sludge wastes and for the incineration of

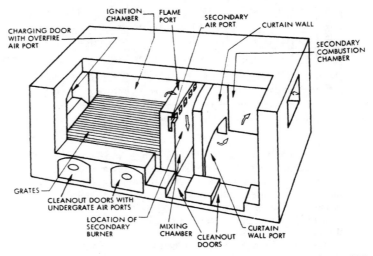

FIGURE 13A.7   Cutaway of an in-line multiple-chamber incinerator. *(Source: U.S. EPA, 1973.)*

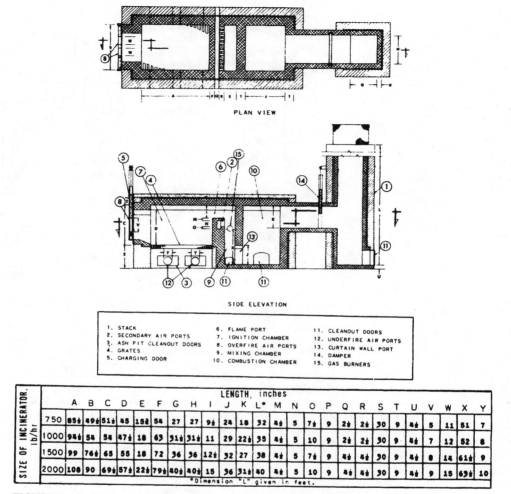

PLAN VIEW

SIDE ELEVATION

1. STACK	6. FLAME PORT	11. CLEANOUT DOORS
2. SECONDARY AIR PORTS	7. IGNITION CHAMBER	12. UNDERFIRE AIR PORTS
3. ASH PIT CLEANOUT DOORS	8. OVERFIRE AIR PORTS	13. CURTAIN WALL PORT
4. GRATES	9. MIXING CHAMBER	14. DAMPER
5. CHARGING DOOR	10. COMBUSTION CHAMBER	15. GAS BURNERS

SIZE OF INCINERATOR, lb/hr	A	B	C	D	E	F	G	H	I	J	K	L*	M	N	O	P	Q	R	S	T	U	V	W	X	Y
750	85½	49½	51½	45	15½	54	27	27	9½	24	18	32	4½	5	7½	9	2½	2½	30	9	4½	5	11	51	7
1000	94½	54	54	47½	18	63	31½	31½	11	29	22½	35	4½	5	10	9	2½	2½	30	9	4½	7	12	52	8
1500	99	76½	65	55	18	72	36	36	12½	32	27	38	4½	5	7½	9	4½	4½	30	9	4½	8	14	61½	9
2000	108	90	69½	57½	22½	79½	40½	40½	15	36	31½	40	4½	5	10	9	4½	4½	30	9	4½	9	15	69½	10

LENGTH, inches
*Dimension "L" given in feet.

FIGURE 13A.8 Design standards for multiple-chamber, in-line incinerators. *(Source: Brunner, 1988a.)*

liquid and gaseous waste. The rotary kiln system has found application in both municipal and industrial waste incineration.

Kiln System.   A rotary kiln system used for waste incineration is shown in Fig. 13A.11. It includes provisions for feeding, air injection, the kiln itself, an afterburner, and an ash collection system. The gas discharge from the afterburner is directed to an air emissions control system. An induced-draft (ID) or exhaust fan is provided within the emission control system to draw gases from the kiln through the equipment line and discharges through a stack to the atmosphere.

As shown in Fig. 13A.11, a rotary kiln system may include a waste heat boiler between the afterburner and the scrubber for energy recovery. The waste heat boiler reduces the temperature of the gas stream sufficiently to allow the use of a fabric filter, or baghouse, for particulate control. The scrubber in this illustration utilizes water and alkali injection. It is used for acid gas control. Dry scrubbing may be used in lieu of the wet scrubbing system shown.

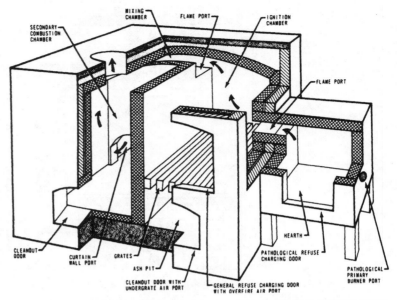

FIGURE 13A.9 Multiple-chamber incinerator with a pathological waste retort. *(Source: Danielson, 1973.)*

There are a number of areas within the kiln system where leakage can occur, as can be seen in each of the figures cited above. The feeding ports cannot be completely sealed, and the kiln seals are areas of potential leakage. The ash system is normally provided with a water seal, but for dry ash collection there will usually be some leakage. To ensure that the leakage is into the system, that no hot, dirty gases leak out of the kiln to the surrounding areas, the kiln is maintained with a negative draft. The ID fan is sized to maintain a negative pressure throughout the system so that leakage is always into, not out of, the kiln system.

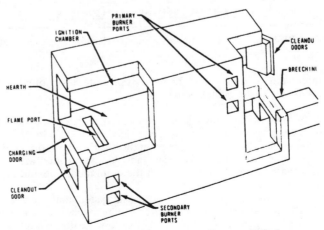

FIGURE 13A.10 Crematory retort. *(Source: Brunner, 1988a.)*

TABLE 13A.7 Comparison between Amounts of Emissions from Single- and Multiple-Chamber Incinerators

Item	Multiple chamber	Single chamber
Particulate matter, gr/st ft^3 at 12% CO_2	0.11	0.9
Volatile matter, gr/st ft^3 at 12% CO_2	0.07	0.5
Total, gr/st ft^3 at 12% CO_2	0.18	1.4
Total, lb/ton refuse burned	3.50	23.8
Carbon monoxide, lb/ton of refuse burned	2.90	197–991
Ammonia, lb/ton of refuse burned	0	0.9–4
Organic acid (acetic), lb/ton of refuse burned	0.22	<3
Aldehydes (formaldehyde), lb/ton of refuse burned	0.22	5–64
Nitrogen oxides, lb/ton of refuse burned	2.50	1
Hydrocarbons (hexane), lb/ton of refuse burned	<1	—

The Primary Combustion Chamber (Rotary Kiln).   The conventional rotary kiln is a horizontal cylinder, lined with refractory, which turns about its horizontal axis. Waste is deposited in the kiln at one end, and the waste burns out to an ash by the time it reaches the other end. Kiln rotational speed is variable, in the range of ¾ to 2½ r/min. The ratio of length to diameter of a kiln used for waste disposal is normally in the range of 2:1 to 5:1.

TABLE 13A.8 Chemical Composition of Pathological Waste and Combustion Data

Ultimate analysis (whole dead animal)		
Constituent	As charged, % by weight	Ash-free combustible, % by weight
Carbon	14.7	50.80
Hydrogen	2.7	9.35
Oxygen	11.5	39.85
Water	62.1	—
Nitrogen	Trace	—
Mineral (ash)	9	—

Dry combustible empirical formula—$C_5H_{10}O_3$

Combustion data (based on 1 lb of dry ash-free combustible)		
Constituent	Quantity, lb	Volume, scf
Theoretical air	7.028	92.40
40% sat at 60°F	7.069	93
Flue gas with theoretical air 40% saturated — CO_2	1.858	16.06
N$_2$	5.402	73.24
H_2O formed	0.763	15.99
H_2O air	0.031	0.63
Products of combustion total	8.054	105.92

Gross heat of combustion—8,820 Btu per lb

Source: Brunner (1988a).

TABLE 13A.9   Design Factors for Pathological Ignition Chamber (Incinerator Cavity, 25 to 250 lb/h)

Item	Recommended value	Allowable deviation, %
Hearth loading	$10 \ lb/(h \cdot ft^2)$	±10
Hearth length-to-width ratio	2	±20
Primary burner design	$\dfrac{10 \ ft^3 \ natural \ gas}{lb \ waste \ burned}$	±10

Source: Brunner (1988a).

TABLE 13A.10   Gas Velocities and Draft (Pathological Incinerators with Hot-Gas Passage below a Solid Hearth)

Item	Recommended values	Allowable deviation, %
Gas velocities		
Flame port at 1600°F, ft/s	20	±20
Mixing chamber at 1600°F, ft/s	20	±20
Port at bottom of mixing chamber at 1550°F, ft/s	20	±20
Chamber below hearth at 1500°F, ft/s	10	±100
Port at bottom of combustion chamber at 1500°F, ft/s	20	±20
Combustion chamber at 1400°F, ft/s	5	±100
Stack at 1400°F, ft/s	20	±25
Draft		
Combustion chamber, in WC	0.25^a	$\left\{ \begin{array}{l} -0 \\ +25 \end{array} \right.$
Ignition chamber, in WC	0.05–0.10	±0

[a] Draft can be 0.20 in WC for incinerators with a cold hearth.
Source: Brunner (1988a).

Most kiln designs utilize smooth refractory on the kiln interior. Some designs, particularly those for the processing of granular material (dirt or powders), may have internal vanes or paddles to encourage motion along the kiln length and to promote turbulence of the feed. Care must be taken in the provision of internal baffles of any kind. With certain material consistencies, such as soil of from 10 to 20 percent moisture content, baffles may tend to retard the movement of material through the kiln.

The kiln is supported by at least two trunnions. One or more sets of trunnion rollers are idlers. Kiln rotation can be achieved by a set of powered trunnion rollers, by a gear drive around the kiln periphery, or through a chain driving a large sprocket around the body of the kiln. The kiln trunnion supports are adjustable in the vertical direction. The kiln is normally supported at an angle to the horizontal, or rake. The rake will normally vary from 2 to 4 percent (¼ to ½ in/ft of length), with the higher end at the feed end of the kiln. Other kiln designs have a zero or slightly negative rake, with lips at the input and discharge ends. These kilns are operated in the slagging mode, as discussed subsequently, with the internal kiln geometry designed to maintain a pool of molten slag between the kiln lips.

TABLE 13A.11 Emissions from Two Pathological-Waste Incinerators with Secondary Burners

Test no.	A	B
Rate of destruction to powdery ash, lb/h	Mixing chamber burner operating	Mixing chamber burner operating
	19.2	99
Type of waste	Placental tissue in newspaper at 40°F	Dogs freshly killed
Combustion contaminants		
gr/st ft$^{3\,a}$ at 12% CO_2	0.200	0.300
gr/st ft^3	0.014	0.936
lb/h	0.030	0.360
lb/ton charged	3.120	7.260
Organic acids		
gr/st ft^3	0.006	0.013
lb/h	0.010	0.050
lb/ton charged	1.040	1.010
Aldehydes		
gr/st ft^3	N.A.[b]	0.006
lb/h	N.A.[b]	0.020
lb/ton charged	N.A.[b]	0.400
Nitrogen oxides		
ppm	42.70	131
lb/h	0.08	0.099
lb/ton charged	8.84	2
Hydrocarbons	Nil	Nil

[a] CO_2 from burning of waste used only to convert to basis of 12% CO_2.
[b] Not available.
Source: Brunner (1988a), p. 462.

A source of heat is required to bring the kiln up to operating temperature and to maintain its temperature during incineration of the waste feed. Supplemental fuel is normally injected into the kiln through a conventional burner or a ring burner when gas fuel is used.

There are a number of variations in kiln design, including the following:

- Parallel flow or counterflow
- Slagging or nonslagging mode
- Refractory or bare wall

The more commonly used kiln design, referred to as the *conventional kiln,* is a parallel-flow system, nonslagging, lined with refractory.

Kiln Exhaust Gas Flow.   When gas flow through the kiln is in the same direction as the waste flow, the kiln is said to have *parallel* or *cocurrent flow.* With countercurrent flow, the gas flows opposite to the flow of waste. The burner(s) is(are) placed at the front of the kiln, the face of the kiln from which air or gas originates.

Generally, a countercurrent kiln is used when an aqueous waste, one with at least 30 percent water content, is to be incinerated. Waste is introduced at the end of the kiln far from the burner. The gases exiting the kiln will dry the aqueous waste, and its temperature will drop. If aqueous waste were dropped into a kiln with cocurrent flow, water would be evaporated at the feed end of the kiln. The feed end would be the end of the kiln at the lowest temperature, and a much longer kiln would be required for burnout of the waste.

TABLE 13A.12 Emissions from Two Pathological-Waste Incinerators without Secondary Burners (Source Tests of Two Pathological-Waste Incinerators)

Test no.	A	B
	Mixing chamber burner not operating	Mixing chamber burner not operating
Rate of destruction to powdery ash, lb/h	26.4	107
Type of waste	Placental tissue in newspaper at 40°F	Dogs freshly killed
Combustion contaminants		
gr/st ft$^{3\ a}$ at 12% CO_2	0.500	0.300
gr/st ft^3	0.017	0.128
lb/h	0.030	0.430
lb/ton charged	2.270	8.040
Organic acids		
gr/st ft^3	0.010	0.034
lb/h	0.020	0.110
lb/ton charged	1.514	2.050
Aldehydes		
gr/st ft^3	0.007	0.010
lb/h	0.013	0.033
lb/ton charged	0.985	0.617
Nitrogen oxides		
ppm	14.700	95
lb/h	0.016	0.082
lb/ton charged	1.210	1.550
Hydrocarbons	Nil	Nil

^{a} CO_2 from burning of waste only used to convert to basis of 12% CO_2.
Source: Brunner (1988a).

Wastes with a light volatile fraction (containing greases, for instance) should utilize a kiln with cocurrent flow. These volatiles will likely be released from the feed immediately upon entering the kiln. Use of a cocurrent kiln provides a higher residence time than use of a countercurrent kiln for the effective burnout of these volatiles.

Slagging Mode. At temperatures in the range of 2000 to 2200°F, ash will start to deform for many waste streams; and as the temperature increases, the ash will fluidize. The actual temperatures of initial deformation and subsequent physical changes to the ash are a function of the chemical constituents present in the waste residual. They are also a function of the presence of oxygen in the furnace. The ash deformation temperatures will vary with reducing vs. oxidation atmospheres, as noted in Table 13A.14. Ash Fusion Temperatures, which lists deformation temperatures for coal and a typical refuse mix. Eutectic properties can be controlled by the use of additives to the molten material.

A kiln can be designed to generate and maintain molten ash during operation. Operation in a slagging mode provides a number of advantages over nonslagging operation. When a kiln is operating in a nonslagging mode, however, and slagging occurs, slagging is undesirable and must be eliminated.

Differences in slagging vs. nonslagging kilns are outlined in Table 13A.15. As noted, the construction of a slagging kiln is more complex than that of a nonslagging kiln, requiring provision of a lip at the kiln exit to contain the molten material. A nonslagging kiln will normally have a smooth transition with no impediments to the smooth discharge of ash.

he WASTE-TO-ENERGY COMBUSTION—INCINERATION TECHNOLOGIES

TABLE 13A.13   Operating Procedures for Crematory

Phase	Duration, 1½ h operation, min	Burner settings	Casket	Body Moisture	Tissue	Bone calcin
Charging[a]	—	Secondary zone on				
Ignition	15	All on	20% burns	—	—	—
Full combustion	30	All on	80% burns	20% evap.	10% burns	—
Final combustion	45	All on	—	80% evap.	90% burns	50%
Calcining	1 to 12 h	All off (or small primary on)	—	—	—	50%
	Duration, 2½ h operation, min					
Ignition	15	All on	20% burns	—	—	—
Full combustion	30	Primary off	60% burns	20% evap.	—	—
Final combustion	15	All on	20% burns	20% evap.	20% burns	50%
	90	All on	—	60% evap.	80% burns	—
Calcining	1 to 12 h	All off (small primary may be on)	—	—	—	—

[a] Charge Casket: 75 lb wood
  Body: 180 lb
  Moisture: 108 lb
  Tissue: 50 lb
  Bone: 22 lb

Source: Brunner (1988a).

Slagging kilns have been designed and operated with a negative rake; e.g., the outer surface of the kiln at the feed end is lower than the kiln surface at the discharge end. This will permit the accumulation of more slag in the kiln than with zero or positive rake. The kiln internal surface must be designed for this operating mode. For instance, as noted previously, an internal refractory lip is required on the kiln feed end.

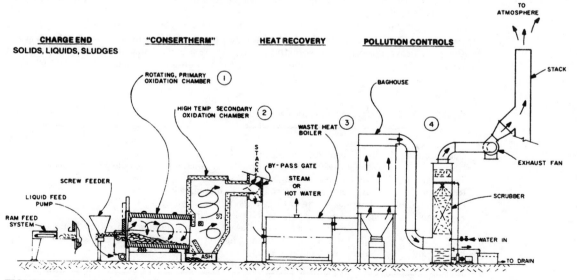

FIGURE 13A.11   Rotary kiln with waste heat boiler. (*Source: R. Rayve, Consertherm, East Hartford, Conn.*)

TABLE 13A.14   Ash Fusion Temperatures

	Reducing atmosphere, °F	Oxidizing atmosphere, °F
Refuse		
Initial deformation	1880–2060	2030–2100
Softening	2190–2370	2260–2410
Fluid	2400–2560	2480–2700
Coal		
Initial deformation	1940–2010	2020–2270
Softening	1980–2200	2120–2450
Fluid	2250–2600	2390–2610

Source: Brunner (1988a).

The slagging kiln can accept metal drums. The ash eutectic properties at the molten slag temperatures will tend to dissolve a ferrous metal drum placed in the kiln. The placing of drums containing waste in a kiln may be undesirable from a safety and maintenance standpoint (even with the tops of the drums removed, localized heating of the drum surface may occur, causing an explosion, and the impact of a dropping drum will eventually damage kiln refractory). However, if drums are to be placed in a kiln (they should be quartered), slagging kilns are able to absorb the drum into a homogeneous residue discharge. The nonslagging kiln can only move the drum through the unit and must include specialty equipment for handling the drum body as it exits the kiln.

Salt-laden wastes will tend to melt in the range of 1300 to 1600°F and can produce severe caking, or deposits, in a nonslagging kiln. Often salt-bearing wastes are prohibited from kilns because they will produce an unacceptable buildup on the kiln surface which can eventually choke off the kiln. In a slagging kiln, however, the temperature is kept high enough to keep the salts in a molten state. The salts combine with the molten ash in the pool at the bottom of the kiln and are maintained in their molten state until quenched. The temperature in a slagging kiln must be sufficiently high to maintain the ash as a molten slag. Temperatures as high as 2600 to 2800°F are not uncommon. A nonslagging kiln will normally operate at temperatures below 2000°F.

The destruction of organic compounds is achieved by a combination of high temperature and residence time. Generally, the higher the temperature, the shorter the residence time required for destruction. Conversely, the higher the residence time, the lower the required temperature. The use of higher temperatures in the slagging kiln reduces the residence time

TABLE 13A.15   Slagging versus Nonslagging Kiln

Factor	Effect
Construction	More complex with slagging kiln
Duty	Slagging kiln can accept drums, salt-laden wastes; nonslagging kiln is limited
Temperature	Higher with slagging kiln
Retention time	Greater residence required in nonslagging kiln
Process control	Thermal inertia or forgiveness in slagging kiln
Emissions	Less particulate, greater NO_x in slagging kiln
Slag	Slagging kiln may require CaO, Al_2O_3, SiO_2 additives; dissolves drums, salts
Ash	Wet, less leachable with slagging; wet or dry with nonslagging kiln
Maintenance	Higher with slagging kiln
Refractory	More critical with slagging kiln

requirements for the off-gas. The afterburner associated with a slagging kiln can often be smaller than that required for a nonslagging kiln.

The molten slag can weigh hundreds or even thousands of pounds. As a concentrated material, a liquid, it represents a significant thermal inertia within the kiln. The molten slag tends to act as a heat sink which provides thermal stability to the system. The slagging kiln is much less subject to temperature extremes than the nonslagging kiln because of the presence of this massive melt. It will maintain a relatively constant-temperature profile under rapid changes in kiln loading. This stability leads to more predictable system behavior. Safety factors employed in the design and operation of downstream equipment (such as an exhaust gas scrubber or the induced-draft fan) can be reduced when a slagging kiln is used.

The tumbling action of a rotating kiln encourages the release of particulate to the gas stream. From 5 to 25 percent of the nonvolatile solids in a feed stream may become airborne with the use of a conventional nonslagging kiln. The presence of the molten slag in a slagging kiln acts much like the fluid ash in a pulverized-coal (PC) burner. The slag will absorb particulate matter from the gas stream and can reduce particulate emissions from the kiln to 25 to 75 percent of the emissions from a nonslagging kiln. However, emissions of NO_x are greater with a slagging than with a nonslagging kiln. The generation of NO_x is generally not significant until the temperature of the process increases above 2000°F. Above this temperature the formation of NO_x will increase substantially. At 2600°F the generation of NO_x is almost 10 times as great as at 1800°F.

A danger in slagging kiln operation is that the melt will solidify. When this happens the kiln will be off-balance. With an eccentric-turning kiln, if rotation of the kiln is not stopped, damage to kiln supports and to the kiln drive may occur. In addition, the incineration process will degrade under a melt freeze. Operating stability will be lost, and demands on downstream equipment (the gas scrubbing system, for instance) may be too severe. One reason for the loss of a molten slag, besides a drop in temperature, is a change in the feed quality. To ensure the maintenance of an adequate melt, additives may have to be employed. These additives may include CaO, Al_2O_3, SiO_2, or another compound or set of compounds, depending on the nature of the waste. Additives will help maintain the eutectic, to ensure that the melt will remain in a molten state.

The molten slag from a slagging kiln is dropped into a wet sump. (The hot slag can "pop" or explode as it contacts the cooling water in the sump.) The slag immediately hardens into a granular material (termed *frit*) with the appearance of gravel or dark glass. The ash from a nonslagging kiln can be collected wet or dry.

Refractory for a slagging kiln will experience more severe duty than that for nonslagging kiln service. The higher operating temperatures will directly affect refractory life, as will the corrosive effect of the melt. In addition, if steel drums are dropped into the kiln, the physical impact of the drum on the kiln surface will be damaging. The molten slag will absorb the steel and ferrous metals, as well as other metals, which are highly corrosive to the refractory. The refractory must resist this corrosive attack, high temperatures, and impact loading. The resulting refractory system will be expensive and will require frequent maintenance.

Operation.   The waste retention time in a kiln can be varied. It is a function of kiln geometry and kiln speed, as shown in the following equation:

$$t = \frac{2.28 \, L/D}{SN}$$

where t = mean residence time, min
L/D = internal length-to-diameter ratio
S = kiln rake slope, in/ft of length
N = rotational speed, rev/min

For a given *L/D* ratio and rake, the solids residence time within the unit is inversely proportional to the kiln speed. By doubling the speed, the residence time will halve. Here is an example of this calculation:

Calculating the residence time for a kiln rotating at $N = 0.75$ rev/min with a 1 percent slope ($S = 0.12$ in/ft of length), with a 4-ft inside diameter and 12-ft length *L*, we get

$$t = \frac{2.28(12/4)}{0.12 \times 0.75} = 76 \text{ min}$$

By inspection, note that a doubling of rotation N would halve the retention time, and halving the rake *S* would double the retention time.

The preceding calculation was for the residence time of solids or other materials within the kiln, not the kiln exhaust gas. The off-gas residence time can be determined by the application of the heat balance and flue gas analyses developed later in this text.

Kiln Seal.   Sealing a kiln is a difficult task. Efficient kiln operation requires that kiln seals be provided and maintained to control the infiltration of unwanted airflow into the system. With too much air, fuel usage increases and process control deteriorates.

The kiln turns between two stationary yokes. The kiln diameter, which can vary from 4 to 20 ft, will have a periphery of from 12 to 60 ft. At 1 ft/min velocity, the kiln surface is moving at a rate of up to 60 ft/min. A seal must close this gap between the yoke and the kiln surface while the kiln is moving at this surface velocity. The kiln surface is not a machined surface and will have variations in texture and dimension, making the task of sealing very difficult. A further problem is that the kiln interior is normally at relatively high temperatures, which tend to encourage wear of the kiln surface.

Two types of seals are illustrated in Fig. 13A.12. The rotating portion of this seal, a T-ring in this illustration, is mounted on the kiln surface. There are as many variations in kiln seal designs as there are kiln and kiln seal manufacturers.

Design Variations.   In an effort to control the air distribution and temperature profile along the length of the kiln, a rotary kiln was developed with air injection ports. This kiln, developed by Universal Energy International, Inc., has a combustion air plenum inserted high through-

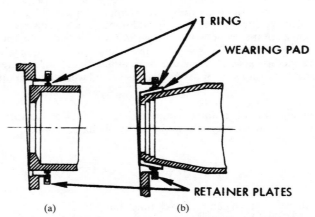

FIGURE 13A.12 Kiln seal arrangements. (a) Single floating-type feed-end air seal. (b) Single floating-type air seal on air-cooled tapered feed end. *(Source: Brunner, 1993.)*

out its length. Combustion air will cool the plenum as well as provide air for feed volatiles. The airflow within the plenum can be directed to any of a number of zones within the kiln. The control of air has been found to allow low- or substoichiometric operation in portions of the kiln or throughout the entire length of the kiln.

The rotary kiln system illustrated in Fig. 13A.13 has been developed for municipal solid waste application. The kiln, or rotary combustor, has no refractory. Without refractory, it is believed that the system maintenance cost is reduced. It is constructed of water tubes which absorb from 25 to 35 percent of the heat generated by the burning waste. Burning begins in the kiln, with air injected through openings in the tube wall construction. Burning of the off-gas is completed within the boiler, which is also constructed of water tubes, with no refractory. A rotary joint in the kiln hot-water circulation system maintains a water seal under the physical motion of the tubes and the relative high-pressure demands of the hot fluid.

Pyrolysis and Controlled Air Incineration

General Description.   Pyrolysis is the destructive distillation of a solid, carbonaceous, material in the presence of heat and in the absence of stoichiometric oxygen. It is an exothermic reaction (i.e., heat must be applied for the reaction to occur).

Ideally a pyrolytic reaction will occur as follows, using cellulose:

$$C_6H_{10}O_5 \xrightarrow{\text{heat}} CH_4 + 2CO + 3H_2O + 3C$$

A gas is produced containing methane, CH_4, carbon monoxide, CO, and moisture. The carbon monoxide and methane components are combustible, providing heating value to the off-gas. The carbon residual, a char, also has heating value. This is an idealized reaction. No oxygen is added, and the original material is pure cellulose, $C_6H_{10}O_5$. In general the initial material is not pure and contains additional components, both organic and inorganic. The off-gas is a mixture of many simple and complex organic compounds. The char is often a liquid which contains minerals, ash, and other inorganics as well as residual carbon or tars.

Pyrolysis as an industrial process has been in use for years, and, although attempts to apply this process to municipal solid waste disposal have been made since the 1960s, it has not met with success in this area in the United States. The pyrolysis process produces charcoal from wood chips, coke and coke gas from coal, fuel gas and pitch from heavy-hydrocarbon still bottoms, etc.

The Pyrolysis System.   An idealized pyrolysis system for disposal of mixed waste is shown in Fig. 13A.14. The waste received is sorted for removal of glass, metal, and cardboard, all of which has possible resale value. The waste stream enters a shredder (grinder), and the shredded material passes through a magnetic separator where residual ferrous metal is removed, for resale.

The balance of the waste stream will be fed into the reactor from a feed hopper. The hopper discharge and the feeder must be provided with air locks to minimize the infiltration of air (oxygen), which will degrade the pyrolysis reaction. Shredding is a necessary step, not only to allow metal removal but also to provide a uniform-size feed of relatively small particles to the reactor. The converter is heated externally, as shown. Other types of pyrolytic reactors are designed to allow sufficient air infiltration to provide some burning within the reactor, generating enough heat internally to sustain the process.

Gas, exiting the reactor, is collected in a storage tank where organic acids and other organic compounds condense and are eventually discharged. Between 30 and 40 percent of the gas is required to heat the pyrolytic reactor; the balance of the gas stream can be used for other processes. In this generalized scheme a significant portion of the heating value of the off-gas is contained within the condensables. If the gas is heated and the discharged char residue is cooled, the condensables will remain in the gaseous state. As the gas cools and the

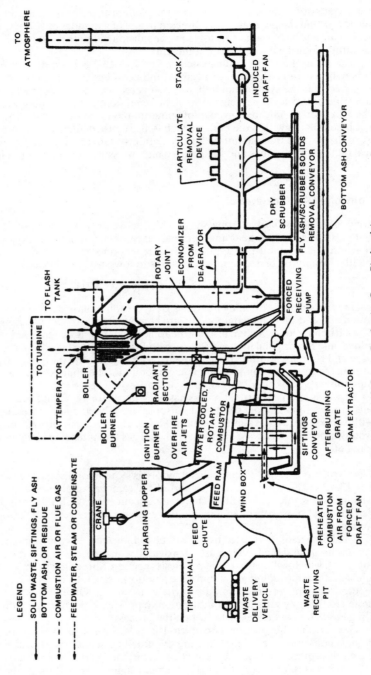

FIGURE 13A.13 Rotary combustor. (Source: Westinghouse/O'Connor Combustion Corp., Pittsburgh.)

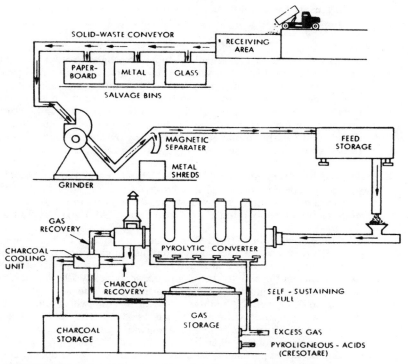

FIGURE 13A.14 Pyrolytic waste conversion. *(Source: Brunner, 1988a.)*

condensables leave the gas stream, the gas heating value will decrease. Therefore, for maximum energy reclamation from the gas, it is important that the gas be kept in a heated state as long as possible—at least long enough for the gas to reach the farthest gas burner. Storage should be minimized because the condensables will leave the gas stream relatively readily in any quiescent area. The residual solid material is termed *charcoal*. This is, ideally, a desired byproduct of this reaction.

A stack is shown immediately downstream of the converter. Upon start-up of the process, when an outside source of heat is required to initiate the reaction (not shown), the initial off-gas is basically composed of steam, carbon dioxide, entrapped air, and trace amounts of carbon monoxide. These components can be vented through the stack until the process stabilizes and pyrolysis gas is produced.

Severe problems have occurred in commercial attempts to develop this technology. It has been found difficult, if not impossible, to clean up the gas exiting the reactor. Although the gas is passed through a secondary combustion chamber and is subject to high-energy scrubbing systems, the organics within the gas stream have not been effectively controlled. Another severe problem lies within the reactor itself. As the waste is heated in the reducing atmosphere of the reactor, the ash, metals, glass, and other noncombustible materials tend to liquefy and form a slag. It is impossible to remove all of the glass or metals from the waste stream, and even the relatively small amount of these materials that may be present in the waste will contribute to the formation of a slag. In a number of the designs of commercial pyrolysis systems, the movement of the slag and the quantity of the slag generated has been impossible to control. Slagging has reached burner ports, and has risen in the reactor to interfere with the pyrolysis reaction itself.

Interest in the development of pyrolysis as an effective means of treatment of municipal solid waste is reviving in Europe, but American firms have abandoned the marketing of this technology.

Related Systems.   There are many studies in progress not far distant from those of the traditional alchemists. Instead of yellow gold the goal is black gold—petroleum and petroleum-derived products. Currently processes and systems are exemplary if they just dispose of waste, generating innocuous residual materials without creating nuisance odor or budget overruns. But new processes will undoubtedly follow the perfectly sound theory of oil from waste (one hydrocarbon from another), and perhaps a workable system will be found to generate the potential of 1 barrel of oil per ton of waste by pyrolysis. A related process, starved air or controlled air combustion has been successfully developed.

Starved Air Incineration.   In the early 1960s a new type of incinerator started gaining in popularity. The *modular combustion unit* has become an economical and efficient system for on-site and central destruction of waste. These incinerators are also known as *controlled air units.* They can be operated as excess air units (EAU) or starved air units (SAU).

Theory of Operation.   The SAU consists of two major furnace components, as shown in Fig. 13A.15, a primary and a secondary combustion chamber. Waste is charged into the primary chamber, and a carefully controlled flow of air is introduced. Only enough air is provided to allow sufficient burning for heating to occur. Typically 70 to 80 percent of the stoichiometric air requirement is introduced into the primary chamber.

The off-gas generated by this starved air reaction will contain combustibles, and this gas is burned in the secondary chamber, which is sized for sufficient residence time to totally destroy organics in the off-gas. As in the primary chamber, a carefully controlled quantity of air is introduced into the secondary chamber, but in this case excess air, 140 to 200 percent of the off-gas stoichiometric requirement, is maintained to effect complete combustion. Gas-cleaning devices such as wet scrubbers or electrostatic precipitators may not be required. The burnout of the off-gas in the secondary chamber is usually sufficient to clean the gas to 0.08 grains/st ft^3.

Figure 13A.16 illustrates the variety of configurations currently marketed for controlled air incineration. They all have a starved air primary section and a secondary or afterburner chamber.

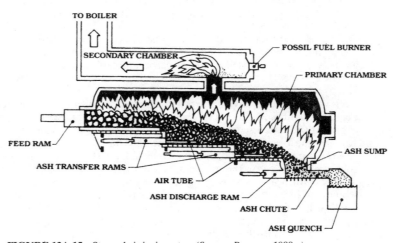

FIGURE 13A.15   Starved air incinerator. *(Source: Brunner, 1988a.)*

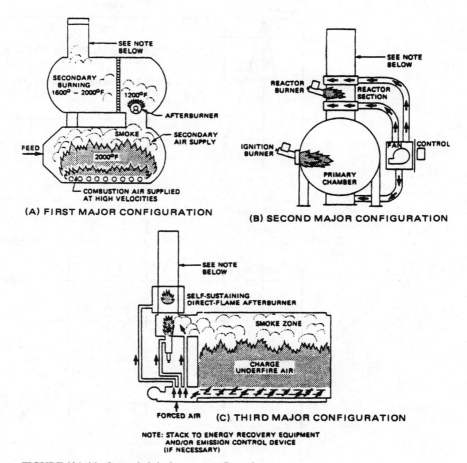

FIGURE 13A.16 Starved air incinerator configurations.

Control.   As can be seen in Fig. 13A.17, the temperature is directly related to the excess air provided. Temperature, therefore, is normally utilized to control airflow in both primary and secondary chambers. Below stoichiometric the temperature of the reaction increases with an increase in airflow. As more air is provided, more combustion will occur, so more heat will be released. This heat release will result in higher temperatures produced. Control of the primary-chamber operation, therefore, where less than complete oxidation is provided, is as follows:

• With higher temperatures, decrease airflow.
• With lower temperatures, increase airflow.

The secondary chamber is designed for complete combustion, greater than stoichiometric air is supplied. At stoichiometric conditions all the combustible material present will combust completely. Additional air will act to quench the off-gas, i.e., will lower the resulting exhaust gas temperature. Therefore, control of the secondary-chamber operation is as follows:

• With higher temperatures, increase airflow.
• With lower temperatures, decrease airflow.

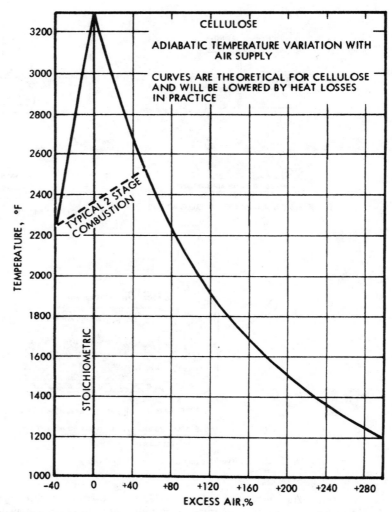

FIGURE 13A.17 Adiabatic temperature variation with air supply. *(Source: Brunner, 1988a.)*

SAUs are often provided with temperature detectors which automatically control fan damper positioning to provide the required chamber airflow.

Incinerable Wastes.   The SAUs were originally developed for the destruction of trash. They are applicable for other solid waste destruction, and their secondary chamber can be used for destruction of gaseous or liquid waste in suspension. It is not applicable for incineration of endothermic materials. The nature of the SAU process is such that turbulence of the waste feed is minimal. Materials requiring turbulence for effective combustion such as powdered carbon or pulp wastes are not appropriate candidates for starved air incineration.

Air Emissions.   Compared to other incineration methods, the airflow in the primary chamber, firing the waste, is low in quantity and is low in velocity. The low velocity and near absence of turbulence of the waste result in minimal amounts of particulate carried along in the gas stream. Complete burning is accomplished in the secondary chamber, and the resulting exhaust gas is clean and practically free of particulate matter (i.e., smoke and soot). The SAU can usually comply with exhaust emissions standards to 0.08 g/st ft^3 without the use of supplemental gas-cleaning equipment such as scrubbers or baghouses.

Waste Charging.   Smaller units, under 750 lb/h, are normally batch-fed. Waste is charged over a period of hours, and after a full load has been placed in the chamber, the chamber is sealed and the waste fired.

Figure 13A.18 illustrates a typical hopper-ram assembly designed to minimize the quantity of air infiltration into the primary chamber when charging. Figure 13A.19 illustrates a double-ram charging system which allows a more continuous feed than the single-ram. Note that the furnace charging door is not opened until the hopper is sealed by the upper ram, preventing air infiltration from the hopper. Larger units are usually provided with a continuous waste charging system, a screw feeder, or a series of moving grates.

Ash Disposal.   As with waste charging, SAUs are provided with both manual and automatic discharge systems. With smaller units, after burnout the chamber is opened and ash residue is manually raked out. With continuous operating units such as that shown in Fig. 13A.15, ash is continually discharged, normally into a wet well, where it is transferred to a container or truck by means of a drag conveyer.

Energy Reclamation.   Waste heat utilization is a viable option provided with SAUs. The hot gas exiting the secondary chamber is relatively clean. Boiler or heat-exchanger surfaces placed within this gas stream will therefore be subject to minimal particulate matter carryover and attendant problems of erosion and plugging.

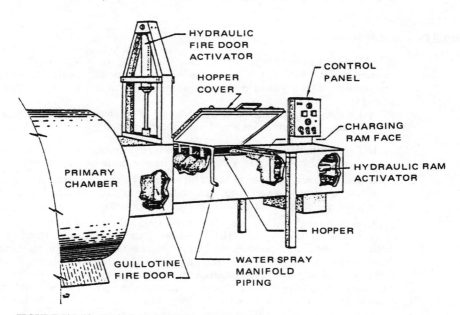

FIGURE 13A.18   Typical standard hopper-ram assembly.

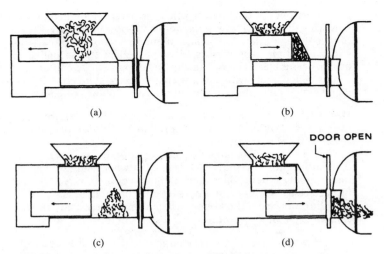

(a)   (b)

DOOR OPEN

(c)   (d)

FIGURE 13A.19   Double-ram type of charging system.

Typical Systems.   Dimensional data of a typical SAU (Morse Boulger, Inc.) system are shown in Fig. 13A.20. Its internal configuration is similar to that of the unit in the upper left of Fig. 13A.16. Its charging system is similar to that of Fig. 13A.18.

Table 13A.16 lists typical SAU systems which are provided with energy generation systems. These are basically "standard" models which are normally modified to the customers' specific waste, heat recovery mode, or other needs.

Central Disposal Systems

The Need for Central Disposal.   Central disposal of waste is a consideration in the disposal of municipal solid waste. In recent years industrial firms with many plants have looked toward incineration of their wastes in a central plant location as an efficient and economical means of disposal.

In Europe, after World War II, the disposal of municipal solid waste at a central location, by incineration, was given impetus by the following factors:

1. Population concentrations and increases required the use of more land for housing and for farming. The use of land for burying refuse was becoming impractical.

2. Technology developed to the point where it became economical to generate energy, i.e., steam and/or electric power, from incineration. Economics of scale dictated that the larger the facility, the more efficient would be its energy generation potential.

3. In general all utilities, including refuse disposal and electric or steam power generating industries, were state-owned. The interests of the electric utility and the refuse disposal authority were therefore common. This conflicts with conditions in the United States, where refuse collection is a public or government function and electric power generation is generally a private-sector function. Cooperation between these agencies in Europe has promoted the development of energy-producing incinerators. The power utility readily purchases energy from an incineration facility, providing revenues for the incinerator operation.

4. The higher cost of fossil fuel, particularly fuel oil, has helped promote energy generation, hence, central disposal facilities.

CAPACITY lbs./hr.	A	B	C	CC*	D	E	F	FF*	G	H	J	JJ*	CHUTE CAP. cu. yd.
100-160	8'-6"	4'-6"	5'-0"	9'-0"	6'-4"	3'-6"	9'0"	13'-0"	1'-4"	11'-0"	2'-0"	6'-0"	0.5
220-350	9'-6"	5'-6"	6'-0"	10'0"	7'-0"	4'-6"	11'-0"	15'-0"	1'-7½"	11'-0"	2'-0"	6'-0"	0.5
320-525	9'-6"	6'-0"	6'-6"	10'-6"	7'-0"	5'-6"	12'-6"	16'-6"	1'-9"	11'-0"	2'-6"	6'-6"	1.0
430-700	10'-0"	6'-6"	7'-0"	11'-0"	8'-0"	5'-6"	13'-0"	17'-0"	2'-2½"	11'-0"	2'-6"	6'-6"	1.0
640-1050	11'-6"	7'-6"	8'-0"	12'-0"	8'-0"	6'-0"	15'-0"	19'-0"	2'-6"	13'-0"	2'-6"	6'-6"	2.0
870-1400	12'-0"	8'-0"	8'-6"	12'-6"	9'-0"	6'-6"	16'-0"	20'-0"	2'-9½"	13'-0"	3'-6"	8'-0"	2.0
1300-2100	12'-6"	9'-6"	10'-0"	14'-0"	9'-6"	8'-6"	19'6"	23'-6"	3'-5"	13'-0"	3'-6"	8'-6"	3.0
1950-3200	14'-6"	10'-6"	11'-0"	15'-0"	10'-0"	9'-6"	21'-6"	25'-6"	3'-11"	13'-0"	4'-0"	9'-0"	3.0
2400-3900	16'-0"	11'-0"	11'-6"	15'-6"	12'-0"	9'-6"	22'-0"	26'-0"	4'-5"	14'-0"	5'-6"	9'-6"	4.0
2900-4700	18'-0"	11'-6"	12'-0"	16'-0"	14'-0"	9'-6"	22'-6"	26'-6"	4'-9"	14'-0"	5'-6"	9'-6"	4.0

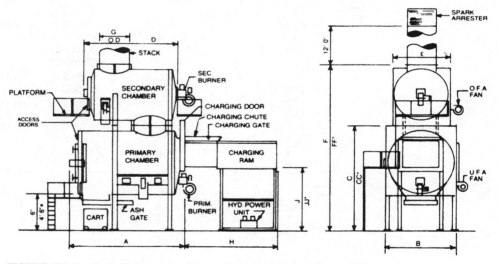

FIGURE 13A.20   Typical SAU system. *(Source: Morse Boulger, Inc., Queens, N.Y.)*

Only in the past decade, when the United States came to the realization that the cost and availability of energy were unreliable and out of its control, has a serious attempt been made to generate energy from waste in central collection and incineration facilities. Table 13A.17 lists information on selected mass burning central disposal facilities in the United States. Cost data are included for reference only.

Municipal Solid Waste.   Central-station incineration is usually applied to municipal solid waste destruction. The average characteristics of refuse and other wastes are listed earlier in this chapter. The actual variation in average waste composition from one country to another is listed in Table 13A.18. The Ash column represents the residual from coal or wood burning for domestic heat in the winter months. For instance, 43 percent of the composition of refuse in the United States was ash due to household coal burning in 1939, whereas 30 years later this component, i.e., ash from coal burning, was absent from the refuse.

TABLE 13A.16   SAU Energy Generation

Burning rate, lb/h	Waste feed, Btu/lb	Energy generation mode	Energy generation rate	Manufacturer
700	6,000	Steam, 100 lb/in² gauge	2,550 lb/h	Morse Boulger
1,000	6,000	Steam, 100 lb/in² gauge	3,850 lb/h	Morse Boulger
1,400	6,000	Steam, 100 lb/in² gauge	5,100 lb/h	Morse Boulger
3,200	6,000	Steam, 100 lb/in² gauge	11,600 lb/h	Morse Boulger
4,700	6,000	Steam, 100 lb/in² gauge	17,000 lb/h	Morse Boulger
1,280	6,285	Steam, 160 lb/in² gauge	2,025 lb/h	George L. Simonds
1,650	8,500	Steam, 150 lb/in² gauge	3,500 lb/h	George L. Simonds
1,050	6,240	Steam, 125 lb/in² gauge	1,900 lb/h	George L. Simonds
650	6,500	Steam, 150 lb/in² gauge	2,500 lb/h	George L. Simonds
1,800/15 gal/h	8,500 trash/ waste oil	Steam, 150 lb/in² gauge	8,000 lb/h	George L. Simonds
1,000	6,500	Steam, 100 lb/in² gauge Hot water, 105° Δt	3,458 lb/h 66 gal/min	Smokatrol
1,500	6,500	Steam, 100 lb/in² gauge Hot water, 105° Δt	5,187 lb/h 86 gal/min	Smokatrol
2,000	6,500	Steam, 100 lb/in² gauge Hot water, 105° Δt	6,916 lb/h 132 gal/min	Smokatrol
2,500	6,500	Steam, 100 lb/in² gauge Hot water, 105° Δt	8,645 lb/h 165 gal/min	Smokatrol
1,250	4,500 7,000	Steam, 150 lb/in² gauge	3,200 lb/h 4,950 lb/h	Consumat
2,100	4,500 7,000	Steam, 150 lb/in² gauge	5,400 lb/h 8,400 lb/h	Consumat
6,250	4,500 7,000	Steam, 150 lb/in² gauge	16,100 lb/h 25,000 lb/h	Consumat
8,400	4,500 7,000	Steam, 150 lb/in² gauge	21,600 lb/h 33,600 lb/h	Consumat

Source: Selected manufacturers' data.

When burning refuse, the generation of dry gas and moisture from combustion can be estimated as follows:

Dry gas   7.5 lb/10,000 Btu fired
Moisture   0.51 lb/10,000 Btu fired

Grate System.   The grate system is one of the most crucial systems within the mass burning incinerator. The grate must transport refuse through the furnace and, at the same time, promote combustion by adequate agitation and good mixing with combustion air. Abrupt tumbling caused by the dropping of burning solid waste from one tier to another will promote combustion. This action, however, may contribute to excessive carryover of particulate matter in the exiting flue gas. A gentle agitation will decrease particulate emissions.

Combustion is largely achieved by injection of combustion air below the grates (i.e., underfire air). Underfire air is also necessary to cool the grates. It is normally provided at a rate of approximately 40 to 60 percent of the total air entering the furnace. Too low a flow of underfire air will inhibit the burning process and will result in high grate temperatures.

Note the ash fusion temperatures listed in Table 13A.14. These temperatures limit the operating temperatures of the grate areas. With insufficient air a reducing atmosphere will result, and the ash deformation temperature can be as low as 1800°F. If the refuse reaches this

TABLE 13A.17 Field-Erected Mass Burn Facilities

Name	Status*	Design tons/day	Original capital costs, $	Year	Capital cost, 1990 $	Capital cost per ton, 1990 $	Additional capital cost	Year	Additional cost, 1990 $	Additional capital cost per ton, 1990 $
Process: MB—refractory										
Energy type—steam										
Betts Avenue	05	1000	5,000,000	65	21,594,896	21,595	36,500,000	89	37,440,152	37,440
City of Waukesha (old plant)	05	175	1,700,000	71	4,856,118	27,749	3,900,000	79	5,806,047	33,177
Davis County	05	400	40,000,000	88	41,693,608	104,234	0		0	0
Average		525	15,566,667		22,714,874	51,193	20,200,000		21,623,100	35,309
Standard deviation		348	17,329,423		15,059,680	37,590	16,300,000		15,817,053	2,132
Energy type—electricity										
McKay Bay Refuse-to-Energy Facility	05	1000	72,700,000	85	81,083,856	81,084	500,000	87	532,861	533
Average		1000	72,700,000		81,083,856	81,084	500,000		532,861	533
Standard deviation		0	0		0	0	0		0	0
Energy type—steam and electricity										
Muscoda	05	125	8,250,000	87	8,792,208	70,338	0		0	0
Average		125	8,250,000		8,792,208	70,338	0		0	0
Standard deviation		0	0		0	0	0		0	0
Process: MB—waterwall										
Energy type—steam										
Brooklyn Navy Yard	02	3000	426,000,000	90	426,000,000	142,000	0		0	0
Hampton/NASA Project Recoup	05	200	10,400,000	78	16,823,896	84,119	2,450,000	87	2,611,020	13,055
Norfolk Naval Station	08	360	3,220,000	67	12,995,170	36,098	5,400,000	87	5,754,899	15,986
Savannah	05	500	35,000,000	85	39,036,240	78,072	0		0	0
Average		1015	118,655,000		123,713,827	85,072	3,925,000		4,182,960	14,521
Standard deviation		1151	177,836,647		174,807,968	37,713	1,475,000		1,571,940	1,466
Energy type—electricity										
Albany (American Ref-Fuel)	02	1500	200,000,000	89	205,151,520	136,768	0		0	0
Alexandria/Arlington R.R. Facility	05	975	75,900,000	85	84,652,896	86,823	2,000,000	89	2,051,515	2,104
Babylon Resource Recovery Project	05	750	85,520,000	85	95,382,272	127,176	0		0	0
Bergen County	02	3000	335,000,000	91	335,000,000	111,667	0		0	0
Bridgeport RESCO	05	2250	211,000,000	85	235,332,768	104,592	0		0	0
Bristol	05	650	58,800,000	85	65,580,888	100,894	0		0	0
Broome County	02	571	77,000,000	90	77,000,000	134,851	0		0	0
Broward County (Northern Facility)	03	2250	216,007,000	90	216,007,000	96,003	0		0	0
Broward County (Southern Facility)	03	2250	277,816,000	90	277,816,000	123,474	0		0	0
Camden County (Foster Wheeler)	03	1050	96,000,000	87	102,309,328	97,437	0		0	0
Camden County (Pennsauken)	03	500	88,000,000	90	88,000,000	176,000	0		0	0
Central Mass. Resource Recovery Project	05	1500	140,000,000	86	152,686,272	101,791	0		0	0

(continued)

13.35

TABLE 13A.17 Field-Erected Mass Burn Facilities (*Continued*)

Name	Status*	Design tons/day	Original capital costs, $	Year	Capital cost, 1990 $	Capital cost per ton, 1990 $	Additional capital cost	Year	Additional cost, 1990 $	Additional capital cost per ton, 1990 $
			Process: MB—waterwall (*Continued*)							
City of Commerce	05	400	35,010,000	86	38,182,472	95,456	1,000,000	89	1,025,758	2,564
Concord Regional S.W. Recovery Facility	05	500	53,500,000	85	59,669,688	119,339	0		0	0
Dakota County	02	800	108,852,000	90	108,852,000	136,065	0		0	0
East Bridgewater (American Ref-Fuel)	02	1500	150,000,000	90	150,000,000	100,000	0		0	0
Eastern-Central Project	02	550	78,000,000	89	80,009,088	145,471	0		0	0
Essex County	03	2277	252,500,000	89	259,003,776	113,748	0		0	0
Fairfax County	05	3000	195,500,000	88	203,777,536	67,926	0		0	0
Falls Township (Wheelabrator)	02	2250	200,000,000	91	200,000,000	88,889	0		0	0
Glendon	02	500	63,500,000	90	63,500,000	127,000	0		0	0
Gloucester County	05	575	60,000,000	90	60,000,000	104,348	0		0	0
Haverhill (Mass Burn)	05	1650	120,000,000	87	127,886,656	77,507	0		0	0
Hempstead (American Ref-Fuel)	05	2505	255,000,000	85	284,406,912	113,536	0		0	0
Hennepin County (Blount)	05	1200	80,000,000	88	83,387,216	69,489	0		0	0
Hillsborough County S.W.E.R. Facility	05	1200	80,500,000	87	85,790,640	71,492	0		0	0
Hudson County	02	1500	179,000,000	89	183,610,592	122,407	0		0	0
Huntington	03	750	153,500,000	90	153,500,000	204,667	0		0	0
Johnston (Central Landfill)	02	750	80,000,000	90	80,000,000	106,667	0		0	0
Lake County	04	528	60,000,000	90	60,000,000	113,636	0		0	0
Lancaster County	03	1200	102,000,000	89	104,627,280	87,189	0		0	0
Lee County	02	1800	146,964,600	90	146,964,600	81,647	0		0	0
Lisbon	02	500	100,000,000	90	100,000,000	200,000	0		0	0
Marion County Solid W-T-E. Facility	05	550	47,500,000	86	51,804,272	94,190	0		0	0
Montgomery County	02	1800	280,000,000	89	287,212,096	159,562	0		0	0
Montgomery County	03	1200	115,000,000	89	117,962,112	98,302	0		0	0
Morris County	02	1340	141,900,000	89	145,555,008	108,623	0		0	0
Energy type—electricity										
New Hampshire/Vermont S.W. Project	05	200	26,500,000	85	29,556,012	147,780	0		0	0
North Andover	05	1500	185,000,000	85	206,334,432	137,556	0		0	0
North Hempstead	02	990	135,000,000	89	138,477,280	139,876	0		0	0
Oklahoma City	08	820	35,000,000	87	37,300,272	45,488	0		0	0
Onondaga County	02	990	132,000,000	90	132,000,000	133,333	0		0	0
Oyster Bay	02	1000	135,000,000	90	135,000,000	135,000	0		0	0
Pasco County	03	1050	90,600,000	89	92,933,632	88,508	0		0	0
Passaic County	02	1434	142,000,000	90	142,000,000	99,024	0		0	0
Pinellas County (Wheelabrator)	05	3150	83,000,000	83	94,280,208	29,930	60,000,000	86	65,436,968	20,774
Portland	05	500	45,500,000	87	48,490,360	96,981	20,600,000	90	20,600,000	41,200

13.36

(continued)

Name	Status*	Design tons/day	Original capital costs, $	Year	Capital cost, 1990 $	Capital cost per ton, 1990 $	Additional capital cost	Year	Additional cost, 1990 $	Additional capital cost per ton, 1990 $
Process: MB—waterwall (Continued)										
Preston (Southeastern Connecticut) S.E. Resource Recovery Facility (SERRF)	03	600	83,000,000	87	88,454,944	147,425	0		0	0
Saugus	04	1380	106,000,000	87	112,966,544	81,860	0		0	0
Spokane	05	1500	33,000,000	74	74,160,992	49,441	95,000,000	90	95,000,000	63,333
Stanislaus County Res Recovery Facility	03	800	82,149,000	87	87,548,016	109,435	0		0	0
Sturgis	05	800	82,200,000	85	91,679,408	114,599	0		0	0
	02	560	0	0	0	0	0	0	0	0
Union County	02	1440	150,000,000	90	150,000,000	104,167	0		0	0
Warren County	05	400	50,300,000	89	51,595,608	128,989	0		0	0
Washington-Warren Counties	03	400	50,000,000	90	50,000,000	125,000	0		0	0
West Pottsgrove Recycling/R.R. Facility	02	1500	150,000,000	91	150,000,000	100,000	0		0	0
Westchester	05	2250	179,000,000	83	203,327,168	90,368	0		0	0
Average		1230	122,359,975		127,837,294	110,691	35,720,000		36,822,848	25,995
Standard deviation		723	69,637,080		70,966,036	32,515	36,537,017		37,301,493	23,547
Energy type—steam and electricity										
Charleston County	05	644	59,000,000	89	60,519,688	93,975	0		0	0
Davidson County	03	210	7,000,000	87	7,460,056	35,524	0		0	0
Harrisburg	05	720	8,300,000	71	23,709,280	32,930	21,300,000	86	23,230,124	32,264
Jackson County/Southern Michigan State Prison	05	200	28,000,000	86	30,537,256	152,686	0		0	0
Kent County	05	625	62,200,000	89	63,802,128	102,083	0		0	0
Nashville Thermal Transfer Corp. (NTTC)	05	1120	24,500,000	74	55,058,920	49,160	36,500,000	85	40,709,224	36,348
Northwest Waste-To-Energy Facility	05	1600	23,000,000	68	86,385,568	53,991	5,000,000	88	5,211,701	3,257
Olmstead County	05	200	30,000,000	87	31,971,664	159,858	0		0	0
Quonset Point	02	710	83,000,000	90	83,000,000	116,901	0		0	0
S.W. Resource Recovery Facility (BRESCO)	05	2250	185,000,000	83	210,142,624	93,397	0		0	0
University City Res. Recovery Facility	05	235	27,000,000	87	28,774,496	122,445	0		0	0
Walter B. Hall Res. Recovery Facility	05	1125	114,000,000	87	121,492,320	107,993	0		0	0
Wayne County	02	300	27,000,000	90	27,000,000	90,000	0		0	0
Average		765	52,153,846		63,834,923	93,149	20,933,333		23,050,350	23,956
Standard deviation		597	48,428,117		52,029,873	39,364	12,862,435		14,492,361	14,731
Process: MB—rotary combustor										
Energy type—steam										
Galax	05	56	2,100,000	85	2,342,175	41,825	160,000	88	166,774	2,978
Average		56	2,100,000		2,342,175	41,825	160,000		166,774	2,978
Standard deviation		0	0		0	0	0		0	0

TABLE 13A.17 Field-Erected Mass Burn Facilities (*Continued*)

Name	Status*	Design tons/day	Original capital costs, $	Year	Capital cost, 1990 $	Capital cost per ton, 1990 $	Additional capital cost	Year	Additional cost, 1990 $	Additional capital cost per ton, 1990 $
			Process: MB—rotary combustor (*Continued*)							
Energy type—electricity										
Auburn (New Plant)	03	200	26,500,000	90	26,500,000	132,500	0		0	0
Delaware County Regional R.R. Project	03	2688	276,000,000	90	276,000,000	102,679	0		0	0
Gaston County/Westinghouse R.R. Center	02	440	42,000,000	89	43,081,824	97,913	0		0	0
MacArthur Energy Recovery Facility	05	518	38,700,000	85	43,162,936	83,326	2,500,000	91	2,500,000	4,826
Mercer County	02	975	117,500,000	88	122,474,976	125,615	0		0	0
Monmouth County	02	1700	220,000,000	90	220,000,000	129,412	0		0	0
Montgomery County (North)	05	300	7,494,000	69	25,688,292	85,628	9,700,000	87	10,337,504	34,458
Montgomery County (South)	02	900	6,150,000	69	21,081,268	23,424	5,000,000	85	5,576,606	6,196
Oakland County	02	2000	172,000,000	90	172,000,000	86,000	0		0	0
San Juan Resource Recovery Facility	02	1040	91,400,000	89	93,754,240	90,148	0		0	0
Skagit County	05	178	14,000,000	87	14,920,112	83,821	0		0	0
Westinghouse/Bay Resource Mgmt. Center	05	510	38,000,000	86	41,443,416	81,262	0		0	0
York County	05	1344	91,200,000	88	95,061,424	70,730	0		0	0
Average		984	87,764,923		91,936,038	91,728	5,733,333		6,138,037	15,160
Standard deviation		737	83,297,756		80,657,588	27,471	2,984,776		3,224,182	13,657
Energy type—steam and electricity										
Dutchess County	05	506	35,000,000	84	39,213,904	77,498	0		0	0
Falls Township (Technochem)	02	70	7,000,000	90	7,000,000	100,000	0		0	0
Monroe County	02	500	100,000,000	91	100,000,000	200,000	0		0	0
Sangamon County	02	450	38,160,000	90	38,160,000	84,800	0		0	0
Sumner County	05	200	9,800,000	81	12,655,414	63,277	5,340,000	90	5,340,000	26,700
Waukesha County (New Plant)	02	600	100,000,000	90	100,000,000	166,667	5,340,000		5,340,000	26,700
Average		388	48,326,667		49,504,886	115,374	5,340,000		5,340,000	26,700
Standard deviation		188	38,326,286		37,635,694	50,187	0		0	0

* Facility status: advanced planning 02, construction 03, shakedown 04, operation 05 through 07, and temporarily shutdown 08.

Source: *Data Summary of Municipal Solid Waste Management Alternatives,* vol. III: *Appendix A—Mass Burn Technologies,* National Technical Information Service, October 1992, NTIS accession number DE92016433.

TABLE 13A.18 A Summary of International Refuse Composition (in Percentages)

	Ash	Paper	Organic matter	Metals	Glass	Miscellaneous
United States (1939)	43.0	21.9	17.0	6.8	5.5	5.8
United States (1970)	0	44.0	26.5	8.6	8.8	12.1
Canada	5	70	10	5	5	5
United Kingdom	40–40	25–30	10–15	5–8	5–8	5–10
France	24.3	29.6	24	4.2	3.9	14
West Germany	30	18.7	21.2	5.1	9.8	15.2
Sweden	0	55	12	6	15	12
Spain	22	21	45	3	4	5
Switzerland	20	40–50	15–25	5	5	—
Netherlands	9.1	45.2	14	4.8	4.9	22
Norway (summer)	0	56.6	34.7	3.2	2.1	8.4
Norway (winter)	12.4	24.2	55.7	2.6	5.1	0
Israel	1.9	23.9	71.3	1.1	0.9	1.9
Belgium	48	20.5	23	2.5	3	3
Czechoslovakia (summer)	6	14	39	2	11	28
Czechoslovakia (winter)	65	7	22	1	3	2
Finland	—	65	10	5	5	15
Poland	10–21	2.7–6.2	35.3–43.8	0.8–0.9	0.8–2.4	—
Japan (1963)	19.3	24.8	36.9	2.8	3.3	12.9

Source: Brunner (1988a).

temperature, slagging will begin, further reducing the air supply by clogging the grates and forming large, unwieldy clinkers. The ash properties of coal are listed for comparison.

Overfire air is injected above the grates. Its main purpose is to provide sufficient air to completely combust the flue gas and flue gas particulate rising from the grates. Numerous injection points are located on the furnace walls above the grates to provide a turbulent over-fire air supply along the furnace length.

Ash and other particles dropping through the grates are termed *siftings,* and they must be effectively removed from the system. Siftings can readily clog grate mechanisms, generate fires, and create housekeeping problems if not attended to. Siftings, due to their small particle size, have been found to be more dense than incinerator ash, approximately 1780 versus 1040 lb/yd^3 for typical incinerator ash.

Grate Design.   A number of different types of grate designs are used in central waste burning facilities. Each grate system manufacturer provides a unique grate feature, attempting to obtain a competitive edge in the marketplace. The grate system manufacturer should be contacted for design and sizing information for a particular grate design. The following listing describes typical grate systems, both generic grate types and grates specific to certain manufacturers:

Traveling Grate.   This type is no longer in common usage. As shown in Fig. 13A.21 it is normally not a single grate but a series of grates which are placed in a manner that separates the drying and burning functions of the incinerator.

Rocking Grate.   As shown in Fig. 13A.22, these grate sections are placed across the width of the furnace. Alternate rows are mechanically pivoted or rocked to produce an upward and forward motion, advancing and agitating the waste. The stroke of the grate sections is 5 to 6 in. This grate will handle refuse on a continuous basis.

Reciprocating Grate.   As shown in Fig. 13A.23, this grate consists of sections stacked above each other similar to overlapping roof shingles. Alternate grate sections slide back and forth while adjacent sections remain fixed. Drying and burning are accomplished on single, short but wide grates. The moving grates are basically bars, stoking bars, which move the waste along and help agitate it.

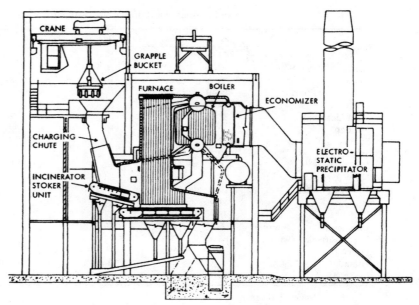

FIGURE 13A.21   Traveling grate system. *(Source: Brunner, 1988a.)*

Rotary Kiln.   As shown in Fig. 13A.24, two traveling grates are initially used for drying the incoming refuse and for initial ignition. The kiln is at the heart of this system. By varying the kiln rotational speed, burnout of the refuse is accurately controlled. The refuse burns out in the kiln, and ash is discharged from the end of the kiln to residue conveyers. Some of the flue gases are diverted for drying the incoming refuse. Flue gas can be passed through a waste heat boiler for energy recovery.

Martin System.   As shown in Fig. 13A.25, this system utilizes reverse reciprocating grates. As the grates move forward and then reverse, there is continuous agitation of the waste. The

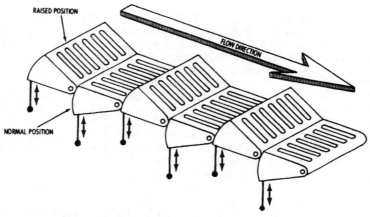

FIGURE 13A.22   Rocking grates.

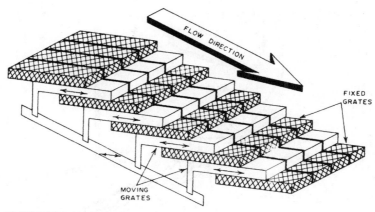

FIGURE 13A.23   Reciprocating grates.

bars making up the grates are hollow, allowing air to circulate within them and keep them relatively cool.

Von Roll System.   As shown in Fig. 13A.26, a series of reciprocating grates is used to move refuse through the furnace. The first grate section dries the refuse, the second is a burning grate, and burnout to ash takes place on the third grate.

VKW System.   Shown in Fig. 13A.27 is a variation of the traveling grate concept. A series of drums is utilized as grates. The drums rotate slowly, agitating the waste and moving it along to subsequent drums. Air passes through openings in these drums, or roller grates, as underfire air. Both speed of rotation on the roller grates and quantity of underfire air per roller grate are variable.

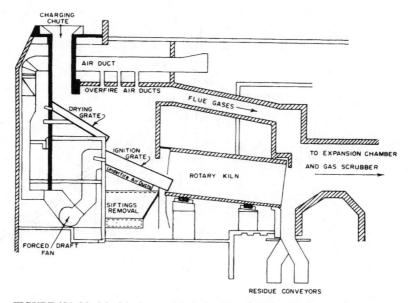

FIGURE 13A.24   Municipal rotary kiln incineration facility. *(Source: Brunner, 1988a.)*

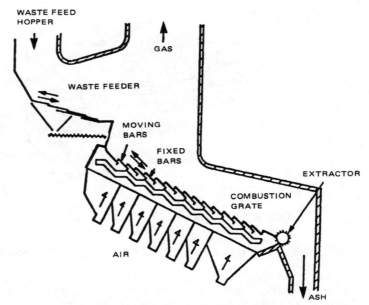

FIGURE 13A.25   Martin system. *(Source: Brunner, 1988a.)*

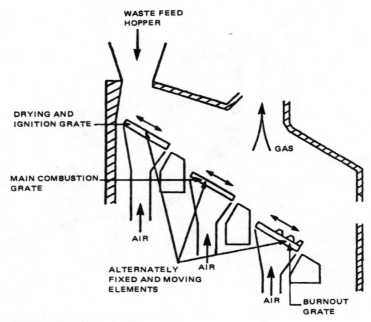

FIGURE 13A.26   Von Roll system. *(Source: Brunner, 1988a.)*

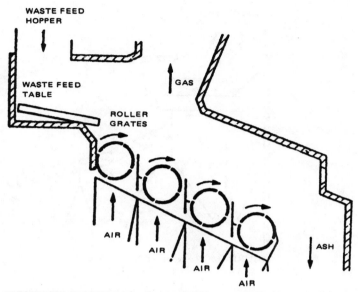

FIGURE 13A.27   VKW system. *(Source: Brunner, 1988a.)*

Alberti System.   As illustrated in Fig. 13A.28, this grate system has a single-section grate constructed of fixed and moving elements arranged, as shown, in a series of steps. Feed is rammed from the feed hopper to the grates. The fixed grate contains the refuse while the moving elements agitate the waste, driving it down to the next grate.

Esslingen System.   As shown in Fig. 13A.29, a traveling grate is used to feed a rocking grate system. It is normally provided with a single-grate section composed of semicircular

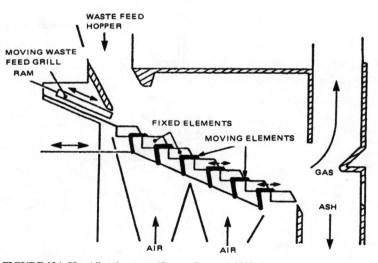

FIGURE 13A.28   Alberti system. *(Source: Brunner, 1988a.)*

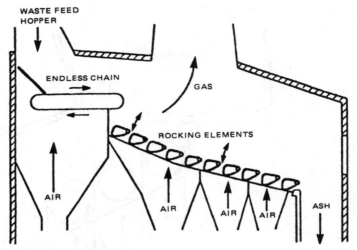

FIGURE 13A.29   Esslingen system. *(Source: Brunner, 1988a.)*

rocking elements. Each movement of these elements promotes transport and agitation of
the waste. Underfire air passes through the rocking elements, keeping them cool and providing for combustion of the waste.

   The Heenan Nichol System.   As shown in Fig. 13A.30, this system utilizes grates composed of three or more sections which are arranged in steps. Each pair of elements moves in
a rocking manner so that at any moment half of the elements are moving. All odd-numbered
elements are linked to each other, as are all the even-numbered elements. The rocking
action moves and agitates the waste.

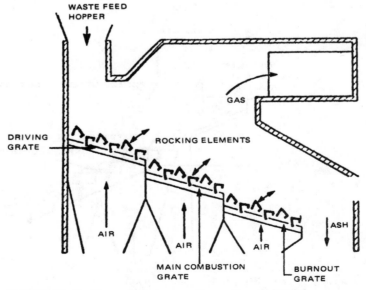

FIGURE 13A.30   Heenan Nichol system. *(Source: Brunner, 1988a.)*

CEC System.   This system, illustrated in Fig. 13A.31, utilizes a single-section grate. Two or more grate sections can be arranged in parallel. The grate is constructed of successive sliding, rocking, and fixed elements. The sliding and rocking elements are synchronized so that the sliding elements move over the rocking elements when the rocking elements are retracted. The sliding elements, therefore, promote transport of the waste while the rocking elements provide the required agitation.

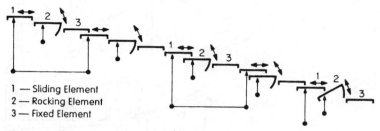

1 — Sliding Element
2 — Rocking Element
3 — Fixed Element

FIGURE 13A.31   CEC system. *(Source: Brunner, 1988a.)*

Bruun and Sorensen System.   As shown in Fig. 13A.32, this system utilizes a series of rollers, up to six in each of its three sections. Odd-numbered rollers turn clockwise while even-numbered rollers turn counterclockwise. Underfire air passes through the rotary grates, or drums. The action of the drums provides good agitation, and the slope of the grate to the horizontal promotes the transport of waste along the grate.

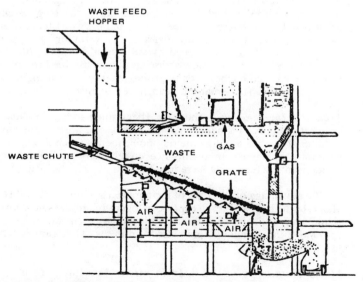

FIGURE 13A.32   Bruun and Sorensen system. *(Source: Brunner, 1988a.)*

Volund System.   This system is shown in Fig. 13A.33. It utilizes a rotary kiln for controlled burning of waste. Reciprocating grates are used for waste drying and initial combustion. Burnout takes place within the kiln.

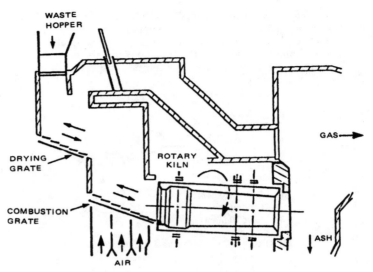

FIGURE 13A.33 Volund System. *(Source: Brunner, 1988a.)*

Associated Disposal Systems.   There are refuse burning systems in use which cannot strictly be classified as grate systems.

Suspension Burning.   An incinerator coupled to a refuse processing system is pictured in Fig. 13A.34. Refuse is shredded and air classified into light and heavy fractions. The light fraction is blown into the boiler through a pneumatic charging system. Figure 13A.35 shows the air distribution within the furnace and around the waste feed. Waste that is not burned in suspension will drop onto the shredder stoker, a variation of the traveling grate, and will burn out. Ash not airborne, produced by suspension burning, will also drop onto the spreader stoker. The stoker moves slowly, discharging its ash load to an ash hopper for ultimate disposal. Heavier components of the refuse that are not incinerated are composed mainly of metals and glass. These materials can, in certain instances, be marketed.

Fluid Bed Incineration.   There have been some attempts to adapt limited European experience with fluid bed incineration of municipal solid waste to the United States. This technology requires that glass and low-melting-point metals (such as aluminum) be removed from the waste stream. These components, in even relatively small quantities, will slag the furnace bed. In addition, the feed must be reduced to uniform size, no larger than 1- to 1½-in mean particle size.

The advantage of this type of incineration system is the ability to add limestone (or other alkali) to the bed, which will capture halogens (chlorides and fluorides) and other compounds, significantly reducing the discharge of acid gases. The effort and resultant high cost required to remove the aluminum and glass from the waste stream, however, restrict the use of this technology.

Sludge Burning in an MSW Incinerator.   Moisture content is the single most important parameter in determining the burning characteristics of a material. The higher the moisture content, the longer it will take that material to burn. When materials of different moisture content are placed in an incinerator at the same time, the lower-moisture-content material will burn off first, while moisture is evaporated from the second material. It will exit the incinerator as it came into the incinerator, but with less moisture.

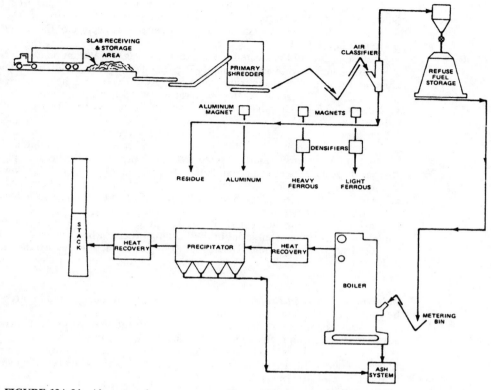

FIGURE 13A.34 Akron recycle energy system. *(Source: Akron Recycle Energy System, Teledyne National, Akron, Ohio.)*

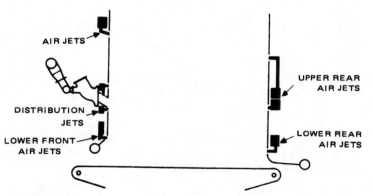

FIGURE 13A.35 Air and feed distribution. *(Source: Babcock & Wilcox, North Canton, Ohio.)*

Municipal solid waste has a moisture content in the range of 20 to 30 percent. The moisture content of sewage sludge is normally in the range of 70 to 80 percent. The only effective method of firing these two waste streams has been by a reduction of the sludge moisture content to that of the MSW.

A refuse incinerator is located at the site of a sewage treatment plant at a New York facility. Sewage sludge, having approximately 75 percent moisture content, is sprayed on top of the refuse as the refuse was entering the incinerator. The refuse has a moisture content of between 20 and 30 percent. At the incinerator ash discharge the refuse is burned out, however, the sludge is unburned. It is found exiting the incinerator smoldering and identifiable as sludge.

In contrast, at a Connecticut facility a refuse incinerator is also located on the site of a wastewater treatment plant. Sludge, as above, is generated at approximately 75 percent moisture content. An extensive sludge drying and conveying system has been employed at this site, however, to reduce its moisture content to from 15 to 20 percent. This allows the sludge to burn in no longer a period of time than it takes the refuse to burn. There have been no problems with burnout at this plant. Sludge is completely fired, with no sludge residual in the ash discharge of the incinerator.

A major reason for the absence of sewage sludge burning facilities at refuse incinerator plants is in the problems inherent in the burning process when materials of these dissimilar moisture fractions are present. The cost of drying equipment to reduce the moisture content of the sludge to that of the refuse is usually prohibitive and the sludge is either taken to its own dedicated incinerator or an alternative sludge disposal method is found.

Incinerator Corrosion Problems.   Severe corrosion has been found in three major areas of the mass burning incinerator system.

Scrubber Corrosion.   The acidic components of the flue gas present corrosion problems in wet scrubbing equipment, which will be discussed later in this chapter.

Corrosion of Grates.   With insufficient airflow, high temperatures and a reducing atmosphere can occur and ash can soften or fluidize. Fluid ash can be exceedingly corrosive, readily attacking cast iron or steel.

TABLE 13A.19   Gas-Phase Corrosion at Elevated Temperatures

Alloy	Temperature, °F	Corrosion rate, mils/month
A106	800	0.9
A106	1000	8
A106	1200	36
T11	800	0.8
T11	1000	6
T11	1200	29

1 mil = 2.54×10^{-2} mm.
Source: Brunner (1988a).

Table 13A.19 illustrates the corrosion rate as a function of temperature for two steel alloys in widespread use in grate construction. Temperature alone greatly increases the corrosive rate, the amount of material lost per month of service. At 1200°F over ⅜ in of material is "wasted" or lost from steel components, for instance.

Fireside Corrosion.   There are two modes of corrosion that affect boiler tubes. Low-temperature or dewpoint corrosion is metal wastage caused by sulfuric or hydrochloric acid condensation. Chlorides and sulfides within the refuse (chlorides are present in plastics) will partially convert to hydrogen chloride, sulfur dioxide, and sulfur trioxide in the exhaust gas stream. These gases will condense at temperatures below 300°F, and their condensate or liquid phase will be hydrochloric and sulfuric acid, both of which will attack steel. It is important, therefore, that the temperature of the boiler tubes, constructed of steel, be kept above 300°F. This is a function of the temperature of the steam or hot water generated. The boiler tubes will be at a temperature close to that of the circulating fluid, and this 300°F rule therefore limits the minimum temperature of the steam or hot water generated.

High-temperature corrosion is a more complex problem. Table 13A.20 lists the steam pressure, temperature, and external tube temperatures for an assortment of incinerator systems burning municipal solid waste. At temperatures exceeding 700°F a complex reaction takes place between the sulfide and chlorine/chloride-bearing flue gas and the steel boiler tube, as illustrated in Fig. 13A.36. Chlorine reacts with the iron in the tube wall to produce ferrous chloride which, upon contact with oxygen in the flue gas, converts to iron oxide. The iron oxide (rust) will leave the surface of the steel, causing wastage of the steel surface. Other components of the refuse which become airborne, such as alkali salts, will promote this corrosion. For incinerators operating above 700°F metal temperatures (most of the incinerator tubes noted in Table 13A.20 operate at temperatures in excess of 700°F), special refractory-lined water walls must be utilized to protect the tubes from this metal wastage.

TABLE 13A.20 Nominal Operating Conditions of Waterwall Incinerators

Location	Steam pressure, lb/in^2 gauge	Steam temp., °F	Metal temp., °F (approx.)
Milan, Italy	500	840	890
Mannheim, Germany	1800	980	1030
Frankfurt, Germany	960	930	980
Munster, Germany	1100	980	1030
Moulineaux, France	930	770	820
Essen Karnap, Germany	—	930	980
Stuttgart, Germany	1100	980	1030
Munich, Germany	2650	1000	1050
Rotterdam, Netherlands	400	680	730
Edmonton, England	625	850	900
Coventry, England	275	415	465
Amsterdam, Netherlands	600	770	820
Montreal, Canada	225	395	445
Chicago (N.W.), Illinois	265	410	460
Oceanside, New York	460	465	515
Norfolk, Virginia	175	375	425
Braintree, Massachusetts	265	410	460
Harrisburg, Pennsylvania	275	460	510
Hamilton, Ontario	250	400	450

Source: Brunner (1988a).

Table 13A.21 indicates the relative performance of various alloys in incinerator fireside areas. Although stainless steels appear to have favorable corrosion resistance, the danger of stainless-steel stress corrosion cracking prohibits its use in pressure vessels such as boilers and high-temperature hot-water heaters. Figure 13A.37 further illustrates the rate of corrosion of carbon steel by chloride attack as a function of metal temperature.

Incinerator Refractory Selection.   Table 13A.22 illustrates the types of problems that can be expected within the various areas of a large incinerator system associated with refractory selection. The nature of the hot gas stream within the incinerator can create significant detrimental effects on grates, walls, ceilings, and other areas within the furnace enclosure. Some of the more common concerns in refractory selection are discussed here.

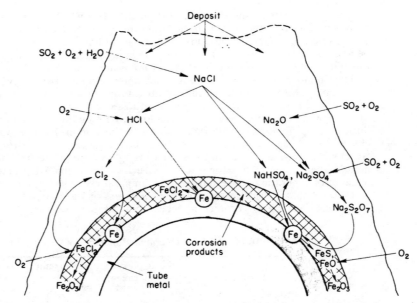

FIGURE 13A.36 Sequence of chemical reactions explaining corrosion on incinerator boiler tube. (*Source: Brunner, 1988a.*)

Abrasion. This is the effect of impact of moving solids within the gas stream or of heavy pieces of materials charged into the furnace upon refractory surfaces. Fly ash also causes abrasive effects. Abrasion is the wearing away of refractory, or any surface, under direct contact with another material with relative motion to its surface.

Slagging. When a portion of charged material, usually ash, metals, or glass within an incinerator, reaches a high enough temperature, deformation of that material will occur. The mate-

TABLE 13A.21 Performance of Alloys in Fireside Areas[a]

	Resistance to wastage		
Alloy	300–600°F	600–1200°F	Moist deposit
Incoloy 825	Good	Fair	Good
Type 446	Good	Fair	Pits
Type 310	Good	Fair	SCC[b]
Type 316L	Good	Fair	SCC
Type 304	Good	Fair	SCC
Type 321	Good	Fair	SCC
Inconel 600	Good	Poor	Pits
Inconel 601	Good	Poor	Pits
Type 416	Fair	Fair	Pits
A106-Grade B (carbon steel)	Fair	Poor	Fair
A213-Grade T11 (carbon steel)	Fair	Poor	Fair

[a] Arranged in approximate decreasing order.
[b] Stress-corrosion cracking.
Source: Brunner (1988a).

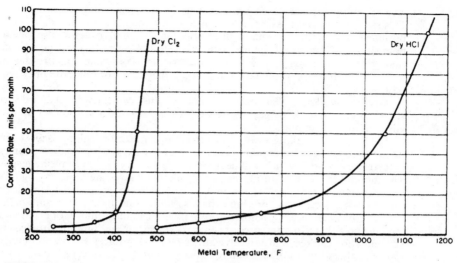

FIGURE 13A.37   Corrosion of carbon steel in chlorine and hydrogen chloride. (*Source: Brunner, 1988a.*)

rial will physically change to a more amorphous state and may begin to flow, as a heavy liquid. When the temperature of the material is then reduced below that required for deformation, the material will solidify into a hard slag. This process can take place when high temperatures are experienced on a grate. When the grate section moves through a lower-temperature zone, a slag may form. Molten ash may become airborne, then attach to a refractory or a metal sur-

TABLE 13A.22   Suggested Refractory Selection for Incinerators

Incinerator part	Temperature (°F) range	Abrasion	Slagging	Mechanical shock	Spalling	Fly ash adherence	Recommended refractory
Charging gate	70–2600	Severe, very important	Slight	Severe	Severe	None	Superduty
Furnace walls, grate to 48 in. above	70–2600	Severe	Severe, very important	Severe	Severe	None	Silicon carbide or superduty
Furnace walls, upper portion	70–2600	Slight	Severe	Moderate	Severe	None	Superduty
Stoking doors	70–2600	Severe, very important	Severe	Severe	Severe	None	Superduty
Furnace ceiling	700–2600	Slight	Moderate	Slight	Severe	Moderate	Superduty
Flue to combustion chamber	1200–2600	Slight	Severe, very important	None	Moderate	Moderate	Silicon carbide or superduty
Combustion chamber walls	1200–2600	Slight	Moderate	None	Moderate	Moderate	Superduty
Combustion chamber ceiling	1200–2600	Slight	Moderate	None	Moderate	Moderate	Superduty
Breeching walls	1200–3000	Slight	Slight	None	Moderate	Moderate	Superduty
Breeching ceiling	1200–3000	Slight	Slight	None	Moderate	Moderate	Superduty
Subsidence chamber walls	1200–1600	Slight	Slight	None	Slight	Moderate	Medium duty
Subsidence chamber ceiling	1200–1600	Slight	Slight	None	Slight	Moderate	Medium duty
Stack	500–1000	Slight	None	None	Slight	Slight	Medium duty

Source: Brunner (1988a).

face within the furnace which is cooler than the air stream. Slag will then form on this surface. This slag can be acidic (as a result of silicon, aluminum, or titanium oxides released from the burning waste), or it can be basic (due to the generation of oxides of iron, calcium, magnesium, potassium, sodium, or chromium), and the selection of refractory must be compatible with these materials to help ensure long refractory life. (For acidic slag, fire-clay or high-alumina and/or silica firebrick would be used. Chrome, magnesite, or forsterite brick would be used for basic slags.)

Mechanical Shock. The impact of falling refuse can cause mechanical shock, as can constant vibration caused by grate bushings or supports and vibration set up by turbulent flow adjacent to air inlet ports.

Spalling. The flaking away of the refractory surface, or spalling, is most commonly caused by thermal stresses or mechanical action. Uneven temperature gradients can cause local thermal stresses in brick which will degrade the refractory surface, causing a spalling condition. The more common type of mechanical spalling is caused by rapid drying of wet brickwork. The steamed water does not have an opportunity to escape the brick surface through the natural porosity of the refractory, but expands rapidly, causing cracking and spalling of the brick.

Fly Ash Adherence. As noted above, fly ash can have fluid properties within the hot gas stream and adhere to refractory or other cooler surfaces within the furnace chamber. Fly ash accumulation can result in corrosive attack on these surfaces. Heavy accumulations will interfere with the normal surface cooling effects within the furnace, resulting in a decrease of furnace refractory life.

As just noted, Table 13A.22 lists various areas within an incinerator and describes the severity of the problems at each of these locations. Normal temperature ranges are noted as well as recommended refractory types.

Heat Recovery or Wasting of Heat. Steam can be generated by utilization of incinerator waste heat, as shown in Table 13A.23, which lists steam production rates. Variations in waste heating value will produce variations in steam generation, as shown in Table 13A.24. But these quantities must be weighed against the overall implications of a boiler installation for waste burning. Note the following comparison between an energy recovery system and an incinerator without provision for heat recovery:

With heat recovery	Without heat recovery
Reduced gas temperatures and volumes due to absorption of heat by heat recovery system	Hotter gas temperatures
Moderate excess air	High excess air required to control furnace temperatures
Moderate-size combustion chamber	Large refractory-lined combustion chamber to handle high gas flow
Smaller air and induced-draft fans required for smaller gas volume	Higher gas volumes due to higher temperatures, requiring larger air and gas flow equipment
Steam facilities including integral water wall, boiler drums, and boiler auxiliary equipment required	No steam facilities required
Operations involve boiler system monitoring, adjustments for steam demand, etc.	Relatively simple operating procedures
Steam tube corrosion is possible as well as corrosion within exhaust gas train	Corrosion possible in the exhaust gas train
Licensed boiler operators are required to operate incinerator	Conventional operators satisfactory
Considerable steam credits possible, including in-plant energy savings in addition to salvage	Only credits are possible salvage of the equipment after its useful life

TABLE 13A.23 Typical Steam Generation

Solid waste type	MSW	MSW	MSW	RDF
Steam temperature, °F	620	500	465	400
Steam pressure, lb/in² gauge	400	225	260	250
Steam production, tons/ton refuse	3.6	1.4–3.0	1.5–4.3	4.2

Note: MSW: Municipal solid waste. RDF: Refuse-derived fuel, i.e., shredded MSW less metals.
Source: Brunner (1988a).

TABLE 13A.24 Steam Production Related to MSW Characteristics

	As-received heating value, Btu/lb				
	6500	6000	5000	4000	3000
Refuse:					
% moisture	15	18	25	32	39
% noncombustible	14	16	20	24	28
% combustible	71	66	55	44	33
Steam generated, tons/ton refuse	4.3	3.9	3.2	2.3	1.5

Source: Brunner (1988a).

Resource Recovery Plant Emissions. A number of states have established regulations specifically governing emissions of central disposal incinerators firing municipal solid waste (resource recovery facilities). Table 13A.25 lists these regulations for six states and the EPA. Most of these criteria are noted as guidelines. A guideline is not necessarily established by statute, but it is used by the regulator as a criterion for the permitting of these facilities.

Biomedical Waste Incineration

Biomedical wastes are generated by hospitals, laboratories, animal research facilities, and other institutional sources. The disposal of these wastes is coming under severe public scrutiny, and regulations are being promulgated to control their disposal. Incineration is a favored method of treating these wastes because it is the only commercially available method of treatment which destroys the organisms associated with this waste completely and effectively.

The Waste Stream. *Biomedical waste* is a term coming into common usage to replace what had been referred to as *pathological* or *infectious wastes* and to include additional related waste streams. Where the term *pathological waste* is used here, it refers to anatomical wastes, carcasses, and similar wastes. Table 13A.26 is a listing of wastes classified as biomedical and includes a description of each waste as well as typical characteristics. The bag designations (red, orange, yellow, blue) are used in Canada. In the United States, generally, most of these wastes are classified as "red bag."

The hospital waste stream has changed significantly in the last few years. Disposable plastics have been replacing glass and clothing in what appears to be, at first look, a means of cutting costs. They represent a greater cost in their disposal, however, since many of these plastics contain chlorine and with the increase in the use of plastics, the increase in chlorine creates the need for additional equipment in the incineration process. The plastics content of the hospital waste stream has grown from 10 percent to over 30 percent in the past 10 years.

TABLE 13A.25 Emissions Limitations for Municipal Waste Incinerators

Pollutant	California guidelines 7% O_2	Illinois guidelines 12% CO_2	New Jersey guidelines 7% O_2	New York guidelines 12% CO_2	Pennsylvania BAT criteria 7% O_2	USEPA guidelines 12% CO	Wisconsin guidelines 12% CO_2
Particulate (below 2 μm), gr/dry st ft³	0.01 0.008	0.01	0.015	0.01	0.015	—	0.015
HCl	30 ppm	30 ppm/1 h or 90% reduction	50 ppm/1 h or 90% reduction	50 ppm/8 h or 90% reduction	30 ppm/1 h or 90% reduction	—	50 ppm
SO_2	30 ppm	50 ppm/1 h or 70% reduction	50 ppm/1 h or 80% reduction	0.2–2.5 lb/(MBtu/h)	50 ppm/1 h or 70% reduction	—	—
NO_x	140–200 ppm	100 ppm/1 h	350 ppm/1 h	BACT	—	—	—
Hydrocarbons	70 ppm	—	70 ppm/1 h	—	—	—	—
CO	400 ppm	100 ppm/1 h	400 ppm/1 h 100 ppm/4 days	—	400 ppm/8 h 100 ppm/4 days	50 ppm/4 h	—
Dioxin	—	—	—	2 ng/Nm³	—	—	3 ng/Nm³
Furnace temp, design, °F	1800 ±200	1800	1800	1800	—	—	1500
Furnace temp, min., °F	—	1500	1500	1500 for 15 min	1800	1800	—
Residence time, min., s	1	1.2	1	1	1	1	1
Lime injection, min., lb/h	—	100	—	—	—	—	—
Baghouse temp., max., °F	—	—	—	300	—	—	250
Combustion efficiency, %	—	99.9/2 h	—	99.9/8 h 99.95/7 days	99.9/4 days	—	—
Minimum O_2, %	—	—	6	—	—	6–12	—
Opacity, max., %	—	10	20	10/6 min	30/3 min/h	—	20

TABLE 13A.26 Characterization of Biomedical Waste

Waste class	Component description	Typical component weight, % (as fired)	HHV dry basis, Btu/lb	Bulk density as fired, lb/ft	Moisture content of component, wt %	Weighted heat value range of waste component, Btu/lb	Typical component heat value of waste as fired, Btu/lb
A1	Human anatomical	95–100	8,000–12,000	50–75	70–90	760–2,600	1,200
(red bag)	Plastics	0–5	14,000–20,000	5–144	0–1	0–1,000	180
	Swabs, absorbents	0–5	8,000–12,000	5–62	0–30	0–600	80
	Alcohol, disinfectants	0–0.2	11,000–14,000	48–62	0–0.2	0–28	20
	Total bag						1,480
A2	Animal infected anatomical	80–100	9,000–16,000	30–80	60–90	720–6,400	1,500
(orange bag)	Plastics	0–15	14,000–20,000	5–144	0–1	0–3,000	420
	Glass	0–5	0	175–225	0	0	0
	Beddings, shavings, paper, fecal matter	0–10	8,000–9,000	20–46	10–50	0–810	600
	Total bag						2,520
A3a	Gauze, pads, swabs, garments, paper, cellulose	60–90	8,000–12,000	5–62	0–30	3,360–10,800	6,400
(yellow bag)	Plastics, PVC, syringes	15–30	9,700–20,000	5–144	0–1	1,440–6,000	3,250
	Sharps, needles	4–8	60	450–500	0–1	3–5	5
	Fluids, residuals	2–5	0–10,000	62–63	80–100	0–11	30
	Alcohols, disinfectants	0–0.2	7,000–14,000	48–62	0–50	0–28	15
	Total bag						9,700
A3b	Plastics	50–60	14,000–20,000	5–144	0–1	6,930–12,000	9,000
(yellow bag)	Sharps	0–5	60	450–500	0–1	0–3	0
Lab waste	Cellulosic materials	5–10	8,000–12,000	5–62	0–15	340–1,200	650
	Fluids, residuals	1–20	0–10,000	62–63	95–100	0–100	30
	Alcohols, disinfectants	0–0.2	11,000–14,000	48–62	0–50	0–28	20
	Glass	15–25	0	175–225	0	0	0
	Total bag						9,700
A3c	Gauze, pads, swabs	5–20	8,000–12,000	5–62	0–30	280–3,600	1,000
(yellow bag)	Plastics, petri dishes	50–60	14,000–20,000	5–144	0–1	6,930–12,000	9,000
R&D	Sharps, glass	0–10	60	450–500	0–1	0–6	0
	Fluids	1–10	0–10,000	62–63	80–100	0–200	100
	Total bag						10,100
B1	Noninfected						
(blue bag)	Animal anatomical	90–100	9,000–16,000	30–80	60–90	810–6,400	1,400
	Plastics	0–10	14,000–20,000	5–144	0–1	0–20,000	1,000
	Glass	0–3	0	175–225	0	0	0
	Beddings, shavings, fecal matter	0–10	8,000–9,000	20–46	10–50	0–810	600
	Total bag						3,000

Source: Ontario Ministry of the Environment (1986).

It is rare to find an incinerator designated for biomedical waste destruction to be fired solely on this type of waste. Generally, particularly in hospitals, installation of an incinerator encourages the disposal of other wastes in the unit. Besides the cost savings this represents in not having to cart away this trash, there is the potential for heat recovery. For example, hospitals generally require steam throughout the year for their laundry, sterilizers, autoclaves, and kitchens. As more waste is fired, more heat is produced and more steam is generated.

Another set of wastes includes those generated in hospital laboratories that are hazardous wastes under the Resource Conservation and Recovery Act (RCRA). Table 13A.27 lists some

TABLE 13A.27 Hazardous Wastes under RCRA Typically Generated by In-Hospital Laboratories

Acetone	Methyl alcohol
Antineoplastics	Methyl cellosolve
Butyl alcohol	Pentane
Cyclohexane	Petroleum ether
Diethyl ether	Tetrahydrofuran
Ethyl alcohol	Xylene

Source: Doyle (1985).

of these wastes. If more than 200 lb/month of these wastes is generated, the incinerator must be permitted under the provisions of RCRA, which are additional to state requirements.

Waste generation rates will vary from one hospital to another, as a function of the number of hospital beds, the number of intensive-care beds, and the presence of other specialty facilities. In the absence of specific generation data, the figures in Table 13A.28 can be used as an estimate of waste generation rates.

TABLE 13A.28 Estimated Waste Generation Rates

Hospital	13 lb/(occupied bed · day)
Rest home	3 lb/(person · day)
Laboratory	0.5 lb/(patient · day)
Cafeteria	2 lb/(meal · day)

Source: Brunner (1987).

Regulatory Issues. Biomedical waste incinerators are generally small, much smaller than the central disposal incinerators that have been in the public eye in many of the densely populated areas of the country. Regulations have addressed the larger municipal solid waste incinerators. Smaller units, such as 2 to 10 tons/day biomedical waste municipators, have generally not been subject to rigorous regulatory attention in the past. The only restriction on their operation in many parts of the country is that they not create a public nuisance. That has meant that no odors are to be generated and that the opacity is to be low, i.e., no greater than Ringleman no. 1 for more than, for instance, 5 min/h. Incinerators have been designed to this standard, which is virtually no standard at all. As public attention is starting to focus on hazardous, dangerous, and toxic wastes, the regulatory attitude toward biomedical waste incinerators is starting to change. These incinerators are not addressed by the federal government yet, but many states are moving in the direction of regulation. In some states these wastes are classified as hazardous; in others they are regulated as a unique waste stream with its own set of regulations; and in still others there is still no regulation of biomedical wastes per se.

Hazardous Waste Incineration. Where hazardous regulations must be complied with, the incinerator design and operation must be subject to the RCRA regulations for handling and disposal. Incineration regulations under RCRA require an extensive analytical and compliance process. In addition to operating requirements, the RCRA incinerator regulations mandate extensive record-keeping and reporting procedures. A detailed, comprehensive operator training program must also be implemented.

Combined Hazardous Waste Systems. Hazardous waste incineration systems require an RCRA permit and strict operating controls and reporting standards. Another significant issue associated with hazardous waste incinerators is that the ash is always considered hazardous. Procedures exist for delisting ash (declaring ash nonhazardous), but this requires extensive testing and administrative activity (filings and petitions) which represent at least 18 months of reporting and review.

If, for example, 1000 lb of biomedical waste were incinerated, approximately 200 lb of ash would be generated. In a state not classifying such waste as hazardous, the ash could be deposited directly in a nonhazardous (municipal waste) landfill. If 100 lb of a hazardous waste were fired in the secondary chamber of this same incinerator, all the ash would be considered hazardous and would have to be deposited in a hazardous waste landfill (unless it were delisted). Where 100 lb of hazardous waste was originally present, now at least 200 lb of hazardous waste must be disposed of. As a general rule, it is impractical and uneconomical to incinerate hazardous and nonhazardous waste in the same incinerator.

Waste Combustion.   Hospital wastes will contain paper and cardboard, plastics, aqueous and nonaqueous fluids, anatomical parts, animal carcasses, and bedding, glass bottles, clothing, and many other materials. Much of this waste is combustible. Lighting a match to a mixed assortment of hospital waste will generally result in a sustained flame, unless it contains a high proportion of liquids, anatomical, or other pathological waste materials.

Thermal treatment technologies include starved air and excess air combustion processes, as described previously in this text. The main advantage of starved air is a low air requirement in the primary chamber. With little air passing through the waste there is less turbulence within the system and less particulate carryover from the burning chamber. This low airflow also results in very low nitrogen oxide generation, although this is not normally a concern in hospital incinerators. Less supplemental fuel is required than with excess air systems, where the entire airflow must be brought to the operating temperature of the incinerator. With the lower airflow, fans, ducts, flues, and air emissions control equipment can be sized smaller than in excess air systems.

In small systems, with less than 100 lb/h throughput, starved air operation is difficult, if not impossible, to achieve for two reasons. It is difficult to control air leakage into the system, and it is not possible to determine an accurate waste heating value on which to base a definition of the stoichiometric air requirement (see Table 13A.29). As listed in Table 13A.29, the stoichiometric air requirement can vary by a factor of 4 depending on the types of materials normally found in a biomedical waste stream. With larger systems the significance of air leakage decreases and the variations in feed characteristics will tend to even out. With units above 1500 lb/h, starved air can be sustained for many medical waste streams.

Waste Destruction Criteria.   Generally, paper waste (cellulosic materials) requires that a temperature of 1400°F be maintained for a minimum of 0.5 s for complete burnout. The temperature-residence time requirement for biomedical waste destruction must be at least equal to the requirements for paper waste; however, the specific relationship between temperature and residence time must be determined for the specific waste. Many states require that a tem-

TABLE 13A.29 Waste Combustion Characteristics

Waste constituent	Btu/lb	lb air/lb waste*
Polyethylene	19,687	16
Polystyrene	16,419	13
Polyurethane	11,203	9
PVC	9,754	8
Paper	5,000	4
Pathological	Will not support combustion	

* Stoichiometric requirement.
Source: Brunner (1988b).

perature of 1800°F be maintained for a minimum of 1 or 2 s, and some states require a 2000°F off-gas temperature.

On the high-temperature side, it is necessary to consider the general nature of much of the biomedical waste stream. It has a high proportion of organic material, including cellulosic waste. The noncombustible portion of this waste (ash) will begin to melt, or at least desolidify, as the temperature increases. Table 13A.18 lists the ash fusion temperature of refuse, which represents the same constituents as much of the biomedical waste stream. Above 1800°F, the ash produced will begin to deform in a reducing atmosphere (i.e., where there is a lack of oxygen in the furnace). When ash starts to deform and then is moved to a cooler portion of the incinerator, or to an area of the furnace where additional oxygen is present, the ash will harden into slag or clinker. This hardened ash can clog air ports, disable burners, corrode refractory, and interfere with the normal flow of material through the furnace. To prevent slagging, the temperature within the incinerator should never be allowed to rise above 1800°F. Higher temperatures also encourage the discharge of heavy metals to the gas stream, which is another reason not to impose an arbitrarily high temperature on the process.

Past Practices.   Modular units have been popular in the past because of their relatively low cost. A major factor contributing to their cost advantage over rotary kilns and other equipment is that they require no external air emissions control equipment to produce a fairly clean stack discharge. When properly designed and operated, they can achieve a particulate emissions rate of 0.08 gr/dry st ft^3 (corrected to 50 percent excess air). As new regulations are promulgated, however, lower emissions limitations will make the use of baghouses or electrostatic precipitators mandatory, and the modular incinerator (with inclusion of control equipment in the system package) will likely lose its price advantage over other systems.

The higher cost of rotary kilns was due to their need for external air emissions control equipment and to inclusion of the drive mechanism necessary for its operation.

Starved Air Process Limitations.   The most important issue associated with starved air combustion is the nature of the waste. (Note that there may be references to pyrolysis in some literature, including sales brochures for such equipment, but this term is usually used in error. The process is starved air combustion.)

For any starved air reaction to occur, the waste must be basically organic and able to sustain combustion without the addition of supplemental fuel, a definition of *autogenous* combustion. Without sufficient heat content to sustain combustion, the concept of substoichiometric burning has no meaning.

A waste with a moisture content in excess of 60 percent will not burn autogenously at 1600°F. As noted in Table 13A.26, the moisture contents of red bag, orange bag, and blue bag wastes are generally in excess of this figure. Starved air combustion will not work when wastes with this moisture fraction are placed in the furnace.

Usually a starved air incinerator is designed to fire a paper waste, and the burners in both the primary and secondary chambers of the incinerator are sized appropriately: for a relatively small supplemental-fuel requirement. When a bag of pathological waste is placed in the primary combustion chamber, the waste will not burn autogenously and supplemental fuel must be added. In most present starved air incinerator designs, the burners have too low a capacity to provide the heat required when organics released by a starved air process are not present. Without unburned organics in the secondary chamber (when starved air combustion has not occurred in the primary combustion chamber), the sizing of the secondary burner is generally inadequate (the burner is too small) to provide the fuel required for complete burnout of the gas stream.

On the other end of the operating range, when a waste with a very high heating value is introduced (a plastic material such as polyurethane or polystyrene, for instance, as noted in Table 13A.29), a good deal of air is required to generate even the substoichiometric requirement necessary to generate heat for the process to advance. This air quantity is often much greater than the airflow present from the fans provided with the unit. If a high-quality paper waste (waste paper, boxes, cartons, cardboard, etc.) is introduced into an incinerator, starved air combustion will generally work, assuming there is good control of infiltration air. The incinerator should be designed, however, not for paper waste, but for the firing of mixed paper and pathological waste, which requires a relatively large heat input and a coordinated increase in airflow (primary and secondary combustion air and burner air fans) in both the primary and the secondary combustion chambers.

Starved air operation of an operating incinerator can be easily checked. Increase the airflow in the primary combustion chamber, and if its temperature increases, the incinerator is operating in the starved air mode. If its temperature decreases, the incinerator is operating as an excess air unit; i.e., starved air operation is not occurring.

One must expect, in the design of biomedical waste incinerators, that although the incinerator charge may be sufficiently large to preclude swings from very high to very low heat value wastes, this is not always the case. It is likely that a single incinerator charge can contain a polystyrene mattress (and very little else) that will start to burn almost at once upon insertion into the incinerator.

Likewise, since much of the waste charged into an incinerator is in opaque bags, and the waste cannot be identified, a charge can consist almost wholly of anatomical waste, or animal carcasses, or liquid (aqueous) materials which have a very low heat content. An incinerator must be designed for this certain variation in waste stream quality. Of the incinerators on the market today, the starved air unit is least able to adapt to changes in waste constituents. As noted previously, this is of particular concern with smaller systems.

Incinerator Analysis

Combustion Properties.   The effectiveness of combustion is related to the combination of three factors, the "three T's": temperature in the furnace, time of residence of the combustion products at the furnace temperature, and turbulence within the furnace.

In general a solid or liquid must be converted to a gaseous phase before burning will occur. (Examine a lit match or a burning log. The flame does not rise directly from the solid. There is a zone immediately above the match, or log, where the gaseous fuel phase has been generated and is mixing with combustion air prior to burning.) The three T's are factors which control the rapidity of conversion of solid and liquid fuel to the gaseous phase.

Furnace Temperature.   Furnace temperature is a function of fuel heating value, furnace design, air admission, and combustion control. The minimum temperature must be higher than the ignition temperature of the waste. The upper temperature limit is normally a function of the enclosure materials and the ash melting temperature. Over 2400°F operation requires use of special refractory materials. The rates of combustion reactions increase rapidly with increased temperatures. Of the three T's (temperature, time, and turbulence), only temperature can be significantly controlled after a furnace is constructed. Time and turbulence are fixed by furnace design and airflow rate and can normally be controlled only over a limited range.

Furnace Temperature Control.   This can be achieved as follows:

Excess air control, i.e., control of the air-fuel ratio. Temperature produced is a direct function of the fuel properties and excess air introduced. Excess air control requires either automatic control or close manual supervision.

Direct heat transfer by the addition of heat-absorbing material within the furnace, such as water-cooled furnace walls. The addition of water sprays in the combustion zone (1000 Btu/lb water evaporated is equivalent to ½ Btu/°F sensible heat in the flue gas) will reduce the furnace temperature. Use of water sprays must be carefully controlled to avoid thermal shock to the furnace refractory.

Furnace Gas Turbulence.   Turbulence is an expression relating the physical relationship of fuel and combustion air in a furnace. A high degree of turbulence, intimate mixing of air and fuel, is desirable. Burning efficiency is enhanced with increased surface area of fuel particles exposed to the air. Fuel atomization maximizes the exposed particle surface. Turbulence helps to increase particle surface area by promoting fuel vaporization. In addition, good turbulence exposes the fuel to air in a rapid manner, helping to promote rapid combustion and maximizing fuel release.

A burner requiring no excess air and producing no smoke is said to have perfect turbulence (a *turbulence factor* of 100 percent). If, for instance, 15 percent excess air is required to achieve a no-smoke condition, the turbulence factor is calculated as follows:

$$\text{Turbulence factor} = \frac{\text{stoichiometric air}}{\text{total air} \times 100\%} = \frac{1.00}{1.15} \times 100\% = 87\%$$

Fuel gas burners can be designed to produce a turbulence factor close to 100 percent.

Retention Time.   Combustion does not occur instantaneously. Sufficient space must be provided within a furnace chamber to allow fuel and combustible gases the time required to fully burn. This factor, termed *dwell time, residence time,* or *retention time,* is a function of furnace temperature, degree of turbulence, and fuel particle size.

Retention time required may be a fraction of a second, as when gaseous waste is burned, or many minutes, as when solid granular waste such as powdered carbon is burned.

Waste Combustion.   The characteristics of an incineration process depend upon the characteristics of the waste incinerated (Table 13A.30). Waste characteristics of interest include the heat release, the air required for combustion of the waste, and the dry gas and moisture generated from waste burning.

Municipal waste will have three components: moisture, ash, and combustibles. Moisture will generally be in the range of 18 to 25 percent, ash (or noncombustibles) will normally be from 10 to 20 percent of the waste, and the balance of the waste will be combustible (or volatiles).

The combustible content of a waste is that portion of the waste that will burn. An equation to estimate heat release Q in Btu/lb, is as follows:

$$Q = 14406C + 67276H_2 - 6187O_2 + 4142S + 2433Cl_2 - 1082N_2$$

where C, H_2, O_2, S, Cl_2 and N_2 are the fractions of carbon, hydrogen, oxygen, sulfur, chlorine, and nitrogen, respectively, in the combustible content of the waste. Table 13A.30 is based on this equation. The heat content of the volatile fraction of municipal waste is normally in the range of 8000 to 12,000 Btu/lb.

TABLE 13A.30   Waste Burning Characteristics for Stoichiometric Combustion

Q, Btu/lb	Air, lb/10 kB	Dry gas, lb/10 kB	H_2O, lb/10 kB	C, %	H_2, %	O_2, %	S, %	N_2, %	Cl_2, %
7,000	7.13	8.00	0.56	47.62	4.41	46.07	0.25	0.82	0.83
8,000	7.01	7.67	0.59	49.36	5.29	43.45	0.25	0.82	0.83
9,000	6.91	7.41	0.61	51.11	6.16	40.84	0.25	0.82	0.83
10,000	6.83	7.20	0.62	52.85	7.03	38.22	0.25	0.82	0.83
11,000	6.76	7.03	0.64	54.59	7.90	35.60	0.25	0.82	0.83
12,000	6.72	6.90	0.65	56.38	8.79	32.93	0.25	0.82	0.83
13,000	6.67	6.77	0.66	58.08	9.65	30.37	0.25	0.82	0.83
14,000	6.63	6.67	0.67	59.83	10.52	27.75	0.25	0.82	0.83
15,000	6.59	6.58	0.68	61.57	11.39	25.14	0.25	0.82	0.83

Note: kB = 1000 Btu (10 kB = 10 kBtu = 10,000 Btu).

Gas Properties.   The gases generated by incineration will include nitrogen, carbon monoxide, oxygen, sulfur dioxide, hydrogen chloride, water vapor, and trace amounts of other gases. These gases can be considered to have a dry component and moisture. In most municipal waste incineration systems from 150 to 250 percent of the stoichiometric (ideal) air requirement is injected into the incinerator as combustion air. This means that from 50 to 150 percent of the air injected into the incinerator remains in the exhaust gas after burning is complete.

The dry gas component of the incinerator off-gas stream will have characteristics similar to those of air because of this relatively large air carryover through the process. In the following calculations, the characteristics of the incinerator off-gas are considered to be the same as that of dry air and moisture. This introduces a slight error, less than 3 percent, in the calculation of incinerator temperature, fuel requirement, scrubber performance, etc. This error is insignificant compared to the error inherent in other waste parameters. For instance, the waste feed rate is not normally known to within 3 percent and the heat content of the waste is generally an estimate. It cannot be determined with an accuracy of less than 5 percent, not just because of the heterogeneous nature of the waste, but because of the difficulty of obtaining a truly representative waste sample for analysis. By assuming that the dry gas characteristics are those of dry air rather than of the individual gases, the calculations are made more expeditious.

Table 13A.31 lists the enthalpy of dry gas and moisture relative to 60°F and 80°F. The properties of a saturated mixture of moisture vapor in air are listed in Table 13A.32.

The Mass Balance.   The flow weight into an incinerator must equal the flow of products leaving the incinerator. Input includes waste fuel, air (including humidity entrained within the air), and supplementary fuel. The flow exiting the incinerator includes moisture and dry gas in the exhaust as well as ash, both in the exhaust as fly ash and exiting as bottom ash.

Table 13A.33, the mass flow table, provides an orderly method of establishing a mass balance surrounding a combustion system. Of initial interest is waste quality: its moisture content, ash content (noncombustible fraction), and heat content. Of prime importance is the generation of moisture and dry flue gas from the combustion process.

For the purpose of this example, the following waste will be assumed, at the indicated firing rate and other combustion parameters:

8000 lb/h of municipal solid waste

5000 Btu/lb as fired

25 percent moisture as fired

20 percent ash as fired

Fired with 100 percent excess air

For the example in Table 13A.33:

Wet feed, as received charging rate is 8000 lb/h.

Moisture, by weight, of the wet feed is 25 percent. The moisture rate is 25 percent of the wet feed rate, 0.25×8000 lb/h $= 2000$ lb/h.

Dry feed equals total wet feed less moisture, 8000 lb/h $-$ 2000 lb/h $= 6000$ lb/h.

Ash is the percentage of total feed that remains after combustion. From the data provided, 20 percent of the total wet feed, 0.20×8000 lb/h $= 1600$ lb/h ash.

Volatile is that portion of feed that is combusted. It is found by subtracting ash from dry feed: 6000 lb/h $-$ 1600 lb/h $= 4400$ lb/h.

Volatile heating value is the Btu value of the waste per pound of volatile matter. The total heating value as charged is 8000 lb/h $\times$ 4500 Btu/lb $= 36,000,000$ Btu/h. With M representing 1 million, the total waste heat value is 36.00 MBtu/h. On a unit volatile basis, with 4400 lb/h volatile charged, the heating value per pound volatile is 36 MBtu/h $\div$ 4400 lb/h $= 8200$ Btu/h.

TABLE 13A.31 Enthalpy of Air, and Moisture

Relative to 60°F			Relative to 80°F	
H_{Air}, Btu/lb	H_{H_2O}, Btu/lb	Temp., °F	H_{Air}, Btu/lb	H_{H_2O}, Btu/lb
21.61	1091.92	150	16.82	1071.91
33.65	1116.62	200	28.86	1096.61
45.71	1140.72	250	40.92	1120.71
57.81	1164.52	300	53.02	1144.51
69.98	1188.22	350	65.19	1168.21
82.19	1211.82	400	77.40	1191.81
94.45	1235.47	450	89.66	1215.46
106.79	1259.22	500	102.00	1239.21
119.21	1283.07	550	114.42	1263.06
131.69	1307.12	600	126.90	1287.11
144.25	1331.27	650	139.46	1311.26
156.87	1355.72	700	152.08	1335.71
169.59	1380.27	750	164.80	1360.26
187.38	1405.02	800	177.59	1385.01
195.26	1430.02	850	190.47	1410.01
208.21	1455.32	900	203.42	1435.31
221.25	1480.72	950	216.46	1460.71
234.36	1506.42	1000	229.57	1486.41
247.55	1532.40	1050	242.76	1512.40
260.81	1558.32	1100	256.02	1538.31
274.15	1584.80	1150	264.36	1564.80
287.55	1611.22	1200	282.76	1591.21
301.02	1638.26	1250	296.23	1618.20
314.56	1665.12	1300	309.77	1645.11
328.17	1692.15	1350	323.38	1672.15
341.85	1719.82	1400	337.06	1699.81
355.58	1747.70	1450	350.82	1727.70
369.37	1775.52	1500	364.58	1755.51
397.17	1832.12	1600	392.33	1812.11
425.08	1890.11	1700	420.29	1870.10
453.24	1948.02	1800	448.45	1928.01
481.57	2007.17	1900	476.78	1987.70
510.07	2067.42	2000	505.28	2047.41
538.72	2128.70	2100	533.93	2108.70
567.52	2189.92	2200	562.73	2169.91
596.45	2252.60	2300	591.66	2232.60
625.52	2315.32	2400	620.73	2295.31
654.70	2377.80	2500	649.91	2357.80
684.01	2443.30	2600	679.22	2423.30
713.42	2511.88	2700	708.63	2491.80

Source: Brunner (1988a).

Dry gas produced from combustion of the waste is 7.60 lb/10 kBtu from Table 11.30. The dry gas flow is this figure multiplied by the Btu released, or $(7.60 \div 10,000)$ lb/Btu $\times$ 36.00 MBtu/h = 27,360 lb/h dry gas.

Combustion H_2O is the moisture generated from burning the waste, 0.60 lb/10 kB from Table 11.30. The combustion moisture flow rate is $(0.60 \div 10,000)$ lb/Btu $\times$ 36.00 MBtu/h = 2160 lb/h moisture.

TABLE 13A.32   Saturation Properties of Dry Air (DA) with Moisture

Temp., °F	Humidity, lb H$_2$O/ lb DA	Enthalpy, Btu mixture/ lb DA	Volume, ft^3 mixture/ lb DA	Temp., °F	Humidity, lb H$_2$O/ lb DA	Enthalpy, Btu mixture/ lb DA	Volume, ft^3 mixture/ lb DA
60	0.01108	0.000	13.329	110	0.05932	65.764	15.725
61	0.01149	0.690	13.363	111	0.06123	68.144	15.796
62	0.01191	1.393	13.398	112	0.06319	70.589	15.869
63	0.01234	2.114	13.433	113	0.06522	73.102	15.944
64	0.01279	2.849	13.468	114	0.06731	75.690	16.020
65	0.01326	3.602	13.504	115	0.06946	78.357	16.098
66	0.01374	4.369	13.540	116	0.07168	81.095	16.178
67	0.01423	5.155	13.680	117	0.07397	83.197	16.259
68	0.01474	5.958	13.613	118	0.07633	86.815	16.343
69	0.01527	6.781	13.650	119	0.07877	89.800	16.428
70	0.01581	7.621	13.688	120	0.08128	92.880	16.515
71	0.01638	8.484	13.762	121	0.08388	96.840	16.603
72	0.01700	9.366	13.764	122	0.08655	99.300	16.695
73	0.01755	10.266	13.803	123	0.08931	102.65	16.789
74	0.01817	11.191	13.842	124	0.09216	106.11	16.885
75	0.01881	12.136	13.882	125	0.09511	109.67	16.983
76	0.01946	13.094	13.922	126	0.09815	113.34	17.084
77	0.02014	14.094	13.963	127	0.10129	117.14	17.187
78	0.02084	15.109	14.004	128	0.10453	121.04	17.293
79	0.02156	16.149	14.046	129	0.10788	125.06	17.402
80	0.02231	17.214	14.088	130	0.11130	129.22	17.514
81	0.02308	18.306	14.131	131	0.11490	133.50	17.628
82	0.02387	19.426	14.175	132	0.11860	137.87	17.746
83	0.02468	20.571	14.219	133	0.12240	142.46	17.867
84	0.02552	21.744	14.264	134	0.12640	147.18	17.991
85	0.02639	22.947	14.309	135	0.13040	152.02	18.119
86	0.02728	24.181	14.355	136	0.13460	157.04	18.251
87	0.02821	25.445	14.402	137	0.13900	162.24	18.386
88	0.02916	26.744	14.449	138	0.14350	167.58	18.525
89	0.03014	28.074	14.497	139	0.14820	173.11	18.669
90	0.03115	29.441	14.547	140	0.15300	178.82	18.816
91	0.03219	30.841	14.597	141	0.15800	184.73	18.969
92	0.03326	32.279	14.647	142	0.16320	190.85	19.126
93	0.03437	33.751	14.699	143	0.16850	197.17	19.288
94	0.03551	35.266	14.751	144	0.17410	203.72	19.454
95	0.03668	36.815	14.804	145	0.17980	210.49	19.626
96	0.03789	38.408	14.854	146	0.18580	217.52	19.804
97	0.03914	40.039	14.913	147	0.19200	224.82	19.987
98	0.04043	41.715	14.968	148	0.19840	232.35	20.176
99	0.04175	43.438	15.025	149	0.20510	240.17	20.374
100	0.04312	45.209	15.083	150	0.22120	248.29	20.576
101	0.04453	47.023	15.142	151	0.21920	256.71	20.786
102	0.04498	48.886	15.202	152	0.22670	265.45	21.004
103	0.04748	50.806	15.263	153	0.23440	274.53	21.229
104	0.04902	52.770	15.325	154	0.24250	283.96	21.462
105	0.05061	54.792	15.389	155	0.25090	293.78	21.704
106	0.05225	56.868	15.453	156	0.25960	303.98	21.955
107	0.05394	59.000	15.519	157	0.26880	314.64	22.216
108	0.05568	61.190	15.587	158	0.27820	325.69	22.487
109	0.05747	63.444	15.655	159	0.28810	337.23	22.769

(continued)

TABLE 13A.32   Saturation Properties of Dry Air (DA) with Moisture (*Continued*)

Temp., °F	Humidity, lb H$_2$O/ lb DA	Enthalpy, Btu mixture/ lb DA	Volume, ft^3 mixture/ lb DA	Temp., °F	Humidity, lb H$_2$O/ lb DA	Enthalpy, Btu mixture/ lb DA	Volume, ft^3 mixture/ lb DA
160	0.29850	349.36	23.063	186	0.87940	1021.0	39.037
161	0.30920	361.79	23.368	187	0.92710	1076.1	40.332
162	0.32050	374.90	23.685	188	0.97900	1135.9	41.737
163	0.33230	388.58	24.017	189	1.03550	1201.0	43.265
164	0.34460	402.89	24.365	190	1.09700	1272.2	44.935
165	0.35750	417.84	24.725	191	1.16500	1350.7	46.764
166	0.37100	433.53	25.102	192	1.24000	1436.3	48.780
167	0.38510	449.95	25.492	193	1.32200	1531.6	51.011
168	0.40000	467.18	25.914	194	1.41400	1637.4	53.488
169	0.41560	485.28	26.347	195	1.51700	1756.2	56.265
170	0.43200	504.29	26.804	196	1.63300	1889.5	59.381
171	0.44930	524.31	27.272	197	1.76500	2040.8	62.918
172	0.46750	545.38	27.787	198	1.91500	2214.0	66.963
173	0.48670	567.61	28.315	199	2.08900	2413.9	71.630
174	0.50700	591.07	28.876	200	2.29200	2648.4	77.102
175	0.52840	615.86	29.465	201	2.53200	2924.2	83.543
176	0.55110	642.11	30.089	202	2.82000	3255.4	91.270
177	0.57520	669.92	30.749	203	3.17300	3660.8	100.750
178	0.60080	699.48	31.449	204	3.61400	4169.8	112.590
179	0.62790	730.86	32.193	205	4.18100	4821.7	127.800
180	0.65600	764.31	32.984	206	4.93900	5694.0	147.600
181	0.68780	800.01	33.829	207	6.00000	6913.0	176.560
182	0.72090	838.20	34.731	208	7.59400	8748.0	219.300
183	0.75630	879.02	35.694	209	10.248	11802	290.440
184	0.79430	922.91	36.728	210	15.54	17887	432.250
185	0.83520	970.11	37.839	211	31.49	36241	859.820

Source: From C. R. Brunner, *Handbook of Hazardous Waste Incineration,* TAB Books Inc., Blue Ridge Summit, Pa., 1989. Derived from O. Zimmerman, I. Lavine, *Psychrometric Tables and Charts,* 1st ed., Industrial Research Service, Dover, N.H.

Dry gas + combustion H$_2$O is the sum of the dry gas and the moisture products of combustion, $27{,}360 + 2160 = 29{,}520$ lb/h. This figure is convenient for obtaining the amount of air required for combustion.

100 percent air is the dry gas and moisture weights that are produced by combustion of the volatile component, which equals the weight of the volatile component plus the weight of the air provided. Likewise, the air requirement is equal to the sum of the dry gas and moisture of combustion less the volatile component. This air requirement is the stoichiometric air requirement, that amount of air necessary for complete combustion of the volatile component of the waste. Using the above figures, the value for 100 percent air is as follows:

$$29520 \text{ lb/h (dry gas + H}_2\text{O)} - 4400 \text{ lb/h (volatile)} = 25120 \text{ lb/h}$$

Total air fraction is the air required for effective combustion. This is basically a function of the physical state of the fuel (gas, liquid, or solid) and the nature of the burning equipment. In this example, 100 percent air is required, providing $1.00 + 1.00 = 2.00$ total air fraction.

Total air is the stoichiometric requirement multiplied by the total air fraction, $25{,}120$ lb/h $\times$ $2.00 = 50{,}240$ lb/h.

TABLE 13A.33 Mass Flow

	Example
Wet feed, lb/h	8,000
Moisture, %	25
lb/h	2,000
Dry feed, lb/h	6,000
Ash, %	20
lb/h	1,600
Volatile	4,400
Volatile htg. value, Btu/lb	8,200
MBtu/h	36.00
Dry gas, lb/10 kBtu	7.60
lb/h	27,360
Comb. H_2O, lb/10 kBtu	0.60
lb/h	2,160
Dry gas + comb. H_2O, lb/h	29,520
100% Air, lb/h	25,120
Total air fraction	2.00
Total air, lb/h	50,240
Excess air, lb/h	25,120
Humid/dry gas (air), lb/lb	0.01
Humidity, lb/h	502
Total H_2O, lb/h	4,662
Total dry gas, lb/h	52,480/(ad)

Excess air provided to the system is the total air less the stoichiometric air requirement, 50,240 lb/h − 25,120 lb/h = 25,120 lb/h.

Humidity/dry gas (air) The humidity of the air entering the incinerator may have a significant effect on the heat balance to be calculated subsequently. In this case, assume a humidity of 0.01 lb of moisture per pound of dry air.

Humidity is the flow of moisture into the system with the air supply. Given the airflow and the fractional humidity, the humidity can be calculated as 50,240 lb/h air × 0.01 lb H_2O/lb air = 502 lb/h moisture.

Total H_2O is that moisture exiting the system. It is equal to the sum of three moisture components: moisture in the feed plus moisture of combustion plus humidity, or 2000 lb/h + 2160 lb/h + 502 lb/h = 4662 lb/h total H_2O.

Total dry gas exiting the system is equal to the sum of the dry gas generated by the combustion of the volatiles, with stoichiometric air, plus the flow of excess air into the system. Thus, 27,360 lb/h dry gas + 25,120 lb/h excess air = 52,480 lb/h total dry gas.

Heat Balance. Heat, like mass, is conserved within a system. The heat exiting a system is equal to the amount of heat entering that system. Table 13A.34 presents a quantitative means of establishing a heat balance for an incinerator. The heat balance must be preceded by a mass flow balance, as discussed previously. The result of a heat balance is a determination of the incinerator outlet temperature, outlet gas flow, supplemental fuel requirement, and total air requirement. The total heat into the incinerator was calculated previously in the mass flow computations. By determining how much of the total heat produced is present in the exhaust gas, the exhaust gas temperature can be calculated. If the calculated exhaust gas temperature is equal to the desired exhaust gas outlet temperature, the process is autogenous and supplemental fuel is not required. If the desired temperature is lower than the actual outlet temperature, additional air must be added (or additional water, if this is possible) to lower the outlet temperature to that desired. This condition is also autogenous burning since supplemental fuel is not added to the system. For the case where the actual outlet temperature is less than the desired outlet temperature, supplemental fuel must be added. The products of combustion must include the products of combustion of the supplemental fuel. Dry flue gas properties are assumed to be identical to those of dry air.

As with Table 13A.33, Table 13A.34 will be described on a step-by-step basis.

TABLE 13A.34   Heat Balance

	Example with fuel oil	Example with gas
Cooling air wasted, lb/h	9,000	
°F	450	
Btu/lb	94	
MBtu/h	0.85	
Ash, lb/h	1,600	
Btu/lb	130	
MBtu/h	0.21	
Radiation, %	1.5	
MBtu/h	0.54	
Humidity, lb/h	502	
Correction (@970 Btu/lb), MBtu/h	−0.49	
Losses, total, MBtu/h	1.11	
Input, MBtu/h	36.00	
Outlet, MBtu/h	34.89	
Dry gas, lb/h	53,480	
H_2O, lb/h	4,662	
Temperature, °F	1,915	
Desired temp., °F	2,000	
MBtu/h	36.41	
Net MBtu/h	1.53	
Fuel oil, air fraction	1.20	1.10
Net Btu/gal	57,578	406
gal/h	26.57	3,768
Air, lb/gal	125.06	0.791
lb/h	3,323	2,980
Dry gas, lb/gal	125.54	0.748
lb/h	3,336	2,818
H_2O, lb/gal	8.75	0.103
lb/h	232	388
Dry gas w/fuel oil, lb/h	55,816	56,298
H_2O w/fuel oil, lb/h	4,894	5,050
Air w/fuel oil, lb/h	53,563	53,220
Outlet MBtu/h	38.61	38.66
Reference temperature t, °F	60	60

Cooling air wasted: assume the incinerator shell is cooled by a flow of 2000 st ft³/min of air. A standard cubic foot of air weighs 0.075 lb/ft³, therefore the mass airflow is 2000 ft³/min × 60 min/h × 0.075 lb/ft³ = 9000 lb/h. It is further assumed that this flow is wasted, i.e., discharged to the atmosphere. The temperature of the air at the discharge point is assumed to be 450°F. From Table 13A.31 the enthalpy of air at 450°F, the cooling air discharge temperature, is 94 Btu/lb. The total heat loss due to the wasted cooling air is the quantity of cooling air discharged multiplied by its enthalpy, 9000 lb/h × 94 Btu/lb = 0.85 MBtu/h.

Ash generated is 1600 lb/h, from the mass flow table. The heating value of ash can be assumed to be 130 Btu/lb based on the equation $Q = c_p(T - 60)$, where Q is the heat value, c_p is the ash specific heat (taken as 0.24 Btu/lb · °F), and T is the ash discharge temperature, at 600°F [0.24(600 − 60) = 130 Btu/lb]. The heat loss through ash discharge is therefore 1600 lb/h ash × 130 Btu/lb = 0.21 MBtu/h.

Radiation: the heat lost by radiation from the incinerator shell can be approximated as a percentage of the total heat of combustion. Table 13A.35 lists typical values of radiation loss. For this case, with a heat release of 36.00 MBtu/h, a radiation loss of 1.5 percent is used. The total loss by radiation is equal to 36.00 MBtu/h released × 1.5 percent, or 0.54 MBtu/h.

TABLE 13A.35 Furnace Radiation Loss Estimates

Furnace rate, MBtu/h	Radiation loss, % of furnace rate
<10	3
15	2.75
20	2.50
25	2
30	1.75
>35	1.50

Humidity is water vapor within the air. The humidity component of the air, 502 lb/h, is taken from the mass flow sheet.

Correction: when one is considering the heat absorbed by the moisture or water within the incinerator, exhaust stream humidity has released its heat of vaporization because it is in the vapor phase. The other moisture components, moisture in the feed and moisture of combustion, enter the reaction as liquid, and the heat of vaporization is released by the reaction. To simplify these moisture calculations, the heat of vaporization of humidity moisture at 60°F, 970 Btu/lb, is added to the total heat capacity of the flue gas. Therefore the correction factor is 502 lb/h humidity × 970 Btu/lb = 0.49 MBtu/h.

Total losses of an incinerator are the sum of the heat discharged as cooling air and the heat lost in the ash discharge and the radiation loss. To this is added the correction for humidity. In this case the total loss is 0.85 MBtu/h + 0.21 MBtu/h + 0.54 MBtu/h − 0.49 Btu/h = 1.11 MBtu/h.

Input to the system is the heat generated from the combustible, or volatile, portion of the feed. From the mass flow table this figure is 36.00 MBtu/h.

Outlet heat content is that amount of heat exiting in the flue gas. The heat left in the flue gas is the heat generated by the feed (input) less the total heat loss, 36.00 MBtu/h − 1.11 MBtu/h = 34.89 MBtu/h.

Dry gas is 52,480 lb/h, from the mass flow sheet.

H_2O is 4662 lb/h, from the mass flow sheet.

Temperature is the temperature of the exhaust gas, X, where the heat content of the dry gas flow plus the heat content of the moisture flow exiting the incinerator equals the outlet MBtu/h. With 52,480 lb/h dry gas and 4662 lb/h moisture at 1900°F and 2000°F, the dry gas enthalpy is 481.57 and 510.07, respectively, and the moisture enthalpy is 2007.17 and 2067.42, respectively (from Table 13A.31). Therefore, the total calculated enthalpy is:

1900°F	34.63 MBtu/h
X°F	34.89 MBtu/h
2000°F	36.41 MBtu/h

By interpolation, the exhaust-temperature X is

$$X = 1900 + (2000 - 1900) \times (34.89 - 34.63) \div (36.41 - 34.63)$$
$$= 1915°F$$

Therefore the exhaust gas temperature is 1915°F.

Desired temperature is the temperature to which it is desired to bring the products of combustion. For this example a temperature of 2000°F will be used. At this temperature the heat content of the wet gas stream is that of the dry gas and that of the moisture at 2000°F or 52,480 lb/h dry gas × 510.07 Btu/lb + 4662 lb/h moisture × 2067.42 Btu/lb = 36.41 MBtu/h.

Net MBtu/h is the amount of heat that must be added to the flue gas to raise its heat content to the desired level. (In this case the desired heat level is evaluated at 2000°F.) The net MBtu/h, therefore, is the desired less the outlet MBtu/h, 36.42 MBtu/h − 34.89 MBtu/h = 1.53 MBtu/h.

Note: this analysis will continue on the basis of using No. 2 fuel oil as supplemental fuel. Table 13A.36 lists fuel oil combustion parameters. Table 13A.37 is a list of combustion

TABLE 13A.36 No. 2 Fuel Oil (139,703 Btu/gal, 7.6 lb/gal)

Total air:	1.1	1.2	1.3
lb air/gal	114.640	125.062	135.483
lb dry gas/gal	115.115	125.537	135.958
lb H₂O/gal	8.615	8.751	8.886

Temp., °F	Heat available, Btu/gal		
200	126,210	125,707	125,206
300	123,016	122,255	121,495
400	119,802	118,780	117,760
500	116,562	115,277	113,995
600	113,283	111,732	110,184
700	109,965	108,146	106,328
800	106,029	103,885	101,742
900	103,197	100,829	98,463
1,000	99,747	97,099	99,747
1,100	96,255	93,325	90,397
1,200	92,721	89,505	86,291
1,300	89,147	85,643	82,140
1,400	85,535	81,738	77,943
1,500	81,887	77,796	73,707
1,600	78,205	73,817	69,431
1,700	74,487	69,799	65,115
1,800	70,746	65,757	60,771
1,832	69,538	64,452	59,369
1,900	66,971	61,679	56,389
2,000	63,175	57,578	51,984
2,100	59,341	53,445	59,349
2,192	55,813	49,628	44,385
2,200	55,507	49,294	43,084
2,300	41,637	45,114	38,594
2,400	47,750	40,916	34,085
2,500	43,852	36,706	29,562
2,600	39,914	32,453	24,995
2,700	35,938	28,162	20,388

TABLE 13A.37 Natural Gas (1000 Btu/st ft³, 0.050 lb/st ft³)

Total air:	1.05	1.10	1.15
lb air/st ft³	0.755	0.791	0.827
lb dry gas/st ft³	0.712	0.748	0.784
lb H₂O/st ft³	0.103	0.103	0.104

Temp., °F	Heat available, Btu/st ft³		
200	861	860	857
300	839	837	834
400	817	814	810
500	794	790	785
600	772	767	761
700	749	743	736
800	722	715	707
900	702	694	685
1000	678	670	660
1100	654	644	633
1200	629	619	607
1300	605	593	580
1400	579	567	553
1500	554	541	526
1600	529	514	498
1700	503	487	470
1800	477	460	442
1832	468	451	433
1900	450	433	414
2000	424	406	385
2100	397	378	356
2192	372	352	329
2200	370	350	327
2300	343	322	298
2400	316	294	269
2500	289	265	239
2600	261	237	210
2700	233	208	179

parameters for natural gas. The incinerator analysis of Tables 13A.38 and 13A.42 includes listings for the use of natural gas as supplemental fuel. The values for natural gas are calculated like those for fuel oil.

Fuel oil, air fraction is the total air fraction required for combustion of supplementary fuel. The total air normally required for efficient combustion of gas fuel is from 1.05 to 1.15 and for light fuel oil is 1.10 to 1.30. In this case assume No. 2 fuel oil is used for supplemental heat, with a total air stoichiometric ratio of 1.20.

Net Btu/gal: when fuel is combusted, the products of combustion must be heated to the desired flue gas temperature. The amount of heat required to heat these combustion products must be subtracted from the total heat of combustion to obtain the effective heating value of the fuel. As the temperature to which the fuel products must be raised increases, the net heat available from the fuel decreases. From Table 13A.36, by bringing the products of combustion of fuel oil, with 1.2 total air, to 2000°F, a net heating value of 57,578 Btu/gal is available. The gallons of fuel oil required to provide the heat required to bring the exhaust gas temperature from its actual temperature, 1915°F, to the desired temperature, 2000°F, are equal to the net MBtu/h required divided by the net Btu/gal available: 1.53 MBtu/h ÷ 57,578 Btu/gal = 26.57 gal/h.

Air required for combustion of fuel oil, with 1.2 total air, from Table 13A.36, is 125.06 lb/gal of fuel oil. The fuel combustion airflow is the unit flow multiplied by the fuel quantity, 125.06 lb/gal × 26.57 gal/h fuel oil = 3323 lb/h air.

Dry gas. From Table 11.36 the dry gas produced from combustion of fuel oil with 1.2 total air is 125.54 lb/gal of fuel oil. The dry gas flow rate is 125.54 lb dry gas/gal × 26.57 gal/h fuel oil = 3336 lb/h.

H_2O produced from combustion of fuel oil with 1.2 total air and 0.013 humidity is 8.75 lb/gal fuel oil from Table 11.36. The moisture flow rate from combustion of fuel oil is 8.75 lb H_2O/gal fuel oil × 26.57 gal fuel oil/h = 232 lb/h.

Dry gas with fuel oil is the total quantity of dry gas exiting the system. It is equal to the dry gas produced from combustion of the waste plus the dry gas produced from fuel combustion: 52,480 lb/h + 3336 lb/h = 55,816 lb/h dry gas.

H_2O with fuel oil is the total quantity of moisture exiting the system, that calculated in the mass flow sheet plus the contribution from combustion of supplementary fuel, 4662 lb/h + 232 lb/h = 4894 lb/h.

Air with fuel oil is the total amount of air entering the incinerator, calculated from the mass flow sheet, plus that needed for supplemental fuel combustion, 50,240 lb/h + 3323 lb/h = 53,563 lb/h.

Outlet MBtu/h, the total heat value of the flue gas exiting the incinerator, is the sum of the heat content of the gas prior to adding supplemental fuel and the heat addition of the supplemental fuel. The supplemental fuel adds 26.57 gal/h × 139,703 Btu/gal = 3.72 MBtu/h to the flue gas. Therefore the flue gas outlet contains 34.89 MBtu/h + 3.72 MBtu/h = 38.61 MBtu/h. As a check on this figure, the outlet temperature will be calculated by using the flue gas flow (55,816 lb/h dry gas and 4894 lb/h moisture):

2000°F   38.59 MBtu/h

X°F   38.61 MBtu/h

2100°F   40.49 MBtu/h

By interpolation,

$$X = 2000 + (2100 - 2000) \times (38.61 - 38.59) \div (40.49 - 38.59) = 2001°F$$

This calculation of flue gas temperature, 2001°F, is in good agreement with the desired gas outlet temperature, 2000°F.

Reference t is the datum temperature for enthalpy. It is the temperature at which feed, supplemental fuel, and air enter the system, 60°F for this example.

Flue Gas Discharge.   To meet the rigorous air pollution codes in effect today, gas scrubbing equipment is often necessary. Table 13A.38, the flue gas discharge table, provides a method of calculating gas flow volumes exiting a wet scrubbing system as well as scrubber flow quantities. This table can also be used when calculating volumetric flow from a dry flue gas system. Table 13A.38 entries are as follows:

Inlet: insert the incinerator outlet temperature, 2000°F, from the heat balance table. This temperature is the inlet temperature of the flue gas processing system.

Dry gas is the flow of dry gas exiting the incinerator. From the heat balance table, this figure is 55,816 lb/h.

Heat is the total heat exiting the incinerator in the flue gas, MBtu/h. From the heat balance table this figure is 38.61 MBtu/h. The heat is calculated in terms of the dry gas component of the flue gas, as Btu/lb dry gas. From the entries for total heat and dry gas flow, the heat is 38.61 MBtu/h ÷ 55,816 lb/h = 692 Btu/lb dry gas.

Adiabatic t: when 1 lb of water evaporates, it absorbs approximately 1000 Btu, without a change in temperature. This heat adsorption is called the *heat of vaporization or latent heat.* Latent heat is opposed to sensible heat, which is the heat required for a change in tempera-

TABLE 13A.38   Flue Gas Discharge

	Example with fuel oil	Example with gas
Inlet, °F	2,000	2,000
Dry gas, lb/h	55,816	56,298
Heat, MBtu/h	38.61	38.66
Btu/lb dry gas	692	687
Adiabatic t_a, °F	178	178
H_2O saturation, lb/lb dry gas	0.6008	0.6008
lb/h	33,534	33,824
H_2O inlet, lb/h	4,894	5050
Quench H_2O, lb/h	28,640	28,774
gal/min	58	58
Outlet temp., °F	120	120
Raw H_2O temp., °F	60	60
Sump temp., °F	148	148
Temp. diff., °F	88	88
Outlet, Btu/lb dry gas	92.880	92.880
MBtu/h	5.18	5.23
Req'd. cooling, MBtu/h	33.43	33.43
H_2O, lb/h	379,886	379,886
gal/min	760	760
Outlet, ft³/lb dry gas	16.515	16.515
ft³/min	15,363	15,496
Fan press., in WC	30	30
Outlet, actual ft³/min	16,586	16,729
Outlet, H_2O/lb dry gas	0.08128	0.08128
H_2O, lb/h	4,537	4,576
Recirc. (ideal), gal/min	58	58
Recirc. (actual), gal/min	464	464
Cooling H_2O, gal/min	760	760

ture without a change in phase. For evaporated water (steam) the sensible heat is approximately 0.5 Btu/lb steam for every rise of 1°F.

The *adiabatic temperature t* of the flue gas is the quench temperature. Quenching of a gas is defined as the use of latent heat of water (or other fluid) to decrease the gas temperature. The process does not involve the addition or removal of heat, only the use of the heat of vaporization of the quench liquid, i.e., water. The term *adiabatic* defines a process where heat is neither added nor removed from a system. Considering the properties of dry flue gas equal to the properties of dry air, listed in Table 13A.32, note that a maximum amount of moisture can be held in dry air at a particular temperature. The table lists saturation moisture quantities, volumes, and enthalpy as a function of temperature. The temperature at which the enthalpy of the saturated flue gas (dry air) is equal to the enthalpy calculated above, Btu/lb dry gas, is the adiabatic temperature of the system. There has been no transfer of heat from the system, only the conversion of latent heat in the quench water to sensible heat in the dry flue gas. In this example the adiabatic temperature is found in Table 11.32 as that temperature where the dry flue gas (saturated mixture) will have an enthalpy of approximately 692 Btu/lb, 178°F.

H_2O saturation: The quenched flue gas, at the adiabatic temperature (178°F in this example) will contain an amount of moisture equal to the maximum amount of moisture that it can hold, saturation. From Table 13A.32, the saturation moisture, in lb H_2O/lb dry gas (air), is read opposite the adiabatic temperature. In this case, for 178°F adiabatic temperature, the saturation moisture is 0.6008 lb H_2O/lb dry gas. The moisture flow is the saturation moisture multiplied by the dry gas flow, 0.6008 lb H_2O/lb dry gas × 55,816 lb dry gas/h = 33,534 lb H_2O/h.

H_2O inlet: The moisture component of the flue gas exiting the incinerator is inserted here from the heat balance table: 4894 lb/h.

Quench H_2O is the moisture required for quenching the incoming flue gas to its adiabatic temperature. This is equal to the saturated moisture content of the flue gas less the moisture initially carried into the system with the flue gas. This figure is equal to H_2O saturation (33,534 lb/h) less H_2O inlet (4894 lb/h), which in this example is equal to 28,640 lb/h. The conversion factor from lb/h of water to gal/min (8.34 lb/gal × 60 min/h) is 500. The quench water required is equal to 28,640 lb/h ÷ 500 = 58 gal/min.

Outlet temperature is that temperature entering the low (negative) pressure side of the induced-draft (ID) fan, or, with no ID fan, the temperature within the stack. This temperature is normally selected in the range of 120 to 160°F. The lower this temperature, the smaller the size of the outlet plume and the lower the volumetric flow rate of flue gas. For this example 120°F was chosen as the outlet temperature. As can be seen below, a lower outlet temperature would require additional amounts of cooling water.

Raw H_2O temperature: this entry is the temperature of the water available for cooling the flue gas from the adiabatic temperature to the outlet temperature. In this example a raw water temperature of 60°F was chosen.

Sump temperature: normally a quantity of water in excess of that calculated for cooling the flue gas is provided for particulate removal. The excess water is generally collected in a sump where a quiescent period is allowed to permit larger particles within the spent water to settle to the sump floor, eventually to be drained. The temperature of the water in the sump must be ascertained to determine the effective cooling rate of the water flow. The sump temperature is a practical impossibility to forecast accurately through detailed calculations, but an empirical relationship has been established. The sump temperature is assumed equal to the adiabatic temperature divided by 1.2. In this example the sump temperature is estimated at 178 ÷ 1.2 = 148°F.

Temperature differential: the temperature differential of note is the difference in temperature between the raw water entering the cooling tower (or scrubber) and the water temperature exiting the tower, the sump temperature. Sump temperature less raw H_2O temperature is, in this example, 148 − 60 = 88°F, the temperature differential.

Outlet: the gas exiting the scrubber system is designed to be at the outlet temperature, 120°F in this case, saturated with moisture. The outlet enthalpy is inserted from Table 13A.32 for the outlet temperature chosen: 92.880 Btu/lb dry gas. The total heat in the outlet flue gas is its enthalpy multiplied by its flow, that is, 92.880 Btu/lb × 55,816 lb dry gas/h = 5.18 MBtu/h.

Required cooling: as noted previously, the flue gas is initially quenched, without heat addition or removal, to its adiabatic temperature. To reduce the temperature to the desired outlet temperature, a supply of cooling water is required. This cooling water must remove the heat content at adiabatic conditions relative to the heat content at outlet conditions. The required cooling is therefore the heat inlet less the outlet MBtu/h. For this example 38.61 MBtu/h heat inlet less 5.18 MBtu/h outlet equals 33.43 MBtu/h required cooling; i.e., with removal of 33.43 MBtu/h from the saturated flue gas stream the flue gas temperature will fall from 178°F to 120°F.

H_2O: the moisture flow referred to is that required to achieve the desired cooling effect. With $Q = WC \Delta t$ or $W = Q/C \Delta t$, where W = cooling water in lb/h, Q = cooling load in Btu/h, C = specific heat of water [1 Btu/(lb · °F)], and Δt = temperature difference of the cooling water across the flue gas stream (sump water temperature less raw water temperature), the required cooling water flow rate can be calculated. For this example:

$$W = 33.43 \text{ MBtu/h} \div 1 \text{ Btu(lb} \cdot \text{°F)} \div 88\text{°F} = 379,886 \text{ lb/h}$$

The flow in gallons per minute is that in lb/h divided by 500, 379,886 ÷ 500 = 760 gal/min.

Outlet: the outlet volumetric flow is obtained with use of Table 13A.32. The specific volume, ft³ mixture/lb dry gas (air), is found in this table for the outlet temperature. For this example, with an outlet temperature of 120°F, the specific volume is 16.515 ft³/lb dry gas. The volumetric flow is equal to the specific volume multiplied by the dry gas flow divided by 60 min/h; that is, 16.515 ft³/lb dry gas × 55,816 lb/h ÷ 60 min/h = 15,363 ft³/min.

Fan pressure: induced draft fans used to clean the incinerator off-gases may require a relatively high pressure. The actual differential pressure value across the fan, inserted here, will be used to modify the volumetric flow rate. For this example the fan pressure is 30 in. WC.

Outlet actual ft³/min. The volumetric flow immediately prior to entering the induced draft fan will experience an expansion because of the fan suction. The volumetric flow correction is as follows: Multiply the value in ft³/min determined above by the ration $407 \div (407 - p)$, where p is the pressure across the fan in inches of water column. The figure 407 is atmospheric pressure (14.7 lb/in² absolute) expressed in inches of water column (14.7 lb/in² absolute ÷ 62.4 lb H_2O/ft³ × 1728 in³/ft³ = 407 inches WC). In this example the corrected, or actual, volumetric flow entering the ID fan is $[407 \div (407 - 30)] \times 15,363$ ft³/min = 16,586 ft³/min actual flow.

Outlet: From Table 11.32, insert the saturation humidity, lb H_2O/lb dry gas (dry air), corresponding to the outlet temperature. For this example, with an outlet temperature of 120°F, the humidity is 0.08128 lb H_2O/lb dry gas. The total moisture exiting the stack is the saturation humidity multiplied by the dry gas flow. For this case, 0.08128 lb H_2O/lb dry gas × 55,816 lb dry gas/h = 4537 lb H_2O/h.

Recirculation (ideal): the recirculation flow is that amount of flow required for quenching, as calculated above (58 gal/min for this example). The temperature of this water flow is not critical. The flow is used adiabatically, where the latent heat (not a temperature change) in the water flow reduces the temperature of the flue gas. It is termed a *recirculation flow* because spent scrubber water from the scrubber sump can be recirculated to the venturi for use at 140°F or greater, instead of a cooler flow of water.

Recirculation (actual): In practice the ideal flow is inadequate to fully clean the gas stream of particulate. Ideal quenching requires intimate contact of each molecule of water with gas, instantaneous evaporation, and instantaneous heat transfer between the moisture and the gas, none of which occurs. To compensate for actual versus ideal conditions, an empirical factor is used. In this case this factor is 8. Therefore, for actual recirculation flow use the ideal flow multiplied by 8: 58 gal/min × 8 = 464 gal/min.

Cooling H_2O: Insert the flow of cooling water calculated above, in this case, 760 gal/min.

Computer Program. The mass flow, heat balance, and flue gas discharge analyses presented in this chapter have been developed into a series of computer programs. The programs accept a number of different waste streams, consider individual gas components, and have afterburner, waste heat boiler, and air emission control systems options, as well as the ability to operate in the starved air or excess air modes. These programs display and print out more comprehensive information than is immediately available from the analysis sheets in this chapter. The programs are available from Incinerator Consultants Incorporated, 11204 Longwood Grove Drive, Reston, VA 22094 [phone: (703) 437-1790, fax: (703) 437-9048].

Energy Recovery

Recovering Heat. Steam is used for incinerator heat recovery far more frequently than hot water or hot air (gas) generation. Steam is more versatile in its application, and 1 lb of steam contains significantly more energy than 1 lb of water or air. In general, while hot water is normally of use only for building heat during winter months or can be used in limited quantities for feedwater heating, steam can be used for process requirements and for equipment loads, which are year-round loads. Further, steam can be converted to hot water or used for air heating when these needs arise. The calculations presented herein are for steam generation.

Approach Temperature. With t the temperature of the heated medium (steam or hot water), t_i the temperature of the entering flue gas, and t_o the temperature of the flue gas exiting the boiler (see Fig. 13A.38), the heat available can be calculated. For any heat exchanger there is an approach temperature t_x. This temperature is the difference between the temperatures of the heated medium (t) and of the exiting flue gas (t_o). Therefore,

$$t_x = t_o - t$$

The more efficient the heat exchanger, the lower the approach temperature. The larger the heat exchanger, the lower t_x until, in the extreme case, with an infinitely large heat exchanger,

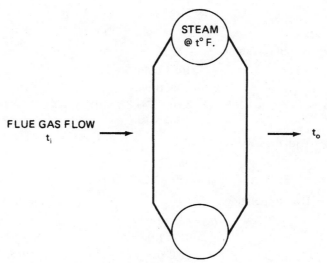

FIGURE 13A.38 Waste heat boiler.

t_x will be zero and the steam (or hot water) will be at the same temperature as the exiting flue gas. In practice, the approach temperature of a waste heat boiler is on the order of 100°F for efficient and 150°F for standard, economical construction.

Available Heat.   The heat available in exhaust or flue gas is equal to that heat at the boiler inlet less the gas heat content at the boiler outlet. With Q the heat available from the flue gas stream, in Btu per pound, note the following:

$$Q = W(h @ t_i - h @ t_o)$$

The flue gas will have a dry and a wet component. Considering the dry gas component to have the properties of air (W_{dg}, h_a) and W_m the moisture component, this equation becomes

$$Q = W_{dg}(h_{ai} - h_{ao}) + W_m(h_{mi} - h_{mo})$$

The inlet temperature t_i is the temperature of the incinerator outlet. The outlet temperature of the heat exchanger (t_o) is defined by the approach temperature (t_x) and the temperature of the heated medium (t):

$$t_o = t_x + t$$

The enthalpy at the outlet of the heat exchanger must be evaluated at t_o.

Example. Consider a gas flow at 1400°F of 15,000 lb/h of dry gas plus 2000 lb/h of moisture. Let the available heat be calculated for the generation of saturated steam at 100, 200, and 400 lb/in² absolute with a 150°F approach temperature (note Table 13A.31 for enthalpy values):

Inlet condition:

$$t_i = 1400°F$$
$$h_a = 341.85 \text{ Btu/lb}$$
$$h_{mi} = 1719.82 \text{ Btu/lb}$$
$$W_{dg} = 15,000 \text{ lb/h}$$
$$W_m = 2000 \text{ lb/h}$$

Outlet condition:
With $p = 100$ lb/in² absolute,

$$t = 328°F$$
$$t_o = 150 + 328 = 478°F$$

By interpolation,

$$h_{ao} = 101 \text{ Btu/lb}$$
$$h_{mo} = 1249 \text{ Btu/lb}$$
$$Q = 15,000(341.85 - 101) + 2000(1719.82 - 1249)$$
$$= 4.554 \text{ MBtu/h}$$

With $p = 200$ lb/in² absolute,

$$t = 382°F$$
$$t_o = 150 + 382 = 532°F$$

By interpolation,

$$h_{ao} = 115 \text{ Btu/lb}$$
$$h_{mo} = 1274 \text{ Btu/lb}$$
$$Q = 15,000(341.85 - 115) + 2000(1719.82 - 1274)$$
$$= 4.294 \text{ MBtu/h}$$

With $p = 400$ lb/in^2 absolute,

$$t = 445°F$$
$$t_o = 150 + 445 = 595°F$$

By interpolation,

$$h_{ao} = 130 \text{ Btu/lb}$$
$$h_{mo} = 1304 \text{ Btu/lb}$$
$$Q = 15,000(341.85 - 130) + 2000(1719.82 - 1305)$$
$$= 4.007 \text{ MBtu/h}$$

These calculations are summarized in Table 13A.39. Steam temperature is listed in Table 13A.40. The inlet is the total heat in the flue gas, related to 60°F:

$$15,000 \times 341.85 + 2000 \times 1714.82 = 8.567 \text{ MBtu/h}$$

TABLE 13A.39

Inlet, MBtu/h	°F	Steam pressure, psia	Steam temperature, °F	Flue gas temperature, °F	Δt, °F	Available heat, MBtu/h	Efficiency, %
8.567	1400	100	328	478	922	4.554	53
8.567	1400	200	382	532	868	4.294	50
8.567	1400	400	445	595	805	4.007	47

TABLE 13A.40 Saturated Steam Properties

Pressure, lb/in^2 absolute	Temperature, °F	Pressure, lb/in^2 absolute	Temperature, °F
14.7	212	45	274
15	213	50	281
16	216	55	287
17	219	60	293
18	222	65	298
19	225	70	303
20	228	75	308
21	231	80	312
22	233	90	320
23	235	100	328
24	238	125	344
25	240	150	358
26	242	175	371
27	244	200	382
28	246	250	401
29	248	300	417
30	250	350	432
32	254	400	445
34	258	450	456
36	261	500	467
38	264	600	486
40	267	700	503

Source: Keenan and Keyes (1957).

The column Δt is the difference in flue gas temperatures entering and leaving the boiler. The efficiency noted is the available heat divided by the total heat in the flue gas entering the boiler.

Of significance is the relationship between available heat and Δt, the temperature difference of the flue gas across the boiler. The available heat is proportional to Δt.

For example, Q at $\Delta t = 805$ °F versus $\Delta t = 922$ °F:

$$Q @ 805 = \frac{805}{922} \times 4.554 = 4.0 \text{ MBtu/h}$$

Q at $\Delta t = 868$ versus $\Delta t = 922$:

$$Q @ 868 = \frac{868}{922} \times 4.554 = 4.3 \text{ MBtu/h}$$

By comparing these values to the calculated values for Q in Table 13A.39, it is clear that the available heat Q is directly proportional to Δt, the temperature loss in the flue gas.

Steam Generation. Given the heat availability, the amount of steam that can be generated can be calculated. Figure 13A.39 shows typical flow through a waste heat boiler producing steam. Makeup water temperature is raised to feedwater temperature by steam flow from the boiler and by the heat contained in return condensate. Condensate is returned to the deaerator.

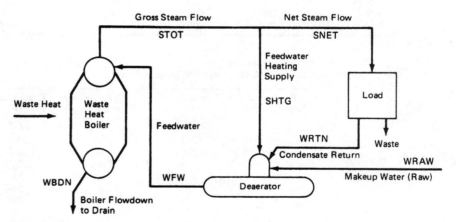

FIGURE 13A.39 Waste heat boiler, steam flow.

Besides raising the feedwater temperature prior to injection into the boiler, the deaerator acts to help release dissolved oxygen from feedwater. Additional feedwater treatment is usually employed to reduce, or prevent, scaling and corrosion of boiler surfaces. Water softeners are used to remove most of the calcium and magnesium hardness from raw water. Chemical addition is also used, typically as follows:

- *Sodium sulfite.* This is an oxygen-scavenging chemical that chemically removes the dissolved-oxygen component not removed in the deaerator. Hydrazine is another oxygen scavenger that is used in high-pressure (over 1200 lb/in^2 absolute) boiler applications.

- *Amine.* There are a number of amines in use for feedwater treatment. They are used for boiler pH or alkalinity control. Excess alkalinity (pH greater than 11) will result in accelerated scale buildup while low pH (below 6) can cause excessive boiler tube corrosion. Normally boiler water pH is maintained in the range of 8.0 to 9.5, slightly alkaline.

- *Phosphates.* This treatment is used to precipitate residual calcium and magnesium hardness remaining in feedwater after softening. Certain phosphates will act as dispersants, preventing adhesion of the precipitate to tube walls.

These chemicals will form a sludge, or mud, which will accumulate in the lower drum of a boiler. The boiler water must have a blowdown on a regular basis to prevent a buildup of mud within the boiler. This blowdown will normally represent from 2 to 5 percent of the boiler steam generation.

Calculating Steam Generation.   By using the steam tables (Table 13A.40), calculations will be performed for obtaining steam, makeup, blowdown, and feedwater flows.

With an approach temperature of 150°F, generating 100 lb/in^2 absolute steam, dry and saturated (h_{stm} = 1187 Btu/lb, h_{bdn} = 298 Btu/lb), the heat available (from Table 13A.39) is 4.554 MBtu/h. For this illustration, blowdown is 4 percent of the feedwater flow, feedwater is provided to the boiler at 220°F (h_{fw} = 188 Btu/lb), and 20 percent of the steam generation is returned as condensate at 170°F (h_{ret} = 138 Btu/lb). In addition, raw water enters the deaerator at 60°F (h_{mu} = 27 Btu/lb), and radiation loss from the boiler is 1 percent of the total boiler input.

The heat available for generating steam is the heat lost by the flue gas less the heat lost by boiler radiation:

$$\text{Waste heat} = 4.554 \text{ MBtu/h} - 0.01 \times 4.554 \text{ MBtu/h} = 4.508 \text{ MBtu/h}$$

Referring to Fig. 13A.39,

$$\text{Waste heat} = \text{heat in steam} + \text{heat in blowdown} - \text{heat in feedwater}$$
$$\text{Heat in blowdown} = \text{WBDN} \times h_{bdn}$$
$$\text{WBDN} = 0.04 \times \text{WFW}$$
$$\text{Heat in blowdown} = 0.04 \times 298 \times \text{WFW} = 11.92\text{WFW}$$
$$\text{Heat in feedwater} = \text{WFW} \times h_{fw}$$
$$= 188\text{WFW}$$
$$\text{Heat in steam} = \text{STOT} \times h_{stm} = \text{STOT} \times 1187 \text{ Btu/lb}$$
$$\text{STOT} = \text{WFW} - \text{WBDN} = \text{WFW} - 0.04\text{WFW} = 0.96\text{WFW}$$
$$\text{Heat in steam} = 0.96\text{WFW} \times 1187 = 1139.52\text{WFW}$$
$$\text{Waste heat} = 4.508 \text{ MBtu/h} = 11.92\text{WFW} + 1139.52\text{WFW} - 188\text{WFW}$$
$$= 963.44\text{WFW}$$

Therefore,

$$\text{WFW} = 4679 \text{ lb/h}$$

and

$$\text{STOT} = 0.96 \times 4679 = 4492 \text{ lb/h}$$

also

$$\text{WBDN} = 0.04 \times 4679 = 187 \text{ lb/h}$$

To calculate steam required for feedwater heating, makeup, and condensate return flows, a material balance and a heat balance must be performed around the deaerator. Material (flow) balance:

$$WFW = SHTG + WRTN + WRAW$$

From above:

$$WRTN = 0.2 \times STOT = 0.2 \times 4492 = 898 \text{ lb/h}$$
$$WFW = 4679 \text{ lb/h}$$

Therefore:

$$SHTG = WFW - WRTN - WRAW = 4679 - 898 - WRAW = 3781 - WRAW$$

Heat balance:

$$WFW \times h_{fw} = SHTG \times h_{stm} + WRTN \times h_{ret} + WRAW \times h_{mu}$$

Therefore,

$$4679 \times 188 = (3781 - WRAW) \times 1187 + 898 \times 138 + WRAW \times 27$$
$$WRAW = 3218 \text{ lb/h}$$
$$SHTG = 3781 - 3218 = 563 \text{ lb/h}$$

Where condensate is not returned to the deaerator, i.e., to the boiler system, the steam required for feedwater heating will increase:
Material balance:

$$SHTG = WFW - WRAW = 4679 - WRAW$$

Heat balance:

$$4679 \times 188 = (4679 - WRAW) \times 1187 + WRAW \times 27$$

Therefore,

$$WRAW = 4030 \text{ lb/h}$$
$$SHTG = 649 \text{ lb/h}$$

In general, with no separate source of heat for feedwater heating (such as returned condensate), 12 to 15 percent of generated steam is required.

Table 13A.41 summarizes the above calculations.

TABLE 13A.41   Calculated Steam Generation

Avail.heat, MBtu/h	STOT, lb/h	WBDN, lb/h	WFW, lb/h	WRET, lb/h	WRAW, lb/h	SHTG, lb/h	WNET, lb/h
4.554	4492	187	4679	898	3218	563	3929
4.554	4492	187	4679	0	4030	649	3843

The net flow of steam (SNET) is that quantity of steam available for useful work. As can be seen, use of condensate return for feedwater heating increases the quantity of steam available for a load.

Waterwall Systems.   Larger incinerators dedicated to destruction of refuse or other paper-type waste materials are often designed with "waterwall" construction, as described previously. These installations can be provided with a variety of features including the following:

- *Convection boiler section.* Boiler tubes are placed perpendicular to the flow of gas as it exits the incinerator. A major portion of available heat is captured by these tubes, producing saturated steam.

- *Economizer.* This is used to heat feedwater by extracting heat from gases as they leave the convection boiler section.

- *Superheater.* A tubular section is normally placed upstream of the convection section. Hot incinerator gases superheat steam generated from the convection section of the boiler.

- *Air preheater.* This is used in lieu of, or directly downstream of, the economizer. It produces heated combustion air from the relatively low-temperature gas flow at this location.

Calculations of steam generation from each of these sections are a complex task and will not be detailed here. Tables 13A.23 and 13A.24 indicate steam generation for typical waterwall incinerators for a variety of waste quality.

To calculate available heat by the methods of this chapter, the exit gas temperature (the temperature of flue gas exiting the boiler sections and entering the air emissions control system) can be assumed to be in the range of 350 to 550°F.

Electric Power Generation.   As discussed above, steam is a useful by-product of the incineration process. The generation of steam from typical facilities is listed in Table 13A.42. Steam rates follow the cost of energy, and can range from $10 to $15 per 1000 lb, which represents a revenue of $70 to $105 per ton, based on a steaming rate of 7000 lb per ton of MSW.

Revenue from steam sales can be generated only if there is a market for steam and if the user is relatively close to the incinerator, where pipeline losses are not significant. In many areas there is no local steam customer, or the potential user does not have a year-round need for steam. If the main use for steam is for winter heating, the sale of steam will go wanting for half the year.

TABLE 13A.42   Typical Steam Generation Rates

MSW quality*	6500	6000	5000	4000	3000
Moisture, %	15	18	25	32	39
Noncombustible, %	14	16	20	24	28
Combustible, %	71	66	55	44	33
Steam generation, lb/ton MSW	8600	7800	6400	4600	3000

* Heat value, Btu/lb as received.

Electric power has a universal market. It is salable practically anywhere in the country. The conversion of steam to electric power, however, results in a loss of energy. Electric power generation requires that steam pass through a turbine, and that the turbine drive a generator to feed a power grid. While 7000 lb of steam per ton of MSW converts to 2050 kWh, this same ton of MSW will generate less than 600 kWh of electrical energy. At a cost of electric power of from $0.09 to $0.15 per kWh, the electric power revenue will bring in from $54 to $90 per ton of MSW. This is less than the revenue generated by steam.

It is usually preferable to find a steam customer. In some areas of the country a crucial factor in the location of an incinerator is its physical proximity to a potential user of steam.

Table 13A.43 lists energy production rates of incinerators throughout the United States. The net kWh listed is the gross generation of electric power less the amount of power that is used to operate the facility. The gross power generation rate will vary from 577 kWh/ton MSW for waterwall units to 350 kWh/ton MSW for smaller, modular systems.

TABLE 13A.43 Energy Production at Mass Burn Facilities in the United States

Facility	State	Design capacity, tons/day	Net power output, MW	Gross power output, MW	Ratio net/gross power output	Net kWh per ton processed	Gross kWh per ton processed	Ratio net/gross kWh/ton	Steam, lb/h	Btu/lb	Start-up year
					Process: Mass burn—waterwall						
Albany (American Ref-Fuel)	NY	1500	40	50	0.80	N/A	N/A	N/A	400,000	5500	88
Alexandria/Arlington R.R. Facility	VA	975	20	22	0.90	470	520	0.90	255,000	4800	89
Babylon Resource Recovery Project	NY	750	14	17	0.82	410	N/A	N/A	185,000	5000	
Bergen County	NJ	3000	80	88	0.91	482	N/A	N/A	808,000	4500	
Bridgeport RESCO	CT	2250	60	67	0.90	640	720	0.89	576,000	5300	88
Bristol	CT	650	14	16	0.84	535	620	0.86	148,000	5000	88
Brooklyn Navy Yard	NY	3000	N/A	N/A	N/A	N/A	N/A	N/A	847,000	N/A	
Broome County	NY	571	15	18	0.83	467	560	0.83	184,000	5200	
Broward County (Northern Facility)	FL	2250	60	67	0.90	638	709	0.90	573,500	5200	
Broward County (Southern Facility)	FL	2250	57	63	0.90	608	676	0.90	576,700	5200	
Camden County (Foster Wheeler)	NJ	1050	21	30	0.70	482	N/A	N/A	260,400	4500	
Camden County (Pennsauken)	NJ	500	10	13	0.78	425	N/A	N/A	110,000	5200	
Central Mass. Resource Recovery Project	MA	1500	36	40	0.90	600	N/A	N/A	336,000	5000	88
Charleston County	SC	644	11	13	0.84	N/A	N/A	N/A	164,000	5000	89
City of Commerce	CA	400	10	12	0.87	630	725	0.87	115,000	5600	87
Concord Regional S.W. Recovery Facility	NH	500	12	13	0.92	470	550	0.85	135,400	5000	89
Dakota County	MN	800	20	23	0.87	550	N/A	N/A	410,000	5000	
Davidson County	TN	210	3	4	0.81	N/A	N/A	N/A	34,000	6000	
East Bridgewater (American Ref-Fuel)	MA	1500	40	50	0.80	N/A	N/A	N/A	400,000	5500	
Eastern-Central Project	CT	550	12	15	0.83	560	N/A	N/A	155,500	5300	
Essex County	NJ	2277	72	76	0.95	N/A	501	N/A	633,000	4500	
Fairfax County	VA	3000	73	85	0.86	540	610	0.89	822,504	4400	90
Falls Township (Wheelabrator)	PA	2250	65	72	0.90	600	N/A	N/A	570,000	5200	
Glendon	PA	500	13	14	0.89	525	N/A	N/A	130,000	5200	
Gloucester County	NJ	575	12	14	0.86	425	475	0.89	135,400	4500	90
Hampton/NASA Project Recoup	VA	200	N/A	N/A	N/A	N/A	N/A	N/A	66,000	N/A	80
Harrisburg	PA	720	5	8	0.64	500	N/A	N/A	170,000	4500	72
Haverhill	MA	1650	41	46	0.89	572	N/A	N/A	396,000	5081	89
Hempstead (American Ref-Fuel)	NY	2505	64	72	0.89	570	N/A	N/A	604,000	4500	90
Hennepin County (Blount)	MN	1200	33	38	0.88	540	700	0.77	350,000	5800	90
Hillsborough County SWER Facility	FL	1200	28	30	0.92	492	N/A	N/A	270,000	4500	87
Hudson County	NJ	1500	38	45	0.85	455	N/A	N/A	410,000	4500	
Huntington	NY	750	21	25	0.84	627	736	0.85	225,000	6000	
Jackson County/Southern MI State Prison	MI	200	2	2	0.85	N/A	N/A	N/A	49,600	4900	87
Johnston (Central Landfill)	RI	750	17	21	0.81	543	N/A	N/A	150,000	5200	
Kent County	MI	625	16	18	0.86	410	N/A	N/A	158,000	5350	
Lake County	FL	528	10	15	0.69	N/A	525	N/A	120,000	5000	90
Lancaster County	PA	1200	30	36	0.83	560	N/A	N/A	291,000	5000	

Facility	State	Design capacity, tons/day	Net power output, MW	Gross power output, MW	Ratio net/gross power output	Net kWh per ton processed	Gross kWh per ton processed	Ratio net/gross kWh/ton	Steam, lb/h	Btu/lb	Start-up year
Lee County	FL	1800	47	50	0.94	630	N/A	N/A	506,250	5000	
Lisbon	CT	500	13	15	0.87	550	600	0.92	135,400	4500	86
Marion County Solid W-T-E Facility	OR	550	11	13	0.84	450	N/A	N/A	133,446	4700	
Montgomery County	MD	1800	69	84	0.83	644	N/A	N/A	512,000	5500	
Montgomery County	PA	1200	29	34	0.85	N/A	460	N/A	269,082	4500	
Morris County	NJ	1340	34	40	0.85	N/A	535	N/A	433,300	5500	
Nashville Thermal Transfer Corp. (NTTC)	TN	1120	3	7	0.40	N/A	N/A	N/A	308,000	4900	74
New Hampshire/Vermont S.W. Project	NH	200	4	5	0.84	N/A	440	N/A	46,200	5400	87
Norfolk Naval Station	VA	360	N/A	N/A	N/A	N/A	N/A	N/A	40,000	N/A	67
North Andover	MA	1500	32	38	0.84	550	N/A	N/A	344,000	5500	85
North Hempstead	NY	990	17	21	0.81	N/A	N/A	N/A	N/A	N/A	
Northwest Waste-To-Energy Facility	IL	1600	N/A	N/A	N/A	N/A	N/A	N/A	330,000	N/A	70
Oklahoma City	OK	820	10	22	0.46	N/A	N/A	N/A	240,000	5200	85
Olmstead County	MN	200	2	3	0.75	N/A	293	N/A	50,000	5500	87
Onondaga County	NY	990	32	38	0.84	640	N/A	N/A	311,646	6000	
Oyster Bay	NY	1000	27	31	0.87	N/A	N/A	N/A	248,000	6000	
Pasco County	FL	1050	29	31	0.94	550	650	0.85	270,900	4800	
Passaic County	NJ	1434	37	45	0.83	625	753	0.83	445,620	5500	
Pinellas County (Wheelabrator)	FL	3150	56	62	0.90	430	N/A	N/A	750,000	4000	83
Portland	ME	500	10	14	0.74	N/A	500	N/A	120,000	5000	88
Preston (Southeastern Connecticut)	CT	600	16	18	0.89	520	N/A	N/A	144,000	5000	
Quonset Point	RI	710	18	21	0.86	455	N/A	N/A	182,000	4750	
S.E. Resource Recovery Facility (SERRF)	CA	1380	30	36	0.83	540	N/A	N/A	351,000	4800	88
S.W. Resource Recovery Facility (BRESCO)	MD	2250	34	60	0.57	350	400	0.88	441,000	5100	85
Saugus	MA	1500	40	50	0.80	550	N/A	N/A	340,000	4500	75
Savannah	GA	500	N/A	N/A	N/A	N/A	N/A	N/A	120,000	N/A	87
Spokane	WA	800	22	26	0.85	497	N/A	N/A	222,600	N/A	
Stanislaus County Res. Recovery Facility	CA	800	17	23	0.76	450	N/A	N/A	201,000	4750	89
Sturgis	MI	560	11	13	0.85	N/A	N/A	N/A	100,000	6000	
Union County	NJ	1440	39	44	0.89	567	670	0.85	360,000	5400	89
University City Res. Recovery Facility	NC	235	4	5	0.75	395	476	0.83	50,000	4500	89
Walter B. Hall Res. Recovery Facility	OK	1125	15	17	0.88	530	600	0.88	240,000	5000	86
Warren County	NJ	400	11	14	0.78	482	N/A	N/A	112,000	4650	88
Washington/Warren Counties	NY	400	11	13	0.85	N/A	N/A	N/A	115,000	5500	
Wayne County	NC	300	4	5	0.85	N/A	N/A	N/A	36,000	N/A	
West Pottsgrove Recycling/R.R. Facility	PA	1500	40	45	0.89	N/A	N/A	N/A	336,000	5200	
Westchester	NY	2250	56	60	0.93	590	N/A	N/A	504,000	4800	84
Numerical average of nonzero values		1138	27	32	0.83	526	577	0.87	291,520	5065	
Standard deviation		754	20	22	0.10	74	115	0.03	199,429	450	

(continued)

13.81

TABLE 13A.43 Energy Production at Mass Burn Facilities in the United States (*Continued*)

Facility	State	Design capacity, tons/day	Net power output, MW	Gross power output, MW	Ratio net/gross power output	Net kWh per ton processed	Gross kWh per ton processed	Ratio net/gross kWh/ton	Steam, lb/h	Btu/lb	Start-up year
				Process: Mass burn—modular							
Agawam/Springfield	MA	360	7	9	0.83	390	N/A	N/A	85,500	4200	88
Barron County	WI	80	0	0	0.26	N/A	N/A	N/A	16,500	4750	86
Batesville	AR	100	N/A	N/A	N/A	N/A	N/A	N/A	6,200	N/A	81
Bellingham	WA	100	1	2	0.67	350	N/A	N/A	23,000	4500	86
Beto 1 Unit (Texas Dept. of Corrections)	TX	25	N/A	N/A	N/A	N/A	N/A	N/A	7,000	N/A	80
Cassia County	ID	50	N/A	N/A	N/A	N/A	N/A	N/A	9,000	N/A	82
Cattaraugus County R-T-E Facility	NY	112	N/A	N/A	N/A	N/A	N/A	N/A	26,000	N/A	83
Center	TX	40	N/A	N/A	N/A	N/A	N/A	N/A	9,000	N/A	86
City of Carthage/Panola County	TX	40	N/A	N/A	N/A	N/A	N/A	N/A	2,500	N/A	86
Cleburne	TX	115	1	1	0.77	N/A	N/A	N/A	18,000	4500	86
Collegeville	MN	50	N/A	N/A	N/A	N/A	N/A	N/A	11,000	N/A	81
Dyersburg	TN	100	N/A	N/A	N/A	N/A	N/A	N/A	20,000	5000	80
Eau Claire County	WI	150	3	3	0.91	263	323	0.81	37,000	5000	
Elk River R.R. Authority (TERRA)	TN	200	N/A	N/A	N/A	N/A	N/A	N/A	50,000	N/A	87
Energy Gen. Facility at Pigeon Point	DE	600	11	13	0.79	532	N/A	N/A	152,000	5500	88
Fergus Falls	MN	94	N/A	N/A	N/A	N/A	N/A	N/A	30,000	N/A	88
Fort Dix	NJ	80	N/A	N/A	N/A	N/A	N/A	N/A	12,000	N/A	86
Fort Leonard Wood	MO	75	N/A	N/A	N/A	N/A	N/A	N/A	8,740	N/A	82
Fort Lewis (U.S. Army)	WA	120	N/A	N/A	N/A	N/A	N/A	N/A	42,000	N/A	
Gatesville (Texas Dept. of Corrections)	TX	13	N/A	N/A	N/A	N/A	N/A	N/A	3,000	N/A	80
Hampton	SC	270	N/A	N/A	N/A	N/A	N/A	N/A	45,000	N/A	85
Harford County	MD	360	N/A	N/A	N/A	N/A	N/A	N/A	75,000	N/A	88
Harrisonburg	VA	100	N/A	N/A	N/A	N/A	N/A	N/A	17,000	N/A	82
Key West	FL	150	2	3	0.85	300	N/A	N/A	42,740	5000	86
Lamprey Regional Solid Waste Cooperative	NH	108	N/A	N/A	N/A	N/A	N/A	N/A	20,000	N/A	80
Lassen Community College	CA	100	1	2	0.78	N/A	N/A	N/A	24,000	6500	84
Lewis County	TN	50	N/A	N/A	N/A	N/A	N/A	N/A	14,000	N/A	88
Long Beach	NY	200	3	5	0.67	N/A	N/A	N/A	58,000	5000	88
Manchester	NH	560	13	14	0.89	425	N/A	N/A	20,000	4500	
Mayport Naval Station	FL	50	N/A	N/A	N/A	N/A	N/A	N/A	N/A	N/A	79
Miami	OK	108	N/A	N/A	N/A	N/A	N/A	N/A	23,000	N/A	82
Miami International Airport	FL	60	N/A	N/A	N/A	N/A	N/A	N/A	15,000	N/A	83
Muskegon County	MI	180	2	3	0.82	373	N/A	N/A	34,000	N/A	
New Hanover County	NC	100	2	4	0.50	N/A	N/A	N/A	54,000	N/A	84
North Slope Borough/Prudhoe Bay	AK	100	N/A	N/A	N/A	N/A	N/A	N/A	N/A	N/A	81
Oneida County	NY	200	1	2	0.55	N/A	N/A	N/A	26,000	N/A	85
Osceola	AR	50	N/A	N/A	N/A	N/A	N/A	N/A	10,000	N/A	80
Oswego County	NY	200	1	4	0.28	275	N/A	N/A	45,000	5000	86
Park County	MT	75	N/A	N/A	N/A	N/A	N/A	N/A	13,000	N/A	82
Pascagoula	MS	150	N/A	N/A	N/A	N/A	N/A	N/A	24,000	N/A	85

Facility	State	Design capacity, tons/day	Net power output, MW	Gross power output, MW	Ratio net/gross power output	Net kWh per ton processed	Gross kWh per ton processed	Ratio net/gross kWh/ton	Steam, lb/h	Btu/lb	Start-up year
Perham	MN	116	N/A	N/A	N/A	N/A	N/A	N/A	23,000	N/A	86
Pittsfield	MA	240	N/A	N/A	N/A	N/A	N/A	N/A	50,000	N/A	81
Polk County	MN	80	N/A	N/A	N/A	N/A	N/A	N/A	21,000	N/A	88
Pope-Douglas W-T-E Facility	MN	80	N/A	N/A	N/A	N/A	N/A	N/A	11,000	N/A	87
Red Wing	MN	72	N/A	N/A	N/A	N/A	N/A	N/A	15,000	N/A	82
Richard Asphalt	MN	57	N/A	N/A	N/A	N/A	N/A	N/A	13,500	N/A	82
Rutland	VT	240	6	7	0.86	470	N/A	N/A	40,000	N/A	88
Salem	VA	100	N/A	N/A	N/A	N/A	N/A	N/A	14,000	N/A	88
St. Croix County	WI	115	1	1	0.58	85	110	0.77	23,500	5000	78
Tuscaloosa Energy Recovery Facility	AL	300	N/A	N/A	N/A	N/A	N/A	N/A	55,880	N/A	89
Wallingford	CT	420	9	11	0.85	384	500	0.77	105,000	4850	84
Waxahachie	TX	50	N/A	N/A	N/A	N/A	N/A	N/A	15,000	N/A	89
Westmoreland County	PA	50	N/A	N/A	N/A	N/A	N/A	N/A	10,000	4500	82
Windham	CT	108	2	2	0.86	N/A	150	N/A	16,800	5000	81
Numerical average of nonzero values		143	4	5	0.71	350	271	0.78	29,651	4920	
Standard deviation		122	4	4	0.19	114	155	0.02	27,108	525	
Process: Rotary combustor											
Auburn	ME	200	4	5	0.76	N/A	N/A	N/A	113,800	5200	N/A
Delaware County Regional	PA	2688	80	90	0.79	600	N/A	N/A	664,972	5200	N/A
Dutchess County	NY	506	9	10	0.92	140	320	0.44	110,000	N/A	88
Falls Township (Technochem)	PA	70	0	1	0.47	130	275	0.47	16,000	4500	
Galax	VA	56	N/A	N/A	N/A	N/A	N/A	N/A	12,000	N/A	86
Gaston County/Westinghouse R.R. Center	NC	440	6	7	0.81	550	N/A	N/A	N/A	N/A	
MacArthur Energy Recovery Facility	NY	518	8	12	0.70	370	N/A	N/A	118,000	4450	89
Mercer County	NJ	975	32	36	0.89	560	655	0.85	314,500	5000	
Monmouth County	NJ	1700	57	63	0.90	N/A	N/A	N/A	N/A	4950	
Monroe County	IN	500	9	11	0.85	N/A	N/A	N/A	110,000	N/A	
Montgomery County (North)	OH	300	6	6	0.95	523	550	0.95	72,000	5000	88
Montgomery County (South)	OH	900	18	19	0.95	482	507	0.95	240,000	5000	
Oakland County	MI	2000	54	62	0.87	645	N/A	N/A	600,000	5200	
San Juan Resource Recovery Facility	PR	1040	22	27	0.81	510	N/A	N/A	254,000	4500	
Sangamon County	IL	450	6	8	0.75	380	N/A	N/A	90,000	N/A	
Skagit County	WA	178	2	2	0.85	345	N/A	N/A	40,000	4500	88
Sumner County	TN	200	0	1	0.86	N/A	N/A	N/A	50,000	5500	81
Waukesha County (New Plant)	WI	600	N/A	N/A	N/A	N/A	N/A	N/A	200,000	N/A	
Westinghouse/Bay Resource Mgmt. Center	FL	510	10	12	0.83	432	480	0.90	136,000	4600	87
York County	PA	1344	30	35	0.86	540	N/A	N/A	330,000	4500	89
Numerical average of nonzero values		759	20	23	0.83	443	465	0.76	192,848	4864	
Standard deviation		680	22	25	0.11	151	131	0.22	181,052	336	

N/A = not available.

Source: *Data Summary of Municipal Solid Waste Management Alternatives,* vol. III: *Appendix A—Mass Burn Technologies,* National Technical Information Service, October 1992, NTIS accession number DE92016433.

REFERENCES

Black, R., and A. Klee (1968) "The National Solid Wastes Survey: An Interim Report," Presented at the 1968 Annual Meeting of the Institute of Solid Wastes of the American Public Works Association, Miami Beach, FL.

Brunner, C. R. (1987) "Biomedical Waste Incineration," Monograph, Presented at the Air Pollution Control Association Annual Conference, New York, NY.

Brunner, C. R. (1988a) *Incineration Systems: Selection and Design,* Incinerator Consultants Incorporated, Reston, VA.

Brunner, C. R. (1988b) Hospital Waste Disposal by Incineration, *Journal of the Air Pollution Control Association,* vol. 38, no. 10, pp. 1297–1309.

Brunner, C. R. (1993) *Handbook of Hazardous Waste Incineration,* McGraw-Hill, New York.

Brunner, C., and S. Schwarz (1983) *Energy and Resource Recovery from Waste,* Noyes, Park Ridge, NJ.

Danielson, J. (1973) *Air Pollution Engineering Manual,* Air Pollution Control District, AP-40, County of Los Angeles, CA.

Doyle, B. W. (1985) The Smoldering Question of Hospital Wastes, *Pollution Engineering.*

Incinerator Institute of America (1972) *Incinerator Standards,* Incinerator Institute of America, New York, 1972.

Keenan, J., and F. Keyes (1957) *Thermodynamic Properties of Steam,* John Wiley & Sons, New York.

Ontario Ministry of the Environment (1986) Incinerator Design and Operating Criteria, vol. II, *Biomedical Waste Incinerators.*

U.S. EPA (1973) *Recommended Methods of Reduction, Neutralization, Recovery or Disposal of Hazardous Waste,* vol. 3, Disposal Process Descriptions, Ultimate Disposal, Incineration and Pyrolysis Processes, 670/2-73-053C, U.S. Environmental Protection Agency, Washington, DC.

U.S. EPA (1980) *Source Category Survey: Industrial Incinerators,* 450 3-80-013, U.S. Environmental Protection Agency, Washington, DC.

CHAPTER 13

WASTE-TO-ENERGY COMBUSTION
Part 13B   Ash Management and Disposal

Floyd Hasselriis

Ash residues from combustion of municipal solid waste generally represent about 25 percent of the incoming waste. These residues are generally in a wet condition when disposed, adding up to 25 to 50 percent to the weight. Consequently, disposal of ash residues imposes a substantial increment to the total cost of operation of a WTE facility. Public apprehensions concerning the environmental effect of ash residues have imposed additional costs of testing and processing.

In the United States, where space for landfills is ample, landfilling of ash residues can be supported. The situation in Europe and Japan is entirely different. Land is scarce and extremely valuable for general uses and agriculture. Hence, in Europe landfilling of ash residues has been restricted, and recycling and beneficial use of these residues has been encouraged, both economically and by favorable regulations.

Ash residues from combustion of MSW need to be disposed of in an environmentally sound and economical manner. Whether placed in landfills or beneficially used, account must be taken of their characteristics and the effect of ash management procedures on their properties, and their environmental impact. The cost of landfill disposal provides an incentive to develop beneficial uses.

13B.1   SOURCES AND TYPES OF ASH RESIDUES

Ash residues are discharged at various locations from the combustion and emission control equipment (see Figs. 13A.1 and 13A.13).

Bottom ash, discharged after the waste has progressed down the stoker, consists of inert residues, glass and metallic objects, and 2 to 10 percent carbon. Bottom ash is usually quenched with water, although it can also be collected in a dry state.

Stoker grate siftings fall through clearances in the grates, and are collected with bottom ash. These may include unburned organic matter.

Boiler ash, carried by combustion gases, consists of flying particles and condensable metal vapors which may attach to refractory and water-cooled walls and be caught by boiler tube surfaces. It may fall onto the stoker into the bottom ash, or it may be collected in hoppers and discharged into the bottom ash.

Fly ash, carried by the combustion gases through the furnace, boiler and scrubber, is collected by the particulate control device. If a wet scrubber is employed the ash will be discharged with the scrubber blowdown. Fly ash collected by an electrostatic precipitator (ESP) or fabric filter may be discharged into the bottom ash, or collected separately. Fly ash can be conditioned (moistened) to prevent dusting and fugitive emissions.

Scrubber reaction products, collected at the bottom of spray-dry or dry lime-injection acid gas scrubbers, include fly ash and reacted or partially reacted alkaline reagent (such as lime) and some carbon.

Mixed ash may contain siftings, bottom ash, boiler deposits, scrubber residues, fly ash, and scrubber products.

13B.2 PROPERTIES OF ASH RESIDUES

Ash residue properties depend upon the municipal solid waste (MSW) burned, the combustion and emission control systems, and the methods of residue collection. What goes in must come out somewhere and in some form.

Composition of Municipal Solid Waste

Unprocessed municipal solid waste (MSW) generally contains about 52 percent combustible matter, 26 percent moisture, and 22 percent ash and noncombustible (inert) materials, as shown in Fig. 13B.1 (Hasselriis, 1984). Recycling of metals and glass reduces but does not eliminate the noncombustible fraction.

Chemical Composition of MSW

The chemical composition of raw MSW is determined after removing (and separately accounting for) the large inert materials (metals, glass, and ceramics), and shredding and performing laboratory analysis on the remaining fraction. Table 13B.1 shows a typical ultimate analysis and analysis of noncombustibles for major and trace metals as performed by ASTM procedures (ASTM, undated). The combustible remainder contains 13.86 percent inherent ash. Major metals constitute 99 percent of this, mainly aluminum, calcium, sodium and potassium, as shown in Fig. 13B.2 (GBB, 1990). Trace metals, totaling about 1 percent of the combustibles, are mainly zinc, tin, and lead, and are shown in Fig. 13B.3.

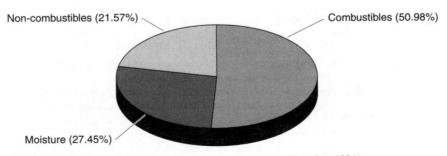

Non-combustibles (21.57%)

Combustibles (50.98%)

Moisture (27.45%)

FIGURE 13B.1 Composition of municipal solid waste. *(Source: Hasselriis, 1984.)*

TABLE 13B.1 Typical Ultimate Analysis and Analysis of Noncombustibles by ASTM Procedures

Ultimate analysis	Percent	Major metals[a]	Percent
Carbon	35.02	Aluminum	33.21
Oxygen	30.16	Calcium	30.20
Hydrogen	11.67	Sodium	17.50
Sulfur	1.88	Potassium	13.12
Chlorine	0.25	Silicon	5.00
Moisture	19.16	Subtotal:	99.04
Ash of combustibles	6.86		
Total	100.00		
		Trace metals	
		Zinc	0.40
		Lead	0.31
		Tin	0.15
		Chromium	0.047
		Nickel	0.020
		Cadmium	0.009
		Copper	0.007
		Subtotal:	0.96
		Total	100.00

[a] Minerals

Source: GBB (1990).

Composition and Quantities of Ash Residues

The composition of bottom ash residues remaining after combustion of unprocessed MSW, determined after separation through a 2-in mesh screen, is described in Table 13B.2.

The typical range in total quantities of residues generated from combustion of MSW and associated emission controls, shown in Table 13B.3, indicates that dry bottom ash and fly ash may be 27 to 39 percent of the weight of MSW, and residues from various types of acid gas

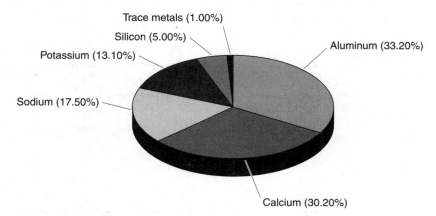

FIGURE 13B.2   Major metal composition of municipal solid waste. *(Source: Hasselriis, 1984.)*

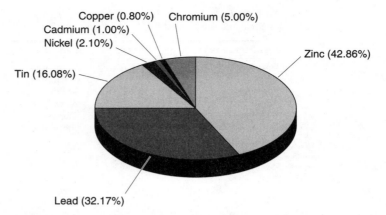

FIGURE 13B.3   Composition of trace metals in municipal solid waste. *(Source: Hasselriis, 1984.)*

emission controls may add 1 to 5 percent to the residues requiring disposal (Thome-Kozmiensky, 1989).

Potentially recoverable are a fraction of the 84 percent mixed metals in the 19 percent over-size fraction and the 22.7 percent ferrous metals and 3.4 percent nonferrous metals in the minus 2-in screened fraction. The fine fly ash component of the bottom ash has been found to contain relatively high levels of toxic metals, serving to contaminate the bottom ash.

Density of Ash Residues

Ash residues from combustion of MSW generally have densities ranging from 65 to 75 lb/ft³ at a water content of 15 to 25 percent. They can be compacted in the ashfill to 135 lb/ft³ (2180 kg/m³) at about 20 percent moisture (Forrester and Goodwin, 1990). This may be compared with portland cement weighing about 94 lb/ft³, and gravel or aggregate at 105 lb/ft³ (1 lb/ft³ = 113.2 kg/m³). Specific gravities of fine and coarse residues from a facility employing a dry lime-injection scrubber and baghouse ranged from 1.9 to 2.5 (120 to 160 lb/ft³).

TABLE 13B.2   Material Fractions of Municipal Solid Waste Ash Residues

Residue	Total (100%)	Plus 2-in, (19.2%)	Minus 2-in, (80.8%)
Metal	111.1	84.0	
Other	1.1	11.4	
Combustibles	4.0	9.7	2.7
Ferrous metal	18.3		22.7
Nonferrous metal	2.7		3.4
Glass	211.2		32.4
Ceramics	8.3		10.3
Minerals and ash	23.0		28.5
	100%	100%	100%

Source: Chesner et al. (1988).

TABLE 13B.3 Quantities of Combustion and Emission
Control Residues

	Quantity of waste, lb/100 lb waste	% of total
Combustion residue		
Bottom ash (slag)	25.0–35.0	90
Filter dust (fly ash)	2.0–4.0	10
Total:	27.0–39.0	100
Additional residues:		
Wet Scrubber Residue	0.8–1.5	3–4
Spray-dry Scrubber Residue	1.6–3.5	6–9
Dry Injection Residue	2.5–4.5	9–12

Source: Thome-Kozmiensky (1989).

Moisture Content

Moisture content of ash residues generally can range widely, from 15 to 57 percent, depending on whether a semidry ash discharger or a water quench tank is used. Excessive moisture content increases disposal costs. Maintaining minimum but sufficient moisture essentially eliminates dust liberation and fugitive dust problems. Landfill density can be optimized by controlling moisture. Dry fly ash can be conditioned with water to eliminate dusting.

Chemical Composition of Ash Residues

The noncombustible components of the MSW appear in the bottom ash and fly ash residues in different fractions, as can be seen in Table 13B.4 (Forrester, 1989).

TABLE 13B.4 Total Metals in Combined Ash, Fly Ash,
and Fly Ash/Scrubber Residues

	(Parts per million parts of ash by weight)		
Metal	Combined ash	Fly ash	Fly ash/ Scrubber
Aluminum	17,857	27,500	12,714
Calcium	33,642	64,857	176,428
Sodium	3828	26,928	1292
Potassium	3071	36,928	8450
Iron	20,428	10,857	3314
Chloride	928	65,428	164,285
Sulfate	7	33	8
Lead	3142	22,143	3257
Cadmium	35	642	160
Zinc	4107	53,500	9143
Manganese	534	649	365
Mercury	ND	3	73

Source: Forrester (1989).

Combined bottom ash (including fly ash) consists mainly of the mineral metals found in common earth. Fly ash has greater proportions of the earth metals, the trace metals, lead, cadmium, and zinc, and the highly soluble chloride and sulfate salts.

Alkaline materials used to scrub acids from the combustion products increase the calcium (or sodium) compounds in the ash residues. Fly ash/scrubber residues from lime-injection acid gas controls contain predominantly calcium and chlorine, mainly calcium chloride salt resulting from the reaction of calcium hydroxide with the hydrogen chloride arising from combustion of chlorine-bearing components in the waste. This high salt content may be the main component that must be considered in the disposal of scrubber reaction products, dry or wet.

The chemical forms of fly ash/scrubber residues, shown in Table 13B.5, consist mainly of calcium, silica, alumina, and iron oxides. This composition is similar to that of portland cement and flue gas desulfurization (FGD) waste (Goodwin, 1982). Vitrified slag from MSW consists primarily of silica, iron, calcium, sodium, and aluminum (Abe, 1984).

Metals Found in Bottom Ash and Fly Ash

The trace (or minor) metals found in combined bottom ash/fly ash, fly ash and combined fly ash/scrubber residues are shown in Table 13B.4 (Forrester, 1989). Lead, cadmium, and zinc have high concentrations in the fly ash since they are volatilized at normal combustion temperatures.

Potentially toxic metals in the fly ash, such as lead, cadmium, chromium, and nickel, are mainly derived from pigments, fillers, and inks used in or on paper and plastic products. Reduction of these sources can reduce the quantities of these metals in the ash residues from combustion of MSW.

Particle (Grain) Size Distribution

Typical size distributions of bottom ash and fly ash, an important factor in disposal and beneficial use of ash residues, are shown in Fig. 13B.4. Typically these distributions plot as straight lines on log-log paper. The knee in the fly ash curve indicates the presence of agglomerated fly ash. The size distribution of bottom ash corresponds well with specifications for aggregate materials (Chesner et al., 1988).

Acidity and Alkalinity

The acidity or alkalinity of ash residues in the presence of water, as measured by pH, generally ranges from neutral (pH = 7) to alkaline (pH = 9). When large amounts of unreacted lime are present, higher alkalinities, up to pH = 11, may be encountered.

TABLE 13B.5 Comparison of MSW APC Waste with Portland Cement and Slag[a]

Component	MSW/APO[a] residue	Portland[b] cement	Utility[b] FGD waste	MSW[c] slag
CaO	62–67	24	37	14
SiO$_2$	18–24	28	24	61
Al$_2$O$_3$	4–8	14	6	5
Fe$_2$O$_3$	1.5–4.5	5	3	

[a] MSW/APC residue is fly ash plus spray-dry acid gas scrubber residue.

[b] *Source:* Goodwin, 1982.

[c] Slag obtained by fusing MSW ash residues (Abe, 1984).

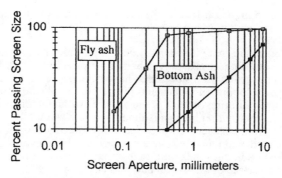

FIGURE 13B.4   Size distributions of particles in fly ash and bottom ash. *(Source: Chesner et al., 1988.)*

Solubility of Metals in Water

Water containing high concentrations of dissolved metal compounds has the potential for contaminating the environment and drinking water supplies. Metals in solid form are not readily dissolved by pure neutral water. The solubility of the trace toxic metals depends on their chemical form and on the acidity or alkalinity (pH) of the water with which they are in contact. Hydroxides, sulfates, and chlorides are more soluble and more readily available for leaching than oxides, silicates, and carbonates. The sulfate and chloride forms of lead and cadmium, found mainly in fly ash as the result of reaction with the sulfur and chlorine in the MSW, are highly soluble.

Most metals are only slightly soluble in water at a pH range from 6 to 9. However, under acidic conditions, such as pH of 5 or less, solubility increases rapidly. While most metals are not soluble under alkaline conditions, lead becomes highly soluble at a pH greater than 10. Since lime has a pH of 12, large quantities of unreacted lime in the fly ash from acid gas control devices can increase the concentration of lead in the solution.

Figure 13B.5 shows the solubility of lead and cadmium under the range from acidity to alkalinity. The limiting concentrations of lead and cadmium prescribed by the U.S. EPA are

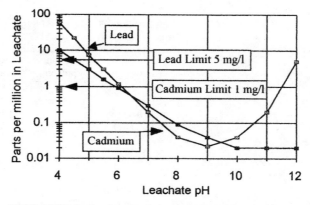

FIGURE 13B.5   Lead and cadmium concentrations in MSW ash residue leachate versus leachate pH, showing U.S. EPA toxic limits and detection limit. *(Source: Donnelly et al., 1987.)*

shown: 5 mg/L for lead and 1 mg/L for cadmium. It is apparent that the concentrations of lead and cadmium are less than these limits within a wide range of pH values (Donnelly et al., 1987).

Lead carbonate is relatively insoluble. The carbonic acid that forms when carbon dioxide in the atmosphere is dissolved in rain may account for the low concentrations of lead found in leachates from ash residues that have been exposed to the atmosphere (Shinn, 1987).

The solubility of metals and the pH, the total acid content of the leaching water, and the alkaline buffering content of the ash are crucial factors in the management of ash residues. If the water is neither highly alkaline nor acidic, the soluble metals will be leached out relatively slowly, and the leachate will contain only low concentrations of the metals. When leached with simulated acid rain, the concentrations of soluble metals in the leachate fall rapidly to a minimum at a liquid/solids ratio of less than 1:1, indicating that most of the leachable metals are on the surface of the fly ash particles (Hjelmar, 1982).

Soluble Salts in Fly Ash

In general, fly ash contains a large fraction of soluble salts, due to the sulfur and chlorine in the waste. Alkaline reagents such as caustic soda and lime, used as scrubbing agents for acid gas control, react with the acid gases to form soluble salts such as sodium and calcium chloride, and various sulfates. Table 13B.6 shows an analysis of the fly ash and scrubber residues from a spray-dry scrubber, and the soluble fraction, which totals 56 percent of the fly ash, primarily chlorides and sulfates (Lebedur et al., 1989).

Permeability of Ash Residues

Permeability is the property of a material which measures the velocity at which water will pass through the material, usually reported as centimeters per second (cm/sec). The permeability of compacted ash has been measured to be from 1×10^{-6} to 1×10^{9} cm/sec. For comparison, landfill liners are typically required to have a permeability of 1×10^{-7} cm/sec. It is possible to prepare ash for use as a landfill liner, thus meeting this specification, by proper compaction, and by adding portland cement and/or lime (Forrester, 1989).

TABLE 13B.6   Solubility of Fly Ash Residues of Spray-Dry Scrubber

Residue	Weight percent	Soluble percent
Fly ash	40	
$CaCl_2$	27	27
$CaSO_3 \cdot 0.5\ H_2O$	20	20
$CaSO_4 \cdot 2\ H_2O$	5	5
CaF_2	2	2
Lime inerts	2	
$CaCO_3$	2	
$MnCl_2$	1	1
Heavy metals	1–2	1
	100	56

Source: Lebedur et al., 1989.

13B.3 *ASH MANAGEMENT*

Equipment

Wet quench systems cool the residues and permit them to be removed from the quench tank by means of drag conveyors. Ash residues commonly have a moisture content of about 50 percent.

Semiwet systems quench the residues with water, but employ mechanical dischargers to push the residues out with a minimum amount of moisture. The heat remaining in the residues serves to drive off moisture so that the discharged residues may have a moisture content close to 25 percent.

Dry removal of the ash residues makes it possible to remove the fine ash component by screening, leaving a useful granular aggregate and a relatively clean ferrous product.

Drag conveyors extract ash residues from water-filled quench tanks. They consist of flytes attached to moving chains carried over sprockets, configured to carry the ash residues up a slope so that they can be conveniently discharged.

Screw conveyors move powderlike dry ash from fly ash hoppers to other locations where they are dropped into water tanks, conditioning (wetting) devices for damp discharge, or into dry collecting containers.

Vibrating conveyors convey damp ash residues from ash dischargers to processing devices or receiving containers.

Vibrating screens separate ash residues into various size fractions, to remove rejects and produce useful fractions.

Grizzley screens remove oversize objects such as wood, bulky appliances, wheels, and miscellaneous metal objects from the ash residue. Vibrating rails separate out the oversize objects.

Trommels are rotary screens that remove oversize objects and clean ash residues to obtain more uniform products.

Magnetic separators, usually placed after grizzley screens to reduce interference by oversize objects, recover ferrous metal.

Pneumatic conveyors transfer fly ash from ash hoppers to remotely located containers. Blower air and fabric filters are used to separate the fly ash from the transport air.

Handling and Storage

Handling and disposal of ash residues must not cause contamination of the environment by fugitive dust or by leaching into the environment or water supplies. Dusting is minimized by keeping the residues in a moist condition (Hahn et al., 1990).

Stored ash residues must be properly managed to prevent unacceptable discharge of dust or leachate. Runoff and leachate must be collected, supervised, and properly disposed of so that soluble metal compounds and salts do not contaminate the environment.

Storing of ash allows chemical reactions to take place which bind the metals, reducing leaching potential. Rainwater percolating through the pile cleans the ash by slowly leaching out the available (soluble) metals.

Ash containers must be drainable to recover leachate, and watertight to prevent leachate from running off into the environment.

Transportation of ash residues must be in covered containers to prevent dust from escaping into the environment. Containers, vehicles, and roads must be washed if contaminated with ash. The wash water must be disposed of properly or recycled back to the ash quench tank.

Processing

Ash residues can be processed at the waste-to-energy facility to reduce the rate of release of contaminants into the environment; facilitate disposal; improve the quality of the residues; remove valuable, useful, or harmful materials; and to prepare portions of the ash for beneficial use. The residues can be treated by washing, chemical treatment, or the use of additives and specific chemicals in order to retain, remove, or immobilize potentially toxic compounds.

Ferrous metal can be separated from ash residues by solid- or electromagnets. As much as 15 percent of the bottom ash from mass-burn facilities is ferrous material which can be extracted magnetically. The quality of the ferrous metal is measured by the amount of contamination with combustibles and fine ash materials. Tumbling of the ferrous product in a trommel can separate contaminants to improve quality, as can washing with water.

Screening processes remove unwanted oversize and undersize components and separate the ash into usable products, including aggregate for use in construction.

Washing processes can provide clean aggregate materials and ferrous metals for beneficial use, and remove the cementitious fly ash. The wash water can be processed and recirculated. The blowdown stream must be treated or evaporated to remove/recover and render harmless the dissolved metals and salts (Hasselriis et al., 1991; Exner et al., 1989).

Treatment

Various methods of treatment may be used to reduce the amount of leachable metal and salt concentrations, and thus render the ash more environmentally acceptable, as well as improve the chemical and physical stability and durability of product so that it can be used for a variety of purposes. Treatment methods include ferrous separation and compaction, and various methods that modify release rates by chemical and physical changes, including solidification, stabilization, and encapsulation; addition of portland cement, phosphates, waste pozzolans, and bituminous materials; washing and chemical treatment, thermal treatment, and vitrification.

After ferrous separation and screening, bottom ash residues have been used for fill and road base under certain conditions.

Encapsulation in asphalt for use in bituminous paving mixtures and in cement, serves to minimize the leaching of metals and salts from the product. Cement blocks and special forms can be made using aggregate processed from ash residues.

Lime and/or portland cement can be mixed with fly ash to encapsulate the toxic metals and/or render them insoluble (Holland et al., 1989).

Phosphate treatment converts soluble lead compounds to insoluble phosphates, thus immobilizing the lead and reducing leaching to acceptable levels (Eighmy et al., 1989).

Carbonic acid absorbed from stack gases can convert soluble lead compounds into relatively insoluble lead carbonates (Shinn, 1987; Wakamura and Nakazato, 1992). Heavy metals can be removed from fly ash by using carbonic acid recovered from flue gases, producing insoluble carbonates.

Washing ash residues can produce clean metals and aggregate suitable for use in concrete and road base, at the expense of physical and chemical processing of the wash water.

Chemical processing systems can process the ash residues while treating the flue gases and minimizing or eliminating water discharges. Calcium hydroxide specifically removes hydrogen chloride (HCl) and mercury; sodium hydroxide removes sulfur dioxide (SO_2); and ammonia (NH_3) can remove nitrogen oxides (NOx) (Fahlenkamp and Hemmer, 1989). Hydrochloric acid used in a primary gas scrubber can be used to remove most of the cadmium and a major portion of the lead content of combined ash. Washing may be especially justified for treatment of fly ash, which can contain as much as 50 percent soluble salts in facilities employing acid gas controls.

Vitrification of bottom ash, mixed ash, or fly ash can be employed to obtain a high-quality glassy frit from which the trace toxic metals such as lead and cadmium do not leach out (DeCesare and Plumley, 1992a, b). Thermal treatment can produce glassy materials, or sintered and ceramic materials (Wakamura and Nakazato, 1992).

13B.4 LANDFILL DISPOSAL

Landfilling

Landfilling of untreated ash is the simplest means of disposal. Ash residues have been codisposed with municipal solid waste (MSW), but for various reasons it may be preferable to place ash residues in separate or dedicated cells. Ash residues have been used to cover MSW as daily cover or as final cover. Placing ash residues in ashfills has the advantage that a solid, relatively impervious mass is created, over which trucks can drive as soon as it is placed. Removing ferrous metal from the ash residues improves the density of the ashfill as well as its stability.

Efficient management of ashfills can increase the density of the ash to as high as 3300 lb/yd^3 (122 lb/ft^3), as compared with about 1800 lb/yd^3 of uncompacted ash. Another benefit of compaction is the potential for reducing the permeability to as low as 1×10^{-6} to 1×10^{-9} cm/sec. Ashfills are so impervious to water that only a small fraction of rain falling on the top surface is able to penetrate the fill; 90 percent or more will run off, without leaching much of the soluble material in the ash. It is important to provide effective runoff collection systems since the water may be significantly contaminated, especially when the ashfill is in the process of being filled, prior to capping (Forrester and Goodwin, 1990).

Ash residues have been used as lining materials for landfills, in lieu of costly clay liners. To prepare the ash for use as a landfill liner, portland cement can be added at the landfill at 6 to 10 percent plus lime at 6 to 7 percent by weight. If the fly ash component of the ash residues contains excess lime from the acid gas scrubber, less lime will be needed.

Codisposal with MSW

There has been concern that when ash residues are codisposed in landfills together with raw MSW, the acids generated by decomposing MSW would increase concentrations of soluble toxic metals in the collected leachate, requiring more stringent containment, leachate treatment, and groundwater monitoring (Francis, 1984). On the other hand, the alkalinity of the ash has the ability to neutralize (buffer) the acids, reducing acid leaching.

Ashfills

Ashfills provide dedicated disposal of ash residues in cells separate from MSW. Ash residues have a high density and low permeability, which minimizes the need for leachate collection and treatment (Fahlenkamp and Hemmer, 1989; Goodwin and Forrester, 1990).

Liners and Containment

Liners are provided in landfills and ashfills to contain the ash residues and to minimize or eliminate leachate penetration into the surroundings, as well as to provide means for collection and removal of leachate and monitoring for indications of leakage. Only a small fraction of the rain falling on an ashfill can percolate through to the leachate collection system. The remainder is runoff which must be collected and properly managed (Forrester, 1989).

Ash residues containing mixed bottom ash, fly ash, and acid gas scrubber residues have cementitious and compaction properties that make them relatively impervious to the penetration of leachate, especially if moisture and lime content are optimized. Completed cells can be covered with plastic liners, ash residues, or other relatively impervious materials to essentially eliminate the generation of leachate after the ashfill cell is closed (Goodwin and Forrester, 1990).

Leachate Disposal and Treatment

The composition of leachate must be known before it can be disposed of or treated. Table 13B.7 shows that actual leachate concentrations measured at various ashfill sites had concentrations of the potentially toxic metals which were far below the EPA toxic limits. In most cases the concentrations of the regulated toxic metals were close to the USEPA drinking water limit (U.S. EPA, 1988; Roffman, 1991; Clark, 1992).

The leachate may be discharged or trucked to wastewater disposal plants if found to be acceptable, or it may require treatment before such disposal. The leachate from fly ash and from mixed bottom ash and fly ash contains substantial amounts (roughly 50 percent) of soluble salts resulting from the removal of acid gases by the emission controls that are now required. It has been described as being similar to salt water (Hjelmar, 1982). The salt content may be more likely to require attention than the low concentrations of soluble metals.

Salinity of Leachate from Ash Residues

The salinity of leachate is measured by electrical conductivity. The effect of highly salty leachate on the environment has been studied. Soils producing leachates that have an electri-

TABLE 13B.7 Ranges of Leachate Concentrations of Inorganic Constituents from Monofills

Constituent	Concentration (CORRE study), mg/L	EP toxicity maximum allowable limit, mg/L	Primary drinking water standard, mg/L
pH	5.2–7.4		
Arsenic	nd–0.400	5.0	0.05
Barium	nd–9.22	100	1.00
Cadmium	nd–0.004	1.0	0.01
Chromium	nd–0.032	5.0	0.05
Copper		1.00	
Iron		0.30	
Lead	nd–0.054	5.0	0.05
Manganese	0.50		
Mercury	nd	0.2	0.002
Selenium	nd–0.340	1.0	0.01
Silver	nd	5.0	0.05
Zinc			5.00
Chloride			250

Source: Roffman, 1991.

cal conductivity greater than 16 mhos/cm are classified as *saline,* causing interference in the uptake of water by plants. Column simulations have shown that after 20 years of leaching, ash residues having an initial leachate conductivity of 21 mhos/cm were reduced by simulated annual acid rainfall to 8 mhos/cm, a level having relatively little impact on most plants (Cundari and Lauria, 1986).

Neutralizing Capacity—Ash/Acid Deposition Mass Balance

The soluble toxic metals are only slowly released due to the presence of alkaline materials that provide powerful buffering against MSW-produced acids and the low quantity of weak acids in acid rain. It has been estimated that acid rain would be resisted for over 1000 years. Long before this time the leachable materials would presumably have been removed (Hartlen and Elander, 1986).

13B.5   *REGULATORY ASPECTS*

The management, disposal, and beneficial use of ash residues and their products is subject to federal and state regulations which require sampling and analysis of the leaching characteristics of residues from combustion of municipal and other types of wastes.

Federal Regulations

Federal regulations broadly classify wastes into hazardous and nonhazardous categories. The Resource Conservation and Recovery Act of 1976 (RCRA) empowered the U.S. EPA to regulate residues from solid waste incinerators. In 1992 the U.S. EPA Administrator sent a memorandum to EPA Regional Administrators stating that the Resource Conservation and Recovery Act completely excludes ash from municipal waste combustors from regulation as a hazardous waste under Subtitle C as long as it is not characterized as toxic, since ash can be managed safely in solid waste landfills under Subtitle D, Section 3001(i) of RCRA. Prior to this statement, states developed various requirements, many requiring that MWC ash be disposed in monofills, for ash only (ashfills), employing single liners, as compared with the double liners required for MSW landfills. Leachate collection and treatment are required in both cases. Landfills for hazardous wastes require much more stringent design and operation due to the greater potential hazards and uncertainty of their leachates.

The U.S. EPA has developed test procedures designed to screen wastes to determine their classification. Various states have developed different regulations as to whether or not ash must be sampled and analyzed prior to disposal. In any case it is generally the responsibility of the producer of the ash to determine whether it exhibits the characteristic of toxicity in accordance with U.S. EPA procedures (U.S. EPA, 1980).

Ash Residue Extraction Leaching Procedures

The following leaching procedures have been applied to determine the characteristics of ash residues under a wide range of conditions to which they might be exposed, and also to discover which procedures might more closely simulate actual conditions of disposal or beneficial use:

- Extraction Procedure Toxicity (EP Tox) Test.
- Acid No. 1. Acetic acid extraction fluid at pH of 4.87 to 5.2.
- Threshold Characteristic Leaching Procedure (TCLP).

- TCLP Fluid No. 1 (Acid No. 2). Similar to EP-Tox.
- TCLP Fluid No. 2 (Acid No. 3). Similar to EP-Tox.
- California Waste Extraction Tests (WET). Uses citric acid.
- Deionized Water (Method SW-924), also known as the Monofill Waste Extraction Procedure (MWEP).
- CO_2 saturated deionized water.
- Simulated acid rain (SAR).
- ASTM Shake Extraction Procedure uses distilled water.
- Leaching Column Tests using simulated acid rain.

The original Extraction Procedure (EP) Toxicity test produced erratic results when testing municipal waste combustion ash residues; hence it was replaced by the Threshold Characteristic Leaching Procedure (TCLP). These screening procedures use an acid leaching medium intended to simulate leaching of ash codisposed in a landfill in a proportion of 15 percent ash to 85 percent MSW. In the TCLP test, a unit sample of ash residue is immersed in 20 units of a specified acetic acid solution. The acidity of the solution is maintained at a pH value between 4.87 to 5.2 and stirred for a 24-h period. The extract (leachate) is analyzed, and the results are compared with the EPA-established limits shown in Table 13B.7. The sample would be characterized as hazardous if any of these limits were exceeded (U.S. EPA, 1980; Francis and Maskarinec, 1987). These limits assume that a 100-time dilution would occur before the leachates could reach drinking water, hence they are established at 100 times the drinking water standard.

Actual leachate from MSW landfills as well as codisposal and ash residue landfills (ashfills) do not generally have pH values as low as 5.0. Extensive testing of actual leachate from these various types of landfills shows that they do not contain significant amounts of the toxic metals lead and cadmium, contrary to the results of laboratory tests employing the TCLP method. In other words, the TCLP test does not simulate actual disposal conditions (Roffman, 1991).

A comparison of the effects of various laboratory leaching procedures on the cadmium and lead concentrations is shown in Table 13B.8 (Francis and Maskarinec, 1987). Actual leachate concentrations from tests sponsored by CORRE/EPA are compared with the EPA Toxic Limits in Table 13B.7. The carbonic acid test was in closest agreement, while the EP Toxicity procedure overestimated leaching by over 100 times (Roffman, 1991).

In general, actual leachates from landfills are more closely simulated by leaching column tests, and tests using simulated acid rain or carbon-dioxide-saturated water and/or deionized

TABLE 13B.8   Metal Concentrations in Extracts of MSW Ash Residues*

Facility:	Chicago	Sumner	Hampton	Auburn
Cadmium:				
WET test	1.6	0.81	1.52	0.18
EP toxicity	0.71	0.24	0.50	0.02
Acetate	0.19	0.52	0.33	0.03
Carbonic acid	0.016	0.012	0.07	0.005
Water	<0.0005	<0.005	<0.005	<0.005
Lead:				
WET test	29.0	35.0	46.0	29.0
EP toxicity	5.8	6.4	10.3	3.15
Acetate	0.5	0.28	1.62	4.20
Carbonic acid	0.025	0.004	0.095	0.012
Water	<0.002	<0.002	<0.002	<0.002

* In mg/L (parts per million)
Source: Francis and Maskarinec (1987).

water. More aggressive leaching tests such as the California WET test and EP or TCLP test, which represent conditions not likely to occur in the environment, serve as the basis for classifying the wastes as being potentially able to produce toxic levels of metals in the leachate.

Leaching column tests show that the soluble metals and salts are gradually removed from the ash as the leachate absorbs and removes them. The larger the quantity of acid in the water the faster the rate of removal. In some cases two to four quantitative washes will have removed essentially all of the lead and cadmium which was soluble at the leaving pH level. Test borings of one ash pile showed that after several years of natural acid rainfall the leachable metals remaining in the ash pile had been reduced to the nondetectable level.

State Regulations

State regulations must be at least as stringent as federal regulations, but may be more detailed and suited to specific state environments. Many states require the collection and analysis of ash residue samples on a periodic basis, and require that these samples, on average, pass the prescribed toxicity tests.

Testing of ash residues may be required to obtain confidence that there will be no harmful effects on the environment after the residues are disposed of or used beneficially.

Samples of ash residues should represent the stream of ash residues from which they are taken. Mixing the fly ash properly with bottom ash avoids unrepresentative "hot spots." Aging samples with normal moisture allows the chemical reactions to take place that would occur under the conditions of disposal, such as converting soluble lead chlorides to insoluble lead carbonates.

Frequency of testing is generally regulated, including extensive testing after start-up of the plant, followed by one or more tests per year in order to assure consistent operation (Fiesinger 1989). If some of the analytical results are found to be critically close to acceptable limits, more samples may be taken to obtain confidence in the average values.

Regularly time-spaced samples will represent the true average characteristic of the ash residues for the period of time over which they are produced, landfilled, or otherwise used.

Daily average samples are collected at uniform time intervals over the entire day. Daily samples are well mixed and coned and quartered to reduce the sample size and provide several identical samples. Weekly samples should include the entire week; monthly samples each week, and annual averages all months. As analyses are accumulated over long periods, greater confidence is established as to the true mean, and fewer analyses are required to assure representation of the residue stream.

Statistical analysis may be needed, especially if a high degree of variability is observed in leaching characteristics of ash residues. While cadmium data are fairly consistent, greater variations have often been found in lead concentrations (Hasselriis, 1994).

Analyses of fly ash samples generally show that untreated fly ash contains enough soluble lead and cadmium to exceed the toxic limit according to the EP or TCLP test. On the other hand, bottom ash generally passes the test, and individual samples of mixed bottom and fly ash samples may occasionally fail due to nonuniform mixing.

13B.6   ACTUAL LEACHING OF MWC ASH

Several points are clear from reviewing the data of several field studies that characterized the leachates from ash monofills, TCLP testing of ash from several waste-to-energy facilities, and leaching of products containing ash.

- No single leach test, including the TCLP, is adequate to fully and accurately predict the potential for an ash or an ash product to release constituents of concern under field disposal and beneficial use conditions.
- Laboratory leach tests routinely overestimate the potential for constituents of concern to leach from an ash and ash product when compared to actual leaching from ash monofills and ash use in field applications.
- Modern WTE facilities routinely pass the requirements of the TCLP.
- Metal concentrations in leachates from ash monofills evaluated over time have routinely met ground water standards and often meet drinking water standards.
- Although leachates from combined ashes in monofills have low concentrations of heavy metals, the total dissolved salts concentration may be several orders of magnitude above drinking water standards.
- If detected at all, levels of dioxins/furans in ash and ash leachates were extremely low and considered not to be a concern when evaluating the environmental and health consequences of using ash.

The Municipal Waste Management Association published results of an analysis of liability issues associated with beneficial use of MWC residues. Results of the analysis demonstrates that local governments that generated MWC ash, which when tested does not exhibit hazardous characteristics, have several levels of protection against environmental liability. This is the case if (1) the generators provide the ash to a bona fide recycling operation; (2) the recycler or the local government treats the ash, if necessary, to satisfy state and federal laws; (3) the recycler has obtained all necessary state and local approvals; (4) the MWC ash is used in products that are introduced into the economic mainstream in a manner that limits the potential for human exposure (Roffman, 1991).

13B.7   TREATMENT OF ASH RESIDUES

Ash Residues Discharged from a WTE Facility

If these residues fail the TCLP test, they have to be sent to a special and more costly landfill. To avoid these costs, and to assure that all products will pass the test, various chemical treatments have been used. The most prominent of these is the WES-PHix™ process, in which the ash residue is sprayed with a phosphate solution while being tumbled in a drum. It has been demonstrated that this treatment reduces the solubility of lead which otherwise increases at the highly alkaline conditions such as 11 or 12 pH which result when excess lime is used for acid gas control. With this treatment, the ash residues may be disposed of in dedicated ashfills, and used for beneficial purposes (Lyons, 1996). Other treatments include control or addition of alkaline agents including portland cement.

Residues from Air Pollution Control Systems

Processes have been developed to recover useful by-products from waste incinerators, and render the remaining residue nonhazardous, thereby eliminating the long-term environmental liability associated with disposing of these materials.

There are two predominant types of air pollution control (APC) systems: (1) wet scrubbers and (2) dry or semidry lime or other alkaline reagent injection systems. Wet scrubbers generally use sodium-based agents, such as $Na(OH)_2$, producing the highly soluble NaCl salt. In dry/semidry lime injection APC systems, the calcium hydroxide ($Ca[OH]_2$) reacts with the acid gases to form significant concentrations of calcium chloride ($CaCl_2$) and calcium sulphate ($CaSO_4$) that comprise the bulk of the APC residue waste stream. The resulting $CaCl_2$ and excess lime present in the APC residues are quickly solubilized upon contact with water. Furthermore, since the stoichiometric ratio of lime addition is greater than 1, the APC residues are highly alkaline, and the potential to solubilize amphoteric metal compounds (such as some Pb compounds) is greatly increased. Consequently, leachates from these residues may contain high concentrations of salts and trace metals, such as Al, Cr, Pb, and Zn.

The fundamental mechanisms of phosphate stabilization of divalent metals in MSW combustion scrubber residues have been studied intensively (Eighmy et al., 1997).

Washing Processes.   One example of washing processes is the APEX technology, which is based on controlled washing, dewatering, and rinsing operations that remove highly soluble amphoteric metals and salts from APC residues. The remaining dewatered residue generated from this treatment process meets the criteria for a nonhazardous material (as determined by the EPA Threshold Characteristic Leaching Procedure [TCLP]). The treated filter cake can be rendered suitable as an amendment for aggregate feedstock, for use in concrete/asphalt manufacturing, and in certain ceramics. The primary by-products of the APEX treatment system include a nonhazardous calcium-enriched solid suitable for construction applications, a highly concentrated lead residue suitable for recycling to smelters, and a commercial grade calcium chloride solution. The trace metals solubilized during the washing stage are then precipitated and separated from the rest of the filtrate. The metal precipitate filtrate can be further processed to generate a concentrated calcium chloride ($CaCl_2$) solution that is commonly used for road construction, dust control, deicing, or as an antifoaming agent in the pulp and paper industry.

In benchmark trials, approximately 25 to 38 percent of the raw APC residue was solubilized during the washing and rinsing process. Consequently, 62 to 75 percent of the residue remained as calcium-enriched filter cake. The filtrate from the filter cake contains mostly chlorides (calcium, sodium, potassium), carbonates, and subpercent levels of soluble hazardous metals such as Cr, Al, Zn, and Pb. It is necessary to remove the metals from the liquid phase. By the APEX treatment process, the concentration of Pb in the precipitate was about 65 percent by weight for most of the test runs. The precipitates contained about 4 percent zinc, 0.3 percent copper, and about 0.26 percent Cr. A series of two filters were used to polish the postmetal precipitation brine stream (Sawell et al., 1999).

13B.8   ENVIRONMENTAL IMPACT OF ASH RESIDUE USE

Risk Assessments

Risk assessments to estimate environmental impact have been carried out for various uses of ash residues: combined ash as landfill daily cover; combined ash as final cover (bottom layer); use of treated ash aggregate (TAA) as a roadway base, as a structural fill, as daily and final landfill covers, as an aggregate substitute in asphalt concrete paving, and reuse and final disposal of paving material containing TAA. Boiler Aggregate™ has been demonstrated for use in producing an asphalt product, as an unregulated fill, and for milling and excavation for reuse. Stockpiling, handling, and transporting operations have also been evaluated, as well as in combined ash as a 30 percent substitution in bituminous pavement.

These assessments have covered specific situations, and generally have evaluated the noncarcinogenic and carcinogenic effects on receptors from exposure to As, Ba, Cd, Cr, Pb, Hg, Ni, Se, Ag, and dioxin and furan congeners. Key receptors included nearby residents, workers, adults and children visiting a site, and those who could be exposed to runoff or fugitive dust.

Direct and indirect exposure pathways evaluated included inhalation of fugitive dust on- and off-site, incidental ingestion and dermal contact with ash and ash products, residential exposure to soils potentially contaminated with particulates and/or leachates from ash and ash products, and similar pathways. In the case of landfill final cover, exposure pathways also included incidental ingestion and dermal contact with surface water and sediment while swimming in a nearby harbor, and consumption of fish from the harbor. Exposure from ingestion of food grown in soil potentially contaminated with TAA was also evaluated. Also included was ingestion of drinking water containing leachate from Boiler Aggregate stockpiles, a road base, and recycled asphalt product pile.

In tests performed in the United States, on a 600-m section of U.S. Route 3 in Laconia, New Hampshire, with MWC bottom ash as 50 percent of the required aggregate in the binder course pavement, roadway runoff, surface water, and groundwater were monitored. Monolith leaching tests have indicated that release rates of chemical constituents are low and occur at levels similar to those for natural aggregates (Musselman et al., 1994, 1995; Eighmy et al., 1993).

At a municipal solid waste WTE facility, the upramp to the tipping floor was paved with a 2-in top course of 5 percent ash-amended asphalt, and the downramp was paved with a 2-in top course of control asphalt. Both materials met a Marshall test, an asphalt content test, a gradation analysis, a specific gravity test, and a compaction test using the Rice Method. To investigate any potential environmental effects, both pavements have been water washed twice a month since October 1998, and soils in the areas where runoff water collects have ben analyzed once a month. Wash water has been analyzed for total suspended solids, dissolved metals (12 metals of interest), chloride, and hardness. Soil has been analyzed for total metals and TCLP metals, chloride, and hardness. Test cores of both pavements have been analyzed for total metals and TCLP-leachable metals, chloride, and hardness. To accelerate the potential effects of weathering, test cores of pavement were broken and subjected to a serial leaching test by placing them in TCLP solution which has been analyzed and replaced monthly. In addition, intact test cores have been subjected to alternating cycles of submersion in TCLP solution and exposure to sunlight, rain, and wind, with leachate tested monthly. Data from the first three months of the environmental testing program are presented in Magee and Hahn et al. (1999).

On numerous occasions the release of chemical constituents to the environment has been demonstrated to not be correlated with the total concentration of the constituents in the material. Thus the total metal analyses do not yield useful information about leaching potential, and specific leaching tests should be performed. Preliminary statistical comparisons of data collected to date demonstrate that the ash-amended asphalt does not leach metals in a manner that is statistically different from normal control asphalt.

Trends in concentrations of metals in ash residues have been investigated over a period of nine years of TCLP testing of ash from the H-Power WTE plant in Honolulu. The data showed that there has been a downward trend in the concentrations of the metals in the ash, except for Ba. The plausible conclusion is that original efforts to reduce Pb and Hg in consumer products, and similar efforts have reduced the amounts of these constituents in the waste going to the WTE facility (Wiles, 1999). It should also be noted that printing inks that originally contained lead, cadmium, and chromium are now largely converted to organic inks.

13B.9   ASH MANAGEMENT AROUND THE WORLD

Today most countries view ash as a resource to be recycled, rather than a waste to be disposed of into a landfill, provided that utilization is protective of the environment. Several countries

TABLE 13B.9   Criteria That Residues in Germany Must Meet Before Landfilling

Parameter	Unit	Landfill Class 1	Landfill Class 2	Limits for bottom ash use in road construction
Loss on ignition, weight	%	3	5	
Total organic carbon, weight	%	1	3	1
Cl	mg/L			250
Cu	mg/L	1	5	0.3
Zn	mg/L	2	5	0.3
Cd	mg/L	0.05	0.1	0.005

have established criteria and procedures for determining acceptable use and disposal options, and all continue to support research and development efforts for improved treatment and use technologies (Wiles, 1999).

In Denmark, laws and regulations define how ash residues may be used, and in what quantities, so that environmental impact will be insignificant (Hjelmar, 1990). In general, in order to produce an environmentally benign bottom ash material by minimizing the quantity of soluble salts, fly ash generally is not mixed with bottom ash.

Ash management practices in Bermuda, Japan, The Netherlands, Denmark, Germany, France, Sweden, and the United Kingdom have been reviewed by a document produced by the U.S. Department of Energy (Wiles, 1999). Germany sets the requirement that residues contain only small quantities of carbon before they can be landfilled (see Table 13B.9), a more realistic view of what can be recycled economically.

The Netherlands has more than a decade of experience with MWC fly ash as a substitute for natural aggregate in asphalt road construction. No differences were found in the leaching of metals between asphalt pieces containing MWC fly ash compared to natural aggregates.

France established requirements as shown in Table 13B.10 to determine when bottom ash is acceptable for utilization. Bottom ash with low leaching characteristics can be used immediately. Bottom ash in Category M can be stored (aged) for as long as 12 months, and its characteristics after storage determine whether it can ultimately be used. Bottom ash in Category L must be landfilled. In tests of plants after nine months of aging, seven of the ashes met requirements of category V; one fell into M; and two into L.

The French concluded that when bottom ash is produced under good combustion conditions and maturation, it is a satisfactory replacement for gravel.

TABLE 13B.10   Categories of Bottom Ash in France Based on Ash Constituents

Constituent	V	M	L
% Unburnt material	<5%	<5%	>5%
Hg[a]	<0.2	in between	>0.4
Cd[a]	<1	in between	>2
Pb[a]	<10	in between	>50
As[a]	<2	in between	>4
CrV[a]I	<1.5	in between	>3
Sulfates[a]	<10,000	in between	>15,000
TOC[a]	<1500	in between	>2,000

[a] mg/kg dry material

13B.10   BENEFICIAL USE OF RESIDUES

Ash residues represent approximately 20 percent of the municipal waste stream. For this reason, reduction in the amount of ash that must be transported and disposed of in ashfills can offer a substantial saving in landfill space and cost (Chesner, 1989; 1993).

MWC ash is widely used in Europe in road construction as compacted road base; structural fill in wind barriers, sound barriers, and highway ramps; and in asphalt applications (Chandler et al., undated; IAWG, 1997). In fact, approximately one-half of the MWC bottom ash generated in Germany is used in road construction.

Similarly, in The Netherlands more than 10 years of experience with MWC fly ash confirms its suitability as a substitute for natural aggregate in asphalt road construction. No differences were found in the leaching of metals between asphalt pieces containing MWC fly ash compared to natural aggregates (Wiles, 1999).

Ash Landfill Operations

The pozzolanic behavior of MSW residues from facilities with acid gas control, due to their high free-lime content and cementlike mineralogy, is beneficial for disposal site management practices. These properties allow the residues to be disposed of as a liner/cap over lifts of MSW and as a final capping material over MSW or other materials. In addition to providing high densities, using more lime or portland cement makes it possible to achieve permeabilities below the liner requirement of 1×10^{-7} cm/s (Forrester, 1989).

Use of Ash Residues for Construction

Ash residues from combustion of MSW have been used as roadway fill and subbase for parking lots, stabilized road base, bituminous paving mixtures, concrete masonry block, and portland cement concrete. Concern about leaching of metals into the environment has led to extensive research into the characteristics and environmental impact of these practices. Table 13B.11 describes the physical and chemical tests that may be used to evaluate the waste material for reuse (Fiesinger, 1992).

TABLE 13B.11   Tests Recommended for Waste Reuse of Granular and Asphaltic Materials

Chemical tests	Physical tests
Elemental composition	Moisture content (ASTM d2216)
Mineralogy	Percent rejected (>¾ in)
Acid neutralizing capacity	Organic content (loss on ignition)
Distilled water leach test	Ferrous content
Bioavailability leach test	Particle size distribution (ASTM C136)
Toxicity characteristics	Absorption and specific gravity (ASTM C127 and C128)
Leaching procedure (TCLP)	Unit weight and voids (ASTM C29)
Lysimeter leach test	Moisture density test (ASTM D1557)
	CBR (ASTM D 1863)
	Sodium sulfate soundness of aggregates (ASTM C-88)
	Los Angeles abrasion test (ASTM C131)
	Unconfined compressive strength (ASTM D2166)
	Marshall stability of asphaltic material (ASTM D1559)

Source: DiPietro et al., 1989.

Percent Available as Aggregate

Recent tests of ash residues from a waste-to-energy facility in Concord, New Hampshire, analyzed residues that combined bottom ash and residues from the dry lime-injection scrubber/baghouse system. The average percent of material smaller than ¾ in, suitable for asphaltic base course, was 65 percent, with a standard deviation of 13 percent. A portion of the remaining 40 percent also has the potential for use in road construction.

Use of Ash Residues in Asphaltic Mixtures

Asphalt has been found to encapsulate ash residues effectively, reducing leaching potential to acceptable levels, so that the asphalt can be safely used for road construction (Lucido, 2000).

Ash residue used as an aggregate in bituminous-base course construction has been studied since the 1970s. Test sections have been continuously evaluated over periods from one to five years. The surface pavement and binder courses using combined bottom ash and fly ash from the WTE facility in Lynn, Massachusetts, constructed in 1980, used 50 percent ash, 2 percent lime, 50 percent natural aggregate, and 13.5 percent asphalt, and were assessed in 1991 to be still performing well. A section built in Washington, D.C., contained 68.5 percent residue, 1.5 percent hydrated lime, 15 percent sand, 15 percent limestone and 9 percent asphalt. Surface asphalt road sections at the SEMASS facility in Rochester, Massachusetts, and in New Jersey, and ash use in road base and subbase structures at SEMASS and in New Hampshire are being evaluated for soil contamination, runoff, and leachate (Wiles, 1999).

Portland Cement Treatment

A patented portland cement–based ash aggregate, McKayanite, using combined ash or bottom ash, has been tested in Florida as landfill cover and as aggregate in road projects, and was found to meet physical standards for road construction materials, showing no adverse effects on groundwater, soil, or ambient air quality. The State of Tennessee Highway Department has developed a standard for acceptable use of ash residues as aggregate in roadbase construction. Progress continues in many other states toward acceptance of ash residues for beneficial use (Wiles, 1999).

At the Commerce WTE facility in Los Angeles, California, fly ash from the spray-dry scrubber/baghouse is mixed with portland cement, then blended with bottom ash in a cement mixer truck. The treated ash-concrete is poured into roll-off containers and stored for 24 hours before transport to the landfill, where it is crushed for use as a subbase for roads at the landfill (Eaton, 1992).

Building Blocks and Other Uses

Use of ash residue to produce aggregate material for use in concrete encapsulates the heavy metals and converts them to chemical forms that are essentially insoluble. Leaching tests of concrete blocks made from bottom ash only and mixed bottom ash/fly ash have shown that the rate of leaching of metals is insignificant in both underwater marine environments and in aboveground applications (Wiles, 1999).

Cement blocks made from MSW combustion in Montgomery County, Ohio, have been used to construct buildings on the county landfill, after testing according to structural testing protocols, including ASTM tests for strength, Underwriter Laboratories (UL) tests for fire resistance, and TCLP tests. The blocks exhibited somewhat higher-than-normal shrinkage indices. The blocks made from coarser bottom ash released more easily from the molds than did blocks containing fly ash as a component (Wiles, 1999).

Processing Ash for 100 Percent Recovery for Use

The American Ash Recycling process, operating at several facilities in the United States, can achieve 100 percent recovery of products; that is, zero discharge to the landfill. The process involves the sorting of ash as it is received; bulk reduction; ferrous separation (three stages); ferrous processing; aggregate wet treatment; air classification to remove unburned product; and nonferrous separation, processing, and sorting. The unburned product is returned to the WTE facility. The aggregate product is treated by the WES-PHix process and sold for various uses in construction. The product is further conditioned as it is stored in piles, awaiting sale (Arcani, 2000).

Figure 13B.6 shows the size distribution of the AAR product. Figure 13B.7 shows the fractions of product after processing of ash residues from the Nashville Thermal WTE facility.

Vitrification

Vitrification of residues from thermal processes produces a dense, grainless, amorphous, glasslike material which contains no organic material and from which inorganic mineral matter does not leach significantly, and which has many beneficial uses.

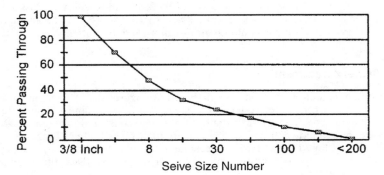

FIGURE 13B.6 Size distribution of American Ash Recycling product. *(Source: Arcani, 2000.)*

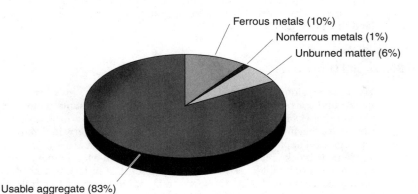

FIGURE 13B.7 Fractions of product after processing ash residues. *(Source: Arcani, 2000.)*

In Japan, where landfills are especially scarce, nonexistent, or costly, vitrification has been applied at many facilities, employing fossil fuels, electric heating, and electric arc as the heat source. Due to the relatively low cost of landfill in the United States, there has been less incentive to pursue this course.

Electric Arc Furnace Vitrification.   An extensive investigation of vitrification was carried out by the American Society of Mechanical Engineers (ASME) and the U.S. Bureau of Mines with the support of governmental agencies and industry (DeCesare and Plumley, 1992a, b). The results of this program provide information on operating parameters and potential constraints of the technology, identified and quantified process residuals, effluents or emissions, identified beneficial uses for the vitrified products, and developed economic data.

Tests were performed on combined furnace bottom ash and fly ash from three mass-burn WTE plants, fly ash from an RDF-fired WTE plant with acid gas scrubber, and combined ash from a multiple-hearth wastewater treatment sludge incinerator.

The tests were performed in a sealed submerged-arc electric furnace having a capacity of 1 ton/h. Fumes from the furnace were cooled and the deposits analyzed. A fabric filter was used to collect particulate for analysis, and an afterburner was provided to burn off any possible organic matter. The stack gases were analyzed for trace organics and metals.

The composition of the furnace feed, vitrified product, metal, matte, and fumes for a typical mass-burn WTE furnace is shown in Table 13B.12. Although the mass balance is not perfect, it is evident that most of the nonvolatile elements reported to the vitreous product and only traces to the fume.

Approximately 83 percent of the feed weight was converted to vitrified product, and 5 percent metal was withdrawn, consisting mostly of about 75 percent Fe, and 8 percent copper, 6 percent phosphorous, and 7 percent silica. The matte, which collects on the top of the charge, contained about 48 percent copper, 15 percent iron, and about 27 percent sulfur. The principal components of the fume solids were NaCl, ZnS, and KCl. They contained 70 percent of the zinc and 37 percent of the lead in the furnace feed, and from 7 to 26 percent silica, varying with the carryover which depended on the nature of the ash residues. Most of the chlorine in the feed material left the system in gaseous form, since essentially none was found in the solid products.

The vitrified product was tested in accordance with TCLP procedures that showed that the vitrified residues were environmentally benign. The levels of leaching from the samples were in all cases 10 to 50 times less than the EPA maximum limit.

The cost of vitrification was estimated to range from $200 per ton of dry ash residue for a 50-ton-per-day facility to $115 per ton for a 300-ton-per-day facility, based on electric power costing $0.051 per kilowatt-hour. The components of the cost of vitrification can be roughly estimated to be 5 percent for drying, 45 percent for electric power, water, and gas, 30 percent for labor and maintenance, and about 20 percent for the cost of capital.

Cold Crown Glass Furnace Vitrification.   Glass furnace technology was investigated as a means of vitrifying WTE facility air pollution control (APC) fly ash residues. In this process, the ash must first be washed to extract chlorine, since the maximum tolerable level is about 5 percent. About 22 percent of the APC residue was dissolved and removed in two extraction steps, during which 25 to 30 percent of the lead was also washed out. Washing removed 96 percent of the chloride, 30 percent of the calcium, 60 percent of the potassium, 20 percent of the lead, and 19 percent of the sulfur, but 99 percent of the volatile heavy metals were retained in the glassy product. The melting temperature ranged from 2200 to 2800°F (1200 to 1550°C). The wash water can be treated to precipitate the metals, especially lead, by raising the pH to about 9. In the second step, the dechlorinated APC residue is blended with glass-forming additives, forming a mixture of about 52 percent residue, 36 percent silica, 9 percent sodium

TABLE 13B.12   Composition of Vitrification Furnace Feed and Products*

Element percent	Feed 93	Vit.prod. 83	Metal 5	Matte 1	Fume 2
Si	186,860	154,018	24	0.09	19
Fe	122,484	85,040	2,028	38	11
Ca	40,409	67,257			21
Al	31,649	48,312	0.28	0.02	8
Na	22,259	14,811			38
Cl	13,089	51			69
C	9,568	372	2.85	0.04	1
Mg	8,458	10,471			2
S	8,343	1,764	93	25	20
K	6,392	4,035			25
Cu	5,842	1,020	202	31	3
Ti	4,131	6,483			1
Pb	3,287	187	4	2	30
Zn	2,718	1,353	2	0.33	59
P	2,399	1,387	75	0.11	7
Bi	1,013	1,766			0.26
Mn	963	1,542	0.06	0.10	0.19
Ba	631	737	0.05		0.11
Sr	306	391			0.04
Cr	280	1,216	1.05	0.03	0.06
Ni	222	53	13	0.22	0.04
Sn	182	138			1.30
B	167	48			0.24
Sb	91	8	3.73	0.14	0.34
Ce	86	69			0.04
Br	63	1			0.97
V	48	59			0.01
Y	43	34			0.02
Mo	43	34	1.02	0.01	0.02
Co	35	21			0.00
As	28	2	0.89	0.01	0.07
Li	19	16			0.02
Cd	18	6	0.12		0.18
Ag	9.51	1.38	0.33	0.16	0.03
Tl	4.32	3.44	0.43		0.00
Hg	2.59	0.69	0.00		0.01
Be	0.86	0.69	0.00		
Se	0.86	0.69	0.53		0.01
Au			0.17		
Total	472,145	402,709	2,449	98	317

* In parts per million parts of residue

Source: DeCesare and Plumley (1992a, b).

carbonate, and 4 percent sodium nitrate. The volume of the resulting product was reduced by 40 percent from the original APC volume. The product was subjected to the TCLP leaching procedure and the accelerated strong acid durability test, neither of which showed appreciable leaching of elements of concern, most being below detectable limits of analysis (Wexerll, 1993; Hnat and Bartone, 1996).

13B.11   ANALYSIS OF ASH RESIDUE TEST DATA

Variability

The variability of analytical test data of ash residues has created serious problems. Individual samples have shown heavy metals concentrations, 5 to 10 times the average, occasionally exceeding the EPA toxic limit values. Early sporadic analyses of ash residues showed that more than half of the samples exceeded the limits, causing ash to acquire the name "toxic ash." As more data became available, and procedures improved, it became apparent that average ash quality is not as highly variable as it had appeared to be, and that bottom ash and properly mixed bottom and fly ash generally do not exhibit the characteristic of toxicity as determined by the TCLP leaching procedure. Pug mills and other devices have been used to assure consistent mixing of fly ash with bottom ash.

The variability of the leaching characteristics of ash residues must be investigated and understood in order to obtain confidence in its quality. The following paragraphs apply statistical methods to an unusually extensive data base in order to illustrate the principles of statistical analysis.

Basic Elements of Ash Sampling and Testing

Ash sampling is carried out in order to obtain sufficient sample to estimate whether specific characteristics of the ash material meet required specifications. The ASTM *Standard Guide for General Planning of Waste Sampling* (D4687) contains guidelines for developing a sampling plan, including sampling procedures, safety plans, quality assurance, general considerations, preservation and containerization, labeling and shipping, and chain-of-custody (ASTM, 1989).

Specification

A material specification generally contains a target range of values to be met, the test methods to be used, and the desired confidence level. The confidence level is the selected degree of confidence that the difference between the mean of the sample and the mean of the population of all possible samples of the material being tested is less than some allowable error. For instance, the U.S. EPA has proposed a 90 percent confidence level in its SW8413 (U.S. EPA, 1993, 1995). Specifically, an ash residue may be classified as exhibiting the characteristic of a hazardous waste if extracted leachate obtained from TCLP leaching procedures exceeds the target of 5 mg/L, on average, for lead, and 1 mg/L for cadmium, the most likely metals to exceed the limits. The method of obtaining the average becomes more critical as the analytical results approach this target.

Selecting Physical Sampling Procedures

Sampling procedures should be related to the purpose of the test program. Performance tests require determination of the average moisture and heating value of ash residues which are representative of the period of days or weeks of testing. Procedures used to determine the leaching characteristics of ash residues destined for landfilling or for constructive use should simulate the ongoing production of ash over the time periods during which the residues are produced, such as monthly or annually.

To obtain an appropriate test-sized sample: (1) take samples or increments for composited samples, (2) combine increments into composite samples, (3) blend or process (screen or crush) the composite, and (4) take subsamples from the composite for laboratory analysis and reference. Representative ash samples can be taken from the full width of a conveyor, or from the conveyor drop-off.

Figure 13B.8 shows the procedure for collecting and dividing ash residue samples in order to obtain laboratory and reference samples over a period of eight hours, during a facility performance test. In this case, 40- to 60-lb samples are collected from the drop-off of a conveyor belt every 10 min. Alternately, a sample could be taken once per hour over a 24-h period. Samples weighing about 10 lb each are sent to the laboratory and the client, and kept for reference in case any samples are lost or damaged.

Coning and quartering reduces the quantity of the sample while retaining representativeness. The composite sample is mixed in a pile, then split into four quarters. Two opposite quar-

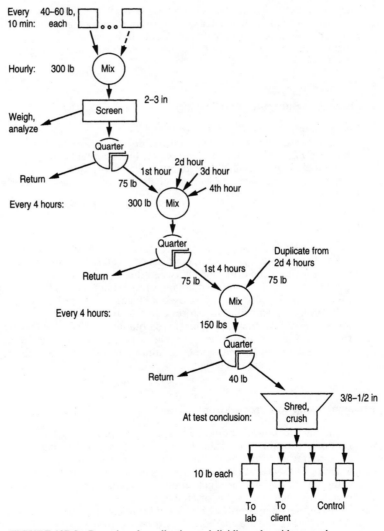

FIGURE 13B.8 Procedure for collecting and dividing ash residue samples.

ters are retained and two are set aside. The retained quarters are combined into a pile: the process is repeated until the desired sample quantity is obtained.

The average characteristics of the ash residue stream leaving the facility can be obtained by taking daily samples for a month, to determine the statistical properties of the residue, after which weekly samples might be taken to sustain a moving average. Unusual results might provoke a return to daily samples until confidence was again restored.

At least 8 samples, and preferably 16 or more samples should be analyzed to obtain the statistical properties of the ash. The following example taken from actual operational tests of a WTE facility illustrates why a large number of samples may be needed to obtain confidence in highly variable data.

Analysis of TCLP Data Obtained from Sampling at a WTE Facility.   *TCLP leaching test data* for lead and cadmium from samples of mixed fly ash and bottom ash collected hourly over a continuous period of 48 h, are plotted sequentially in Fig. 13B.9, revealing the degree of variation in individual samples. These data are listed in Table 13B.13 (Feder and Mika, 1982). It is apparent that a series of spikes occurred in the lead analysis, and to a smaller extent in the cadmium analysis, at roughly six-hour intervals. These spikes were attributed to boiler cleaning, in this case by tube rapping. The deposits that formed on the tubes fell onto the stoker, or were conveyed externally to the ash quench tank. Analysis of the tube deposits showed high concentrations of lead and cadmium, which, volatilized in the combustion process, condensed on the particulate that adhered to the tubes. The lead spikes were two to three times the EP limit of 5 mg/L, although the average of 48 analyses was only 2.85 mg/L. Cadmium concentrations were far less than the EPA limit of 1 mg/L.

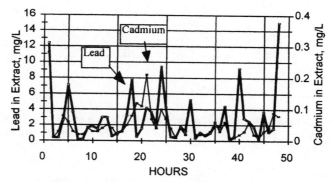

FIGURE 13B.9   Sequential plot of TCLP extract concentrations for Pb and Cd over 48 hourly samples. *(Source: Feder and Mika, 1982.)*

The distribution of the logarithms of the *combined* ash data of Fig. 13B.9 is shown in Fig. 13B.10. The data plots on a line which is nearly straight except at the ends, and therefore the data may be described as log-normal. The lead and cadmium EP data plotted on a logarithmic scale fall on a fairly straight line for 41 points, representing the log-normal distribution which is typical of natural variability, and which indicates that there is a uniform probability of the individual readings to occur. The upper seven lead data points, which exceeded 5 mg/L, represent an unusual condition, in this case spikes due to tube rapping. They must be considered to be a "different population," which distorts the normal distribution. Three high cadmium points also represent spikes. It is interesting to note that the *combined (mixed)* fly ash and bottom ash had an average leachate concentration of about 2 mg/L, over the two-week period; hence, the average for the combined ash passed the test, contrary to the separated bottom and fly ash.

TABLE 13B.13   EP Extracts of Lead and Cadmium in Bottom Ash*

Sample number	Lead, mg/L	Cadmium, mg/L	Sample number	Lead, mg/L	Cadmium, mg/L
1	12.40	0.28	25	4.27	0.07
2	0.41	0.01	26	0.51	0.04
3	0.38	0.03	27	0.40	0.01
4	2.62	0.08	28	1.79	0.04
5	7.11	0.06	29	1.26	0.02
6	3.65	0.03	30	5.28	0.12
7	0.18	0.02	31	0.34	0.02
8	0.15	0.02	32	1.10	0.02
9	1.70	0.04	33	0.65	0.02
10	1.81	0.03	34	0.98	0.03
11	1.52	0.03	35	2.04	0.06
12	3.04	0.04	36	1.60	0.03
13	3.01	0.05	37	4.43	0.06
14	0.57	0.04	38	0.08	0.01
15	0.98	0.02	39	0.59	0.01
16	1.19	0.03	40	9.21	0.02
17	3.23	0.05	41	2.87	0.03
18	7.93	0.08	42	2.54	0.06
19	0.53	0.12	43	1.81	0.02
20	1.56	0.11	44	0.21	0.01
21	4.22	0.21	45	3.66	0.03
22	3.69	0.07	46	1.15	0.04
23	2.11	0.04	47	1.58	0.09
24	9.44	0.10	48	15.08	0.08
Avg. of 48	2.59	0.05	Avg. of 16	2.55	0.05
Std. dev.	2.69	0.05	Std. dev.	3.07	0.06
90% UCL	3.46	0.05	90% UCL	3.21	0.05
Avg. of 24	3.06	0.06	Avg. of 8	3.18	0.06
Std. dev.	3.10	0.06	Std. dev.	3.90	0.08
90% UCL	3.68	0.06	90% UCL	4.04	0.06

* Collected from June 3 to June 5, 1982

Source: Feder and Mika, 1982.

Separated fly ash and bottom ash TCLP extraction data for lead samples collected daily on two separate weeks are shown in Fig. 13B.11. Over the wide range, the bottom ash data for each week's data generally tend to fall on nearly straight lines in this plot on logarithmic coordinates. The fly ash curves are not as straight, indicating that a mixture of several components is present.

Eighty percent of the first week's samples of bottom ash exceeded the limit of 5 mg/L. Only one of the samples of fly ash taken during the second week passed. The separated fly ash exhibited much higher levels of lead as indicated by the TCLP test. Thus, a large number of representative samples of the stream of ash generated by the plant will be required to obtain confidence in the true value of the average. The wide range during and between the two weeks illustrates why a large number of samples is needed to obtain confidence in the true value of the average. Eighty percent of one week's samples of bottom ash passed the limit of 5 mg/L.

Histograms show how the data are distributed. A histogram of the *lead concentrations* in mixed ash as listed in Table 13B.11 and shown in Fig. 13B.9 is presented in Fig. 13B.12. Characteristic of log normal data, this distribution has a high peak which tails off with higher concentrations of lead. It is difficult to ascertain the average from this diagram.

A histogram of the *logarithms* of the lead concentrations is shown in Fig. 13B.13. This graph exhibits the familiar bell curve of the normal probability distribution, indicating that the *logarithms* of the data are normally distributed. This graph, called a log-normal distribution, is characteristic of ash properties, as it is of a wide range of natural phenomena, such as the breakage of coal.

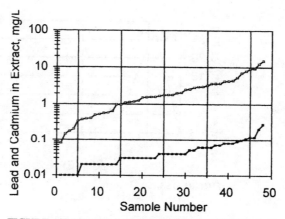

FIGURE 13B.10   Distribution of logarithms of TCLP extract concentrations for Pb and Cd over 48 hourly samples. *(Source: Feder and Mika, 1982.)*

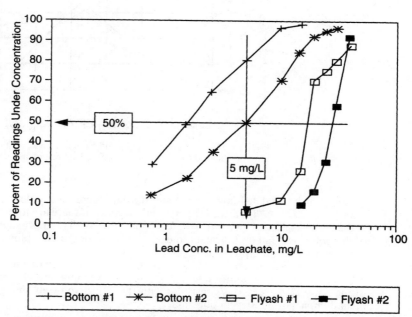

FIGURE 13B.11   Distribution of TCLP extract concentrations for Pb in *separated* bottom ash and fly ash samples taken during two different weeks. *(Source: Feder and Mika, 1982.)*

The average concentration of lead in the EP extract is not easily determined from the linear graph shown in Fig. 13B.12, whereas the distribution of the histogram of the logarithm of the data exhibits the logarithmic mean, wherein all points have an equal probability of happening. The point that has the same weight on either side appears to be about 1.1, the antilog of which is about 3.0 mg/L.

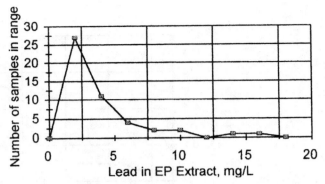

FIGURE 13B.12 Histogram of lead concentrations of leachate from TCLP test of lead samples of Table 13B.13.

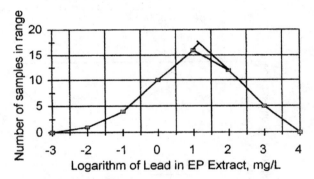

FIGURE 13B.13 Histogram of logarithms of lead concentrations in leachate obtained by TCLP test of combined bottom ash and fly ash.

To confirm that data is log-normal, sort the data, increasing or decreasing, and plot at equal intervals on *probability versus log* paper. For instance, if there are four data points, place them at the 20, 40, 60, and 80 percent "less than" locations. With log-normally distributed data most of the points will fall on a straight line. Even if most of the data plot along a straight line on log-log paper they are probably log-normal.

Numerical methods are used to analyze data. The true mean of a set of data which has been collected in samples of uniform samples weight at even intervals of time is the simple average of the data. This average is consistent with the objective of obtaining the average environmental impact of the stream of residue.

For a first approximation it is useful to know that for a set of normally distributed data, about 85 percent of the data points will have values less than the mean plus one standard deviation $(x + s)$, 95 percent will be less than $x + 2s$, and 98 percent will be less than $x + 3s$. A pocket calculator can be used to determine the mean (x-bar) and the standard deviation (s) of a set of normally distributed data.

What is important to an operator of a facility is how many samples must be collected and analyzed in order to be confident that 90 percent of the samples will be below the regulatory limit. The upper bound of this data range is called the *upper confidence limit* (UCL) at 90 percent confidence level.

Confidence Limits

The confidence interval is a criterion which can be used to find the range of data which will include 90 percent (or alternately, 95 percent) of the datum points in a set of data. In other words, 90 percent of the data will not exceed the upper confidence limit.

The upper confidence limit (UCL) as set forth by U.S. EPA (1993) is:

$$UCL = mean + sampling error = x + t_n s_x = x + t_n s/(n)^{0.5}$$

where: UCL = upper confidence limit below which the actual mean of the characteristic being tested will be found at the specified confidence level

x = mean of the data set

t_n = probability factor (Student's t) corresponding to the desired level of confidence and number of samples

s_x = standard error of the mean, calculated by dividing the standard deviation(s) of the data set by the square root of the number of analyses in the data set

s = standard deviation of the data set

n = number of datum points in the data set

Determining the Number of Samples Needed

Example: The number of samples required can be found by trial and error procedures, sharpening the numbers by stages. Data from other sources or from preliminary tests may be used as a first estimate of the number of samples needed to obtain the desired confidence that the upper limit will be below the target.

Table 13B.14 represents a series of trials to determine how many samples must be taken to be certain that the upper confidence limit is below the target, for lead, of 5 mg/L. The data set in Table 13B.13 has been used as the basis. The mean and standard deviation were calculated for different numbers of consecutive samples, as if the sampling were stopped at different points. For each number of samples analyzed, n, the degrees of freedom, $(n-1)$ are used to select the value of the Student's t. The standard error is added to the mean to obtain the upper confidence limit (UCL). The closer the UCL is to the target of 5.0 mg/L, the more samples would be needed.

The UCL for 16 samples would be:

$$UCL = 2.55 + 1.34 * 3.07/(16)^{0.5} = 3.58 \text{ mg/L}$$

This is within about 30 percent of the target of 5.0 mg/L. With 4 and 8 samples, the UCL exceeds the limit of 5 mg/L. With 30 and 48 samples, the UCLs are almost the same.

When there are cyclical factors in the flow of ash, as seen in Fig. 13.7, the time during a shift at which the samples were collected can severely affect the data. This is illustrated by Table

TABLE 13B.14 Calculation of Sampling Error Versus Number of Samples

Number of samples	Degrees of freedom $(n-1)$	Student t	Std. deviation s	Mean (avg) x	Std. error $s_x = s/n^{0.5}$	UCL = mean + std error
4	3	1.68	4.96	3.95	4.17	8.12
8	7	1.415	4.10	3.36	2.05	5.41
16	15	1.341	3.07	2.55	1.03	3.58
30	29	1.31	2.91	2.90	0.70	3.59
48	47	1.31	3.21	2.85	0.61	3.46

TABLE 13B.15   Analysis of 48 Samples, Taking One from Each Shift

	First hour in shift			Seventh hour in shift	
Hour	Lead, mg/L	Cadmium, mg/L	Hour	Lead, mg/L	Cadmium, mg/L
1	12.4	0.28	7	0.18	0.08
9	1.70	0.02	15	0.98	0.04
17	3.23	0.03	23	2.11	0.11
25	4.27	0.10	31	0.34	0.04
33	0.65	0.02	39	0.59	0.02
41	2.87	0.02	47	1.58	0.01
Average	4.19	0.08		0.96	0.054
Std.dev.	3.85	0.09		0.69	0.034
90% UCL	4.33	0.13		2.12	0.074

13B.15, which shows the average, standard deviation and 90 percent UCL of samples taken during six 8-hour shifts over a period of 48 h. The UCL based on samples taken during the first hour (after cleaning of the boiler tubes) is 4.33 mg/L, compared with 2.12 mg/L based on samples collected during the seventh hour (prior to cleaning the tubes). The average of these two UCL values is close to the average of the UCL found for 48 hourly samples, 3.46 mg/L. Taking consecutive hourly samples produced much more reliable results than taking one sample per shift at the same hour. This would be true even if the samples were composited and one analysis performed per composite sample.

The number of samples needed may be calculated directly from the following equation:

$$n_r = t^2 s^2 / e^2$$

where n_r = number of samples required to be analyzed to show that the specified target range of values has been met at the desired confidence level

t = the confidence factor for a designed confidence level: Student's "t" values are used, based on the desired confidence level and the actual or planned number of samples analyzed. For a confidence limit of 90 percent, t is 1.31 for 30 samples, 1.38 for 10 samples, and 1.64 for 4 samples

s^2 = the variance (square of the standard deviation), calculated for the data set, or estimated from prior sampling and testing

e^2 = the allowable error, or absolute value of the difference from the specified target value which defines the maximum acceptable target value and the mean of the corresponding data set

For example, assuming an error (e^2) of 30 percent = 1.5 and a mean of 2.5, using a trial (e^2) of 3.07 and (t) of 1.34 for 16 samples, we nearly confirm this assumption:

$$n_r = t^2 s^2 / e^2 = (1.34)^2 \times (3.07)^2 / (1.5)^2 = 14 \text{ samples}$$

REFERENCES

Abe, S. (1984) Melting Treatment of Municipal Waste, *Recycling International,* p. 173 ff.

Arcani, G. B. (2000) "Ash Recycling in Nashville, Tennessee," Pub. No. GR-WTE-0108, SWANA, 8th Annual North American Waste-to-Energy Conference, Nashville, TN.

ASTM (undated) *Standards Developed by Committee E-38-01 Energy,* American Society for Testing and Materials, Philadelphia, PA.

ASTM (1989) *Standard Guide for General Planning of Waste Sampling* (D4687), American Society for Testing and Materials.

Chandler, A. J. et al. (undated) Municipal Solid Waste Incinerator Residue, *Studies in Environmental Science 67,* The International Ash Working Group (IAWG), Elsevier, Amsterdam, the Netherlands.

Chesner, W. A. (1989) "Aggregate-Related Characteristics of MSW Combustion Residue," Resource Recovery Report, Proceedings of the 2d International Conference on MSW Combustor Ash Utilization, Washington, DC.

Chesner, W. A. (1993) Working Toward Beneficial Use of Waste Combustor Ash, *Solid Waste & Power,* vol. VII, no. 5.

Chesner, W. H., R. J. Collins, and T. Fung (1988) *Assessment of the Potential Suitability of Southwest Brooklyn Incinerator Residue in Asphaltic Concrete Mixes,* NYSERDA Report 90-15, State of New York, Albany, NY.

Clark, R. A. (1992) "Summary of the Field Leachate Generated from Northern States Power Company's Red Wing RDF Ash Disposal Facility," Paper No. 92-47.02, 85th Annual Meeting of the Air and Waste Management Association, Kansas City, MO.

Cundari, K. L., and J. M. Lauria (1986) "The Laboratory Evaluation of Expected Leachate Quality from a Resource Recovery Ashfill," 1986 Triangle Conference for Environmental Technology, Chapel Hill, NC.

DeCesare, R., and A. Plumley (1992a) "Results from the ASME/US Bureau of Mines Investigation—Vitrification of Residue from Municipal Waste Combustion," 1991 ASME National Waste Processing Conference, Detroit, MI.

DeCesare, R., and A. Plumley (1992b) "Results from the ASME/US Bureau of Mines Investigation—Vitrification of Residue from Municipal Waste Combustion," Paper 92-47.03, 85th Annual Meeting of the Air and Waste Management Association, Kansas City, MO.

Donnelly, J. E., E. Jons, and P. Mahoney (1987) "By-product Disposal from MSW Flue Gas Cleaning Systems," Paper #87-94A.3., 1987 APCA Annual Meeting, New York, NY.

Eaton, M. (1992) "The Commerce Refuse-to-Energy Facility Ash Handling and Treatment System," Chesner Engineering, 5th International Conference on MSW Combustor Ash Utilization.

Eighmy, T. T. et al. (1989) "Theoretical and Applied Methods of Lead and Cadmium Stabilization in Combined Ash and Scrubber Residues," Resource Recovery Report, Proceedings of the 2d International Conference on MSW Combustor Ash Utilization, Washington, DC.

Eighmy, T. T. et al. (1991, 1993) The Laconia, NH, Bottom Ash Paving Project, Environmental Research Group, University of New Hampshire, Durham, NH.

Eighmy, T. T. et al. (1997) "Fundamental Mechanisms of Phosphate Stabilization of Divalent Metals in MSW Combustion Scrubber Residues," Pub. No. GR-WTE 0105, SWANA, 5th Annual North American Waste-to-Energy Conference.

Exner, R. et al. (1989) Thermal Effluent Treatment for Flue Gas Treatment Systems in Refuse Incineration Plants from the Point of View of Residue Minimization by Recycling, *Recycling International,* pp. 1372–1392.

Feder, W. A., and J. S. Mika (1982) *Summary Update of Research Projects with Incinerator Bottom Ash Residue,* Executive Office of Environmental Affairs, Commonwealth of Massachusetts.

Fiesinger, T. (1989) "Sampling of Incinerator Ash," Resource Recovery Report, p. 51, Proceedings of the 2d International Conference on MSW Combustor Ash Utilization, Washington, DC.

Fiesinger, T. (1992) *Incinerator Ash Management: Knowledge and Information Gaps to 1987,* Report 92-6, New York State Energy Research and Development Authority, Albany, NY.

Forrester, K. (1989) "State-of-the-Art in Refuse-to-Energy Facility Ash Residue Characteristics, Handling, Reuse, Landfill Design and Management—A Summary," Resource Recovery Report, pp. 167ff, Proceedings of the 2nd International Conference on MSW Combustion Ash Utilization, Washington, DC.

Forrester, K. E., and R. W. Goodwin (1990) MSW Ash Field Study: Achieving Optimal Disposal Characteristics, *Journal of Environmental Engineering,* Paper No. 25094, vol. 116, no. 5.

Francis, C. W. (1984) *Leaching Characteristics of Resource Recovery Ash in Municipal Waste Landfills,* DOE ERD-83-289, No. 2456, Oak Ridge National Laboratory, Oak Ridge, TN.

Francis, C. W., and M. P. Maskarinec (1987) *Leaching of Metals from Alkaline Wastes by Municipal Waste Leachate,* ORNL/T-10050, Pub. No. 2846, Oak Ridge National Laboratory, Oak Ridge, TN.

GBB (1990) *Studies of Waste Composition,* Gephman, Brickner & Bratton, Inc., Falls Church, VA.

Goodwin, R. W. (1982) "Chemical Treatment of Utility and Industrial Waste," ASCE National Conference on Environmental Engineering, Minneapolis, MN.

Goodwin, R. W., and K. E. Forrester (1990) MSW Ash Field Study: Achieving Optimal Disposal Characteristics, Paper No. 25094, vol. 116, no. 5, pp. 880–889, American Society of Civil Engineers, *Journal of the Environmental Engineering Division.*

Hahn, J., R. G. Rumba, G. T. Hunt, and J. Wadsworth (1990) "Fugitive Particulate Emissions Associated with MSW Ash Handling—Results of a Full Scale Field Program," 83d Annual Meeting of the Air and Waste Management Association, Pittsburgh, PA.

Hartlen, J., and P. Elander (1986) *Residues from Waste Incineration—Chemical and Physical Properties,* SGI Varia 172, Swedish Geotechnical Institute, Linkoping, Sweden.

Hasselriis, F. (1984) *Processing Refuse-Derived Fuel,* Ann Arbor Science, available from author.

Hasselriis, F. (1994) "Analysis of Data Obtained from an Historic Ash Residue Leaching Investigation," ASME National Waste Processing Conference, New York, NY.

Hasselriis, F., D. Drum et al. (1991) "The Removal of Metals by Washing of Incinerator Ash," 84th Annual Meeting of the Air and Waste Management Association, Vancouver, BC.

Hjelmar, O. (1982) "Leachate from Incinerator Ash Disposal Sites," International Workshop on Municipal Waste Incineration, Montreal, Canada.

Hjelmar, O. (1990) "Regulatory and Environmental Aspects of MSW Ash Utilization in Denmark," Resource Recovery Report, 3d International Conference on Ash Utilization and Stabilization (ASH III), Washington, DC.

Hnat, J., and L. Bartone (1996) "Recycling of Boiler and Incinerator Ash into Value Added Glass Products," pp. 129ff, Proceedings of the 17th Biennial ASME Waste Processing Conference.

Holland, P. et al. (1989) "Evaluation of Leachate Properties and Assessment of Heavy Metal Immobilization from Cement and Lime Amended Incinerator Residues" vol. 1, Proceedings of the International Conference on Municipal Waste Combustion, Hollywood, FL.

IAWG (1997) Municipal Solid Waste Incinerator Residues, International Ash Working Group, Elsevier, the Netherlands.

Lebedur, H. et al. (1989) Emission Reduction in the Second Domestic Waste Incinerator Plant in the RZR Herten, *Recycling International,* p. 1220.

Lucido, S. P. (2000) "The Use of Municipal Waste Combustor Ash as a Partial Replacement of Aggregate in Bituminous Paving Material," Pub. No. GR-WTE-0108, SWANA, 8th Annual North American Waste-to-Energy Conference, Nashville, TN.

Lyons, M. R. (1996) The WES-PHix Ash Stabilization Process, pp. 165ff, 17th Biennial ASME SWPD.

Magee, B. H., J. Hahn et al. (1999) "Environmental Testing of Municipal Solid Waste Ash-Amended Asphalt," pp. 103ff, North American Waste-to-Energy Conference.

Musselman, C. N. et al. (1994) "The New Hampshire Bottom Ash Paving Demonstration, Laconia, New Hampshire," ASME National Waste Processing Conference, New York, NY.

Musselman, C. N. et al. (1995) "The New Hampshire Bottom Ash Paving Demonstration, Laconia, New Hampshire," Proceedings of the 8th International Conference on MSW Combustor Ash Utilization.

Roffman, H. K. (1991) Major Findings of the U.S. EPA/CORRE MWC-Ash Study, A&WMA VIP-19, *Municipal Waste Combustion.*

Sawell, S. E., S. A. Hetherington, and F. C. Mitchell (1999) "Treatment of Residues from Air Pollution Control Systems," Pub. No. GR-WTE 0107, SWANA, 7th Annual North American Waste-to-Energy Conference, Tampa, FL.

Shinn, C. E. (1987) *Toxicity Characteristic Leaching Procedure (TCLP), Extraction Procedure Toxicity (EPTox) and Deionized Water Leaching Characteristics of Lead from Municipal Waste Incinerator Ash,* Department of Environmental Quality, Portland, OR.

Thome-Kozmiensky, K. (1989) Measures to Reduce Incinerator Emissions, *Recycling International,* p. 1009.

U.S. EPA (1980) *Hazardous Waste Management System: Identification and Listing of Hazardous Waste, Toxicity Test Extraction Procedure (EP),* Federal Register, 45(98), 33063-33285, Washington, DC.

U.S. EPA (1988) National Secondary Drinking Water Regulations, 40-CFR 143, U.S. Environmental Protection Agency, Washington, DC.

U.S. EPA (1993) Test Methods for Evaluating Solid Waste Physical/Chemical Methods, 3d ed., SW-846, *Sampling and Testing Methods for Solid Waste,* U.S. Environmental Protection Agency, Washington, DC.

U.S. EPA (1995) *Guidance for the Sampling and Analysis of Municipal Waste Combustion Ash for the Toxicity Characteristic,* U.S. Environmental Protection Agency, Washington, DC.

Wakamura, Y., and K. Nakazato (1992) "Technical Approach for Flyash Stabilization in Japan," Paper No. 92-20.08, 85th Annual Meeting of the Air and Waste Management Association, Kansas City, MO.

Wexerll, D. R. (1993) "Cold Crown Vitrification of Municipal Waste Combustor Flyash," Paper No. 93-RA-117.02, 86th Annual Meeting of the Air and Waste Management Association, Denver, CO.

Wiles, C. (1999) *Beneficial Use and Recycling of Municipal Waste Combustion Residues—A Comprehensive Resource Document,* NREL/BK-570-25841, NREL, Boulder, CO.

CHAPTER 13
WASTE-TO-ENERGY COMBUSTION
Part 13C   Emission Control

Floyd Hasselriis

13C.1   INTRODUCTION

Combustion of wastes has long been recognized as a "final" disposal solution, because the organic matter is destroyed and only solid residues remain. By comparison, land-filling is a solution that amounts to storage, with the continuing risk of unwanted consequences (Taylor, 1992; Jones, 1994).

As of the year 2000, over 90 percent of municipal waste is combusted in Japan, 75 percent in Europe, where landfill of organic matter is essentially prohibited. In the United States, only 15 percent is combusted although in some states, it approaches 50 percent: the low cost competition of landfills has been a major factor in limiting combustion. Moreover, in the United States properly designed and maintained landfills are accepted by regulatory authorities. Emissions from landfills can be higher than those from waste combustion facilities, and collection of landfill gases has increasingly become a regulatory requirement (IWSA, undated).

Waste combustion results in discharge of gaseous and particulate matter to the atmosphere and causes public concern for health and the environment. In order to take advantage of combustion technology, great efforts and continuous evolution have been applied to minimize negative affects. In addition, it is necessary to dispose of the solid residues of combustion which have the potential for harm if not properly managed, mainly due to the solubility of metals, and the risk that they potentially impose on the environment. Management of the ash residues is treated in Sec. 13C.2.

To gain and maintain public acceptance for combustion of wastes, it has been necessary to reduce the emissions from waste combustion below the levels of concern to health and the environment. Acceptable levels are regulated by law, based on standards developed by health authorities. These reductions can be achieved by proper treatment of the flue gases before they are discharged from the stack.

This chapter reviews the emission levels that have already been achieved and projects those which are anticipated as the presently promulgated U.S. Environmental Protection Agency (U.S. EPA) regulations come into effect and existing facilities are brought up to these standards. These standards are summarized in the text that follows.

The nature of the critical pollutants (PM, SO_2, HCl, NOx, metals, and organics) and the emission control devices which are used to remove or convert them are described. The functions of the control devices are reviewed and performance data are presented. The variability of emissions, a significant factor from a regulatory standpoint, is discussed and evaluated.

After the gases leave the stack, they are dispersed, resulting in dilution by 10 thousand to 1 million times before they reach ground level. The impact on health and the environment of typical modern waste combustion facilities is estimated based on evaluation of the ground level concentrations of pollutants as well as deposition of particulate matter. The studies cited show that the risk of additional cancer cases is less than 10 in one million, and that the Hazard Index is far less than 1, the U.S. EPA regulatory levels of concern.

Inventory of MWCs in the United States

Table 13C.1 lists the municipal waste combustion (MWC) facilities, the number of individual units, and the total combustion capacity in the United States. This capacity peaked in 1993 and has stabilized at about 100,000 tons per day since 1999, due to closing of some plants, and construction of new facilities in accordance with more stringent regulations for combustion of wastes (IWSA, undated).

TABLE 13C.1 Number of MWC Facilities, Units, and Total Combustion Capacity

Year	Number of MWC facilities	Number of MWC units	Total combustion capacity (tons per day)
1990	126	302	93,274
1993	125	299	104,676
1996	123	286	103,335
1999	108	259	100,102
2000	106	256	99,082
2005	106	256	99,082

Note: The emissions, from both large MWC units regulated under 40 CFR Part 60, subpart Cb (greater than 250 tons per day capacity) and small MWC units regulated under subpart BBBB (35 to 250 tons per day capacity) were estimated based on the extensive data bases available.

Source: IWSA (undated).

Reductions in Emissions from MWCs. As the result of the evolution of treatment methods, emissions of the critical pollutants have been greatly reduced, based on the emission levels permitted in 1990. These reductions have already substantially impacted the environment, as indicated by the reduction of lead and cadmium in cigarettes grown in Canada (Ricket and Kaiserman, 1994). The critical pollutants and the percent reductions that are expected to result by the year 2000 and 2005 from regulations now in place are shown in Table 13C.2. How these reductions have been achieved is described in this chapter.

Annual Emissions. Based on all MWCs operating or expected to be operating in the United States, annual emissions have been estimated for the critical pollutants, based on the time range from 1990 to 2005 and on compliance dates for both large MWC units (year 2000) and small

TABLE 13C.2 Estimated Percent Reductions in Emissions Expected for Years 2000 and 2005

Pollutant	Percent reduction from 1990—permitted levels by year 2000	Percent reduction from 1990—permitted levels by year 2005
Dioxins, Toxic Equivalent (TEQ)	99	99+
Cadmium	68	78
Lead	77	94
Mercury	88	92
PM	71	94
HCl	88	95
SO_2	72	77
NOx	24	26

Source: IWSA (undated).

TABLE 13C.3   Total Annual Emissions from MWC Units in the United States

| | Grams/year | | Tons per year | | | | | | |
Year	Total Dioxins	TEQ Dioxins	PM	HCl	SO$_2$	NOx	Cd	Pb	Hg
1990	209,000	4170	17,000	52,000	34,500	58,500	8.59	148	51
1993	200,000	4000	12,500	45,000	31,900	64,700	6.36	105	30
1996	28,900	577	8320	24,000	21,500	62,000	5.5	73	24
1999	18,300	366	7000	18,600	17,200	56,300	4.38	50	17
2000	2030	41	4870	6000	9700	44,500	2.78	34	6
2005	601	12	1040	2380	8050	43,600	1.88	9	4
Reduction	99.7%	99.7%	94%	95%	77%	25%	78%	94%	92%

Source: IWSA (undated).

MWC units (year 2005). The emission estimates on which Table 13C.2 is based are derived from the most current available information and are considered to be more accurate than earlier estimates. Emissions were calculated based on currently available test data and emission factors.

Actual and Anticipated Emissions from MWCs

From the 1995 inventory of MWC units, a database of large and small MWC units was developed for the years from 1990 to 2005, taking into account the regulations that would apply during this period. Using these inventories, emissions were calculated for the eight critical pollutants using currently available compliance test data and U.S. EPA AP-42 emission factors (U.S. EPA, 1996a). A summary of the estimates of annual emissions is presented in Table 13C.3. A substantial reduction in MWC unit emissions has occurred since 1990 and additional reductions are projected in future years, the result of retrofit of air pollution control devices (APCD) on existing MWC units; retirement of several existing MWC units; and special actions, most notably EPA's dioxin initiative and voluntary mercury reduction by battery manufacturers. A summary of percent reductions, based on regulatory standards, over the years 2000 and 2005 for the eight pollutant emissions was given in Table 13C.2. The graphical presentations such as Figs. 13C.1 and 13C.2 are more striking than figures listed in tables.

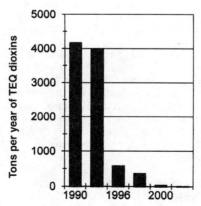

FIGURE 13C.1   Estimated annual emissions of TEQ dioxins from all U.S. MWCs. (*Source: IWSA, undated.*)

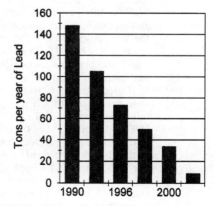

FIGURE 13C.2   Total tons per year of lead emissions from U.S. MWCs versus year. (*Source: IWSA, undated.*)

13C.2 EMISSIONS FROM COMBUSTION

Emission control systems are designed to control the various pollutants contained in the products of combustion, especially those of major concern for environment and health: particulate matter (PM), SO_2, HCl, NOx, metals, and organics. Each pollutant is subject to different mechanisms of control, however, and specific emission control techniques have different impacts on individual pollutants.

Pollutants of Concern

The pollutants of concern that may be found in the combustion products, prior to the air pollution control device (APCD) and/or in the stack emissions, along with the controls used to reduce their discharge to the atmosphere, are discussed in the text that follows. (Kilgroe and Licata, 1996).

Particulate Matter.   The quantity and concentration of particulate matter (PM) exiting the furnace of a waste combustion system depends on the waste characteristics, and the design and operation of the combustion system. While most of the inorganic, noncombustible fraction of municipal solid waste (MSW) and biomedical waste (BMW) will be discharged as bottom ash, a substantial fraction will be formed from combustion and released into the flue gas. Generally 99 percent or more of this particulate matter is captured by the APCD and is not emitted to the atmosphere. Opacity monitors, previously required to be operated in the stack, are not effective at these low emissions levels.

Particulate matter varies in particle diameters from less than 1 micron (μm) to hundreds of microns. Fine particulates, having diameters less than 10 μm (known as PM10). Particulates smaller than 2.5 μm are of concern because of the greater potential for inhalation and passage of these fine particles into the pulmonary region of the lungs. Also, acid gases, metals, and toxic organics preferentially adsorb onto particulates in this size range, so they can be absorbed within the lungs.

The physical properties of the waste being fed, the method of feeding, and the quantity and distribution of overfire and underfire air influence PM concentrations in the flue gas. The concentration of PM emissions at the inlet of the APCD will depend on the combustor design, air distribution, and waste characteristics. The higher the underfire/overfire air ratio or the excess air levels, the greater the entrainment of PM in the flue gases, and the higher the PM levels at the APCD inlet. Combustors with boilers that change the direction of the flue gas flow may remove a significant portion of the PM prior to the APCD. For instance, RDF (refuse-derived fuel) furnaces typically have higher PM carryover due to the suspension firing of the RDF, and starved-air furnaces have substantially lower carryover than excess-air furnaces due to the low gas velocities in the primary furnace.

Metals.   Metals are present throughout municipal and medical wastes (WASTE, 1993). The metals emitted as components of PM (e.g., arsenic [As], cadmium [Cd], chromium [Cr], and lead [Pb]) and as vapors such as Hg are highly variable and are essentially independent of the combustor type. Most of the metals are vaporized during combustion and condense onto particulates in the flue gas as its temperature is reduced; hence the metal can be removed effectively by the PM control device, at efficiencies greater than 99 percent (Sorum et al., 1997). Mercury, on the other hand, still has a high vapor pressure at typical APCD operating temperatures, and capture by the PM control device is highly variable. A high level of carbon in the fly ash, or the injection of activated carbon greatly enhances Hg adsorption onto the particles removed by the PM control device.

Acid Gases.   Combustion of wastes produces hydrogen chloride (HCl) and sulfur dioxide (SO_2), as well as hydrogen fluoride (HF), hydrogen bromide (HBr), and sulfur trioxide (SO_3) at much lower concentrations. Concentrations of HCl and SO_2 in uncontrolled flue gases are related directly to the chlorine and sulfur contents in the waste, which vary considerably based on seasonal and local waste variations. The major sources of chlorine in MSW are paper, food, and plastics. Sulfur in MSW may derive from asphalt shingles, gypsum wallboard, and tires. The presence of PVC plastics in BMW results in relatively high HCl concentrations in uncontrolled emissions. Controlled stack emissions of SO_2 and HCl depend partly on the chemical form of sulfur and chlorine in the waste, as well as the availability of alkali materials in combustion-generated fly ash that act as sorbents, added reagents, and the type of emission control system used (Licata et al., 1994).

Carbon Monoxide.   Carbon monoxide emissions are the result of incomplete oxidation of the carbon in the waste to carbon dioxide (CO_2). High levels of CO indicate that the combustion gases were not held at a sufficiently high temperature in the presence of sufficient oxygen (O_2) for a long enough time and with sufficient mixing of the gases to convert CO to CO_2. In the first stages of combustion in a fuel bed, waste first releases CO, hydrogen (H_2), and unburned hydrocarbons. Additional air converts these gases to CO_2 and H_2O. However, adding too much air to the combustion zone can lower the local gas temperature and quench (retard) the oxidation reactions. Conversely, if too little air is added, the probability of incomplete mixing increases, allowing greater quantities of unburned hydrocarbons to escape the furnace, thus increasing CO emissions. Because O_2 levels, air distribution, and the effectiveness of mixing vary among combustor types, CO levels also vary substantially. For example, semi-suspension-fired RDF units generally have higher CO levels than do mass burn units, due to the effects of carryover of incompletely combusted materials into low-temperature portions of the furnace, and, in some cases, due to instabilities that result from fuel feed characteristics and distribution over the fuel bed. Likewise, two-chamber starved-air systems usually have very low CO emissions due to the inherently effective mixing that they can achieve.

Carbon monoxide concentration is a direct indicator of combustion effectiveness and efficiency, and is an important indicator of instabilities and nonuniformities in the combustion process. During unstable combustion conditions when more carbonaceous material is available, serving as precursors for the formation of trace organics, chlorinated dibenzo-dioxins (commonly called PCDD or CDD), chlorinated dibenzofurans (commonly called PCDF or CDF), and organic hazardous air pollutant levels are likely to exist or be produced so as to appear in the products of combustion. The relationship between emissions of CDD/CDF and carbon monoxide (CO) indicates that high levels of CO (several hundred parts per million by volume [ppmv]), resulting from poor combustion conditions generally correlate with high CDD/CDF emissions. When CO levels are low, however, simple correlations between CO and CDDs/CDFs may not be found due to the fact that many mechanisms contribute to CDD/CDF formation.

Nitrogen Oxides.   Nitrogen oxides are produced by all combustion processes using air as a source of oxygen due to nitrogen present in the fuel, and also due to the nitrogen in the combustion air. Nitric oxide (NO) is the primary component of NOx, but nitrogen dioxide (NO_2) and nitrous oxide (N_2O) are also formed in smaller amounts. The combination of these compounds is referred to as NOx. Nitrogen oxides are formed during combustion through oxidation at relatively low temperatures (less than 1090°C [2000°F]), and fixation of atmospheric nitrogen occurs at higher temperatures. Because of the relatively low temperatures at which municipal medical and hazardous waste furnaces operate, 70 to 80 percent of NOx is associated with nitrogen in the waste. Acrylic plastics are a major source of nitrogen in MSW and BMW. Hazardous wastes may contain nitrogen in many forms.

Organic Compounds.   Organic compounds, including chlorinated dibenzodioxins and chlorinated dibenzofurans (CDDs/CDFs), chlorobenzenes (CB), polychlorinated biphenyls (PCBs), chlorophenols (CPs), and polyaromatic hydrocarbons (PAHs), are present in municipal solid waste (MSW) and biomedical waste (BMW) and can also be formed during the combustion and postcombination processes. Organics in the flue gas exist in the vapor phase or may be condensed or absorbed on fine particulates. Organics are controlled by proper design and operation of both the combustor and the air pollution control devices (APCDs). Activated carbon injection has been found to be effective in adsorbing CDD/CDFs as well as other trace organic compounds (Licata et al., 1994).

Due to their relatively high toxicity levels, emphasis is placed on levels of CDDs/CDFs in the tetra- through octa- homolog groups and specific isomers within those groups that have chlorine substituted in the 2, 3, 7, and 9 positions. The U.S. EPA New Source Performance Standards (NSPS) and emission guidelines for MWCs BMCs regulate the total tetra- through octa-CDDs/CDFs. The rest of the world focuses on the Toxic Equivalent (TEQ) of 2,3,7,8 TCDD, a factor less than the total CDD/CDF by a nominal factor of 60 (used by the U.S. EPA), but ranging from 40 to 100 (U.S. EPA, 1999). Because the main effect of dioxins on humans is on the Ah receptor, methods of measuring the toxicity of dioxins and similar compounds have been developed.

13C.3   EMISSION STANDARDS AND GUIDELINES

Technology-Based Standards versus Risk-Based Standards

The trend of regulations has been from risk-based standards toward technology-based standards. Before present federal regulations were in place for municipal waste combustion, many states took the initiative and wrote standards based on the health risk associated with the emissions at ground level, and required risk assessments to be prepared on the basis of modeling. U.S. EPA regulations for hazardous waste combustion (RCRA and TSCA) include the requirement for carrying out risk assessments based on trial burn stack test results and environmental modeling (U.S. EPA, 1999).

The Clean Air Act Amendments of 1990 required that Maximum Available Control Technology (MACT) standards be developed which would take into account the "best performing" combustion systems, requiring that new facilities meet these more stringent emission standards, and that existing facilities be upgraded to higher standards (U.S. EPA, 1996b). In spite of having to comply with the new federal standards, and more stringent state standards, permits may continue to be subject to public scrutiny on a one-by-one basis, which means that environmental impact will still be considered for pollutants which are not numerically regulated by federal and state regulations. New Source Performance Standards (NSPS) are firm for new facilities, and less stringent guidelines are set for existing units. Note that *unit* means a single line of equipment, not the entire facility.

Guidelines

In addition to stack emission requirements, guidelines have also been issued for the design and operation of waste combustion systems. Guidelines are used for review, but need not necessarily be followed if the performance and compliance tests can be met.

Combustion temperatures are to be maintained at 1800°F for one (or two) seconds after the last injection of secondary combustion air.

To confirm the probability of a given system being able to comply with these regulations, a number of calculations may be made. Some of these will be described and carried out for typical systems in the following sections.

Regulatory Standards

The following summaries greatly simplify the actual requirements, but are presented to point out the differences in general approaches to the various types of systems. More detailed summaries are presented in Tables 13C.4 to 13C.8.

Municipal Waste Combustor (MWC) emission limits have been established by the U.S. EPA for existing and new units, and for two size categories: between 38.6 tons per day (TPD) and 248 TPD, and greater than 248 TPD, as shown in Table 13C.4 (U.S. EPA, 1996b).

TABLE 13C.4 New Source Performance Standards and Emission Guidelines for Municipal Waste Combustors

Pollutant (Test method)	Units	Existing units >38.6 T/D	Existing units >248 T/D	New units >38.6 T/D	New units >248 T/D
Particulates (EPA method 5 or 29)	mg/dscm (gr/dscf)	70 (0.03)	27 (0.012)	24 (0.010)	24 (0.010)
Opacity	6 min. avg.	10%	10%	10%	10%
CO (EPA method 10 or 108)	ppmv	100	40	20	20
Dioxins/furans (EPA method 23)	ng/dscm Total	125	60 (ESP) 30 (FF)	13	13
HCl (EPA method 26)	ppmv or % reduction	250 or 50%	31 or 95%	25 or 95%	25 or 95%
SO₂	ppmv or % reduction	80 or 50%	31 or 75%	30 or 80%	30 or 80%
NOx	ppmv	N/A	200	N/A	150
Lead (EPA method 29)	µg/dscm	1,600	490	200	200
Cadmium (EPA method 29)	µg/dscm	100	40	20	20
Mercury (EPA method 29)	µg/dscm or % reduction	80 or 85%	80 or 85%	80 or 85%	80 or 85%

Ref. EPA 40 CFR Part 60, Subparts Eb, Cb, AAAA, and BBBB, December 19, 1995.
Source: U.S. EPA (1996b).

MWCs are required to demonstrate the following in periodic compliance tests:

Particulate matter (PM) emissions are limited to 0.015 grains/dscf at 7 percent oxygen

Carbon monoxide (CO) emissions are not to exceed 50 ppmv on a rolling average, with exemptions of 150 ppmv for refuse-derived fuel (RDF) systems.

Hydrogen chloride (HCl) is limited to 30 ppmv at 7 percent oxygen or 95 percent control.

Sulfur dioxide (SO₂) is limited to 50 ppmv at 7 percent oxygen or 80 percent control

Nitrogen oxides (NOx) are limited to 150 ppmv at 7 percent oxygen

Mercury, lead, and cadmium emissions are limited quantitatively

The system is permitted to operate at a maximum feed rate, usually specified as tons per day (T/D) of defined wastes having a specified reference heating value.

Hospital/Medical/Infectious Waste Incinerators (HMIWIs): Emission limits have been established by the U.S. EPA for four existing categories and three categories of new units (Strong and Copland, 1998). Special limits have been set for existing small, remote HMIWI units in rural areas, which have little environmental impact and cannot support the costs of emission controls from an economic point of view. Other existing and new units have three size categories: small units, <200 pounds per hour (lb/h); medium units, 200–500 lb/h; and large units, >500 lb/h. Emission limits for existing HMIWI are shown in Table 13C.5, and for new HMIWIs in Table 13C.6.

TABLE 13C.5 Emission Guidelines for Existing Hospital/Medical/Infectious Waste Incinerators

Pollutant (Test method)	Small units (<200 lb/h)	Medium Units (200–500 lb/h)	Large units (>500 lb/h)
Particulates (EPA Method 5 or 29)	115 mg/dscm (0.05 gr/dscf)	69 mg/dscm (0.03 gr/dscf)	34 mg/dscm (0.015 gr/dscf)
CO (EPA Method 10 or 108)	40 ppmv	40 ppmv	40 ppmv
Dioxins/furans (EPA Method 23)	125 ng/dscm total CDD/CDF (2.3 ng/dscm TEQ)	125 ng/dscm total CDD/CDF (2.3 ng/dscm TEQ)	125 ng/dscm total CDD/CDF (2.3 ng/dscm TEQ)
HCl (EPA Method 26)	100 ppmv or 93% reduction	100 ppmv or 93% reduction	100 ppmv or 93% reduction
SO_2 (testing not required)	55 ppmv	55 ppmv	55 ppmv
NOx (testing not required)	250 ppmv	250 ppmv	250 ppmv
Lead (EPA Method 29)	1.2 mg/dscm or 70% reduction	1.2 mg/dscm or 70% reduction	1.2 mg/dscm or 70% reduction
Cadmium (EPA Method 29)	0.16 mg/dscm or 65% reduction	0.16 mg/dscm or 65% reduction	0.16 mg/dscm or 65% reduction
Mercury (EPA Method 29)	0.55 mg/dscm or 85% reduction	0.55 mg/dscm or 85% reduction	0.55 mg/dscm or 85% reduction

(1998) EPA 40 CFR Part 60 Subpart Ce (U.S. EPA, 1998).
Source: New Source Performance Standards for Hospital/Medical/Infectious Waste Incinerators, EPA August 1997.

TABLE 13C.6 New Source Performance Standards for New Hospital/Medical/Infectious Waste Incinerators

Pollutant (Test method)	Small units (<200 lb/h)	Medium units (200–500 lb/h)	Large units (>500 lb/h)
Particulates (EPA method 5 or 29)	69 mg/dscm (0.03 gr/dscf)	34 mg/dscm (0.015 gr/dscf)	34 mg/dscm (0.015 gr/dscf)
CO (EPA method 10 or 108)	40 ppmv	40 ppmv	40 ppmv
Dioxins/furans (EPA method 23)	125 ng/dscm total CDD/CDF (2.3 ng/dscm TEQ)	25 ng/dscm total CDD/CDF (0.6 ng/dscm TEQ)	25 ng/dscm total CDD/CDF (0.6 ng/dscm TEQ)
HCl (EPA method 26)	15 ppmv or 99% reduction	15 ppmv or 99% reduction	15 ppmv or 99% reduction
SO_2 (testing not required)	55 ppmv	55 ppmv	55 ppmv
NOx (testing not required)	250 ppmv	250 ppmv	250 ppmv
Lead (EPA method 29)	1.2 mg/dscm or 70% reduction	0.07 mg/dscm or 98% reduction	0.07 mg/dscm or 98% reduction
Cadmium (EPA method 29)	0.16 mg/dscm or 65% reduction	0.04 mg/dscm or 90% reduction	0.04 mg/dscm or 90% reduction
Mercury (EPA method 29)	0.55 mg/dscm or 85% reduction	0.55 mg/dscm or 85% reduction	0.55 mg/dscm or 85% reduction

EPA 40 CFR Part 60 Subpart Ce (Strong and Copland, 1998).
Source: Final New Source Performance Standards for Hospital/Medical/Infectious Waste Incinerators, U.S. EPA August 1998.

New HMIWIs are required to demonstrate the following:

Particulate matter (PM) emissions are limited to 0.03 to 0.015 grains/dscf at 7 percent oxygen

Carbon monoxide (CO) emissions are not to exceed 40 ppmv on a rolling average

Hydrogen chloride (HCl) is limited to 15 ppmv at 7 percent oxygen or 95 percent control. Sulfur dioxide and nitrogen oxides (NOx) are not limited.

Mercury, lead, and cadmium emissions are limited quantitatively

The system is permitted to operate at a maximum feed rate, usually specified as pounds per hour or tons per day of defined wastes.

Hazardous Waste Combustor (HWC) emission limits have been established by the U.S. EPA, as shown in Table 13C.7, for existing and new units, with specific limits for hazardous waste incinerators (HWIs), cement kilns (CKs), and lightweight aggregate kilns (LWAKs). The metals are divided into Hg, and semi- and low-volatile groups. Averaging times are different for different pollutants. All data are corrected to 7 percent oxygen, dry basis (U.S. EPA, 1999).

HWIs must demonstrate the following in trial burns and periodic compliance tests:

Principal organic hazardous compounds (POHCs) are required to demonstrate a 99.99 percent Destruction and Removal Efficiency (DRE)

Hydrogen chloride (HCl) emissions to be less than 5 lb/h or 99 percent removed

Particulate matter (PM) emissions are limited to 0.03 grains/dscf at 7 percent oxygen.

Carbon monoxide (CO) emissions are not to exceed 150 ppmv on a rolling average

Mercury, lead, and cadmium emissions are limited quantitatively.

TABLE 13C.7   National Emission Standards for Hazardous Waste Combustors*

Pollutant	Averaging time	Units	Incinerator		Cement kiln		Lightweight aggregate kiln	
			Existing	New	Existing	New	Existing	New
PM	CEM 2 h	mg/dscm (gr/dscf)	69 (0.030)	69 (0.03)	69 (0.030)	69 (0.030)	69 (0.030)	69 (0.030)
CDD/DF	Stack	ng/dscm	0.20	0.20	0.20	0.20	0.20	0.20
HC	CEM	ppmv	12	12	20	20	14	14
CO	CEM	ppmv	100	100			100	100
HCl + C$_2$	CEM-h	ppmv	280	67	630	67	450	62
Hg	CEM 10 h	µg/dscm	50	50	50	50	72	72
Semivol Pb,Cd, (sum)	CEM 10-h	µg/dscm	270	62	57	60	60	60
Low vol. As,Be,Cr,Sb (sum)	Stack or CEM 10-h	µg/dscm	210	80	130	80	340	80

* Final Rule, June 19, 1998.
EPA 40 CFR Part 63 Subpart EEE.
Source: U.S. EPA (1999).

Critical operating parameters are to be maintained within limits established by the trial burn tests, and may require waste feed shutoffs when these limits are exceeded.

European Countries impose hazardous waste incinerator emissions as outlined in Table 13C.8. Notably different from the U.S. EPA limits are the averaging times (i.e., 24 h average, maximum hour, weekly mean, and 24 h maximum). Also notable is the division of heavy metals into three classes, each class being the sum of a group of like-behaving metals. Note that the Federal Republic of Germany requires correcting the data only when the oxygen level exceeds 11 percent oxygen, whereas the European Community Directive calls for correcting all data to 11 percent oxygen (Schüttenhelm et al., 2000).

TABLE 13C.8   Waste Incinerator Emission Guidelines for Some European Countries

Measurement	Federal Republic of Germany 17.BImSch V, Nov. 1990 All Waste Plants mg/Nm^3 dry at >11% Oxygen		European Community Directive for New Facilities mg/Nm^3 at 11% O_2	
	24 h avg.	max. h	Weekly mean	24 h max.
HCl	10	60	50	65
SO ($SO_2 + SO_3$)	50	200	300	390
HF	1	4	2	2.7
NOx (NO_2)	200	400	—	—
CO	50	100	100	130
C (organic)	10	20	20	26
Particulates	10	30	30	39
Heavy metals				
Class I	Cd + Tl í = 0.05 (>0.5 hr)		Cd + Hg = 0.2	
Class II	Hg = 0.05 (>0.5 h)		Ni + As = 1.0	
Class III	Sb,As,Pb,Co,Cr,Cu,Mn,V,Sn,Ni í = 0.5		Pb,Cu,Cr,Mn í = 5.0	
PCDD/PCDF	0.1 ng/Nm^3 I-TEQ >500 min		0.1 ng/Nm^3 I-TEQ (8 hr. avg.)	
Combustion temperature	850°C >2 sec for MSW; MWI & Sludge >6% O_2, 1200°C >2 sec. Others		850°C at >6% O_2 >2 sec.	

Source: Schüttenhelm et al. (2000).

Good Combustion Practice

U.S. EPA Standards for Good Combustion Practice (GCP) have been developed, which apply to all waste combustors: Essentially they require that no waste be fed until the furnace temperature is at least 1600°F, and that the gaseous products of combustion be retained for at least one second at 1800°F or higher for most wastes, or for two seconds at 2000°F for wastes containing large quantities of halogenated compounds that are more difficult to destroy.

Hazardous waste incinerators have been subjected to stringent regulation because of the potentially toxic emissions from a wide range of organic and inorganic compounds that may be in the wastes, and which define them as hazardous (U.S. EPA, 1999). These regulations require that trial burns be performed as a permit condition, during which it is demonstrated that the system can achieve 99.99 percent destruction and removal efficiency (DRE) while burning POHCs and heavy metals. The assumption underlying these tests was that the emissions would be related to the feed materials. Hence the feed must be analyzed in normal operation, and be anticipated as the basis for the trial burns. In view of the fact that modern combustion and emission control systems that can meet present regulatory standards actually destroy or remove all but traces of the target substances, the tendency has been to focus on the actual emissions and assure that they will not exceed the new, stringent standards.

Stack Testing and Monitoring

An essential part of the focus on actual stack emissions, which applies to all waste combustors, is the requirement that monitoring instruments be installed and operated to provide a continuous record of compliance. Continuous measurement of carbon monoxide (CO) and/or hydrocarbons (HCs) and oxygen provides assurance of good combustion. These are useful surrogates for trace organic compounds including dioxins and furans. Measurement of HCl, SO_2, and NOx may be required, where applicable. Opacity measurements provide continuous supervision of the combustion process as well as the emission control system.

Ash Residue Management

Ash residues from combustion of wastes generally have to be tested for toxicity and managed appropriately. If they fail the Threshold Characteristic Leaching Procedure (TCLP) test, they may have to be treated or disposed in suitable landfills. The source of toxicity is generally heavy metals, specifically lead and cadmium, which are not destroyed by the combustion process, and then, end up in the fly ash or bottom ash residues. The use of alkaline reagents for control of the acid gases adds another component to the disposal problem, as excess alkalinity can increase the solubility of these metals in the ash residues; and carbon, which absorbs dioxins and mercury. Special chemical treatments have been found to reduce the solubility of metals.

Operator Certification and Training

In addition to requiring good combustion practice and regulating emissions, an essential component of the newly promulgated regulations is the requirement for operator training and certification, applied appropriately to municipal, medical, and hazardous waste combustion systems. The American Society of Mechanical Engineers (ASME) provides certification of operators of waste combustion systems.

Continuous Emissions Monitoring

Continuous monitoring of emissions provides the public and the regulators with confidence that the facility is being operated properly and within permit limits. Continuous opacity monitors were once required and relied upon for this purpose. However, the low emissions levels of modern systems cannot be read by these monitors: they serve only to give an alarm indicating extreme conditions. Continuous monitors for CO are required for essentially all waste combustors, generally with a scale of 0 to 200 ppmv, to cover the permitted range. Continuous SO_2 monitors are required for municipal combustors, but not for medical waste combustors, because sulfur levels are low in the latter. On the other hand, continuous HCl monitors are generally required on larger medical and all hazardous waste combustion facilities, but not generally for MWCs.

13C.4 EMISSION CONTROL DEVICES

The following devices, used as components in complete emission control systems, are currently available for control of emissions.

- Cyclone separators
- Quench venturi
- Wet venturi scrubber (WVS)
- Packed-bed scrubber (PBS)
- Plate scrubber (PS)
- Electrostatic precipitator (ESP)
- Wet electrostatic precipitator (WESP)
- Dry venturi (DV)
- Dry sorbent injection (DSI)
- Reactor tower (RT)
- Conditioning (cooling) tower
- Spray-dry reactor (SDR)
- Fabric filter (FF)
- SNCR
- SCR

For MSW combustion systems, regulatory requirements in the United States are different for different categories (e.g., large or small mass-burn, small modular, large and small RDF. Large and small are divided by the limit of 250 tons per day capacity per unit for small units.

For medical waste systems, there are four capacity sizes—small, medium, and large, and a special category for smaller systems in rural hospitals that may not need emission controls to meet the less stringent standards. The main choices in general use for emission control components are: venturi/wet scrubbers; spray-dry reactors; gas cooling towers; dry injection reactor towers; fabric filters, final wet scrubbers, and activated carbon injection.

For hazardous waste incinerators, the regulations treat all sizes alike. The main components of emission control systems may be: electrostatic precipitators, venturi/wet scrubbers; spray-dry reactors; gas cooling towers; dry injection reactor towers; fabric filters, final wet scrubbers, and activated carbon injection.

The various components of pollution control systems are described in the text that follows, and later, the complete systems. Table 13C.9 lists the types and numbers of air pollution control devices (APCDs) employed at operating U.S. WTE facilities. Different sizes and types of plants have different equipment (IWSA, undated).

Particulate Control

Cyclone Separators. Cyclone separators use the centrifugal forces attained by forcing the gases through a tangential entry to a cylinder to cause solid particles to follow the walls, and drop out of the gas stream, while the gas stream is diverted away without those collected particles. Cyclones are useful for collecting particles larger than 10 μm in diameter, and are still often used to knock out the large particles in gas streams entering scrubbing towers and electrostatic precipitators (ESPs), but since they are not effective on smaller particles, we will concentrate on the devices that actually collect the fine PM that is of environmental concern.

TABLE 13C.9   Types of Air Pollution Control Devices on Operating U.S. WTE Plants*

Technology	Number	ESP[†]	Spray dry	Wet scrubber	Cyclone	Dry sorbent injection	Fabric filter	SNCR for NOx	Carbon injection
Large mass burn	47	13	42	0	0	2	36	21	25
Small mass burn	22	9	9	1	0	7	13	4	3
Small modular	13	7	2	3	1	2	3	0	2
Large RDF	10	4	9	1	1	0	7	3	2
Small RDF	3	1	0	0	1	1	2	0	0
Large RDF combustor	4	1	2	0	0	1	3	0	0
Small RDF combustor	4	1	0	0	1	1	2	0	0
Total large	62	18	53	1	1	3	46	24	27
Total small	41	17	11	5	2	11	19	4	5

* 1999.
[†] electrostatic precipitation.
Source: IWSA (undated).

Venturi Scrubber/Packed Tower Scrubbers.   Medical waste and hazardous waste combustors are often served by venturi scrubbers followed by packed tower scrubbers due to their effectiveness in removing acid gases and organic vapors, and where solid, inorganic particulate matter is not a major contaminant in the flue gases.

In single-stage scrubbers, the flue gas reacts with an alkaline scrubber liquid to simultaneously remove HCl and SO_2. In two-stage scrubbers, a low-pH water scrubber for HCl removal is installed upstream of the alkaline SO_2 scrubber. The alkaline solution, typically containing calcium hydroxide ($Ca[OH]_2$), reacts with the acid gas to form salts, which are generally insoluble and may be removed by sequential clarifying, thickening, and vacuum filtering. The dewatered salts or sludges must then be disposed of in suitable landfills.

Quench Sections.   Quench sections are used to cool combustion products rapidly down to the temperatures at which emission control devices must operate to be effective. Water sprays are used for this process, evaporating the water and cooling the gases to close to the wet bulb temperature. Gases entering a secondary chamber exit at a temperature of 1800°F and are cooled to about 180°F, whereas those which have been cooled by a heat recovery boiler or heat exchanger will be cooled to about 130°F. Scrubbers that follow a quench section cool the gases close to the wet bulb temperature by providing more time and surface contact.

Venturi Scrubbers.   Venturi scrubbers use a converging duct section followed by a diverging section to accelerate and then decelerate the gas stream, while at the same time spraying water into the converging section of the scrubber. As the gases pass through the diverging section, most of the pressure drop lost in the converging section is recovered. The permanent pressure loss must be overcome by a fan that moves the gases through the system.

The water droplets, moving at a slower velocity than the gases, take significant time to pass through the venturi, while becoming targets for the particulate matter carried by the gases. The droplets thus absorb fine particulate, at the same time absorbing some acid gases, such as HCl and HF. As the gases pass through the diverging section, they slow up and the accelerated droplets of water continue to capture dust particles, while agglomerating into larger water droplets which can be dropped out and collected beyond the venturi, typically by cyclonic action.

The effectiveness of venturi scrubbers in collecting particulate matter depends mainly on the difference between the entering velocity and the throat gas velocity, which in turn depends upon the relative cross-sectional areas of the duct and the throat. The pressure drop measured between entrance and throat is directly dependent upon these areas and the corresponding velocities achieved.

Venturi scrubbers are usually followed by wet scrubbers or absorbing towers, of either packed tower or plate types. They usually have demisters to remove most of the droplets entrained by the gases, as seen in Fig. 13C.3. To meet more stringent standards, various add-on devices have been employed. A condensing section, using cold or cooled water, agglomerates the droplets and condenses them. HEPA (high-efficiency particle arrester) filters have been used to remove ultrafine particles (under 1 µm in diameter), especially salt and metal fumes.

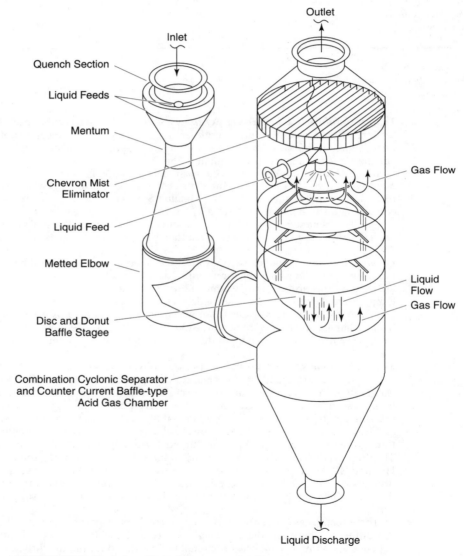

FIGURE 13C.3   Venturi scrubber with tray-type absorbing tower.

The diameters of wet scrubbers are sized so as to reduce the gas velocities to levels that allow droplets of water to fall. The depth of packing is determined by the desired approach to theoretical zero emissions. For removal of absorbable gases or vapors, the concept of "number of transfer units" is used, equivalent to adding heat transfer surface to heat exchangers to bring the exit temperature as close to the entering temperature of the cooling medium. The concentration of salts in the circulating liquid is a limiting factor, hence the rate of circulation, and the percentage of blowdown are essential design factors.

Collection Efficiency

Cyclones, venturis, wet scrubbers, and other particle collection devices are evaluated by collection efficiency, defined as the percentage of incoming particles that is removed. Collection efficiency varies with the size of the particles to be collected. Each particle size group has a different efficiency. Hence, to obtain the overall collection efficiency and the actual emissions quantity, a size distribution of the incoming particles is needed. It is necessary to use the efficiency for each size group, multiply this by the fraction of the incoming particulate in each size range, and then sum the quantities of each size in the emissions to obtain the total emissions. This process is illustrated in the following text.

The typical collection efficiencies of cyclones, venturi scrubbers, electrostatic precipitators (dry and wet), and fabric filters for a series of average particle sizes are listed in Table 13C.10.

TABLE 13C.10   Collection Efficiency of Particulate Matter Control Devices

| Particle diameter (microns) | Efficiency of particulate controls | | | |
	Cyclone	Venturi scrubber	Electrostatic precipitator (ESP)	Fabric filter
0.05			99	99.9
0.1			97	99.8
0.2			96	99.6
0.5		1	94	99.5
1		80	94	99.4
2	0.05	95	95	99.5
5	5	99	99	99.9
10	60	99.5	99.5	99.91
20	90	99.9	99.8	99.95

Source: Achternbosch and Richters (2000).

Collection Efficiency of Venturi Scrubbers.   The efficiency depends upon the particle size and the delta P, or pressure drop, as seen in Fig. 13C.4. At 10 in delta P the efficiencies fall severely with the smaller particle sizes. At 40 in the efficiency is seen to be 80 percent even with 0.25-μm particles. However, these scrubbers are not effective with smoke and fumes, including metal fumes, which are in the less-than-1-μm range. The U.S. EPA has placed emphasis on minus-2.5 μm due to the fact that they are readily inhaled. Because the efficiency is dependent on particle size, it is necessary to obtain the particle size distribution of the PM and perform a fractional efficiency analysis in order to determine the overall efficiency of the venturi for the specific size distribution.

Table 13C.11 shows a penetration calculation (penetration is efficiency minus 1) for a specific size distribution of particulate emissions from combustion. For each average particle size diameter measured by the instrument (a series of cyclones), there is a fraction of the total par-

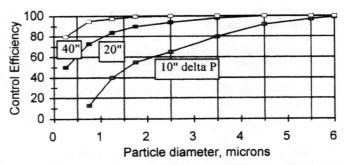

FIGURE 13C.4   Collection efficiency versus particle diameter and pressure differential, Δ P of venturi scrubbers.

ticle mass. For instance, 0.063 (6.3 percent) of the particles average 6 μm in diameter. For this size, the penetration is zero for delta P from 10 to 40 in. However, at a delta P of 10 in for particles of 1.5 to 2 μm, which are 16.5 percent of the total, the penetration is 35 percent. Adding the penetrations, for the 10-in case, the total is 0.259, for an efficiency of 100* (1 − 0.259) or 74.07 percent.

Emissions.   Assuming that the particulate concentration discharged by a combustion device or at the boiler exit is 1 grain per dry standard cubic foot (gr/dscf), the 74 percent efficiency of the 10-in venturi would result in an emission of (1 − .74)* 1.0 = 0.26 gr/dscf. The venturi with a 40-in pressure drop, however, would reduce the emissions to (1 − 989)* 1.0 = 0.011 gr/dscf, well within the limit of 0.015 gr/dscf, but would not be counted on to meet the standard of 0.010 gr/dscf.

Wet Scrubbers.   Wet scrubbers are highly effective in absorbing HCl and SO_2 from the gas stream. They employ a vertical bed of surface-generating shapes, or trays, through which the gases to be scrubber pass vertically, while water or recirculated solution enters from the top and contacts the gases. Their effectiveness depends upon the temperature at which the scrub-

TABLE 13C.11   Determination of Penetration and PM at Outlet of Venturi Scrubber

Particle diameter (μm)	Fraction in size range		Fractional efficiency = Pt (Dp)		
			Dp = 10 in	Dp = 20 in	Dp = 40 in
6	0.063	x	0 = 0	0 = 0	0 = 5–6
5–6	0.042	x	0.01 = 0.00042	0 = 0	0 = 0
4–5	0.077	x	0.03 = 0.00231	0 = 0	0 = 0
3–4	0.138	x	0.08 = 0.1104	0.02 = 0.00276	0 = 0
2–3	0.245	x	0.20 = 0.049	0.06 = 0.0174	0 = 0
1.5–2	0.165	x	0.35 = 0.05775	0.10 = 0.0165	0.005 = 0.0008
1.0–1.5	0.173	x	0.45 = 0.07785	0.16 = 0.02768	0.025 = 0.0043
0.5–1.0	0.087	x	0.60 = 0.0522	0.27 = 0.2349	0.05 = 0.00435
0–0.5	0.010	x	0.87 = 0.0087	0.50 = 0.005	0.20 = 0.002
Sum Pt* Dp *dPp = total penetration =			0.25927	0.09013	0.0115
1-total penetration = collection =			0.7407	0.91	0.989

Note: Pt stands for penetration.

ber is operated (usually close to the saturation or wet bulb temperature of the solution), the gas velocity, liquid to gas ratio, packing height (Buonicore and Davis, 1992).

Performance calculations of packed-bed are based on tests which determine the performance factors. These are best obtained from the manufacturer who performed the tests and who will guarantee the performance. Given a specific scrubber, with a set of performance conditions, it is possible to extrapolate its performance under different conditions of operation. For instance, when the number of transfer units (NTU) built into the scrubber is known versus a removal efficiency, the efficiency under different conditions can be calculated, or, the change in efficiency which would result from an increase in the NTU.

A scrubber used to remove SO_2 from a given concentration of 300 mg/cm to 30 mg/cm has a NTU as follows:

$$NTU = \ln [Y1(inlet)/Y2(outlet)] = \ln (300/30) = 2.3$$

The NTU required to reduce the SO_2 from 3000 to 30 would be 4.6.

Wet scrubbers perform as a function of the *difference* in concentration between the entering gas and leaving absorbing liquid (Y1), and the leaving gas and the entering absorbing liquid (Y2). This means that the concentration in the liquid is an important factor: the liquid absorbs the acid, accumulating it and increasing its concentration. To control concentration, a portion of the liquid must be "blown down" and discharged from the system. Achieving a high removal efficiency is obtained at the cost of blowing down and probably having to treat the blowdown to make it suitable for discharge to the sewer or publically operated treatment works (POTW). The ideal method would be to remove the products in a dry benign form, rather than diluted with water, increasing the cost or limitations of disposal.

Blowdown. *Blowdown* is needed to maintain stable concentrations.

Makeup = x = vapor + blowdown

Blowdown = y

Makeup concentration = 250 ppm

Blowdown concentration = 1000 ppm

If vapor = 1 gpm, $x = 1 + y$

Concentration balance: $x(250) = 1(0) + y(1000)$

hence $x = 4\ y$, or $y = x/4$

Then: makeup = $x = 1 + y = 1 + 0.25x$
    $x - 0.25\ x = 1 = (0.75)\ x = 1$

Then: $x = 1 / 0.75 = 1.33$ gpm
    $y = 1.33 / 4 = 0.33$ gpm

Thus, with a makeup of 1.33 gallons per minute (gpm), the blowdown would be 0.33 gpm, or 25 percent of the feed. The blowdown should be controlled by total solids content, so that it maintains a constant concentration in the scrubber water.

Electrostatic Precipitators. ***Electrostatic precipitators*** (ESPs) remove dry dust particles from gas streams by employing a series of high-voltage discharge electrodes to charge particulate matter, and attracting the charged particles to grounded metal plates parallel to the direction of gas flow, where they are collected and periodically cleaned by rapping the plates, causing the particles to fall off (see Fig. 13C.5). Although most of the dust is removed in the first stage, upon rapping, some of the collected PM becomes reentrained in the flue gas, so it

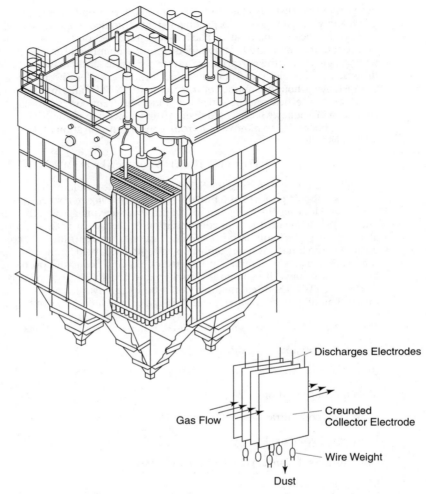

Discharges Electrodes

Gas Flow

Creunded
Collector Electrode

Wire Weight

Dust

FIGURE 13C.5   Three-field electrostatic precipitator, showing schematic of discharge electrode wires and grounded collector electrodes. (*Courtesy Research Cottrell.*)

is necessary to have two or more sets of charging wires and plates in series so that the next set can pick up the re-entrained particles. To meet present standards for PM emissions, ESPs are provided with four or more sets in series. The voltage which may be applied to each bank of the ESP is limited, since sparking, which occurs at some point, varies with the amount and nature of the particulate on the plates. The time intervals between rapping cycles, and the voltages which can be maintained for optimum performance, are variables which greatly affect the overall collection efficiency of the ESP (Buonicore and Davis, 1992).

Small particles have lower migration velocities than large particles and are therefore more difficult to collect. As compared to pulverized coal fired combustors, in which only 1 to 3 percent of the fly ash is generally smaller than 1 μm, 20 to 70 percent of the fly ash at the inlet of the PM control device for MWCs is reported to be smaller than 1 μm, requiring greater collection areas and lower flue gas velocities than required for many other combustion types.

The specific collection area (SCA) of an ESP is used as an indicator of collection efficiency. The greater the collection plate area, the greater the ESP's PM collection efficiency. The SCA is expressed as square feet per cubic feet per minute (square meters per cubic meter per minute) of flue gas, in effect giving the gas velocity. Most recent ESPs have SCAs in the range of 400 to 600 ft^2/1000 $ft^3 \cdot$ min. This area corresponds to a very low gas velocity of 0.4 to 0.6 ft/min, requiring very large-volume units. (See Table 13C.12.)

TABLE 13C.12   Electrostatic Precipitator Design Parameters

	Particulate	Acid-gas control
Particulate loading, gr/actual cubic foot (acf)	0.5–9	
Required efficiency, %	98–99.9	
Number of fields	3–4	
SCA, ft^2/1000 acfm	400–550	
Average secondary voltage, kV	35–55	
Average secondary current, mA/1000 ft^3	30–50	
Gas velocity, ft/sec	3.0–3.5	
Flue gas temperature, °F	350–450	230–300
Flue gas moisture, % vol.	8–16	12–20
Ash resistivity, ohm-cm	10^9–10^{12}	10^8–10^9

ESP Efficiency.   The efficiency of an ESP, which varies with the particle size, can be estimated by the Deutsch-Anderson equation:

$$\eta = 1 - \exp{(-Aw/Q)}$$

where η = Fractional collection efficiency
   A = Area of the collection plates, m^2
   w = Drift velocity of the charged particles, commonly ranging from 0.02 to 0.2 m/s, depending upon the particle diameter.
   Q = Flow rate of the gas stream, m/s

Typical ESP. A 400-ton-per-day MSW combustor would have an actual gas flow through the ESP of about 245,000 lb/h. At a temperature of 450°F, the volume may be estimated by multiplying the 13.5 ft^3/lb at standard conditions by the temperature correction [(450 + 460)/(460 + 70) = 1.717] to get 23 ft^3/lb. Thus, the actual gas flow would be 245,000[lb/h]*23[ft^3/lb]/60[min/h] = 93,900 ACFM. If the ESP has a specific area of 550 ft^2/1000 ft^3, the plate area would be 51,600 ft^2. This plate area would be divided into three or more fields.

Fabric filters (FF), also known as *baghouses,* are widely used for PM, metals, and acid gas control. They remove particulate matter by passing the flue gas through a large number of porous cylindrical fabric bags hanging vertically from a tube sheet. Particulate matter is collected on one of the bag surfaces, from which it is periodically removed and collected in hoppers. The cleaning process may use either reverse air, with the bags off-line, or pulse cleaning, with the bags either on- or off-line.

A fabric filter consists of 4 to 16 individual compartments that can be operated independently, and taken off-line for maintenance and/or cleaning. The collected particulate builds up on the bag, forming a filter cake. As the thickness of the filter cake increases, the pressure drop across the bag also increases. Once pressure drop across the bags in a given compartment reaches a set limit, that compartment is subjected to cleaning, either online (pulse cleaning), or off-line.

A fabric filter (baghouse) is shown in Fig. 13C.6. Dirty gases enter the enclosure containing the suspended bags, and flows through the bags to exit as clean gases, leaving the dust on

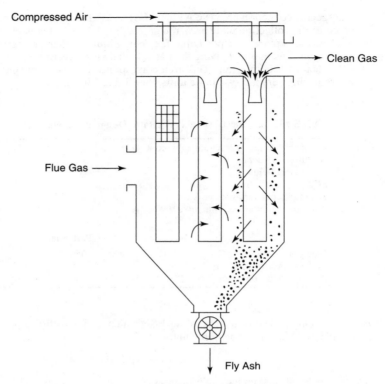

FIGURE 13C.6   Baghouse (fabric filter) module, showing tubular filter bags held by bag retainers; dust hopper; and sealing (rotary) valve through which the dust is discharged.

the outside of the bags. The dust is dropped from the bags after being loosened by pulses of air applied internally to the bags. Table 13C.13 lists fabric filter design parameters.

In *reverse-air* fabric filters, flue gas flows through the filter bags, leaving the particulate on the bags. Once the preset pressure drop across the filter cake is reached, air is blown through the filter in the opposite direction; the filter bag collapses; and the filter cake falls off and is collected in a hopper below.

TABLE 13C.13   Fabric Filter Design Parameters

	Type of fabric filter	
	Reverse air	Pulse jet
Operating temperature, °F	230–450	
Type of fabric	Woven fiberglass	
Fabric coating	10% Teflon B or acid resistant	
Fabric weight, oz/yd^2	9.5	16 or 22
Bag diameter, inches	8	6
Net air-to-cloth ratio	1.5–2.0:1	3.5–4.0:1
Minimum number of compartments	6	4
Overall pressure drop, in. w.g.	4–6	8–10
Estimated bag life, years	3–4	1.5–2

Source: Buonicore and Davis (1992); Kenna and Turner (1989).

In a *pulse-jet* fabric filter, compressed air is used, being pulsed through the inside of the filter bags to remove the particulate filter cake. The filter bag expands and collapses to its pre-pulsed shape, and the filter cake falls off and is collected in the hopper.

The bags are usually 6 to 8 in in diameter, with lengths ranging from 10 to 25 ft. Baghouses are generally provided with a weather-protection house above the tube sheet so that the bags can be removed and replaced. The air-to-cloth ratio determines the cost of the bags for a given gas flow. A range of A/C of 1.5 to 4.5 applies to baghouses used with municipal waste combustors. This ratio represents the velocity of the gas passing through the bags (i.e., 1.5 to 4.5 ft/sec). At least one baghouse module is provided so that one module can be taken off-line for cleaning and/or maintenance. At least four modules are generally used so that the transfer from four to three modules will not too severely affect the gas flow controls.

Filtering Area.   The total filtering area required for a given application is:

$$A_f = Q/v_f$$

where A_f = total filtering area, m^2
Q = Volumetric flow rate of gas stream, m^3/min
v_f = Filtering velocity (air-to-cloth ratio), m/min

For cylindrical bags, $A_b = \pi dh$

where A_b = Filtering area for each bag, m^2
π = 3.1416
d = Diameter of bag, m
h = Length of bag, m

The number of bags required is: $N = A_f/A_b$

Pressure Drop.   The pressure drop for a baghouse filter is best determined by experience. Normally design pressure drops range from 4 to 8 in of water.

Filter Cloth Area.   Given that the volume of dry gases at 300°F entering the baghouse is 32,000 CFM on a dry basis. If the moisture content of the gases is 13.6 percent, by volume, the actual volume is (1+.136)* 32,000 = 36,352 ACFM for each unit. At an air/cloth ratio of 4.5, the baghouse area is determined as follows:

Volume of contaminated air stream Q = 36,352 ft^3/min

Air/cloth ratio: A/c = 4.0 ft^3 min air / ft^2 bag area = 4.0 ft/min

Area of bags = 36,352 ft^3/min/4.0 ft^2 = 9100 ft^2

Area of each bag, 8-in diameter by 20 ft long = (8/12) in * 20 = 42 ft^2

Number of bags = 9000 / 42 = 220 bags. On 12-in/12-in pitch, 20 ft × 11 ft tube sheet

Acid Gas Control

Spray dryers (SD) are frequently used as the acid gas control technology for waste combustors. When used in combination with an ESP or FF, the system can control CDD/CDF, PM (and metals), SO_2, and HCl emissions. Spray dryer/fabric filter systems have become favored over SD/ESP systems due to more efficient metals removal. In the spray drying process, lime slurry is injected into the SD through either a rotary atomizer or through dual-fluid nozzles using steam or air for atomization. The water in the slurry evaporates to cool the flue gas, but before evaporating, the droplets absorb the acid gases where the acids react with the lime, to form calcium salts that can be removed by the PM control device. The SD is designed to provide sufficient contact and residence time to produce a dry product before leaving the SD adsorber vessel. The residence time in the adsorber vessel is typically 10 to 15 sec, resulting in very large vessel diameters. The particulate leaving the SD contains fly ash plus calcium salts, water, and unreacted hydrated lime (Teller, 1994).

The SD outlet temperature and lime-to-acid gas stoichiometric ratio (SR) are the key design and operating parameters that significantly affect SD performance. The term *stoichiometric* refers the chemically necessary (ideal) quantity of reagent needed to completely react with the acid in question. The outlet temperature must be high enough to ensure that the slurry and reaction products are adequately dried prior to collection in the PM control device. For MWC flue gas containing significant chlorine, the SD outlet temperature must be higher than about 115°C (240°F) to control agglomeration of PM and sorbent by calcium chloride. To provide a necessary safety margin, the outlet gas temperature from the SD is usually kept around 140°C (285°F). This temperature is controlled by the quantity of water sprayed into the gases. The acid gas concentrations are controlled by the quantity of lime added to the slurry.

A measure of performance of spray-dry scrubbers is the stoichiometric ratio, the molar ratio of calcium in the lime slurry fed to the SD divided by the theoretical amount of calcium required to completely react with the inlet HCl and SO_2. At a ratio of 1.0, the moles of calcium are equal to the moles of incoming HCl and SO_2. More than the theoretical amount of lime is generally fed to the SD, because of mass transfer limitations, incomplete mixing, and differing rates of reaction (SO_2 reacts more slowly than HCl). The stoichiometric ratio used in SD systems varies depending on the level of acid gas reduction required, the temperature of the flue gas at the SD exit, and the type of PM control device used. Lime is fed in quantities sufficient to react with the peak acid gas concentrations expected without severely decreasing performance. See Fig. 13C.7. The lime content in the slurry is generally about 10 percent by weight, but cannot exceed approximately 30 percent by weight without clogging the lime slurry feed system and spray nozzles.

Duct sorbent injection (DSI), involves injecting dry alkali sorbents into flue gas downstream of the combustor boiler outlet and upstream of the PM control device, and is effective in the control of acid gas as well as CDD/CDF and PM emissions from MWCS (Teller, 1994).

In DSI, powdered sorbent is pneumatically injected into either a separate reaction vessel or a section of flue gas duct located downstream of the boiler's economizer, or quench tower

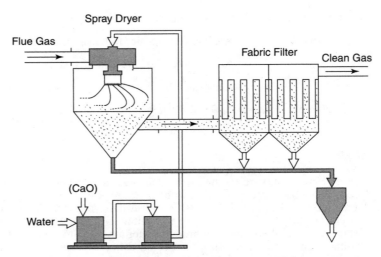

FIGURE 13C.7 Spray-dry absorber tower with baghouse, lime slurry supply system, fabric filter, and ash collection.

if no boiler is present. See Fig. 13C.8. Alkali in the sorbent (generally calcium, or sodium hydroxides, or sodium bicarbonate) reacts with HCl, HF, and SO_2 to form alkali salts (e.g., calcium chloride [$CaCl_2$], calcium fluoride [CaF_2], and calcium sulfite [$CaSO_3$]). By lowering the acid content of the flue gas, and not evaporating water into the gas stream, downstream equipment can be operated at reduced temperatures while minimizing the potential for acid corrosion of equipment. Solid reaction products, fly ash, and unreacted sorbent are collected with either an ESP or FF.

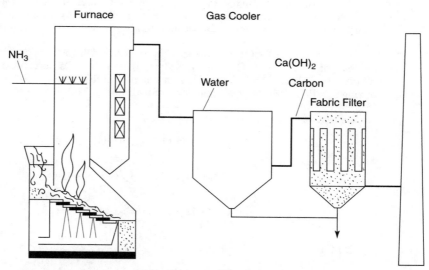

FIGURE 13C.8   Wet gas cooling followed by dry reagent injection with baghouse, lime slurry supply system, fabric filter, and ash collection.

Acid gas removal efficiency with DSI depends on the method of sorbent injection, flue gas temperature, sorbent type and feed rate, and the extent of sorbent mixing with the flue gas. Flue gas temperature at the point of sorbent injection can range from about 150 to 320°C (300 to 600°F) depending on the sorbent being used, the means of cooling the gases, and other aspects of the process. Sorbents that have been successfully used include hydrated lime ($Ca[OH]_2$), soda ash ($Na_2 CO_3$), and sodium bicarbonate ($NaHCO_3$). DSI systems can achieve removal efficiencies comparable to SD systems. Recirculation towers that increase residence time may be required to achieve comparable acid gas and sorbent efficiencies. Flue gas cooling by dry heat exchange or water injection, combined with DSI makes it possible to increase CDD/CDF and acid gas removal through a combination of vapor condensation and adsorption onto the sorbent surface (Teller, 1994).

Furnace injection has been employed to achieve a degree of acid gas control. The basic chemistry of furnace sorbent injection (FSI) is similar to DSI. Both use a reaction of sorbent with acid gases to form alkali salts. By injecting sorbent directly into the furnace (at temperatures of 870 to 1200°C (1600 to 2200°F) limestone can be calcined in the combustor to form more reactive lime, thereby allowing use of less expensive limestone as a sorbent. At these temperatures, SO_2 and lime react in the combustor, thus providing a mechanism for effective removal of SO_2. By injecting sorbent into the furnace rather than into a downstream duct, additional time is available for mixing and reaction between the sorbent and acid gases. Removing a significant portion of the HCl before the flue gas exits the combustor can reduce the formation of CDD/CDF in later sections of the flue gas breaching.

Alkaline Reagents

Various lime and related alkaline products are used in spray dry and dry injection scrubbers to react with the acid gases, HCl. SO_2, and HF to convert them into salts which can be collected by the particulate filter (Kilgroe and Licata, 1996).

Calcium oxide (CaO) is called "pebble lime" or "quick lime." Hydrated lime [$Ca(OH)_2$] is made from CaO by adding 32 percent by weight of water in a hydrator. $Ca(OH)_2$ is a powder having a mean particle size of 5 µm, and is highly reactive.

CaO is not very reactive with acid gases at the temperatures and conditions that exist in waste combustion facilities, and has to be converted to the hydrate form to reactive in scrubbing systems. CaO has been demonstrated to absorb acid gases in high temperature applications such as furnace injection.

CaO converts to $Ca(OH)_2$ in the slaking process, in which four parts of water are added to one part of CaO to form $Ca(OH)_2$ in a slurry that is about 25 percent solids. This conversion requires two phases that takes place in a slaker. The first phase is to convert the hydrate by mixing 3.96 lb of free water with one part of hydrate (1.32 lb) that results in a 25 percent slurry (5.28 lb).

The reactions are as follows:

$$CaO + H_2O \Longrightarrow Ca(OH)_2 + Heat$$
$$56 \qquad\qquad 18 \qquad\qquad 74$$

$$Ca(OH)_2 + SO_2 \Longrightarrow CaSO_3 + H_2O$$
$$74 \qquad 64 \qquad\quad 120 \qquad 18$$

$$Ca(OH)_2 + 2HCl \Longrightarrow CaCl_2 + 2\,H_2O$$
$$74 \qquad 73 \qquad\quad 111 \qquad 36$$

The capture ratio is 74 / 73 = 1.104

$$1 \text{ lb of CaO yields } 1.32 \text{ lb of } Ca(OH)_2$$

$$1.156 \text{ lb } Ca(OH)_2 \text{ captures } 1.0 \text{ lb } SO_2$$

$$1.04 \text{ lb } Ca(OH)_2 \text{ captures } 1.0 \text{ lb HCl}$$

The following emission factors are listed for a typical MWC facility:

$$SO_2 = 5.03 \text{ lb/ton (212 ppmv @ 7\% } O_2)$$

$$HCl = 7.03 \text{ lb/ton (532 ppmv @ 7\% } O_2)$$

The characteristic stoichiometric reaction of $Ca(OH)_2$ is:

$$5.03 \text{ lb } SO_2/\text{ton MSW} \times 1.156 \text{ lb } Ca(OH)_2/\text{lb } SO_2 = 5.815 \text{ lb } Ca(OH)_2/\text{ton MSW}$$

$$7.03 \text{ lb HCl/ton MSW} \times 1.104 \text{ lb } Ca(OH)_2 \text{ HCl} = 7.761 \text{ lb } Ca(OH)_2/\text{ton MSW}$$

$$\text{Total} = 13.575 \text{ lb/ton}$$

Pebble Lime

$$CaO + H_2O \Longrightarrow Ca(OH)_2$$
$$56 \quad 18 \qquad\qquad 74$$

$$CaO + H_2O \Longrightarrow Ca(OH)_2$$
$$56 \quad 18 \qquad\qquad 74$$

The characteristic stoichiometric reaction of $Ca(OH)_3$ is:

$$5.03 \text{ lb } SO_2/\text{ton MSW} \times 0.875 \text{ lb CaO/lb } SO_2 = 4.401 \text{ lb CaO/ton MSW}$$

$$7.03 \text{ lb HCl/ton MSW} \times 0.767 \text{ lb CaO/lb HCl} = 5.392 \text{ lb CaO/ton MSW}$$

$$\text{Total} = 9.793 \text{ lb/ton}$$

Since CaO contains about 7 percent unreactive material and inerts that are lost in the slaking process, the usage is adjusted to compensate: The adjustment is $1.07 \times 9.793 = 10.5$.

Due to the inability to provide absolute contact between the lime and the acid gases, more lime is required in the process. In addition, several unwanted chemical reactions take place that also use some undefined portion of the lime. For example, lime will react with carbon dioxide in the flue gas as follows:

$$Ca(OH)_2 + CO_2 = CaCO_2 + H_2O$$

The ratio of the actual amount of lime used to the theoretical amount required is called the *stoichiometric ratio*. A typical MWC equipped with a spray dryer and an ESP will require about 35 lb pebble lime per ton of MSW, while a MWC with a spray-dry baghouse will require about 20 lb/ton to meet the NSPS standards of 25 ppmv of HCl (a 93.5 percent reduction) and 30 ppm of SO_2 (an 85.8 percent reduction). The stoichiometric ratio for a plant with an ESP would be:

$$35 + 10.5 = 45.5 \text{ lb/ton MSW}$$

SO_2 removal using dry lime injection. The removal efficiency achieved at various gas temperatures at the dry scrubber, at SRs from 1.0 to 2.5, is shown in Fig. 13C.9. It is apparent that reducing the gas temperature and/or increasing the SR increase the removal efficiency. Meeting an efficiency of 80 percent requires reducing the temperature to 285°F at a SR of 1.5, for instance. Using heating surface rather than water injection produces a substantial increase in heat recovery (Finnis, 1998).

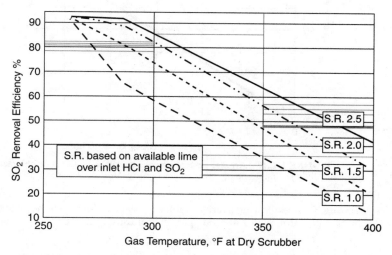

FIGURE 13C.9 SO_2 removal efficiency by dry scrubbing at various gas temperatures and stoichiometric ratios (S.R.). (*Source: Finnis, 1998.*)

Sodium Bicarbonate Injection After a Municipal Waste Combustor.   The refuse combustion unit tested consisted of two identical lines working in parallel, which merge before the common electrofilter. Each line consisted of a combustor, a tubular reactor (residence time 5 sec), and a cyclone. The bicarbonate was pulverized and then distributed equally into the two lines. The bicarbonate was injected into the exit of a venturi tube placed at the entrance to the reactor. The solid residues were trapped by the electrofilter (Maziuk, 1998).

The waste combustion plant treating 5 t/h of refuse produced 85 m³/h water at 180°C under a 15-bar pressure, corresponding to a power of 8.6 MW. Flue gas flow was 28,000 Nrn³/h. The heterogeneous nature of the fuel prevented determining in advance the exact bicarbonate requirements. The injection rate varied between 50 and 150 kg/h.

Stoichiometric Ratios.   For removal of SO and HCl, the following global reactions apply:

$$NaHCO_3 + HCl \rightarrow NaCl + H_2O + CO_2$$

$$2\,NaHCO_3 + SO_2 + \frac{1}{2}\,O_2 \rightarrow Na_2SO_4 + H_2O + 2\,CO_2$$

In other words, $84/36.5 = 2.30$ kg of $NaHCO_3$ are required to remove 1 kg of HCl, and $2 \times 84/64 = 2.625$ kg of $NaHCO_3$ are required to remove 1 kg of SO_2. Depending upon the percent of acid gas to be removed, an amount more than the stoichiometric reagent is needed. Figure 13C.10 shows the percent removal versus stoichiometric ratio for HCl and SO_2, based on tests in a facility employing dry injection of sodium bicarbonate (nahcolite).

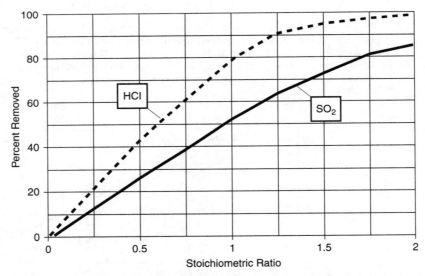

FIGURE 13C.10   Average percent removal of acid gases versus stoichiometric ratio, from test of dry injection of sodium bicarbonate. (*Source: Maziuk, 1998.*)

Testing.   The following average values for the two lines together were measured at the bicarbonate injection points at a flue gas temperature of 225°C:

HCl content: 880 mg/Nm³ (dry) Å 28%

SO₂ content: 143 mg/Nm³ (dry) Å 34%

HF content: 6 mg/Nm³ (dry) Å 36%

During tests the composition of the flue gases to be purified fluctuated widely. A range of stoichiometric ratios between 0.9 and 2.03 was covered. On leaving the reactors and cyclones, the gases were already 75 percent purified compared with levels at the electro-filter outlet.

Other alkaline reagents have been used, including trona, containing calcium and magnesium carbonates.

Nitrogen Oxides Control

Nitrogen oxides emissions derive from two sources: the fuel, and conversion of nitrogen in the air. The conversion from nitrogen takes place in the flame, and depends upon the flame temperature (McDonald et al., 1993).

The control of NOx emissions can be accomplished through either control of the combustion process, injection of reactants, or the use of add-on controls. Combustion controls include use of refractory furnaces (without waterwall cooling), staged combustion, low excess air (LEA), and flue gas recirculation (FGR). Add-on controls which have been used on MWCs include selective noncatalytic reduction (SNCR), selective catalytic reduction (SCR), and natural gas reburning.

Combustion controls involve the control of temperature or O_2 to reduce NOx formation. With LEA, less air is supplied, which lowers the supply of O_2 that is available to react with N_2 in the combustion air. In staged combustion, the amount of underfire air is reduced, which generates a starved-air region.

In FGR, cooled flue gas and ambient air are mixed to become the combustion air. This mixing reduces the O_2 content of the combustion air supply and lowers combustion temperatures. Due to the lower combustion temperatures present in MWCS, especially in refractory furnaces, most NOx is produced from the oxidation of nitrogen present in the fuel.

Selective noncatalytic reduction (SNCR), also known as Thermal deNOx, is the most common method used for reduction of NOx. Injection of ammonia or urea into the furnace in a region having the optimum gas temperatures achieves the optimum reduction. With SNCR, ammonia (NH_3) or area is injected into the furnace along with chemical additives to reduce NOx to NO_2 without the use of catalysts. MWCs equipped with SNCR, have achieved NOx reductions of about 45 percent. Figure 13C.11 shows a typical installation with three levels of injection in the furnace.

With selective catalytic reduction (SCR), ammonia (NH_3) is injected into the flue gas *downstream* of the boiler where it mixes with NOx in the flue gas and passes through a catalyst bed, where NOx is reduced to NO_2 by a reaction with NH_3. Reductions of up to 80 percent have been observed, but problems with catalyst poisoning and deactivation reduce performance over time.

Natural gas reburning involves limiting combustion air to produce a low excess air (LEA) zone. Recirculated flue gas and natural gas are then added to this LEA zone to produce a fuel-rich zone that inhibits NOx formation and promotes reduction of NOx to NO_2. Natural gas reburning has achieved NOx reductions of 50 to 60 percent.

Table 13C.14 compares the NOx emissions from three California MSW combustion systems equipped with thermal deNOx.

At the Commerce WTE facility, a study evaluated the two injection levels installed, carrier air injection pressure and ammonia injection rate. The study concluded that optimum performance was achieved by injection of an NH_3-to-NOx mole ratio of about 1.5 through the upper elevation of nozzles. Even when there was substantial ammonia slip at the economizer exit, the level at the stack due to the spray dryer-baghouse was held to less than 5 ppm. The lower levels achieved by SERRF are due to a higher rate of ammonia injection as compared with Commerce and Stanislaus. The higher ammonia injection rate also explains the higher ammonia slip numbers, which become the limiting factor in NOx control by this method (Kilgroe and Licata, 1996; U.S. EPA, 1989).

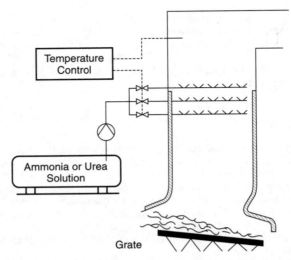

FIGURE 13C.11 Diagram of nitrogen oxide (NOx) control system having SNCR, using ammonia or urea injection at three levels.

Mercury and Dioxins Control

Dry activated carbon (DAC) has been proven to be effective in removing dioxins and mercury from the flue gases from municipal waste and medical waste combustion systems. Test data gives an indication of the quantity of DAC required in relation to the concentrations of mercury and/or dioxins present upstream of the point of injection. This quantity is affected by the temperature of the gases and the retention time before the gases enter the particulate collection device, such as ESP or fabric filter. DAC may be injected into the flue gas duct as dry powder, separately or combined with dry lime or other reagents (Teller, 1994; Licata and Hartenstein, 1998).

Sodium sulfide has been used for mercury control, using DSI/FF as the APC (Andersson and Weimer, 1991). Aqueous sodium sulfide will react with mercury to form solid HgS that can be collected in the PM control device. In this process, a dilute Na_2S solution is injected into cooled gas ($<400°F$) prior to injection of hydrated lime used to control the acid gases. Mercury in the gas initially is absorbed by the solution droplet. As the droplet evaporates, HgS and sodium salts precipitate. Feed rates of Na_2S vary from 0.05 to 0.5 kg/Mg (0.1 to 1 lb/ton) of MSW, depending on site-specific conditions such as the amount of mercury in the flue gas, the level of control required, and the level of carbon in fly ash (which also absorbs mercury).

TABLE 13C.14 NOx Emissions from MWCs with Thermal DeNOx

Emissions	Commerce	Stanislaus		SERRF		
ppm @ 7% Oxygen		Unit 1	Unit 2	Unit 1	Unit 2	Unit 3
Uncontrolled NOx	128–217	298	318	—	210	259
Controlled NOx	104	93	112	49	72	54
Ammonia slip	~2	3.7	5	—	—	35

Source: (McDonald et al., 1993).

Tests of four WTE facilities with different types of emission control systems, shown in Table 13C.14, provide comprehensive data obtained when the input and output concentrations were measured simultaneously along with operating conditions. Such data are normally only available from governmentally-sponsored tests.

Sodium tetrasulfide (Na_2S_4), now in use as an additive for inexpensive control of mercury emissions, offers an opportunity to bring most MWCs into compliance with Hg emissions limits (Licata et al., 2000). In February 1999, Germany reduced the Hg emissions standard for MWCs to a daily average of 30 $\mu g/Nm^3$ at 11 percent O_2, based on the use of continuous emissions monitors (CEMs). In the United States, several states have proposed or adopted 28 $\mu g/Nm^3$ or 85 percent reduction, whichever is less restrictive. In view of the concern that these standards might not be achievable solely by the use of activated carbon injection, alternatives were investigated, and Na_2S_4 was found to offer promise, since it can capture both ionic $HgCl_2$ and Hg° in accordance with the following simplified reactions:

$$Na_2S_4 + HgCl_2 \leftrightarrow HgS + 2NaCl + 3\ S^\circ$$

and

$$S^\circ + Hg^\circ \leftrightarrow HgS$$

Since the Na_2S_4 solution can be injected into the flue gas duct, it can easily be retrofitted to an existing flue gas cleaning plant. The Na_2S_4 reacts with the mercury to form mercury sulfide (HgS). The red allotrope is known as cinnabar. It is a nonpoisonous insoluble salt that is thermally state up to 400°C, and thus effectively immobilizes the mercury by chemical binding. Pilot plant tests showed that at a dose rate of 80 mg/Nm^3 of Na_2S_4 and an average inlet Hg of 148 $\mu g/dscm$ at 7 percent O_2 the average outlet was 26 $\mu g/Nm^3$, at 82.4 percent removal efficiency. At a dose rate of 120, and an inlet of 360, the outlet was 24, at 93 percent removal. With simultaneous injection of activated carbon, the total activated carbon feed rate could be reduced from 230 to 6/57 mg/dscm resulting in a reduction in activated carbon usage to 18 lb/h. The author calculated that for an 800 TPD MWC, the annual cost of activated carbon would be $320,000. The Na_2S_4 solution would cost $176,000 per year, saving $144,000. These tests were at a facility using an ESP. For facilities having FF, the technology should be even more effective due to the additional retention and contact time (Licata et al., 2000).

Dioxin Control and Reduction by Catalytic Filter.   The latest development in dioxin control is the catalytic fabric filter system developed by W. L. Gore, called Remedia (Fritsky et al., 2000). The new system consists of a Gore-Tex membrane laminated to a catalytically active felt. The felt is composed of chemically active fibers containing a variety of specially produced catalysts. As gases pass through the felt, a catalytic reaction is induced and dioxins/furans are decomposed into harmless gaseous components. The temperature range required for catalytic reaction is as low as 285°F (140°C) to 500°F (260°C). A minimum temperature of 356°F (180°C) is preferred. These temperatures can be achieved at the boiler outlet, without the need for evaporative water cooling.

Tests at a MSW combustor in Roeselare, Belgium, having an ESP followed by dry lime injection and a baghouse, showed that the particulate level was at or below the detection level of 0.2 mg/Nm^3 at 11 percent O_2. The existing baghouse system using powdered activated carbon had a total release of PCDD/F of 69 g/ton municipal waste, whereas with the catalytic filter (no carbon) it was 4.7 g/ton. The catalyst filter system not only decreased the gaseous dioxin/furan emissions by more than 99 percent but also decreased the particulate phase dioxin/furan emissions by more than 93 percent. In other words, the PCDD/F were largely destroyed, and not found in the fly ash, essentially eliminating concern about depositing dioxin-laden fly ash in a landfill.

Tests at a 135-ton/day medical waste combustor in the United States showed that the raw gas from the boiler had a PCDD/F concentration of 2.57 ng/Nm^3 TEQ at 11 percent oxygen, whereas the outlet concentration was 0.042 ng/Nm^3, a 98 percent reduction. A 100-ton/day municipal waste combustor in Japan having a dry lime scrubber and baghouse reported reduction of PCDD/F from 3.536 ng/Nm^3 to 0.011, (99.7 percent reduction) after seven months of operation, and with particulate emissions below 1 mg/Nm^3.

Wet Scrubber with Granular Carbon Bed.   Due to the fact that medical waste incinerators are often operated intermittently, which is unfavorable for fabric filters, wet scrubbers may be more desirable. In order to meet regulatory requirements, it may be necessary to add downstream controls which can capture the fine particulate matter, dioxins, and toxic metals. The New East Carolina University MWI has installed a granular carbon bed for this purpose (Sanders et al., 2000).

The requirement for more stringent mercury control has resulted in achieving further reductions in dioxin levels, generally far below regulatory standards.

Uncontrolled mercury levels prior to APCDs ranged from as low as 75 to as high as 1500 µg/dscm, and averaged roughly 650 µg/dscm. The high variability of mercury levels entering and leaving the APC make it difficult to obtain anything better than a range of removal efficiency. (See Table 13C.15.)

Achievable Emission Limits and Averaging Times

Due to the high variability of mercury in the waste and hence in the uncontrolled and controlled emissions, it is necessary to analyze the data statistically in order to make estimates of the future performance of the control system (White et al., 1992), and to decide on the amount of carbon which should be injected on a continuous basis. On the basis of extensive tests at Stanislaus, it appears that the decision of whether to assess compliance based on the average of multiple one-hour tests or on one multiple-hour test is arbitrary from a statistical viewpoint (White et al., 1992).

TABLE 13C.15   Summary of APC Performance in Removing Mercury

Facility	Carbon addition rate (mg/dscm)	Operating temperature (°F)	Inlet Hg level (µg/dscm)	Outlet Hg level (µg/dscm)	Removal efficiency (%)
Stanislaus MWC (SD/FF)	183672	285	400–700	300–550	20–40
			500–650	100–200	65–85
			350–1200	40–140	70–90
			500–1200	30–60	92–96
	Sodium sulfide injection rate (kg/h)		Inlet Hg level (µg/dscm)	Outlet Hg level (µg/dscm)	Removal efficiency (%)
Burnaby MWC (DSI/FF)	236	400	670	84	87
			1200–1500	470–750	50–60
			660–780	90–105	85–90
			Inlet Hg level (µg/dscm)	Outlet Hg level (µg/dscm)	Removal efficiency (%)
Fergus Falls, MN—MWC, Wet Scrubber			600	6–50	92–99
Basel, Switzerland MWC—Wet Scrubber	Unit 1 Unit 2		170–510 75–360	15–20 <15–30	90–96 88–94

Source: Gleiser et al. (1993).

Complete Combustion/Emission Control Systems

Waste combustion systems consist of the combustion system plus the emission control system, operating as a unit, operated by a centralized control system, and having the required instrumentation and continuous emissions monitors. Figure 13C.12 shows a cross section of a modern waste-to-energy system. Following the grate furnace is an SNCR injection point where ammonia or urea is sprayed into the combustion chamber, after which the gases pass through the boiler, reactor, or cooling tower (not numbered), fabric filter, and fan. Not shown is the activated carbon (or coke) injection used to control dioxins and mercury. Up to this point, this facility is typical of WTE systems installed in recent years in the United States. In European systems, additional acid gas controls are typically added. The gases pass through HCl and SO_2 scrubbers before passing to the stack. The stack gases are monitored by the continuous monitors located in the emission control room. European systems are typically located in downtown heavily populated areas where essentially complete removal of the acid gases is required. They differ from systems which meet the requirements of the U.S. EPA, which control HCl and SO_2 adequately without the need for final reduction of these acid gases.

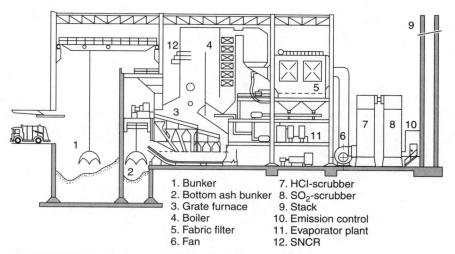

1. Bunker	7. HCl-scrubber
2. Bottom ash bunker	8. SO_2-scrubber
3. Grate furnace	9. Stack
4. Boiler	10. Emission control
5. Fabric filter	11. Evaporator plant
6. Fan	12. SNCR

FIGURE 13C.12 Emission control system for municipal waste combustor with selective noncatalytic NOx reduction (SNCR) control in furnace, spray-dry scrubber with lime slurry injection, fabric filter, shown with additional HCl and SO_2 wet scrubbers typically employed in European facilities. (*Source: Achternbosch and Richters, 2000.*)

Figure 13C.13 shows a similar WTE system in which a water-spray cooling tower and injection of dry alkaline reagent ($Ca(OH)_2$, sodium bicarbonate or trona) is used to react with HCl and SO_2 rather than a spray-dry reactor after the gases are cooled. Dry activated carbon is used before the fabric filter, to remove dioxins and mercury (WASTE, 1993).

Figure 13C.14 shows an upgraded municipal solid waste combustion system with more complex wet flue gas cleaning system reflecting older systems originally having only an ESP. The retrofit added a spray-dry scrubber, an additional ESP to protect downstream devices from particulate matter, wet HCl and SO_2 scrubbers, ammonia injection before the selective catalytic reduction tower, and a coke adsorber tower. The ESPs are needed to reduce particulate matter that would contaminate the towers.

1. Maneuvering apron
2. Receiving hall
3. Maintenance bay
4. Refuse bunker
5. Grabbing crane
6. Feed chute
7. Grate
8. Ash discharger

9. Ash bunker
10. Boiler
11. Superheater
12. Economizer
13. Conditioning tower
14. Reactor (lime injection)
15. Fabric filters
16. Stack

FIGURE 13C.13   Municipal solid waste combustion system with selective noncatalytic NOx reduction (SNCR) control in furnace, water-spray cooling tower after the boiler, followed by dry reagent injection, reactor tower, and fabric filter.

Comparison of Different Systems.   Different flue gas cleaning systems for MSW combustion systems have been compared by Achternbosch and Richters (2000). Their focus was on chlorine, sulfur, and mercury, and the investment costs for alternate emission control systems. The amounts of residues are compared. Their conclusion is that the least costly systems are those common in the United States, employing spray-dry scrubbers and fabric filters, as com-

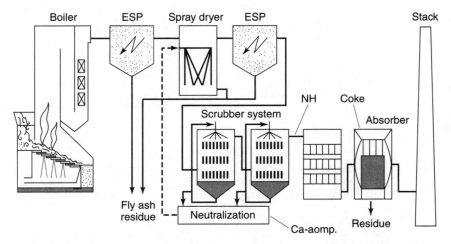

FIGURE 13C.14   WTE facility originally equipped with an ESP, upgraded by adding a spray-dryer, HCl and SO_2 scrubbers, ammonia injection prior to tower (SCR) to remove NOx, and an adsorber using coke to remove mercury and dioxins. (*Source: Achternbosch and Richters, 2000.*)

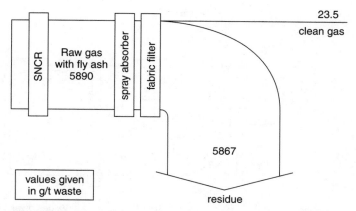

FIGURE 13C.15   Chlorine balance of the plant in Fig. 13C.12. (*Source: Achternbosch and Richters, 2000.*)

pared with the systems in Germany, employing additional wet scrubbers for control of HCl and SO$_2$. For the U.S. systems to meet the more stringent acid gas control requirements of European countries, wet scrubbers may be preferred, since without them the consumption of lime or other reagents is higher due to less efficient use of the reagent, and a corresponding increase in the cost of reagent and disposal of residues. In other words, both systems can meet the same standards of chlorine and sulfur control, but the simpler U.S.-type system may be less expensive in capital cost, but more in operating costs due to increased use of reagent chemicals, and the cost of disposal of the increased residues. A major difference is that in Europe landfills are scarcer and disposal is more costly, justifying the use of the chemically more efficient wet scrubbers for sulfur and chlorine acid gases. Figures 13C.15 to 13C.17 show chlorine balances for three plants for comparison.

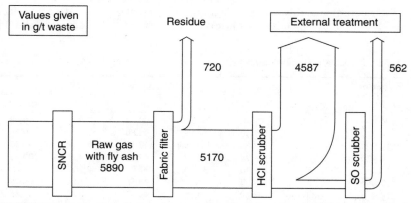

FIGURE 13C.16   Chlorine balance of the plant in Fig. 13C.13. (*Source: Achternbosch and Richters, 2000.*)

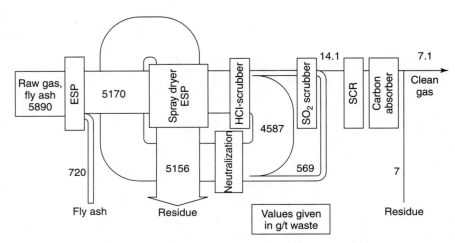

FIGURE 13C.17   Chlorine balance of the facility in Fig. 13C.14. (*Source: Achternbosch and Richters, 2000.*)

13C.5   CONTROLLED AND UNCONTROLLED EMISSION FACTORS

Emission Factors

The emission factors (EF), defined as the pounds of a pollutant per ton of waste, in pounds per million tons, or grams/tonne, or pounds per million Btu, etc., provides convenient reference numbers for tabulating stack test data, and hence for use in predicting emissions from facilities having similar combustion and emission control configurations.

Table 13C.16 is a compilation published by the U.S. EPA in AP-42, providing EFs for uncontrolled emissions, and emissions controlled by ESPs, spray-dryer + ESP, dry sorbent injection + FF, and spray-dry scrubber + FF (U.S. EPA, 1996a). The numbers in this table have been calculated by assuming uncontrolled emissions considered to be typical and controlled emissions from selected facilities. It is important to note that the uncontrolled emissions have not been measured at the same facility as the controlled emissions. While the efficiencies determined this way may be useful for general guidance, data from facilities where input and output of the emission controls were measured at the same time should be more reliable. The

TABLE 13C.16   Particulate Matter, Metals and Acid Gas Emission Factors—AP-42 (Pounds per million tons of MSW)

	No control	ESP control		SD/ESP control		DSI/FF control		SD/FF control	
	E.F.*	E.F.	Effy. %[†]	E.F.	Effy. %	E.F.	Effy. %	E.F.	Effy. %
PM	25,100,000	210,000	99.16	70,300	99.72	17,900	99.93	62,000	99.75
As	4,370	21.7	99.50	13.7	99.69	10.3	99.76	4.2	99.90
Cd	10,000	646	93.54	75.1	99.25	23.4	99.77	27.1	99.73
Cr	8,970	113	98.74	259	97.11	200	97.77	30	99.67
Hg	4,790	6,620	−38.20	3,260	31.94	2,200	54.07	2,200	54.07
Ni	7,850	112	98.57	270	96.56	143	98.18	52	99.34
Pb	213,000	3,000	98.59	915	99.57	297	99.86	261	99.88

* EF = emission factor, pounds per million tons of MSW.
[†] Control efficiencies calculated from uncontrolled emissions of various other WTE facilities.
Source: U.S. EPA (1996).

measurements and efficiency calculations shown in Tables 13C.19 and 13C.21 were obtained at the same facility, hence are more reliable.

AP-42 also expresses emission factors alternatively in µg/standard cubic meter. The conversion factor is 8.06 lb/million tons per µg/dscm, based on the assumption that the heating value of the waste was 4500 Btu/lb. The value in µg/cm was based on actual stack measurements, converted to 7 percent oxygen or 12 percent CO_2. This calculation is, therefore, for illustration purposes only. Individual comparisons, while more accurate, would show the same general tendencies. This calculation is, of course, for illustration purposes only. Individual comparisons, while more accurate, would show the same general tendencies, from which the weight units are calculated.

Table 13C.17 shows, for general guidance, uncontrolled emission factors obtained by averaging a number of EPA tests of then-existing medical waste incinerators.

TABLE 13C.17   Uncontrolled Emission Factors for General Medical Waste

Pollutant	Uncontrolled emission factors (µg or mg/kg waste)
µg/kg waste	
Dioxin/furan	32
Cd	2000
Pb	28,600
Hg	25,500
Cr(total)	422
Cr (VI)	32
Ni	<124
Fe	4780
Mn	245
As	118
mg/kg waste	
CO	2500
NOx	1350
SO_2	566
PM	3,000
HCl	11,000
Benzene	1300

Source: (Walker and Cooper, 1992).

Tracing Metals from Waste to Emissions

It is important to have a clear perspective regarding the fate of air pollutants, especially the heavy metals, starting from concentrations in the waste, in the combustion process, before and after the air pollution control systems, and as concentrations in the stack gases, as they are diluted in the atmosphere before the metals reach the ground. From such analyses the metals concentrations in the stack may be compared with regulatory standards, to see whether they comply with the standards, and with what safety factor. Finally, after dilution before they reach ground level, as estimated by modeling studies, comparisons may be made with health standards and with "acceptable ground level concentrations" established by health authorities and amended by state regulatory authorities.

Reduction and Partitioning of Metals.   Only a few comprehensive studies have been made to measure the "partitioning" of metals from the waste to the various streams leaving the process. Such studies are costly and difficult to run, especially since the samples taken for analysis should be taken as near simultaneously as possible. Three studies are cited in the paragraphs that follow, the first for a European facility having an ESP, followed by a scrubbing tower, the second for a Canadian facility having a dry-sorbent injection system and baghouse, and the third for a California facility having a spray-dry scrubber and baghouse.

Example 1:   A study by Sorum et al. analyzed the MSW, bottom ash, ESP ash, scrubber filter cake, and drain water, as well as the flue gas leaving the stack. This facility has an ESP followed by a washing tower (wet scrubber). Table 13C.18 summarizes the results for lead, cadmium, and mercury (Sorum et al., 1997).

TABLE 13C.18   Distribution of Metals Discharges from MSW Combustion System

Sampling point	Annual emissions, kg/yr		
	Lead	Cadmium	Mercury
Waste feed	38,060	434	139
Bottom ash	359,508	272	9
ESP fly ash	1871	104	3
Scrubber filter cake	146	28	121
Stack gas	92	33	7
Drain water	0.80	0.02	0.80
Stack concentration:	0.162 mg/m^3	0.057 mg/m^3	0.06 mg/m^3
Background conc.	0.1 µg/m^3	0.01 µg/m^3	
Ratio: stack/background	620	175	

Based on ground-level concentrations measured in the New York/New Jersey metropolitan area.

Source: NYSDEC (1993).

The bottom ash contained 35,950/38,060 = 94.4 percent of the lead. The emissions were 0.2 percent of the MSW, or, in other words, the lead in the MSW was reduced by 99.76 percent, or by a factor of 413. Likewise, the 62.6 percent of the cadmium stayed in the bottom ash, 24 percent was removed in the ESP ash, 6 percent was in the filter cake, and 7.6 percent was emitted from the stack. Most of the mercury (87 percent) was collected in the scrubber filter cake, and only 5 percent was emitted.

The flow diagram of the annual lead emissions can be shown as follows:

38,000 kg/year [MSW] → [Combustor] → [ESP] → [Scrubber] → Stack: 92 kg/year

Lead   ↓   ↓   ↓   Emissions

35,950   1871   146 kg/year

Ash residue   Fly ash   Filter cake

The stack emissions for lead are therefore 92/38,000 or 1/413 of the input, or, stated another way, the removal efficiency is 99.76 percent.

Example 2:   The complete mass-balance of the Burnaby 240 tonnes per day WTE facility having a dry lime injection scrubber and baghouse. Table 13C.19 gives a numerical demonstration of the factors by which the metals are reduced, and the efficiency of capture by the emission controls, as measured in extensive tests of the Burnaby facility. With the exception of the highly volatile mercury, of which over 50 percent reached the stack, the other metals were reduced by factors from 20 times for vanadium to as much as 5000 times for lead. Overall reduction of metals from 4,700,000 to only 4720: a 99.4 percent reduction. Individual APC

TABLE 13C.19 Reduction and Partitioning of Metals in Municipal Waste Measured at Burnaby

Metal	Metal in waste (lb/Mt)	Boiler emissions (lb/Mt)	Boiler reduction waste/boiler out	Stack emissions (lb/Mt)	Overall reduction waste/stack	APC reduction boiler out/stack	APC control efficiency (%)
Mercury	3630	3630	1	1,934	2	2	46.7
Boron	222,000	7496	30	1,370	162	5	81.7
Zinc	3,746,000	249,860	15	725	5167	345	99.7
Lead	326,000	21,681	15	363	898	60	98.3
Nickel	33,000	1612	20	105	314	15	93.5
Chromium	185,000	2821	66	97	1907	29	96.6
Tin	98,000	1120	88	31	3161	36	97.2
Cadmium	27,000	5723	5	18	1500	318	99.7
Arsenic	15,800	1048	15	11	1436	95	99.0
Selenium	9600	81	119	10	960	8	87.7
Vanadium	40	40	1	2	20	20	95.0
Copper	28,400	14,508	2	54	526	269	99.6
Totals	4,694,470	309,620		4720			98.5

Source: Rigo and Chandler (1994).

control efficiencies for the critical metals, lead, and cadmium were 98.3 and 99.7 percent, respectively. Of the 4,694,470 lb/Mt of the metals measured in the waste, 94 percent was collected as ash residue; only 309,620 or 6 percent passed through the boiler to the emission control; and only 4720 or 1.5 percent of this were emitted to the stack (Rigo and Chandler, 1994).

Example 3: Tests of the 400 ton-per-day Commerce WTE facility, having a spray-dry scrubber and fabric filter (baghouse) provided data for Table 13C.20. Stack emissions are compared with boiler outlet concentrations to obtain control efficiencies ranging from 91 percent for mercury to 99.99 percent for lead. The lead entering the emission controls were 1133 times the stack emissions. The emission factors, in pounds per million tons, are also shown. One million tons represents about seven years of operation. The removal efficiency of the emissions of measured metals was found to be $(684,756 - 5,962)/684,756 = 99.1$ percent. The boiler emissions are thus 115 times the stack emissions.

Emissions from WTE versus Fossil Fuels

Table 13C.21 gives a comparison between the emissions from waste-to-energy facilities and those from fossil-fuel-fired utility boilers. Note that based on equivalent electric power generation, WTE facilities generally have much lower emissions than those of the fossil fuels (not including distillate oil and gas). This calculation is, of course, for illustration purposes only. Individual comparisons, while more accurate, would show the same general tendencies.

Recycling and Pollution Prevention

Recycling and pollution prevention can reduce the quantity of metals in the waste stream, and reduce the discharge of these metals to the ash residue and to the stack. The first step in investigating the affect of reductions in the waste is to obtain an analysis of the metals content of all of the components in the waste, after which the degree of reduction which can be expected can be determined. Table 13C.22 lists the components of MSW analyzed at Burnaby, and the concentrations of the metals in each component (Rigo and Chandler, 1994).

TABLE 13C.20   Heavy Metals Collected and Emitted by Commerce Resource to Energy Facility

	Boiler emissions µg/Nm³	Stack emissions µg/Nm³	Control effy %	Collected lb/Mton MSW	Range of Emitted lb/Mton MSW	Range of AP-42 lb/MTon MSW
Magnesium	89,933	270	>99.70	89,663	<2,160	
Barium	4695	117	97.51	4578	936	
Silicon	1860	66	96.45	1794	528	
Calcium	193,000	56	99.97	192,944		448
Copper	8818	54	99.39	8764	<432	9–153
Iron	84,167	54	99.94	84,113	<432	
Mercury	475	41	91.28	434	331	113–3460
Zinc	90,933	38	99.96	90,895	308	90–420
Aluminum	178,000	16	>99.99	177,984	<130	
Molybdenum	522	12	>97.61	510	<100	
Nickel	4240	6	99.85	4234	50	2–258
Selenium	84	2.7	>96.76	81	<22	1–8
Chromium	3620	2.3	99.94	3618	19	1–210
Tin	800	2	>99.75	798	<16	
Cadmium	1680	2	99.88	1,678	16	3–145
Lead	18,133	2	99.99	18,131	16	8–230
Manganese	3235	1	99.97	3234	8	4–129
Cobalt	111	0.3	99.69	111	3	
Antimony	822	0.3	>99.96	822	<2	1–23
Beryllium	7	0.2	>97.24	7	<2	0.01–4
Bismuth	31	0.16	>99.49	31	<1	
Arsenic	78	0.16	>99.79	78	<1	
Vanadium	257	0.09	99.96	257	1	
Total	685,501	745.2	99.89	684,756	5962	

Source: Teller (1994).

TABLE 13C.21   Comparison of Emissions from WTE Facilities with Those from Fossil Fuels*

	Residual Oil	Bituminous coal (pulverized)	Lignite coal (pulverized)	Waste-to-energy (mass burn/refuse derived fuel)
Arsenic (As)	0.22	0.46	0.91	<0.033
Beryllium (Be)	0.06	0.03	0.06	<0.017
Cadmium (Cd)	0.18	0.10	0.11	0.063
Chromium (Cr)	0.24	4.56	570	<0.19
Copper (Cu)	3.19	2.28	3.42	0.43
Mercury (Hg)	0.04	0.23	0.23	0.17
Nickel (Ni)	1436	3.42	3.42	0.84
Lead (Pb)	0.34	0.87	0.11	0.44
Selenium (Se)	NR	0.29	0.29	<0.022
Vanadium (V)	3.4	4.0	4.0	0.025
Zinc (Zn)	0.47	8.0	8.0	1.23
Particulate	1,030	440	440	150

* lb/1000 megawatt-hours (MWh).
Source: Getz (1993).

TABLE 13C.22 Contribution of Components of MSW to Metals

		Percent in MSW	Parts per million		Parts of MSW	
			Cd	Cr	Hg	Pb
Paper	fine	2.09	0.002	0.07	0.006	0.09
	books	0.24	0.001	0.02	0.000	0.00
	magazines	glued	0.88	0.000	0.15	0.003	0.00
		not glued	0.93	0.003	0.05	0.003	0.05
	laminates	wax/plastic	1.66	0.005	0.05	0.002	0.12
		foil	0.30	0.000	0.13	0.000	0.28
	newsprint	glued	0.29	0.000	0.00	0.000	0.01
		not—b&w	4.55	0.005	0.17	0.014	0.33
		color	1.32	0.001	2.84	0.038	0.08
	browns	corrugate	9.19	0.009	0.17	0.028	0.35
		kraft	1.86	0.002	0.09	0.002	0.17
		box	1.68	0.003	0.09	0.008	0.20
	mixed paper		13.52	0.230	4.46	0.027	30.96
Plastic	film	color	3.13	0.207	3.60	0.013	11.33
		flexible	2.51	0.070	2.16	0.005	7.00
		rigid	0.3	0.112	0.36	0.001	0.10
	food	pete	0.015	0.001	0.00	0.000	0.01
		hdpe	0.182	0.005	0.03	0.000	0.11
		pvc	0.001	0.000	0.00	0.000	0.02
		dpe	0.001	0.000	0.00	0.000	0.00
		pp	0.026	0.000	0.01	0.000	0.02
		ps	0.006	0.000	0.00	0.000	0.00
		misc	0.684	0.542	0.30	0.003	1.08
	housewares	clear	0.064	0.001	0.00	0.000	0.04
		white	0.262	0.007	1.56	0.001	0.11
		blue	0.039	0.113	0.00	0.000	0.03
		yellow	0.049	0.001	0.63	0.000	1.21
		other	0.663	0.670	2.38	0.002	4.29
	toys etc.		0.257	0.195	0.59	0.000	0.00
	video tape		0.001	0.022	0.00	0.000	0.01
Organics	yard	lawn	10.87	0.652	10.98	0.152	16.74
		branches	2.46	0.027	0.59	0.010	1.53
	food	organic	6.76	0.066	0.75	0.010	2.39
	wood	finished	3.29	0.036	3.72	0.007	18.52
		unfinished	6.06	0.002	3.51	0.024	19.63
	textiles		4.4	0.123	19.36	0.048	5.63
	footwear		0.65	0.077	11.90	0.001	0.87
Metals	ferrous	beer cans	0.015	0.009	0.05	0.005	0.03
		soft drinks	0.012	0.007	0.04	0.004	0.03
		food	1.26	0.543	3.64	0.071	4.33
		band	0.06	0.009	0.30	0.000	0.36
	non-ferrous	beer	0.058	0.002	0.55	0.000	0.04
		soft drink	0.182	0.011	0.16	0.001	0.06
		food	0.016	0.000	0.03	0.000	0.02
		manufactured	0.40	0.022	5.42	0.001	0.38
		foil	0.326	0.166	0.44	0.003	0.00
		other	0.001	0.000	0.00	0.000	0.00
Glass	combined	clear	1.52	0.073	0.43	0.003	1.67
		green	0.12	0.000	1.13	0.000	0.02
		brown	0.13	0.002	0.06	0.001	0.13
		other	0.02	0.001	0.02	0.000	0.02

TABLE 13C.22   Contribution of Components of MSW to Metals (*Continued*)

		Percent in MSW	Parts per million		Parts of MSW	
			Cd	Cr	Hg	Pb
Inorganic light	dirt, rock	0.60	0.120	1.12	0.002	9.27
construction	drywall	0.09	0.002	0.01	0.000	0.03
	fiberglass	0	0.050	14.10	1.100	40.80
	other	0.87	0.400	34.00	0.100	30.10
Small appliances	plastic	0.15	0.005	0.38	0.000	0.99
Household batteries	carbon	0.011	0.003	0.00	0.002	0.00
	ni-cad	0.007	8.400	0.00	0.000	0.01
	alkaline	0.012	0.233	0.01	0.029	0.02
Fines		7.6	0.334	8.74	0.106	19.68
Total percent:		93.24				
Total parts per million:			13.5	93.5	0.73	163.40

Source: Rigo and Chandler (1994).

The sources and fates of mercury are of special interest. It can be seen that the highest concentrations of mercury were found in certain paper fractions, plastic film, lawn waste, unfinished wood, textiles, food, fiberglass, and as expected, batteries. The use of mercury as an antifungal agent in corrugated cardboard has essentially ceased. The mercury (and other metals) in printing inks has also been reduced if not eliminated by use of organic colors. Mercury in batteries has been phased out. It has been found that the levels measured at Burnaby were declining at the time of the tests, and have subsequently declined further. Therefore these emission factors are no longer valid. In any case, concentrating on elimination of batteries would not have resulted in a substantial reduction: The total of 0.73 ppm of MSW would be reduced by only 0.03 (4 percent) to 0.70 ppm if the battery fraction were eliminated, assuming that there was a direct relationship. It is not known whether the mercury in the batteries reports to the ash residues or to the stack gases. On the other hand, it was found that most of the lead goes to the ash residues (Rigo et al., 1993).

13C.6   VARIABILITY OF EMISSIONS

Stack emissions of pollutants vary considerably with time. Compliance tests performed semiannually over a period of years give some indication of the variability that is taking place from hour to hour, and from day to day. Only continuous monitors could reveal actual variations. However, the principles of statistics allow us to interpret data sets taken over years to obtain the inherent variability of pollutants. First, we should note that HCl and SO_2 are controlled, hence do not vary much. However, PM, and the metals contained in the particulate matter, do vary. Figure 13C.18 shows the variability over a period of four years at a single facility. Note that dioxins (TEQ) exhibited a similar variability. Table 13C.23 shows the statistical reduction of the data, from which the standard deviations are found. It is seen that the mean plus two standard deviations encompasses most of the data, and the mean plus three standard deviations includes the highest readings reported.

The significance of this is that from a regulatory point of view, while the average for the year is the environmental impact, any single test may show readings which significantly exceed the average. When regulators require that certain stack numbers not be exceeded, it is necessary to take variability into account. The practical way to deal with this problem is to set permit conditions at the level at which 95 percent or even 90.9 percent of the data will not be exceeded.

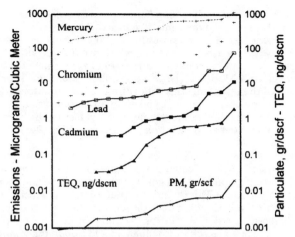

FIGURE 13C.18   Distribution of test data from a single WTE facility with dry lime-injection baghouse. Metals, particulate (PM), and dioxins (TEQ) show a similar range of variation. (*Source: Hasselriis, 1995.*)

13C.7   DISPERSION OF POLLUTANTS FROM STACK TO GROUND

Dispersion

The environmental impact and health risk to humans resulting from waste combustion depends on the waste, the combustion and emission control system, and the burning capacity of the incinerator. While emission controls typically reduce PM and metals emissions by 99.9 percent or more (with the exception of mercury) as the gases leave the stack of a combustion system (emitting only 1 out of 1000 units entering the emission control system), the gases are subjected to dispersion after they leave the stack, resulting in a much larger reduction in concentrations, factors typically 20 thousand to 1 million times before they reach the ground

TABLE 13C.23   Compliance Test Stack Emissions Measured over Four+ Years* WTE Facility with Dry Lime Injection and Fabric Filter

	TSP gr/dscf	Cadmium µg/dscm	Lead µg/dscm	Mercury µg/dscm	Chromium µg/dscm	Nickel µg/dscm
Low	0.0009	0.100	2.89	67.3	2.00	1.80
Median	0.0025	0.902	11.90	350	4.62	3.67
Maximum	0.0200	11.60	621.00	1170	78.75	23.70
Avg.	0.0045	2.07	74.50	460.89	12.24	8.04
Std. deviation	0.0047	3.11	153.43	287.45	19.01	6.88
Avg. + 3SD	0.0185	11.39	534.78	1321	69.27	28.68
Ratio SD/avg	1.04	1.50	2.06	0.62	1.55	0.86
Max./median	8.00	12.86	52.00	3.34	17.00	64.58
Max./avg.	4.44	5.60	8.33	2.24	6.43	2.95

 * WTE facility with dry lime injection and fabric filter.
 Source: Hasselriis (1995).

level. The degree of dispersion before the gases reach the ground depends upon the temperature of the gases and their initial velocity. The presence of buildings near the stack also influences downwash. The terrain influences the dispersion: Rural terrain allows the gases to loft more freely than urban conditions, which provide turbulence and bring the gases closer to the ground. Computer modeling is carried out to determine the points of maximum concentration on the ground. Models arrive at a maximum ground level concentration in $\mu g/m^3$ per gram per second of the pollutant. This ratio, often called the *dispersion factor* or *unit dispersion factor,* unfortunately is meaningless, giving no direct indication of the degree of dispersion. For this reason, it is useful to define a dimensionless factor which directly shows the degree of dispersion, and define it as the *dilution factor* (Hasselriis, 1995).

Dilution Factor. The factor by which the pollutant concentration leaving the stack is diluted by dispersion can be calculated by comparing the stack concentration (in g/m^3) with the ground level concentration GLC (in g/m^3) at the point of maximum impact resulting from the same 1 g/sec emission rate. This defines the dilution factor, which directly evaluates the phenomenon of dilution per se, resulting from dispersion. The stack concentrations obtained by stack testing are then multiplied by this factor to obtain the GLC. The information needed to do this calculation is the actual stack flow volume in addition to the grams per second of pollutant.

Dilution factors (DF) are calculated for a given facility in accordance with the following equations:

$$\text{Dilution factor} = \frac{\text{Concentration in the stack } (g/m^3)}{\text{Maximum ground level concentration } (g/m^3)}$$

$$\text{Dilution factor} = \frac{(g/sec)/(m^3/sec)}{(g/m^3)/(g/sec)}$$

m^3/sec = actual stack volumetric flow rate = (meters/second) × (square meters stack area)

$$\frac{g/m^3}{g/sec} = \text{grams per cubic meter per g/sec, by modeling}$$

Effect of Stack Height and Burning Capacity on Dilution Factor

Waste combustors having greater burning capacities usually have higher stack heights. Stack heights are usually set by good engineering practice (GEP), usually 2.5 times the height of the building. Large municipal waste-to-energy facilities have stack heights of 90 m or greater, whereas BMCs commonly have short stacks less than 10 m in height. Dilution factors which have been calculated from modeling data performed for a wide range of combustion facilities are plotted, in Fig. 13C.19, against the stack height upon which the modeling was based. The stack concentrations in micrograms per cubic meter ($\mu g/m^3$), corresponding to a unit emission rate of 1 g/sec, have been divided by the maximum annual average ground level concentrations in $\mu g/m^3$, in order to obtain the dilution factors for each facility. Most of these modeling studies were performed for the California Air Resources Board (CARB) as part of their cadmium and dioxin studies (Fry et al., 1990).

This data set includes medical and commercial waste incinerators, municipal sludge incinerators, biomass (wood) burners, and municipal waste incinerators, yet the range of DFs for each category is remarkably similar. The DFs range from 20 thousand to almost 1 million. The small MWCs with short stacks produced DFs that were similar to the municipal waste combustors with high stacks. In spite of the fact that the larger MSW combustors have capacities 50 to 100 times greater than the MWCs, their higher stacks result in only achieving the same

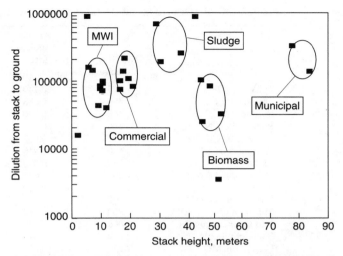

FIGURE 13C.19 Dilution factors achieved by various combustors versus stack height. (*Source: Hasselriis, 1995.*)

DFs as the small MWCs. It is important to recognize this when considering the impact of small medical waste incinerators, especially those located in rural areas (Hasselriis et al., 1992).

Table 13C.24 shows the range of unit dispersion factors (per g/sec) and annual average dilution factors resulting from three specific combustion sources, obtained by detailed modeling of the sites. The facilities selected illustrate the fact that two similar MWCs exhibited radically different dispersion and dilution factors. On the other hand, one MWC showed a dilution factor higher than a large municipal solid waste (MSW) combustor.

TABLE 13C.24 Dispersion and Dilution Factors for Medical and Municipal Waste Sources

Facility (all in California)	Burn rate (lb/h)	Unit dispersion ($\mu g/m^3$ per g/s)	Dilution factor
Cedars Sinai [MWI, 2-m stack]	980	27	16,000
Kaiser Permanente [MWI, 7.3-m stack]	980	5	145,000
St. Stanislaus [Municipal waste]	57,000	0.117	133,500

Source: Hasselriis et al. (1992).

Table 13C.25 lists dilution factors calculated from unit annual dispersion factors for 16 U.S. WTE facilities, for the capacity in tons per day of each facility. This calculation is based on the 5780 m^3/ton of waste burned, at a stack temperature of 300°F, wet basis, and 7 percent oxygen, that is, the actual stack emission volume. The equation is simply: DF = 14.9 ∗ E + 6/[(ton/day) ∗ (unit dispersion factor)]. This table shows that while the unit dispersion factors vary from 0.12 to 0.23 ($\mu g/m^3$) per (g/s), a two-time range, the dilution factors range from 71,000 to

TABLE 13C.25   Dispersion and Dilution Factors versus Plant Capacity

Plant	Tons/day capacity	Annual dispersion factor $(\mu g/m^3)/(g/s)$	Annual dilution factor
Tulsa Co., OK	375	0.140	283,810
Ocean City, NJ	400	0.088	423,295
Portland, ME	400	0.230	161,957
Pennsauken, NJ	500	0.031	961,290
Gloucester Co., NJ	575	0.068	381,074
Broome Co., NY	600	0.210	118,254
Johnston, RI	750	0.055	361,212
Montgomery Co., PA	1,200	0.035	354,762
Camden, NJ	1,400	0.012	886,905
Los Angeles, CA	1,600	0.130	71,635
Montgomery Co., MD	1,800	0.032	258,681
Mid Connecticut, CT	2,000	0.049	152,041
Philadelphia, PA	2,250	0.044	150,505
Dickerson, MD	2,250	0.026	252,757
Delaware Co., PA	2,688	0.033	167,974
Detroit, MI	3,000	0.049	101,361

Source: Hasselriis et al. (1992).

961,000, a range of 13.5. The dilution factors show no trend with capacity, indicating that site conditions for each plant are highly variable. However, it can be concluded that the ground-level concentrations are consistently low compared with stack concentrations. The estimation of health risk at ground level is a highly complex process, as will be discussed.

Acceptable Ground-Level Concentrations

Acceptable ground-level concentrations of many pollutants have been determined by various health agencies, such as OSHA. Exposure is expected to take place during eight-hour working days. Standards based on health effects use annual averages (with some exceptions such as PM, where a 24-h averaging time is also considered). Models provide corrections for other exposure times.

Health risk of cancer due to exposure to dioxins and furans, as well as toxic metals such as cadmium, are based on the annual average, since cancer-based risk involves exposure of the most effected individual (MED) for 70 years at that location.

The health risk impact of waste combustion emissions should include not only inhalation, but also other pathways resulting from deposition of particulate matter and its entry into soil, water, and crops, fish, and animals which may be eaten. Some studies have concluded that total risk is five times greater than inhalation alone (Hahn and Sofaer, 1990). It should not be overlooked that on-site MWCs are generally operated eight hours per day or less, thus contributing only one-third the emissions that their capacity implies. The shorter operating time of on-site MWCs may reduce the total risk to a factor of two times inhalation alone.

Health risks due to cancer are established on the basis of *unit risks*. The unit risk is the number of additional cancer cases per million that might be caused by exposure to an airborne concentration of 1 $\mu g/m^3$. The unit risk factors for the major cancer-causing or promoting pollutants are listed in Table 13C.26. For dioxins, a concentration of 3×10^{-8} $\mu g/m^3$, or 30 femtograms/m^3, has been adopted by some states.

TABLE 13C.26   Acceptable Workplace Concentrations and Unit Risk Values for Pollutants

	Ground level concentration[a] ($\mu g/m^3$)	Unit risk value for inhalation cases per million[b,c] at 1 $\mu g/m^3$
Pollutant		
Total suspended particulate:		
24-hour	150	
Annual	7	
Hydrochloric acid:		
3-minute	140	
annual	50	
Arsenic	0.00023	0.004
Beryllium	0.00042	0.003
Cadmium	0.00056	0.002
Chromium VI	0.000083	0.012
Lead	0.150–1.5	
Mercury	0.012–0.80	
Nickel	0.0033	0.0005
Benzene	0.00042–0.000120	
Dioxin Equivalent	3.0×10^{-8}	33 @ 1 pg/m[3c]

[a] New York State DEC "Air Guide 1," NYSDEC also regulations of Pennsylvania and North Carolina.
[b] Health Risk Assessment, St. Lawrence County, NY, RamTrac Corp.
[c] Unit risk is posed by 70-year exposure to an airborne concentration of 1 $\mu g/m^3$ of each substance except Dioxin equivalent 2,3,7,8-TCDD, for which the applicable unit is picograms per cubic meter of air (pg/m^3) where 1 pg = 10^{-6} μg.
Source: Hasselriis (1992).

13C.8   RISK ASSESSMENT

The risk assessment process consists of the following components (Kelly, 1997):

- Hazard identification: identifying the chemical substances of concern and compiling, reviewing, and evaluating data relevant to toxic properties of these substances
- Dose-response evaluation: assessing the relationship between dose and response for each chemical of potential concern
- Exposure assessment: identification of potential exposure pathways, the fate and transport of chemicals in the environment (including dispersion modeling), and the estimation of the magnitude of chemical exposure for the potential exposure pathways
- Risk characterization: calculating numerical estimates of risks for each substance through each route of exposure using the dose-response information and the exposure estimates

The general approach used by the EPA guidance provides estimates of:

- Individual risks based on exposure within defined subareas surrounding the facility, expressed both as averages across the subareas and at the location of maximum chemical concentrations within each subarea
- Risks to potentially more highly exposed or susceptible subgroups, such as young children, within the general population

- Risks associated with specific activities that may result in elevated exposures, such as subsistence fishing
- Individual risks based on high-end exposure to subgroups of the population that are believed to be potentially more highly exposed
- Cumulative risks to the population in the vicinity of the incinerator as a result of stack emissions

This approach allows for the estimation of risk to specific segments of the population, taking into account site-specific activity patterns, the numbers of individuals in each subgroup, and actual locations of individuals within these subgroups.

Toxicity Assessment.   Consistent with U.S. EPA guidance, potential carcinogenic and noncarcinogenic effects are evaluated separately, assuming potentially carcinogenic substances to pose a finite cancer risk at all exposure levels, therefore a "no-threshold" assumption, based on a 70-year lifetime exposure. The U.S. EPA uses a linearized multistage model to develop the cancer *slope factor* (SF), which is generally believed to overpredict the true potency of a chemical.

Noncancer effects assume that a minimum threshold level of exposure must be reached before the effect will occur. The estimated level of daily human exposure below which it is unlikely that adverse effects will result is known as the *reference dose* (RfD) and *reference concentration* (RfC). The noncarcinogenic effects of certain chemicals are typically derived from experimental animal studies, incorporating uncertainty factors to extrapolate from the high dose exposures in the animal experiments to the low doses likely to be received by humans from environmental sources, and taking into account individuals who are likely to be more susceptible than the general population to the chemical.

For incinerator emissions, exposure to individuals living and working in the vicinity of a facility is evaluated for both inhalation and indirect, multipathway routes of exposure, specifically:

- Inhalation of air
- Ingestion of and dermal contact with soil
- Consumption of meat, dairy products, and eggs from locally raised livestock
- Consumption of locally grown vegetables
- Ingestion of and dermal contact with surface water during swimming

Fate and Transport Modeling.   The ISC-COMPDEP model is used to estimate chemical concentrations in air associated with the routine emissions from the facility. The results of this modeling can be used directly to assess inhalation exposures, and at the starting point for evaluating exposures through indirect pathways, including the result of wet and dry deposition of particulate matter and vapor onto soil and vegetation, followed by ingestion by livestock, and ingestion of vegetables or livestock grown or raised locally.

Health Risk Analyses

Health risk assessments have been prepared for many WTE facilities. The one prepared for Spokane Regional Solid Waste System WTE Facility may be of special interest since it was performed a second time, using actual rather than assumed emissions, and using the latest U.S. EPA protocol for multipathway risk assessments. Table 13C.27 shows the carcinogenic and noncarcinogenic risk summary for this 800-ton-per-day facility for the maximum affected adult and child residents, based on 1995 actual stack concentrations (Kelly, 1997).

TABLE 13C.27 Summary of Cancer Risks and Noncancer Hazard Indices—Spokane WTE Facility

	Additional carcinogenic risk			Noncarcinogenic hazard		
	Actual	Threshold	Ratio	Actual	Hazard index	Ratio
Adult resident	$2 \times$ E-7	$1 \times$ E-5	50	0.02	1.0	50
Child resident	$5 \times$ E-8	$1 \times$ E-5	200	0.05	1.0	20

Source: Delta Toxicology (1998).

This calculation of cancer risks ranged from 0.2×10^{-6} for the theoretical adult, based on inhalation, ingestion of soil and vegetables, and dermal contact with soil from pasture, agricultural, forest, garden, and watershed areas. Vegetable exposure includes garden vegetables as well as produce grown within the limits. In addition, children were assumed to imbibe breast milk. The upper 95 percent confidence limit was used as is conventional, to be even more conservative. *The U.S. EPA considers these risk levels to be of no significance.* It was estimated that of the total risk, 83 percent was due to inhalation, and 17 percent was due to eating vegetables grown in the area.

Excess cancer risk was calculated to be $2 \times$ E-7, or 0.2 per million exposed individuals. Thus the maximum adult exposure, is ¹⁄₅₀, compared with the U.S. EPA criterion of 10 additional cancer cases per million exposed individuals as being protective of human health.

It is important to understand the implication of "additional cancer risk." Assuming that the risk of contracting cancer is about 300,000 in one million persons, (half of whom die of cancer), the risk of 10 additional cases becomes 300,010 cases of cancer. In this case, the risk is increased to 300,000.2 per million cases.

Hazard Index.   The Hazard Index or quotient is used to evaluate noncancer risks, based on the toxicity of metals and chemicals. It is calculated by dividing the estimated exposure concentration or dose by the appropriate toxicological benchmark value. Hazard quotients exceeding one indicate potentially moderate to high magnitude risks (the magnitude of the hazard quotient indicating the relative magnitude of risk) and hazard quotients of one or less indicate that risks are low to negligible.

For the Spokane facility, as seen in Table 13C.27 the noncarcinogenic hazard based on actual emissions from the stack, related to ground level conditions, was 0.03 for a child resident, hence the ratio of the Hazard Index of 0.03 to the criterion of 1.0 is 33. The Hazard Indices were 0.02 to 0.05. (See Table 13C.24.) These estimates are extremely conservative worst-case estimates.

Distribution of Risk.   An assessment performed for the Onondaga WTE facility in New York State in 1990 wherein the risk analysis was based on anticipated emissions (later found to be very conservative), calculated that inhalation contributed 67 percent to total risk, locally caught fish 17.5 percent, and beef 4.5 percent (Holstein, 1990).

Cancer Risks and Health Indices for a Hazardous Waste Incinerator.   The complex analysis performed for the Waste Technologies Incinerator (WTI) estimated cancer risks to range from 1×10^{-6} for the theoretical subsistence farmer to 6×10^{-7} for the farmer adult or child. *The U.S. EPA considers these risk levels to be of no significance.* The Hazard Indices were all less than 0.1.

Risk Assessment for Hazardous Waste Incinerator.   The comprehensive risk assessment performed for the hazardous waste incinerator (WTI) due to exposure from direct and indirect pathways arrived at cancer risks of 0.2 additional cancer cases per resident adult, 0.4 per resident child, 0.6 per farmer adult or child, and 1.0 per subsistence farmer or child. Hazard indices ranged from 0.01 to 0.07 for these groups (U.S. EPA, 1997).

Environmental Monitoring.   Ultimately, the public wanted to know what impact a new WTE facility would have on the environment by comparing preconstruction with postoperation of the facility, to monitor changes, if any, in environmental media, due to the operation of the facility. In a study done in Montgomery County, Maryland, dioxins/furans, polycyclic-aromatic hydrocarbons (PAHs), and polychlorinated-biphenls (PCBs), and trace metals arsenic, beryllium, cadmium, chromium, nickel, lead, and mercury were measured in the environment. The media sampled were air, soil, earthworms, garden vegetables, surface water, fish and sediment from the farm ponds, dairy milk, and hay. The monitoring was carried out from February 1996 to February 1997.

The findings show that distributions of 24-hour and 12-day air samples analyzed for dioxin/furans were significantly lower in the operational phase than in the preoperational phase. The same applies to chromium, lead, mercury, and nickel analyses of air samples: post operational samples were in all cases lower than preoperational and background samples, and in all cases the mean values were far below levels of concern (Rao, 1998).

13C.9   CALCULATION OF MUNICIPAL WASTE COMBUSTOR EMISSIONS

Typical composition of municipal solid waste:

Component	Percent in MSW
Carbon	26.4
Hydrogen	3.74
Oxygen	18.19
Nitrogen	0.40
Chlorine	0.31
Fluorine	0.01
Sulfur	0.20
Moisture	30.0
Ash/Inert	20.5

Higher heating value (HHV) = 5000 Btu/lb.

Net heating value (NHV) = 4354.

Moisture and ash-free HHV (MAFHHV) = 10,000 Btu/lb.

Waste feed rate at 750 tpd = 31.3 tph = 62,550 lb/h.

Heat released by waste = 62,550 [lb/h] * 5000 [Btu/lb] = 312,752,000 Btu/h.

The heat and mass balance determines that a furnace temperature of 2200°F requires the use of 88 percent excess air. The gaseous products are 7.38 lbp/lbf, for a mass flow of 62,550 [lb/h] * [7.38 lbp/lbf] = 465,630 lb/h.

The heat recovered by the boiler, Q_b assuming the boiler exit temperature is 450°F and an average specific heat of 0.30 Btu/lb-°F, is:

$Q_b = wc(T_{in} - T_{out}) = 465,630$ [lbp/h] * 0.30 [Btu/lb °F] * $(2200 - 450)$[°F] = 244,455,000 Btu/h.

The boiler efficiency (heat recovered by boiler/heat supplied in fuel) = 244,455,000/(62,550 lb/h * 5000 Btu/lb) = 78.5 percent.

Steam generation at 1200 Btu/lb steam = 244,455,000/1200 [Btu/lb] = 20,371 lb/h.

Power generation at 11,000 Btu/kWh = 244,455,000 [Btu/h]/11,000 [Btu/kWh] = 22.2 MW.

Water evaporated to cool gases to 300°F: 465,630 [lbp/h] $*$ 0.3 [Btu/lb-°F] $*$ (450 − 300) [°F] = 20,925,000 Btu/h/[1000 Btu/lb] = 20,925 lb/h (41.8 gal/min).

Gas flow entering fabric filter and handled by the induced-draft fan: 465,630 + 20,925 = 486,000 lb/h.

The specific volume of the gases at 300°F: calculated from 13.59 std. ft^3/lb

$$\text{Volume} = 13.59 \text{ [std. ft}^3\text{/lb]} * [(460 + 300)/530] * 468,000 \text{ [lb/h]}/60 \text{ [min/h]}$$
$$= 157,852 \text{ ACFM}$$

Volumetric composition of the stack gases from heat and mass balance:

	Wet	Dry Basis	
$CO_2 =$	8.47%	9.80%	$[CO_2 + O_2 = 19.73\%]$
$O_2 =$	8.58	9.93	
$N_2 =$	69.30	80.18	
$H_2O =$	13.58	0	
	100%	100%	

$$HCl = 396 \text{ ppmv (parts per million by volume)}$$

$$HF = 25 \text{ ppmv}$$

$$SO_2 = 406 \text{ ppmv}$$

Factor needed to correct to 7 percent oxygen, dry

$$\text{Ratio} = \frac{20.5 - (O_2)_{\text{actual}}}{20.5 - (O_2)_{\text{standard}}} = \frac{20.5 - 9.93}{20.5 - 7} = 10.57 * 100/13.5 = 0.78$$

Using the EPA factor to correct from the higher oxygen content (more excess air) to the "reference" 7 percent oxygen increases the concentration (note that while the actual sum of $CO_2 + O_2$ is 19.73 percent in this case, the "standard" factor used in EPA calculations is 20.5 percent:

$$(HCl)\text{corrected} = 393/0.78 = 504 \text{ ppm}$$

$$(SO_2)\text{corrected} = 403/0.78 = 517 \text{ ppm}$$

To comply with 25 ppmv HCl, the removal efficiency must be (500 − 25)/500 = 95%.

To comply with 30 ppmv SO_2, the removal efficiency must be (517 − 30)/517 = 94%.

The alternative is 80 percent removal, reducing 517 to [517 − (0.80 $*$ 517)] = 103 ppmv.

Alkaline reagent needed to control SO_2 and HCl to regulatory limits:

Stoichiometric requirements for HCl input is 0.31 lb/100 lb, or 62,550 [lbw/h] $*$ 0.0031 [Cl] = 194 lb/h $*$ [35 + 1]/35 = 200 lb/h.

From Table 13C.28, at 300°F, the $Ca(OH)_2$ required would be 1.6 $*$ 200 = 319 lb/h.

Sulfur input is 0.20 lb/100 lb or 62,550 $*$ 0.002 = 125 lb/h [S] $*$ [44/12] = 459 lb/h.

From Table 13C.28, at 300°F, the $Ca(OH)_2$ required would be 3.7 $*$ 459 = 1697 lb/h.

In practice, the lime that must be added is greater than the stoichiometric quantity, depending upon the percent removal required. These numbers may have to be increased by a factor of at least 50 percent, depending upon the effectiveness of the APC system itself.

TABLE 13C.28   Calculated Theoretical Reagent Requirements per Tonne of Gaseous Pollutants

Pollutant	Gas temperature		Stoichiometric ratio Ca(OH)$_2$	Consumption in tonnes (94%)
	°C	°F		
SO$_2$	120–160	250–320	3.0	3.7
	160–220	320–428	3.5	4.3
	220–280	428–536	5.0	6.2
HCl	120–160	250–320	1.5	1.6
	160–220	320–428	1.8	2.0
	220–280	428–536	2.0	2.2
HF	120–280	250–536	1.0	2.1

Metals in stack gases after emission controls of MSW facility:

Using the data from the tests of the 750 tpd combustor at Burnaby, a facility with dry lime injection with reactor tower, and fabric filter:

Waste flow = 750 * 2000/24 = 62,500 lb/h = 31 t/h.

Stack gas flow = 926 dscm/min × 60 = 55,560 dry std. m^3/h.

Cadmium Emissions.   Cadmium in the waste at Burnaby was estimated to be 27,000 lb/million tons (Mt). (This is 13.5 ppm weight of MSW. Assuming ash is 20 percent of MSW, this is 13.5/.2 = 67.5 ppm in the ash if all goes to ash, or, if 80 percent goes to ash, 54 ppm cadmium in the dry ash.) Leaving the boiler it was measured to be 5723 lb/Mt, a reduction factor of 5. This means that 80 percent of the cadmium reported to the ash residues, and 20 percent to the gases entering the emission controls. From the measured stack emissions of 18 lb/Mt, the control efficiency is calculated to be 99.7 percent.

Cadmium mass flow in stack was 18 lb/Mt * 27 t/h * 454 g/lb/1,000,000 lb/Mt = 0.221 g/h.

Cadmium concentration in stack = 0.221 g/h/49,800 dry std. m^3/h = 4.43 µg/dscm.

This is 22 percent of the U.S. EPA standard of 20 µg/dscm for new MWCs. If the control efficiency had been 99.00 instead of 99.70, the result would be 15 µg/dscm.

European Guideline: 200 µg/dscm corrected to 11 percent for hg + cd

U.S. EPA Cd standard:

for small existing MSW units: 100 µg/dscm corrected to 7 percent oxygen

for large existing MSW units: 40 µg/dscm corrected to 7 percent oxygen

for new MSW units: 20 µg/dscm corrected to 7 percent oxygen

Lead Emissions.   Lead in the stack was 363 lb/Mt * 27 t/h * 454 g/lb/1000000 lb/Mt = 4.45 g/h.

Lead concentration in stack = 4.45 g/h/49,800 m^3/h = 89.4 µg/dscm.

This is 44 percent of the U.S. EPA standard of 200 µg/dscm for new MWCs.

If the control efficiency had been 97.00 instead of 98.3, this would result in 158 µg/dscm.

Mercury Emissions.   Mercury entering the APC was 3630 lb/Mt * 27 t/h * 454 g/lb/1000000 lb/Mt = 44.5 g/h.

Mercury concentration in stack = 44.5 g/h/49,800 m^3/h = 894 µg/dscm.

This exceeds the standard of 20 µg/dscm for new units and 100 µg/dscm for small existing units. Sodium sulfide was used at Burnaby to meet the existing mercury standard.

13C.10 *CONVERSIONS AND CORRECTIONS*

Correction of Emission Factors for Heating Value

The EPA has listed emission factors for MSW in AP-42, using a higher heating value (HHV) of 4500 Btu/lb as a reference value. When the heating value is different than 4500 Btu/lb, a correction must be made, as follows:

Example: AP-42 lists emissions of nitrogen oxides (NOx) as 2.86 kg/mg of fuel, and as 358 ppmdv at 4500 Btu/lb.

Correct these emissions to the actual higher heating value of 5000 Btu/lb:

Calculate lb/ton and lb/million Btu:

   Multiply 2.86 kg/mg by 5000/4500 *to get* 3.18 kg/mg at 5000 Btu/lb

   Multiply 3.18 kg/mg by 2 *to get* 6.36 lb/ton

Convert from lb/ton to lb/million Btu:

   1 ton at 4500 Btu/lb × 2000 lb/ton = 9 million Btu/ton

   2.86 kg/mg @ 4500 Btu/lb *n.ultiplied by* 2 = 5.72 lb/ton

   5.72 lb/ton divided by 9 MBtu = 0.64 lb/MBtu

   1 ton at 5000 Btu/lb contains 2000 lb = 10 million Btu/ton

   6.36 lb/ton divided by 10 Mbtu/ton = 0.64 lb/MBtu

Correct ppmv at 4500 Btu/lb to 5000 Btu/lb:

   Multiply 358 ppmdv @4500 Btu/lb by 5000/4500 *to get* 398 ppmdv.

Conversions from Volumetric to Other Bases

The U.S. EPA has developed the volumetric emission factors, in grams per cubic meter based on a standardized F factor of 9570 ft^3/MBtu, for stoichiometric combustion (zero excess air or zero oxygen). This F factor is reasonably accurate for combustion of wastes and fuels, but a more precise number may be calculated from the actual composition of the waste.

Convert μg/dscm to (lb pollutant)/(million Btu of waste):

$$1 \frac{\mu g}{m^3} @ 7\% \ O_2 \times \frac{21\text{-}0}{21\text{-}7} \times \frac{m^3}{35.3 \ ft^3} \times \frac{g}{10^6 \ \mu g} \times \frac{9570 \ ft^3}{10^6 \ Btu} \times \frac{lb}{454 \ g} = 0.90 \ \frac{lb}{10^6 \ Btu}$$

Convert ppmv to lb/million Btu

$$1 \ ppmv @ 7\% \ O_2 \times MW \ \frac{lb}{lb\text{-}mol} \times \frac{21\text{-}0}{21\text{-}7} \times \frac{1}{35.3} \ \frac{lb\text{-}mol}{ft^3} \times 9570 \ \frac{ft^3}{10^6 Btu} = \frac{1}{2,682} \ \frac{lb}{10^6 \ Btu}$$

Convert ppmv at 7% oxygen to mg/m^3

$$ppm \times \frac{1}{106} \times MW \ \frac{lb}{lb\text{-}mol} \times \frac{lb\text{-}mol}{385 \ ft^3} \times \frac{35.3 \ ft^3}{m^3} \times \frac{454 \ g}{lb} \times \frac{1000 \ mg}{g} \times \frac{21}{21\text{-}7} = \frac{MW}{16} \ mg/m^3$$

Example: 100 ppm SO$_2$ (MW = 44) = 100 × 44/16 = 274 mg/m^3 at 7% oxygen and 70°F.

TABLE 13C.29   Conversion of ppmv at 0 percent Oxygen to lb/million Btu and mg/m^3

Gas	Molecular weight (MW)	lb/million Btu	mg/m^3
		100 ppmv at 0% $O_2 =$	100 ppmv at 0% O_2
CO_2	28	0.104	116.5
SO_2	44	0.163	183.1
NOx	46	0.172	191.3
HCl	36.5	0.136	151.8
HF	20	0.746	83.2
CH_4	16	0.060	66.6
NH_3	31	0.063	112.3

Corrections for Excess Air

Emissions tests performed in the stack of combustion devices are measured at the actual wet gas flow, but must be reported at dry reference conditions, such as 7 percent oxygen or 12 percent CO_2, in order to standardize all reported data and relate it to regulated emission standards. Correction to dry conditions is made based on measured moisture in the stack. Reference standards vary from 3 percent oxygen, used in California, to 11 percent oxygen, used in Europe.

Calculate excess air:

$$\%EA = \frac{O_2 - 0.5\ CO}{0.266\ (N_2 - .5\ N_f) - O_2} \times 100\%$$

Where: O_2, CO_2, and CO are the molar or volume fractions of the gases in the flue gas as determined by an Orsat or equivalent analysis.

N_f is the mole fraction of fuel nitrogen in the combined waste and fuel feeds to the combustor, determined from the ultimate analysis. This value is usually negligible.

Normally N_f and CO can be neglected, yielding this simpler expression:

$$\%EA = \frac{O_2}{0.266\ (N_2) - O_2} \times 100\%$$

Correct from (Concentration)$_{actual}$ to (Concentration)$_{standard}$

Since the sum of O_2 plus CO_2 in the dry products of combustion is about 20.5, corrections for excess air are usually approximated by this expression:

$$Ratio = \frac{20.5 - (O_2)_{actual}}{20.5 - (O_2)_{standard}} \times 100\%$$

The factor 20.5 is a standardized approximation. The actual sum of $CO_2 + O_2$ varies somewhat with the fuel.

Useful Relationships

Stoichiometric combustion air needed per million Btu can be calculated for typical moisture- and ash-free MSW and similar wastes as follows, using consistent values of HHV from a heat and mass balance calculation:

$$\text{lb combustion air / million Btu} = \frac{1,000,000 \ [\text{Btu}] * 5.105 \ [\text{lb}_a]}{7457 \ [\text{Btu/lb}_f] \ [\text{lb}_f]} = 684.6 \ \text{lb}_a/\text{MBtu}$$

Pounds of stoichiometric products per million Btu would be:

$$\text{lb products / million Btu} = \frac{1,000,000 \ [\text{Btu}] * 6.105 \ [\text{lb}_a]}{7457 \ [\text{Btu/lb}_f] \ [\text{lb}_f]} = 819 \ \text{lb}_p/\text{mmBtu}$$

With 50 percent excess air, we get pounds combustion air per million Btu:

$$\text{lb air / million Btu} = \frac{1,000,000 \ [\text{Btu}][\text{lb}_f] * 7.66 \ [\text{lb}_a]}{7457 \ [\text{Btu/lb}_f] \ [\text{lb}_f]} = 713 \ \text{lb}_a/\text{mmBtu}$$

Pounds of products at 50 percent excess air would be:

$$\text{lb products / million Btu} = \frac{1,000,000 \ [\text{Btu}] * 8.66 \ [\text{lb}_a]}{7457 \ [\text{Btu/lb}_f] \ [\text{lb}_f]} = 1161 \ \text{lb}_a/\text{mmBtu}$$

Weight and volume of products per million Btu at standard temperature and pressure:
   Volume of 1 mol of ideal gas = 387 ft^3 / lb-mol @ 70°F (20°C), 1 atmosphere.
   1 lb-mol of dry products (see above) weighs 29.51 lb.
   1 lb-mol of products weighs about 29.75 lb at zero excess air.

Therefore, the volume of wet products corrected to 70°F (70 + 460 = 530°R) is:
V_{wp} = 387/29.75 = 13.0 ft^3/lb$_f$
V_{wp} = 6.105[lb$_p$/lb$_f$] * 13 ft^3/lb$_f$ = 79.416 ft^3/lb$_f$

The volume of dry products is calculated when reporting emissions.

To get the volume of dry products we subtract 0.555 [H$_2$O] from 6.105 to get:
V_{dp} = (6.105 − 0.555)[lb$_{dp}$/lb$_f$] * 13.0 ft^3/lb$_f$ = 72.15 ft^3/lb$_f$.

Common Conversion Factors

To convert from	to	Multiply by
Milligrams/m^3	Micrograms/m^3	1000
	Micrograms/liter	1.0
	ppm by volume (20°C)	(24.04/M)
	ppm by weight	0.8347
	lb/ft^3	62.43 × 10^{-9}

Micrograms/m³	Milligrams/m³	0.001
	Micrograms/liter	0.001
	ppm by volume (20°C)	(0.02404/M)
	ppm by weight	834.7×10^{-6}
	lb/ft³	62.43×10^{-12}
Micrograms/liter	Milligrams/rn³	1.0
	Micrograms/rn³	1000
	ppm by volume (20°C)	(24.04/M)
	ppm by weight	0.8347
	lb/ft³	62.43×10^{-9}
ppm by volume (20°C)	Milligrams/m³	(M/24.04)
	Micrograms/m³	(M/0.02404)
	Micrograms/liter	(M/24.04)
	ppm by weight	(M/28.8)
	lb/ft³	$(M/385.1 \times 10^{6})$
ppm by weight	Milligrams/m³	1.198
	Micrograms/rn³	1.198×10^{-3}
	Micrograms/liter	1.198
	ppm by volume (20°C)	(28.8/M)
	lb/ft³	7.48×10^{-6}
lb/ft³	Milligrams/rn³	16.018×10^{6}
	Micrograms/rn³	16.018×10^{9}
	Micrograms/liter	16.018×10
	ppm by volume (20°C)	$(385.1 \times 10^{6}/M)$
	ppm by weight	133.7×10^{3}

Note:
- cm = 0.0328 ft
- gal (US) = 0.1337 ft³
- Liter = 0.03532 ft³ = 0.001 rn³
- Microgram = 0.000001 g
- Micron = 0.0000394 in = 0.001 mm
- Milligram = 0.001 g
- lb = 7,000 grains = 453.6 g
- M = Molecular weight

REFERENCES

Achternbosch, M., and U. Richters (2000) "Material Flows and Investment Costs of Flue Gas Cleaning Systems of MSWI," International Thermal Treatment Technologies Conference, Portland, OR.

Andersson, C., and B. Weimer (1991) "Mercury Emission Control—Sodium Sulfide Dosing at the Hogdalen Plant in Stockholm," pp. 664–674, Proceedings of the 2d Annual Conference on Municipal Waste Combustion, Air and Waste Management Association, Tampa, FL.

Buonicore, A. J., and W. T. Davis (1992) *Air Pollution Control Engineering Manual,* Air and Waste Management Association, Van Nostrand Reinhold, New York.

Delta Toxicology (1998) *Multipathway Risk Assessment for the Spokane Regional Solid Waste System Waste-to-Energy Facility, Spokane, Washington,* Delta Toxicology, Inc., Crystal Bay, NV.

Finnis, P. (1998) "Heat Recovery Dry Injection Scrubbers for Acid Gas Control," Proceedings of the ASME Asian–North American Solid Waste Management Conference (ANACON), Los Angeles, CA.

Fritsky, K., J. Kumm, and M. Wilken (2000) "Combined PCDD/F Destruction and Paniculate Control in a Baghouse: Experience with a Catalytic Filter System at a Medical Waste Incineration Plant," International Thermal Treatment Technologies Conference, Portland, OR.

Fry, B. et al. (1990) *Technical Support Document to Proposed Dioxins and Cadmium Control Measure for Medical Waste Incinerators,* California Air Resources Board, Sacramento, CA.

Getz, N. (1993) How Does Waste-to-Energy "Stack Up?" *Municipal Waste Combustion,* pp. 951–965.

Gleiser, R., K. Nielsen, and K. Felsvang (1993 Control of Mercury from MSW Combustors by Spray Dryer Absorption Systems and Activated Carbon Injection, *Municipal Waste Combustion,* pp. 106–120.

Hahn, J. L., and D. S. Sofaer (1990) "A Comparison of Health Risk Assessments for Three Ogden Martin Systems Inc. Resource Recovery Facilities Using Estimated (Permitted) and Actual Emission Levels," ASME National Waste Processing Conference, Long Beach, CA.

Hasselriis, F. (1995) "Variability of Metals and Dioxins in Stack Emissions of Three Types of Municipal Waste Combustors over Four Year Period," Paper No. 95-RP147B.03, Air and Waste Management Association, San Antonio, TX.

Hasselriis, F., D. Corbus, and R. Kasinathan, "Environmental and Health Risk Analysis of Medical Waste Incinerators Employing State-of-the-Art Emission Controls," Paper No. 91-30.3, 84th Annual Meeting of the Air and Waste Management Association, Kansas City, MO.

Holstein, E. (1990) *Health Risk Assessment for the Onondaga County Resource Recovery Facility,* Environmental Health Associates, Cambridge, MA.

IWSA (undated) Summary of the National Emission Estimates for Municipal Waste Combustion Units, Prepared by Eastern Research Group for the Integrated Waste Services Association, Washington, DC.

Jones, K. (1994) Comparing Air Emissions from Landfills and WTE Plants, *Solid Waste Technologies,* March/April.

Kelly, K. E. (1997) "Comparative Assessment of Estimated vs. Actual Emissions and Associated Health Risks from a Modern Municipal Waste-to-Energy Facility," International Incineration Conference, University of California.

Kenna, J. D., and J. H. Turner (1989) *Fabric Filter-Baghouses I: Theory, Design and Selection,* ETS International, Inc.

Kilgroe, J., and A. Licata (1996) "Control of Air Pollution Emissions from Municipal Waste Combustors," 17th Biennial ASME Waste Processing Conference, Atlantic City, NJ.

Licata, A., M. Babu, and L-P. Nethe (1994) "Acid Gases, Mercury and Dioxin from MWCs," Proceedings of the ASME National Waste Processing Conference, Boston, MA.

Licata, A., and H. Hartenstein (1998) Mercury and Dioxin Control for Municipal Waste Combustors, Asian–North American Solid Waste Management Conference (ANACON), Los Angeles, CA.

Licata, A., W. Schüttenhelm, and M. Klein (2000) "Mercury Control for MWCs Using the Sodium Tetrasulfide Process," 8th Annual North American Waste-to-Energy Conference, Nashville, TN.

Maziuk, J. (1998) *Results of Emission Testing at a Medical Waste Incinerator Using Dry Injection/Fabric Filter APC Technology,* Church & Dwight, Princeton, NJ.

McDonald, B., G. Fields, and M. McDannel (1993) *Selective Non-Catalytic Reduction (SNCR) Performance of Three California WTE Facilities,* Carnot, Tustin, CA.

NYSDEC (1993) *Incineration 2000 Phase II Report,* New York State Department of Environmental Conservation.

Rao, R. K. (1998) "Communicating Health Risks to the Community from a State-of-the-Art Waste-to-Energy Resource Recovery Facility through Multimedia Environmental Monitoring Programs," 6th Annual Waste-to-Energy Conference, Miami Beach, FL.

Ricket, W. S., and M. Kaiserman, (1994) Levels of Lead, Cadmium and Mercury in Canadian Cigarette Tobacco as Indicators of Environmental Change: Results from a 21-Year Study (1968–1988), *Environmental Science Technology,* vol. 28, no. 5.

Rigo, H. G., and J. Chandler (1994) "Metals in MSW—Where Are They and Where Do They Go in an Incinerator?," ASME National Waste Processing Conference, Boston, MA.

Rigo, H. G., J. Chandler, and S. Sawell (1993) Debunking Some Myths About Metals, *Municipal Waste Combustion,* pp. 609–627.

Sanders, D. L. et al. (2000) "The New East Carolina University Medical Waste Incinerator: Combining a Wet Scrubber with Granular Carbon Bed," International Thermal Treatment Technologies Conference, Portland, OR.

Schüttenhelm, W., R. Holste, and A. Licata (2000) "New Trends in Flue Gas Cleaning Technologies for European and Asian Waste Incineration Facilities," SWANA Pub. No. GR-WTE-0108, 8th Annual North American Waste-to-Energy Conference, Nashville, TN.

Sorum, Lars, M. Fossum, E. Evensen, and J. E. Hustad (1997) "Heavy Metal Partitioning in a Municipal Solid Waste Incinerator," Proceedings of the North American Waste-to-Energy Conference.

Strong, B., and R. Copland (1998) "Summary of Final New Source Performance Standards and Emission Guidelines for New and Existing Hospital/Medical/Infectious Waste Incinerators," U.S. EPA 40 CFR Part 60, 91st Annual Meeting of the Air and Waste Management Association.

Taylor, H. F. (1992) "Potential Greenhouse Gas Emissions from Disposal of MSW in Sanitary Landfills vs. Waste-to-Energy Facilities," Paper No. 9216.05P, 85th Annual Meeting of the Air and Waste Management Association.

Teller, A. (1994) Emission Control, p. 11.159, *in Handbook of Solid Waste Management,* McGraw-Hill, New York.

U.S. EPA (1989) "Emissions Test Results from the Stanislaus County, California, Resource Recovery Facility," International Conference on Municipal Waste Combustion, Hollywood FL, U.S. Environmental Protection Agency, Washington, DC.

U.S. EPA (1996a) *Compilation of Air Pollution Emission Factors AP-42,* 5th ed., vol. 1, Stationary Point and Area Sources, U.S. Environmental Protection Agency, Washington, DC.

U.S. EPA (1996b) *Draft Technical Support Document for MWC MACT Standards,* U.S. Environmental Protection Agency, OSWER, Washington, DC.

U.S. EPA (1997) *Risk Assessment for the Waste Technologies Industries (WTI) Hazardous Waste Incineration Facility (East Liverpool, Ohio),* EPA-905-R97-002a, U.S. Environmental Protection Agency Region 5, Chicago, IL.

U.S. EPA (1999) NESHAPS: *Final Standards for Hazardous Air Pollutants for Hazardous Waste Combustors,* 40 CFR Parts 60 to 271, U.S. Environmental Protection Agency, Washington, DC.

Walker, B. L., and C. D. Cooper (1992) Air Pollutant Emission Factors for Medical Waste Incinerators, *Journal of the Air and Waste Management Association,* vol. 42, no. 6, pp. 784–791.

WASTE (1993) *Waste Analysis, Testing and Evaluation: The Fate and Behavior of Metals in Mass Burn Incineration,* A. J. Chandler Associates Ltd. et al., Willowdale, ON.

White, D. M. et al. (1992) "Parametric Evaluation of Activated Carbon Injection for Control of Mercury Emissions from a Municipal Waste Combustor," Paper No. 92-40.06, Annual Air and Waste Management Association Meeting, Kansas City, MO.

CHAPTER 14
LANDFILLING*

Philip R. O'Leary
George Tchobanoglous

The safe and reliable disposal of municipal solid waste (MSW) and solid waste residues is an important component of integrated waste management. Solid waste residues are waste components that are not recycled, that remain after processing at a materials recovery facility, or that remain after the recovery of conversion products and/or energy. Historically, solid waste has been placed on or in the surface soils of the earth or deposited in the oceans. Ocean dumping of municipal solid waste was officially abandoned in the United States in 1933. *Landfill* is the term used to describe the physical facilities used for the disposal of solid wastes and solid waste residuals in the surface soils of the earth. Since the turn of the last century, the use of landfills, in one form or another, has been the most economical and environmentally acceptable method for the disposal of solid wastes, both in the United States and throughout the world. Today, landfill management incorporates the planning, design, operation, environmental monitoring, closure, and postclosure control of landfills.

Although many landfills have been constructed in the past with little or no thought for the long-term protection of public health and the environment, the focus of this chapter is modern landfilling practice. In the past 20 years, practices have changed substantially so that recently constructed landfills have overcome the problems formerly associated with "dumps." The major topics covered in this chapter include:

1. A description of the landfill method of solid waste disposal, including environmental concerns, regulatory requirements, and siting considerations

2. Generation, composition, control, and management of landfill gases

3. Formation, composition, and management of leachate

4. Intermediate and final landfill cover

5. Landfill structural characteristics and settlement

6. Landfill design considerations

7. Development of landfill operation plan

8. Environmental quality monitoring

9. Landfill closure, postclosure care, and remediation

---

* Adapted from G. Tchobanoglous, H. Theisen, and S. A. Vigil, *Integrated Solid Waste Management, Engineering Principles and Management Issues,* McGraw-Hill, New York, 1993, and P. O'Leary and P. Walsh, *Solid Waste Landfills Correspondence Course,* University of Wisconsin–Madison, 1992.

Additional details on the subjects covered in this chapter may be found in Bagchi (1990), Crawford and Smith (1985), Pfeffer (1992), and Tchobanoglous et al. (1993).

14.1 THE LANDFILL METHOD OF SOLID WASTE DISPOSAL

Landfilling is the term used to describe the process by which solid waste and solid waste residuals are placed in a landfill. In the past, the term *sanitary landfill* was used to denote a landfill in which the waste placed in the landfill was covered at the end of each day's operation. Today, *sanitary landfill* refers to an engineered facility for the disposal of MSW designed and operated to minimize public health and environmental impacts. Landfills for individual waste constituents such as combustion ash, asbestos, and other similar wastes are known as *monofills*. Landfills for the disposal of hazardous wastes are called *secure landfills*. Those places where waste is dumped on or into the ground in no organized manner are called *uncontrolled land disposal sites* or *waste dumps*.

In developing countries, the implementation of improved land disposal practices is progressing at varying rates dependent upon the available resources and national regulatory standards. The need to improve land disposal practices is being forced along by consolidating populations, where rural residents are moving to cities resulting in rapid urban population growth. This has created an ever increasing need for better solid waste disposal practices. In a number of instances, less than adequate disposal practices have resulted in accidents that have led to loss of life. Large, uncontrolled dumps in urban areas are also a significant source of air pollution and water contamination and, as such, over time these facilities will need to be closed and replaced with landfills that meet conventional standards.

Definition of Terms

The general features of a sanitary landfill are illustrated in Fig. 14.1. Some terms commonly used to describe the elements of a landfill are defined as follows. The term *cell* is used to describe the volume of material placed in a landfill during one operating period, usually 1 day (see Fig. 14.1*b*). A cell includes the solid waste deposited and the daily cover material surrounding it. *Daily cover* usually consists of 6 to 12 in of native soil or alternative materials such as compost, foundry sand, or auto shredder fluff that are applied to the working faces of the landfill at the end of each operating period. Historically, daily cover was to prevent rats, flies, and other disease vectors from entering or exiting the landfill. Today, daily landfill cover is used primarily to control the blowing of waste materials, to reduce odors, and to control the entry of water into the landfill during operation. A *lift* is a complete layer of cells over the active area of the landfill (see Fig. 14.1*b*). Typically, landfills comprise a series of lifts. A *bench* (or *terrace*) is typically used where the height of the landfill will exceed 50 to 75 ft. Benches are used to maintain the slope stability of the landfill, for the placement of surface water drainage channels, and for the location of landfill gas recovery piping. The *final lift* includes the landfill cover layer.

Landfill liners are materials (both natural and man-made) that are used to line the bottom area and below-grade sides of a landfill (see Fig. 14.1*a*). Liners usually consist of successive layers of compacted clay and/or geosynthetic material designed to prevent migration of landfill leachate and landfill gas. The final *landfill cover* layer is applied over the entire landfill surface after all landfilling operations are complete (see Fig. 14.1*c*). Landfill covers consist of successive layers of compacted clay and/or geosynthetic material designed to prevent the migration of landfill gas and to limit the entry of surface water into the landfill.

The liquid that forms at the bottom of a landfill is known as *leachate*. In general, leachate is a result of the percolation of precipitation, uncontrolled runoff, and irrigation water into the landfill. Leachate will also include water initially contained in the waste. Leachate contains a

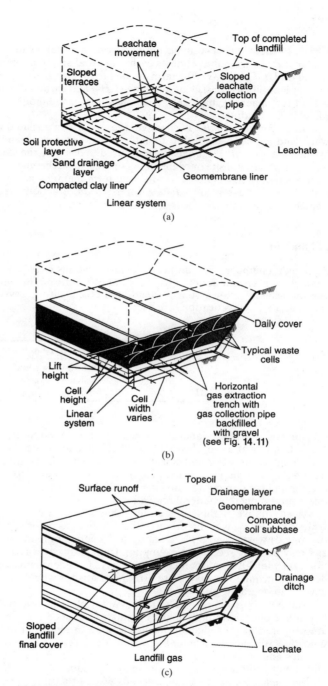

FIGURE 14.1 Cutaway views of a sanitary landfill: (*a*) after geomembrane liner has been installed over compacted clay layer and before drainage and soil protective layers have been installed; (*b*) after two lifts of solid waste have been completed; and (*c*) completed landfill with final cover installed.

variety of chemical constituents derived from the solubilization of the materials deposited in the landfill and from the products of the chemical and biochemical reactions occurring within the landfill. *Landfill gas* is the term applied to the mixture of gases found within a landfill. The bulk of landfill gas consists of methane (CH_4) and carbon dioxide (CO_2), the principal products of the anaerobic biological decomposition of the biodegradable organic fraction of the MSW in the landfill.

Environmental monitoring involves the activities associated with collection and analysis of water and air samples used to monitor the movement of landfill gases and leachate at the landfill site. *Landfill closure* is the term used to describe the steps that must be taken to close and secure a landfill site once the filling operation has been completed. *Postclosure care* refers to the activities associated with the long-term maintenance of the completed landfill (typically 30 to 50 years). *Remediation* refers to those actions necessary to stop and clean up unplanned contaminant releases to the environment.

Classification of Landfills

Although a number of landfill classification systems have been proposed over the years, the classification system adopted by the state of California in 1984 is perhaps the most widely accepted classification system for landfills. In the California system, as reported in the following table, three classifications are used.

Class	Type of waste
I	Hazardous waste
II	Designated waste
III	Municipal solid waste (MSW)

The majority of the landfills throughout the United States are designed for commingled MSW. In many of these Class III landfills, limited amounts of nonhazardous industrial wastes and sludge from water and wastewater treatment plants are also accepted. In many states, treatment plant sludges are accepted if they are dewatered to a solids content of 51 percent or greater and contain no free-flowing liquids. The acceptance of liquid wastes into MSW landfills is now banned by federal regulations.

An alternative method of landfilling that is being tried in several locations throughout the United States involves shredding of the solid wastes before placement in a landfill. Shredded (or milled) waste can be placed at up to 35 percent greater density than unshredded waste, and may possibly receive an exemption from daily cover requirements in some state regulations. Blowing litter, odors, flies, and rats have not been significant problems. The use of shredders has declined but may be reintroduced at sites where more rapid waste decomposition may be an operating goal.

Another approach is to bale the MSW for placement in the landfill. This method has the advantage of easier handling and eliminates the need for compaction equipment. The bales are prepared at a production facility located in either an off-site transfer station or at an unloading station on the landfill property. The bales are moved to the working face on flatbed vehicles and stacked with forklifts or similar equipment. Cover is applied as a lift is completed, but daily covering may not always be required.

Designated wastes are nonhazardous wastes that may release constituents in concentrations that are in excess of applicable water quality objectives established by various state and federal agencies. Combustion ash, asbestos, and other, similar wastes often identified as designated wastes are typically placed in lined monofills to isolate them from materials placed in municipal landfills.

Landfilling Methods

The principal methods used for the landfilling of MSW may be classified as (1) excavated cell/trench, (2) area, and (3) canyon. The principal features of these types of landfills, illustrated in Fig. 14.2, are described as follows. Landfill design details are presented later in the chapter.

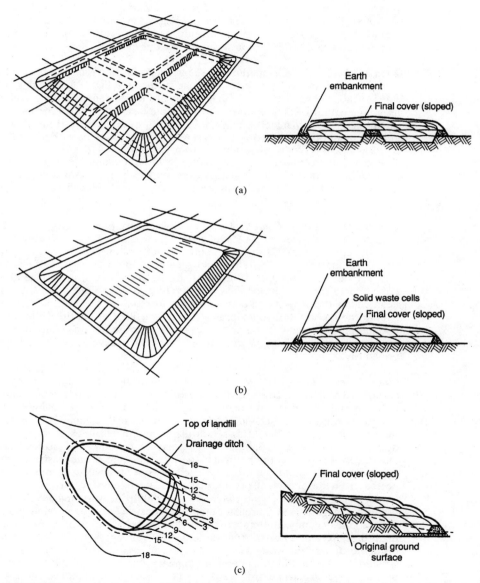

FIGURE 14.2   Commonly used landfilling methods: (*a*) excavated cell/trench; (*b*) area; (*c*) canyon/depression.

Excavated Cell/Trench Method.   The *cell/trench* method of landfilling (see Fig. 14.2*a*) is ideally suited to areas where an adequate depth of cover material is available at the site and where the water table is not near the surface. Typically, solid wastes are placed in cells or trenches excavated in the soil (see Fig. 14.2*a*). The soil excavated from the site is used for daily and final cover. The excavated cells or trenches are lined with synthetic membrane liners, low-permeability clay, or a combination of the two to limit the movement of both landfill gases and leachate. Excavated cells are typically square, up to 1000 ft in width and length, with side slopes of 2:1 to 3:1. Trenches vary from 200 to 1000 ft in length, 3 to 10 ft in depth, and 15 to 50 ft in width. A variation of this method is the *artesian* or *zone of saturation* landfill (Adams et al., 1998). These landfills are constructed below the naturally occurring groundwater table surface. Drainage systems control the entry of groundwater into the landfill cell. Both lined and unlined sites have been constructed using this method.

Area Method.   The area method is used when the terrain is unsuitable for the excavation of cells or trenches in which to place the solid wastes (see Fig. 14.2*b*). High groundwater conditions, such as those that occur in many parts of Florida and elsewhere, necessitate the use of area-type landfills. Site preparation includes the installation of a liner and leachate management system. Cover material must be hauled in by truck or earthmoving equipment from adjacent land or from borrow-pit areas. As noted, in locations with limited material that can be used as cover, compost produced from yard wastes and MSW, foundry sand, and auto shredder fluff have been used successfully as intermediate cover material. Other techniques include the use of movable temporary cover materials such as soil and geosynthetics. Soil and geosynthetic blankets, placed temporarily over a completed cell, can be removed before the next lift is begun.

Canyon/Depression Method.   Canyons, ravines, dry borrow pits, and quarries have been used for landfills (see Fig. 14.2*c*). The techniques to place and compact solid wastes in canyon/depression landfills vary with the geometry of the site, the characteristics of the available cover material, the hydrology and geology of the site, the type of leachate and gas control facilities to be used, and the access to the site. Control of surface drainage often is a critical factor in the development of canyon/depression sites. Typically, filling starts at the head end of the canyon and ends at the mouth, so as to prevent the accumulation of water behind the landfill. Canyon/depression sites are filled in multiple lifts, and the method of operation is essentially the same as previously described. If a canyon floor is reasonably flat, the initial landfilling may be carried out using the excavated cell/trench method discussed previously.

Other Types of Landfills.   Various other configurations of landfills are constructed to meet specialized objectives. These include construction and demolition waste landfills that receive only materials that are the result of tearing down buildings and removing roadways. Other specialized landfills are those associated with receiving high volumes of industrial waste such as that from paper mills, foundries, power plants, and mines. Each of these landfills has unique design considerations. The landfills may or may not contain all of the conventional design elements, depending upon the particular specialized nature of the waste. For example, a power plant ash landfill would not have a gas recovery system since no decomposition of waste is expected, given the fact that all organic matter had been removed during the combustion process.

An emerging technology for more quickly stabilizing waste in conventional landfills is the *bioreactor* (Fig. 14.3). A bioreactor landfill is constructed and operated in a manner that will enhance the decomposition rate of the organic material within municipal solid waste. Operating procedures are adjusted from those used at conventional landfills to quickly initiate the decomposition of the waste. Gas collection facilities are installed immediately upon the construction of the landfill cell so that methane gas can be recovered. To accelerate the decomposition rate, the leachate withdrawn from the base of the landfill is recycled and, in addition, other sources of moisture, such as sewage sludge, may be added to the waste profile. Bioreac-

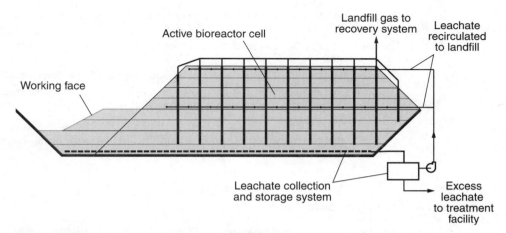

FIGURE 14.3   Bioreactor landfill with leachate recirculation and landfill gas recovery. *(Adapted from Solid and Hazardous Waste Education Center, University of Wisconsin–Madison, 2000.)*

tor landfills are being viewed as an option that will reduce the long-term care period of landfills after they are closed by quickly stabilizing the waste. In addition, some designs have as their goal reducing the waste volume to the maximum extent possible and in the shortest period of time so that more waste material can be disposed of on the original landfill site. The methods for lining and covering bioreactor landfills are still under consideration. Design issues that are currently being evaluated are those associated with slope stability, landfill liner leakage, methods for collecting landfill gas in a partially opened cell, and constructing leachate recirculation systems that will be effective in inclement weather and will minimize odors (Pohland and Kim, 2000).

Reactions Occurring in Landfills

Solid wastes placed in a sanitary landfill undergo a number of simultaneous and interrelated biological, chemical, and physical changes. The most important biological reactions occurring in landfills are those related to the conversion of the organic material in MSW, leading to the evolution of landfill gases and, eventually, leachate. Important chemical reactions that occur within the landfill include dissolution and suspension of landfill materials and biological conversion products in the liquid percolating through the waste, evaporation and vaporization of chemical compounds and water into the evolving landfill gas, sorption of volatile and semivolatile organic compounds into the landfilled material, dehalogenation and decomposition of organic compounds, and oxidation-reduction reactions affecting metals and the solubility of metal salts. Among the more important physical changes in landfills is the settlement caused by consolidation and decomposition of landfilled material. The reactions occurring in landfills are discussed in greater detail in Secs. 14.2 and 14.3.

Concerns with the Landfilling of Solid Wastes

Concerns with the landfilling of solid waste are related to the following:

• The uncontrolled release of landfill gases that might migrate off-site and cause odor and other potentially dangerous conditions

- The impact of the uncontrolled discharge of landfill gases on the greenhouse effect in the atmosphere
- The uncontrolled release of leachate that might migrate to underlying groundwater or to surface streams
- The breeding and harboring of disease vectors in improperly managed landfills
- The health and environmental impacts associated with the release of the trace gases found in landfills arising from the hazardous materials that were often placed in landfills in the past

The goal for the design and operation of a landfill is to eliminate or minimize the impacts associated with the aforementioned concerns.

Federal and State Regulations for Landfills

In planning for the implementation of a new landfill, special attention must be paid to the many federal and state regulations that have been enacted to improve the performance of sanitary landfills. The principal federal requirements for municipal solid waste landfills are contained in Subtitle D of the Resource Conservation and Recovery Act (RCRA) and in Environmental Protection Agency (EPA) Regulations on Criteria for Classification of Solid Waste Disposal Facilities and Practices (40 CFR Parts 257 and 258) (U.S. EPA, 1991). The final version of Part 258—Criteria For Municipal Solid Waste Landfills (MSWLFs) was signed on September 11, 1991. The subparts of Part 258 deal with the following areas:

Subpart A	General
Subpart B	Location Restrictions
Subpart C	Operating Criteria
Subpart D	Design Criteria
Subpart E	Groundwater Monitoring and Corrective Action
Subpart F	Closure and Postclosure Care
Subpart G	Financial Assurance

Additional details on the implementation requirements for these subparts are summarized in Table 14.1. Many state environmental protection agencies have parallel regulatory programs that deal specifically with their unique geologic and soil conditions and environmental and public policy issues. Landfill owners, operators, and persons contemplating siting landfills must study their state's regulations carefully and become aware of the public policy issues affecting landfill regulation. It should also be noted that the aforementioned landfill regulations necessitate extensive record keeping to document compliance.

The Clean Air Act also contains provisions dealing with the air emissions from landfills. In addition to the federal government, many of the states have also adopted regulations governing the design, operation, closure, and long-term maintenance of landfills. In many cases, the regulations that have been adopted by the individual states have been more restrictive than the federal requirements.

One approach for reducing the generation of leachate and the emission of decomposition by-products from a landfill is to limit the amount of biodegradable waste that enters the landfill. In the European Union, guidelines have been set that, when implemented, will greatly reduce the amount of biodegradable material that is allowed to be placed within a landfill. It is expected that the implementation of these types of standards will cause waste system operators to consider aggressive source segregation and composting, incineration, and the implementation of bioreactor-type landfills.

TABLE 14.1 Summary of U.S. Environmental Protection Agency Regulations for Municipal Solid Waste Landfills

Item	Requirement
Applicability	All active landfills that receive municipal solid waste (MSW) after October 9, 1993. Certain requirements also apply to landfills which received MSW after October 9, 1991, but closed within 2 years. Certain exemptions for very small landfills. Some requirements are waived for existing landfills. New landfills and landfill cells must comply with all requirements.
Location requirements	Airport separation distances of 5000 and 10,000 ft, and in some instances greater than 6 mi are required. Landfills located on floodplains can operate only if flood flow is not restricted. Construction and filling on wetlands is restricted. Landfills over faults require special analysis and possibly construction practices. Landfills in seismic impact zones require special analysis and possibly construction practices. Landfills on unstable soils require special analysis and possibly construction practices.
Operating criteria	Landfill operators must conduct a random load-checking program to ensure exclusion of hazardous waste. Daily cover with 6 in of soil or other suitable materials is required. Disease vector control is required. Permanent monitoring probes are required. Probes must be tested every 3 months. Methane concentrations in occupied structures cannot exceed 1.25 percent. Methane migration off-site must not exceed 5 percent at the property line. Clean Air Act criteria must be satisfied. Access must be limited by fences or other structures. Surface water drainage run-on to the landfill and runoff from the working face must be controlled for 25-year rainfall events. Appropriate permits must be obtained for surface water discharges. Liquid wastes or wastes containing free liquids cannot be landfilled. Extensive landfill operating records must be maintained.
Liner design criteria	Geomembrane and soil liners or equivalent are required under most new landfill cells. Groundwater standards may be allowed as the basis for liner design in some states.
Groundwater monitoring	Groundwater monitoring wells must be installed at many landfills. Groundwater monitoring wells must be sampled at least twice per year. A corrective action program must be initiated where groundwater contamination is detected.
Closure and postclosure care	Landfill final cover must be in place within 6 months of closure. The type of cover is soil or geomembrane and must be less permeable than the landfill liner. Postclosure care and monitoring of the landfill must continue for 30 years.
Financial assurance	Sufficient financial reserves must be established during the site operating period to pay for closure and postclosure care amounts.

Source: 40 CFR Parts 257 and 258, 1991.

Landfill Siting Considerations

One of the most difficult tasks faced by public agencies and private waste management firms in implementing an integrated waste management program is the siting of new landfills. Factors that must be considered in evaluating potential sites for the long-term disposal of solid waste include:

- Haul distance
- Location restrictions

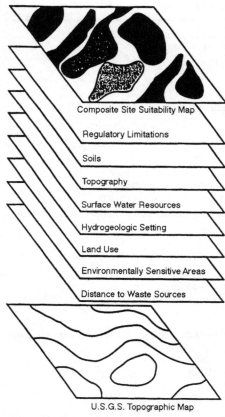

Composite Site Suitability Map

Regulatory Limitations

Soils

Topography

Surface Water Resources

Hydrogeologic Setting

Land Use

Environmentally Sensitive Areas

Distance to Waste Sources

U.S.G.S. Topographic Map

FIGURE 14.4   Overlay maps of various site criteria used in the screening of potential landfill sites. *(From Barlaz et al., 1989.)*

- Available land area
- Site access
- Soil conditions and topography
- Climatalogical conditions
- Surface-water hydrology
- Geologic and hydrogeologic conditions
- Existing land use patterns
- Local environmental conditions
- Potential ultimate uses for the completed site

Final selection of a disposal site usually is based on the results of a detailed site survey, results of engineering design and cost studies, the conducting of one or more environmental impact assessments, and the outcome of public hearings. An overlay procedure for assembling and displaying the relevant site selection information is illustrated in Fig. 14.4. A site scoring procedure that compares proposed sites to an ideal site provides a method for rating sites with a wide range of attributes (Baldasano et al., 1999). The list of technically feasible sites can be refined using input from the public (Thomas and Barlaz, 1999). Landfills are often viewed as LULUs (locally undesirable land uses). With this in mind the public's viewpoints must be incorporated into the landfill development process (Blight and Fourie, 1999). State or local regulations may specify special procedures for interacting with the public when siting a landfill. Often an extensive public information and negotiation process must be conducted concurrently with the technical development activities to site a new landfill successfully. An example is shown in Fig. 14.5. The public's challenges to landfills during the siting process are understandable, but economic impact studies of landfills generally do not show widespread reduction of property values.

14.2   GENERATION AND COMPOSITION OF LANDFILL GASES

A solid waste landfill can be conceptualized as a biochemical reactor, with solid waste and water as the major inputs, and with *landfill gas* and *leachate* as the principal outputs. Material stored in the landfill includes partially biodegraded organic material and the other inorganic waste materials originally placed in the landfill. Landfill gas control systems are employed to prevent unwanted movement of landfill gas into the atmosphere. The recovered landfill gas can be used to produce energy or flared under controlled conditions to eliminate the discharge of harmful constituents to the atmosphere. These topics are considered in greater detail in subsections that follow.

Generation of Landfill Gases

The generation of the principal landfill gases (CO_2 and CH_4), the variation in their rate of generation with time, and the sources of trace gases in landfills are considered in the following discussion.

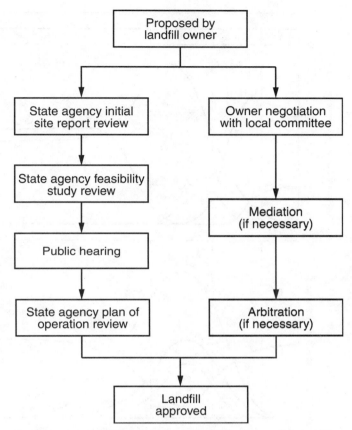

FIGURE 14.5   State Landfill Approval Regulatory Process (*Solid and Hazardous Waste Education Center, University of Wisconsin–Madison, Shannon Morris, 2001.*)

Principal Landfill Gases.   The generation of principal landfill gases is thought to occur in five more or less sequential phases, as illustrated in Fig. 14.6. Each of these phases is described here; additional details may be found in Christensen and Kjeldsen (1989), Emcon Associates (1980), Farquhar and Rovers (1973), Parker (1983) and Pohland (1987, 1991). Additional details on the anaerobic digestion process may be found in Akesson and Nilsson (1998), Baldwin et al. (1998), Barlaz et al. (1989), Holland et al. (1987), and Manna et al. (1999).

Phase I—Initial Adjustment.   Phase I is the *initial adjustment phase,* in which the organic biodegradable components in municipal solid waste begin to undergo bacterial decomposition soon after they are placed in a landfill. In Phase I, biological decomposition occurs under aerobic conditions because a certain amount of air is trapped within the landfill. The principal source of both the aerobic and the anaerobic organisms responsible for waste decomposition is the soil material that is used as a daily and final cover. Digested wastewater treatment plant sludge, disposed of in many MSW landfills, and recycled leachate are other sources of organisms.

Phase II—Transition Phase.   In Phase II, identified as the *transition phase,* oxygen is depleted and anaerobic conditions begin to develop. As the landfill becomes anaerobic, nitrate and sulfate, which can serve as electron acceptors in biological conversion reactions, are often reduced to nitrogen gas and hydrogen sulfide. Measuring the oxidation/reduction potential can monitor the onset of anaerobic conditions. Reducing conditions sufficient to

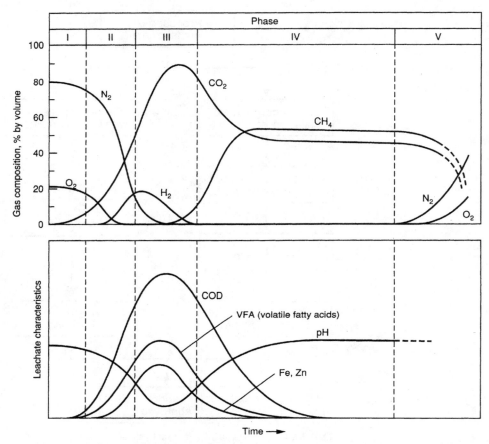

FIGURE 14.6 Generalized phases in the generation of landfill gases (I—Initial Adjustment, II—Transition Phase, III—Acid Phase, IV—Methane Fermentation, and V—Maturation Phase). *(Adapted from Farquhar and Rovers, 1973; Parker, 1983; Pohland, 1987; and Pohland, 1991.)*

bring about the reduction of nitrate and sulfate occur at about −50 to −100 mV. The production of methane occurs when the oxidation/reduction potential values are in the range from −150 to −300 mV. As the oxidation/reduction potential continues to decrease, members of the consortium of microorganisms responsible for the conversion of the organic material in MSW to methane and carbon dioxide begin the three-step process in which the complex organic material is converted to organic acids and other intermediate products, as described in Phase III. In Phase II, the pH of the leachate, if formed, starts to drop due to the presence of organic acids and the effect of the elevated concentrations of CO_2 within the landfill (see Fig. 14.6).

Phase III—Acid Phase. In Phase III, known as the *acid phase*, the bacterial activity initiated in Phase II is accelerated with the production of significant amounts of organic acids and lesser amounts of hydrogen gas. The first step in the three-step process involves the enzyme-mediated transformation (hydrolysis) of higher-molecular-mass compounds (e.g., lipids, organic polymers, and proteins) into compounds suitable for use by microorganisms as a source of energy and cell carbon. The second step in the process (acidogenesis) involves the bacterial conversion of the compounds resulting from the first step into lower-molecular-weight intermediate compounds as typified by acetic acid (CH_3COOH) and small concentrations of fulvic and other more complex organic acids. CO_2 is the principal gas generated during Phase III. Smaller amounts of hydrogen gas (H_2) will also be produced. The microorganisms involved in this conversion, described collectively as nonmethanogenic, consist of facultative and obligate anaerobic bacteria. These microorganisms are often identified in the literature as *acidogens* or *acid formers*.

Because of the acids produced during Phase III, the pH of the liquids held within the landfill will drop. The pH of the leachate, if formed, will often drop to a value of 5 or lower because of the presence of the organic acids and the effect of the elevated concentrations of CO_2 within the landfill. The biochemical oxygen demand (BOD_5), the chemical oxygen demand (COD), and the conductivity of the leachate will increase significantly during Phase III due to the dissolution of the organic acids in the leachate. Also, because of the low pH values in the leachate, a number of inorganic constituents, principally heavy metals, will be solubilized during Phase III. Many essential nutrients are also removed in the leachate in Phase III. If leachate is not recycled, the essential nutrients will be lost from the system. It is important to note that if leachate is not formed, the conversion products produced during Phase III will remain within the landfill as sorbed constituents and in the water held by the waste as defined by the field capacity (see Sec. 14.4).

Phase IV—Methane Fermentation Phase. In Phase IV, known as the *methane fermentation phase,* a second group of microorganisms, which converts the acetic acid and hydrogen gas formed by the acid formers in the acid phase to methane (CH_4) and CO_2, becomes more predominant. In some cases, these organisms will begin to develop toward the end of Phase III. The bacteria responsible for this conversion are strict anaerobes and are called *methanogenic.* Collectively, they are identified in the literature as *methanogens* or *methane formers.* In Phase IV, both methane and acid fermentation proceed simultaneously, although the rate of acid fermentation is considerably reduced.

Because the acids and the hydrogen gas produced by the acid formers have been converted to CH_4 and CO_2 in Phase IV, the pH within the landfill will rise to more neutral values in the range of 6.8 to 8. In turn, the pH of the leachate, if formed, will rise, and the concentration of BOD_5 and COD and the conductivity value of the leachate will be reduced. With higher pH values, fewer inorganic constituents are solubilized; as a result, the concentration of heavy metals present in the leachate will also be reduced.

Phase V—Maturation Phase. Phase V, known as the *maturation phase,* occurs after the readily available biodegradable organic material has been converted to CH_4 and CO_2 in Phase IV. As moisture continues to migrate through the waste, portions of the biodegradable material that were previously unavailable will be converted. The rate of landfill gas generation diminishes significantly in Phase V, because most of the available nutrients have been removed with the leachate during the previous phases and the substrates that remain in the landfill are slowly biodegradable. The principal landfill gases evolved in Phase V are CH_4 and CO_2. Depending on the landfill closure measures, small amounts of nitrogen and oxygen may also be found in the landfill gas. During the maturation phase, the leachate will often contain higher concentrations of humic and fulvic acids, which are difficult to process further biologically.

Duration of Phases. The duration of the individual phases in the production of landfill gas will vary depending on the distribution of the organic components in landfill, the availability of nutrients, the moisture content of waste, moisture routing through the waste material, and the degree of initial compaction. For example, if several loads of brush are compacted together, the carbon/nitrogen ratio and the nutrient balance may not be favorable for the production of landfill gas. The generation of landfill gas will be retarded if sufficient moisture is not available. Increasing the density of the material placed in the landfill will decrease the availability of moisture to some parts of the waste and thus reduce the rate of bioconversion and gas production. Typical data on the percentage distribution of principal gases found in a newly completed landfill as a function of time are reported in Table 14.2.

Volume of Gas Produced. The general anaerobic transformation of the organic portion of the solid waste placed in a landfill can be described by the following equation.

TABLE 14.2   Typical Percentage Distribution of Landfill Gases during the First 48 Months

Time interval since cell completion, months	Average percent by volume		
	Nitrogen, N_2	Carbon dioxide, CO_2	Methane, CH_4
0–3	5.2	88	5
3–6	3.8	76	21
6–12	0.4	65	29
12–18	1.1	52	40
18–24	0.4	53	47
24–30	0.2	52	48
30–36	1.3	46	51
36–42	0.9	50	47
42–48	0.4	51	48

Source: Merz and Stone (1970)

$$\text{Organic matter} + H_2O + \text{nutrients} \rightarrow \text{new cells} + \text{resistant organic matter}$$
$$+ CO_2 + CH_4 + NH_3 + H_2S + \text{heat} \tag{14.1}$$

Assuming methane, carbon dioxide, and ammonia are the principal gases that are produced, Eq. (14.1) can be represented with the following equation (Rich, 1963):

$$C_aH_bO_cN_d \rightarrow nC_wH_xO_yN_z + mCH_4 + sCO_2 + rH_2O + (d - nx)NH_3 \tag{14.2}$$

where $s = a - nw - m$ and $r = c - ny - 2s$. The terms $C_aH_bO_cN_d$ and $C_wH_xO_yN_z$ are used to represent (on a molar basis) the composition of the organic material present at the start and the end of the process, respectively. If it is assumed that the biodegradable portion of the organic waste is stabilized completely, the corresponding expression is

$$C_aH_bO_cN_d + \left(\frac{4a - b - 2c + 3d}{4}\right)H_2O \rightarrow \left(\frac{4a + b - 2c - 3d}{8}\right)CH_4$$

$$+ \left(\frac{4a - b + 2c + 3d}{8}\right)CO_2 + dNH_3 \tag{14.3}$$

An important point to note is that the reaction given by Eq. (14.3) requires the presence of water. Landfills lacking sufficient moisture content have been found in a "mummified" condition, with decades-old newsprint still readable. Hence, although the total amount of gas that will be produced from solid waste derives straightforwardly from the reaction stoichiometry, the rate and the period of time over which that gas production takes place will vary significantly with local hydrologic conditions and landfill operating procedures.

The volume of the gases released during anaerobic decomposition can be estimated in a number of ways. For example, if the individual organic constituents found in MSW (with the exception of plastics) are represented with a generalized formula of the form $C_aH_bO_cN_d$, then the total volume of gas can be estimated by using Eq. (14.3). In general, the organic materials present in solid wastes can be divided into two classifications: (1) those materials that will decompose rapidly (3 months to 5 years) and (2) those materials that will decompose slowly (up to 50 years or more). The rapidly decomposable components of the organic fraction of MSW include food waste, newspaper, cardboard, and a portion of the yard wastes. The slowly decomposable components of the organic fraction of MSW include rubber, leather, the woody portions of yard waste, and wood. The theoretical amount of gas that would be expected under optimum conditions from the conversion of the rapidly and slowly biodegradable organic wastes in a landfill will vary from 12 to 15 and 14 to 16 ft³/lb of biodegradable organic solids destroyed, respectively. However, because the biodegradable fraction of the organic waste depends to a large extent on the lignin content of the waste, not all of the organic matter will be degraded at the same rate. Widely varying rates have been observed in the field, with the typical values ranging between 1 and 4 ft³/lb of MSW.

Variation in Gas Production with Time.   The overall rate at which the organic material in a landfill will be decomposed biologically will, as noted previously, depend on the distribution of the organic components in landfill, the availability of nutrients, the moisture content of waste, the routing of moisture through the fill, and the degree of initial compaction. Under normal conditions, the rate of decomposition of mixed organic wastes deposited in a landfill, as measured by gas production, reaches a peak within the first 2 years and then slowly tapers off, continuing in many cases for periods up to 25 years or more. If moisture is not added to the wastes in a well-compacted landfill, it is not uncommon to find materials in their original form years after they were buried.

The variation in the rate of gas produced from the anaerobic decomposition of the rapidly (5 years or less—some highly biodegradable wastes are decomposed within days of being placed in a landfill) and slowly (5 to 50 years) biodegradable organic materials in MSW can be modeled as shown in Fig. 14.7. As shown in Fig. 14.7, the yearly rates of decomposition for rapidly and slowly decomposable material are based on a triangular gas production model in which the peak rate of gas production occurs in 1 and 5 years, respectively, after gas production starts. Gas production is assumed to start at the end of the first full year of landfill

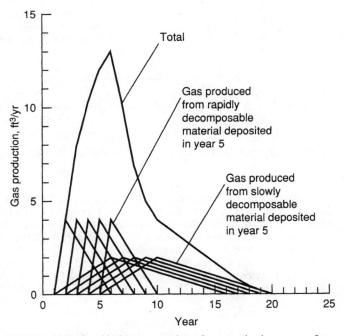

FIGURE 14.7 Graphical representation of gas production over a 5-year period from the rapidly and slowly decomposable organic materials placed in a landfill.

operation. Although a triangular gas production model is used in Fig. 14.7, it should be noted that a variety of different models have been used, including a first-order model.

The total rate of gas production from a landfill in which wastes were placed for a period of 5 years is obtained graphically by summing the amount of gas produced from the rapidly and slowly biodegradable portions of the MSW deposited each year (see Fig. 14.7). The total amount of gas produced corresponds to the area under the rate curve. As noted previously, in many landfills the available moisture is insufficient to allow for the complete conversion of the biodegradable organic constituents in the MSW. The optimum moisture content for the conversion of the biodegradable organic matter in MSW is on the order of 45 to 60 percent. Also, in many landfills, the moisture that is present is not distributed uniformly. When the moisture content of the landfill is limited, the gas production curve is more flattened out and is extended over a greater period of time. An example of the effect of reduced moisture content on the production of landfill gas is illustrated in Fig. 14.8. The production of landfill gas over extended periods of time is of great significance with respect to the management strategy to be adopted for postclosure maintenance. The goal of leachate recirculation is to enhance the rate of gas production and thus reduce the time required to stabilize the biodegradable organic matter in the landfill.

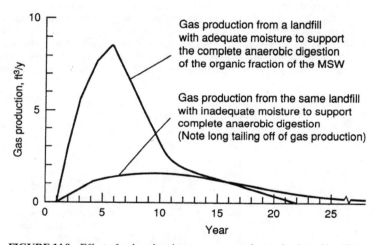

FIGURE 14.8   Effect of reduced moisture content on the production of landfill gas.

Variations in temperature, landfill cell depth, and waste density also will influence the amount of gas and timing of gas generation. Findings in studies (El-Fadel et al., 1997; Manna et al., 1999) showed that the more refined estimates of gas generation rates are possible by incorporating mathematical models of landfill density, layering, biochemical feedback, and temperature into gas generation models.

Landfills are frequently cited (Baldasano and Soriano, 2000; Baldasano et al., 1999; and Irving et al., 1999) as a source of greenhouse gases. The Global Warming Potential (GWP) of MSW is estimated to be 2.323 tons of CO_2 per ton of waste landfilled. Specifically,

CH_4 emission expressed in terms of CO_2
$= 0.085$ tons CH_4 per ton MSW $\times 25$ tons CO_2 per ton CH_4
$= 2.13$ tons CO_2 per ton MSW

CO_2 emission $= 0.193$ tons CO_2 per ton MSW

CO_2 total equivalent GWP $= 2.323$ tons CO_2 per ton MSW

Procedures for preparing national estimates of methane emissions from landfilling practices are described in Kmet et al. (1981). Recent research has begun to investigate the oxidation of methane as it moves through the landfill cover soil (Lang et al., 1987). Oxidation of 14 to 24 percent of the methane moving through the landfill cover soil has been observed at two North American landfills. In an effort to control methane emissions, research is ongoing to identify the soil conditions that will optimize methane oxidation in the landfill cover (Lang and Tchobanoglous, 1989; Thomas and Barlaz, 1999).

Sources of Trace Gases.   Trace constituents in landfill gases have two basic sources. They may be brought to the landfill with the incoming waste or they may be produced by biotic and abiotic conversion reactions occurring within the landfill. Trace compounds mixed with the incoming waste are typically in liquid form, but tend to volatilize. As noted previously, the occurrence of significant concentrations of volatile organic compounds (VOCs) in landfill gas is associated with older landfills, which accepted industrial and commercial wastes that contained VOCs and other organic compounds from which VOCs can be derived. In newer landfills, where the disposal of hazardous waste has been banned, the concentrations of VOCs in the landfill gas have been reduced significantly. Laboratory studies verify this result (Thomas and Barlaz, 1999).

Composition of Landfill Gas

Landfill gas comprises a number of gases that are present in large amounts (the principal gases) and in very small amounts (the trace gases). The principal gases are produced from the decomposition of the biodegradable organic fraction of MSW. Trace gases, although present in small percentages, may be toxic and could present risks to public health.

Principal Landfill Gas Constituents.   Gases found in landfills include ammonia (NH_3), carbon dioxide (CO_2), carbon monoxide (CO), hydrogen (H_2), hydrogen sulfide (H_2S), methane (CH_4), nitrogen (N_2), and oxygen (O_2). The typical percentage distribution of the gases found in the landfill is reported in Table 14.3. Data on the molecular weight and density are presented in Table 14.4. As shown in Table 14.3, methane and carbon dioxide are the principal gases pro-

TABLE 14.3   Typical Constituents Found in and Characteristics of Landfill Gas

Component	Percent (dry volume basis)
Methane	45–60
Carbon dioxide	40–60
Nitrogen	2–5
Oxygen	0.1–1.0
Ammonia	0.1–1.0
Sulfides, disulfides, mercaptans, etc.	0–1.0
Hydrogen	0–0.2
Carbon monoxide	0–0.2
Trace constituents	0.01–0.6

Characteristic	Value
Moisture content	Saturated
Specific gravity	1.02–1.06
Temperature, °F	100–160
High heating value, Btu/std ft^3	475–550

Source: Adapted in part from Ham et al. (1979), Lang et al. (1987), and Parker (1983).

TABLE 14.4 Molecular Weight and Density of Gases Found in
Sanitary Landfill at Standard Conditions (0°C, 1 atm)

Gas	Formula	Molecular weight	Density g/L	Density lb/ft³
Air		28.97	1.2928	0.0808
Ammonia	NH_3	17.03	0.7708	0.0482
Carbon dioxide	CO_2	44.00	1.9768	0.1235
Carbon monoxide	CO	28.00	1.2501	0.0781
Hydrogen	H_2	2.016	0.0898	0.0056
Hydrogen sulfide	H_2S	34.08	1.5392	0.0961
Methane	CH_4	16.03	0.7167	0.0448
Nitrogen	N_2	28.02	1.2507	0.0782
Oxygen	O_2	32.00	1.4289	0.0892

Note: For ideal gas behavior, the density is equal to mp/RT where m is the
molecular weight of the gas, p is the pressure, R is the universal gas constant, and
T is the temperature.
Source: Adapted from Perry et al. (1984).

duced from the anaerobic decomposition of the biodegradable organic waste components in
MSW. When methane is present in the air in concentrations between 5 and 15 percent, it is
explosive. Because only limited amounts of oxygen are present in a landfill when methane concentrations reach this critical level, there is little danger that the landfill will explode. However,
methane mixtures in the explosive range can be formed if landfill gas migrates off-site and is
mixed with air. The concentration of these gases that may be expected in the leachate will
depend on their concentration in the gas phase in contact with the leachate.

Trace Landfill Gas Constituents. Summary data on the concentration of trace compounds
found in landfill gas samples from 66 landfills are reported in Table 14.5. In another study conducted in England, gas samples were collected from three different landfills and analyzed for
154 compounds. A total of 116 organic compounds were found in landfill gas (Young and
Heasman, 1985). Many of the compounds found would be classified as VOCs. The data presented in Table 14.5 are representative of the trace compounds found at most MSW landfills.
The presence of these gases in the leachate that is removed from the landfill will depend on
their concentration in the landfill gas in contact with the leachate. It should be noted that the
occurrence of significant concentrations of volatile organic compounds in landfill gas is associated with older landfills, which accepted industrial and commercial wastes that contained
VOCs. In newer landfills in which the disposal of hazardous waste has been banned, the concentrations of VOCs in the landfill gas have been extremely low. Even at these low concentrations, some landfill operators must install emission control facilities for VOCs to achieve
compliance with air quality protection standards imposed by health-based risk assessment
(Deipser and Stegmann, 1994; Pohland et al., 1993; Pohland et al., 2000; Reinhart, 1993;
Thomas and Barlaz, 1999).

Air Pollution Considerations. The emission of nonmethane hydrocarbons from landfills is
being regulated in the United States. These emission standards apply to landfills over 2.5 million m³ in size, and specify that the quantity of nonmethane hydrocarbons be limited. The typical control mechanism is to install a landfill gas recovery system and an accompanying energy
recovery unit. The collection of the landfill gas in this manner not only results in reduced
emissions, but also recovery of energy. When the methane content falls below an economical
threshold for energy recovery, the landfill gas will then be flared. Other techniques are also
being investigated for reducing emissions by routing the landfill gas through earthen filters
constructed on the top of the landfill.

TABLE 14.5 Typical Concentrations of Trace Compounds Found in
Landfill Gas at 66 California MSW Landfills

Compound	Concentration, ppb by volume		
	Median	Mean	Maximum
Acetone	0	6,838	240,000
Benzene	932	2,057	39,000
Carbon dioxide	330,000,000	10,000,000	534,000,000
Chlorobenzene	0	82	1,640
Chloroform	0	245	12,000
1,1-dichloroethane	0	2,801	36,000
Dichloromethane	1,150	25,694	620,000
1,1-dichloroethene	0	130	4,000
Diethylene chloride	0	2,835	20,000
1,2-trans-dichloroethane	0	36	850
2,3-dichloropropane	0	0	0
1,2-dichloropropane	0	0	0
Ethylene bromide	0	0	0
Ethylene dichloride	0	59	2,100
Ethylene oxide	0	0	0
Ethyl benzene	0	7,334	87,500
Hydrogen sulfide	0	0	0
Hydrogen	0	0	4
Methane	440,000,000	70,000,000	740,000,000
Nitrogen	12	26	98
Oxygen	1	2	17
1,1,2-trichloroethane	0	0	0
1,1,1-trichloroethane	0	615	14,500
Trichloroethylene	0	2,079	32,000
Toluene	8,125	34,907	280,000
1,1,2,2-tetrachloroethane	0	246	16,000
Tetrachloroethylene	260	5,244	180,000
Vinyl chloride	1,150	3,508	32,000
Methyl ethyl ketone	0	3,092	130,000
Styrenes	0	1,517	87,000
Vinyl acetate	0	5,663	240,000
Xylenes	0	2,651	38,000

Source: Adapted from CIWMB (1988).

Movement of Landfill Gases

Under normal conditions, gases produced in soils are released to the atmosphere by means of molecular diffusion. In the case of an active landfill, the internal pressure is usually greater than atmospheric pressure and both convective (pressure-driven) flow and diffusion will release landfill gas. Other factors influencing the movement of landfill gases include the sorption of the gases into liquid or solid components and the generation or consumption of a gas component through chemical reactions or biological activity.

Movement of Principal Gases.   Although most of the methane escapes to the atmosphere, both methane and carbon dioxide have been found at concentrations up to 40 percent each, at lateral distances of up to 400 ft from the edges of unlined landfills. Methane concentrations over 5 percent have been measured at a distance of 1000 ft. The movement of landfill gas in unconsolidated soils is controlled by several mechanisms (Williams et al., 1999). Diffusion and the pressure of the gas in the landfill combine to move the gas away from the landfill. Both aerobic and anaerobic processes, resulting in faster declines in methane concentrations that can be accounted for by dispersion alone, oxidize methane in the gas plume. Geologic variations further complicate prediction of gas movement.

If methane is allowed to migrate underground in an uncontrolled manner, it can accumulate (because its specific gravity is less than that of air) below buildings or in other enclosed spaces at, or close to, a sanitary landfill. With proper venting, methane, with the exception of the fact that it is a greenhouse gas, should not pose a problem by itself. Odorous compounds and VOCs mixed with the methane may lead to odor complaints and the need for emission controls.

Upward Migration of Landfill Gas.   The principal gases, methane and carbon dioxide, can be released through the landfill cover into the atmosphere by convection and diffusion. The diffusive flow through the cover can be estimated by using Eq. (14.4):

$$N_A = D\alpha^{4/3} \frac{(C_{A_{atm}} - C_{A_{fill}})}{L} \tag{14.4}$$

where N_A = gas flux of compound A, g/cm^2 · s (lb-mol/ft^2 · d)
  D = effective diffusion coefficient, cm^2/s (ft^2/d)
  α = total porosity, cm^3/cm^3 (ft^3/ft^3)
  $C_{A_{atm}}$ = concentration of compound A at the surface of the landfill cover, g/cm^3 (lb-mol/ft^3)
  $C_{A_{fill}}$ = concentration of compound A at bottom of the landfill cover, g/cm^3 (lb-mol/ft^3)
  L = depth of the landfill cover, cm (ft)

Typical values for the coefficient of diffusion for methane and carbon dioxide are 0.20 cm^2/s (18.6 ft^2/d) and 0.13 cm^2/s (14.1 ft^2/d), respectively (Lang and Tchobanoglous, 1989). It is also common to assume dry soil conditions, thus $\alpha_{gas} = \alpha$. Assuming dry soil conditions introduces a safety factor in that any infiltration of water into the landfill cover will reduce the gas-filled porosity and thus reduce the vapor flux from the landfill. Typically, porosity values for different types of clay vary from 0.010 to 0.30.

Downward Migration of Landfill Gas.   Ultimately, carbon dioxide, because of its density, can accumulate in the bottom of a landfill. If a clay or soil liner is used, the carbon dioxide can move from there downward primarily by diffusive transport through the liner and the underlying formation until it reaches the groundwater (note that the movement of carbon dioxide can be limited with the use of a geomembrane liner). Because carbon dioxide is readily soluble in water, it usually lowers the pH, which in turn can increase the hardness and mineral content of the groundwater through solubilization.

Movement of Trace Gases.   In a manner similar to that outlined previously for the principal gases, the movement of trace gases due to diffusion can be estimated using the following equation:

$$N_i = \frac{D\alpha^{4/3}(C_{i(s)}W_i)}{L} \tag{14.5}$$

Estimated values of the diffusion coefficient D for 12 trace compounds are reported in Table 14.6 for temperatures varying from 0 to 50°C. Porosity values typically vary from 0.010 to 0.30 for different types of clay. The term $C_{i(s)}W_i$ corresponds to the concentration of the compound in question just below the cover at the top of the landfill. If the value of the term $C_{i(s)}W_i$ is to be estimated in the field, measurements should be taken by inserting a gas probe through the landfill cover, to a point just beyond the bottom of the cover, and recording both the concentration of the compound and the temperature at this point in the landfill. By obtaining actual

TABLE 14.6 Selected Physical Properties for 12 Trace Compounds Found in Landfills

Compound	0°C			10°C			20°C			30°C			40°C			50°C		
	D^*	$vp^\dagger$	$C_s^\ddagger$	D	vp	C_s	D	vp	C_s	D	vp	C_s	D	vp	C_s	D	vp	C_s
Ethyl benzene	0.052	2.0	12.48	0.055	3.9	23.47	0.059	7.3	42.44	0.062	13	73.08	0.066	22	119.7	0.069	36	189.9
Toluene	0.056	6.7	36.26	0.060	12	62.65	0.064	22	110.9	0.068	37	180.4	0.073	59	278.5	0.077	92	420.9
Tetrachloroethene	0.053	4.1	39.95	0.057	7.9	74.27	0.061	15.6	127.1	0.065	24	210.7	0.069	40	340.0	0.073	63	581.9
Benzene	0.066	27	123.9	0.070	47	208.1	0.075	76	325.0	0.081	122	504.6	0.086	185	740.7	0.091	274	1063
1,2-Dichloroethane	0.063	24	139.6	0.068	41	230.0	0.072	62	363.0	0.077	107	560.7	0.082	164	831.9	0.088	243	1194
Trichloroethene	0.059	20	154.5	0.063	36	268.4	0.067	60	424.8	0.072	94	654.5	0.077	146	984.1	0.082	217	1417
1,1,1-Trichloroethane	0.058	36	282.2	0.062	61	461.3	0.067	100	715.9	0.071	153	1061	0.076	231	1580	0.081	338	2240
Carbon tetrachloride	0.058	32	289.3	0.062	54	470.9	0.066	90	741.2	0.071	138	1124	0.075	209	1648	0.080	308	2353
Chloroform	0.065	61	427.9	0.070	100	676.7	0.075	160	1026	0.080	240	1517	0.085	354	2166	0.090	508	3012
1,2-Dichloroethene	0.077	110	626.7	0.082	175	961.8	0.087	269	1428	0.092	399	2048	0.097	576	2862	0.102	810	3901
Dichloromethane	0.074	155	773.6	0.080	242	1165	0.085	349	1702	0.091	536	2410	0.097	763	3322	0.103	1060	4472
Vinyl chloride	0.080	1280	4701	0.085	1810	6413	0.091	2548	8521	0.098	3350	11090	0.104	4410	14130	0.110	5690	17660

* Diffusion coefficient, cm²/s.
† Vapor pressure, mm Hg.
‡ Saturation vapor concentration, g/m³.

Source: From Herrera et al. (1989).

field measurements, an estimate of the average emission rate can be obtained very quickly. If field measurements are not available, the value of the term $C_{i(s)}W_i$ can be estimated using the data given in Table 14.6 for $C_{i(s)}$ and a value of 0.001 as an estimate for W_i.

Trace organics contained within the landfill gases that are moving through the nonsaturated zone of the soil profile may go into solution when they contact liquid water. This could occur either as infiltration through the soil profile moves downgradient or at the groundwater table surface interface. Trace organics from landfill gas have been shown to be a source of significant groundwater contamination. The fact that this does occur further complicates the monitoring of groundwater quality traditional approaches of searching for groundwater contamination downgradient from the landfill. The movement of landfill gas that is not influenced by groundwater gradient further complicates this. If the landfill gas moves in a direction opposite from the groundwater gradient, it is possible that groundwater contamination from the trace organics will actually occur upgradient from the landfill.

Active and Passive Control of Landfill Gases

The release of landfill gases is controlled to reduce atmospheric emissions, to minimize the release of odorous emissions, to minimize subsurface gas migration in unlined landfills, and to allow for the recovery of energy from methane. Control systems can be classified as active or passive. In active gas control systems, energy in the form of an induced vacuum is used to control the flow of gas. In passive gas control systems, the pressure of the gas, which is generated within the landfill, serves as the driving force for the movement of the gas. For both the principal and trace gases, passive control during times when the principal gases are being produced at a high rate can be achieved by providing paths of lower permeability to guide the gas flow in the desired direction. A gravel-packed trench, for example, can serve to channel the gas to a flared vent system. When the production of the principal gases is limited, passive controls are not very effective because the weaker molecular diffusion mechanism will be the primary transport mechanism. The additional consideration is that the cover soils and earth surrounding the landfill may have a significant gas permeability relative to a vent trench. However, at this stage in the life of the landfill it may not be as important to control the residual emission of the methane in the landfill gas. Control of VOC emissions, however, may necessitate the use of both active and passive gas control facilities. It is generally recommended that the use of passive vents be limited to those areas where the chance of methane entering structures via underground pathways is minimal. In areas where buildings are in close proximity to landfills, active venting systems are often recommended.

Active Control of Landfill Gas.   Both vertical and horizontal gas wells have been used for the extraction of landfill gas from within landfills. In some installations both types of wells have been used. The management of the condensate that forms when landfill gas is extracted is also an important element in the design of gas recovery systems.

Vertical Gas Extraction Wells.   A typical gas recovery system using vertical gas extraction wells is illustrated in Fig. 14.9. The wells are spaced so that their radii of influence overlap. For completed landfills, the radius of influence for gas wells is sometimes determined by conducting gas drawdown tests in the field. Typically, an extraction well is installed along with gas probes at regular distances from the well, and the vacuum within the landfill is measured as a vacuum is applied to the extraction well. Both short-term and long-term extraction tests can be conducted. Because the volume of gas produced will diminish with time, some designers prefer to use a uniform well spacing and to control the radius of influence of the well by adjusting the vacuum at the well head. For deep landfills, with a composite cover containing a geomembrane, a 150- to 200-ft spacing is common for landfill gas extraction wells. In landfills with clay and/or soil covers, a closer spacing (e.g., 100 ft) may be required to avoid pulling atmospheric gases into the gas recovery system. The entry of air introduces oxygen into the landfill, which may affect the methane-producing bacteria, and can, by spontaneous combustion, result in the development of an internal landfill fire.

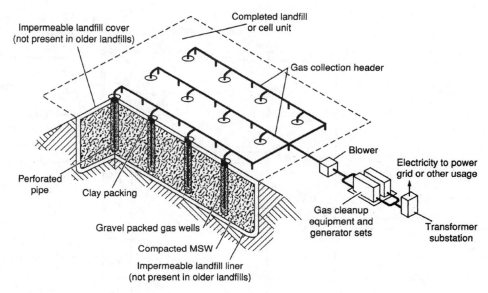

FIGURE 14.9 Landfill gas recovery system using vertical wells.

Vertical gas extraction wells are usually installed after the landfill or portions of the landfill have been completed. In older landfills, vertical wells are installed both to recover energy and to control the movement of gases to adjacent properties. The typical extraction well design consists of 4- to 6-in pipe casing—usually polyvinylchloride (PVQ) or polyethylene (PE)—set in an 18- to 36-in borehole (see Fig. 14.10). The bottom third to half of the casing is perforated and set in a gravel backfill. The remaining length of the casing is not perforated and is backfilled with soil and sealed with a clay (SCS Engineers, Inc., 1989b). Landfill gas recovery wells are typically designed to penetrate 80 percent of the depth of the waste in the landfill, because their radius of influence will extend to the bottom of the landfill. However, to allay the public's fear concerning the escape of landfill gas, some designers now place gas recovery wells all the way to the bottom of the landfill. In instances where effective wells cannot be developed inside the landfill due to well clogging or small radii of influence, wells may be placed in the ground immediately adjacent to the landfill. The available vacuum in the collection manifold at the wellhead is typically 10 in of water.

Horizontal Gas Extraction Wells.   An alternative to vertical gas recovery wells is horizontal wells. The use of horizontal wells was pioneered and developed by the County Sanitation Districts of Los Angeles County (see Figs. 14.11 and 14.12). The use of vertical perimeter wells in conjunction with horizontal gas extraction wells is also illustrated in Figs. 14.11 and 14.14. Horizontal wells are installed after two or more lifts have been completed. The horizontal gas extraction trench is excavated in the solid waste by a backhoe. The trench is then backfilled halfway with gravel, and a perforated pipe with open joints is installed (see Fig. 14.13). The trench is then filled with gravel and capped with solid waste. By using a gravel-filled trench and a perforated pipe with open joints, the gas extraction trench remains functional even with the differential settling that will occur in the landfill with the passage of time (see Fig. 14.13b). The horizontal trenches are installed at approximately 80-ft vertical intervals and on 200-ft horizontal intervals. (Stahl et al., 1982).

Condensate Management.   Condensate forms when the warm landfill gas is cooled as it is transported in the header leading to the blower. Gas collection headers are usually installed with a minimum slope of 3 percent to allow for differential settlement. Because headers are constructed in sections that slope up and down throughout the extent of the landfill, conden-

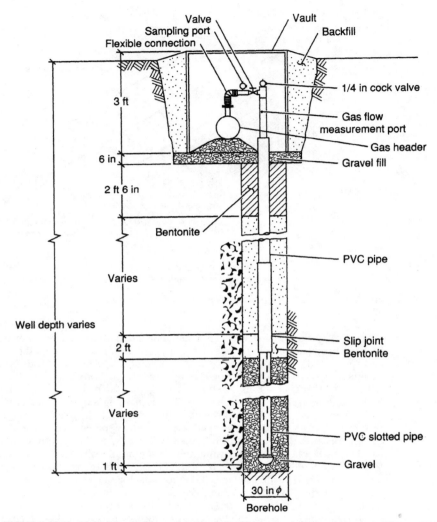

FIGURE 14.10 Typical landfill gas extraction well. (*Courtesy of California Integrated Waste Management Board.*)

sate traps are installed at the low spots in the line (see Fig. 14.9). A typical condensate trap in which the condensate is collected in a holding tank is shown in Fig. 14.14. Condensate from the holding tanks is pumped out periodically and recirculated with leachate to the landfill, transported to an authorized disposal facility, treated on-site prior to disposal, or discharged to a local sewer. In some states, the direct return of condensate to the landfill is allowed.

Passive Control of Landfill Gas.   One of the most common passive methods for the control of landfill gases is based on the fact that relieving gas pressure within the landfill interior can reduce the lateral migration of landfill gas. For this purpose, vents are installed through the final landfill cover extending down into the solid waste mass (see Fig. 14.15). Gas moves through the vent system to the landfill exterior. Due to relatively low gas pressures, many landfills, equipped with passive vents and capped with soil covers, have experienced vegetative

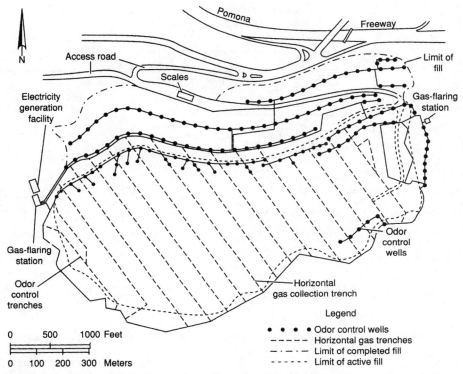

FIGURE 14.11   Plan view of gas collection facilities at Puente Hills landfill. *(Courtesy County Sanitation Districts of Los Angeles County.)*

stress on the landfill cover or underground gas migration outside the landfill, indicating that only a portion of the gas is flowing through the passive vents. These field observations are consistent with mathematical models that predict that passive vents are not effective in controlling gas movement under normal field conditions (Williams et al., 1999). Where landfills are located near occupied buildings, active control systems are usually necessary to achieve adequate migration control.

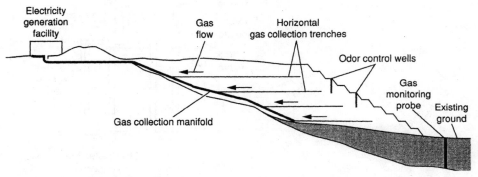

FIGURE 14.12   Sectional view through Puente Hills landfill showing horizontal gas collection trenches. *(Courtesy County Sanitation Districts of Los Angeles County.)*

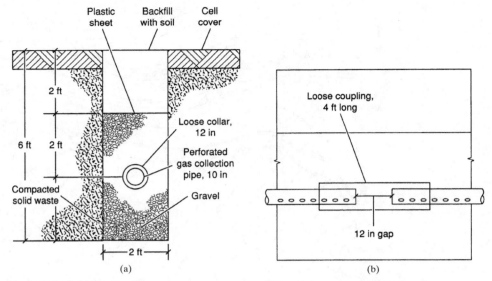

FIGURE 14.13   Details of horizontal gas extraction trench: (*a*) section through trench; (*b*) side view. (*Courtesy County Sanitation Districts of Los Angeles County.*)

If the methane in the venting gas is of sufficient concentration, several vents can be connected and equipped with a gas burner (see Fig. 14.16). Where waste gas burners are used, it is recommended that the well penetrate into the upper waste cells. The height of the waste burner can vary from 10 to 20 ft above the completed fill. The burner can be ignited either by hand or by a continuous pilot flame. To derive maximum benefit from the installation of a waste gas burner, a pilot flame is recommended. It should be noted, however, that passive vents with burners may not achieve the VOC and odor destruction efficiencies that are

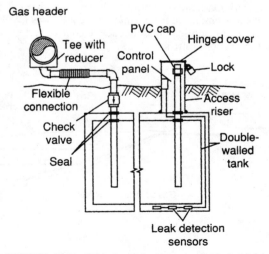

FIGURE 14.14   Typical condensate trap with holding tank.

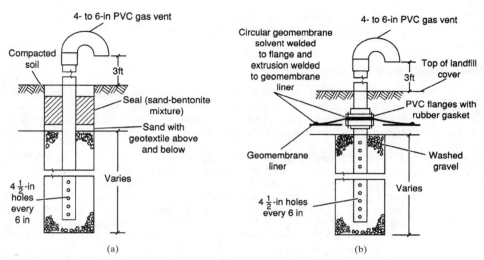

FIGURE 14.15   Typical gas vents used in the surface of a landfill for the passive control of landfill gas: (*a*) gas vent for landfill with a cover that does not contain a geomembrane liner; (*b*) gas vent for a landfill with a cover that contains a synthetic membrane liner. (*Courtesy County Sanitation Districts of Los Angeles County.*)

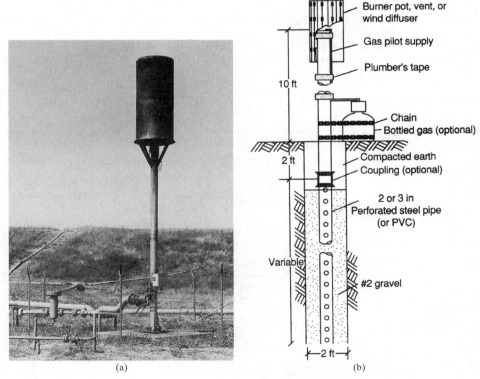

FIGURE 14.16   Typical candlestick-type waste gas burner used to flare landfill gas from a well vent or several interconnected well vents: (*a*) without pilot flame; (*b*) with pilot flame.

required by many urban air quality control agencies, and, thus, their use is not considered good practice. Gas burners are considered later in this section.

Management of Landfill Gas

Typically, landfill gases that have been recovered from an active landfill are either flared or used for the recovery of energy in the form of electricity, or both. More recently, the separation of the carbon dioxide from the methane in landfill gas has been suggested as an alternative to the production of heat and electricity.

Flaring of Landfill Gases.   A common method of treatment for landfill gases is thermal destruction, in which the methane and any other trace gases (including VOCs) are combusted in the presence of oxygen to CO_2, sulfur dioxide (SO_2), oxides of nitrogen, and other related gases. The thermal destruction of landfill gases is usually accomplished in a specially designed flaring facility (see Figs. 14.17 and 14.18). Because of concerns over air pollution, modern flaring facilities are designed to meet rigorous operating specifications to ensure effective destruction of VOCs and other, similar compounds that may be present in the landfill gas. For example, a typical requirement might be a minimum combustion temperature of 1500°F and a residence time of 0.3 to 0.5 s, along with a variety of controls and instrumentation in the flaring station. Typical requirements for a modern flaring facility are summarized in Table 14.7. Where the landfill gas contains less than 15 percent methane, supplemental natural gas or propane may need to be supplied to the flare to sustain combustion. Installation of a carbon filter is an alternative approach to flaring for control of VOCs.

Landfill Gas Energy Recovery Systems.   Landfill gas is usually converted to electricity (see Fig. 14.19). In smaller installations, it is common to use dual-fuel internal combustion piston engines (see Fig. 14.19*a*). In larger installations, the use of turbines is common (see Fig. 14.19*b*). Where piston-type engines are used, the landfill gas must be processed to remove as

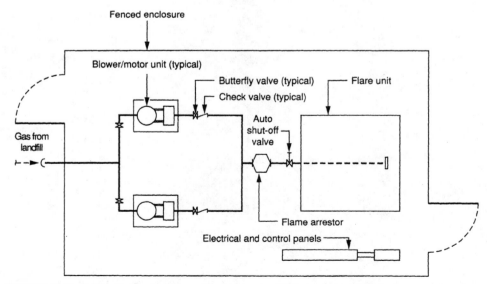

FIGURE 14.17   Schematic layout of blower/flare station for the flaring of landfill gas. (*Courtesy of California Integrated Waste Management Board.*)

FIGURE 14.18 Large array of ground effects flares used to flare landfill gas.

much moisture as possible to limit damage to the cylinder heads. If the gas contains hydrogen sulfide (H_2S), the combustion temperature must be controlled carefully to avoid corrosion problems. Alternatively, the landfill gas can be passed through a scrubber containing iron shavings, or other proprietary scrubbing devices, to remove the H_2S before the gas is combusted.

Combustion temperatures will also be critical where the landfill gas contains VOCs released from wastes placed in the landfill before the disposal of hazardous waste was banned in municipal landfills. The typical service cycle for dual-fuel engines running on landfill gas varies from 3000 to 10,000 h before the engine must be overhauled. In most installations, low-Btu landfill gas is compressed under high pressure so that it can be used more effectively in the gas turbine. The typical service cycle for gas turbines running on landfill gas is approximately 10,000 h.

Other energy recovery methods are also available or under development. Landfill gas can be used to fuel utility boilers at institutional or industrial facilities located near the landfill. After scrubbing, the landfill gas is piped directly from the landfill to the boiler. Another option implemented by some municipalities is to operate vehicles in their service fleet with compressed landfill gas. Fuel cell technology is being developed in an effort to achieve higher conversion efficiencies and lower emissions. A 37 percent energy efficiency was demonstrated with a phosphoric acid fuel cell generating 120 kW (MacKay et al., 1985).

Gas Purification and Recovery.   Where there is a potential use for the CO_2 contained in the landfill gas, the CH_4 and CO_2 in landfill gas can be separated. The separation of the CO_2 from the CH_4 can be accomplished by physical adsorption, chemical adsorption, or membrane separation. In physical and chemical adsorption, one component is adsorbed preferentially by a suitable solvent. Membrane separation involves the use of a semipermeable membrane to remove the CO_2 from the methane. Semipermeable membranes have been developed that

TABLE 14.7   Important Design Elements for Enclosed Ground-Level Landfill Gas Flares

Item	Comments
Automatically controlled combustion air louvers	Used to control the amount of combustion air and the temperature of the flame.
Automatic pilot restart system	To ensure continuous operation.
Failure alarm with an automatic isolation system	The alarm and isolation system is used to isolate the flare from the landfill gas supply line, shut off the blower, and notify a responsible party of the shutdown.
Heat shield	A heat shield should be provided around the top of the flare shroud for use during source testing.
Source test ports with adequate and safe access provided	Test ports used for sampling.
Temperature indicator and recorder	Used to measure and record gas temperature in the flare stack. Whenever the flare is in operation, a temperature of 1500°F or greater must be maintained in the stack as measured by the temperature indicator 0.3 s after passing through the burner.
View ports	A sufficient number of view ports must be available to allow visual inspection of the temperature sensor location within the flare.

Source: Adapted from SCS Engineers, Inc. (1989b).

allow CO_2, H_2S, and H_2O to pass while the CH_4 molecule is retained. Membranes are available as flat sheets or as hollow fibers.

14.3   FORMATION, COMPOSITION, AND MANAGEMENT OF LEACHATE

Leachate may be defined as liquid that has percolated through solid waste and has extracted dissolved or suspended materials. In most landfills, leachate is composed of the liquid that has entered the landfill from external sources, such as surface drainage and rainfall and the liquid produced from the decomposition of the wastes, if any.

Formation of Leachate in Landfills

Preparing a water balance on the landfill can assess the potential for the formation of leachate (Fenn et al., 1975). The water balance involves summing the amounts of water entering the landfill and subtracting the amount of water consumed in chemical reactions and the quantity leaving as water vapor. The potential leachate quantity is the quantity of water in excess of the moisture-holding capacity of the landfill material.

Preparation of Landfill Water Balance.   The components that make up the water balance for a landfill cell are illustrated in Fig. 14.20. As shown in the figure, the principal components involved in the water balance are (1) the water entering the landfill cell from above, the moisture in the solid waste, the moisture in the cover material, and the moisture in the sludge, if the disposal of sludge is allowed, and (2) the water leaving the landfill as part of the landfill gas, as saturated water vapor in the landfill gas, and as leachate.

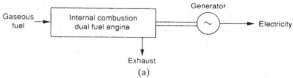

(a)

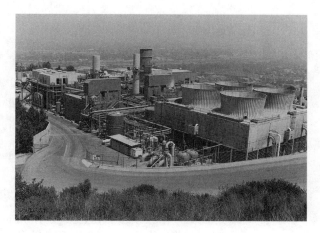

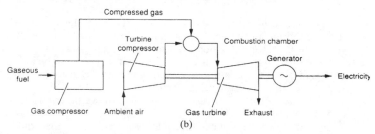

(b)

FIGURE 14.19   Schematic flow diagrams for the recovery of energy from gaseous fuels: (*a*) using internal combustion engine; (*b*) using a gas turbine.

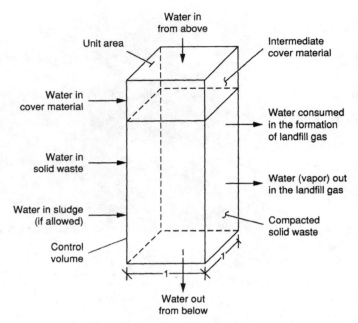

FIGURE 14.20 Definition sketch for water balance used to assess leachate formation in a landfill.

The terms that constitute the water balance can be put into equation form:

$$\Delta S_{SW} = W_{SW} + W_{TS} + W_{CM} + W_{A(R)} - W_{LG} - W_{WV} - W_E + W_{B(L)} \qquad (14.6)$$

where ΔS_{SW} = change in the amount of water stored in solid waste in landfill, lb/yd^3

W_{SW} = water (moisture) in incoming solid waste, lb/yd^3

W_{TS} = water (moisture) in incoming treatment plant sludge, lb/yd^3

W_{CM} = water (moisture) in cover material, lb/yd^3

$W_{A(R)}$ = water from above (for upper landfill layer water from above corresponding to rainfall, or water from snowfall), lb/yd^2

W_{LG} = water lost in the formation of landfill gas, lb/yd^3

W_{WV} = water lost as saturated water vapor with landfill gas, lb/yd^3

W_E = water lost due to surface evaporation, lb/yd^2

$W_{B(L)}$ = water leaving from bottom of element (for the cell placed directly above a leachate collection system; water from bottom corresponds to leachate), lb/yd^3

Water in Solid Waste.   Water entering the landfill with the waste materials is the moisture that is inherent in the waste material and moisture that has been absorbed from the atmosphere or from rainfall where the storage containers are not sealed properly. In dry climates, some of the inherent moisture contained in the waste can be lost, depending on the conditions of the storage. The moisture content of residential and commercial MSW varies from about 15 to 35 percent, depending on the season.

Water in Cover Material.   The amount of water entering with the cover material will depend on the type and source of the cover material and the season of the year. The maximum amount of moisture that can be contained in the cover material is defined by the field capacity (FC) of the material. The field capacity is defined as the liquid that remains in the pore space subject to the pull of gravity. Typical values for soils range from 6 to 12 percent for sand to 23 to 31 percent for clay loams (see Table 14.12).

Water from Above.   For the upper layer of the landfill, the water from above corresponds to the precipitation that has percolated through the cover material. For the layers below the upper layer, water from above corresponds to the water that has percolated through the solid waste above the layer in question. In landfills with leachate recirculation, the water from above will also include the recirculated leachate. One of the most critical aspects in the preparation of a water balance for a landfill is to determine the amount of the rainfall that actually percolates through the landfill cover layer. Where a geomembrane is not used, the amount of rainfall that percolates through the landfill cover can be determined using the latest version of the Hydrologic Evaluation of Landfill Performance (HELP) model (Schroeder et al., 1984a,b). A simplified method for estimating the amount of percolation that can be expected is presented in Sec. 14.5.

Water Lost in the Formation of Landfill Gas.   Water is consumed during the anaerobic decomposition of the organic constituents in MSW. The amount of water consumed by the decomposition reaction can be estimated by using Eq. (14.3). The amount of water consumed per cubic foot of gas produced is typically in the range from 0.012 to 0.015 lb H_2O/ft^3 of gas.

Water Lost as Water Vapor.   Landfill gas usually is saturated in water vapor. The quantity of water vapor escaping the landfill is determined by assuming the landfill gas is saturated with water vapor. The numerical value for the mass of water vapor contained per cubic foot of landfill gas at 90°F is about 0.0022 lb H_2O/ft^3 landfill gas.

Water Lost Due to Evaporation.   There will be some loss of moisture to evaporation as the waste is being landfilled. The amounts are not large and are often ignored. The decision to include these variables in the water balance analysis will depend on local conditions.

Water Leaving from Below.   Water leaving from the bottom of the *first* cell of the landfill is termed *leachate.* As noted previously, water leaving the bottom of the second and subsequent cells corresponds to the water entering from above for the cell below the cell in question.

Field Capacity of Solid Waste.   Water entering the landfill that is not consumed and does not exit as water vapor may be held within the landfill or may appear as leachate. Both the waste material and the cover material are capable of holding water against the pull of gravity. The quantity of water that can be held against the pull of gravity is referred to as *field capacity* (FC). The potential quantity of leachate is the amount of moisture within the landfill in excess of the landfill field capacity. The field capacity, which varies with the overburden weight, can be estimated using the following equation (Huitric, 1979, 1980):

$$FC = 0.6 - 0.55 \left(\frac{W}{10,000 + W} \right) \tag{14.7}$$

where FC = field capacity (i.e., the fraction of water in the waste based on the dry weight of
        the waste)
      W = overburden mass calculated at the midheight of the waste in the lift
        in question, lb

The landfill water balance is prepared by adding the mass of water entering a unit area of a particular layer of the landfill during a given time increment to the moisture content of that layer at the end of the previous time increment, and subtracting the mass of water lost from the layer during the current time increment. The result is referred to as the available water in the current time increment for the particular layer of the landfill. To determine whether any leachate will form, the field capacity of landfill is compared to the amount of water that is present. If the field capacity is less than the amount of water present, then leachate will be formed.

In general, it has been found that the quantity of leachate is a direct function of the amount of external water entering the landfill. In fact, if a landfill is constructed properly, the production of leachate can be reduced substantially. When wastewater treatment plant sludge is added to the solid wastes to increase the amount of methane produced, leachate control facilities must be provided. In those landfills where leachate treatment may be required, the most common approach is to transport the leachate to a municipal wastewater treatment plant.

Composition of Leachate

When water percolates through solid wastes that are undergoing decomposition, both biological materials and chemical constituents are leached into solution. Typical data on the characteristics of leachate are reported in Table 14.8 for both new and mature landfills. Because the range of the observed concentration values for the various constituents reported in Table 14.8 is rather large, especially for new landfills, great care should be exercised in using the typical values that are given.

Variations in Leachate Composition.   It should be noted that the chemical composition of leachate will vary greatly depending on the age of landfill and the history of events preceding the time of sampling. For example, if a leachate sample is collected during the acid phase of decomposition (see Fig. 14.6), the pH value will be low and the concentrations of BOD_5, TOC,

TABLE 14.8 Typical Data on the Composition of Leachate from New and Mature Landfills

Constituent	Value, mg/L*		Mature landfill (greater than 10 years)
	New landfill (less than 2 years)		
	Range[†]	Typical[‡]	
BOD_5 (5-day biochemical oxygen demand)	2,000–30,000	10,000	100–200
TOC (total organic carbon)	1,500–20,000	6,000	80–160
COD (chemical oxygen demand)	3,000–60,000	18,000	100–500
Total suspended solids	200–2,000	500	100–400
Organic nitrogen	10–800	200	80–120
Ammonia nitrogen	10–800	200	20–40
Nitrate	5–40	25	5–10
Total phosphorus	5–100	30	5–10
Ortho phosphorus	4–80	20	4–8
Alkalinity as $CaCO_3$	1,000–10,000	3,000	200–1000
pH	4.5–7.5	6	6.6–7.5
Total hardness as $CaCO_3$	300–10,000	3,500	200–500
Calcium	200–3,000	1,000	100–400
Magnesium	50–1,500	250	50–200
Potassium	200–1,000	300	50–400
Sodium	200–2,500	500	100–200
Chloride	200–3,000	500	100–400
Sulfate	50–1,000	300	20–50
Total iron	50–1200	60	20–200

* Except pH, which is unitless.
[†] Representative range of values. Higher maximum values have been reported in the literature for some of the constituents.
[‡] Typical values for new landfills will vary with the metabolic state of the landfill.

Source: Developed from Bagchi (1990), County of Los Angeles and Engineering Science, Inc. (1969), Ehrig (1989), SWPCB (1954), and SWRCB (1967).

COD, nutrients, and heavy metals will be high. If, on the other hand, a leachate sample is collected during the methane fermentation phase (see Fig. 14.6), the pH will be in the range from 6.5 to 7.5, and the BOD_5, TOC, COD, and nutrient concentration values will be significantly lower. Similarly the concentrations of heavy metals will be lower because most metals are less soluble at neutral pH values. The pH of the leachate will depend not only on the concentration of the acids that are present, but also on the partial pressure of the CO_2 in the landfill gas that is in contact with the leachate.

The biodegradability of the leachate will also vary with time. Checking the BOD_5/COD can monitor changes in the biodegradability of the leachate. Initially, the BOD_5/COD ratios will be around 0.5. Ratios in the range from 0.4 to 0.6 are taken as an indication that the organic matter in the leachate is readily biodegradable. In mature landfills, the BOD_5/COD ratio is often in the range of 0.05 to 0.2. The reason that the BOD_5/COD ratio drops is that the leachate from mature landfills typically contains humic and fulvic acids, which are not readily biodegradable.

Because of the variability in the characteristics of leachate, the design of leachate treatment systems is complicated. For example, the type of treatment plant designed to treat a leachate with the characteristics reported for a new landfill would be quite different from one designed to treat the leachate from a mature landfill. The problem of analysis is complicated further by the fact that the leachate that is being generated at any point in time is a mixture of leachate derived from solid waste of different ages.

Trace Compounds. The presence of trace compounds (some of which may pose health risks) will depend on the concentration of these compounds in the gas phase within the landfill. The expected concentrations can be estimated using Henry's law. It is interesting to note that, as more communities and operators of landfills institute programs to limit the disposal of hazardous wastes with MSW, the quality of the leachate from new landfills is improving with respect to the presence of trace constituents. A study of 48 landfills found a clear differentiation between leachate characteristics from hazardous, codisposal, and MSW landfills.

Movement of Leachate in Unlined Landfills

Under normal conditions, leachate is found in the bottom of landfills. From there, its movement in unlined landfills is through the underlying strata, although some lateral movement may also occur, depending on the characteristics of the surrounding material. As leachate percolates through the underlying strata, many of the chemical and biological constituents originally contained in it will be removed by the filtering and adsorptive action of the material composing the strata. In general, the extent of this action depends on the characteristics of the soil, especially the clay content. Because of the potential risk involved in allowing leachate to percolate to the groundwater, best practice calls for its elimination or containment.

Control of Leachate in Landfills

Landfill liners are now commonly used to limit or eliminate the movement of leachate and landfill gases from the landfill site. To date (2000), the use of clay as a liner material has been the favored method of reducing or eliminating the seepage (percolation) of leachate from landfills. Clay is favored for its ability to adsorb and retain many of the chemical constituents found in leachate and for its resistance to the flow of leachate. However, the use of combination composite geosynthetic and clay liners is gaining in popularity, especially because of the resistance afforded by geomembranes to the movement of both leachate and landfill gases and the implementation of U.S. EPA and similar standards in Western Europe. Typical specifications for geomembrane liners are given in Table 14.9.

TABLE 14.9 Performance Tests Used to Measure Properties of Synthetic Liners and Typical Values

Test	Test method	Typical values
Chemical resistance:		
Resistance to chemical waste mixtures	EPA method 9090	10% tensile strength change over 120 days
Resistance to pure chemical reagents	ASTM D543	10% tensile strength change over 7 days
Durability:		
Carbon black percent	ASTM D1603	2%
Carbon black dispersion	ASTM D3015	A-1
Accelerated heat aging	ASTM D573, D1349	Negligible strength change after 1 month at 110°C
Strength category:		
Tensile properties	ASTM D638, Type IV;	2400 lb/in^2
Tensile strength at yield	dumbbell 2 in/min	4000 lb/in^2
Tensile strength at break		15%
Elongation at yield		700%
Stress cracking resistance:		
Environmental stress crack resistance	ASTM D1693, condition C	1500 h
Toughness:		
Tear resistance initiation	ASTM D1004 die C	45 lb
Puncture resistance	FTMS 101B, method 2031	230 lb
Low-temperature brittleness	ASTM D746, procedure B	−94°F

Source: Adapted from Bagchi (1990) and World Waste (1986).

Liner Systems for MSW.   The objective in the design of landfill liners is to minimize the infiltration of leachate into the subsurface soils below the landfill to substantially reduce the potential for groundwater contamination. A number of liner designs have been developed to minimize the movement of leachate into the subsurface below the landfill. Some of the many types of liner designs that have been proposed are illustrated in Fig. 14.21. In the multilayer landfill liner designs illustrated in Fig. 14.21, each of the various layers has a specific function. For example, in Fig. 14.21a the clay layer and the geomembrane serve as a composite barrier to the movement of leachate and landfill gas. The sand layer serves as a collection and drainage layer for any leachate that may be generated within the landfill. The geotextile layer is used to minimize the intermixing of the soil and sand layers. The final soil layer is used to protect the drainage and barrier layers. A modification of the liner design shown in Fig. 14.21a involves the installation of leachate collection pipes in the leachate collection layer. Composite liner designs employing a geomembrane and clay layer provide more protection and are hydraulically more effective than either type of liner alone.

In Fig. 14.21b, a specifically designed open-weave plastic mesh (geonet) and geotextile filter cloth are placed over the geomembrane, which, in turn, is placed over a compacted clay layer. A protective soil layer is placed above the geotextile. The geonet and the geotextile function together as the drainage layer to convey leachate to the leachate collection system. The permeability of the liner system composed of a drainage layer and filter layer is equivalent to that of coarse sand or gravel (see Table 14.12). When preparing a liner system design, the long-term reliability of manufactured materials for drainage media must be compared to the characteristics of soils with regard to biofouling and clogging.

In the liner system shown in Fig. 14.21c, two composite liners, commonly identified as the primary and secondary composite liners, are used. The primary composite liner is used for the

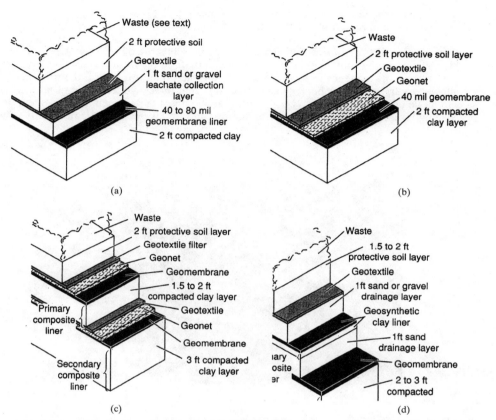

FIGURE 14.21 Typical landfill liners: (*a, b*) single composite barrier types; (*c, d*) double composite barrier types. Note in the double liner systems the first composite liner is often identified as the primary liner or as the leachate collection system while the second composite liner is identified as the leachate detection layer. Leachate detection probes are normally placed between the first and second liners.

collection of leachate, while the secondary composite liner serves as a leak detection system and a backup for the primary composite liner. A modification of the liner system, shown in Fig. 14.21*c*, involves replacing the sand drainage layer with a geonet drainage system, as shown in Fig. 14.21*b*. The two-layer composite design shown in Fig. 14.21*d* is the same as the liner shown in Fig. 14.21*c*, with the exception that the clay layer below the first geomembrane liner is replaced with a geosynthetic clay liner (GCL). A manufactured product, the GCL is made from a high-quality bentonite clay (from Wyoming) and an appropriate binding material. The bentonite clay is essentially a sodium montmorillonite mineral that has the capacity to absorb as much as 10 times its weight in water. As the clay absorbs water, it becomes putty-like and very resistant to the movement of water. Permeabilities as low as 10^{-10} cm/s have been observed. Available in large sheets (12 to 14 by 100 ft), GCLs are overlapped in the construction of a liner system.

Liner Systems for Monofills.   Liner systems for monofills are usually composed of two geomembranes, each provided with a drainage layer and a leachate collection system (see Fig. 14.21*c* and *d*). A leachate detection system is placed between the first and second liners as well as below the lower liner. In many installations, a thick (3- to 5-ft) clay layer is used below the two geomembranes for added protection.

Construction of Clay Liners. In all the liner designs illustrated in Fig. 14.21, great care must be exercised in the construction of the clay layer. Perhaps the most serious problem with the use of clay is its tendency to form cracks due to desiccation. It is critical that the clay not be allowed to dry out as it is being placed. To ensure that the clay liner performs as designed, it should be laid in 4- to 6-in layers, with adequate compaction between the placement of succeeding layers (see Fig. 14.22b). Laying the clay in thin layers limits the possibility of leaks

(a)

(b)

FIGURE 14.22 Preparation of compacted clay layer before geomembrane liner is placed:.(*a*) spreading and compacting clay layer subbase and (*b*) maintaining moisture content of clay layer during compaction.

resulting from the alignment of clods that could occur if the clay layer is applied in a single pass. Another problem that has been encountered when clays of different types have been used is cracking due to differential swelling. To avoid differential swelling, the same type of clay must be used in the construction of the liner. Detailed specifications usually are prepared to describe the construction and testing procedures necessary to achieve a high-quality clay liner.

Construction of Geomembrane Liners. Special care must be exercised when installing geosynthetic lining (and cover) systems. The chemical compatibility of the lining material with the waste materials must be ensured to avoid liner failure (Stessel et al., 1998). Physical failure of the material must also be considered (Kodikara, 2000). A summary of the defects found in a survey of landfill studies is presented in Table 14.10. Various tests can be conducted to determine the strength of the geosynthetic materials. These tests are important for properly specifying the design of materials for lining sideslope liners and covers. Installation of geosynthetics must be accompanied by a quality control and quality assurance program. It is general practice to leak-test all seams in a landfill geomembrane liner. In addition, samples of the lining material should be collected and tested for compliance with the construction specifications.

Leachate Collection Systems

A leachate collection system comprises the landfill liner, the leachate collection system, the leachate removal facilities, and the leachate holding facilities.

TABLE 14.10 Typical Defects and Possible Causes

Stage	Type of defect	Possible cause/comment
Manufacture	Pinholes, excessive thickness changes, poor stress crack resistance	Unusual now for procedures with good quality control; poor resin
Delivery	Scuffing, cuts, brittle cracks, tears, punctures	Unloading with unsuitable plant or lifting equipment; impact; poorly prepared storage areas
Placement	Scratches, cuts, holes, tears, crimps	Dragging sheet along ground, trimming of panels, rough subgrade, use of equipment on top of sheet without protection layer, wind damage, large wrinkles, folds, damage by lifting bars
Welding	Cuts, overheating, scoring, poor adhesion, crimping	Careless edge trimming, welding speed or temperature incorrect, excessive grinding, dirt or damp in weld area, excessive roller pressure
Cover placement	Tears, cuts and scratches, holes, stress in membrane	Action of earthmoving plant, insufficient cover during placement, careless probing of cover depth; contraction of sheet due to ambient temperature reduction
Postinstallation	Holes, tearing, slits, cracks	Puncture from drainage materials, puncture by items of deposited waste, opening of partial depth cuts, pulling apart of poor-quality welds, downdrag stresses caused by settling waste, differential settlement in the base

Source: From Kodikara (2000).

Landfill Liner.   The type of landfill liner selected will depend to a large extent on the local geology and environmental requirements of the landfill site. For example, in locations where there is no groundwater, a single compacted clay liner has been sufficient. In locations where both leachate and gas migration must be controlled, the use of a composite liner composed of a clay liner and a geosynthetic liner with an appropriate drainage and soil protection layer will be necessary. Federal regulations for MSW landfills now mandate either the construction of some type of liner that is equivalent to a geomembrane and clay composite liner, or placing the landfill over a soil formation that will severely restrict leachate movement to protect groundwater quality. New cells added to existing landfills must also comply with this standard. Special wastes are regulated by state standards.

Leachate Collection Facilities.   Collection of the leachate that accumulates in the bottom of a landfill is usually accomplished by using a series of sloped terraces and a system of collection pipes. As shown in Fig. 14.23a, the terraces are sloped so that the leachate that accumulates on the surface of the terraces will drain to leachate collection channels. Perforated pipe, placed in each leachate collection channel (see Fig. 14.23b), is used to convey the collected leachate to a central location, from which it is removed for treatment or reapplication to the surface of the landfill.

The cross slope of the terraces is usually 1 to 5 percent, and the slope of the drainage channels is 0.5 to 1.0 percent. The configuration and slope of the drainage system can be analyzed using the equations developed by Wang (Kmet et al., 1981). The cross slope and flow length of

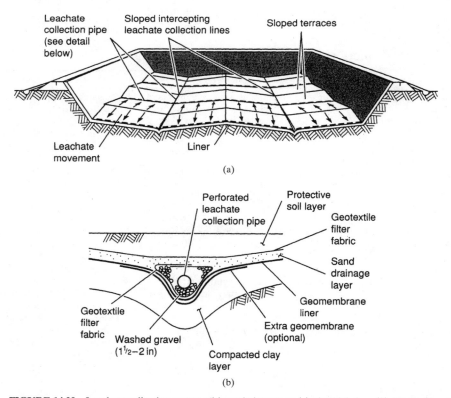

FIGURE 14.23   Leachate collection system with graded terraces: (*a*) pictorial view; (*b*) detail of typical leachate collection pipe.

the terraces determine the depth of leachate above the liner. Flatter and longer slopes result in higher head buildup. The design objective is not to allow the leachate to pond in the bottom of the landfill so as to create a significant hydraulic head on the landfill liner (less than 1 ft at the highest point, as specified in the federal Subtitle D landfill regulations). The depth of flow in the perforated drainage pipe increases continually from the upper reaches of the drainage channel to the lower reaches. In very large landfills, the drainage channels will be connected to a larger cross-collection system.

Leachate Removal and Holding Facilities.   Two methods have been used for the removal of leachate, which accumulates within a landfill. In Fig. 14.24a, the leachate collection pipe is passed though the side of the landfill. Where this method is used, great care must be taken to ensure that the seal where the pipe penetrates the landfill liner is sound. An alternative method used for the removal of leachate from landfills involves the use of an inclined collection pipe located within the landfill (see Fig. 14.24b). Leachate collection facilities are used where the leachate is to be recycled from or treated at a central location. A typical leachate collection access vault is shown in Fig. 14.25a. In some locations, the leachate removed from the landfill is

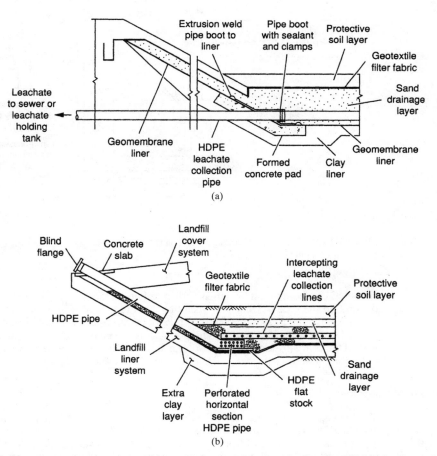

FIGURE 14.24   Typical systems used to collect and remove leachate from landfills: (a) leachate collection pipe passed through side of landfill; (b) inclined leachate collection pipe and pump located within landfill.

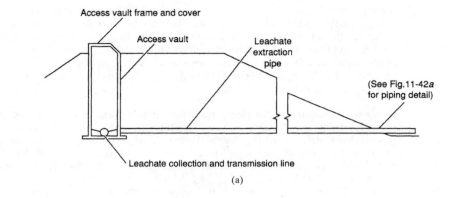

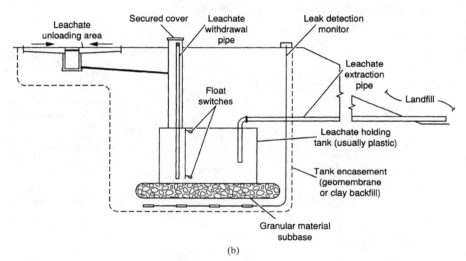

FIGURE 14.25   Examples of leachate collection facilities: (*a*) leachate collection and transmission access vault; (*b*) leachate holding tank.

collected in a holding tank such as shown in Fig. 14.25*b*. The capacity of the holding tank will depend on the type of treatment facilities that are available and the maximum allowable discharge rate to the treatment facility. Typically, leachate-holding tanks are designed to hold from 1 to 3 days' worth of leachate production during the peak leachate production period. Double-walled tanks are preferred because of the added safety they afford compared to a single-walled tank. Tank materials must be carefully selected to resist corrosion.

Leachate Management

The management of leachate—when and if it forms—is key to the elimination of the potential for a landfill to pollute underground aquifers. A number of alternatives have been used to manage the leachate collected from landfills, including (1) leachate recycling, (2) leachate evaporation, (3) treatment followed by spray disposal, (4) wetlands treatment, and (5) discharge to municipal wastewater collection systems. These options are discussed briefly as follows.

Leachate Recycling.   An effective method for the treatment of leachate is to recirculate the leachate through the landfill (see Fig. 14.26).

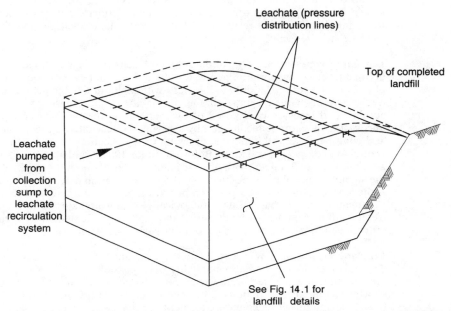

FIGURE 14.26 Schematic of leachate recirculation system used to apply leachate to landfill for treatment. In addition to the leachate distribution system placed on top of the final lift of the landfill as shown, leachate distribution systems are often placed on intermediate lifts.

The rate at which leachate is recirculated will influence the decomposition process (Watts and Charles, 1999). A European study found that higher recirculation rates improve solubilization of the fresh wastes and more quickly establish methanogenic conditions. The benefits of higher recirculation rates must be balanced against the operational concerns associated with the hydraulic limitations that may be experienced at the higher rates.

The design parameters for leachate recirculation systems are not yet fully developed (Koerner and Daniel, 1992). Hydraulic routing through the landfill is the most important consideration. Research shows that it is not possible to view a landfill as either homogeneous or isotropic. Channeling, variable waste permeability, and the effect of soil layers must be considered when recirculating leachate. System failures have occurred where leachate has flowed through the landfill sideslope covers and/or has accumulated on the surface. The design engineer must provide flexible operating systems and provisions for leachate storage outside the landfill. During site operation, low permeable daily and intermediate cover must be avoided or removed before another waste layer is placed. Infiltration into the site and storage of water within the landfill must be appropriately managed to add water to the site when desired and limit water content when moisture content becomes excessive (Reinhart, 1996).

During the early stages of landfill operation, the leachate will contain significant amounts of total dissolved solids (TDS), BOD_5, COD, nutrients, and heavy metals (see Sec. 14.3). When the leachate is recirculated, the constituents are attenuated by the biological activity and by other chemical and physical reactions occurring within the landfill. For example, the simple organic acids present in the leachate will be converted to CH_4 and CO_2. Because of the rise in pH within the landfill when CH_4 is produced, metals will be precipitated and retained within the landfill. An additional benefit of leachate recycling is the recovery of landfill gas that contains CH_4.

Typically, the rate of gas production is greater in leachate recirculation systems. To avoid the uncontrolled release of landfill gases when leachate is recycled for treatment, the landfill should be equipped with a gas recovery system. Ultimately, it will be necessary to collect, treat, and dispose of the residual leachate. In large landfills it may be necessary to provide leachate storage facilities.

Leachate Evaporation.   One of the simplest leachate management systems involves the use of lined leachate evaporation ponds. Leachate that is not evaporated is sprayed on the completed portions of the landfill. In locations with high rainfall, the lined leachate storage facility is covered with a geomembrane during the winter season to exclude rainfall. The accumulated leachate is disposed of by evaporation during the warm summer months by uncovering the storage facility and by spraying the leachate on the surface of the operating and completed landfill. Odorous gases that may accumulate under the surface cover are vented to a compost or soil filter (Bohn and Bohn, 1988; Tchobanoglous et al., 2003). Soil beds are typically 2 to 3 ft deep, with organic loading rates of about 0.1 to 0.25 lb/ft^3 of soil. During the summer when the pond is uncovered, surface aeration may be required to control odors. If the storage pond is not large it can be left covered year round. Another example involves treatment of the leachate (usually biologically) with winter storage and spray disposal of the treated effluent on nearby lands during the summer. If enough land is available, spraying of effluent can be carried out on a continuous basis, even when it is raining.

Leachate Treatment.   Where leachate recycling and evaporation is not used, and the direct disposal of leachate to a treatment facility is not possible, some form of pretreatment or complete treatment will be required. Because the characteristics of the collected leachate can vary so widely, a number of options have been used for the treatment of leachate. The principal biological and physical/chemical treatment operations and processes used for the treatment of leachate are summarized in Table 14.11. The treatment process or processes selected will depend to a large extent on the contaminant(s) to be removed. Typical examples of the types of biological and physical/chemical processes that have been used for the treatment of leachate are shown in Fig. 14.27. Design details on the treatment options reported in Table 14.11 may be found in McQuade and Needham, 1999.

The type of treatment facilities used will depend primarily on the characteristics of the leachate and secondarily on the geographic and physical location of the landfill. Leachate characteristics of concern include TDS, COD, SO_4^{-2}, and heavy metals, as well as nonspecific toxic constituents. Leachate containing extremely high TDS concentrations (e.g., >50,000 mg/L) may be difficult to treat biologically. High COD values favor anaerobic treatment processes, as aerobic treatment is expensive. High sulfate concentrations may limit the use of anaerobic treatment processes due to the production of odors from the biological reduction of sulfate to sulfide. Heavy metal toxicity is also a problem with many biological treatment processes. Another important question is how large should the treatment facilities be. The capacity of the treatment facilities will depend on the size of the landfill and the expected useful life. The presence of nonspecific toxic constituents is often a problem with older landfills that received a variety of wastes, before environmental regulations governing the operation of landfills were enacted.

Wetland Treatment.   Constructed artificial wetlands take advantage of natural processes to reduce leachate pollutant concentrations. Organic matter is most effectively removed with other constituents being degraded or retained within the wetland to a lesser degree. The design parameters for a particular system must be derived from experimental data that takes into account both the leachate characteristics and the soil conditions (Mulamoottil et al., 1999).

TABLE 14.11 Commonly Used Leachate Treatment Processes

Treatment process	Application	Comments
Biological processes		
Activated sludge	Removal of organics from leachate	Defoaming additives may be necessary; separate clarifier needed
Sequencing batch reactors	Removal of organics	Similar to activated sludge, but no separate clarifier needed; applicable only to relatively low flow rates
Aerated stabilization basins	Removal of organics	Requires large land area
Fixed film processes (trickling filters, rotating biological contactors)	Removal of organics	Commonly used on industrial effluents similar to leachates, but untested on actual landfill leachates
Anaerobic lagoons and contactors	Removal of organics	Lower power requirements and sludge production than aerobic systems; requires heating; greater potential for process instability; slower than aerobic systems
Nitrification/ denitrification	Removal of nitrogen	Nitrification/denitrification can be accomplished simultaneously with the removal of organics
Physical/chemical		
Sedimentation/flotation	Removal of suspended matter	Of limited applicability alone; may be used in conjunction with other treatment processes
Filtration	Removal of suspended matter	Useful only as a polishing step
Air stripping	Removal of ammonia or volatile organics	May require air pollution control equipment
Steam stripping	Removal of volatile organics	High energy costs; condensate steam requires further treatment
Adsorption	Removal of organics	Proven technology; variable costs depending on leachate
Ion exchange	Removal of dissolved inorganics	Useful only as a polishing step
Ultrafiltration	Removal of bacteria and high-molecular-weight organics	Subject to fouling; of limited applicability to leachate
Reverse osmosis	Dilute solutions of inorganics	Costly; extensive pretreatment necessary
Neutralization	pH control	Of limited applicability to most leachates
Precipitation	Removal of metals and some anions	Produces a sludge, possibly requiring disposal as a hazardous waste
Oxidation	Removal of organics; detoxification of some inorganic species	Works best on dilute waste streams; use of chlorine can result in formation of chlorinated hydrocarbons
Evaporation	Where leachate discharge is not permissible	Resulting sludge may be hazardous; can be costly except in arid regions
Wet air oxidation	Removal of organics	Costly; works well on refractory organics

Source: Adapted from SCS Engineers, Inc. (1989a) and Tchobanoglous et al. (2003).

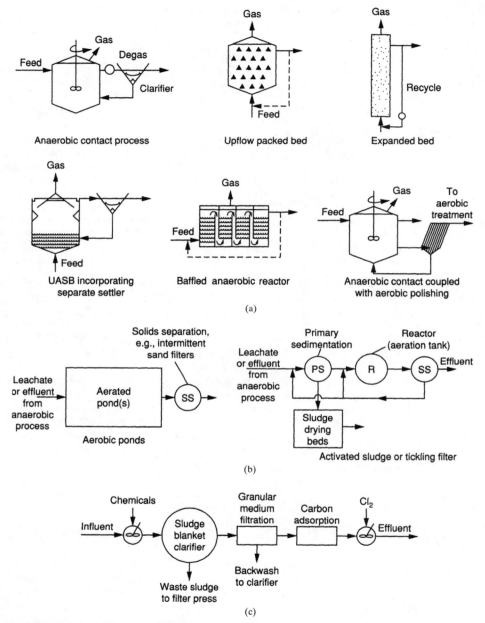

FIGURE 14.27 Typical processes used for the treatment of leachate: (*a*) anaerobic processes; (*b*) aerobic processes; (*c*) chemical treatment process for the removal of heavy metals and selected organics. Additional details on these and other treatment processes may be found in Tchobanoglous et al. (2003).

Discharge to Wastewater Treatment Plant.   In those locations where a landfill is located near a wastewater collection system or where a pressure sewer can be used to connect the landfill leachate collection system to a wastewater collection system, leachate is often discharged to the wastewater collection system. In many cases pretreatment, using one or more of the methods reported in Table 14.11, or leachate recirculation may be required to reduce the organic content before the leachate can be discharged to the sewer. In locations where sewers are not available and evaporation and spray disposal are not feasible, complete treatment followed by surface discharge may be required.

In developing countries where uncontrolled waste practices have resulted in the construction of large waste disposal sites, the amount of leachate being generated is leading to significant concerns. As these facilities are being converted to conventional landfills, it has been found that the quantities of leachate that now require management and collection are often underestimated. This has resulted in more leachate needing treatment than was originally anticipated, overloading of leachate treatment facilities, and, in some instances, significant contamination of surface water resources.

14.4   *INTERMEDIATE AND FINAL LANDFILL COVER*

The use and type of intermediate cover and the design and performance of the final landfill cover are critical issues in the implementation of the landfill method of waste disposal.

Intermediate Cover

Intermediate cover layers are used to cover the wastes placed each day to enhance the aesthetic appearance of the landfill site, to limit the amount of surface infiltration, and to eliminate the harboring of disease vectors. The greatest amount of water that enters a landfill and ultimately becomes leachate enters during the period when the landfill is being filled. Some of the water, in the form of rain and snow, enters while the wastes are being placed in the landfill. Water also enters the landfill by first infiltrating and subsequently percolating through the intermediate landfill cover. Thus, the materials and method of placement of the intermediate cover can limit the amount of surface water that enters the landfill. The question of whether an intermediate cover layer is even needed, or should be required, is currently the subject of renewed debate, especially in landfills designed to recirculate leachate (Wright, 1986).

Materials Used for Intermediate Cover Layers.   The types of materials that have been used as intermediate landfill cover include a variety of native soils, composted MSW, composted yard waste, yard waste mulch, agricultural residues, old carpets, synthetic foam, geomembranes, and construction and demolition waste. Of the materials listed, only the native soils and geomembranes are the most effective in limiting the entry of surface water into the landfill. In general, synthetic foam works well, except when it rains. To be effective, the intermediate cover, using the materials cited here, must be sloped properly to enhance surface water runoff. In some landfill operations, a very thick layer of soil (3 to 6 ft) is placed temporarily over the completed cell. Any rainfall that infiltrates the intermediate cover layer is retained by virtue of its field capacity. When a second lift is to be placed over the first lift, the soil is removed and stockpiled before filling begins. By using the operating technique of temporarily storing additional cover material over a completed cell, the amount of water entering the landfill can be limited significantly.

Intermediate Cover Layers Using Compost.   Where the amount of native soil available for use as intermediate cover material is limited, alternative waste materials have been used for the purpose. Suitable materials that can be used as a substitute for native soil include compost and mulch produced from yard wastes and compost produced from MSW (see Fig. 14.28).

(a)

(b)

FIGURE 14.28   Yard waste used as intermediate landfill cover: (*a*) size of yard waste is reduced using a tub grinder; (*b*) ground-up waste is applied to face of landfill.

An important advantage of using compost and mulch produced from MSW is that the landfill volume that would have been occupied by the soil used for intermediate cover is now available for the disposal of waste materials. In locations where the amount of cover material is limited, the use of composted MSW can increase the capacity of the landfill significantly. Excess compost produced at the landfill site can be stored until it is needed. Cured compost placed on the MSW deposited in the landfill also serves as an odor filter. The use of composted MSW for intermediate cover is expected to increase significantly in the coming years, as the conservation of landfill capacity becomes a more important issue (Moshiri, 1993).

Final Landfill Cover

The primary purposes of the final landfill cover are (1) to minimize the infiltration of water from rainfall and snowfall after the landfill has been completed, (2) to limit the uncontrolled release of landfill gases, (3) to suppress the proliferation of vectors, (4) to limit the potential for fires, (5) to provide a suitable surface for the revegetation of the site, and (6) to serve as the central element in the reclamation of the site. To meet these purposes the landfill cover must do the following (Hatheway et al., 1987; Koerner and Daniel, 1992):

- Be able to withstand climatic extremes (e.g., hot/cold, wet/dry, and freeze/thaw cycles).
- Be able to resist water and wind erosion.
- Have stability against slumping, cracking and slope failure, and downslope slippage or creep.
- Resist the effects of differential landfill settlement caused by the release of landfill gas and the compression of the waste and the foundation soil.
- Resist failure due to surcharge loads resulting from the stockpiling of cover material and the travel of collection vehicles across completed portions of the landfill.
- Resist deformations caused by earthquakes.
- Withstand alterations to cover materials caused by constituents in the landfill gas.
- Resist the disruptions caused by plants, burrowing animals, worms, and insects.

The cover must be configured in such a manner that it can be maintained efficiently and be amenable to relatively easy repair. It is important to note that under current legislation all of these purposes and attributes must continue to be satisfied far into the future. Federal regulations also establish minimum standards for MSW landfill covers, specifying permeability and construction materials. The general features of a landfill cover, some typical types of landfill cover designs, and the long-term performance requirements for landfill covers are considered next.

General Features of Landfill Covers.   A modern landfill cover, as shown in Fig. 14.29, is made up of a series of layers, each of which has a special function. The subbase soil layer is used to contour the surface of the landfill and to serve as a subbase for the barrier layer. In some cover designs, a gas collection layer is placed below the soil layer to transport landfill gas to gas management facilities. The barrier layer is used to restrict the movement of liquids into the landfill and the release of landfill gas through the cover. The drainage layer is used to transport rainwater and snowmelt that percolates through the cover material away from the barrier layer and to reduce the water pressure on the barrier layer. The protective layer is used to protect the drainage and barrier layers. The surface layer is used to contour the surface of the landfill and to support the plants that will be used in the long-term closure design of the landfill.

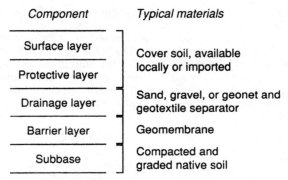

Component	Typical materials
Surface layer	Cover soil, available locally or imported
Protective layer	
Drainage layer	Sand, gravel, or geonet and geotextile separator
Barrier layer	Geomembrane
Subbase	Compacted and graded native soil

FIGURE 14.29   Typical components that constitute a landfill cover.

It should be noted that not all of the layers will be required in each location. Sometimes the subbase layer can also be used as the gas collection layer. Of the layers identified in Fig. 14.29, the barrier layer is the most critical for the reasons cited here (Hatheway et al., 1987; Koerner and Daniel, 1992). Although clay has been used in many existing landfills as the barrier layer, a number of problems are inherent with its use. For example, clay is difficult to compact on a soft foundation, compacted clay can develop cracks due to desiccation, clay can be damaged by freezing, clay will crack due to differential settling, the clay layer in a landfill cover is difficult to repair once damaged, and, finally, the clay layer does not restrict the movement of landfill gas to any significant extent. As a consequence, use of a geomembrane in combination with clay or two or more geomembranes is recommended over the use of clay alone as a barrier layer in landfill covers.

An alternative approach for covering landfills in those regions where evapotranspiration exceeds rainfall is to layer selected soil materials in such a fashion that the total infiltration through the cover is less than or equal to the amount of infiltration through geomembrane constructed covers (Chadwick et al., 1999; Coons et al., 2000). Alternate earthen covers are constructed by selecting soil materials that will moderately reduce the rate of percolation through the soil and, at the same time, store the moisture within the soil cover until such time that evapotranspiration will remove it. This has the effect of relying upon natural conditions to build landfill covers that are equivalent to those covers constructed from geomembranes and clay barrier soils (Chadwick et al., 1999; Coons et al., 2000).

Typical Landfill Cover Designs.   Some of the many types of cover designs that have been proposed and used are illustrated in Fig. 14.30. In Fig. 14.30*a,* the geotextile filter cloth is used to limit the intermixing of the soil with the sand layer. If the available topsoil at the landfill site is not suitable for plant growth, a suitable topsoil must be brought to the site or the available topsoil should be amended to improve its characteristics for plant growth. The use of a composite barrier design comprising a geomembrane and a clay layer is illustrated in Fig. 14.30*b.* In the cover design illustrated in Fig. 14.30*c,* a 6- to 10-ft-thick layer of soil is used as the cover layer. Functionally, the soil layer is sloped adequately to maximize surface runoff. The depth of soil is used to retain rainfall that does not run off and infiltrates into the soil cover. The flexible membrane liner is used to limit the release of landfill gases. In another design, the waste is first covered with a base layer of old carpets or similar materials. A flexible membrane liner is placed over the base layer. A layer of astroturf is placed over the flexible membrane liner. Use of the astroturf is advantageous because the amount of maintenance required is minimized.

Long-Term Performance and Maintenance of Landfill Covers.   Regardless of the design of the final landfill cover, the following question must be considered: How will the integrity and performance of the landfill cover be maintained as the landfill settles, as a result of the loss of

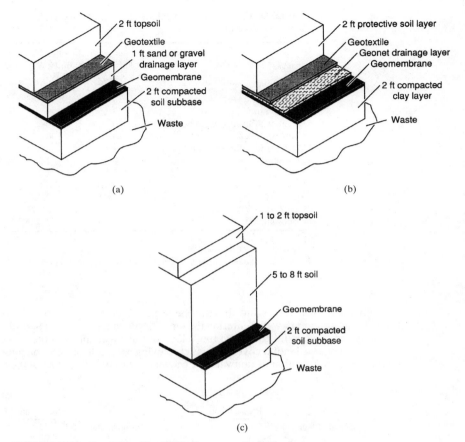

FIGURE 14.30 Examples of landfill final cover configurations.

weight due to the production of landfill gas and by long-term consolidation? For example, how will a composite liner be repaired to maintain adequate drainage? Typically, if settlement occurs, the landfill cover material is stripped back, soil or composted waste is added to adjust the grade, and the various layers replaced. Where a thick soil cover is used, regrading the cover layer may restore proper surface drainage. In drier climates, where vegetation is planted on the soil cover layer, a sprinkler system may be required to sustain the vegetation during the summer. In landfills where astroturf is used, when the turf starts to fall apart, the landfill cover is opened, the used turf is placed in the landfill, the flexible membrane is repaired, and a new astroturf layer is added to the top.

Percolation Through Intermediate and Final Cover Layers

If it is assumed (1) that the cover material is saturated, (2) that a thin layer of water is maintained on the surface, and (3) that there is no resistance to flow below the cover layer, then the theoretical amount of water expressed in gallons that could enter the landfill per unit area in a 24-h period for various cover materials is given in Table 14.12 in column 2. Clearly, these data

TABLE 14.12   Typical Permeability Coefficients for Various Soils (Laminar Flow)

Material	Coefficient of permeability, K	
	ft/d	gal/ft² · d
Uniform coarse sand	1333	9970
Uniform medium sand	333	2490
Clean, well-graded sand and gravel	333	2490
Uniform fine sand	13.3	100
Well-graded silty sand and gravel	1.3	9.7
Silty sand	0.3	2.2
Uniform silt	0.16	1.2
Sandy clay	0.016	0.12
Silty clay	0.003	0.022
Clay (30 to 50% clay sizes)	0.0003	0.0022
Colloidal clay	0.000003	0.000022

Note: ft/day × 0.3048 = m/d
gal/ft² · day × 0.0408 = m³/m² · d

Source: Adapted from Davis and DeWiest (1966) and Salvato et al. (1971).

are only theoretical values, but they can be used in assessing the worst possible situation. In practice, the amount of water entering the landfill will depend on local hydrological conditions, the design of the landfill cover, the final slope of the cover, and whether vegetation has been planted. In general, landfill cover designs employing a flexible membrane liner are designed to eliminate the percolation of rainwater or snowmelt into the waste below the landfill cover.

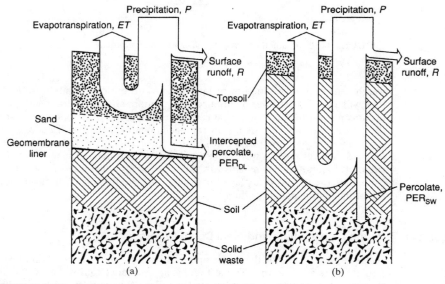

FIGURE 14.31   Definition sketch for water balance for landfill: (*a*) for landfill cover containing a drainage layer and geomembrane liner; (*b*) for landfill with no drainage layer or geomembrane liner.

Estimation of the percolation of rainwater or snowmelt through the soil layer above the drainage layer (see Fig. 14.31*a*) or through a cover layer composed of soil only (see Fig. 14.31*b*) is usually accomplished by using one of the many available hydrologic simulation programs. Perhaps the best known is the Hydrologic Evaluation of Landfill Performance (HELP) model, which is being revised continuously (Schroeder et al., 1984a, 1984b). Percolation through the landfill cover layer can also be estimated using a standard hydrological water balance. Referring to Fig. 14.31, the water balance for a soil landfill cover is given by the following expression:

$$\Delta S_{LC} = P - R - \text{ET} - \text{PER}_{SW} \tag{14.8}$$

where ΔS_{LC} = change in the amount of water held in storage in a unit volume of landfill cover, in
P = amount of precipitation per unit area, in
R = amount of runoff per unit area, in
ET = amount of water lost through evapotranspiration per unit area, in
PER_{SW} = amount of water percolating through unit area of landfill cover into compacted solid waste, in

The total amount of water that can be stored in a unit volume of soil will depend on the field capacity (FC) and the permanent wilting percentage (PWP). The FC is defined as the amount of water that is retained in a soil against the pull of gravity. Soil moisture tension at FC is typically between $\frac{1}{10}$ and $\frac{1}{3}$ atm (Hansen et al., 1979). The PWP is defined as the amount of water left in a soil when plants are no longer able to extract any more. Soil moisture tension at PWP is approximately 15 atm (Hansen et al., 1979). The difference between the FC and the PWP represents the amount of water that can be stored in a soil. Typical FC and PWP values for representative soils are given in Table 14.13. If a layered landfill cover is used, the field capacity of each layer must be considered in the analysis. Typical runoff coefficients for completed landfill covers are given in Table 14.14. Monthly precipitation and evapotranspiration data are site specific, but local weather bureau data are usually acceptable.

TABLE 14.13 Typical Field Capacity (FC) and Permanent Wilting Point (PWP) Values for Various Soil Classifications

	Value, %			
	Field capacity		Permanent wilting point	
Soil classification	Range	Typical	Range	Typical
Sand	6–12	6	2–4	4
Fine sand	8–16	8	3–6	5
Sandy loam	10–18	14	4–8	6
Fine sandy loam	0	0	0	0
Loam	18–26	22	8–12	10
Silty loam	0	0	0	0
Light clay loam	0	0	0	0
Clay loam	23–31	27	11–15	12
Silty clay	27–35	31	12–17	15
Heavy clay loam	0	0	0	0
Clay	31–39	35	15–19	17

Source: Adapted from Hansen et al. (1979), Linsley et al. (1958), and the U.S. Army Corps of Engineers (1956).

TABLE 14.14   Typical Runoff Coefficients for Storms of 5- to 10-Year Frequency

| Type of cover | Slope, % | Runoff coefficient | | | |
| | | With grass | | Without grass | |
		Range	Typical	Range	Typical
Sandy loam	2	0.05–0.10	0.06	0.06–0.14	0.10
	3–6	0.10–0.15	0.12	0.14–0.24	0.18
	7	0.15–0.20	0.17	0.20–0.30	0.24
Silt loam	2	0.12–0.17	0.14	0.25–0.35	0.30
	3–6	0.17–0.25	0.22	0.35–0.45	0.40
	7	0.25–0.36	0.30	0.45–0.55	0.50
Tight clay	2	0.22–0.33	0.25	0.45–0.55	0.50
	3–6	0.30–0.40	0.35	0.55–0.65	0.60
	7	0.40–0.50	0.45	0.65–0.75	0.70

Source: Developed in part from Frevert et al. (1963), Linsley et al. (1958), and WPCF and ASCE (1969).

14.5   *STRUCTURAL AND SETTLEMENT CHARACTERISTICS OF LANDFILLS*

The structural characteristics and settlement of the landfill must be considered in the design of gas collection and surface water drainage facilities, during filling operations, and before a decision is reached on the final use to be made of a completed landfill.

Structural Characteristics

When solid waste is initially placed in a landfill, it behaves in a manner that is quite similar to other fill material. The nominal angle of repose for waste material placed in a landfill is approximately 1.5:1. Because solid waste has a tendency to slip when the slope angle is too steep, the slopes used for the completed portions of a landfill will vary from 2.5:1 to 4:1, with 3:1 being the most common.

As landfills have increased in size, the need to consider slope stability has become much more important. In some cases uncontrolled dumping has resulted in the placement of waste at considerably steep slopes, and at heights in excess of 100 ft. There have been major slope failures at a number of large uncontrolled sites that have resulted in a significant loss of life. In the United States, there has been a major slope failure every few years since 1980 (see Fig. 14.32). To avoid slope failure, it is necessary to do a slope stability analysis as part of the design process.

Slope Stability.   Problems have often occurred where the landfill had more waste placed in it than had been originally anticipated. This can result in either a foundation-type failure or slopes becoming too steep. One sign that failure is about to occur is that cracks may open within the waste, indicating that slight undersurface movement has begun to take place. These cracks need to be carefully monitored, and if the cracks are filled with materials and then reappear, it is a danger signal that a failure may be imminent. Slope stability of a landfill can be determined by conducting a soil and waste mechanics analysis. This would normally take place at any site where the side slopes exceed 3:1. The analysis considers the arrangement of waste placement, the angle of repose of the waste, the stress-strain characteristics of the liner and cover materials, and the ability of the foundation soils to support the landfill. Procedures

FIGURE 14.32   Landfill slope failure. *(Tim Stark, University of Illinois, 1998.)*

for conducting a landfill stability analysis are in Stark (1999), Stark et al. (1998), and Stark et al. (2000).

Because of the problems with slippage, many landfills are benched (see Fig. 14.2c) where the height of the landfill will exceed 50 ft. In addition to helping to maintain slope stability, benches are also used for the placement of surface water drainage channels and for the location of landfill gas recovery piping.

Seismic Protection.   Coupled with the slope stability analysis is seismic protection of landfills. In seismically active areas, it is very important that a seismic analysis be conducted to determine the critical design factors associated with preventing slope failure during an earthquake. Failures have occurred but are less common than the slope stability failures previously described. Seismic failures are associated with ground motion being transmitted through the waste, resulting in a portion of the landfill experiencing a slope failure or the cover materials becoming displaced from the waste material. This seismic analysis requires specialized technical skills. More information is provided in Hashash et al. (2001).

Settlement of Landfills

As the organic material in landfill is decomposed and weight is lost as landfill gas and leachate components, the landfill settles. Settlement also occurs as a result of increasing overburden mass as landfill lifts are added and as water percolates into and out of the landfill. Landfill settlement results in ruptures of the landfill surface and cover, breaks and misalignments of gas recovery facilities, cracking of manholes, and interference with subsequent use of the landfill after closure.

In general, the construction of permanent facilities on completed landfills is not recommended because of the uneven settlement characteristics, varying bearing capacity of the upper layers of the landfill, and the potential problems that can result from gas migration, even with the use of gas collection facilities. When the final use of the landfill is known before waste placement begins, it is possible to control the deposition of certain materials during the operation of the landfill. For example, relatively inert materials such as construction and demolition wastes can be placed in those locations where buildings and/or other physical facilities are to be placed in the future. Recent regulatory trends have further limited the placement of structures on completed landfills.

Effect of Waste Decomposition.   Once placed in a landfill, the organic components of the waste will decompose, resulting in loss of as much as 30 to 40 percent of the original mass. The rate of decomposition is directly related to the moisture content of the waste, with wet waste decomposing the fastest. The loss of mass results in a loss of volume, which becomes available for refilling with new waste. The volume that is lost is usually filled in when higher lifts are subsequently placed over the initial lifts. Weight and volume will also be lost after a landfill is closed.

Effect of Overburden Pressure (Height).   The density of the material placed in the landfill will increase with the weight of the material placed above it, so that the average specific weight of waste in a lift depends on the depth of the lift. The maximum specific weight of solid waste residue in a landfill under overburden pressure will vary from 1750 to 2150 lb/yd³ (Huitric, 1979; Huitric et al., 1980; and Pfeffer, 1992). The following relationship can be used to estimate the increase in the specific weight of the waste as a function of the overburden pressure:

$$D_{W_p} = D_{W_i} + \frac{p}{a + bp} \tag{14.9}$$

where D_{W_p} = specific weight of the landfill material at pressure p, lb/yd³
D_{W_i} = initial compacted specific weight of the waste, lb/yd³
p = overburden pressure, lb/ft²
a = empirical constant, yd³/ft²
b = empirical constant, yd³/lb

Typical specific weights versus applied pressure curves for compacted solid waste for several initial specific weights are shown in Fig. 14.33. The increase in the specific weight of the waste

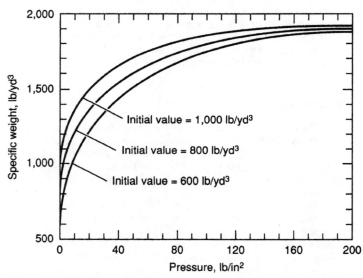

FIGURE 14.33   Specific weight of solid waste placed in landfill as function of the initial compacted specific weight of the waste and the overburden pressure.

material in the landfill is important in (1) determining the actual amount of waste that can be placed in a landfill up to a given grade limitation and (2) determining the degree of settlement that can be expected in a completed landfill after closure.

Extent of Settlement.   The extent of settlement depends on the initial compaction, the characteristics of wastes, the degree of decomposition, the effects of consolidation when water and air are forced out of the compacted solid waste, and the height of the completed fill. Representative data on the degree of settlement to be expected in a landfill as a function of the initial compaction are shown in Fig. 14.34. It has been found in various studies that about 90 percent of the ultimate settlement occurs within the first 5 years.

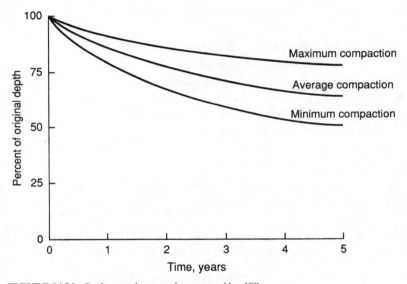

FIGURE 14.34   Surface settlement of compacted landfills.

Several methods have been proposed for predicting settlement rates in landfills that account for waste decomposition, in addition to the weight of the material (Edgers et al., 1992; Edil et al., 1990; El-Fadel et al., 1999; Watts and Charles, 1999). A one-dimensional model (El-Fadel et al., 1999) for landfill settlement is described by the following:

Intermediate secondary settlement:

$$S_{(t)} = H_0 C_{\alpha 1} \log \left(\frac{t}{t_{\text{initial}}} \right) \quad t_{\text{initial}} < t < t_2$$

Long-term secondary settlement:

$$S_{(t)} = H_0 C_{\alpha 2} \log \left(\frac{t}{t_2} \right) \quad t_2 < t < t_{\text{final}}$$

where $C_{\alpha 1}$ = coefficient of intermediate secondary compression (varies from 0.015 to 0.035)
$C_{\alpha 2}$ = coefficient of long-term secondary compression (varies from 0.132 to 0.25)
t_{initial} = end of initial settlement period (16 days)
t_{final} = end of field experiment observations (1576 days)
t_2 = time at which slope of stress-strain curve changes (day)

14.6 LANDFILL DESIGN CONSIDERATIONS

Among the important topics that must be considered in the design of a landfill (though not necessarily in the order given) are the following:

1. Layout of landfill site
2. Types of wastes that must be handled
3. The need for a convenience transfer station
4. Estimation of landfill capacity
5. Evaluation of the local geology and hydrogeology of the site
6. Selection of leachate management facilities
7. Selection of landfill cover
8. Selection of landfill gas control facilities
9. Surface water management
10. Aesthetic design considerations
11. Development of landfill operation plan
12. Determination of equipment requirements
13. Environmental monitoring
14. Public participation
15. Closure and postclosure care

The development of an operational plan for a landfill and the determination of equipment requirements are considered in following sections. Environmental monitoring is considered in Sec. 14.8. Closure and postclosure care is considered in Sec. 14.9. Important factors that must be considered in the design of landfills are reported in Table 14.15. Throughout the development of the engineering design report, careful consideration must be given to the final use or uses to be made of the completed site. Land reserved for administrative offices, buildings, and parking lots should be filled with dirt only. Buildings should be protected from the underground gas migration and should be sealed. Protection can be accomplished with membrane seals or soil gas extraction systems.

Layout of Landfill

In planning the layout of a landfill site, the location of the following must be determined:

1. Access roads
2. Equipment shelters
3. Scales, if used
4. Office space
5. Location of convenience transfer station, if used
6. Storage and/or disposal sites for special wastes
7. Identification of areas to be used for waste processing (e.g., composting)
8. Definition of the landfill areas and areas for stockpiling cover material
9. Drainage facilities
10. Location of landfill gas management facilities
11. Location of leachate treatment facilities, if required

14. Location of monitoring wells

13. Placement of barrier berms or structures to limit sight lines into the landfill

14. Plantings

A typical layout for a landfill disposal site is shown in Fig. 14.35. Because site layout is specific for each case, Fig. 14.35 is meant to serve only as a guide. However, the items identified on Fig. 14.35 can be used as a checklist of the areas that must be addressed in the preliminary layout of a landfill.

TABLE 14.15 Important Factors That Must Be Considered in the Design of Landfills

Factors	Remarks
Access	Paved all-weather access roads to landfill site; temporary roads to unloading areas.
Cell design and construction	Each day's wastes should form one cell; cover at end of day with 6 in of earth or other suitable material; typical cell width, 10–30 ft; typical lift height including intermediate cover, 10–14 ft; slope of working faces, 3:1.
Completed landfill characteristics	Finished slopes of landfill, 3:1; height to bench, if used, 50–75 ft; slope of final landfill cover, 3–6%.
Environmental requirements	Install vadose zone gas- and liquid-monitoring facilities; install up- and downgradient groundwater monitoring facilities; locate ambient air monitoring stations.
Equipment requirements	Number and type of equipment will vary with the type of landfill and the capacity of the landfill.
Final cover	Use multilayer design; slope of final landfill cover, 3–6%.
Fire prevention	Water on site; if nonpotable, outlets must be marked clearly; proper cell separation prevents continuous burn-through if combustion occurs.
Groundwater protection	Divert any underground springs; if required, install perimeter drains, well point system, or other control measures.
Intermediate cover material	Maximize use of on-site soil materials; other materials such as compost produced from yard waste and MSW can also be used to maximize the landfill capacity; typical waste-to-cover ratios vary from 5:1 to 10:1.
Land area	Area should be large enough to hold all community wastes for a minimum of 5 years, but preferably 10 to 25 years; area for buffer strips or zones must also be included.
Landfill gas management	Develop landfill gas management plan including extraction wells, manifold collection system, condensate collection facilities, the vacuum blower facilities, and flaring facilities and or energy production facilities. Operating vacuum at well head, 10 in of water.
Landfill liner	Single clay layer (2–4 ft) or multilayer design incorporating the use of a geomembrane. Cross slope for terrace-type leachate collection systems, 1–2%; maximum flow distance over terrace, 100 ft; slope of drainage channels, 0.5–0.75%. Slope for piped type leachate collection system, 1–2%; size of perforated pipe, 4 in.; pipe spacing, 20 ft.
Landfilling method	Landfilling method will vary with terrain and available cover; most common methods are excavated cell/trench, area, canyon.
Leachate collection	Determine maximum leachate flow rates and size leachate collection pipe and/or trenches; size leachate pumping facilities; select collection pipe materials to withstand static pressures corresponding to the maximum height of the landfill.
Leachate treatment	On basis of expected quantities of leachate and local environmental conditions, select appropriate treatment process.
Surface drainage	Install drainage ditches to divert surface water runoff; maintain 3–6% grade on finished landfill cover to prevent ponding; develop plan to divert stormwater from lined but unused portions of landfill.

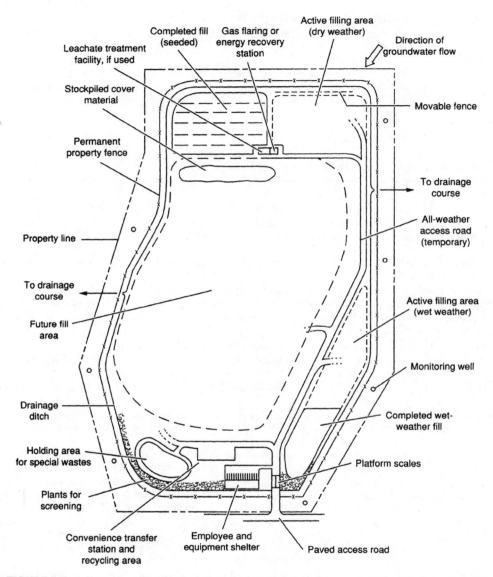

FIGURE 14.35 Typical layout of a landfill site showing all of the elements involved in the implementation of a new landfill (see Figure 14.51 for completed landfill).

Types of Wastes

Knowledge of the types of wastes to be handled is important in the design and layout of a landfill, especially if special wastes are involved. It is usually best to develop separate disposal sites or monofills for designated and special wastes such as asbestos or incinerator ash because, under most conditions, special treatment of the site will be necessary before these wastes can be landfilled. The associated disposal costs are often significant, and it is wasteful to use this landfill capacity for wastes that do not require special precautions. If significant quantities of demolition wastes are to be handled, it may be possible to use them for embankment stabilization. In some cases, it may not be necessary to cover demolition wastes on a daily basis.

Need for a Convenience Transfer Station

Because of safety concerns and the many new restrictions governing the operation of landfills, many operators of landfills have constructed convenience transfer stations at the landfill site for the unloading of wastes brought to the site by individuals and small-quantity haulers (see Fig. 14.36). By diverting private individuals and small-quantity haulers to a separate transfer facility, the potential for accidents at the working face of the landfill is reduced significantly. Excluding small vehicles also enhances operating efficiency at the working face. In some locations, transfer facilities are also used for the recovery of recyclable materials. Waste materials are usually emptied into large transfer trailers, each of which is hauled to the disposal site, emptied, and returned to the transfer station. The need for a convenience transfer station will depend on the physical characteristics and the operation of the landfill and whether there is a separate location where the public can be allowed to dispose of waste safely.

FIGURE 14.36   Small direct-load convenience transfer station located at landfill.

Estimation of Landfill Capacity

The nominal volumetric capacity of a proposed landfill site is determined by first laying out several different landfill configurations, taking into account appropriate design criteria, including the planned thickness of the liner and final cover systems (see Figs. 14.35 and 14.51). The next step is to determine the surface area for each lift. The nominal volume of the landfill is determined by multiplying the average area between two adjacent contours by the height of the lift and summing the volume of successive lifts. If the cover material will be excavated from the site, then the computed volume corresponds to the volume of solid waste that can be placed in the site. If the cover material has to be imported, then the computed capacity must be reduced by a factor to account for the volume occupied by the cover material. For example, if a cover-to-waste ratio of 1:5 is adopted, then the capacity reported must be multiplied by a factor of 0.833 (%).

The nominal volumetric capacity of the landfill is used as a preliminary estimate of landfill capacity for estimating purposes. The actual total capacity of the landfill to accept waste on a weight basis will depend on the initial density at which the residual solid waste is placed in the landfill, on the subsequent compaction of the waste material due to overburden pressure, and loss of mass as a result of biological decomposition. The impacts of these factors on the capacity of the landfill are considered in the following subsections.

Impact of Compactibility of Solid Waste Components.   The initial density of solid wastes placed in a landfill varies with the mode of operation of the landfill, the compactibility of the individual solid waste components, and the percentage distribution of the components. If the waste placed in the landfill is spread out in thin layers and compacted against an inclined surface, a high degree of compaction can be achieved. With minimal compaction, the initial density will be somewhat less than the compacted density in a collection vehicle. In general, the initial density of solid waste placed in a landfill will vary from 550 to 1200 lb/yd^3, depending on the degree of initial compaction given to the waste. Typical compactibility data for the components found in MSW are reported in Table 14.16. Volume-reduction factors are given for both normally compacted and well-compacted landfills.

TABLE 14.16   Typical Compaction Factors for Various Solid Waste Components Placed in Landfills

| | Compaction factors for components in landfills* | | |
Component	Range	Normal compaction	Well compacted
Food wastes	0.2–0.5	0.35	0.33
Paper	0.1–0.4	0.2	0.15
Cardboard	0.1–0.4	0.25	0.18
Plastics	0.1–0.2	0.15	0.10
Textiles	0.1–0.4	0.18	0.15
Rubber	0.2–0.4	0.3	0.3
Leather	0.2–0.4	0.3	0.3
Garden trimmings	0.1–0.5	0.25	0.2
Wood	0.2–0.4	0.3	0.3
Glass	0.3–0.9	0.6	0.4
Tin cans	0.1–0.3	0.18	0.15
Nonferrous metals	0.1–0.3	0.18	0.15
Ferrous metals	0.2–0.6	0.35	0.3
Dirt, ashes, brick, etc.	0.6–1.0	0.85	0.75

* Compaction factor = V_f/V_i where V_f = final volume of solid waste after compaction and V_i = initial volume of solid waste before compaction.

Impact of Cover Material.   Federal regulations mandate that some type of permanent or temporary daily cover always be placed over MSW. The cover material, typically soil, is incorporated into a landfill at each stage of its construction. Alternative materials, such as foams or blankets, may also be used. Daily cover, consisting of 6 in to 1 ft of soil, is applied to the working faces of the landfill at the close of operation each day to stop material from blowing from the working face and to control disease vectors such as insects and rats. Certain clay and silt soils, when used as daily cover, function satisfactorily as cover. However, later, when new MSW is landfilled on top of the cover, downward movement of leachate is impeded. If the waste cell is located near the outside edge of the landfill, hydrostatic pressure may result in the leachate leaking through the side of the landfill. To overcome this problem, some operators remove and stockpile the daily cover before placing new MSW. This saves landfill space, conserves cover soil, and prevents leachate seeps through the side of the land-fill.

Interim cover is a thicker layer of daily cover material applied to areas of the landfill that will not be worked for some time. Final covers usually are 3 to 6 ft thick and include several layers, as discussed previously, to enhance drainage and support surface vegetation. The quantity of cover material necessary for operation of the landfill is an important factor in determining the capacity of a landfill site. Usually, daily and interim cover needs are expressed as a waste/soil ratio, defined as the volume of waste deposited per unit volume of cover provided. Typically, waste/soil ratios range from 4:1 to 10:1.

Impact of Waste Decomposition and Overburden Height. The loss of mass through biological decomposition results in a loss of volume, which becomes available for refilling with new waste. In the preliminary assessment of site capacity, only compaction due to overburden is considered. At later stages of landfill design, the loss of landfill material to decomposition should be considered. Timing is also important. An area that is filled rapidly to the topmost height allowed by a regulatory permit will not have sufficient time to achieve maximum settlement and forestall the opportunity to refill before closure.

Evaluation of Local Geology and Hydrogeology

To evaluate the geologic and hydrogeological characteristics of a site that is being considered for a landfill, core samples must be obtained. Sufficient borings should be made so that the geologic formations under the proposed site can be established from the surface to (and including) the upper portions of the bedrock or other confining layers. At the same time, the depth to the surface water table should be determined along with the piezometric water levels in any bedrock or confined aquifers that may be found. The resulting information is then used to determine:

1. The general direction of groundwater movement under the site
2. Whether any unconsolidated or bedrock aquifers are in direct hydraulic connection with the proposed landfill site
3. The type of liner system that will be required
4. The suitability of soils available at the site for use as liner and cover materials

A portion of the borings are usually converted to permanent monitoring wells, from which samples are collected periodically and tested to determine variations in background groundwater quality. One year's worth of data is frequently the minimum required. Geophysical data collection techniques can be used to characterize large areas quickly as a preliminary reconnaissance step before commencing borings. Electromagnetic, resistivity, and seismic techniques can provide the designer with preliminary information that can be used to reduce the number of sites under consideration before engaging in subsurface borings.

The presence of wetlands should also be noted during site evaluation in preparation for application to the U.S. Army Corps of Engineers for a wetland exemption. Permits for constructing landfills in wetland areas can be controversial and mitigation may be required. The operators of landfills located over faults, in seismic impact zones, or over unstable soils are required, by federal regulation, to demonstrate that the landfill can operate continuously in an environmentally safe manner. Maps from state highway departments and the U.S. Geological Survey identify faults and seismic areas. Unstable soils must be identified through borings.

Selection of Leachate Management Facilities

The principal leachate management facilities required in the design of a landfill include the landfill liner, the leachate collection system, and the leachate treatment facilities. To provide assurances to the public that leachate will not contaminate underground waters, most states under federal mandates now require some type of liner for all new landfill cells. The current trend is toward the use of composite liners including a geomembrane and clay layer. In extremely arid areas where no possibility exists of contaminating the groundwater, it may be possible to develop a landfill without a liner. In arid regions, the amount of leachate generated may not justify the cost of installing a liner (Boltze and de Freitas, 1997). The U.S. EPA standards allow a liner exemption under certain arid climate and geologic conditions. Nevertheless, the use of a liner system is a critical factor in siting new landfills. Further, the relative cost of a liner system is not great, considering the potential environmental benefits. Multi-million-dollar costs can result where groundwater pollution occurs. To determine the size of the leachate collection and treatment facilities, the quantity of leachate must be estimated using the methods outlined in Sec. 14.3. As noted previously, the most common alternatives that have been used to manage the leachate collected from landfills include (1) leachate recycling, (2) treatment followed by disposal, and (3) discharge to municipal wastewater collection systems. The particular option used will depend on local conditions.

Selection of Landfill Cover

As discussed previously, a landfill cover usually comprises several layers, each with a specific function (see Fig. 14.29). The use of a geomembrane liner as a barrier layer is becoming more common to limit the entry of surface water and to control the release of landfill gases. The specific cover configuration selected will depend on the location of the landfill and the local climatalogical conditions. For example, to allow for regrading, some designers favor the use of a deep layer of soil. To ensure the rapid removal of rainfall from the completed landfill and to avoid the formation of puddles, the final cover should have a minimum slope of about 3 to 5 percent. Cover slope stability must be considered during design when the slope is greater than 25 percent.

Selection of Gas Control Facilities

Because the uncontrolled release of landfill gas, especially methane, contributes to the greenhouse effect, and because landfill gas can migrate laterally underground to potentially cause explosions or kill vegetation and trees, most new landfills are equipped with gas collection and treatment facilities. To determine the size of the gas collection and processing facilities, the quantity of landfill gas must first be estimated using the methods outlined in Sec. 14.3. Because the rate of gas production varies, depending on the operating procedures (e.g., without or with leachate recycle), several rates should be analyzed. The next step is to determine the rate of gas production with time. The decision to use horizontal or vertical gas recovery wells depends on the design and capacity of the landfill. The decision to flare or to recover energy from the landfill gas depends on the capacity of the landfill site and the opportunity to sell power produced from the conversion of landfill gas to electrical energy or the availability of utility boilers for energy recovery. In many small landfills located in remote areas, gas collection equipment is not used routinely.

Surface Water Management

Elimination or reduction of the amount of surface water that enters the landfill is of fundamental importance in the design of a sanitary landfill because surface water is the major contributor to the total volume of leachate. Storm water runoff from the surrounding area must not be allowed to enter the landfill, and surface water runoff (from rainfall) must not be allowed to accumulate on the surface of the landfill. The proper management of storm water runoff that flows away from the landfill is also important. The total and peak runoff flow rates will be increased significantly when a relatively impermeable and sloping cover is placed over a landfill that is located on previously level land. Increased sediment loading to nearby surface water bodies may also occur.

Federal and many state regulations restrict the location and operation of landfills on floodplains and floodways. Provisions must be made to minimize obstruction of floodwater flow and reduction in floodplain storage capacity by landfilling activities. Landfills not capable of excluding floodwaters from entry may be required to close under certain circumstances.

Surface Water Drainage Facilities.   An important step in the design of a landfill is to develop an overall drainage plan for the area that shows the location of storm drains, culverts, ditches, and subsurface drains as the filling operation proceeds. In those locations where storm water runoff from the surrounding areas can enter the landfill (e.g., landfills located in canyons), the site must be graded appropriately and properly designed drainage facilities must be installed (see Fig. 14.37). The drainage facilities may be designed to remove the runoff from the surrounding area only, or from the surrounding area as well as the surface of the landfill. In current federal regulations, a 25-year storm event must be used as the basis for design. In locations where the entire leachate liner system is installed at one time, the design of the liner must allow for the diversion of storm water not falling on the wastes being landfilled. In locations where only the surface water from the top of the landfill must be removed, the drainage facilities should be designed to limit the travel distance of the surface water. In many designs, a series of interceptor ditches is used. Flow from the interceptor ditches is routed to a larger main ditch for removal from the site.

Storm Water Storage Basins.   Depending on the location and configuration of the landfill and the capacity of the natural drainage courses, it may be necessary to install a storm water storage or retention basin. In many cases, it may be necessary to construct storm water storage basins to contain the diverted storm water flows so as to minimize downstream flooding. Typically, storm water must be collected from the completed portions of the landfill as well as from areas yet to be filled. An example of a large storm water retention/storage basin is illustrated in Fig. 14.38. Standard hydrological procedures are followed in sizing the storm water basins (Hjelmfelt and Cassidy, 1975; Linsley et al., 1991; Linsley et al., 1958). Discharges from storm water facilities must have a federal or state discharge permit.

Environmental Monitoring Facilities

Monitoring facilities are required at new landfills and at selected existing landfills for (1) gases and liquids in the vadose zone, (2) groundwater quality both upstream and downstream of the landfill site, and (3) air quality at the boundary of the landfill and from any processing facilities (e.g., flares). The specific number of monitoring stations will depend on the configuration and size of the landfill and the requirements of the local air and water pollution control agencies. Environmental monitoring is considered in greater detail in Sec. 14.8.

Aesthetic Design Considerations

Aesthetic design considerations relate to minimizing the impact of the landfilling operation on nearby residents as well as on the public that may be passing by the landfill.

Screening of Landfilling Areas.   Screening of the daily landfilling operations from nearby roads and residents with berms, plantings, and other landscaping measures is one of the most important examples of an aesthetic design consideration (see Fig. 14.39a). Screening of the

FIGURE 14.37 Typical drainage facilities used at landfills: (*a*) trapezoidal lined ditch; (*b*) vee lined ditch; (*c*) shaped-vee lined ditch. Note the trapezoidal ditch cross section is expandable to accommodate a wide range of flows.

FIGURE 14.38 View of large storm water retention/storage basin at a large landfill. The size of the basin can be estimated from the size of the vehicles parked in the bottom of the basin.

active areas in the landfill must be taken into account in the preliminary design and layout of the landfill.

Control of Birds.   Birds at the landfill site are not only a nuisance; they can cause serious problems if the landfill site is located near an airport. Federal regulations limit landfill development within 6 mi of an airport, and in some instances greater separation distances are required. Techniques to control birds at landfill sites include the use of noisemakers, recordings of the sounds made by birds of prey, and overhead wires. The control of birds at reservoirs and fishponds with overhead wires dates back to the early 1930s (Amling, 1981; McAtee, 1936). The use of overhead wires to control seagulls at landfills was pioneered by the County Sanitation Districts of Los Angeles County in the early 1970s (see Fig. 14.39*b*). Because seagulls descend in a circular pattern when landing, it appears that the wires may interfere with the birds' guidance system. The poles are typically spaced 50 to 75 ft apart, with line spans from 500 to 1200 ft (Mathias, 1984). Crisscrossing improves the effectiveness of the wire system. Typically, 100-lb-test monofilament fish line is used, although stainless steel wire has also been used.

Control of Blowing Materials.   Depending on the location, windblown paper, plastics, and other debris can be a problem at some landfills. The most common solution is to use portable screens near the operating face of the landfill (see Fig. 14.39*c*). To avoid problems with vectors, the material accumulated on the screens must be removed daily. Prompt pickup of paper that is not retained by portable screens is important for maintaining an image of good landfill operation.

Open-topped vehicles hauling waste to the landfill should be covered with a tarpaulin to prevent paper and dust from falling onto the highway. Waste that does fall out of a truck

FIGURE 14.39   Aesthetic considerations in landfill design: (*a*) view of landscaped landfill in which filling operations are not visible from nearby freeway; (*b*) overhead wire system used to control seagulls at landfills; (*c*) wire screen used to control blowing papers and plastic; (*d*) daily cover used to control vectors at landfills.

should be picked up promptly by either the vehicle operator or the landfill personnel. Periodic collection of all litter along the roads leading to the landfill is also a good approach to help the landfill operator enhance community relations.

Control of Pests and Vectors.   The principal vectors of concern in the design and operation of landfills are pests, including mosquitoes and flies, and rodents, such as rats and other burrowing animals. Flies and mosquitoes are controlled by the placement of daily cover and by the elimination of standing water. Standing water can be a problem in areas where white goods and used tires are stored for recycling. The use of covered facilities for the storage of these materials will eliminate most problems. Rats and other burrowing animals are controlled by the use of daily cover (see Fig. 14.39*d*).

Public Participation

Landfill development is often controversial. For privately owned sites there is no mandate to discuss development with representatives of the public until permit applications are submitted and hearings are conducted. Often this approach leads to very time-consuming, protracted hearings, after which a ruling may or may not allow landfill construction. Usually legal appeals are filed, which result in extended delays.

Boards and councils that must decide how to proceed often discuss the development of publicly owned sites. This approach, while more open to public scrutiny, can also become embroiled in public controversy. A better approach is to establish a protocol that involves the

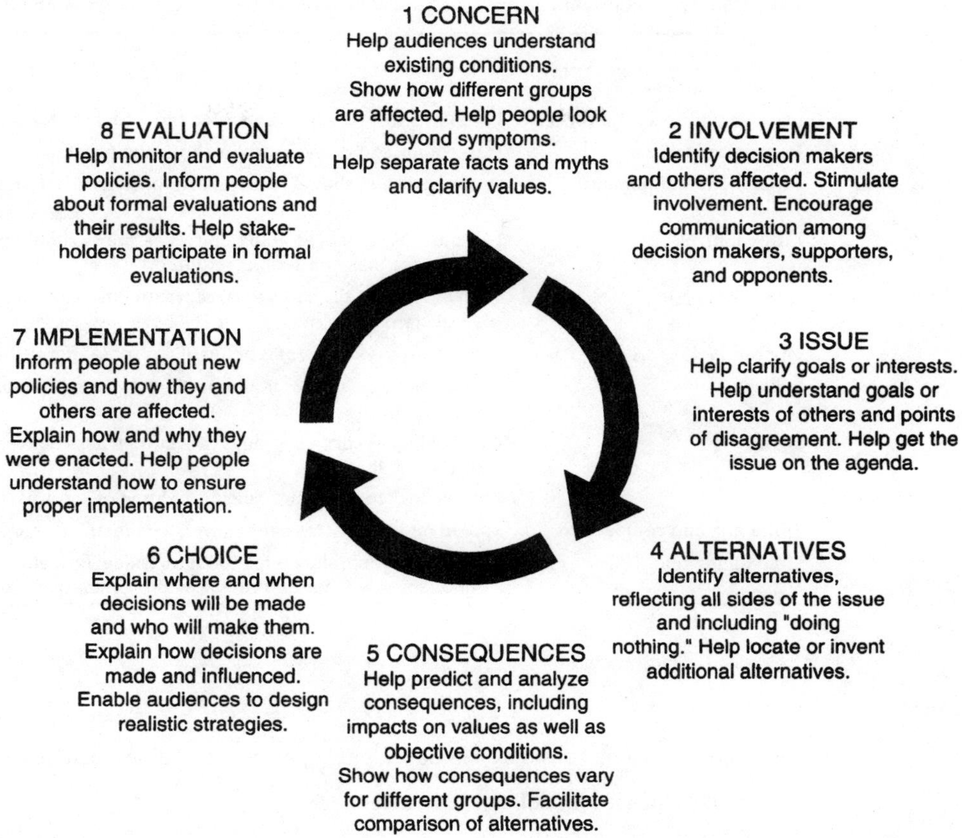

1 CONCERN
Help audiences understand existing conditions. Show how different groups are affected. Help people look beyond symptoms. Help separate facts and myths and clarify values.

8 EVALUATION
Help monitor and evaluate policies. Inform people about formal evaluations and their results. Help stakeholders participate in formal evaluations.

2 INVOLVEMENT
Identify decision makers and others affected. Stimulate involvement. Encourage communication among decision makers, supporters, and opponents.

7 IMPLEMENTATION
Inform people about new policies and how they and others are affected. Explain how and why they were enacted. Help people understand how to ensure proper implementation.

3 ISSUE
Help clarify goals or interests. Help understand goals or interests of others and points of disagreement. Help get the issue on the agenda.

6 CHOICE
Explain where and when decisions will be made and who will make them. Explain how decisions are made and influenced. Enable audiences to design realistic strategies.

5 CONSEQUENCES
Help predict and analyze consequences, including impacts on values as well as objective conditions. Show how consequences vary for different groups. Facilitate comparison of alternatives.

4 ALTERNATIVES
Identify alternatives, reflecting all sides of the issue and including "doing nothing." Help locate or invent additional alternatives.

FIGURE 14.40   Issue evolution/education intervention model. *(From House and Young, 1989.)*

various public interests in an educated decision-making process. The protocol shown in Fig. 14.40 defines an issue evolution and education intervention model that seeks to enhance understanding and decision making.

14.7   *LANDFILL OPERATION*

The development of a workable operating schedule, a filling plan for the placement of solid wastes, an estimate of the equipment requirements, development of landfill operating records and billing information, a load inspection for hazardous waste, traffic control on highways leading to the landfill, and a site safety and security program are important elements of a landfill operation plan. An ongoing community relations program is also a part of managing a landfill. Other factors that must be considered in the operation of a landfill are reported in Table 14.17.

TABLE 14.17 Important Factors That Must Be Considered in the Operation of Landfills

Factors	Remarks
Communications	Telephone for emergencies.
Days and hours of operation	Usual practice is 5 to 6 days/week and 8 to 10 h/day.
Employee facilities	Rest rooms and drinking water should be provided.
Equipment maintenance	A covered shed should be provided for field maintenance of equipment.
Litter control	Use movable fences at unloading areas; crews should pick up litter at least once per month or as required.
Operation plan	With or without the codisposal of treatment plant sludges and the recovery of gas.
Operational records	Tonnage, transactions, and billing if a disposal fee is charged.
Salvage	No scavenging; salvage should occur away from the unloading area; no salvage storage on site.
Scales	Essential for record keeping if collection trucks deliver wastes; capacity to 100,000 lb.
Security	Provide locked gates and fencing, lighting of sensitive areas.
Spreading and compaction	Spread and compact waste in layers less than 2 ft thick.
Unloading area	Keep small, generally under 100 ft on a side; operate separate unloading areas for automobiles and commercial trucks.

Landfill Operating Schedule

Factors that must be considered in developing operating schedules include the following:

- Arrival sequences for collection vehicles
- Traffic patterns at the site
- The time sequence to be followed in the filling operations
- Effects of wind and other climatic conditions
- Commercial and public access

For example, because of heavy truck traffic early in the morning, it may be necessary to restrict public access to the site until later in the morning. Also, because of adverse winter conditions, the filling sequence should be established so that the landfill operations are not impeded by unusual weather conditions. If it is not possible to control blowing paper during high-wind conditions, closing the landfill when winds exceed, for example, 35 mi/h may be necessary.

Solid Waste Filling Plan

Once the general layout of the landfill site has been established, it will be necessary to select the placement method to be used and to lay out and design the individual solid waste cells. The specific method of filling will depend on the characteristics of the site, such as the amount of available cover material, the topography, and the local hydrology and geology. Details on the various filling methods were presented in Sec. 14.1. To assess future development plans, it will be necessary to prepare a detailed plan for the layout of the individual solid waste cells. A typical example of such a plan is shown in Fig. 14.41.

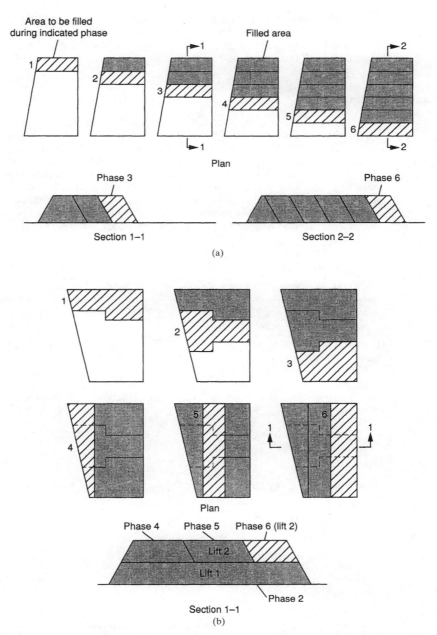

FIGURE 14.41 Typical examples of solid waste filling plans: (*a*) filling plan for single-lift landfill; (*b*) filling plan for a multilift landfill.

On the basis of the characteristics of the site or the method of operation (e.g., gas recovery), it may be necessary to incorporate special features for the control of the movement of gases and leachate from the landfill. These might include the use of horizontal and vertical gas extraction wells, composite liners, and special extraction facilities.

Equipment Requirements

The type, size, and amount of equipment required will depend on the size of the landfill and the method of operation. The types of equipment that have been used at sanitary landfills include crawler tractors, scrapers, compactors, draglines, and motorgraders (see Fig. 14.42). Of these, crawler tractors are most commonly used. Properly equipped tractors can be used to perform all the necessary operations at a sanitary landfill, including spreading, compacting, covering, trenching, and even hauling cover materials (Brunner and Keller, 1972). The size and amount of equipment will depend primarily on the size of the landfill operation. Local site conditions will also influence the size of the equipment. Average equipment requirements that may be used as a guide for landfill operations are reported in Table 14.18.

TABLE 14.18   Typical Equipment Requirements for Sanitary Landfills

Approximate population	Daily wastes, tons	Equipment Number	Equipment Type	Size, lb	Accessory*
0–20,000	0–50	1	Tractor, crawler	10,000–30,000	Dozer blade Front-end loader (1–2 yd^3) Trash blade
20,000–50,000	50–150	1	Tractor, crawler	30,000–60,000	Dozer blade Front-end loader (2–4 yd^3) Bullclam Trash blade
		1	Scraper or dragline		
		1	Water truck		
50,000–100,000	150–300	1–2	Tractor, crawler	30,000+	Dozer blade Front-end loader (2–5 yd^3) Bullclam Trash blade
		1	Scraper or dragline†		
		1	Water truck		
100,000	300‡	1–2	Tractor, crawler	45,000+	Dozer blade Front-end loader Bullclam Trash blade
		1	Steel wheel compactor		
		1	Scraper or dragline†		
		1	Water truck		
		*	Road grader		

* Optional, depends on individual needs.
† The choice between a scraper or dragline will depend on local conditions.
‡ For each 500-ton increase add one each of each piece of equipment.

FIGURE 14.42 Views of equipment used at landfills: (*a*) crawler tractor with dozer blade; (*b*) high track crawler tractor with trash blade; (*c*) steel wheel compactor with trash blade; (*d*) self-loading scraper; (*e*) water wagon; (*f*) dragline.

Landfill Operating Records

To determine the quantities of waste that are disposed, an entrance scale and gatehouse will be required. Personnel who are responsible for weighing the incoming and outgoing trucks would use the gatehouse. The sophistication of the weighing facilities will depend on the number of vehicles that must be processed per hour and the size of the landfill operation. In some larger landfills, weigh stations are equipped with radiation detectors to detect the presence of radioactive substances in the incoming wastes. Many weigh stations are monitored with continuously recording video systems. Some examples of weighing facilities are shown in Fig. 14.43. If the

(a)

(b)

FIGURE 14.43   Typical truck-weighing facilities: (*a*) at large landfill; (*b*) at small landfill.

weight of the solid wastes delivered is known, then the in-place specific weight of the wastes can be determined and the performance of the operation can be monitored. The weight records would also be used as a basis for charging participating agencies and private haulers for their contributions.

Load Inspection for Hazardous Waste

Load inspection is the term used to describe the process of unloading the contents of a collection vehicle near the working face or in some designated area, spreading the wastes out in a thin layer, and manually inspecting the wastes to determine whether any hazardous wastes are present (see Fig. 14.44). Federal standards mandate randomly selecting MSW loads for inspection. The presence of radioactive wastes can be detected with a handheld radiation-measuring device or at the weigh station, as previously described. If hazardous wastes are found, the waste collection company is responsible for removing the hazardous materials or is billed for their removal. At some landfills, if a company is caught bringing in hazardous wastes

(a)

(b)

FIGURE 14.44 Inspection of solid waste unloaded at landfill for the presence of hazardous wastes at the Frank R. Bowerman landfill in Orange County, Calif.: (*a*) residential load; (*b*) commercial load.

a second time, a high fine is levied. If caught a third time, the company is banned from discharging wastes at the landfill.

Public Health and Safety

Public health and safety issues are related to worker health and safety and to the health and safety of the public.

Health and Safety of Workers.   The health and safety of the workers at landfills are critical in the operation of a landfill. The types of accidents that have occurred at landfills include puncture wounds from sharp objects, equipment rollovers, laborers being run over, personnel falling into holes, and asphyxiation in confined spaces. The federal government, through the Occupational Safety and Health Administration (OSHA) and state-instituted OSHA-type programs, has established requirements for a comprehensive health and safety program for the workers at landfill sites. Because the requirements for these programs change continually, the most recent regulations should be consulted in the development of worker health and safety programs. Depending on the activities at the landfill, careful attention must be given to the types of protective clothing and boots, air-filtering headgear, and punctureproof gloves supplied to the workers.

Safety of the Public.   As noted previously, safety concerns and the many new restrictions governing the operation of landfills have forced landfill operators to reexamine past operational practices with respect to public safety and site security. As a result, the use of a convenience transfer station at the landfill site to minimize the public contact with the working operations of the landfill is gaining in popularity.

Site Safety and Security

The increasing number of lawsuits over accidents at landfill sites has caused landfill operators to improve security at landfill sites significantly. Most sites now have restricted access and are fenced and posted with "no trespassing" and other warning signs. In some locations, television cameras are used to monitor landfill operations and landfill access.

Community Relations

An important and often overlooked aspect of landfill operation is sustaining good community relations. The landfill manager must maintain a dialogue with neighbors, municipal leaders, community activists, and state government representatives in an effort to build trust through honest communications. While community relations activities do not guarantee continued support for the landfilling operation, poor relations almost certainly will result in complaints and problems.

Landfill Operator Training

It is increasingly being recognized that a carefully planned and executed employee training program must accompany the operation of landfills. Landfills contain many technical components of a specialized nature, which, if not properly managed, will fail to achieve the desired results. The operation of a landfill is an expensive activity. If it is improperly managed, the result can be even higher costs and loss of investment as the result of system failures. Training programs for operators must be focused on the environmental objectives, the design elements of the landfill, the equipment operation, environmental regulations, and health and safety protection. Many governmental units require that the landfill manager and, possibly, the equipment operators each have a valid license that is issued after a specified training program has been completed. In some jurisdictions, continuing education is also required on an ongoing basis.

14.8 ENVIRONMENTAL QUALITY MONITORING AT LANDFILLS

Environmental monitoring is conducted at sanitary landfills to ensure that no contaminants that may affect public health and the surrounding environment are released from the landfill. The monitoring required may be divided into three general categories: (1) vadose zone monitoring for gases and liquids, (2) groundwater monitoring, and (3) air quality monitoring. Environmental monitoring involves the use of both sampling and nonsampling methods. Sampling methods involve the collection of a sample for analysis, usually at an off-site laboratory. Nonsampling methods are used to detect chemical and physical changes in the environment as a function of an indirect measurement such as a change in electrical current. Representative devices that have been used to monitor landfill sites are listed in Table 14.19. The typical instrumentation of a landfill for environmental monitoring is illustrated in Fig. 14.45.

Vadose Zone Monitoring

The *vadose zone* is defined as that zone from the ground surface to where the permanent groundwater is found. An important characteristic of the vadose zone is that the pore spaces are not filled with water, and that the small amounts of water that are present coexist with air. Vadose zone monitoring at landfills involves both liquids and gases.

Liquid Monitoring in the Vadose Zone.   Monitoring for liquids in the vadose zone is necessary to detect any leakage of leachate from the bottom of a landfill. In the vadose zone, moisture held in the interstices of the soil particles or within porous rock is always held at pressures below atmospheric pressure. To remove the moisture it is necessary to develop a negative pressure or vacuum to pull the moisture away from the soil particles. Because suction must be applied to draw moisture out of the soil in the vadose zone, conventional wells or other open cavities cannot be used to collect samples in this zone. The sampling devices used for sample extraction in the unsaturated zone are called *lysimeters*. The commonly used classes of lysimeters are (1) the ceramic cup, (2) the hollow fiber, (3) the membrane filter, (4) pressure, (5) wick, and (6) pan (Bagchi, 1990; Emcon Associates, 1980).

The most commonly used device for obtaining samples of moisture in the vadose zone is the ceramic cup sampler (see Fig. 14.46), consisting of a porous cup or ring made of ceramic material, which is attached to a short section of nonporous tubing (e.g., PVC). When the cup is placed in the soil, the pores become an extension of the pore space of the soil. Soil moisture is drawn in through the porous ceramic element by the application of a vacuum. When a sufficient amount of water has collected in the sampler, the collected sample is pulled to the surface through a narrow tube by the application of a vacuum or is pushed up by air pressure.

The wick lysimeter relies on the fabric media to extract water from the soil profile by capillary action. The pan lysimeter, which has limited capability at low soil moisture contents, is used to detect leakage in a landfill liner. The water is recovered in a manner similar to that of the suction lysimeter.

Gas Monitoring in the Vadose Zone.   Monitoring for gases in the vadose zone is necessary to detect the lateral movement of any landfill gases. A typical example of a vadose zone gas-monitoring probe is illustrated in Fig. 14.47. In many monitoring systems, gas samples are col-

TABLE 14.19 Representative Devices Used to Monitor Landfill Gases and Leachate at Landfills

Type	Application/description
	Sampling methods*
Air quality:	
Evacuated flask	Collection of air grab samples for analysis.
Gas syringe	Collection of air grab samples for analysis.
Air collection bag	Collection of air grab samples for analysis.
Active air sampler	Continuous collection and analysis of gas samples.
Groundwater:	
Monitoring wells; single- and multiple-depth	Used to collect groundwater samples. Multiple extraction wells are used to collect samples from different depths.
Piezometers	Used to collect groundwater samples.
In landfills:	
Piezometers	Used to collect leachate samples. Piezometers can be installed before filling of the landfill is initiated or after the landfill has been completed.
Vadose zone:	
Collection lysimeter	Used to collect liquid samples below landfill liners.
Soil gas probes; single- and multiple-depth	Used to monitor landfill gases and volatile organic compounds in the soil. The gas may be analyzed in situ using a portable gas chromatograph or tested in a laboratory after absorption in charcoal.
Suction cup lysimeter	Used to obtain liquid samples from the vadose zone.
	Nonsampling methods†
Groundwater:	
Conductivity cells	Used to monitor changes in groundwater conductivity. Conductivity cells are often located in or near monitoring wells.
In landfills:	
Piezometer	Used to measure the depth of leachate in landfills.
Temperature blocks	Used to measure temperature.
Temperature probes	Used to measure temperature.
Vadose zone:	
Electrical probes	Used to determine the salinity of the vadose zone. A four-probe array is installed so that conductivity of the soil can be measured.
Electrical resistance blocks	Used to measure changes in water content of the vadose zone. Electrode blocks embedded in porous material are installed in the soil. Electrical properties of the blocks change with the changing water content of the vadose zone.
Gamma ray attenuation probes	Used for detecting changes in moisture content of the vadose zone. Based on gamma ray transmission and scattering. In the transmission method, two wells are installed at a known distance apart. A single well is used in the scattering method. Usually limited to shallow depth because of difficulties in installing parallel wells.
Heat dissipation sensors	Used to monitor water content of the vadose zone by measuring the rate of heat dissipation from the block to the surrounding soil.
Neutron moisture meter	Used to obtain a profile of the moisture content of the soil below the landfill. Meter can be installed below a landfill or moved through a borehole next to the landfill.
Salinity sensors	Used to monitor soil salinity. Electrodes attached to a porous ceramic cup are installed in the soil.
Thermocouple psychrometers	Used to detect changes in moisture content. Operation is based on cooling of a thermocouple junction by the Peltier effect. Wet bulb and dew point. The dew point method is used more commonly in landfill monitoring.
Tensiometers	Used to measure the matric potential of soil. Tensiometers measure the negative pressure (capillary pressure) that exists in unsaturated soil.
Time-domain reflectometry (TDR)	Based on the difference in dielectric properties of water and soil. Bandwidth and short-pulse length, which are sensitive to the high-frequency electrical properties of the material, are measured.
Wave-sensing devices	Use of seismic or acoustic wave propagation properties for leak detection. In the seismic wave technique, the difference in travel time of Rayleigh waves between the source and geophones is used to detect leaks. In the acoustic emission monitoring (AEM) technique, sound waves generated by flowing water from a leak are utilized in leak detection.

* Methods involving the collection of samples for subsequent laboratory analysis.
† Methods involving physical and electrical measurements.

14.78

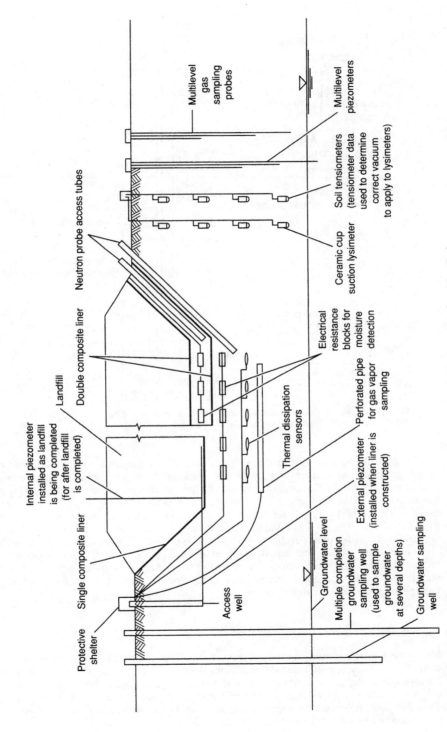

FIGURE 14.45 Instrumentation of a landfill for the collection of environmental monitoring data. Not all of the devices and instrumentation shown would be used at an individual landfill.

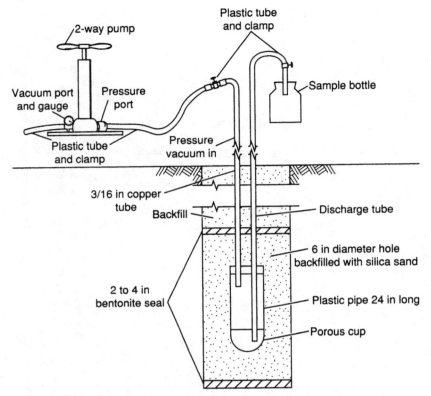

FIGURE 14.46   Porous cup suction lysimeter for the collection of liquid samples from the vadose zone. *(Courtesy of California Integrated Waste Management Board.)*

lected from multiple depths in the vadose zone. Where landfills are located near occupied buildings, testing as frequently as twice per week has been necessary to monitor landfill gas migration adequately. The accuracy to which these probes can measure the total quantity of gas migrating from the site must be recognized. A detailed study of 18 landfill sites (Boltze and deFreitas, 1997) resulted in recommendations that where a continuous, permanent system of gas monitoring is not available the monitoring program consist of (1) low-frequency (i.e., weekly) testing of probes combined with (2) short periods of intensive monitoring. The intensive monitoring should be timed to observe periods when high gas migration rates are expected. Planning the intensive monitoring should be based on waste decomposition, weather, and soil conditions observed during the low-frequency testing that is associated with changing gas migration rates.

Groundwater Monitoring

Monitoring of the groundwater is necessary to detect changes in water quality that may be caused by the escape of leachate and landfill gases. Both down- and upgradient wells are required to detect any contamination of the underground aquifer by leachate from the landfill. An example of a well used for the monitoring of groundwater is illustrated in Fig. 14.48. To obtain a representative sample, the same type of equipment should be used each time and the well must be purged prior to sample collection.

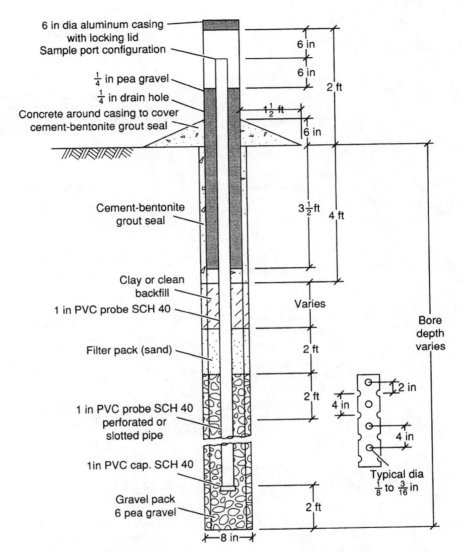

FIGURE 14.47 Vadose zone gas monitoring probe. *(Courtesy Waste Management, Inc.)*

By federal regulation, all new MSW landfills must install groundwater monitoring facilities. Existing sites have a number of years to implement the 1993 requirements for monitoring. There are also extensive regulations for sample collection, testing, and data analysis.

Samples collected from groundwater monitoring wells must be tested in a manner that ensures consistent, reliable, and precise analytical results. Special procedures are necessary to prevent interference of sample collection equipment with the chemical analytical results. The standards for testing groundwater samples are contained in Stark et al., 2000. It is very important that the personnel responsible for collecting and handling samples maintain protocols for ensuring that the water samples do not degrade while they are being transported to the laboratory. This includes keeping the water samples cold, delivering the samples to the

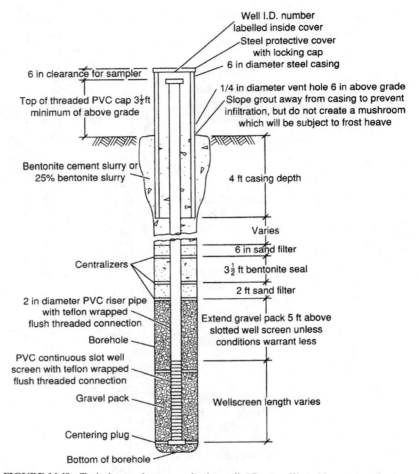

6 in clearance for sampler

Top of threaded PVC cap 3½ft
minimum of above grade

Bentonite cement slurry or
25% bentonite slurry

Centralizers

2 in diameter PVC riser pipe
with teflon wrapped
flush threaded connection

Borehole

PVC continuous slot well
screen with teflon wrapped
flush threaded connection

Gravel pack

Centering plug

Bottom of borehole

Well I.D. number
labelled inside cover

Steel protective cover
with locking cap

6 in diameter steel casing

1/4 in diameter vent hole 6 in above grade
Slope grout away from casing to prevent
infiltration, but do not create a mushroom
which will be subject to frost heave

4 ft casing depth

Varies

6 in sand filter

3½ ft bentonite seal

2 ft sand filter

Extend gravel pack 5 ft above
slotted well screen unless
conditions warrant less

Wellscreen length varies

FIGURE 14.48   Typical groundwater monitoring well. *(Courtesy Waste Management, Inc.)*

laboratory according to the specified procedures, and properly handling the samples once they reach the laboratory. In addition, the laboratory equipment must be calibrated using standard methods.

To determine whether a landfill is affecting the groundwater quality, it is necessary to apply statistical methods to the analytical results. Examples of these statistical methods can be found in SWPCB (1954). A box plot is a simple method for displaying groundwater monitoring data (Fig. 14.49). It is relatively expensive to maintain a groundwater monitoring system; however, it is important that the necessary protocols be implemented in order to achieve reliable results.

The results of the groundwater monitoring tests are used to determine whether a landfill is significantly degrading the environment. In a number of countries, groundwater quality standards have been established which specify that the amount of allowable contamination cannot exceed drinking water standards or some specified limit that is based upon maintaining groundwater quality to a sufficient level to protect human health. In order to have an early warning of potentially harmful contamination, Preventive Action Limits (PALs) have been

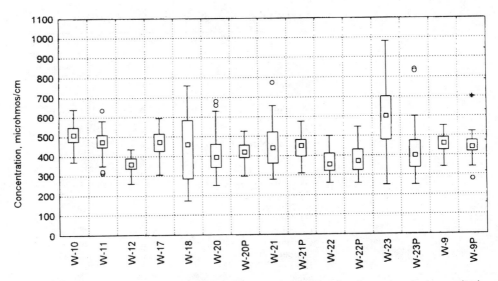

FIGURE 14.49   Box plot of specific conductance analysis on samples taken from groundwater monitoring wells at a landfill. *(SWRCB, 1967)*

established for a number of parameters. An example of the enforcement limits and PALs is shown in Table 14.20.

Landfill Air Quality Monitoring

Air quality monitoring at landfills involves (1) the monitoring of ambient air quality at and around the landfill site, (2) the monitoring of landfill gases extracted from the landfill, and (3) the monitoring of the offgases from any gas processing or treatment facilities.

Monitoring Ambient Air Quality.   Ambient air quality is monitored at landfill sites to detect the possible movement of gaseous contaminants from the boundaries of the landfill site. Gas sampling devices can be divided into three categories: (1) passive, (2) grab, and (3) active. Passive sampling involves the collection of a gas sample by passing a stream of gas through a collection device in which the contaminants contained in the gas stream are removed for subsequent analysis. Passive sampling was commonly used in the past, but is seldom used today. Grab samples are collected using an evacuated flask, a gas syringe, or an air collection bag made of a synthetic material (see Fig. 14.50). An active sampler involves the collection and analysis of a continuous stream of gas.

Monitoring Extracted Landfill Gas.   Landfill gas is monitored to assess the composition of the gas and to determine the presence of trace constituents that may pose a health or environmental risk.

Monitoring Offgases.   Monitoring offgases from treatment and energy recovery facilities is done to determine compliance with local air pollution control requirements. Both grab and continuous sampling have been used for this purpose.

TABLE 14.20   Example of Groundwater Quality Standards

Substance	Enforcement standard (mg/L)	Preventive action limit (mg/L)
Chloride	250	125
Iron	0.3	0.15
Manganese	0.05	0.025
Sulfate	250	125
Zinc	5	2.5
Nitrate + Nitrite (as N)	10	2
Phenol	6	1.2
	(μg/L)	(μg/L)
Acetone	1000	200
Arsenic	50	5
Benzene	5	0.5
Cadmium	5	0.5
Chromium	100	10
Copper	1300	130
Lead	15	1.5
Nickel	100	20
1,1,1-Trichloroethane	200	40
1,1,2-Trichloroethane	5	0.5
Vinyl chloride	0.2	0.02

Source: *Wisconsin Department of Natural Resources Administrative Code NR 140, March 2000.*

14.9   *LANDFILL CLOSURE, POSTCLOSURE CARE, AND REMEDIATION*

Landfill closure and *postclosure care* are the terms used to describe what is to happen to completed landfill in the future. To ensure that completed landfills will be maintained 30 to 50 years into the future, many states and the federal government have passed legislation that requires the operator of a landfill to put aside enough money so that when the landfill is completed, the facility can be closed, maintained, and monitored properly for 30 to 40 years. Greater long-term care periods are being considered in some European countries.

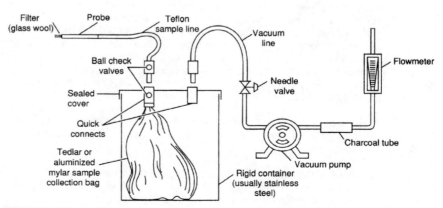

FIGURE 14.50   Sampling apparatus for the collection of air grab samples at landfills.

Development of Long-Term Closure Plan

Perhaps the most important element in the long-term maintenance of a completed landfill is the availability of a closure plan in which the requirements for closure are delineated clearly. A closure plan must include a design for the landfill cover and the landscaping of the completed site. Closure must also include long-term plans for runoff control, erosion control, gas and leachate collection and treatment, and environmental monitoring. The closure plan for the landfill layout given in Fig. 14.35 is presented in Fig. 14.51.

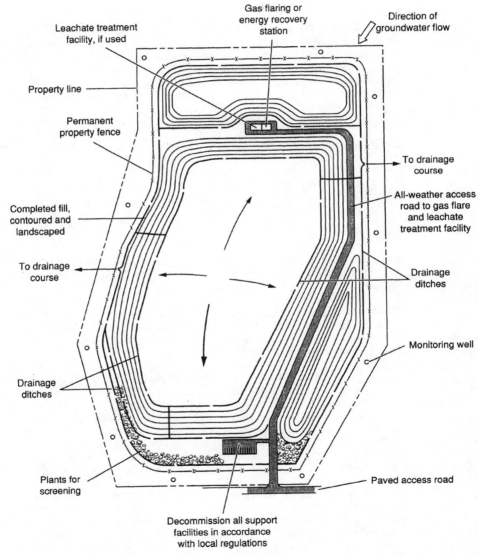

FIGURE 14.51   Plan view of completed landfill showing all of the elements involved in closure and postclosure care.

Cover and Landscape Design.   The landfill cover must be designed to divert surface runoff and snowmelt from the landfill site and to support the landscaping design selected for the landfill. Increasingly, the final landscaping design is based on local plant and grass species as opposed to nonnative plant and grass species. In many water-short locations in the southwest, a desert type of landscaping is favored.

Control of Landfill Gases.   The control of landfill gases is a major concern in the long-term maintenance of landfills. Because of the concern over the uncontrolled release of landfill gases, most modern landfills have some sort of gas control system installed before the landfill is completed. Older completed landfills without gas collection systems are being retrofitted with gas collection systems.

Collection and Treatment of Leachate.   As with the control of landfill gas, the control of leachate discharges is another major concern in the long-term maintenance of landfills. Again, most modern landfills have some sort of leachate control system, as previously discussed. Older completed landfills without leachate collection systems are being retrofitted with leachate collection systems. These retrofitted collection systems are similar in construction to vertical gas wells in which leachate pumps are installed. The leachate head wells, as they are commonly referred to, can be difficult to install due to obstructions encountered during drilling. Poor hydraulic flow conditions through the waste may also limit their effectiveness.

Environmental Monitoring Systems.   To be able to conduct long-term environmental monitoring after a landfill has been completed, it will be necessary to install monitoring facilities. The monitoring required at completed landfills usually involves (1) vadose zone monitoring for gases and liquids, (2) groundwater monitoring, and (3) air quality monitoring. The required facilities have been described previously.

Postclosure Care

Postclosure care involves the routine inspection of the completed landfill site, maintenance of the infrastructure, and environmental monitoring. These subjects are considered briefly as follows.

Routine Inspections.   A routine inspection program must be established to monitor continually the condition of the completed landfill. Criteria must be established to determine when corrective action must be taken. For example, how much settlement will be allowed before regrading must be undertaken?

Infrastructure Maintenance.   Infrastructure maintenance typically involves the continued maintenance of surface water diversion facilities; landfill surface grades; the condition of liners in covers, where used; revegetation; and maintenance of landfill gas and leachate collection equipment. The amount of regrading that will be required will depend on the amount of settlement (see Fig. 14.52). In turn, the rate of settlement will depend on the rate of gas formation and the degree of initial compaction achieved in the placement of the waste materials in the landfill. The amount of equipment that must be available at the site will depend on the extent of the landfill and the nature of the facilities that must be maintained.

Environmental Monitoring Systems.   Long-term environmental monitoring is conducted at completed landfills to ensure that there is no release of contaminants from the landfill that may impact health or the surrounding environment. The monitoring required at completed landfills usually involves (1) vadose zone monitoring for gases and liquids, (2) groundwater monitoring, and (3) air quality monitoring. The number of samples collected for analysis and the frequency of collection will usually depend on the regulations of the local air pollution

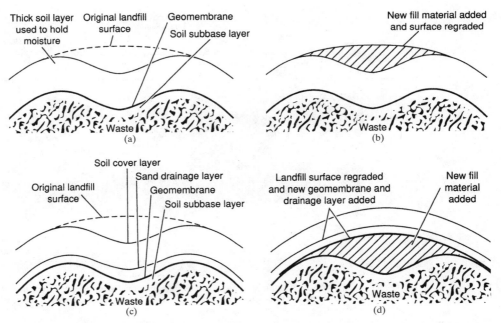

FIGURE 14.52   Schematic representation of the repair of a landfill cover employing a geomembrane to restore drainage: (*a*) landfill after closure and settlement; (*b*) landfill repair procedure; (*c*) landfill employing a drainage layer and a geomembrane after closure and settlement; (*d*) landfill after repair to restore surface drainage.

and water pollution control agencies. The EPA has developed a baseline procedure for sampling of groundwater that should be reviewed (40 CFR 258).

Remediation

Remedial actions may be necessary if unacceptable levels of environmental emissions are detected in the postclosure monitoring program. Remedial actions may be the result of landfill gas migration, toxic air emissions, leachate polluting the groundwater, or some other unforeseen event. The severity of the problem will determine the intensity of the remedial action and the long-term cost.

Migration Control.   Federal regulations specify that methane concentrations cannot exceed 5 percent methane at the property boundary of the MSW landfill. Some states require even lower concentrations. Landfill gas migration may unexpectedly extend into areas on which there are occupied buildings. Emergency measures to secure the area and evacuate buildings are the first steps that must be initiated without delay. Local fire departments usually have the appropriate equipment to measure for the presence of methane in buildings. Wells are then usually installed not only on or adjacent to the landfill to stop gas movement away from the site, but also in the vicinity of the buildings to remove the gas from the ground. The wells in or adjacent to the landfill will likely be operated for years until the concentration of methane being generated is determined to not be a threat. The wells located near the occupied buildings will temporarily operate, usually on the order of months, until the methane in the vadose zone is reduced to safe levels.

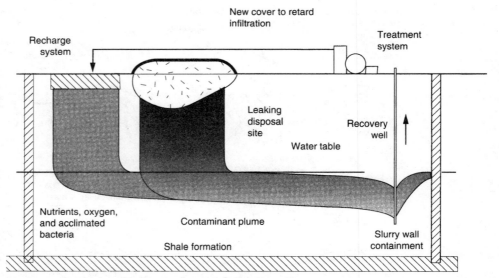

FIGURE 14.53 Typical example of a groundwater remediation system involving the use of slurry wall containment, recovery wells, treatment of contaminated groundwater, and groundwater recharge with nutrient addition to achieve in situ remediation.

Toxic Air Emissions.   A number of landfill operators have unexpectedly found it necessary to install landfill gas control and recovery systems to limit the release of toxic compounds into the atmosphere. The technology and system configurations are described earlier in this chapter. The necessary duration for operating these systems is unknown.

Groundwater Remediation.   Unlined landfills and landfills without leachate collection systems are the most likely to have a deleterious effect on groundwater quality. A remedial action program is instituted under federal or state regulations whereby contamination is detected in groundwater monitoring wells. As shown in Fig. 14.53, the first remediation step is placing a new, highly impermeable cap over the landfill to reduce water draining through the waste. Subsequent measures are designed to limit or cut off the movement of contaminated groundwater away from the landfill by the installation of bentonite slurry walls or the operation of recovery wells which control subsurface hydraulics. Contaminated groundwater in the aquifer surrounding the site is treated in aboveground facilities and either reinjected, sprayed onto nearby land, or discharged to a surface water. In situ bioremediation techniques may also be utilized to remove contaminants from the groundwater. Remediation by natural attenuation, where advective, dispersive, adsorptive, and biodegradation processes are relied upon to cleanse groundwater contamination, has gained favor, due to reduced long-term costs. Computer models simulating natural remediation are used to predict the time necessary to achieve an acceptable level of groundwater quality improvement. Complete restoration by any of the remediation procedures will likely take years or decades to complete.

REFERENCES

Adams, B., B. Karney, C. Cormier, and A. Lai (1998) "Artesian Landfill Liner System: Optimization and Numerical Analysis," *Journal of Water Resources Planning and Management,* vol. 124, no. 6.

Akesson, M., and P. Nilsson (1998) "Material Dependence of Methane Production Rates in Landfills," *Waste Management and Research,* vol. 16, no. 2.

Amling, W. (1981) "Exclusion of Gulls from Reservoirs in Orange County, California," *Progressive Fish Culturist,* vol. 43, no. 3.

Bagchi, A. (1990) *Design, Construction, and Monitoring of Sanitary Landfill,* John Wiley & Sons, New York.

Baldasano, J. M., and C. Soriano (2000) "Emission of Greenhouse Gases from Anaerobic Digestion Processes: Comparison with Other Municipal Solid Waste Treatments," *Water Science and Technology,* vol. 14, no. 3.

Baldasano, J., C. Soriano, and L. Boada (1999) "Emission Inventory for Greenhouse Gases in the City of Barcelona, 1987–1996," *Atmospheric Environment,* vol. 33.

Baldwin, T., J. Stinson, and R. Ham (1998) "Decomposition of Specific Materials Buried Within Sanitary Landfills," *Journal of Environmental Engineering,* vol. 124, no. 12.

Barlaz, M. A., D. M. Schaefer, and R. K. Ham (1989) "Bacterial Population Development and Chemical Characteristics of Refuse Decomposition in a Simulated Sanitary Landfill," *Applied and Environmental Microbiology,* vol. 55, no. 1.

Blight, G. E., and A. B. Fourie (October, 1999) "Leachate Generation in Landfills in Semi-Arid Climates," *Proceedings of the Institution of Civil Engineers, Geotechnical Engineering,* United Kingdom.

Bohn, H., and R. Bohn (1988) "Soil Beds Weed Out Air Pollutants," *Chemical Engineering,* vol. 95, no. 6.

Boltze, U., and M. de Freitas (1997) "Monitoring Gas Emissions from Landfill Sites," *Waste Management and Research,* vol. 15.

Brunner, D. R., and D. J. Keller (1972) *Sanitary Landfill Design and Operation,* U.S. Environmental Protection Agency, Publication SW-65ts, Washington, DC.

Chadwick, D. G., M. Ankeny, L. M. Greer, C. V. Mackey, and M. E. McClain (1999) "Field Test of Potential RCRA-Equivalent Covers at the Rocky Mountain Arsenal, Colorado."

Chanton, J., C. M. Rutkowski, and B. Mosher (1999) "Quantifying Methane Oxidation from Landfills Using Stable Isotope Analysis of Downwind Plumes," *Environmental Science & Technology,* vol. 23.

Christensen, T. H., and P. Kjeldsen (1989) "2.1. Basic Biochemical Processes in Landfills," in T. H. Christensen, R. Cossu, and P. Stegmann (eds): *Sanitary Landfilling: Process, Technology and Environmental Impact,* Academic Press, Harcourt Brace, Jovanovich, London.

Chugh, S., W. Clarke, P. Pullammanappallil, and V. Rudolph (1998) "Effect of Recirculated Leachate Volume on MSW Degradation," *Waste Management and Research,* vol. 16, no. 6.

CIWMB (1988) *Landfill Gas Characterization,* California Integrated Wast Management Board, State of California, Sacramento, CA.

Coons, L. M., M. Ankeny, and G. M. Bulik (2000) "Alternative Earthen Final Covers for Industrial and Hazardous Waste Trenches in Southwest Idaho," *Proceedings of the 3rd Annual Arid Climate Symposium,* SWANA, Albuquerque, NM.

County of Los Angeles, and Engineering-Science, Inc. (1969) *Development of Construction and Use Criteria for Sanitary Landfills, An Interim Report,* Department of County Engineer, Los Angeles, U.S. Department of Health, Education, and Welfare, Public Health Service, Bureau of Solid Waste Management, Cincinnati, OH.

Crawford, J. F., and P. G. Smith (1985) *Landfill Technology,* Butterworths, London.

Davis, S. N., and R. J. M. DeWiest (1966) *Hydrogeology,* John Wiley & Sons, New York.

De Visscher, A., D. Thomas, P. Boeckx, and O. Van Cleemput (1999) "Methane Oxidation in Simulated Landfill Cover Soil Environments," *Environmental Science & Technology,* vol. 33.

Deipser, A., and R. Stegmann (1994) "The Origin and Fate of Volatile Trace Components in Municipal Solid Waste Landfills," *Waste Management and Research,* vol. 12.

Edgers, L., J. J. Noble, and E. Williams (1992) "A Biologic Model for Long-term Settlement in Landfills. in: M. Usmen, and Y. B. Acar, *Environmental Geotechnology, Proceedings of the Mediterranean Conference on Environmental Geotechnology,* 1992.

Edil, T. B., V. J. Ranguette, and W. W. Wuellner (1990) "Settlement of Municipal Refuse," in: A. Landva, and D. Knowles, *Geotechnics of Waste Fills—Theory and Practice,* 1990.

Ehrig, H. J. (1989) "Leachate Quality," in T. H. Christensen, R. Cossu, and P. Stegmann (eds): *Sanitary Landfilling: Process, Technology and Environmental Impact,* Academic Press, Harcourt Brace, Jovanovich, London.

El-Fadel, M., A. N. Findikakis, and J. O. Leckie (1997) "Numerical Modelling of Generation and Transport of Gas and Heat in Sanitary Landfills III. Sensitivity Analysis," *Waste Management and Research,* vol. 15.

El-Fadel, M., S. Shazbak, E. Saliby, and J. Leckie (1999) "Comparative Assessment of Settlement Models for Municipal Solid Waste Landfill Applications," *Waste Management and Research,* vol. 17.

Emcon Associates (1980) *Methane Generation and Recovery from Landfills,* Ann Arbor Science, Ann Arbor, MI.

Farquhar, C. J., and F. A. Rovers (1973) "Gas Production During Refuse Decomposition," *Water, Air and Soil Pollution, Vol. 2,* pp. 483–495.

Fenn, D. G., K. J. Hanley, and T. V. DeGeare (1975) *Use of Water Balance Method for Predicting Leachate Generation from Solid Waste Disposal Sites,* EPA/530/SW-168, U.S. Environmental Association, Washington DC.

Fisher, S. R., and K. W. Potter (1998) *Evaluation of the Use of DUMPSTAT to Detect the Impact of Landfills on Groundwater Quality,* Wisconsin Department of Natural Resources.

Frevert, R. K., G. O. Schwab, T. W. Edminster, and K. K. Barnes (1963) *Soil and Water Conservation Engineering,* John Wiley & Sons, Inc., New York, NY.

Gibbons, R. D., D. G. Dolan, H. May, K. O'Leary, and R. O'Hara (1999) "Statistical Comparison of Leachate from Hazardous, Codisposal, and Municipal Solid Waste Landfills," *Ground Water Monitoring and Remediation.*

Gilbert, R. B., S. G. Wright, and E. Liedtke (1998) "Uncertainty in Back Analysis of Slopes: Kettleman Hills Case History," *Journal of Geotechnical and Geoenvironmental Engineering,* vol. 124, no. 12.

Givens, R. D. (1997) "Evaluation of Statistical Procedures for Groundwater Monitoring at Municipal Landfills," M.S. Thesis, Department of Civil and Environmental Engineering, Madison, WI.

Ham, R. K., et al. (1979) Recovery, Processing and Utilization of Gas from Sanitary Landfills, EPA-600/2-79-001.

Hansen, V. E., O. W. Israelsen, and G. E. Stringham (1979) *Irrigation Principles and Practices,* John Wiley & Sons, New York, NY.

Hashash, M. A. Y., T. D. Stark, and A. Abdulamit (2001) "Equivalent Linear Dynamic Response Analysis of Geosynthetic Lined Landfills," *Proceedings of Geosynthetics 2001, Portland,* Industrial Fabrics Association International, St. Paul, MN.

Hatheway, A. W., P. Geol, and C. C. McAneny (1987) "An In-Depth Look at Landfill Covers," *Waste Age,* vol. 17, no. 8.

Herrera, T. A., R. Lang, and G. Tchobanoglous (1989) "A Study of the Emissions of Volatile Organic Compounds Found in Landfills," *Proceedings of the 43rd Annual Purdue Industrial Waste Conference,* Lewis Publishers, Inc., Chelsea, MI, pp. 229–238.

Hilger, H., A. G. Wollum, and M. Barlaz (2000) "Landfill Methane Oxidation Response to Vegetation, Fertilization, and Liming," *Journal of Environmental Quality,* vol. 29.

Hjelmfelt, A. T., Jr., and J. J. Cassidy (1975) *Hydrology for Engineers and Planners,* Iowa State University Press, Ames, IA.

Holland, K. T., J. S. Knapp, and J. G. Shoesmith (1987) *Anaerobic Bacteria,* Chapman and Hall, New York.

House, V. W., and A. A. Young (1989) *Working with our Publics, Module 6, Education for Public Decisions,* Agricultural Extension Service and Department of Adult and Community College Education, North Carolina State University, Raleigh, NC.

Huitric, R. L. (1979) "In-Place Capacity of Refuse to Absorb Liquid Wastes," Presented at the Second National Conference on Hazardous Material Management, San Diego, CA.

Huitric, R. L., S. Raksit, and R. T. Haug (1980) *Moisture Retention of Landfilled Solid Waste,* County Sanitation Districts of Los Angeles County, Los Angeles, CA.

IPCC (1996) "Wastes," Module 6, *Revised 1996 IPCC Guidelines for National Greenhouse Gas Inventories,* Intergovernmental Panel on Climate Change.

Irving, W., J. Woodbury, M. Gibbs, D. Pape, and V. Bakshi (1999), "Applying a Correction Factor to the IPCC Default Methodology for Estimating National Methane Emissions from Solid Waste Disposal Sites," *Waste Management and Research,* vol. 17.

Khire, M., C. Benson, and P. Bosscher (1997) "Water Balance Modeling of Earthen Landfill Covers," *Journal of Geotech. and Geoenvironmental Engineering,* vol. 123, no. 8.

Kmet, P., K. J. Quinn, and C. Slavik (1981) "Analysis of Design Parameters Affecting the Collection Efficiency of Clay Lined Landfills," presented at the Fourth Annual Madison Conference of Applied Research and Practice on Municipal and Industrial Waste, September 28–30, 1981, University of Wisconsin-Extension, Madison, Wisconsin.

Kodikara, J. (2000) "Analysis of Tension Development in Geomembranes Placed on Landfill Slopes," *Geotextiles and Geomembranes,* vol. 18.

Koerner, R. M., and D. E. Daniel (1992) "Better Cover-Ups," *Civil Engineering,* vol. 62, no. 5.

Lang, R. J., and G. Tchobanoglous (1989) *Movement of Gases in Municipal Solid Waste Landfills: Appendix A, Modelling the Movement of Gases in Municipal Solid Waste Landfills,* prepared for the California Waste Management Board, Department of Civil Engineering, University of California, Davis, Davis, CA.

Lang, R. J., T. A. Herrera, D. P. Y. Chang, G. Tchobanoglous, and R. G. Spicher (1987) *Trace Organic Constituents in Landfill Gas,* prepared for the California Waste Management Board, Department of Civil Engineering, University of California, Davis, Davis, CA.

Lang, R. J., W. M. Stallard, L. C. Stiegler, T. A. Herrera, D. P. Y. Chang, and G. Tchobanoglous (1989) *Summary Report: Movement of Gases in Municipal Solid Waste Landfills,* prepared for the California Waste Management Board, Department of Civil Engineering, University of California, Davis, Davis, CA.

Linsley, R. K., J. B. Franzini, D. Fryberg, and G. Tchobanoglous (1991) *Water Resources Engineering,* 4th ed., McGraw-Hill, New York.

Linsley, R. K., M. A. Kohler, and J. H. Paulhus (1958) *Hydrology for Engineers,* McGraw-Hill, New York.

Mackay, K. M., P. V. Roberts, and J. A. Cherry (1985) "Transport of Organic Contaminants in Groundwater," *Environmental Science and Technology,* vol. 19, no. 5, pp. 384–392.

Manna, L., M. C. Zanetti, and G. Genon (1999) "Modeling Biogas Production at Landfill Site," *Resources, Conservation, and Recycling,* vol. 26.

Mathias, S. L. (1984) "Discouraging Seagulls: The Los Angeles Approach" *Waste Age,* vol. 15, no. 11.

McAtee, W. L. (1936) "Excluding Birds from Reservoirs and Fishponds," Leaflet 120, U.S. Department of Agriculture, Washington, DC.

McCreanor, P., and D. Reinhart (1999) "Hydrodynamic Modeling of Leachate Recirculating Landfills," *Waste Management and Research,* vol. 17.

McQuade, S. J., and A. D. Needham (October, 1999) "Geomembrane Liner Defects—Causes, Frequency, and Avoidance," *Proceedings of the Institution of Civil Engineers, Geotechnical Engineering,* United Kingdom.

Merz, R. C., and R. Stone (1970) *Special Studies of a Sanitary Landfill,* U.S. Department of Health Education and Welfare, Washington, DC.

Moshiri, G. A., and C. C. Miller (1993) "An Integrated Solid Waste Facility Design Involving Recycling, Volume Reduction, and Wetlands Leachate Treatment," in G. A. Moshiri (ed.) *Constructed Wetlands for the Water Quality Improvement,* Lewis Publishers, Boca Raton, FL.

Mulamoottil, G., E. A. McBean, and F. Rovers (1999) *Constructed Wetlands for the Treatment of Landfill Leachates,* Lewis Publishers, Boca Raton, FL.

NAS (2000) *Natural Attenuation for Groundwater Remediation,* National Academy of Science, Washington, DC.

NTIS (1955) *Hydrologic Evaluation of Landfill Performance Model,* National Technical Information Service, Springfield, VA.

O'Leary, P., and P. Walsh (1992) *Solid Waste Landfills Correspondence Course,* University of Wisconsin-Madison, Madison, WI.

Parker, A. (1983) Chapter 7: "Behaviour of Wastes in Landfill-Leachate," and Chapter 8: "Behaviour of Wastes in Landfill-Methane Generation," in J. R. Holmes (ed.) *Practical Waste Management,* John Wiley & Sons, Chister, England.

Perry, R. H., D. W. Green, and J. O. Maloney (eds.) (1984) *Chemical Engineers' Handbook,* 6th ed., McGraw-Hill, New York.

Pfeffer, J. T. (1992) *Solid Waste Management Engineering,* Prentice Hall, Englewood Cliffs, NJ.

Pohland, F. G., W. H. Cross, J. P. Gould, and D. R. Reinhart (1993) *Behavior and Assimilation of Organic and Inorganic Priority Pollutants Codisposed with Municipal Refuse,* U.S. Environmental Protection Agency, Washington, D.C.

Pohland, F. G. (1987) *Critical Review and Summary of Leachate and Gas Production from Landfills,* EPA/600/S2-86/073, U.S. EPA, Hazardous Waste Engineering Research Laboratory, Cincinnati, OH.

Pohland, F. G. (1991) "Fundamental Principles and Management Strategies for Landfill Codisposal Practices," *Proceedings Sardinia 91 Third International Landfill Symposium,* vol. II, pp. 1445–1460, Grafiche Galeati, Imola, Italy.

Pohland, F. G., and J. C. Kim (2000) "Microbially Mediated Attenuation Potential of Landfill Bioreactor Systems," *Water Science and Technology,* vol. 41, no. 3.

Popov, V., and H. Power (2000) "Numerical Analysis of Efficiency of Landfill Venting Trenches," *Journal of Environmental Engineering,* vol. 126.

Reddy, K., E. S. Motan, and C. Oliver (1999) "Parametric Seismic Evaluation of Landfill Liner and Final Cover Slopes," *Journal of Solid Waste Technology and Management,* vol. 26, no. 1.

Reinhart, D. R. (1993) "A Review of Recent Studies on the Sources of Hazardous Compounds Emitted from Solid Waste Landfills: a U.S. Experience," *Waste Management & Research,* vol. 11.

Reinhart, D. (1996) "Full-Scale Experiences with Leachate Recirculating Landfills: Case Studies," *Waste Management & Research,* vol. 14.

Rich, L. G. (1963) Unit Processes of Environmental Engineering, John Wiley & Sons, New York.

Salvato, J. A., W. G. Wilkie, and B. E. Mead (1971) "Sanitary Landfill-Leaching Prevention and Control," *Journal WPCF,* vol. 43, no. 10, pp. 2084–2100.

Schroeder, P. R., et al. (1984a) *The Hydrologic Evaluation of Landfill Performance (HELP) Model, User's Guide for Version I,* EPA/530/SW-84-009, 1, U.S. EPA Office of Solid Waste and Emergency Response, Washington, DC.

Schroeder, P. R., et al. (1984b) *The Hydrologic Evaluation of Landfill Performance (HELP) Model, Documentation for Version I,* EPA/530/SW-84-010, 2, U.S. EPA Office of Solid Waste and Emergency Response, Washington, DC.

SCS Engineers, Inc. (1989a). *Procedural Guidance Manual for Sanitary Landfills: Volume I, Landfill Leachate Monitoring and Control Systems,* California Waste Management Board, Sacramento, CA.

SCS Engineers, Inc. (1989b) *Procedural Guidance Manual for Sanitary Landfills: Volume II, Landfill Gas Monitoring and Control Systems,* California Waste Management Board, Sacramento, CA.

Spiegel, R. J., J. L. Preston, and J. C. Trocciola (1999) "Fuel Cell Operation on Landfill Gas at Penrose Power Station," *Energy,* vol. 24.

Stahl, J. F., M. Moshiri, and R. Huitric (1982) *Sanitary Landfill Gas Collection and Energy Recovery,* County Sanitation Districts of Los Angeles County, Los Angeles, CA.

Stark, T. D. (February, 1999) "Stability of Waste Containment Facilities," *Proceedings of Waste Tech '99,* National Solid Wastes Management Association, New Orleans, LA, pp. 1–24.

Stark, T. D., D. Arellano, D. Evans, V. Wilson, and J. Gonda (1998) "Unreinforced Geosynthetic Clay Liner Case History," *Geosynthetics International Journal,* Industrial Fabrics Association International (IFAI), vol. 5, no. 5, pp. 521–544.

Stark, T. D., W. D. Evans, and V. Wilson (March, 2000) "An Interim Slope Failure Involving Leachate Recirculation," *Proceedings of Waste Tech '00, National Solid Wastes Management Association,* Orlando, FL.

Stessel, R. I., W. M. Barrett, and X. Li (1998) "Comparison of the Effects of Testing Conditions and Chemical Exposure on Geomembranes Using the Comprehensive Testing System," *Journal of Applied Polymer Science,* vol. 70.

SWPCB (1954) *Report on the Investigation of Leaching of a Sanitary Landfill,* California State Water Pollution Control Board, publication 10, Sacramento, CA.

SWRCB (1967) In-Situ Investigation of Movement of Cases Produced from Decomposing Refuse, Final Report, California State Water Resources Control Board, The Resource Agency, publication 35, Sacramento, CA.

Tchobanoglous, G., F. L. Burton, and H. D. Stensel (2003) *Wastewater Engineering: Treatment and Reuse,* 4th ed., McGraw-Hill, New York, pp. 1–1848.

Tchobanoglous, G., H. Theisen, and S. A. Vigil (1993) *Integrated Solid Waste Management, Engineering Principles and Management Issues,* McGraw-Hill, New York.

Thomas, C., and M. Barlaz (1999) "Production of Non-methane Organic Compounds During Refuse Decomposition in a Laboratory-scale Landfill," *Waste Management & Research,* vol. 17.

U.S. Army Corps of Engineers (1956) *Snow Hydrology,* North Pacific Division, Portland, OR.

U.S. EPA (1991) *Solid Waste Facility Critoria: Final Rule,* 40 CFR Parts 257 and 258, U.S. Environmental Protection Agency, Published in Federal Register, October 9, 1991.

U.S. EPA (1995) *Test Methods for Evaluating Solid Waste,* U.S. Environmental Protection Agency, Springfield, VA.

Visvanathan, C., D. Pokhrel, W. Cheimchaisri, J. P. A. Hettiaratchi, and J. S. Wu (1999) "Methanotrophic Activities in Tropical Landfill Cover Soils: Effects of Temperature, Moisture Content, and Methane Concentration," *Waste Management,* vol. 17.

Watts, K. S., and J. A. Charles (October, 1999) "Settlement Characteristics of Landfill Wastes," *Proceedings of the Institution of Civil Engineers, Geotechnical Engineering,* United Kingdom.

Williams, G. M., R. S. Ward, and D. J. Noy (1999), "Dynamics of Landfill Gas Migration in Unsolidated Sands," *Waste Management & Research,* vol. 17.

World Wastes (1986) *Equipment Catalog,* Communication Channels, Inc., Atlanta, GA.

WPCF and ASCE (1969) *Design and Construction of Sanitary and Storm Sewers,* Water Pollution Control Federation and American Society of Civil Engineers, WPCF Manual of Practice No 9, Washington DC.

Wright, T. D. (1986) "To Cover or Not to Cover?" *Waste Age,* vol. 17, no. 3.

Young, P. J., and L. A. Heasman (1985) "An Assessment of the Odor and Toxicity of the Trace Components of Landfill Gas," *Proceedings of the GRCDA 8th International Landfill Gas Symposium.*

CHAPTER 15
SITING MUNICIPAL SOLID WASTE FACILITIES

David Laws
Lawrence Susskind
Jason Corburn

15.1   INTRODUCTION

Because solid waste management agencies will continue to utilize recycling, composting, land-filling, and incineration for the foreseeable future, new facilities will have to be built. When the siting of such facilities is not viewed as part of a process of building consensus on solid waste management goals, public opposition can lead to long delays and often to the rejection of proposed facilities. Continued failure to site needed solid waste management facilities will compromise our national ability to meet basic environmental protection and resource management needs.

The difficulties that plague siting efforts usually stem from the failure to take adequate account of the concerns that affected groups have whenever new solid waste management facilities are proposed. People bordering on the site are likely to resist siting efforts because they are afraid that their property values or quality of life could be adversely affected. Local business leaders are likely to resist new facilities because they may increase taxes or operating costs. Environmental groups will push for source reduction and extensive recycling before they support additional landfills, and they are likely to oppose siting incinerators on the grounds that they pose potential risks to human health and safety. There may well be other groups with additional concerns, depending on the type, scale, location, and cost of proposed facilities.

This chapter reviews the political, technical, economic, and ethical aspects of siting that are prominent sources of public concern, discusses how they arise and are usually handled in the series of choices involved in a typical siting process, and, finally, presents a siting credo that responds effectively to public concerns and embodies the best practical advice available for anyone who will manage or participate in an effort to site a solid waste management facility (National Workshop on Facility Siting, 1989, 1990).

15.2   UNDERSTANDING THE SOURCES OF PUBLIC CONCERN

Someone charged with selecting a site for a solid waste management facility ought to expect opposition. While the sources of resistance are often characterized as NIMBYism, a not-in-my-backyard attitude, this characterization assumes that opposition to new facilities is based solely on selfish desires to shunt the burden of public responsibility elsewhere, but this is often not the case. While there may be residents who would prefer no change of any kind, the most vigorous opposition usually comes from those who have legitimate concerns or favor alternative methods of solid waste management, other "more appropriate" sites, or different ways of making decisions.

Anyone reading in the newspaper one morning that an empty field at the end of the street was being considered for an incinerator or a landfill would have concerns. What risks will the facility pose to family and neighbors? How should such risks be calculated? Will the operators of the proposed facility meet their obligations? Will the technology they have chosen work as expected? What will happen if something goes wrong? What kinds of impacts will the proposed facility have on the immediate area? Will neighbors move away? What about those who have no choice about moving? Could the facility affect groundwater quality? Who will pay if it does? What kind of traffic and noise will the facility generate? What might it do to property values? Will it affect schools? Why should this particular neighborhood be asked to bear the burden so that others can dispose of their trash? Will wastes from other communities be trucked in? Is someone going to make a profit at the expense of others? How was this site selected? These concerns and others might readily lead potential neighbors to conclude that their best strategy is to be first and loudest in raising objections in order to increase the chances that another site will be selected instead.

Solid waste management facilities represent long-term commitments of public resources that can dramatically alter the quality of life in a community. Thus, neighbors of proposed facilities and members of the public have legitimate grounds for concern. They also have reason to expect that every effort will be made to ensure that wise decisions are made.

Political Aspects of the Siting Process

Siting decisions hinge on the trust that citizens have in government, technology, and business. To the extent that corruption, closed decision-making processes, and highly publicized accidents involving advanced technologies (like Three Mile Island, the *Challenger,* or the Exxon *Valdez*) have undermined public confidence, siting decisions have become that much more difficult. For the public official, corporate representative, or technical consultant involved in the siting process, this general lack of trust often translates into skepticism, an unwillingness to accept assertions at face value, requirements that extra margins of safety be met, and demands for risk reduction or compensation.

Advocacy groups and citizen leaders have developed considerable sophistication in putting forward such demands. There are national networks they can tap for advice and legal assistance. While they rarely have the power to veto a siting decision, they can often mount legal and political challenges that will tie up a project indefinitely.

Healthy skepticism is not necessarily inappropriate and should perhaps even be encouraged when decisions of great significance must be made, but when profound distrust coupled with unreasonable demands produces political gridlock, everyone is hurt. Without some willingness to "suspend disbelief" and engage in a community-wide dialogue, unwise decisions are likely to result.

Technical Aspects of the Siting Process

There are several types of technical analysis that are important to making wise siting decisions, although they are rarely precise enough to produce definitive answers. Good technical analysis can clarify operating assumptions, specify areas of uncertainty, highlight data deficiencies, and spell out the sensitivity of key findings to slight variations in underlying assumptions. The key technical analyses required to evaluate solid waste management options involve forecasts of demand, site suitability analysis, impact and mitigation analysis, risk assessment, and specification of monitoring and management standards. Because nonobjective judgments play an important role in all of these analyses, it is imperative that groups concerned about the impacts of proposed solid waste management facilities become involved in the decision-making process.

Economic Aspects of the Siting Process

From an economic perspective, a solid waste management facility represents a stream of benefits and costs. The benefits flow from the ability to dispose of waste, although for most people such a service is often taken for granted. Benefits of this sort are usually distributed fairly evenly across all users of a facility. Costs include disposal fees and taxes, indirect costs such as traffic, noise, odor, changes in property values, and elevated risks to human health and safety. Except for certain out-of-pocket costs, these are likely to be distributed *unevenly*. They tend to fall hardest on those closest to a facility, and drop off across a gradient that correlates roughly with distance from it.

The obvious economic imperative is to find a technological and locational option that is efficient: one that provides the greatest level of *net* benefit. As with technical analysis, however, we suggest that the definition of costs and benefits hinges on a great many subjective judgments. Again, unless concerned groups are involved in preparing such analyses, they will be likely to question the legitimacy of the results.

Moreover, even if a facility is efficient, and the overall benefits to the "gainers" greatly outweigh the costs, resistance can be anticipated. The potential losses faced by the relatively small group of "losers," while modest in aggregate terms, are likely to be significant to individuals and provide them with a substantial (even pressing) incentive to act. And these individuals will face only minimal organizational costs if they wish to act collectively, since the losers are likely to be concentrated geographically and may even know each other. The beneficiaries, on the other hand, cannot be expected to support a proposal since, though numerous, they receive only modest individual benefits and face large organizational costs if they wish to act as a group (O'Hare et al., 1983).

Ethical Aspects of the Siting Process

The siting of solid waste management facilities raises ethical questions of various kinds. The distribution of costs and benefits may not be fair. When solid waste facilities serve regional needs or where facilities accept waste hauled in from long distances for high fees, questions of fairness to the host community are almost always raised. If disproportionate burdens fall on those who appear to be targeted because they are poor, people of color, or disadvantaged in another way, issues of fairness are often framed as questions about justice or the abridgement of rights. When historical patterns of siting appear to repeatedly burden particular communities, legal and political charges of discrimination will probably be raised. These have been some of the motivating concerns behind calls for environmental justice (EJ), which have become prominent and should be anticipated in efforts to site solid waste management facilities. The prominent legal cases suggest how environmental justice concerns can shape facility siting.

The siting of waste management facilities has been challenged in the courts using both the Equal Protection clause of the Fourteenth Amendment and Title VI of the Civil Rights Act of 1964. In a landmark case brought under the Equal Protection clause, *Bean v. Southwestern Waste Management Corporation*, a solid waste disposal facility was challenged as a discriminatory siting because the facility was to be sited next to a predominantly African-American public school. The case ultimately failed to meet the strict standard of discriminatory intent, but it initiated the use of demographic data to analyze the distributional fairness of facility siting (*Bean v. Southwestern*, 1979; Schwartz, 1995).

Another solid waste facility siting challenged in the courts as environmental discrimination was *Chester Residents Concerned for Quality Living v. Seif.* In this case, residents of the predominantly African-American community of Chester, Pa., challenged as discriminatory a decision by the State of Pennsylvania Department of Environmental Protection to issue an operating permit for a waste facility in their community. The Chester residents received injunctive relief from the district court, but this ruling was later overturned by the court of appeals. The case eventually made it to the U.S. Supreme Court, where the brief on the case was declared moot after the Department of Environmental Protection revoked the original permit (*Chester Residents,* 1997).

The case law over environmental justice claims is presently in an indeterminate state, but will surely continue to shape solid waste facility siting efforts in the future. The U.S. Environmental Protection Agency (EPA) recently published guidance in the Federal Register for investigating Title VI complaints used to challenge facility permits (U.S. EPA, 2000). Despite the slowly emerging legal guidelines, environmental justice has emerged as a powerful grassroots movement that can frustrate and derail waste facility siting. Behind most environmental justice claims is the notion that locally unwanted land uses (LULUs), including solid waste facilities, have historically been sited in communities with the least economic and political clout, chiefly low-income communities and communities of color. These same communities see themselves continuing to bear unfair burdens without access to the processes by which siting decisions are made. While it is often difficult to define justice in a way that will satisfy all stakeholding groups in every situation, steps can be taken to ensure that siting processes are viewed as fair by as many groups as possible, and especially by populations who have historically shouldered disproportionate burdens.

At a minimum, procedural guarantees can be put in place to ensure that local stakeholders, especially people of color, play a role in making siting decisions. These guarantees should ensure that all stakeholders have the opportunity to participate in open deliberations that affirm their right and ability to understand issues, ask questions, voice concerns, and propose ways of resolving those concerns. In effect, siting processes need to be publicly transparent and decisions should be taken only after there has been sufficient opportunity to explore alternatives that might meet the interests of all stakeholding parties.

15.3   A TYPICAL SITING CHRONOLOGY

A typical siting process begins with a determination that a facility of some sort is needed. Then, technological options are either explicitly or implicitly considered. These two steps can occur either before alternative sites are considered or at the same time that specific sites are reviewed. In either case, the next step is to forecast and assess ways of mitigating potential impacts. This is generally undertaken as part of an environmental impact process. Finally, operational guidelines for the management of the facility are prepared, usually as the product of all the prior steps.

Determining Need

Need is the pivotal question in any siting process. If a community accepts the need for a facility, a siting process has a much better chance of succeeding. If doubts linger about the way need was determined, those opposed—for whatever reasons—are likely to find allies to join in blocking actions. While it is clear that judgments about need should be based on an analysis of past practice and future requirements, it is also clear that determination of need rests heavily on a great many nonobjective judgments, such as estimates of the size and composition of future waste streams, which are always debatable.

An assessment of need involves forecasting population growth (including both rates and absolute levels), consumption levels, and the prospect of compliance with new regulations. Forecasts hinge on assumptions about how people will live in the future and how much waste they will generate. They also require judgments about the changing composition of the waste stream. Economists, demographers, civil engineers, sociologists, and others can offer advice on how to make such estimates, but it is clear that judgments, more than facts, dominate.

A needs assessment must also take account of what disposal costs are likely to be, and how the behavior of individuals and firms will be influenced by price. The not-in-anybody's-backyard movement is a challenge to the way in which the need for new waste disposal facilities has traditionally been determined. This movement focuses on blocking all new facilities, not to push construction to other locations, but rather to increase the pressure for source reduction and recycling. Proponents argue that source reduction and recycling goals will be undermined if new facilities are built. In short, they question the need for any new solid waste management facilities.

Choosing a Technology

Suppose that the parties involved in a siting decision are able to agree that, indeed, a new facility of some sort is required. This may include a shared understanding that some reduction and recycling goals must also be met. Given such an agreement—and assuming that the volume and composition of the waste stream have been specified—it may be tempting to think that the choice of technology will be obvious. Unfortunately, this is not the case. The choice of a best technology is as open to challenge as the determination of need. The magnitude of the impacts (in terms of costs and risks) that alternative technologies might have, as well as probable impacts of alternative technologies on ecology, human health, safety, and welfare, are likely to all be quite sensitive to assumptions that are open to challenge.

Some technological impacts are relatively easy to forecast, such as the increase in traffic that will occur during the construction and operation of a new facility. The level of noise associated with the given volume of traffic can also be extrapolated. The tolerance of individuals to different traffic and noise levels is, however, much harder to forecast with certainty. Assumptions must be made about whether traffic and noise will be constant or intermittent, what the hours of operation will be, and how sensitive different households will be. While ballpark estimates may be feasible, the individuals involved will demand the final word.

It may be possible to forecast the likely impacts the various technologies will have on human health. Estimates of the level of toxic metals in stack emissions from the incineration of municipal solid waste may be feasible, for example (Washburn et al., 1989). In many cases, however, the absence of historical or baseline data makes this very difficult. Moreover, the relevance of historical or comparable situations will be questioned by those who see things differently (Ozawa, 1991).

Forecasting the mobility of chemicals in the ash from a municipal solid waste incinerator requires the use of a leaching test. Several such tests are available. The forecast of how hazardous the ash from a particular facility may be depends on which test is used, as well as the standard for extrapolation. It also depends on assumptions about the content of the waste stream. Estimates of the costs (and health risks) associated with disposing of the ash depend on how federal regulations governing household waste are presumed to apply. Moreover, assessments of the health effects of alternative disposal technologies require a detailed study of possible exposure pathways. Some of these are not yet well understood. Estimating the magnitude of a human dose requires assumptions about body weight, rates of respiration, consumption of food and water, incidental soil ingestion, and the extent of dermal absorption (Washburn et al., 1989). Variations in these assumptions substantially influence estimates of health effects. The use of average or standard figures, often substituted for empirical data, can obscure differences important to certain segments of the local population.

Even when the risks associated with alternative technologies can be estimated with reasonable certainty, the acceptability of such risks usually varies among different parts of the population. This may be because of the character of the risks (e.g., unfamiliar, involuntary, undetectable risks are usually less acceptable than familiar, voluntary, and detectable risks) or the way they are distributed across the population (e.g., risks to children may be treated more seriously than risk to adults) (Sandman, 1987). Any of these items is sufficient to alter the assessment of a particular technology.

Site Selection

Many aspects of site suitability can be assessed only with a particular technology in mind. The converse is also true. The question that is most often asked in evaluating sites is, "How suitable is this particular site for the activity we have in mind?" Those responsible for selecting a site may deem some locations inappropriate on the basis of exclusionary criteria (related to health and safety standards) that do not require contingent analysis. Asking the question this way, however, leaves a variety of sites that may be acceptable, although each will present a different combination and distribution of costs and risks. Thus, there is a strong temptation (as with the choice of technology) to ask which is the best or optimal site. Unfortunately, answers to this question will probably reveal more about the perspective that the analyst adopts than about the objective characteristics of the situation.

Any number of factors can be taken into account in assessing the suitability of a site. Transportation access, soil capability, adjacent land uses, and land ownership patterns are usually given consideration in assessing sites. If a landfill is the technology of choice, the community will want to consider hydrology and geology. If an incinerator is planned, air circulation patterns around the site will be important. Impacts on flora and fauna will need to be considered. Historical patterns of siting may be brought up, as well as the location of other facilities that pose possible health threats. Thus, there is no all-purpose list of factors that can substitute for good judgment exercised with respect to knowledge of local conditions.

As more factors are added, it becomes increasingly difficult to amalgamate all the concerns identified. As the analysis becomes more comprehensive, complexities and tradeoffs make it more difficult to justify a final decision. Unfortunately, there is no generally accepted method for drawing together all these considerations into a single metric that allows for sites to be compared or ranked.

It is also unlikely that there will be agreement on how the many characteristics of sites ought to be weighted in making a final decision. Different groups are likely to have their own opinions about the relative importance of each consideration. There is no objective system for weighting the relative importance of various features of alternative sites. This explains why the task of identifying a best or optimal site is so difficult. The only way that this challenge can be met is to build a consensus among all the stakeholding parties that synthesizes all siting considerations.

Site selection is also haunted by the political polarization that accompanies the actual designation of candidate sites. Once sites have been announced, the majority of residents in a region breathe a sigh of relief. Those who have been designated to host a facility, however, often feel stigmatized. They may feel enormous pressure to mount whatever political resistance they can muster in the hope of coming off the list quickly. A willingness to participate in reasoned deliberations may be interpreted as a sign of weakness. While a careful analysis of alternative sites would surely benefit from the inclusion of numerous options for purposes of comparison, the political pressure to eliminate sites quickly will be enormous. If sites under consideration are not announced—in an effort to avoid premature political battles—neighbors of those sites that remain after site comparisons are complete will challenge the legitimacy of the analysis on the grounds that they had no chance to contribute crucial information that only those in the area could possibly have.

Assessing and Mitigating Impacts

The assessment of impacts is an integral part of site selection. A thorough understanding of prospective sites can only be gained by studying the likely impacts of the proposed activity and the prospects of mitigating them. It is rare, however, for impacts to be studied before a favored site has been selected. Impact assessments are more commonly undertaken to comply with federal or state regulations *after* a site has been selected. In this context, such assessments tend to be narrowly focused in ways that cast the favored site and technology in a positive light (Barzok, 1986; Susskind, 1978).

The models available for forecasting environmental impacts are usually derived and calibrated using data from other times and places. It is sometimes unclear whether these are appropriate to new conditions and how they should be recalibrated. The costs of building new models or recalibrating old ones usually exceed the resources available for most impact assessments. Longitudinal data that would help in determining prospective sites are usually unavailable.

Environmental impact assessments are expected to enrich our understanding of the trade-offs associated with each solid waste management alternative. Yet, like the analyses involved in the earlier steps of the siting process, they are unlikely to be conclusive (Elliott, 1981). The sources of uncertainty are too numerous, the data too sparse or coarse (if available at all), and the analytic methods themselves dependent on choices that cannot be made solely on objective grounds.

Environmental impact assessment becomes even more complex when an attempt is made to assess possible mitigation measures. Suppose a landfill is considered for a site adjacent to a wetland. The functions of that wetland are likely to be complex (and not completely understood). The importance of the wetland for the survival of neighboring plant and animal populations may not be well understood. Finally, it is difficult to judge (much less allocate responsibility for) the cumulative impacts on a wetland that the proposed facility will have, given other changes that might also occur in the area (Contant and Wiggins, 1991; Dickert and Tuttle, 1985). There is a possibility that beyond a certain threshold a process of catastrophic, irreversible change will occur, but, prior to that point, impacts will be minimal. If the threshold is breached, however, mitigation will not be possible. The boundary that separates environmentally divergent outcomes may be difficult to predict or, in the worst case, recognizable only after it has been passed.

If a portion of a wetland is adversely affected by a new facility, is rehabilitating an adjacent wetland or creating a new one some distance away a suitable response? Without a clear sense of the wetland's functions, it is difficult to answer such a question. What if there is also a chance that groundwater will be contaminated by a proposed landfill? Is securing an alternative supply of water a satisfactory mitigation measure? What does securing mean? Is an insurance policy satisfactory, or should an actual supply of water be set aside? And for how long should a supply be secured? Groundwater contamination, even if stopped, can persist for an extended period.

Even when mitigation measures are straightforward and relatively easy to quantify (such as providing guarantees that property values will not decline), the acceptability of such measures is not necessarily clear. Some residents may be able to move. But dislocating families from their homes may produce stresses that are difficult to predict, much less to value. Are cash payments that permit residents to move an appropriate mitigation measure? Such payments, like artificial wetlands, offset some of the effects of change, but they cannot guarantee that the future will be like the past, and they may have very different meaning to some of the affected parties.

Managing the Facility

Assumptions about operating standards must be made before a forecast of the impacts of a facility can be completed. Thus, clarifying how a facility will be managed is an integral part of the siting process. Yet it is difficult to develop reliable or precise estimates of future organizational performance because many of the factors most likely to affect it—operational procedures, management structures, and individual behavior—are difficult to quantify and forecast (Elliott, 1984).

Perceptions of the risk associated with each type of waste management option will eventually form the basis for voluntary consent or imposition of a new facility. In the former case, technology, site, and impact assessments will permit reasonable people to feel secure. In the latter (which we do not favor), assessments will provide a justification that elected officials or the court can use to impose a choice. The expectations that people have about how a facility will be operated, and about the effectiveness of monitoring and response systems, will also play an influential role in the judgments they make about their security and the acceptability of a facility. In an atmosphere of minimal trust (or even healthy skepticism) neighbors and other affected parties may demand guarantees that provide for independent monitoring and give the host community a substantial role in overseeing management performance. Communities may even wish to bind a facility operator through covenants that require incremental improvements in performance as new knowledge becomes available over time.

It is important that management arrangements satisfy the concerns of affected citizens and build trust in the operator and the siting process. For a facility's neighbors, such arrangements are the only guarantee against threats to their health, safety, and welfare. If an operator believes in its ability to manage a facility within specified standards, then the risks associated with providing contractual guarantees on performance ought to be minimal.

15.4   *BUILDING CONSENSUS ON SITING CHOICES*

Our analysis of the siting process highlights its indeterminate character. Each of the series of questions that must be resolved to site and operate a solid waste management facility hinges on nonobjective judgments and raises concerns about justice and fairness as well as technical appropriateness and efficiency. Issues of need, health, and security will be as much at issue as economic and technical questions about the appropriateness of sites and the cost of competing designs. Opponents of facilities are sophisticated and know how to bring out this characteristic of siting with analysis and political action. They may raise questions about how the problem was framed, what knowledge is valid and sufficient to address disputed questions, and how ongoing responsibilities for protecting public health, safety, and welfare will be met. The siting of solid waste management facilities is inescapably a political process.

Too often, proponents respond to these characteristics by making siting "political in the wrong way." They may adopt a "decide-announce-defend" strategy, stressing the need for strong leadership and internal demands to get something done. Moving down this path, they are likely to rely on the guidance of a small group of professionals with pertinent expertise. This group, in turn, draws heavily on current ideas about best practice, responds to current legal requirements (such as health and safety regulations), and tries to take account of political realities. They provide few opportunities for public participation or even comment prior to deciding what they think ought to be done. In the eyes of many elected and appointed officials, opening up the decision process before the experts have announced their proposed strategy will only undermine the "objective" character of the decisions. Once the agency and its experts (in consultation with industry) reach a decision about a site and a technology, they may well allow hearings or other occasions for unhappy groups to sound off.

Proposals are typically presented as the technically best way to meet a need or respond to a crisis. Once the announcement has been made, agency personnel shift into a defensive mode to weather the storm of political protest. They seek to justify the proposal on grounds that (1) they used the best available evidence and analyses, (2) they followed all required procedures, and (3) the outcome is fair because it was based on an objective process and criteria. Interactions with angry or dissatisfied groups are likely to be polarized, with communication limited to legalistic exchanges as everyone prepares for lawsuits.

Proponents of facilities may feel that they are acting responsibly and providing leadership by getting good advice, making tough choices, and standing by their decisions. Indeed, this view periodically leads to calls for preemptive legislation as a way to avoid siting problems. While such proposals may be well intentioned, avoiding procedural safeguards and opportunities for public scrutiny and deliberation seems inconsistent with the complexity of siting decisions and the influence they can have on people's lives. And efforts to preempt public involvement may only prompt affected parties to recast their challenges or move them to other forums. Short-circuiting established review processes or providing government agencies with the ability to site by fiat is only likely to erode trust further and, in the end, make siting decisions more difficult. Even well-intentioned efforts to involve the public can backfire when they polarize the process into a discussion between those who understand and those who need to be educated. The combination of technical and moral complexity that characterizes siting decisions calls for a process of mutual learning between those responsible for siting and operating solid waste management facilities and those whose lives will be affected.

This concluding section outlines an alternative approach that tries to make siting "political in the right way." This approach starts from the presumption that siting decisions can rarely (if ever) be justified on technical grounds alone and involve issues about which reasonable people can disagree, even if disagreements have not always been reasonable historically. The organizing insight, backed by a growing body of practical experience, is that the institutional commitments shape the way parties engage issues and each other. Changing the organization can make a process more constructive without masking disagreements. The siting process can be made "political in the right way" by moving nonobjective judgments and questions of fairness to center stage and involving stakeholding groups in open discussions about the tensions, tradeoffs, and choices that are involved. This process of engagement is best understood, and organized, as a process of building consensus among the individuals and groups most directly affected by, interested in, and responsible for the siting and operation of solid waste management facilities.

Consensus building works in such contexts by initiating and sustaining a conversation among stakeholders that takes as its subject the design of a practical plan of action, and makes progress by clarifying and engaging disagreements rather than suppressing or avoiding them. The growing body of experience with consensus building suggests that it is not only an attractive, but also a practical, way to deal with problems like siting of solid waste management facilities. The following sections provide an overview of consensus building by discussing the progression of steps and problems that characterize the process. This should not be interpreted as a manual for consensus building, and those interested should consult other sources or seek professional assistance.* This overview is organized in four sections: identifying and convening stakeholders, managing the conversation, getting agreement, and anticipating implementation.

Identifying and Convening Stakeholders

The goal of a consensus-building effort is to reach agreement on a plan of action that is acceptable to stakeholders and is acknowledged as legitimate by the broader public. One of the key questions is always, "Who participated in the development of the consensus?" An outcome carries more credibility when the citizens whose welfare is at stake shape analysis and participate in making choices about sites, alternative designs, and management conditions. Moreover, sustained interaction, if managed and organized appropriately, can raise the likelihood that the members of a diverse group will negotiate an agreement that each is willing to stand behind as fair, efficient, stable, and wise. This assumes that those in charge of the siting process can identify relevant stakeholders and get them to agree to participate. This often requires more than sending out invitations. It may be difficult to identify or engage the full range of parties who are affected or have an interest. Some obvious groups of stakeholders may be difficult to engage because there is no organization through which they can be contacted. Others may be historically disenfranchised and therefore disinterested or resistant to participating in a political process. Convenors may need to help such groups coalesce and develop sufficient organizational capacity to choose representatives and keep their constituents informed. It may seem counter-

---

* The most comprehensive guide is the *Consensus-Building Handbook* (Susskind, Mackearnan, and Thomas-Larmer, 1999).

intuitive to help organize groups that may offer opposition, but this is seen as preferable to trying to engage an amorphous group that can only react and has a limited capacity to learn, negotiate, or commit.

Those groups that are organized may view the invitation to participate skeptically. They may have reservations about committing to a process they do not understand or they may be jaded by previous experience. They may be wary of suspending effective lines of action, such as protest, that are well understood. They may be suspicious that the invitation is insincere and will turn out to be an effort to placate or "educate" them. They may not want to devote resources until other groups have committed to participate.

These challenges make the convenor's job difficult and present a persistent rationale for settling for a group composed of organized, willing, and manageable parties. It is important to keep in mind the pitfalls this approach may create down the road. Critics looking to discredit the process or the outcome will look first at who was involved. If significant groups were excluded or not invited, they will have easy grounds on which to mount a challenge. (This is also why it is often preferable to deal with the practical problems of managing a large group rather than the political problems raised by limiting invitations.) An early effort to solicit participation can also prevent problems down the road when new groups seek to join the process in midstream. It is often not feasible to exclude such groups, yet incorporating them can add delays and disrupt working patterns the group has labored to establish. These observations argue for a robust effort to secure broad initial participation. The risks and demands of inviting broad participation are tempered by its self-regulating quality. Those who elect to participate will find that the demands of consensus building are significant; if they do not have a real stake in the outcome they will often drop out.

Convening is a difficult responsibility. The tensions that convenors must confront often make it difficult for project sponsors to play this role. Professional facilitators may be better able to fill the demands of organizing parties, provide the kind of guarantees that will get skeptics to temporarily suspend disbelief, and solve the chicken-and-egg game of getting enough parties to commit to participate to make it worth anyone's while to commit. These professionals have experience meeting with groups, know how to address their concerns, and understand how to sequence commitments to build a group that can pass the test of public legitimacy.

Formulating an Agenda and Ground Rules

For all the reasons cited here, the most proponents should expect of the stakeholders who come to the table is a temporary suspension of disbelief. This is several steps removed from the level of commitment necessary to build consensus among a diverse group on the difficult and divisive issues involved in choosing a site and technology. It is imperative that whoever is managing the process begin immediately to build credibility and commitment among the parties around the table.

The first opportunity to transform skepticism into commitment comes in the constitutional phase of consensus building in which norms, goals, and formal rules are established and an agenda is specified. All participants will scrutinize this phase closely because it provides the first direct evidence of what the process will be like and of the competence and intentions of the organizers. Three pieces of evidence they will find salient are how the goals for the process are established, how these goals are translated into an agenda, and how the ground rules shape the roles of citizens, professionals, sponsors, and consultants.

Proponents seeking to act responsibly may feel a need to establish goals. Handing this responsibility over to others may appear to be an abdication of their professional or statutory obligations. Even a well-intentioned and grounded effort to translate formal mandates or technical analysis directly into goals is likely to be viewed by stakeholders as a betrayal of the commitments that made them willing to suspend judgment and come to the table. Without a say in determining what the conversation is about, the invitation to participate may appear hollow.

The analysis that leads proponents to set goals confuses the process of consultation with the legislative and administrative responsibilities. To provide sufficient scope in goals for participants to regard the process as legitimate, it is often necessary to constitute consensus-building groups as advisory bodies. Proponents may also feel pressure to set goals because they fail to recognize that, as participants, they retain a veto over recommendations that fail to respond to concerns they have raised, particularly when they can back their concerns with reasons or the recommendations requiring them to take action. Whatever institutional arrangement is decided upon, experience suggests it must offer sufficient scope for participants to shape the goals if there is to be any hope that they will endorse and support the outcome. Critical or skeptical parties, in particular, may want goals that preserve their right to say no when they can supply good reasons.

The next step is to translate these goals into an agenda. The power of agenda setting will be clear to participants, as will the fact that broad or encompassing goals cannot be pursued through a narrow agenda. Efforts to shape or preempt discussion by dictating or limiting the agenda will threaten any progress that has been made in building trust and legitimacy. This may mean that controversial questions about need, health impacts, and facility management will have to be opened to further discussion, even when the responsible parties feel they have been settled by analysis. Proponents may be reluctant to open themselves to criticism or fragile consensus to dispute. They should consider, however, how commitments that seem reasonable to them may look to parties who did not participate and whose welfare is on the line.

The interaction through which the goals and agenda are pursued is shaped by a set of procedural ground rules. These guidelines express the recognition of equality and mutual respect that underpins the commitment to reason, to be open to reason, and to seek a plan that is acceptable to all, and provide the foundation for consensus building. They begin with the effort to be inclusive, but also depend on the offer to participate, having the kind of fair value that is guaranteed by symmetrical rights to ask questions, offer evidence, make proposals, and offer counterproposals. The legitimacy of the process and the viability of the outcome hang by these procedural threads. Responsibility for upholding them cannot be delegated to any single party. All participants must play a role in shaping these commitments and ensuring that they are upheld.

As with the convening stage it may be difficult for proponents of a siting process to bring their particular perspective to the debate and simultaneously manage the dialogue. These demands provide another rationale for using trained facilitators. These professionals understand the relationship between goals, agenda, and process. They also know the importance of tying their role to their ongoing ability to uphold ground rules in a way that helps stakeholders participate and sustain their support. This is not quite the high-wire act that it may sound like, but it often entails explicit promises to keep the conversation open and a commitment to step aside if the participants feel the discussion is being directed rather than managed in an open and ethically justifiable way.

Managing the Conversation

Managing a siting process in this way implies more than formulating consensus on an agenda and upholding a set of procedural conditions. The information pooling that takes place during consensus building continually generates new possibilities. Good management requires a persistent effort to keep the conversation focused and on track. Two central problems are managing consultation on technical issues and engaging the diversity of perspectives around the table. Both involve a central dynamic in consensus building. Parties who have a stake in siting decisions are involved in an ongoing effort to make sense of a complex and changing world. The series of choices they face in siting a solid waste management facility are not controversial just because some people are shortsighted, uninformed, or pursuing narrow interests. Legitimate differences arise from diverging interests, from technical information that is uncertain and open to interpretation, and from different reasonable views on the nature and scope of public obligations to promote public health, safety, and welfare.

A common and reasonable reaction to this complexity is to seek the guidance of experts. While the assistance of technical experts is often essential to understanding the issues that solid waste management involves, their participation raises new challenges. First, the selection of experts is not a trivial matter. The areas where the group seeks advice may be precisely those areas that are controversial. Experts in these areas may have affiliations and commitments that make them appear controversial to some stakeholders. These commitments may be what make them worth consulting in the first place, but may also shape the way they interpret evidence and render areas of agreement and disagreement. The net effect may look unacceptably biased to some stakeholders around the table.

One implication is that the responsibility for selecting experts must be held by the stakeholders. Moreover, stakeholders' responsibilities do not end with the choice of experts, but extend to understanding the complex evidence that weighs on the choices at hand. They may find it is not possible to understand controversial issues by consulting a single expert. As soon as they confer with more than one consultant, however, they raise a new tension. Conversation will threaten to devolve into a contest between competing experts. The tension between these poles and the responsibilities it creates are persistent and must be managed on an ongoing basis.

The participation of experts raises a second challenge. The practices and norms that characterize participatory forums may be foreign to them. Experts may find it difficult to express their understanding in terms that can become part of a common conversation or be brought into relationship with other considerations. They may have difficulty engaging problems in the terms in which stakeholders have framed them. Technical experts may feel it is their responsibility to report a scientific consensus or interpret technical information into recommendations about how to act. Consensus building creates a distinctive challenge for experts. It asks them to provide participants with a complex story that may highlight ambiguities and tensions in their own research. This kind of consultation may violate well-established professional norms.

The interdisciplinary character of solid waste management and the significance of local knowledge create a third challenge: the need for knowledge integration. Solid waste management raises questions about engineering, economics, epidemiology, toxicology, and geology. Experts may not be used to working across the boundaries between these disciplines. Local residents may have knowledge to contribute that shapes the implications of technical insights and constraints. The tensions between these relevant forms of knowledge will often be heightened by the fact that siting is an exercise in prospective reasoning. Much of the most relevant knowledge may draw on projections and scenario modeling that different consultants and stakeholders treat differently. Experts and stakeholders must find ways to integrate these forms of knowledge by relating them to the problem at hand.

All of this goes to say that consultation with experts is unavoidable if stakeholders wish to develop agreements that they can endorse as morally apt *and* technically sound. Yet involving experts raises distinctive problems for managers and participants. The challenge is to fit experts into the broader framework of consensus building rather than adapt this framework to their expectations and patterns of practice. Stakeholders should help formulate questions, identify what kinds of evidence would address these questions, and review and interpret findings. Even when organizers and stakeholders can express their expectations clearly and experts acknowledge them, they should expect that habits and practices learned over a long time may be reasserted as experts try to put the goals they have embraced into practice. The only real guarantee stakeholders have is their sustained ability to monitor and reshape the conversation.

Difficulties in managing the conversation are not limited to interactions between experts and citizens. Cultural and value differences can also impair efforts to understand and reach agreement. Values, like interests, are often plural, but the differences are less easily addressed. Parties may be unwilling to accept the kind of trades and packages that are used to address divergent interests. Values may engage identity in a way that makes it difficult for losses in one area to be compensated for by gains in another. Still, consensus-building processes can often proceed in the face of conflicts over deeply held values. Skilled facilitators and participants know how to build understanding by providing opportunities for informal exchange, by sharing stories, and by working on practical solutions that address needs without asking participants to compromise values.

Differences in cultural norms about communication will also raise a challenge when a diverse group engages in face-to-face negotiation. Efforts to site solid waste management facilities often involve ethnic groups, recent immigrants, and historically disadvantaged populations, such as low-income populations and people of color. The environmental justice movement has highlighted the different communicative resources and norms that these populations bring to public dialogues, as well as the distributive burdens they experience. Once again, a mediator can help ensure that the process allows for different types of communication and provides assistance for the diverse group of stakeholders around the table.

Finally, managing the conversation requires persistent attention to time commitments. On the one hand, participants must be willing to make a substantial enough investment of time to permit a full airing of views, substantial debate, and a careful exploration of the sources of disagreement. On the other hand, consensus-building processes require realistic deadlines. Parties who want to participate, but have limited financial resources, must be able to gauge the resources they will need to commit. If the costs of participating are being underwritten by a public agency, that entity will also need a time horizon in order to make cost estimates. Without an agreed-upon timetable, the group may be held hostage by a minority that uses discussion as a way of blocking progress. Finally, without deadlines, even well-intentioned parties may postpone making choices out of a commitment to resolve persistent ambiguities and find terms that are more advantageous for all. Good management must find a way to get agreement on deadlines that allow sufficient time for careful deliberation, but also ensure that milestones are reached at regular intervals.

Throughout these steps there is a persistent potential for conversation to polarize into a clash of absolutes. Positional rhetoric is self-perpetuating and can easily escalate. Either trend can gut the possibility for constructive engagement and effective negotiation. The chances for this are particularly high when members of a community have been singled out as the potential host. They may feel so threatened that they will doubt even responsible analysis and feel compelled to take a strong position and hold it tenaciously. Even reasoned responses may trigger an escalation of demands, a reinterpretation of evidence, and a search grounds on which to mount a legal challenge. Under such circumstances, the sponsor may feel under attack and seek to protect itself. Once interaction moves in this direction, it becomes difficult to distinguish aggressive moves from defensive ones. Suspicion will color even the most reasonable requests for information and invitations to talk and undercut the possibility of constructive negotiation.

One way to minimize these problems is to keep multiple options under consideration for as long as possible, ideally until the very end of a siting process. When parties on both sides feel that they are being treated fairly and openly, communication may improve, producing a more creative and acceptable outcome. If the costs of keeping multiple options open seem onerous, they should be compared with the costs of picking a site quickly and then having to justify the choice in protracted legal and political battles.

Getting Agreement on a Site and Design

If the conversation is managed well, the stakeholders will reach a point at which they must try to reach agreement on the choice of a site and a technology. The practical test of their efforts to build consensus will come when the facilitator asks them, "Can you live with the proposal that is on the table?" While it is important to strive for unanimity in response to this question, it may be necessary to settle for overwhelming support. A vivid image of this moment should be created and sustained and then counterbalanced by an equally vivid image of the moment at which the members of the group announce their decision to the public and face concerns about the veracity and legitimacy of the choices they have made.

The diptych these images create should help temper the appeal of strategies to push for agreement without full consideration of the issues and tradeoffs involved. It can be frustrating to uphold rights to raise questions, offer evidence, and introduce counterproposals. They acquire a sudden significance, however, at the moment the facilitator asks the question, "Can you live with this proposal?" Residual doubts and latent skepticism that agreement at earlier stages has been manufactured will often surface at this stage as opposition to a proposal. Parties may withhold support if they feel those who oppose a proposal have been excluded from questioning or challenging it. It will be difficult to generate support if some parties are being asked to make compromises that others cannot see themselves making. Misgivings may also be inarticulate and surface as opposition because "something doesn't feel right." The test of public legitimacy will be hard to pass if every effort has not been made to address the interests of parties who hold out or if these parties can offer reasons that others find compelling.

These rights imply responsibilities that become important in the effort to reach agreement. Participants who want to be heard are also obliged to try to craft proposals that will meet the interests of other parties as well as their own. This commitment is expressed practically in efforts participants make to invent options and create mutual gains. Mutual gains challenge the notion that progress involves compromise and that for any party to "win," another must lose. They also open the scope for negotiation. Creative negotiators might, for example, find ways to reconcile the interest of community members in monitoring performance with the operators' desires to manage a facility in an economically viable way. If they are approached in the right way, differences in interests and priorities can provide a rich source of joint gains.

The effort to build agreement on the choice of a site and technology will be facilitated if the following procedural commitments are met. As with the other stages of consensus building, the effort to get agreement will be greatly enhanced by the involvement of facilitators or mediators. These professionals are familiar with the issues raised by the commitments outlined as follows and provide a procedural means to handle some of the most difficult problems involved in building agreement.

Get Agreement That the Status Quo Is Unacceptable.   The cornerstone of agreement on a solution is the belief that the solid waste problem requires action. Efforts to site new facilities must demonstrate that they address a social need. This will undoubtedly require public explanation of the costs associated with doing nothing as well as with alternative methods of addressing waste management needs. Unsettled doubts will resurface as opposition throughout the process. Until there is agreement that a new facility is needed, it does not make sense to move ahead.

Guarantee That Stringent Health and Safety Standards Will Be Met.   The second piece of the foundation for getting agreement is a guarantee that concerns about health and safety will be addressed in an acceptable way. Neighbors of a proposed facility may not want to discuss other issues until they are satisfied that their health and safety concerns will be met. At a minimum, this requires that facilities meet all applicable federal and state standards. Residents often wish to add local performance standards.

Efforts to create packages (see below) often founder because they fail to clearly distinguish health and safety concerns from costs and benefits which may more appropriately be traded against each other. Asking parties, even indirectly, to exchange health and safety for other benefits or forms of compensation will inevitably undermine support for the outcome and the credibility of the process.

Deliberations about health and safety should be informed by joint investigations of the risks associated with alternatives, including the risks posed by taking no action. Consideration should also be given to possible mitigation measures. Risks that can be mitigated (such as arrangements for alternative supplies of water in case of contamination) may be differentiated from other, more problematic risks. Management provisions that provide community oversight and control (e.g., through enforceable shutdown provisions) may assuage fears tied to doubts about an operator's commitment or ability to meet health and safety standards.

Separate Inventing from Committing.   The effort to generate mutual gains will be enhanced if the process of inventing options can be separated from making commitments. The kind of free exchange that facilitates invention will be curtailed if participants feel that by suggesting something they are agreeing to it. By clearly distinguishing a phase of inventing, participants can play games like "What if?" that open new options and create space for negotiation.

Use a Single Text.   The preceding steps assume there is a package on the table. One of the best ways to bring focus and move a discussion along is to provide a preliminary proposal. This proposal should address the items of the agenda and, initially, include the widest possible range of ideas and options. It can be prepared by the facilitators or by subcommittees working on specific issues. Once this single text is available, it can be improved incrementally. The process of improving the agreement should alternate between brainstorming and efforts to consolidate improvements by revising the text. The latter is best done by the facilitator or another neutral party to avoid attribution or the appearance of authorship, which might interfere with the ability of parties to support a proposal—even one that makes sense.

Use Packages and Contingent Agreements.   One way to achieve joint gains is to package agreements where tradeoffs are made across issues. The idea is to marry complementary issues into an agreement package. Operators may be willing to meet neighbors' demands for additional traffic control measures, for instance, if they are permitted to configure the footprint of a landfill in an advantageous way. By packaging these issues that present low costs to one party and high value to another, mutual gains can be created. The risk is that once issues become linked it can be difficult to decouple them. Complex packages can create a kind of gridlock and get in the way of agreement. So the idea is to link issues but not make the entire process hinge on the final package.

Contingent agreements provide a way to allow decisions to go forward even if stakeholders do not completely agree. Disagreements about the management of a facility can sometimes be resolved by contingent agreements that spell out what will be done in case of accidents, interruptions of service, changes in standards, or the emergence of new scientific information about risks or impacts. Contingent agreements can also be used to specify the conditions under which a facility will be shut down temporarily or permanently, responsibilities for taking action, and the means of guaranteeing that agreements will be kept at no additional cost to those adversely affected. Contingent agreements are similar to short-term commitments that are automatically reopened for evaluation and reauthorization by triggers fixed at an earlier point in time. This enables parties to try out agreements, test their efficacy or fairness, and return to them after learning takes place. Contingent agreements are often used when the science or technical analysis is highly uncertain, but parties agree that action is still necessary. Contingent agreements often allow skeptical parties to share in the monitoring of facility performance. For these agreements to be legitimate, parties must ensure that an enforceable mechanism is established to reopen agreements when one of the contingencies is triggered.

Use Compensation and Benefit Sharing to Make the Host Community Better Off.   Even after all health and safety concerns have been attended to, and all impacts have been mitigated to the fullest extent possible, there may be impacts that cannot be handled to the satisfaction of all affected parties. Moreover, the balance between the costs and benefits of a facility is likely to appear unfair to those most immediately affected and likely to voice opposition. Traffic, fears about property values, odors, and other threats to the quality of life are easily identified by affected parties. Benefits, on the other hand, are often distributed widely and are easy to underestimate, particularly when waste disposal costs have been low and services are taken for granted. When a facility responds to real social needs, however, the level of benefits should be more sufficient to offset local costs and impacts that are difficult to mitigate. Compensation provides a mechanism to share the social benefits of waste management facilities in a more equitable way. It also facilitates the development of packages and creation of mutual gains.

Compensation must be treated carefully, however. Advocates of using compensation are careful to distinguish it as a form of benefit sharing, rather than a payment for compromised health and safety measures. Any confusion over this distinction is likely to trigger fears that residents are being bribed and suggest connotations of an underhanded process that are at odds with the goals and organization of consensus building. Appropriate forms of compensation might include insurance policies to guarantee property values or services, such as drinking water, that are threatened by the new facility. Taxes or waste disposal fees might be reduced and environmental or public benefit offsets created that preserve or rehabilitate environmentally sensitive areas. Public amenities, such as parks, may be constructed. In certain cases, cash payments to selected individuals may even be deemed appropriate. The net effect should be to make the host community better off accepting the facility than rejecting it.

Anticipating Implementation

Concerns about implementation cast a long shadow over efforts to build consensus around the choice of site, design, and operating conditions for solid waste management facilities. Stakeholders may be skeptical that others will adhere to their side of the bargain. When agreements need to be ratified by parties and institutions that are not at the negotiating table, the shadow implementation casts may grow even larger. We suggest that all agreements should include a mechanism to hold parties to the commitments and agreements they make. These implementation clauses and responsibilities must be part of the agreement that is negotiated. Often, joint teams of participants can work collaboratively to oversee implementation. Alternatively, individual parties can agree to undertake certain tasks and periodically report to the larger group on their progress.

Some Legal Issues. The consensus-building process must often take into account legal issues and procedures. These rules might include local siting regulations or mandatory public meeting requirements commonly called *sunshine laws.* In some instances, legal mechanisms may be required to ensure proper implementation of agreements. These all create a strong rationale for seeking professional advice. Legal mechanisms may be used to create enforceable agreements by legally binding an organization to carry out the consensually reached agreement. The ability of public and private parties to bind themselves in this way is likely to vary widely. Additionally, parties should ensure that the consensus-building process is consistent with any other legally enforceable procedural requirements, potential restrictions on governmental representatives participating in the process, and disclosure and confidentiality protections.*

15.5   CONCLUSIONS

This chapter presented a view of facility siting that takes account of the complex substantive and procedural concerns that are involved. Good technical advice is critical to making wise choices about the need for and location of new solid waste management facilities, but it is equally important to recognize the significance of the political and ethical concerns that drive many parties' involvement in solid waste management decisions. The consensus-building approach we have outlined provides a practical procedural framework for responding to these concerns.

---

* For a comprehensive discussion of potential legal issues in a consensus-building process, see D. Golann and E. E. Van Loon, "Legal Issues in Consensus Building," in Susskind, Mackearnan, and Thomas-Larmer, *The Consensus-Building Handbook,* 1999.

The rationale for this view is enhanced if siting and solid waste management are situated in a broader political and institutional context. Here they can be seen as complex social problems that raise controversies about the responsibilities of the state and the legitimate use of public authority. They are part of a family of controversial issues that challenges the ability of our democratic institutions to respond to complex and pressing social problems. The approach we have outlined addresses the need for action and gives positive expression to concerns about democratic legitimacy. It creates a context for effective decision making that also provides a central role for citizens in collective decisions about the goals, means, and implementation of strategies for solid waste management. Consensus building provides a practical procedure for reconciling the imperatives of action and democratic legitimacy that engages the moral and technical complexity that solid waste management entails.

Acknowledgments

We wish to thank our colleagues Howard Kunreuther, Professor of Decision Sciences at the University of Pennsylvania's Wharton School of Business and Director of the Wharton Risk and Decision Processes Center, and Tom Aarts and Kevin Fitzgerald, doctoral candidates at the Wharton School, for their continued collaboration in developing and promulgating the Facility Siting Credo. Portions of this chapter appeared in an earlier form in Laws and Susskind (1991).

REFERENCES

Barzok, L. (1986) The Role of Impact Assessment in Environmental Decision-Making in New England: A Ten Year Retrospective, *Environmental Impact Assessment Review*, vol. 6, no. 2.

Bean v. Southwestern Waste Management Corp. (1979) 482 F. Supp. 673 (S.D. Tex. 1979). Aff'd without opinion, 782 F. 2d 1038 (5th Cir. 1986).

Chester Residents Concerned for Quality Living v. Seif (1997) 132 F. 3d 925.

Contant, C. K., and L. L. Wiggins (1991) Defining and Analyzing Cumulative Environmental Impacts, *Environmental Impact Assessment Review*, vol. 11, no. 4.

Dickert, T. G., and A. E. Tuttle (1985) Cumulative Impact Assessment in Environmental Planning: A Coastal Watershed Example, *Environmental Impact Assessment Review*, vol. 5, no. 1.

Elliott, M. L. (1981) Pulling the Pieces Together: Amalgamation in Environmental Impact Assessment, *Environmental Impact Assessment Review*, vol. 2, no. 1.

Elliott, M. L. (1984) Improving Community Acceptance of Hazardous Facilities Through Alternative Systems for Mitigating and Managing Risk, *Hazardous Waste*, vol. 1, no. 3.

Laws, D., and L. Susskind (1991) Changing Perspectives on the Siting Process, *Maine Policy Review*, vol. 1, no. 1.

National Workshop on Facility Siting (1989) Massachusetts Institute of Technology, Cambridge, MA.

National Workshop on Facility Siting (1990) The Wharton School, University of Pennsylvania, Philadelphia, PA.

O'Hare, M., L. S. Bacow, and D. Sanderson (1983) *Facility Siting and Public Opposition,* Van Nostrand Reinhold, New York.

Ozawa, C. (1991) *Recasting Science: Consensual Procedures in Public Policy Making,* Westview, Boulder, CO.

Sandman, P. M. (1987) Getting to Maybe: Some Communications Aspects of Siting Hazardous Waste Facilities, in R. W. Lake (ed.) *Resolving Locational Conflict,* Rutgers Center for Urban Policy Research, New Brunswick, NJ.

Schwartz, A. (1995) The Law of Environmental Justice, *Environmental Law Reporter*, vol. 25, pp. 10543–10552.

Susskind, L. E. (1978) It's Time to Shift Our Attention from Impact Assessment to Strategies for Resolving Environmental Disputes, *Environmental Impact Assessment Review 2.*

U.S. EPA (2000). *Draft Title VI Guidance for EP Assistance Recipients Administering Environmental Permitting Programs,* U.S. Environmental Protgection Agency Federal Register, vol. 65, no. 124, Washington, DC.

Washburn, S. T., J. Brainard, and R. H. Harris (1989) Human Health Risks of Municipal Solid Waste Incineration, *Environmental Impact Assessment Review,* vol. 9, no. 3.

CHAPTER 16
FINANCING AND LIFE-CYCLE COSTING OF SOLID WASTE MANAGEMENT SYSTEMS

Nicholas S. Artz
Jacob E. Beachey
Philip O'Leary

The concept of integrated solid waste management is being applied in many communities in the United States. There are numerous waste-to-energy (WTE) facilities, material recovery facilities (MRFs), and other waste management facilities in operation and the number is growing rapidly. Most states have some form of regulations or incentives to encourage recycling; many states promote composting and waste minimization as well.

Changes in solid waste management have had a significant effect on public works operations and will continue to have an impact for years to come. The waste management technologies become more complex as we move away from the traditional method of simply collecting the waste in packer trucks and disposing of it in the municipal landfill. With the increasing complexity in waste management technology comes an increased complexity in the requirements for financing the new programs. Not only is there a need for greater capital expenditures, which usually means financing through borrowed funds, but also there is a need to finance multiple facilities in addition to the traditional MSW landfill. For example, communities choosing to implement an integrated solid waste management system may include recyclables processing, composting, incineration, and landfilling in their system. The need for multiple facilities in such an integrated system often leads to system financing rather than individual facility financing.

The increased complexity of integrated solid waste management has also resulted in a movement toward privatization of services. Municipalities do not wish to become involved in operations where they lack experience and often contract with private firms that specialize in such services. This desire for limited public involvement is also a factor in the financing of solid waste management facilities and systems.

The purpose of this chapter is to help public works officials deal with some of the changes in solid waste management as they relate to financing alternatives. The following sections summarize the options available in the 1990s for financing integrated solid waste management systems, review some of the issues involved in selecting the best financing mechanism for the local situation, and present a list of steps typically needed for securing system financing. In addition, the last section describes, and presents examples of, life-cycle costing (LCC) analysis.

16.1   *FINANCING OPTIONS*

A number of options are available for financing solid waste management facilities. Choosing from among these options will involve consideration of several issues discussed later in this chapter. The "best" financing option is obviously not the same for all communities. The discussions that follow describe the more prominent financing options from which choices may be made and provide a basic understanding of their structure and applicability. Clearly, not all of the options described are available for every financing need. Also, combinations of these options are often used to finance solid waste management projects.

Private Equity

Privately owned facilities may be financed in total or in part by the use of equity—i.e., the owner's cash. The owner may be the vendor who builds and operates the facility or a third party who contributes equity in anticipation of a sound investment return.

In general, privately owned solid waste management facilities have been financed with a combination of equity and tax-exempt project revenue bonds (Chen et al., 1992). The equity is often for that portion of the facility that does not qualify for tax-exempt debt, which may be 10 to 20 percent of the facility cost. For example, that portion of a waste-to-energy facility used to produce an energy product cannot normally be financed with tax-exempt revenue bonds.

Owners are often hesitant to provide more than a minimal amount of equity because some of the investment returns are fixed (i.e., they do not increase as the equity increases above the minimum) (Turbeville, 1990). The owner is allowed the tax benefit of an accelerated depreciation schedule on the full value of the facility even though the amount of equity may be only 10 to 20 percent of the facility's cost. Also, the owner retains the residual value of the facility after the debt is retired.

Third-party investors have a favorable effect on a project in some instances. Such investors sometimes have greater potential for maximizing use of the tax benefits generated by the project and/or may require a lower rate of return on their investment than the project vendor. The net effect can be a less expensive project.

Some solid waste management facilities are financed entirely by owner equity. This avoids the time and expense of obtaining debt financing and is the easiest means of introducing a new technology. It is often the choice for financing less capital-intensive operations including small recyclables-processing facilities.

In other instances, the amount of private owner equity in a solid waste project may be set in the service contract. For example, municipalities sometimes require an owner-operator to post considerable equity in a facility or system to ensure the continued interest of the owner in meeting the terms of a service agreement.

Traditional Loans

Solid waste management facilities may be financed through traditional loans between the borrower and lending institutions. Different lenders will market loans for different periods of time or for different phases of a project (Chen et al., 1992; Lee and Ashdown, 1992).

Commercial banks, finance companies, and thrift institutions generally provide construction loans on a project. These lenders are interested in short-term loans of 1 to 3 years and do not usually participate in the permanent financing of a project.

Permanent lenders provide financing after the project is operational. These long-term lenders include insurance companies, pension funds, and other financial institutions with long-term sources of cash. Permanent lenders may provide project financing for 20 years or longer, depending on the expected life of the facility (or facilities) included in the financing.

Construction loans and permanent loans can be structured and committed back-to-back if project sponsors, investors, or construction lenders are unwilling to risk refinancing at the end of construction. In the solid waste industry, construction and permanent financing commitments are usually required at the beginning of the project. This avoids the risk of the permanent financing being too costly or unavailable when needed. Typically, under this arrangement, the permanent lender repays the construction loan when the project is at a pre-agreed acceptance stage (Chen et al., 1992).

Traditional loans may be used to finance solid waste projects where tax-exempt financing is not readily available. Such loans must generally be accompanied by owner equity as part of the loan collateral. Traditional loan financing is more common with private ownership than public ownership.

Tax-Exempt Bonds

Tax-exempt bonds issued by a governmental agency are an alternative to taxable debt on some solid waste management projects. Because the interest paid on funds raised from these bonds is exempt from federal taxation, the interest rate may be 2 or 3 percentage points lower than that on taxable bonds (MacCarthy, 1991).

Two basic types of tax-exempt bonds may be issued by a state or local government to finance solid waste projects: general obligation bonds and project revenue bonds. Each of these bonds and their various forms is discussed.

General Obligation Bonds.   General obligation (GO) bonds are tax-exempt certificates of indebtedness that may be used by local governments to finance their capital expenditures. The local government pledges its full faith and credit and taxing power as the security behind the debt service on the bonds. GO bonds are generally considered the most secure form of debt which, coupled with their tax-exempt status, results in the lowest interest rate on a project.

The use of GO bonds requires voter approval and is limited by the general obligation debt capacity of the municipality or other governmental unit choosing to use them. They are not typically used to finance large solid waste management projects because of the need to preserve a community's GO debt capacity and the availability of other financing mechanisms. Also, public ownership of the project is required when GO bonds are used.

Project Revenue Bonds.   Revenue bonds are more commonly used than GO bonds to finance solid waste management projects. These bonds are also tax-exempt, but are not as secure as GO bonds, and, therefore, generally have higher interest rates. As the name implies, revenue bonds are largely secured by the revenues from the project being financed. A project mortgage and other guarantees may be pledged as well, but the credit and taxing power of a local government is not included. Two types of tax-exempt project revenue bonds exist as a result of the Tax Reform Act of 1986: government-purpose bonds (GPBs) and private activity bonds (PABs).

Government-Purpose Bonds.   While the defining characteristics of GPBs are somewhat complex, the basic criteria for their use in solid waste management projects are as follows (Chen et al., 1992; MacCarthy, 1991; Ollis, 1992):

- The project must be publicly owned.
- Limitations on the sale of project outputs to private business must be met.
- Private operations of any part of the project must not exceed 5 years and may be canceled after 3 years.

GPBs may be beneficial in financing publicly owned and operated solid waste management projects. They generally carry a lower interest rate than PABs, since PAB interest is included in the calculations of the alternative minimum tax for individuals and corporations (Ollis, 1992). However, the restrictions on the use of GPBs, particularly with respect to private sector involvement, result in more solid waste projects financed with PABs.

Private Activity Bonds.   PABs are also subject to certain restrictions, but allow private ownership and/or long-term private operation of a solid waste project. Privately owned projects desiring to use PAB financing must obtain a portion of the state's annual allotment of PABs. The annual state ceiling is equal to $50 multiplied by the state's population or $150 million—whichever is greater.

Competition with other projects for a PAB allocation may lead to public ownership, which is exempted from the state's allocation cap. However, PABs are the only means of obtaining tax-exempt financing for privately owned projects. PABs for private use must be issued through a public agency and the funds from the bonds passed on to the private owner through a loan or other ancillary agreement (Horning, 1991).

Whether PABs are used for publicly or privately owned solid waste projects, they allow much greater private involvement than GPBs. The ability to enter into long-term service agreements with a vendor allows a local government to share project risks and responsibilities in a manner not available with GPB financing.

PABs may not be used for certain expenditures in solid waste projects such as the energy-generating equipment in a waste-to-energy facility. This factor, plus demand for equity as additional debt service security, normally results in PABs being used in conjunction with other funds to finance solid waste projects.

Taxable Bonds

Taxable bonds—in particular, taxable municipal bonds—may be used for all or partial financing of solid waste projects. Taxable municipal bonds (TMBs) are commonly used to finance costs that do not qualify for PAB financing in a publicly owned project (Chen et al., 1992). They may be used for that purpose in privately owned projects, as well, when the nonqualifying costs are not all covered by equity. In some instances, TMBs may be substituted for PABs when the tax-exempt bond allocation for private use is not available.

Although TMBs require higher interest rates than tax-exempt bonds, they afford a private owner more favorable depreciation periods on solid waste equipment. This benefit has the effect of at least partially offsetting the higher interest costs.

Federal/State Grants and Loans

State and federal sources of financial assistance to solid waste projects are limited and vary over time. However, money in the form of grants or loans has periodically become available for projects that can show a demonstration or research function.

In cases where state or federal funding may be available, local funding may also be required at some level.

Public Funds

A local government (i.e., county or municipality) may sometimes use general or special reserve funds it possesses to pay for a publicly owned project. This form of equity financing may reduce some or all of the numerous steps necessary to obtain debt financing.

Public funds are typically used to finance projects that are less capital-intensive or portions of projects not qualifying for PABs (Chen et al., 1992; Horning, 1991). Material recovery facilities used to process recyclables and composting operations are examples of solid waste facilities that might be financed in total with public funds. These facilities are generally lower in capital cost than waste-to-energy facilities, for example, and are more difficult to finance with debt because of the uncertainty in prices for their products.

The availability of public funds for solid waste project financing may depend on whether a means of collecting money specifically for solid waste management services is available. In many areas, various forms of surcharges for services are being assessed to provide public funds for expanded or new solid waste projects.

16.2 ISSUES IN FINANCING CHOICES

Choosing between financing options may involve a variety of project issues. These issues are addressed as follows, along with their potential effects on financing solid waste management projects.

Facility/System Financing

As previously indicated, solid waste management systems are becoming more complex; they often include several types of facilities to accomplish the necessary or desired waste processing and disposal. Historically, waste-to-energy facilities and landfills have been financed individually with debt payable from revenues derived through tipping fees and energy sales. The recycling and composting facilities included in many solid waste management systems today, however, are not as amenable to individual facility financing. The uncertainties of markets for recovered materials and compost coupled with difficulties in predicting waste composition make the economic feasibility of these facilities difficult to demonstrate.

The movement toward integrated solid waste management and the importance of demonstrating economic viability to attract capital for solid waste facilities have resulted in more system financings. System financings rely on the strength and diversity of all facilities in the system to secure the repayment of debt or equity. If one facility in the system does not meet expectations, another may take up the slack. For example, if a material recovery facility is not paying for itself, revenues from another facility in the system (e.g., the landfill) will, it is hoped, cover the deficit.

In general, long-term debt financing of recycling/composting facilities will require a system financing structure if revenues are the principal means of securing the debt (Horning, 1991). It will be necessary to provide assurances that a shortage in revenues from the recycling/composting facility can be covered by revenues from another element in the system, such as a landfill or waste-to-energy facility. Without a system financing structure, an MRF or composting facility will probably be excluded from project revenue bond financing.

Ownership

Ownership of solid waste management facilities may be either public or private. Public ownership is usually through a municipal government unit, authority, or agency. Private ownership can be through a private corporation, partnership, or sole proprietorship.

The choice between public and private ownership affects not only financing choices, but also project implementation, including options for procurement and operation (Artz, 1990). Features of solid waste management projects under public versus private ownership are shown in Table 16.1.

In the past, private ownership of capital-intensive waste management facilities was sometimes chosen to avoid public agency involvement and risk in an unfamiliar area. Also, the private ownership tax benefits were substantial prior to the Tax Reform Act of 1986 and were often judged to result in a lower-cost project.

TABLE 16.1   Features of Public versus Private Ownership of Solid Waste Management Facilities

	Public ownership	Private ownership
Procurement options	A/E Turnkey Full service	Full service
Financing options	General obligation bonds (GO) Government-purpose bonds (GPB) Private activity bonds (PAB) Taxable municipal bonds Traditional loans Federal/state grants Public funds	Private activity bonds (PAB) Taxable bonds Private equity Traditional loans
Operation	Public (typically) with A/E Public/private with turnkey Private with full service	Private
Public risk	Similar*	Similar*
Implementation time	Less than with private ownership	Greater than with public ownership

* Applies primarily to facilities/systems financed with large bond issues.
Source: Franklin Associates, Ltd.

Currently, public ownership of highly capitalized waste management facilities is frequently recommended as the most cost-effective and practical approach. Publicly owned projects are reported to require less time to finance and implement and may involve little, if any, additional public risk. Recent comparisons suggest that risk allocation between the public and private sectors in a solid waste project is virtually irrespective of ownership.

Tax-exempt debt financing is easier to obtain with public ownership and is one of the reasons public ownership is more often used than in the past. Whereas PABs issued for private use are limited, no such limit is set for public use, and PABs are frequently used to finance publicly owned solid waste projects. For public projects with limited private sector involvement operationally or otherwise, GPBs provide even lower-cost financing than PABs. GO bonds provide the lowest-cost form of public debt financing, but, as noted previously, limitations on their use has resulted in infrequent use of GO bonds to finance solid waste projects.

Other financing options with public ownership include the use of public funds, federal/state grants and loans (as available), taxable municipal bonds, and traditional loans from lending institutions. These options may be used in combination with each other or with the tax-exempt bonds previously described.

Private ownership financing options for solid waste projects include private equity, traditional loans, taxable bonds, and PABs. Some combination of these options is typically used. In some instances, taxable municipal bonds may be issued to assist in private project financing. Private financing without the use of tax-exempt bonds (i.e., PABs) allows the owner more favorable equipment depreciation periods for tax purposes. This added tax benefit without tax-exempt debt must be considered in comparing financing options under private ownership.

Procurement and Operation

Three basic forms of procurement are used for solid waste management projects:

- Architectural/engineering (A/E)
- Turnkey
- Full service

The A/E procurement method is the standard approach that governmental bodies use to build most public facilities. A consulting engineer is retained by the governmental entity to prepare the facility design and a contractor is hired through a bidding process to build the facility. The facility is publicly owned and publicly operated in most cases.

With a turnkey arrangement, a single contractor will have responsibility for both designing and building the facility. Turnkey procurements usually involve public ownership. The completed facility may be operated either publicly or privately. The turnkey contractor, who is intimately familiar with the facility design and construction, is often hired to operate the facility.

A full-service procurement involves one private entity accepting project responsibility for design, construction, and operation. This form of procurement is normally considered mandatory for private ownership of a capital-intensive waste management facility, but may be used with public ownership as well.

Most waste management facility or system procurements follow one of the three basic options described here or close variations thereof. Any of these options may be used with public ownership, while full service would usually be the only acceptable procurement for private ownership.

Procurement and operation of waste management facilities are related to financing insofar as they affect the choice of public or private ownership. For example, public operation is incompatible with private ownership, although private operation and public ownership are compatible with a turnkey or full-service procurement. An A/E procurement will require public ownership and, in general, public operation.

Risk Allocation

Financing solid waste projects generally requires that the risks be allocated between the public and private participants (Chen et al., 1992). Lenders, including bond investors, are interested in obtaining maximum security on their investment. The credit rating of a project will determine the availability of lenders and the interest rate.

Most solid waste project financings require similar allocations of risks regardless of whether the project is publicly or privately owned. The vendor accepts the completion and technical risks of construction and the responsibility of operating the facility properly to meet certain performance standards. The local government/public agency guarantees the waste supply, including payment for any shortfalls. Generally, the public entity also assumes the risk of force majeure events and the risk of changes in laws that affect operation.

In addition to project revenues backed by waste stream guarantees, lender/bondholder security may include a project mortgage, letter of credit, bond insurance, and a company guaranty, if the company is sufficiently strong (Ollis, 1992). A financially strong vendor may be willing to assume some risks normally borne by the governmental unit, but a substantial price will usually be charged.

A governmental unit assumes the highest level of risk when it issues GO bonds for a project. This gives bondholders the highest degree of security because the full taxing power of the local government is pledged to the repayment of the bonds.

Implementation Time

In general, the time required for financing a project is least when no debt is required and is greatest when debt is used.

Solid waste projects may be expected to require more time to implement with private ownership than public ownership. Long negotiation periods are often associated with arranging private ownership. In addition, tax-exempt PAB financing is more time-consuming when issued for private use. PABs for private use require obtaining a governmental issuer and obtaining a portion of the state's annual allotment of PABs. If the needed allotment is not available in the year requested, the request would not be allowed until, at least, the following year.

Issuing tax-exempt bonds for public use is often less complex and may, therefore, be less time-consuming. However, GO bonds cannot be issued without a public vote, which adds to the time requirement. Further, if the GO bond issue fails, other financing must be arranged.

Cost of Financing

Since the loss of most of the tax benefits of private ownership following the Tax Reform Act of 1986, public ownership financing is often recommended for solid waste projects. However, where tax-exempt PABs are available for private use, there may be no cost advantage with public ownership.

Clearly, the least expensive financing usually involves the use of tax-exempt bonds, which often carry interest rates of 2 or 3 percentage points below that of taxable debt. GO bonds are the least expensive because of their comparatively low risk. Of the project revenue bonds, GPBs carry lower interest rates than PABs. Interest on PABs is subject to the alternative minimum tax calculations for individuals and corporations and must, therefore, offer somewhat higher rates than GPBs.

For solid waste projects financed over 20 years, debt service payments may be 10 to 20 percent lower with tax-exempt bonds than with taxable debt. This advantage must be compared with the tax benefits of private ownership where tax-exempt debt is *not* used. The shorter equipment depreciation periods for tax purposes (5 to 7 years versus 10 years with tax-exempt financing) are of some added benefit in lowering project costs. In most cases, however, they will probably not be worth giving up the cost savings from tax-exempt bonds.

16.3   STEPS TO SECURE SYSTEM FINANCING

Once a decision to proceed with implementation of a facility for solid waste management has been made, financing must be considered. The steps necessary to secure financing will vary, depending on the financing options chosen and the party obtaining the financing. In general, PAB financing on behalf of a private company involves the greatest effort, and the steps necessary to secure this form of financing are emphasized in this section. The assistance of technical consultants, bond counsel, and an investment banking firm usually will be needed in the financing process.

The steps generally required in the more complex financing processes are summarized as follows. They are presented in an order in which they might normally occur, but specific financings may dictate variations in this order. References 3 and 5 were used in developing the process descriptions.

Decisions on Issues and Options in Financing

The first step in financing involves choosing between financing options. This will be done in view of the issues attendant to financing choices, as described previously. Decisions on ownership, procurement, operation, cost, etc., may enter into the final determination of what financing option or combination of options will be chosen. Quite often, pending the availability of grants, equity, or public funds, tax-exempt bonds are chosen for the permanent financing of most of a proposed system/facility. If GO bonds are chosen, in conjunction with public ownership, their use will require voter approval. However, project revenue bonds—PABs, usually—are chosen far more frequently to finance solid waste projects.

Feasibility Study and Plan

A study and plan providing details on project feasibility and the role of the proposed facility will also be needed early in the financing process. Bond counsel will use the description of the facility to be financed to make an initial determination that the project will qualify under the Internal Revenue Code for the tax-exempt financing desired. The feasibility of the project will also be closely studied by the investment banking firm and others prior to preparing the financing documents.

Determine Issuer of Bonds

If the project is determined to qualify for tax-exempt PAB funding, bond counsel must then determine what state agency or local government can act as issuer of the bonds and which state statutes apply to the financing. A private company seeking financing should contact the chosen governmental issuer to officially apply for assistance with the financing. The company will need to convince the prospective issuer to perform this service.

Prepare and Adopt Bond Resolution

Once an issuer for the bonds has been identified, bond counsel will draft a bond resolution to be adopted by the governmental issuer. A final resolution signifying the governmental issuer's intention to issue bonds for the project will need to be adopted. Major project expenditures that are incurred prior to the resolution's adoption may not be paid for by bond proceeds. Thus, it is important to obtain adoption of this resolution as early as possible.

Structure Bond Security

The investment banking firm selected to underwrite the bonds will suggest sources of repayment or security for the bondholders. In addition to pledging the project revenues from tipping fees and the sale of recovered materials/energy products, other potential sources of collateral to back the bonds include:

- Project mortgage
- Company guarantee, if the company is financially strong
- Flow control guarantees through contracts or ordinances that ensure waste delivery and tipping fees to the project
- Letter of credit, surety bond, or bond insurance from an institution with a high credit rating

The greater the security of the bonds, the easier they will be to market. At minimum, guarantees of waste delivery will generally need to accompany project revenue pledges when revenues are the principal source of bond security. Without waste delivery guarantees, the revenues from a solid waste project can be very uncertain and revenue bonds may not be marketable.

Prepare Financing Agreements

After further review of project feasibility, bond counsel may begin drafting the agreements needed to issue the bonds; for PABs issued on behalf of a private company these will include:

- The agreement between the governmental issuer and the company
- The indenture stating the terms under which the bonds will be issued
- The bond purchase agreement providing for the sale of the bonds

Prepare Project Report

Factors relevant to obtaining financing for the project should be addressed in this report. The purpose, costs, and function of each portion of the project should be described. Bond counsel should determine which elements of the project qualify for bond financing and the amount of financing needed. The status of permits needed to construct and operate the project and other factors necessary to the project should be reviewed, as well.

Obtain PAB Volume Cap Allocation

Private company ownership of the project will require obtaining an allocation from the state's annual allotment of PABs if tax-exempt bonds are to be used. Bond counsel may need to assist the company and the governmental issuer of the bonds in the timing and method for obtaining the allocation.

Provide Official Statement

The investment banking firm (underwriter) will use an official statement or other disclosure documents to offer the bonds to the public or to private investors. The official statement summarizes the project, the financing arrangements, the forms of security offered to bondholders, etc. Substantial detail on the bonds being offered is provided in the statement. A bond issue rating may be included if obtained from a national rating agency.

Final Execution

Before PABs are offered for sale on behalf of a private company, a public hearing must be conducted and final documents and closing papers must be executed by all parties.

16.4   LIFE-CYCLE COSTING

This section provides an overview of the basic concepts of life-cycle costing (LCC) analysis. LCC is a method of comparing new projects by taking into account relevant costs over time, including the project's initial investment, future replacement costs, operation and maintenance costs, project revenues, and salvage or resale values. All the costs and revenues over the life of the project are adjusted to a consistent time basis and combined to account for the time value of money. This analysis method provides a single cost-effectiveness measure that makes it easy to compare projects directly.

Time Value of Money

The value of money changes, depending on when it is spent or received. There are two reasons for this: inflation and the opportunity cost of money. Inflation erodes the buying power of money over time, and results in dollars spent today buying fewer goods and services than they did a few years ago. The opportunity cost reflects the fact that money invested has the opportunity to yield a return over time, even in the absence of inflation.

Since the value of money changes with time, cash flows from one year cannot be combined directly with flows from another in a meaningful way, but must first be "discounted" to a common year, usually the first year of the project. These discounted values can then be summed to obtain the total life-cycle cost, which can be compared with the total life-cycle cost of an alternative project that may have different proportions of initial costs and net annual operating costs.

Discount Factors

The formula for discounting a future value F to a present value P is

$$P = F \times \frac{1}{[(1 + d/100)]^n} = F \times \text{PWF}\ (d, n)$$

where d is the discount rate expressed in percent and n is the number of years in the future. The effect of discounting is to reduce the costs of the future to today's values. The present worth factors PWF (d, n), that convert future year values into present values for various discount rates and years have been calculated and are shown in Table 16.2.

If all investments yielded the same rate of return, then all future cash flows would be discounted at that rate. However, since different investments yield different rates, the choice of rate to use is sometimes difficult to determine. The discount rate commonly used is the cost of capital, which is the weighted average rate at which the borrowing agency is financed.

TABLE 16.2   Single Present-Worth Factors PWF (d, n)*

Year n	Discount rate d, %							
	3	4	5	6	7	8	9	10
1	0.9709	0.9615	0.9524	0.9434	0.9346	0.9259	0.9174	0.9091
2	0.9426	0.9246	0.9070	0.8900	0.8734	0.8573	0.8417	0.8264
3	0.9151	0.8890	0.8638	0.8396	0.8163	0.7938	0.7722	0.7513
4	0.8885	0.8548	0.8227	0.7921	0.7629	0.7350	0.7084	0.6830
5	0.8626	0.8219	0.7835	0.7473	0.7130	0.6806	0.6499	0.6209
6	0.8375	0.7903	0.7462	0.7050	0.6663	0.6302	0.5963	0.5645
7	0.8131	0.7599	0.7107	0.6651	0.6227	0.5835	0.5470	0.5132
8	0.7894	0.7307	0.6768	0.6274	0.5820	0.5403	0.5019	0.4665
9	0.7664	0.7026	0.6446	0.5919	0.5439	0.5002	0.4604	0.4241
10	0.7441	0.6756	0.6139	0.5584	0.5083	0.4632	0.4224	0.3855
11	0.7224	0.6496	0.5847	0.5268	0.4751	0.4289	0.3875	0.3505
12	0.7014	0.6246	0.5568	0.4970	0.4440	0.3971	0.3555	0.3186
13	0.6810	0.6006	0.5303	0.4688	0.4150	0.3677	0.3262	0.2897
14	0.6611	0.5775	0.5051	0.4423	0.3878	0.3405	0.2992	0.2633
15	0.6419	0.5553	0.4810	0.4173	0.3624	0.3152	0.2745	0.2394
16	0.6232	0.5339	0.4581	0.3936	0.3387	0.2919	0.2519	0.2176
17	0.6050	0.5134	0.4363	0.3714	0.3166	0.2703	0.2311	0.1978
18	0.5874	0.4936	0.4155	0.3503	0.2959	0.2502	0.2120	0.1799
19	0.5703	0.4746	0.3957	0.3305	0.2765	0.2317	0.1945	0.1635
20	0.5537	0.4564	0.3769	0.3118	0.2584	0.2145	0.1784	0.1486

* The factor for finding the present value P worth of a future amount F, is $[1 + (d/100)]^{-n}$.

Source: Franklin Associates, Ltd.

Capital Recovery Factors

The cost of a waste management system is generally made up of two parts: the capital required to purchase land, buildings, and equipment and the annual costs to operate the system. Capital investments are costs incurred at the beginning of the project. These costs are frequently financed with borrowed funds. The borrowed money and accrued interest are repaid with income received later in the project from the sale of energy or materials, from tipping fees, or from taxes. The constant annual payment required to repay the financed amount is determined by multiplying the borrowed amount by a capital recovery factor CRF (d, n), which is calculated by

$$ \text{CRF}\ (d, n) = \frac{d}{1 - (1 + d)^{-n}} $$

where d is the interest rate expressed as a decimal and n is the number of interest periods. Table 16.3 lists capital recovery factors per thousand dollars as a function of interest rate and length of financing term.

LCC Case Studies

In this section, two example analyses are shown to illustrate the LCC method. While every project has unique features, these generalized examples illustrate the methodology. Two types of systems are examined: a privately owned waste-to-energy (WTE) system for MSW and a privately owned material recovery facility for recyclables separately collected from households. Both systems are assumed to be financed with private activity bonds.

The WTE facility for this analysis has a capacity of 1000 tons per day, and generates revenue from the sale of electricity. Revenue from the 200-ton/day MRF is derived from the sale of processed recyclables. In both cases, the revenues are supplemented by tipping fees or taxes

TABLE 16.3   Capital Recovery Factors CRF (d, n)*

Interest	Years										
	3	4	5	6	7	8	9	10	15	20	30
5.0	367.21	282.01	230.97	197.02	172.82	154.72	140.69	129.50	96.34	80.24	65.05
5.5	370.65	285.29	234.18	200.18	175.96	157.86	143.84	132.67	99.63	83.68	68.81
6.0	374.11	288.59	237.40	203.36	179.14	161.04	147.02	135.87	102.96	87.18	72.65
6.5	377.58	291.90	240.63	206.57	182.33	164.24	150.24	139.10	106.35	90.76	76.58
7.0	381.05	295.23	243.89	209.80	185.55	167.47	153.49	142.38	109.79	94.39	80.59
7.5	384.54	298.57	247.16	213.04	188.80	170.73	156.77	145.69	113.29	98.09	84.67
8.0	388.03	301.92	250.46	216.32	192.07	174.01	160.08	149.03	116.83	101.85	88.83
8.5	391.54	305.29	253.77	219.61	195.37	177.33	163.42	152.41	120.42	105.67	93.05
9.0	395.05	308.67	257.09	222.92	198.69	180.67	166.80	155.82	124.06	109.55	97.34
9.5	398.58	312.06	260.44	226.25	202.04	184.05	170.20	159.27	127.74	113.48	101.68
10.0	402.11	315.47	263.80	229.61	205.41	187.44	173.64	162.75	131.47	117.46	106.08
10.5	405.66	318.89	267.18	232.98	208.80	190.87	177.11	166.26	135.25	121.49	110.53
11.0	409.21	322.33	270.57	236.38	212.22	194.32	180.60	169.80	139.07	125.58	115.02
11.5	412.78	325.77	273.98	239.79	215.66	197.80	184.13	173.38	142.92	129.70	119.56
12.0	416.35	329.23	277.41	243.23	219.12	201.30	187.68	176.98	146.82	133.88	124.14

* The constant annual payment, in dollars, required to repay a present amount of $1000, as a function of the compound interest rate and number of years shown.

Source: Franklin Associates, Ltd.

to pay for the facilities. The estimated costs used for these examples are thought to be typical for the central United States, but may not apply to any specific community.

The two analyses differ in the way collection of incoming material is handled. Since a WTE facility does not require a separate collection system (i.e., the vehicles that collect the waste for disposal simply deliver the waste to a new site), collection costs are not included in that analysis. However, the haul distance will be affected if the distances to the new and old facilities are different. Adding an MRF to an existing system, on the other hand, generally requires additional equipment and staff for collecting recyclables and delivering them to the MRF. These additional costs may be partially offset by avoided MSW collection and disposal costs; however, the avoided costs are usually not proportional to the reduction in quantities disposed and are often quite small.

The life-cycle cost of a project is determined by annualizing the capital costs and then summing all discounted annual capital and operating costs for the life of the system. This life-cycle costing approach is a particularly useful tool for comparing total costs of alternative waste management scenarios over a 20-year period, where one scenario has higher capital requirements than the other.

A listing of typical capital cost elements for a financed waste-to-energy facility, MRF, composting facility, or landfill is shown in Fig. 16.1. The costs over and above the direct construction costs may increase the total bond issue requirement for a large WTE facility by 50 percent or more. An explanation of these additional costs is as follows:

* *Start-up costs* are funds used to operate the facility during the testing and shakedown period, before revenues are routinely generated. The start-up time depends on the type and complexity of the system. For a large WTE facility, the start-up time is typically 6 months to a year. The time for getting an MRF into commercial operation can vary from 1 month or less for a manual sorting station to more than 6 months for a large mechanized processing facility.

* *Interest during construction and start-up* is money included in the bond issue to pay the interest costs during construction and start-up of the facility, when there may be reduced or no revenues. The construction time depends on the complexity of the facility, ranging from 1 or 2 months for a manually operated MRF with minimal equipment to 2 years or more for a large WTE facility.

* Direct construction costs
  Land
  Site development
  Buildings, with utilities
  Process equipment
  Mobile equipment
  Design and engineering
  Delivery and installation
  Construction supervision
  Contingencies/profit
* Interest during construction
* Start-up costs
* Legal and financial fees
* Debt service reserve fund

FIGURE 16.1 Typical capital cost components for solid waste management facilities (when financed with borrowed funds).

- *Legal and financing fees* are for legal counsel and financial advice. These costs are typically in the neighborhood of 4 percent of the total bond issue.
- *The debt service reserve fund* is money set aside to pay for unanticipated problems. It is more likely to be required for the more complex or unproven technologies, and may amount to a year's debt service payment.

The 20-year life-cycle cost analyses for the two case studies are shown in Tables 16.4 and 16.5. The assumptions used for the analysis are listed as footnotes. The MRF costs were derived from a recent survey of 10 operating MRFs (NSWMA, 1992). The present values of the net costs are developed for each year of the 20 years assumed in the analysis. An 8 percent annual cost of capital is used for discounting. This is the same rate assumed for bond interest.

TABLE 16.4   Twenty-Year Life-Cycle Cost Analysis of Waste-to-Energy Facility
(Production of electricity for sale) 1000 ton/day capacity

Year	Costs, $1000 Capital (debt service)	Operation and maintenance	Residue disposal	Total cost	Revenues Energy sales, $1000	Net cost (tipping fee required) $1000 per year	$ per ton	Present value of net cost $1000 per year	$ per ton
1	14,259	13,000	2250	29,509	9,540	19,969	66.56	19,969	66.56
2	14,259	13,455	2329	30,043	9,874	20,169	67.23	18,675	62.25
3	14,259	13,926	2410	30,595	10,219	20,376	67.92	17,469	58.23
4	14,259	14,413	2495	31,167	10,577	20,590	68.63	16,345	54.48
5	14,259	14,918	2582	31,759	10,947	20,812	69.37	15,297	50.99
6	14,259	15,440	2672	32,372	11,331	21,041	70.14	14,320	47.73
7	14,259	15,980	2766	33,005	11,727	21,278	70.93	13,409	44.70
8	14,259	16,540	2863	33,662	12,138	21,524	71.75	12,559	41.86
9	14,259	17,119	2963	34,341	12,562	21,778	72.59	11,766	39.22
10	14,259	17,718	3067	35,043	13,002	22,041	73.47	11,026	36.75
11	14,259	18,338	3174	35,771	13,457	22,314	74.38	10,336	34.45
12	14,259	18,980	3285	36,524	13,928	22,596	75.32	9,691	32.30
13	14,259	19,644	3400	37,303	14,416	22,888	76.29	9,089	30.30
14	14,259	20,331	3519	38,110	14,920	23,189	77.30	8,527	28.42
15	14,259	21,043	3642	38,944	15,442	23,502	78.34	8,002	26.67
16	14,259	21,780	3770	39,808	15,983	23,826	79.42	7,511	25.04
17	14,259	22,542	3901	40,703	16,542	24,160	80.53	7,052	23.51
18	14,259	23,331	4038	41,628	17,121	24,507	81.69	6,623	22.08
19	14,259	24,147	4179	42,586	17,720	24,866	82.89	6,223	20.74
20	14,259	24,993	4326	43,577	18,341	25,237	84.12	5,848	19.49

Total life-cycle cost in discounted dollars — 229,737

Average life-cycle cost in discounted dollars — 38.29

Assumptions:

Total capital required	$140,000,000	O&M cost (year 1)	$13,000,000
PAB interest rate	8%	Residue quantity	75,000 tons/year
Inflation rate	3.5%	Residue disposal cost (year 1)	30 dollars/ton
Discount rate	8%	Salable electricity	530 kWh/ton
Facility financing period	20 years	Electricity revenue (year 1)	6 cents/kWh
MSW throughput	300,000 tons/year	O&M cost (year 1)	43 dollars/ton
		Financing cost (year 1)	48 dollars/ton

Source: Franklin Associates, Ltd.

TABLE 16.5 Twenty-Year Life-Cycle Cost Analysis of Material Recovery Facility
(Including collection of commingled recyclables) 200 tons/day capacity

	Collection costs			MRF costs				Total costs	Revenues	Net cost[e]			Net present value	
Year	Capital[a] debt service	Operation and maintenance	Total collection	Capital[b] debt service	Operation and maintenance	Residue disposal	Total MRF	Collection and MRF	Material sales	Dollars/ year	Dollars/ ton	Dollars/ hh/month	Dollars/ year	Dollars/ ton
1	921,800	3,992,400	4,914,200[f]	875,000	1,368,400	56,900	2,300,300[d]	7,214,500	1,206,200	6,008,300	142.71	2.33	6,008,300	142.71
2	921,800	4,132,100	5,053,900	875,000	1,416,300	58,900	2,350,200	7,404,100	1,248,400	6,155,700	146.22	2.39	5,699,700	135.39
3	921,800	4,276,700	5,198,500	875,000	1,465,900	61,000	2,401,900	7,600,400	1,292,100	6,308,300	149.84	2.45	5,408,400	128.46
4	921,800	4,426,400	5,348,200	875,000	1,517,200	63,100	2,455,300	7,803,500	1,337,300	6,466,200	153.59	2.51	5,133,100	121.93
5	921,800	4,581,300	5,503,100	875,000	1,570,300	65,300	2,510,600	8,013,700	1,384,100	6,629,600	157.47	2.57	4,873,000	115.75
6	921,800	4,741,600	5,663,400	875,000	1,625,300	67,600	2,567,900	8,231,300	1,432,500	6,798,800	161.49	2.64	4,627,100	109.91
7	921,800	4,907,600	5,829,400	875,000	1,682,200	70,000	2,627,200	8,456,600	1,482,700	6,973,900	165.65	2.70	4,394,700	104.39
8	1,124,400	5,079,400	6,203,800	982,200	1,741,100	72,500	2,795,800	8,999,600	1,534,600	7,465,000	177.32	2.89	4,355,800	103.46
9	1,124,400	5,257,200	6,381,600	982,200	1,802,000	75,000	2,859,200	9,240,800	1,588,300	7,652,500	181.77	2.97	4,134,400	98.20
10	1,124,400	5,441,200	6,565,600	982,200	1,865,100	77,600	2,924,900	9,490,500	1,643,900	7,846,600	186.38	3.04	3,925,300	93.24
11	1,197,300	5,631,600	6,828,900	982,200	1,930,400	80,300	2,992,900	9,821,800	1,701,400	8,120,400	192.88	3.15	3,761,300	89.34
12	1,197,300	5,828,700	7,026,000	982,200	1,998,000	83,100	3,063,300	10,089,300	1,761,600	8,328,300	197.82	3.23	3,571,900	84.84
13	1,197,300	6,032,700	7,230,000	982,200	2,067,900	86,000	3,136,100	10,366,100	1,822,600	8,543,500	202.93	3.31	3,392,700	80.59
14	1,197,300	6,243,800	7,441,100	982,200	2,140,300	89,000	3,211,500	10,652,600	1,886,400	8,766,200	208.22	3.40	3,223,300	76.56
15	1,455,100	6,462,300	7,917,400	1,118,600	2,215,200	92,100	3,425,900	11,343,300	1,952,400	9,390,900	223.06	3.64	3,197,200	75.94
16	1,455,100	6,688,500	8,143,600	1,118,600	2,292,700	95,300	3,506,600	11,650,200	2,020,700	9,629,500	228.73	3.73	3,035,600	72.10
17	1,455,100	6,922,600	8,377,700	1,118,600	2,372,900	98,600	3,590,100	11,967,800	2,091,500	9,876,300	234.59	3.83	2,882,800	68.48
18	1,455,100	7,164,900	8,620,000	1,118,600	2,456,000	102,100	3,676,700	12,296,700	2,164,700	10,132,000	240.67	3.93	2,738,400	65.04
19	1,455,100	7,415,700	8,870,800	1,118,600	2,542,000	105,700	3,766,300	12,637,100	2,240,400	10,396,700	246.95	4.03	2,601,800	61.80
20	1,455,100	7,675,200	9,130,300	1,118,600	2,631,000	109,400	3,859,000	12,989,300	2,318,900	10,670,400	253.45	4.14	2,472,500	58.73

Total life-cycle cost in discounted dollars   79,437,300
Average life-cycle cost in discounted dollars   94.34

Assumptions:

Material throughput	162 tons/day
	42,100 tons/year
MRF operation	260 days/year
Total households in collection area	215,000
Participation rate	75%
Setout rate	50%
MRF building cost (year 1)	$4,725,000
MRF equipment cost (year 1)	$2,050,000
Collection trucks and recycling bins (year 1)	$5,306,200
PAB interest rate	8%
Inflation rate	3.5%
Discount rate	8%
Residue quantity	1895 tons/year
Residue disposal cost (year 1)	30 dollars/ton
Material sales revenue (year 1)	30 dollars/ton

[a] Financing period for trucks is 7 years and recycling bins, 10 years.
[b] Financing period for building is 20 years and MRF equipment, 7 years.
[c] First year collection costs are $117 per ton.
[d] First year MRF costs are $55 per ton processed.
[e] The net cost may be partially offset by avoided MSW collection and disposal costs.
[f] Net cost distributed to all households (hh) in the collection area (not just the participants).

16.15

As shown in Table 16.4, the first-year cost of the WTE facility is $66.56 per ton. This cost is higher than that of landfilling in most communities, and analyzed on a first-year basis, one may conclude that WTE is a much more costly option. However, since a rather large component of the WTE cost is capital investment debt service, which remains fixed for the 20 years, the average discounted life-cycle cost is much lower ($38.29 per ton). This value is a better number to compare with other systems over a 20-year life cycle. Usually, a life-cycle cost analysis of the continuation of the existing system will also be conducted, including capital and operation and maintenance (O&M) costs. Then the life-cycle costs can be compared directly.

The MRF costs shown in Table 16.5 show that the first-year costs (including collection of recyclables) are about $143 per ton, or about $2.33 per household per month. The annual cost in the twentieth year, discounted to present-value dollars, becomes $59 per ton. These are net costs after subtracting the revenues from the sale of recyclables (based on 1993 market prices). The MRF costs would be expected to be at least partially offset by the savings experienced in the existing system collection and disposal costs. The average life-cycle cost for the recycling/MRF operation in discounted dollars is $94 per ton.

16.5 SUMMARY

The requirements for financing solid waste management projects can be substantial when tax-exempt PABs are used. Other forms of financing—particularly those where little, if any, debt is included—can be easier to arrange. The steps described in this chapter provide a general description of the process required when PABs are issued for private use. They are generic in nature and the specifics of a given project may result in more or less effort than indicated. Professional assistance will be needed with most solid waste management project financings. With the advent of system versus facility financings, the complexities of financing solid waste management projects are even greater than before.

The last section of this chapter describes the process of life-cycle cost analysis. The tables provide hypothetical examples of life-cycle costs over 20 years for waste-to-energy and recovery of materials for recycling. The importance of life-cycle costs in comparing solid waste management alternatives is demonstrated.

REFERENCES

Artz, N. S. (1990) "Integrated Solid Waste Planning for a Regional Area," Franklin Associates, Ltd., presented at the First U.S. Conference on Municipal Solid Waste Management, Washington, DC.

Chen, P. M., G. D. France, and S. A. Sharpe (1992) "Financing Solid Waste Disposal Projects in the 1990s," presented by S. E. Howard at the National Conference of State Legislatures, Kansas City, MO.

Horning, C. (1991) "Laws Give New Shape to Solid Waste Contracts and Finance," *Solid Waste & Power,* vol. 5, no. 7.

Lee, W. B., and E. T. Ashdown (1992) "Financing Waste Facilities during the Credit Crunch," *World Wastes,* vol. 35, no. 3.

MacCarthy, R. N. (1991) "Financing Recycling Facilities," *Waste Age,* vol. 22, no. 3.

NSWMA (1992) *The Cost to Recycle at a Materials Recovery Facility,* National Solid Wastes Management Association.

Ollis, R. W. (1992) "Financing Recycling Programs," *Waste Age,* vol. 23, no. 3.

Turbeville, W. C. (1990) "Cutting Waste Facility Finance Costs," *Waste Age,* vol. 21, no. 5.

APPENDIX A
GLOSSARY

Absorption   Penetration of one substance into or through another.

Acid gas scrubber   Device that removes particulate and gaseous impurities from a gas stream. This generally involves the spraying of an alkaline solid or liquid, and sometimes the use of condensation or absorbent particles.

Activated carbon   Highly absorbent form of carbon used to remove odors and toxic substances from gaseous emissions or to remove dissolved organic material from wastewater.

Adhesion   Molecular attraction that holds the surfaces of two substances in contact, such as water and rock particles.

Adsorption   Attachment of the molecules of a liquid or gaseous substance to the surface of a solid.

Aeration   Process of exposing bulk material, such as compost, to air. *Forced aeration* refers to the use of blowers in compost piles.

Aerobic   Biochemical process or environmental condition occurring in the presence of oxygen.

Aerobic digestion   Utilization of organic waste as a substrate for the growth of bacteria that function in the presence of oxygen to stabilize the waste and reduce its volume. The products of this decomposition are carbon dioxide, water, and a remainder consisting of inorganic compounds, undigested organic material, and water.

Aerosol   Particle of solid or liquid matter that can remain suspended in the air because of its small size.

Agricultural solid wastes   Wastes produced from the raising of plants and animals for food, including manure, plant stalks, hulls, and leaves.

Air-cooled wall   Refractory wall with a lane directly behind it through which cool air flows.

Air emissions   Solid particulates (such as unburned carbon) and gaseous pollutants (such as oxides of nitrogen or sulfur) or odors. These can result from a broad variety of activities, including exhaust from vehicles, combustion devices, landfills, compost piles, street sweepings, excavation, demolition, and so on.

Air pollutant   Dust, fumes, smoke, and other particulate matter, vapor, gas, odorous substances, or any combination thereof. Also, any air pollution agent or combination of such agents, including any physical, chemical, biological, radioactive substances, or matter that is emitted into or otherwise enters the ambient air.

Air pollution   Presence of unwanted material in the air in excess of standards. The term *unwanted material* here refers to material in sufficient concentrations, present for a sufficient time to interfere significantly with health, comfort, or welfare of persons, or with the full use and enjoyment of property.

Ambient air   Portion of the atmosphere external to buildings to which the general public has access.

Anaerobic digestion   Utilization of organic waste as a substrate for the growth of bacteria that function in the absence of oxygen to reduce the volume of waste. The bacteria consume the carbon in the waste as their energy source and convert it to gaseous products. Properly controlled, anaerobic digestion will produce a mixture of methane and carbon dioxide, with a sludge remainder consisting of inorganic compounds, undigested organic material, and water.

Ash   Residue that remains after a fuel or solid waste has been burned. (See also *bottom ash* and *fly ash*.)

At-site time   Time spent unloading and waiting to unload the contents of a collection vehicle or loaded container at a transfer station, processing facility, or disposal site.

Avoided costs   Cost savings resulting from a recycling, incineration, or energy conservation program. A cost saving can be avoided disposal fees.

Backyard composting   Controlled biodegradation of leaves, grass clippings, and/or other yard wastes on the site where they were generated.

Bacteria   Single-cell, microscopic organisms with rigid cell walls. They may be aerobic, anaerobic, or facultative anaerobic; some can cause disease; and some are important in the stabilization and conversion of solid wastes.

Baffles   Deflector vanes, guides, grids, grating, or other similar devices constructed or placed in air or gas flow systems, flowing water, or slurry systems to effect a more uniform distribution of velocities; absorb energy, divert, guide, or agitate fluids; and check eddies.

Bagasse   Agricultural waste material consisting of the dry pulp residue that remains after juice is extracted from sugar cane or sugar beets.

Baghouse   Air pollution abatement device used to trap particulates by filtering gas streams through large fabric bags usually made of cloth or glass fibers.

Baler   Machine used to compress recyclables into bundles to reduce volume. Balers are often used on newspaper, plastics, and corrugated cardboard.

Biodegradable   Substance or material that can be broken down into simpler compounds by microorganisms or other decomposers such as fungi.

Biodegradable volatile solids (BVS)   Portion of the volatile solids of the organic matter in MSW that is biodegradable.

Biological waste   Waste derived from living organisms.

Biomass   Amount of living matter in the environment.

Blowdown   Minimum discharge of recirculating water for the purpose of discharging materials contained in the process, the further buildup of which would cause concentrations or amounts exceeding limits established by best engineering practice.

Bottle bill   Legislation requiring deposits on beverage containers; appropriately called beverage container deposit law (BCDL).

Bottom ash   Nonairborne combustion residue from burning fuel or waste in a boiler. The material falls to the bottom of the boiler and is removed mechanically. Bottom ash constitutes the major portion (about 90 percent) of the total ash created by the combustion of solid waste.

British thermal unit (Btu)   Unit of measure for the amount of energy a given material contains (e.g., energy released as heat during the combustion is measured in Btu). Technically, 1 Btu is the quantity of heat required to raise the temperature of 1 lb of water 1°F.

Bulky waste   Large wastes such as appliances, furniture, some automobile parts, trees and branches, palm fronds, and stumps.

Burning rate   Volume of solid waste incinerated or the amount of heat released during incineration. The burning rate is usually expressed in pounds of solid waste per square foot of burning area per area or in British thermal units per cubic foot of furnace volume per hour.

Buy-back recycling center   Facility that pays a fee for the delivery and transfer of ownership to the facility of source-separated materials for the purpose of recycling or composting.

Capital costs   Those direct costs incurred in order to acquire real property assets such as land, buildings, and machinery and equipment.

Carbonaceous matter   Pure carbon or carbon compounds present in solid wastes.

Carbon dioxide (CO_2)   Colorless, odorless, nonpoisonous gas that forms carbonic acid when dissolved in water. It is produced during the thermal degradation and microbial decomposition of solid wastes and contributes to global warming.

Carbon monoxide (CO)   Colorless, poisonous gas that has an exceedingly faint metallic odor and taste. It is produced during the thermal degradation and microbial decomposition of solid wastes when the oxygen supply is limited.

Carcinogenic   Capable of causing the cells of an organism to react in such a way as to produce cancer.

Centrifugal collector   Mechanical system using centrifugal force to remove aerosols from a gas stream.

Chain grate stoker   Stoker with a moving chain as a grate surface. The grate consists of links mounted on rods to form a continuous surface that is generally driven by a shaft with sprockets.

Charcoal   Dark or black porous carbon prepared from vegetable or animal substances (as from wood by charring in a kiln from which air is excluded).

Charge   Amount of solid waste introduced into a furnace at one time.

Classification   Separation and rearrangement of waste materials according to composition (e.g., organic or inorganic), size, weight, color, shape, and the like, using specialized equipment.

Clean Air Act   Act passed by Congress to have the air "safe enough to protect the public's health" by May 31, 1975. Required the setting of National Ambient Air Quality Standards (NAAQS) for major primary air pollutants.

Clean Water Act   Act passed by Congress to protect the water resources of the nation. Requires the U.S. EPA to establish a system of national effluent standards for major water pollutants, requires all municipalities to use secondary sewage treatment by 1988, sets interim goals of making all U.S. waters safe for fishing and swimming, allows point-source discharges of pollutants into waterways only with a permit from the EPA, requires all industries to use the best practicable technology (BPT) for control of conventional and nonconventional pollutants and to use the best available technology (BAT) that is reasonable and affordable.

Coal refuse   Waste products of coal mining, cleaning, and coal preparation operations and containing coal, matrix material, clay, and other organic and inorganic material.

Cocollection   Collection of ordinary household garbage in combination with special bags of source-separated recyclables.

Coding   In the context of solid waste, coding refers to a system to identify recyclable materials. The coding system for plastic packaging utilizes a three-sided arrow with a number in the center and letters underneath. The number and letters indicate the resin from which each container is made: 1 = polyethylene terephthalate (PETE), 2 = high-density polyethylene (HDPB), 3 = vinyl (V), 4 = low-density polyethylene (LDPE), 5 = polypropylene (PP), 6 = polystyrene (PS), and 7 = other/mixed plastics. Noncoded containers are recycled through mixed-plastics processes. To help recycling sorters, the code is molded into the bottom of bottles with a capacity of 16 oz or more and other containers with a capacity of 8 oz or more.

Codisposal   Burning of municipal solid waste with other material, particularly sewage sludge: the technique in which sludge is combined with other combustible materials (e.g., refuse, refuse-derived fuel, coal) to form a furnace feed with a higher heating value than the original sludge.

Cofiring or coburning   Combustion of MSW along with other fuel, especially coal.

Cogeneration   Production of electricity as well as heat from one fuel source.

Collection routes   Established routes followed in the collection of commingled and source-separated wastes from homes, businesses, commercial and industrial plants, and other locations.

Collection systems   Collectors and equipment used for the collection of commingled and source-separated waste. Waste collection systems may be classified from several points of view, such as the mode of operation, the equipment used, and the types of wastes collected. In this text, collection systems have been classified according to their mode of operation in two categories: (1) hauled container systems and (2) stationary container systems.

Collection, waste   Act of picking up wastes at homes, businesses, commercial and industrial plants, and other locations, loading them into a collection vehicle (usually enclosed), and hauling them to a facility for further processing or transfer to a disposal site.

Combustible   Various materials in the waste stream that are burnable, such as paper, plastic, lawn clippings, leaves, and other organic materials; materials that can be ignited at a specific temperature in the presence of air to release heat energy.

Combustion   Chemical combining of oxygen with a substance, which results in the production of heat.

Combustion air   Air used for burning a fuel.

Combustion gases   Mixture of gases and vapors produced by burning.

Commercial sector   One of the four sectors of the community that generates garbage. Designed for profit.

Commercial solid wastes   Wastes that originate in wholesale, retail, or service establishments, such as office buildings, stores, markets, theaters, hotels, and warehouses.

Commercial waste   All types of solid wastes generated by stores, offices, restaurants, warehouses, and other nonmanufacturing activities, excluding residential and industrial wastes.

Commingled recyclables   Mixture of several recyclable materials in one container.

Commingled waste   Mixture of all waste components in one container.

Compaction   Unit operation used to increase the specific weight (density in metric units) of waste materials so that they can be handled, stored, and transported more efficiently.

Compactor   Any power-driven mechanical equipment designed to compress and thereby reduce the volume of wastes.

Compactor collection vehicle   Large vehicle with an enclosed body having special power-driven equipment for loading, compressing, and distributing wastes within the body.

Component separation   Separation or sorting of wastes into components or categories.

Composite liner   Liner composed of both a plastic and soil component for a landfill.

Composition   Set of identified solid waste materials, categorized into waste categories and waste types.

Compost   Relatively stable mixture of organic wastes partially decomposed by an aerobic and/or anaerobic process. Compost can be used as a soil conditioner.

Composting   Controlled biological decomposition of organic solid waste materials under aerobic or anaerobic conditions. Composting can be accomplished in windrows, static piles, and enclosed vessels (known as *in-vessel composting*).

Concentration   Amount of one substance contained in a unit of another substance.

Conservation   The planned management of a natural resource to prevent exploitation, destruction, or neglect.

Construction and demolition waste   Waste building materials, packaging, and rubble resulting from construction, remodeling, and demolition operations on pavements, houses, commercial buildings, and other structures. The materials usually include used lumber, miscellaneous metal parts, packaging materials, cans, boxes, wire, excess sheet metal, and other materials.

Consumer waste   Materials used and discarded by the buyer, or consumer, as opposed to wastes created and discarded in-plant during the manufacturing process.

Consumption   Amount of any resource (material or energy) used.

Container   Receptacle used for the storage of solid wastes until they are collected.

Controlled-air incinerator   Incinerator with excess or starved air having two or more combustion chambers in which the amounts and distribution of air are controlled. The U.S. EPA prefers to use the term *combustor* instead of *incinerator*.

Conversion   Transformation of wastes into other forms; for example, transformation by burning or pyrolysis into steam, gas, or oil.

Conversion products   Products derived from the first-step conversion of solid wastes, such as heat from combustion and gas from biological conversion.

Corrosive   Defined for regulatory purposes as a substance having a pH level below 2 or above 12.5, or a substance capable of dissolving or breaking down other substances, particularly metals, or causing skin burns.

Corrugated container   According to SIC Code 2653, a paperboard container fabricated from two layers of kraft linerboard sandwiched around a corrugating medium. *Kraft linerboard* means paperboard made from wood pulp produced by a modified sulfate pulping process, with basis weight ranging from 18 to 200 lb, manufactured for use as facing material for corrugated or solid-fiber containers. Linerboard also may mean that material that is made from reclaimed paper stock.

Cost-effective   Measure of cost compared with an unvalued output (e.g., the cost per ton of solid waste collected) such that the lower the cost, the more cost-effective the action.

Cover material   Soil or other material used to cover compacted solid wastes in a sanitary landfill.

Crusher   Mechanical device used to break secondary materials such as glass bottles into smaller pieces.

Cullet   Clean, generally color-sorted, crushed glass used in the manufacture of new glass products.

Curbside collection   Collection of recyclable materials at the curb, often from special containers, to be brought to various processing facilities. Collection may be both separated and/or mixed wastes.

Curbside separation   To separate commingled recyclables prior to placement in individual compartments in truck providing curbside collection service; this task is performed by the collector.

Cyclone separator   Separator that uses a swirling airflow to sort mixed materials according to the size, weight, and density of the pieces.

Decomposition   Breakdown of organic wastes by bacterial, chemical, or thermal means. Complete chemical oxidation leaves only carbon dioxide, water, and inorganic solids.

Decontamination or detoxification   Processes that will convert pesticides into nontoxic compounds or the selective removal of radioactive material from a surface or from within another material.

Degradable plastics   Plastics specifically developed for special products that are formulated to break down after exposure to sunlight or microbes. By law, six-pack rings are degradable; however, they degrade only gradually, causing litter and posing a hazard to birds and marine animals.

Degradation   (Also biodegradation) Natural process that involves assimilation or consumption of a material by living organisms.

Deinking   Removal of ink, filler, and other nonfibrous material from printed waste paper.

Demolition wastes   Wastes produced from the demolition of buildings, roads, sidewalks, and other structures. These wastes usually include large, broken pieces of concrete, pipe, radiators, ductwork, electrical wire, broken-up plaster walls, lighting fixtures, bricks, and glass.

Densification   Unit operation used to increase the density of waste materials so that they can be stored and transported more efficiently.

Densified refuse-derived fuel (d-RDF)   Refuse-derived fuel that has been compressed or compacted through such processes as pelletizing, briquetting, or extruding, causing improvements in certain handling or burning characteristics.

Deposit   Matter deposited by a natural process; a natural accumulation of iron ore, coal; money paid as security.

Dewatering   Removal of water from solid wastes and sludges by various thermal and mechanical means.

Digestion, anaerobic   Biological conversion of processed organic wastes to methane and carbon dioxide under anaerobic conditions.

Dioxin   Generic name for a group of organic chemical compounds formally known as polychlorinated dibenzo-*p*-dioxins. Heterocyclic hydrocarbons that occur as toxic impurities, especially in herbicides.

Discards   Include the MSW remaining after recovery for recycling and composting. These discards are usually combusted or disposed of in landfills, although some MSW is littered, stored, or disposed of on site, particularly in rural areas.

Dispersion technique   Use of dilution to attain ambient air quality levels, including any intermittent or supplemental control of air pollutants varying with atmospheric conditions.

Disposal   Activities associated with the long-term handling of (1) solid wastes that are collected and of no further use and (2) the residual matter after solid wastes have been processed and the recovery of conversion products or energy has been accomplished. Normally, disposal is accomplished by means of sanitary landfilling.

Disposal facility   Collection of equipment and associated land area that serves to receive waste and dispose of it. The facility may incorporate one or more disposal methods.

Diversion rate   Measure of the amount of material now being diverted from landfilling for reuse and recycling compared with the total amount of waste that was thrown away previously.

DOT   Department of Transportation.

Draft   Pressure difference between an incinerator (or combustor) and the atmosphere.

Drag conveyer   Conveyer that uses vertical steel plates fastened between two continuous chains to drag material across a smooth surface.

Drop-off center   Location where residents or businesses bring source-separate recyclable materials. Drop-off centers range from single-material collection points (e.g., easy-access "igloo" containers) to staffed, multimaterial collection centers.

Dump   Site where mixed wastes are indiscriminately deposited without controls or regard to the protection of the environment. Dumps are now illegal.

Ecosystem   System made up of a community of living things and the physical and chemical environment with which they interact.

Eddy-current separation   Electromagnetic technique for separating aluminum from a mixture of materials.

Effluent   Waste materials, usually waterborne, discharged into the environment, treated or untreated; the liquid leaving wastewater treatment systems.

Electrostatic precipitator (ESP)   Gas-cleaning device that collects entrained particulates by placing an electrical charge on them and attracting them onto oppositely charged collecting electrodes. They are installed in the back end of the incineration process to reduce air pollution.

Embedded energy   Sum of all the energy involved in product development, transportation, use, and disposal.

Emission rate   Amount of pollutant emitted into atmospheric circulation per unit of time.

Encapsulation   Complete enclosure of a waste in another material in such a way as to isolate it from external effects such as those of water or of air.

Endemic plant   Plant species that is confined to a specific location, region, or habitat.

Energy   Ability to do work by moving matter or by causing a transfer of heat between two objects at different temperatures.

Energy recovery   Conversion of solid waste into energy or a marketable fuel. A form of resource recovery in which the organic fraction of waste is converted to some form of usable energy, such as burning processed or raw refuse, to produce steam.

Environment   Water, air, land, and all plants and human and other animals living therein, and the interrelationships that exist among them.

Environmental impact statement (EIS)   Document, prepared by the EPA or under EPA guidance (generally a consultant hired by the applicant and supervised by EPA), which identifies and analyzes in detail the environmental impacts of a proposed action. Individual states also may prepare and issue an EIS as regulated by state law. Such state documents may be called *environmental impact reports* (EIRs).

Environmental quality   Overall health of an environment determined by comparison with a set of standards.

Evaporation   Physical transformation of a liquid to a gas.

Exhaust system   System comprising a combination of components that provide for enclosed flow of exhaust gas from the furnace exhaust port to the atmosphere.

External costs   Cost relating to, or connected with, outside expenses.

Facility operator   Full-service contractors or other operators of a part of a resource recovery system.

Fee   Dollar amount charged by a community to pay for services (e.g., tipping fee at a landfill).

Ferrous metals   Metals composed predominantly of iron. In the waste materials, these metals usually include tin cans, automobiles, refrigerators, stoves, and other appliances. In resource recovery, often used to refer to materials that can be removed from the waste stream by magnetic separation.

Filter   Membrane or porous device through which a gas or liquid is passed to remove suspended particles or dust.

Firebrick   Refractory brick made from fireclay.

Fireclay   Sedimentary clay containing only small amounts of fluxing impurities, high in hydrous aluminum, and capable of withstanding high temperatures.

Fixed grate   Grate without moving parts, also called a *stationary grate*.

Flammable waste   Waste capable of igniting easily and burning rapidly.

Flash point   Minimum temperature at which a liquid or solid gives off sufficient vapor to form an ignitable vapor-air mixture near the surface of the liquid or solid.

Flow control   Legal or economic means by which waste is directed to particular destinations. For example, an ordinance requiring that certain wastes be sent to a combustion facility is waste flow control.

Flow diagram of a process   Diagram that shows the assemblage of unit operations, facilities, and manual operations used to achieve a specified waste separation goal.

Flue   Any passage designed to carry combustion gases and entrained particulates.

Flue gas   Products of combustion, including pollutants, emitted to the air after a production process or combustion takes place.

Fluidized bed combustion   Oxidation of combustible material within a bed of solid, inert (noncombustible) particles that, under the action of vertical hot airflow, will act as a fluid.

Fly ash   All solids, including ash, charred papers, cinders, dusty soot, or other matter that rise with the hot gases from combustion rather than falling with the bottom ash. Fly ash is a minor portion (about 10 percent) of the total ash produced from combustion of solid waste, is suspended in the flue gas after combustion, and can be removed by pollution control equipment.

Food wastes   Animal and vegetable wastes resulting from the handling, storage, sales, preparation, cooking, and serving of foods; commonly called *garbage.*

Forced draft   Positive pressure created by the action of a fan or blower that supplies the primary or secondary combustion air in an incinerator.

Fossil fuel   Natural gas, petroleum, coal, and any form of solid, liquid, or gaseous fuel derived from such materials for the purpose of creating useful heat.

Front-end loader   (1) Solid waste collection truck that has a power-driven loading mechanism at the front; (2) vehicle with a power-driven scoop or bucket at the front, used to load secondary materials into processing equipment or shipping containers.

Front-end recovery   Salvage of reusable materials, most often the inorganic fraction of solid waste, prior to the processing or combusting of the organic fraction. Some processes for front-end recovery are grinding, shredding, magnetic separation, screening, and hand sorting.

Front-end system   Those processes used for the recovery of materials from solid wastes and the preparation of individual components for subsequent conversion process (e.g., composting, waste to energy, etc.).

Fuel   Any material that is capable of releasing energy or power by combustion or other chemical or physical means.

Full material recovery facility (MRF)   Process for removing recyclables and creating a compostlike product from the total of full mixed municipal solid waste (MSW) stream. Differs from a "clean" MRF, which processes only commingled recyclables. (See *waste recovery facility.*)

Functional element   Used in this text to describe the various activities associated with the management of solid wastes from the point of generation to final disposal. In general, a functional element represents a physical activity. The six functional elements used throughout this book are waste generation, onside storage, collection, materials processing and recovery, transfer and transport, and disposal.

Furnace   Combustion chamber; an enclosed structure in which heat is produced.

Garbage   Solid waste consisting of putrescible animal and vegetable waste materials resulting from the handling, preparation, cooking, and consumption of food, including waste materials from markets, storage facilities, handling and sale of produce, and other food products. Generally defined as *wet food waste,* but not synonymous with *trash, refuse, rubbish,* or *solid waste.* (See *food wastes.*)

Gas control system   System at a landfill designed to prevent explosion and fires due to the accumulation of methane concentrations and damage to vegetation on final cover of closed portions of a landfill or vegetation beyond the perimeter of the property on which the landfill is located and to prevent objectionable odors off-site.

Gas scrubber   Device where a caustic solution is contacted with exhaust gases to neutralize certain combustion products, primarily sulfur oxides (SO) and secondary chlorine (Cl).

Gaseous emissions   Waste gases released into the atmosphere as a by-product of combustion.

Generation rate   Total tons diverted, recovered, and disposed per unit of time divided by the population. The annual per capita generation rate is the total tons generated in 1 year divided by the population of residents.

Generation   Refers to the amount (weight, volume, or percentage of the overall waste stream) of materials and products as they enter the waste stream and before material recovery, composting, or combustion takes place.

Generator   Any person, by site or location, whose act or process produces a solid waste; the initial discarding of a material.

Grain loading   Rate at which particles are emitted from a pollution source, in grains per cubic foot of gas emitted (7000 gr = 1 lb).

Grate Device used to support the solid fuel or solid waste in a furnace during drying, ignition, or combustion. Openings are provided for passage of combustion air.

Gravity separation Separation of mixed materials based on the differences of material size and specific gravity.

Gross national product (GNP) Total market value of all the goods and services produced by a nation during a specified time period.

Groundwater Water beneath the surface of the earth and located between saturated soil and rock. It is the water that supplies wells and springs.

Growth rate Estimation of progressive development; the rate at which a population or anything else grows.

Hammermill Type of crusher used to break up waste materials into smaller pieces or particles, which operates by using rotating and flailing heavy hammers.

Haul distance Distance a collection vehicle travels (1) after picking up a loaded container (hauled container system) or from its last pickup stop on collection route (stationary container system) to a materials recovery facility, transfer station, or sanitary landfill, and (2) distance the collection vehicle travels after unloading to the location where the empty container is to be deposited or to the beginning of a new collection route.

Haul time Elapsed or cumulative time spent transporting solid wastes between two specific locations.

Hauled container system Collection systems in which the containers used for the storage of wastes are hauled to the disposal site, emptied, and returned to either their original location or some other location.

Haulers Those persons, firms, or corporations or governmental agencies responsible (under either oral or written contract, or otherwise) for the collection of solid waste within the geographic boundaries of the contract community(ies) or the unincorporated county and the transportation and delivery of such solid waste to the resource recovery system as directed in the plan of operations.

Hazard Having one or more of the characteristics that cause a substance or combination of substances to qualify as a hazardous material.

Hazardous waste Waste, or combination of wastes, that may cause or significantly contribute to an increase in mortality or an increase in serious irreversible or incapacitating illness or that poses a substantial present or potential hazard to human health or the environment when improperly treated, stored, transported, disposed of, or otherwise managed. Hazardous wastes include radioactive substances, toxic chemicals, biological wastes, flammable wastes, and explosives.

Heat balance Accounting of the distribution of the heat input and output of an incinerator or boiler, usually on an hourly basis.

Heavy metals Hazardous elements, including cadmium, mercury, and lead, which may be found in the waste stream as part of discarded items such as batteries, lighting fixtures, colorants, and inks.

Hierarchy of integrated waste management Source reduction, recycling, waste transformation, and disposal. It should be noted that the EPA uses the term *combustion* instead of *transformation*. Further, the U.S. EPA does not make a distinction between waste transformation (combustion) and disposal as both are viewed as viable components of an integrated waste management program. A distinction is made between transformation and disposal in California and other states.

High-density polyethylene (HDPE) Recyclable plastic, used for items such as milk containers, detergent containers, and base cups of plastic soft drink bottles.

High-grade paper Relatively valuable types of paper, such as computer printout, white ledger, and tab cards. Also used to refer to industrial trimmings at paper mills that are recycled.

Household hazardous waste collection Program activity in which household hazardous wastes are brought to a designated collection point where the household hazardous wastes are separated for temporary storage and ultimate recycling, treatment, or disposal.

Household hazardous waste Those wastes resulting from products purchased by the general public for household use, which, because of their quantity, concentration, or physical, chemical, or infectious characteristics, may pose a substantial known or potential hazard to human health or the environment when improperly treated, disposed, or otherwise managed.

Hydrocarbon Any of a vast family of compounds containing carbon and hydrogen in various combinations, found especially in fossil fuels.

Hydrogen sulfide (H₂S)   Poisonous gas with the odor of rotten eggs that is produced from the reduction of sulfates in, and the putrefaction of, a sulfur-containing organic material.

Ignition temperature   Lowest temperature of a fuel at which combustion becomes self-sustaining.

Impermeable   Restricts the movement of products through the surface.

Incineration   Engineered process involving burning or combustion to thermally degrade waste materials. Solid wastes are reduced by oxidation and will normally sustain combustion without the use of additional fuel. Incineration is occasionally referred to as *combustion* in this text.

Industrial unit   Site zoned for an industrial business and that generates industrial solid wastes.

Industrial waste   Materials discarded from industrial operations or derived from industrial operations or manufacturing processes, all nonhazardous solid wastes other than residential, commercial, and institutional. Industrial waste includes all wastes generated by activities such as demolition and construction, manufacturing, agricultural operations, wholesale trade, and mining. A distinction should be made between scrap (those materials that can be recycled at a profit) and solid wastes (those that are beyond the reach of economical reclamation).

Infectious waste   Waste containing pathogens or biologically active material that, because of its type, concentration, or quantity, is capable of transmitting disease to persons exposed to the waste.

Infrastructure   Substructure or underlying foundation; those facilities upon which a system or society depends; for example, roads, schools, power plants, communication networks, and transportation systems.

Inorganic   Not composed of once-living material (e.g., minerals); generally, composed of chemical compounds not principally based on the element carbon.

Integrated solid waste management   Management of solid waste based on a combination of source reduction, recycling, waste combustion, and disposal. The purposeful, systematic control of the functional elements of generation; waste handling, separation, and processing at the source; collection; separation and processing and transformation of solid waste; transfer and transport; and disposal associated with the management of solid wastes from the point of generation to final disposal.

Integrated waste management   Management of solid waste based on a consideration of source reduction, recycling, waste transformation, and disposal arranged in a hierarchical order. The purposeful, systematic control of the functional elements of generation, onside storage, collection, transfer and transport, processing and recovery, and disposal associated with the management of solid wastes from the point of generation to final disposal.

Intermediate processing center (IPC)   Usually refers to a facility that processes residentially collected mixed recyclables into new products for market; often used interchangeably with *materials recovery facility (MRF)*. A facility where recyclables that have been separated from the rest of the waste are brought to be separated and prepared for market (crushed, baled, etc.). An IPC can be designed to handle commingled or separated recyclables or both.

Internal costs   Expenses of, relating to, or occurring within the confines of an organized structure.

Investment tax credit   Reduction in taxes permitted for the purchase and installation of specific types of equipment and other investments.

Jurisdiction   City or county responsible for preparing any one or all of the following: the countywide integrated waste management plan or the countywide siting element.

Kraft paper   Comparatively coarse paper noted for its strength and used primarily as a wrapper or packaging material.

Landfill, sanitary   Engineered method of disposing of solid wastes on land in a manner that protects human health and the environment. Waste is spread in thin layers, compacted to the smallest practical volume, and covered with soil or other suitable material at the end of each working day, or more frequently, as necessary.

Large-quantity generator   Sources, such as industries and agriculture, that generate more than 1000 kg of hazardous waste per month.

Leachate   Liquid that has percolated through solid waste or another medium and has extracted, dissolved, or suspended materials from it, which may include potentially harmful materials. Leachate collection and treatment is of primary concern at municipal waste landfills.

Liner   Impermeable layers of heavy plastic, clay, and gravel that protect against groundwater contamination through downward or lateral escape of leachate. Most sanitary landfills have at least two plastic liners or layers of plastic and clay. Also refers to the material used on the inside of a furnace wall to ensure that a chamber is impervious to escaping gases.

Litter   That highly visible portion of solid wastes that is generated by the consumer and carelessly discarded outside the regular disposal system. Litter accounts for only about 2 percent of the total solid waste volume.

Locally unwanted land use (LULU)   For example, landfills.

Long-term impact   Future effect of an action, such as an oil spill.

Low-grade paper   Less valuable types of paper, such as mixed office paper, corrugated paperboard, and newspaper.

Magnetic separation   Use of magnets to separate ferrous materials from commingled waste materials in MSW.

Magnetic separator   Equipment usually consisting of a belt, drum, or pulley with a permanent or temporary electromagnet and used to attract and remove magnetic materials from other materials.

Mandatory recycling   Programs that, by law, require consumers to separate trash so that some or all recyclable materials are not burned or dumped in landfills.

Manual separation   Separation of wastes by hand. Sometimes called *hand picking* or *hand sorting*, manual separation is done in the home or office by keeping food wastes separate from newspaper, or in a materials recovery facility by picking out large cardboard and other recoverable materials.

Market development   Method of increasing the demand for recovered materials so that end markets for the materials are established, improved, or stabilized and thereby become more reliable.

Mass burn   Controlled combustion of unseparated commingled MSW.

Mass combustion   Burning of as-received, unprocessed, commingled refuse in furnaces designed exclusively for solid waste disposal and energy recovery. Sometimes referred to as *mass burn*.

Mass-burn facility   Type of incinerator (or combustor) that burns solid waste without any attempt to separate recyclables or process waste before burning.

Material recovery   Extraction of materials from the waste stream for reuse or recycling. Examples include source separation, front-end recovery, in-plant recycling, postcombustion recovery, leaf composting, and so on.

Materials balance   Accounting of the weights of materials entering and leaving a processing unit, such as an incinerator, usually on an hourly basis.

Materials recovery facility (MRF)   Physical facilities used for the further separation and processing of wastes that have been separated at the source and for the separation of commingled wastes.

Materials recovery/transfer facilities (MR/TFs)   Multipurpose facilities that may include the functions of a drop-off center for separated wastes, a materials recovery facility, a facility for the composting and bioconversion of wastes, a facility for the production of refuse-derived fuel, and a transfer and transport facility.

Mechanical separation   Separation of waste into various components using mechanical means, such as cyclones, trommels, and screens.

Metal   Mineral source that is a good conductor of electricity and heat, and that yields basic oxides and hydroxides.

Methane (CH_4)   Odorless, colorless, flammable, and asphyxiating gas that can explode under certain circumstances and that can be produced by solid wastes undergoing anaerobic decomposition. Methane emitted from municipal solid waste landfills can be used as fuel.

Microorganisms   Microscopically small living organisms, including bacteria, yeasts, simple fungi, actinomycetes, some algae, slime molds, and protozoans, that digest decomposable materials through metabolic activity. Microorganisms are active in the composting process.

Mixed paper   Waste type that is a mixture, unsegregated by color or quality, of at least two of the following paper wastes: newspaper, corrugated cardboard, office paper, computer paper, white paper, coated paper stock, or other paper wastes.

Mixed refuse   Garbage or solid waste that is in a fully commingled state at the point of generation.

Mixed-waste processing facility   Facility that processes mixed solid waste to remove recyclables and, sometimes, refuse-derived fuel and/or a compost substrate.

Mixing chamber   Chamber usually placed between the primary and secondary combustion chamber and in which the products of combustion are thoroughly mixed by turbulence that is created by increased velocities of gases, checkerwork, or turns in the direction of the gas flow.

Mixture   Any combination of two or more chemical substances if the combination does not occur in nature and is not, in whole or in part, the result of a chemical reaction.

Modular incinerator   Smaller-scale waste combustion units prefabricated at a manufacturing facility and transported to the municipal waste combustor facility site.

Moisture content   Weight loss (expressed in percent) when a sample of solid wastes is dried to a constant weight at a temperature of 100 to 105°C.

Mulch   Any material, organic or inorganic, applied as a top-dressing layer to the soil surface. Mulch is also placed around plants to limit evaporation of moisture and freezing of roots and to nourish the soil.

Municipal incinerator (or combustor)   A privately or publicly owned incinerator (or combustor) primarily designed and used to burn residential and commercial solid wastes within a community.

Municipal solid waste (MSW)   Includes all of the wastes that are generated from residential households and apartment buildings, commercial and business establishments, institutional facilities, construction and demolition activities, municipal services, and treatment plant sites.

Municipal solid waste composting   Controlled degradation of municipal solid waste, including after some form of preprocessing to remove noncompostable inorganic materials.

National Ambient Air Quality Standards (NAAQS)   Federal standards that limit the concentration of particulates, sulfur dioxide, nitrogen dioxide, ozone, carbon monoxide, and lead in the atmosphere.

Native plant   General term referring to plants that grow in a region.

Natural resource   Material or energy obtained from the environment that is used to meet human needs; material or energy resources not made by humans.

Nitrogen   A tasteless, odorless gas that constitutes 78 percent of the atmosphere by volume. One of the essential ingredients of composting.

Nonferrous metals   Any metal scraps that have value and that are derived from metals other than iron and its alloys in steel, such as aluminum, copper, brass, bronze, lead, zinc, and other metals, and to which a magnet will not adhere.

Non-point source   Undefined wastewater discharges, such as runoff from urban, agricultural, or strip-mined areas, which do not originate from a specific point.

Nonrecyclable   Not capable of being recycled or used again.

Nonrenewable (resource)   Not capable of being naturally restored or replenished; resources available in a fixed amount (stock) in the earth's crust; they can be exhausted either because they are not replaced by natural processes (copper) or because they are replaced more slowly than they are used (oil and coal).

Not in my back yard (NIMBY)   Refers to the fact that people want the convenience of products and proper disposal of the waste generated by their use of products, provided the disposal area is not located near them.

Odor threshold   Lowest concentration of an airborne odor that a human being can detect.

Office wastes   Discarded materials that consist primarily of paper waste, including envelopes, ledgers, and brochures.

Off-route time   All time spent by the collectors on activities that are nonproductive from the point of view of the overall collection operation.

Oil   Oil of any kind or in any form, including, but not limited to, petroleum, fuel oil, sludge, oil refuse, and oil mixed with wastes other than dredged spoil.

Old newspaper (ONP)   Any newsprint that is separated from other types of solid waste or collected separately from other types of solid waste and made available for reuse and that may be used as a raw material in the manufacture of a new paper product.

Onside handling, storage, and processing   Activities associated with the handling, storage, and processing of solid wastes at the source of generation before they are collected.

Opacity   Degree of obscuration of light (e.g., a window has zero opacity, while a wall has 100 percent opacity).

Operational costs   Those direct costs incurred in maintaining the ongoing operation of a program or facility. Operational costs do not include capital costs.

Organic materials   Chemical compounds containing carbon, excluding carbon dioxide, combined with other chemical elements. Organic materials can be of natural or anthropogenic origin. Most organic compounds are a source of food for bacteria and are usually combustible.

Organic soil amendment   Plant and animal residues added to mineral soil to improve soil structure and enhance nutritional content of the soil.

Oscillating-grate stoker   Stoker whose entire grate surface oscillates to move the solid waste and residue over the grate surface.

Packaging   Any of a variety of plastics, papers, cardboard, metals, ceramics, glass, wood, and paperboard used to make containers for food, household, and industrial products.

Packed tower   Pollution control device that forces dirty gas through a tower packed with crushed rock, wood chips, or other packing while liquid is sprayed over the packing material. Pollutants in the gas stream either dissolve in or chemically react with the liquid.

Paper   Term for all kinds of matted or felted sheets of fiber. Made from the pulp of trees, paper is digested in a sulfurous solution, bleached, and rolled into long sheets. Acid rain and dioxin are standard by-products in this manufacturing process. Specifically, as one of the two subdivisions of the general term, *paper* refers to materials that are lighter in basic weight, thinner, and more flexible than *paperboard,* the other subdivision.

Paperboard   Type of matted or sheeted fibrous product. In common terms, paperboard is distinguished from paper by being heavier, thicker, and more rigid. (See also *special wastes.*)

Partially allocated costs   Costs of adding a recycling program to an existing operation, such as a waste-hauling company or public works department. Also known as *incremental costs.*

Participant   Any household that contributes any materials at least once during a specified tracking period.

Participation rate   Measure of the number of people participating in a recycling program or other similar program, compared with the total number of people that could be participating.

Particulate matter (PM)   Tiny pieces of partially incinerated matter, resulting from the combustion process, that can have harmful health effects on those who breathe them. Pollution control at municipal waste combustor facilities is designed to limit particulate emissions.

Pathogen   Organism capable of causing disease. The four major classifications of pathogen found in solid waste are (1) bacteria, (2) viruses, (3) protozoans, and (4) helminths.

Permeable   Having pores or openings that permit liquids or gases to pass through.

Permits   Official approval and permission to proceed with an activity controlled by the permitting authority. Several permits from different authorities may be required for a single operation.

Petroleum   Mineral resource that is a complex mixture of hydrocarbons, an oily, flammable bituminous liquid, occurring in many places in the upper strata of the earth.

Photodegradable   Refers to plastics that will decompose if left exposed to sunlight.

Pickup time   For a hauled container system, it represents the time spent driving to a loaded container after an empty container has been deposited, plus the time spent picking up the loaded container and the time required to redeposit the container after its contents have been emptied. For a stationary container system, it refers to the time spent loading the collection vehicle, beginning with the stopping of the vehicle prior to loading the contents of the first container and ending when the contents of the last container to be emptied have been loaded.

Plant community   Assemblage of plants coexisting together in a common habitat or environment.

Plastics   Synthetic materials consisting of large molecules, called *polymers,* derived from petrochemicals (compared with natural polymers such as cellulose, starch, and natural rubbers).

Point of generation   Physical location where the generator discards material (mixed refuse and/or separated recyclables).

Point source   Specific, identifiable end-of-pipe discharges of wastes into receiving bodies of water (e.g., municipal sewage treatment plants, industrial wastewater treatment systems, and animal feedlots).

Pollutant   Dredged spoil, solid waste, incinerator residue, sewage, garbage, sewage sludge, munitions, chemical wastes, biological materials, radioactive materials, heat, wrecked or discarded equipment, rock, sand, cellar dirt, and industrial, municipal, and agricultural waste discharged into the environment. Any solid, liquid, or gaseous matter that is in excess of natural levels or established standards.

Pollution   Presence of matter or energy whose nature, location, or quantity produces undesired environmental effects. Also, the artificial or human-introduced alteration of the chemical, physical, biological, and radiological integrity of water.

Polyethylenes   Group of resins created by polymerizing ethylene gas. The two major categories are (1) high-density polyethylene and (2) low-density polyethylene.

Polyethylene terephthalate (PET)   Plastic resin used to make packaging, particularly soft drink bottles.

Polyvinyl chloride (PVC)   Plastic made by polymerization of vinyl chloride with peroxide catalysts. A typically insoluble plastic used in packaging, pipes, detergent bottles, wraps, and so on.

Porosity   Ratio of the volume of pores of a material to the volume of its mass.

Postconsumer recycling   Reuse of materials generated from residential and commercial waste, excluding recycling of material from industrial processes that has not reached the consumer, such as glass broken in the manufacturing process.

Precycling   Activities such as source and size reduction, material selection when shopping, and reducing toxicity of products in manufacturing prior to recycling, which helps reduce the amounts of municipal solid wastes generated.

Primary materials   Virgin or new materials used for manufacturing basic products. Examples include wood pulp, iron ore, and silica sand.

Primary standard   Natural air emissions standard intended to establish a level of air quality that, with an adequate margin of error, will protect public health.

Privatization   Assumption of responsibility for a public service by the private sector, under contract to local government or directly to the receivers of the service.

Process waste   Any designated toxic pollutant that is inherent to or unavoidable resulting from any manufacturing process, including that which comes into direct contact with or results from the production or use of any raw material, intermediate product, finished product, by-product, or waste product.

Processing   Any method, system, or other means designated to change the physical form or chemical content of solid wastes.

Program   Full range of source reduction, recycling, composting, special waste, or household hazardous waste activities undertaken by or in the jurisdiction or relating to management of the jurisdiction's waste stream to achieve the objectives identified in the source reduction, recycling, composting, special waste, and household hazardous waste components, respectively.

Public Utilities Regulatory Policies Act (PURPA) of 1978   Federal law whose key provision mandates private utilities to buy power commissions equal to the "avoid cost" power production to the utility. The act is intended to guarantee a market for small producers of electricity at rates equal or close to the utilities' marginal production costs.

Pulp   Moist mixture of fibers from which paper is made.

Putrescible   Subject to biological and chemical decomposition or decay. Usually used in reference to food wastes and other organic wastes.

Pyrolysis   Way of breaking down burnable waste by combustion in the absence of air. High heat is usually applied to the wastes in a closed chamber, and all moisture evaporates and materials break down into various hydrocarbon gases and carbonlike residue.

Rack collection   Collection of old newspapers at the same time as residential waste collection. The waste paper is placed in a side or front rack attached to the waste collection truck.

Radioactive   Substance capable of giving off high-energy particles or rays as a result of spontaneous disintegration of atomic nuclei.

Rate structure   That set of prices established by a jurisdiction, special district (as defined in Government Code Sec. 56036), or other rate-setting authority to compensate the jurisdiction, special district, or rate-setting authority for the partial or full costs of the collection, processing, recycling, composting, and/or transformation or landfill disposal of solid wastes.

Raw materials   Substances still in their natural or organic state, before processing or manufacturing; or the starting materials for a manufacturing process.

Reactive   For regulatory purposes, defined as a substance that tends to react spontaneously with air or water, to explode when dropped, or to give off toxic gases.

Rear-end system   Those chemical, thermal, and biological systems and related ancillary facilities used for the transformation (conversion) of processed solid wastes into various products.

Reclamation   Restoration to a better or more useful state, such as land reclamation by sanitary land-filling, or the extraction of useful materials from solid wastes.

Recoverable resources   Materials that still have useful physical or chemical properties after serving a specific purpose and can therefore be reused or recycled for the same or other purposes.

Recovery   Refers to materials removed from the waste stream for the purpose of recycling and/or composting. Recovery does not automatically equal recycling and composting, however. For example, if markets for recovered materials are not available, the materials that were separated from the waste stream for recycling may simply be stored or, in some cases, sent to a landfill or combustor. The extraction of useful materials or energy from waste.

Recycled material   Material that is used in place of a primary, raw, or virgin material in manufacturing a product and consists of material derived from postconsumer waste, industrial scrap, material derived from agricultural wastes, and other items, all of which can be used in the manufacture of new products. Also referred to as *recyclables*.

Recycling   Separating a given waste material (e.g., glass) from the waste stream and processing it so that it may be used again as a useful material for products that may or may not be similar to the original.

Recycling program   Should include the following: types of collection equipment used, collection schedule, route configuration, frequency of collection per household, whether curbside setout containers are provided by the program, publicity and educational activities, and budget, financial evaluation (costs, revenues, and savings), processing and handling procedures, market prices, ordinances, and enforcement activities.

Refuse   All solid materials that are discarded as useless. A term often used interchangeably with the term *solid waste*.

Refuse-derived fuel (RDF)   Combustible, or organic, portion of municipal waste that has been separated out and processed for use as fuel.

Renewable resources   Naturally occurring raw material or form of energy, such as the sun, wind, falling water, biofuels, fish, and trees, derived from an endless or cyclical source, where, through management of natural means, replacement roughly equals consumption (sustained yield).

Request for bid (RFB)   Mechanism for seeking bidders to supply recycling goods and services or to purchase secondary materials.

Request for proposal (RFP)   Mechanism for seeking qualified firm or individuals to supply recycling goods or services.

Request for qualifications (RFQ)   Mechanism for determining the experience, skills, financial resources, or expertise of a potential bidder or proposer.

Residential wastes   Wastes generated in houses and apartments, including paper, cardboard, beverage and food cans, plastics, food wastes, glass containers, and garden wastes.

Residual oil   General term used to indicate a heavy viscous fuel oil.

Residual wastes   Those solid, liquid, or sludge substances from human activities in the urban, agricultural, mining, and industrial environments remaining after collection and necessary treatment.

Residue   Solid or semisolid materials remaining after processing, incineration, composting, or recycling have been completed. Residues are usually disposed of in landfills.

Resource conservation   Reduction of the amounts of wastes generated, reduction of overall consumption, and utilization of recovered resources.

Resource Conservation and Recovery Act (RCRA) of 1976   Requires states to develop solid waste management plans and prohibits open dumps; identifies lists of hazardous wastes and sets the standards for their disposal. This law amends the Solid Waste Disposal Act of 1965 and expands on the Resource Recovery Act of 1970 to provide a program to regulate hazardous waste.

Resource recovery   Describes the extraction of economically usable materials or energy from wastes. The concept may involve recycling or conversion into different and sometimes unrelated uses.

Reusability   Ability of a product or package to be used more than once in its same form.

Reuse   Use of a waste material or product more than once.

Reverse vending machine   Machine that accepts empty beverage containers (or other items) and rewards the donor with a cash refund.

Rotary kiln stoker   Cylindrical, inclined device, utilized for the combustion of materials at high temperatures, that rotates, thus causing the solid waste to move in a slow cascading and forward motion.

Rubbish   General term for solid wastes—excluding food wastes and ashes—taken from residences, commercial establishments, and institutions.

Sanitary landfill   Engineered method of disposing of solid wastes on land in a manner that protects human health and the environment. Waste is spread in thin layers, compacted to the smallest practical volume, and covered with soil or other suitable material at the end of each working day.

Scrap   Products that have completed their useful life, such as appliances, cars, construction materials, ships, and postconsumer steel cans; also includes new scrap materials that result as by-products when metals are processed and products are manufactured. Steel scrap is recycled in steel mills to make new steel products.

Screening   Unit operation that is used to separate mixtures of materials of different sizes into two or more size fractions by means of one or more screening surfaces.

Scrubber   Device for removing unwanted dust particles, liquids, or gaseous substances from an airstream by spraying the airstream with a liquid (usually water or a caustic solution) or forcing the air through a series of baths; common antipollution device that uses a liquid or slurry spray to remove acid gases and particulates from municipal waste combustion facilities flue gases.

Secondary burner   Burner installed in the secondary combustion chamber of an incinerator to maintain a minimum temperature and to complete the combustion of incompletely burned gas.

Secondary combustion air   Air introduced above or below the fuel (waste) bed by a natural, induced, or forced draft.

Secondary material   Material that is used in place of a primary or raw material in manufacturing a product.

Secure landfill   Landfill designed to prevent the entry of water and the escape of leachate by the use of impermeable liners.

Separation   To divide wastes into groups of similar material, such as paper products, glass, food wastes, and metals. Also used to describe the further sorting of materials into more specific categories, such as clear glass and dark glass. Separation may be done manually or mechanically with specialized equipment.

Setout   Quantity of material placed for collection. Usually a setout denotes one household's entire collection of recyclable materials, but in urban areas, where housing density makes it difficult to identify ownership of materials, each separate container or bundle is counted as a setout. A single household, for example, may have three setouts: commingled glass, metals, and newspapers.

Sewage sludge   Semiliquid substance consisting of settled sewage solids, combined with varying amounts of water and dissolved materials.

Shredder   Machine used to break up waste materials into smaller pieces by cutting, tearing, shearing, and impact action.

Shredding   Mechanical operations used to reduce the size of solid wastes.

Shrinkage   Difference in the purchase weight of a secondary material and the actual weight of the material when consumed.

SIC code   Standards published in the *U.S. Standards Industrial Classification Manual* (1987).

Silo   Storage vessel, generally tall relative to its cross section, for dry solids; materials are fed into the top and withdrawn from the bottom through a control mechanism.

Size reduction, mechanical   Mechanical conversion of solid wastes into small pieces. In practice, the terms *shredding, grinding,* and *milling* are used interchangeably to describe mechanical size reduction operations.

Sludge   Any solid, semisolid, or liquid waste generated from a municipal, commercial, or industrial wastewater treatment plant, water supply treatment plant, or air pollution control facility, or any other such waste having similar characteristics and effects. Must be processed by bacterial digestion or other methods, or pumped out for land disposal, incineration, or composting.

Slurry   Pumpable mixture of solids and fluid.

Small-quantity generator   Sources such as small businesses and institutions that generate less than 1000 kg of hazardous waste per month.

Smoke   Particles suspended in air after incomplete combustion of materials containing carbon.

Soil liner   Landfill liner composed of compacted soil used for the containment of leachate.

Solid waste disposal facility   Any solid waste management facility that is the final resting place for solid waste, including landfills and incineration facilities that produce ash from the process of incinerating municipal solid waste.

Solid waste management   See *integrated solid waste management.*

Solid wastes   Any of a wide variety of solid materials, as well as some liquids in containers, which are discarded or rejected as being spent, useless, worthless, or in excess, including contained gaseous material resulting from industrial, commercial, mining, and agricultural operations, and from community activities. (See also *commercial, construction and demolition, hazardous, industrial, municipal,* and *residential wastes.*)

Source reduction   Reduction of the amount of materials entering the waste stream by voluntary or mandatory programs to eliminate the generation of waste. The design, manufacture, acquisition, and reuse of materials so as to minimize the toxicity of the waste generated.

Source-separated materials   Waste materials that have been separated at the point of generation. Source-separated materials are normally collected separately.

Source separation   Separation of waste materials from other commingled wastes at the point of generation.

Special wastes   Special wastes include bulky items, consumer electronics, white goods, yard wastes that are collected separately, hazardous wastes, concrete, batteries, used oil, asphalt, and tires. Special wastes are usually handled separately from other residential and commercial wastes.

Spray chamber   Chamber equipped with water sprays that cool and clean the combustion products passing through it.

Stack   Any chimney, flue, vent, roof monitor, conduit, or duct arranged to discharge emissions to the ambient air.

Stack emissions   Air emissions from combustion facility stacks.

Stationary container systems   Collection systems in which the containers used for the storage of wastes remain at the point of waste generation, except for occasional short trips to the collection vehicle.

Statistically representative   Those representative and random samples of units that are taken from a population sample. For the purpose of this definition, population sample includes, but is not limited to, a sample from a population of solid waste generation sites, solid waste facilities and recycling facilities, or a population of items of materials and solid wastes in a refuse load of solid waste.

Stoichiometric air   Amount of air theoretically required to provide the exact amount of oxygen for total combustion of a fuel. Municipal solid waste incineration technologies make use of both substoichiometric and excess air processes.

Styrofoam   Also known as *polystyrene,* a synthetic material consisting of large molecules called *polymers,* which are derived from petrochemicals. Experts agree that styrofoam will never decompose.

Subtitle C   Hazardous waste section of the Resource Conservation and Recovery Act (RCRA).

Subtitle D   Solid, nonhazardous waste section of the Resource Conservation and Recovery Act (RCRA).

Subtitle F   Section of the Resource Conservation and Recovery Act (RCRA) requiring the federal government to actively participate in procurement programs fostering the recovery and use of recycled materials and energy.

Superfund   Common name for the Comprehensive Environmental Response, Compensation, and Liability Act (CERCLA) to clean up abandoned or inactive hazardous waste dump sites.

Tare   Weight of extraneous material, such as pallets, strapping, bulkhead, and sideboards, that is deducted from the gross weight of a secondary material shipment to obtain net weight.

Thermal efficiency   Ratio of heat used to total useful energy generated.

Threshold dose   Minimum application of a given substance to produce a measurable effect.

Tipping fee   Fee, usually dollars per ton, for the unloading or dumping of waste at a landfill, transfer station, recycling center, or waste-to-energy facility. Also called a *disposal* or *service fee.*

Tipping floor   Unloading area for wastes delivered to an MRF, transfer station, or waste combustor.

Tire-derived fuel (TDF)   Form of fuel consisting of scrap tires shredded into chips.

Ton   Unit of weight in the U.S. customary system of measurement, an avoirdupois unit equal to 2000 lb. Also called a *short ton* or *net ton;* equals 0.907 metric tonnes.

Toxic   Defined for regulatory purposes as a substance containing poison and posing a substantial threat to human health and/or the environment.

Transfer   Act of transferring wastes from the collection vehicle to larger transport vehicles.

Transfer station   Place or facility where wastes are transferred from smaller collection vehicles (e.g., compactor trucks) into larger transport vehicles (e.g., over-the-road and off-road tractor trailers, railroad gondola cars, or barges) for movement to disposal areas, usually landfills. In some transfer operations, compaction or separation may be done at the station.

Transformation, waste   (See *waste transformation.*)

Transport   Transport of solid wastes transferred from collection vehicles to a facility or disposal site for further processing or action.

Trash   Wastes that usually do not include food wastes but may include other organic materials, such as plant trimmings. Generally defined as dry waste material, but in common usage, it is a synonym for *rubbish* or *refuse.*

Treatment process sludges   Liquid and semisolid wastes resulting from the treatment of domestic wastewater and industrial wastes.

Trommel   Perforated, rotating, horizontal cylinder that may be used in resource recovery facilities to break open trash bags, to remove glass and such small items as stone and dirt, and to remove cans from incinerator residue.

Tub grinder   Machine used to grind or chip wood and yard wastes foe mulching, composting, or for the use as a biomass fuel.

Turbidity   Cloudiness of a liquid.

Unacceptable waste   Motor vehicles, trailers, comparable bulky items of machinery or equipment, highly inflammable substances, hazardous waste, sludges, pathological and biological wastes, liquid wastes, sewage, manure, explosives and ordinance materials, and radioactive materials. Also includes any other material not permitted by law or regulation to be disposed of at a landfill, unless such landfill is specifically designed, constructed, and licensed or permitted to receive such material. None of such material constitutes either processable waste or unprocessable waste.

Unprocessable waste   That portion of the solid waste stream that is predominantly noncombustible and therefore should not be processed in a mass-burn resource recovery system; includes, but is not limited to, metal furniture and appliances, concrete rubble; mixed roofing materials; noncombustible building debris; rock, graver, and other earthen materials; equipment; wire and cable; and any item of solid waste exceeding 6 ft in any one of its dimensions or being in whole or in part of a solid mass, the solid mass portion of which has dimensions such that a sphere with a diameter of 8 in could be contained within such solid mass portion, and processable waste (to the extent that it is contained in the normal unprocessable waste stream); excludes unacceptable waste.

U.S. Environmental Protection Agency (U.S. EPA)   Federal agency created in 1970 and charged with the enforcement of all federal regulations having to do with air and water pollution, radiation and pesticide hazard, ecological research, and solid waste disposal.

Vapor   Gaseous phase of substances that are liquid or solid at atmospheric temperature and pressure (e.g., steam).

Vibrating screen   Mechanical device that sorts material according to size.

Virgin material   Any basic material for industrial processes that has not previously been used (e.g., wood-pulp trees, iron ore, silica sand, crude oil, and bauxite. (See also *primary materials, secondary material.*)

Vitrification   Process whereby high temperatures effect permanent chemical and physical change in a ceramic body.

Volatile solid (VS)   Portion of the organic material that can be released as a gas when organic material is burned in a muffle furnace at 550°C (1022°F).

Volume   Three-dimensional measurement of the capacity of a region of space or a container. Volume is commonly expressed in terms of cubic yards or cubic meters. Volume is not expressed in terms of mass or weight.

Volume reduction   Processing of wastes so as to decrease the amount of space they occupy. Reduction is presently accomplished by three major processes: (1) mechanical, which used compaction techniques (baling, sanitary landfills, etc.) and shredding; (2) thermal, which is achieved by heat or combustion (incineration) and can reduce volume by 80 to 90 percent; and (3) biological, in which the organic waste fraction is degraded by bacterial action (composting, etc.).

Volume-based rates   System of charging for garbage pickup that charges the waste generator rates based on the volume of waste collected, so that the greater the volume of waste collected, the higher the charge. Pay-by-the-bag systems and variable-can rates are types of volume-based rates.

Voluntary separation   Willing participation in waste recycling as opposed to mandatory recycling.

Waste   Unwanted materials left over from manufacturing processes, or refuse from places of human or natural habitation.

Waste categories   Grouping of solid wastes with similar properties into major solid waste classes, such as grouping together office, corrugated, and newspaper as a paper waste category, as identified by a solid waste classification system, except where a component-specific requirement provides alternative means of classification.

Waste composition   Relative amount of various types of materials in a specific waste stream.

Waste diversion   To divert solid waste, in accordance with all applicable federal, state, and local requirements, from disposal at solid waste landfills or transformation facilities through source reduction, recycling, or composting.

Waste generation   Act or process of generating solid wastes.

Waste generator   Any person whose act or process produces solid waste, or whose act first causes solid waste to become subject to regulation.

Waste minimization   Action leading to the reduction of waste generation, particularly by industrial firms.

Waste recovery facility (WRF)   Facility for separating recyclables and creating a compostlike material from the total of full mixed municipal solid waste stream.

Waste reduction   The prevention or restriction of waste generation at its source by redesigning products or the patterns of production and consumption.

Waste sources   Agricultural, residential, commercial, and industrial activities, open areas, and treatment plants where solid wastes are generated.

Waste stream   Describes the total flow of solid waste from homes, businesses, institutions, and manufacturing plants that must be recycled, burned, or disposed of in landfills; or any segment thereof, such as the *residential waste stream* or the *recyclable waste stream*. The total waste produced by a community or society, as it moves from origin to disposal.

Waste transformation   The transformation of waste materials involving a phase change (e.g., solid to gas). The most commonly used chemical and biological transformation processes are combustion and aerobic composting.

Wastewater   Water carrying dissolved or suspended solids from homes, farms, businesses, institutions, and industries.

Water table   Level below the earth's surface at which the ground becomes saturated with water. Landfills and composting facilities are designed with respect to the water table in order to minimize potential contamination.

Waterwall furnace   Furnace constructed with walls of welded steel tubes through which water is circulated to absorb the heat of combustion. These furnaces can be used as incinerators. The stream of hot water thus generated may be put to a useful purpose or simply used to carry the heat away to the outside environment.

Waterwall incinerator   Incinerator whose furnace walls consist of vertically arranged metal tubes through which water passes and absorbs the radiant energy from burning solid waste.

Weight-based rates   System of charging for garbage pickup that charges based on weight of garbage collected, so that the greater the weight collected, the higher the charge. The logistics of implementing this system are currently being experimented with.

Wet scrubber   Antipollution device in which a lime slurry (dry lime mixed with water) is injected into the flue gas stream to remove acid gases and particulates.

Wetland   Area that is regularly wet or flooded and has a water table that stands at or above the land surface for at least part of the year. Coastal wetlands extend back from estuaries and include salt marshes, tidal basins, marshes, and mangrove swamps. Inland freshwater wetlands consist of swamps, marshes, and bogs. Federal regulations apply to landfills sited at or near wetlands.

White goods   Large worn-out or broken household, commercial, and industrial appliances, such as stoves, refrigerators, dishwashers, and clothes washers and dryers.

Windrow   Large, elongated pile of composting material.

Yard waste   Leaves, grass clippings, prunings, and other natural organic matter discarded from yards and gardens. Yard wastes may also include stumps and brush, but these materials are not normally handled at composting facilities.

APPENDIX B

FACTORS FOR THE CONVERSION OF U.S. CUSTOMARY UNITS TO THE INTERNATIONAL SYSTEM (SI) OF UNITS

Multiply the U.S. customary unit		By	To obtain the corresponding SI unit	
Name	Abbreviation		Name	Symbol
acre	ac	40476.8564	square meter	m^2
acre	ac	0.4047	hectare	ha
atmosphere	atm	1.0133×10^5	pascals	Pa (N/m^2)
British thermal unit	Btu	1.0551	kilojoule	kJ
British thermal unit	Btu	0.2931	kilowatt per hour	kW · h
British thermal units per cubic foot	Btu/ft^3	37.259	kilojoules per cubic meter	kJ/m^3
British thermal units per hour per square foot	Btu/h·ft	23.158	joules per second per square meter	J/s · m^2
British thermal units per square foot per hour	Btu/ft^2·h	3.1525	kilowatt per meter square per second	kW/m^2 · s
British thermal units per kilowatthour	Btu/kWh	1.0551	kilojoules per kilowatthour	kJ/kWh
British thermal units per pound	Btu/lb	2.326	kilojoules per kilogram	kJ/kg
British thermal units per pound mass per degree Fahrenheit	Btu/lb$_m$ · °F	4.187	joules per kilogram per Kelvin	J/kg · K
British thermal units per ton	Btu/ton	1.16×10^{-3}	kilojoules per kilogram	kJ/kg
degree Celsius	°C	plus 273	Kelvin	K
calorie	C	4.187	joule	J (W · s)
cubic foot	ft^3	0.0283	cubic meter	m^3
cubic foot	ft^3	28.3168	liter	L
cubic feet per minute	ft^3/min	4.7190×10^{-4}	cubic meters per second	m^3/s
cubic feet per minute	ft^3/min	0.4719	liters per second	L/s
cubic feet per second	ft^3/s	2.8317×10^{-4}	cubic meters per second	m^3/s
cubic yard	yd^3	0.7646	cubic meter	m^3
day	d	86.4000	kilosecond	ks
degree Fahrenheit	°F	0.555(°F − 32)	degree Celsius	°C
foot	ft	0.3048	meter	m

(*Continues*)

(*Continued*)

Multiply the U.S. customary unit		By	To obtain the corresponding SI unit	
Name	Abbreviation		Name	Symbol
feet per minute	ft/min	5.0800×10^{-3}	meters per second	m/s
feet per second	ft/s	0.3048	meters per second	m/s
feet of water	ft H_2O	2.989×10^{-2}	pascal	Pa (N/m^2)
gallon	gal	3.7854×10^{-3}	cubic meter	m^3
gallon	gal	3.7854	liter	L
gallons per minute	gal/min	6.3090×10^{-2}	liters per second	L/s
grain	gr	0.0648	gram	g
horsepower	hp	0.7457	kilowatt	kW
horsepower-hour	hp-h	2.6845	megajoule	MJ
inch	in	2.5400	centimeter	cm
inch	in	2.5400×10^{-2}	meter	m
inches of mercury	in Hg	3.367	pascal	Pa (N/m^2)
kilowatthour	kWh	3.600	megajoule	MJ
pound force	lb$_f$	4.448	newton	N
pound mass	lb$_m$	0.4536	kilogram	kg
pound mass per hour	lb$_m$/h	0.4536	kilogram per second	kg/s
pounds per capita per day	lb/capita · d	0.4536	kilograms per capita per day	kg/capita · d
pounds per cubic foot	lb/ft^3	16.0181	kilograms per cubic meter	kg/m^3
pounds per cubic yard	lb/yd^3	0.5933	kilograms per cubic meter	kg/m^3
pounds per square foot	lb/ft^2	47.8803	newtons per square meter	N/m^2
pounds per square inch	lb/in^2	6.8948	kilonewtons per square meter	kN/m^2
million gallons per day	Mgal/d	4.3813×10^{-2}	cubic meters per second	m^3/s
miles	mi	1.6093	kilometer	km
miles per hour	mi/h	1.6093	kilometers per hour	km/h
miles per hour	mi/h	0.4470	meters per second	m/s
miles per gallon	mpg	0.425	kilometers per liter	km/L
ounce	oz	28.3495	gram	g
square foot	ft^2	9.2903×10^{-2}	square meter	m^2
square inch	in^2	6.452×10^{-4}		
square mile	mi^2	2.5900	square kilometer	km^2
square yard	yd^2	0.8361	square meter	m^2
ton (2000 pounds mass)	ton (2000 lb$_m$)	907.2	kilogram	kg
watthour	Wh	3.6000	kilojoule	kJ
yard	yd	0.9144	meter	m

INDEX